HVDC TRANSMISSION+1

VSC HVDC based MMC topology in Power Systems

HVDC TRANSMISSION+1

VSC HVDC based MMC topology in Power Systems

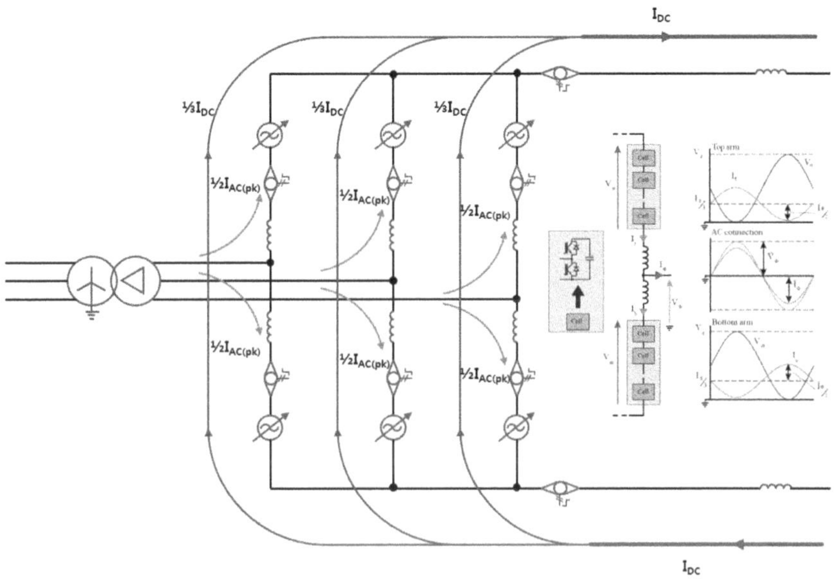

Chan-Ki Kim
Korea Electric Power Research Institute, South Korea

Seung-Il Moon
Seoul University, South Korea

Kyeon Hur
Yonsei University, South Korea

Jang-Mok Kim
Pusan University, South Korea

Gilsoo Jang
Korea University, South Korea

World Scientific

NEW JERSEY • LONDON • SINGAPORE • BEIJING • SHANGHAI • HONG KONG • TAIPEI • CHENNAI • TOKYO

Published by

World Scientific Publishing Co. Pte. Ltd.
5 Toh Tuck Link, Singapore 596224
USA office: 27 Warren Street, Suite 401-402, Hackensack, NJ 07601
UK office: 57 Shelton Street, Covent Garden, London WC2H 9HE

British Library Cataloguing-in-Publication Data
A catalogue record for this book is available from the British Library.

HVDC TRANSMISSION +1
MMC based VSC Technology in Power Systems

ISBN 978-981-121-229-1 (hardcover)
ISBN 978-981-121-358-8 (ebook for institutions)
ISBN 978-981-121-359-5 (ebook for individuals)

For any available supplementary material, please visit
https://www.worldscientific.com/worldscibooks/10.1142/11607#t=suppl

Desk Editors: Amanda Yun/Britta Ramaraj

Typeset by Diacritech Technologies Pvt. Ltd.
Chennai - 600106, India

FOREWORD

Now that the world is entering "*the era of energy transition,*" countries around the world are taking measures to expand renewable energy such as solar and wind power while reducing their dependency on conventional fossil fuels.

These global trends are capable of bidirectional power transmission, consumers and producers are the same "prosumer" (producer + consumer), of the new horizontal power grid.

The new power grid means a combined network of distributed power systems interconnecting micropower systems, and is a concept opposite to existing large-scale AC power systems.

I am pleased that this book has been published in time with these trends, and I am very proud to have written the foreword of this book, as the president of Korea Electric Power Research Institute (KEPRI).

The core technology of these new power grids is "Voltage Source Converter High-Voltage Direct Current (VSC HVDC) technology." VSC HVDC technology can control reactive power and active power simultaneously, and is self-protecting, self-healing, and self-recovering without being affected by frequency and phase. Therefore, VSC HVDC technology further strengthens the existing power system networks leading the change for tomorrow's world.

This book is written by highly qualified engineers who have spent long careers in the field of HVDC technology. It explains not only the important theoretical concept of VSC HVDC operation and control, but also goes into great depths explaining practical considerations for HVDC station design, station equipment, and HVDC transmission circuits. This book contains contents both useful for industry and academia, and not doubt meets the world's top standards.

I would like to pay my tribute to the researchers who created this material and the famous researchers in Korea and overseas who helped create this book.

Finally, I would like to share the philosophy of those who are dedicated to HVDC system research:

"We believe that electrical energy will make more contributions to humanity"

Korea Electric Power Corporation (KEPRI)
KEPRI President
Kim, Suk-Cheol

PREFACE

In the past couple of years, the development of DC transmission has gained significant momentum as a result of two key drivers: first, the development of modern conversion technology (VSC: voltage source converters) and second, the need for the transmission of offshore wind power. Plans for the interconnection of multiple renewable power sources, loads, and AC grids through DC technologies are leading to an exciting transmission concept: high-voltage direct current (HVDC) grids. HVDC grids and supergrids have sparked so much interest that researchers and engineers across the globe are talking about them, studying them, supporting them, or questioning them.

One of the important means of this energy transit policy is "**HVDC technology!**" This technology has the advantage of being able to control power flow regardless of the frequency, phase, and magnitude of the AC network.

The objectives of this book are to answer why HVDC is favorable over AC technologies for power transmission; what the key technologies and challenges are for developing an HVDC grid; how an HVDC grid will be designed and operated; and how future HVDC grids will evolve. The book also devotes significant attention to non-technical aspects such as the influence of energy policy and regulatory frameworks. As such, this book is the first to discuss all relevant topics within the scope of HVDC for offshore grids and the supergrid. It provides valuable information for researchers, industry, and policy makers.

This book describes the technical contents of the HVDC system in accordance with this world trend. The book contains theoretical content based on mathematical analysis and the practical implementation of the system. It will be beneficial for graduate students majoring in power electronics and power systems.

The new modular multi-level converter (MMC) technology adopted by the industry in recent years in particular has demonstrated clear advantages with inherent properties such as built-in redundancy, higher efficiency, and lower harmonic content. State-of-the-art MMCs are core components in flexible AC transmission systems (FACTS) equipment and VSC HVDC schemes.

The fast speed of MMC technology development has left a knowledge gap between the technology specialists, VSC HVDC project developers, and key personnel involved in those projects. There is a need for field engineers, project developers, those involved in the development and management of these projects, and graduate students to gain insights into the design, control, and application of this new MMC technology. In this respect, what is needed is a reference book to fill the gap between field engineers and HVDC specialists.

This book is a result of collaboration between experts from industry and academia. It provides theoretical insights into the design and control of MMC technology and investigates practical aspects of the project planning, design, manufacturing, implementation, and commissioning of MMC-HVDC and multi-terminal HVDC transmission technologies.

The authors hope the reader will find the book useful and stimulating and wish the reader an interesting journey into the world of MMCs and VSC HVDC systems.

Chan-Ki Kim
Seung-il Moon
Kyeon Hur
Jang-Mok Kim
Gil-Soo Jang

ACKNOWLEDGMENTS

We would like to thank the following experts for their comments and suggestions:

- Seung-Ki Sul—Professor, Seoul National University, South Korea, IEEE Fellow
- Byung-Moon Han—Professor, Myungji University, South Korea
- A. Gole—Professor, University of Manitoba, Canada, IEEE Fellow
- Shawn S.J. Chen—President, NR Electric Co. Ltd., China
- Sung-Arm Kim—Vice-President, Korea Electric Power Corporation (KEPCO)
- Jae-Seung Kim—Director General, Korea Electric Power Corporation (KEPCO)
- TangGwang Fu—General Director, State Grid, China
- R. Kuffel—Chief Engineer, RTDS Technologies, Canada
- Liu-Jing Hui—President, LJ Power, China
- N. Mocleod—Ex-director, General Electric (GE), UK
- L. Travin—IEC HVDC WG Secretary, Russia
- Tim Stott—Director, GE, UK
- Tae-Kyun Kim—Director General, KEPCO
- Ji-Won Kang—Director General, KEPCO
- C. Barker—Director, GE, UK
- Mike Li—Director, GE, UK
- C. Oates—Ex-Director, Alstom, UK
- Dong-Wook Ryu—Vice-President, Korea Electrotechnology Research Institute (KERI)
- Seok-Joo Kim—Vice-President, KERI
- Peter Lips—Ex-Director, Siemens, IEEE Fellow, Germany
- Timothy Green—Professor, Imperial College, UK
- Seung-Yong Park—Vice-President, Hyosung Heavy Industries, South Korea
- Hak-Sung Lee—Vice-President, LSIS, South Korea
- Jung-Ho Lee—Vice-President, Korea Electrotechnology Research Institute (KERI)
- Sakai—General Director, TMEIC, Japan
- Jae-Jung Jung—Professor, Kyungbok University, South Korea
- Hee-Jin Kim—Doctor, Yonsei University, South Korea
- S. Kobayashi—Chief Engineer, TMEIC, Japan
- Joon-Hee Lee—Student, Seoul National University, South Korea
- Shenghui Cui—Doctor, RWTH Aachen University, China
- Kyu-Sup Lee—Student, Seoul National University, South Korea
- Hong-Joo Jung—Senior Manager, Hyosung Heavy Industries, South Korea
- Yong-Ho Jung—Executive Director, LSIS, South Korea
- Do-Hoon Kwon—Researcher/Doctor, KERI, Seoul National University, South Korea

ABOUT THE AUTHORS

Chan-Ki Kim obtained his MSc and PhD in Electrical Engineering from Chung-Ang University, South Korea, in 1993 and 1996, respectively. Since 1996, he has been with Korea Electric Power Research Institute (KEPRI), the R&D center of KEPCO (Korea Electric Power Corporation). His research interests are high-voltage direct current (HVDC) and Power Electronics. He developed the HVDC simulator, HVDC commissioning technology, and HVDC control algorithms. Related to these developments, to date he has published over 76 technical papers in widely read journals, including journals by KIEE and IEEE, and submitted 58 patents and programs, and has published a book on HVDC entitled, *HVDC Transmission—Power Conversion Application in Power System*—Published by Wiley and IEEE Press in 2009. He received the Technical Award from the Ministry of Science and Technology of the Korean Government and Excellent Paper Awards from KIEE in 2002, 2004, and 2014, respectively. He is a senior member of IEEE and KIEE.

Seung-Il Moon has been a Professor in the School of Electrical and Computer Engineering, Seoul National University since 1997. His research interests include power system operation, power quality, flexible AC transmission systems, HVDC control, smart grid, and renewable energy. From 1993 to 1997, he was an assistant professor in the School of Electrical Engineering, Chonbuk National University, South Korea. He has participated government's energy policies and organizations. He was the president of Korea Electrical Engineering and Science Research Institute (KESRI), the chairman of the Electric Power Policy Council of the Ministry of Trade, Industry, and Energy, and a member of the National Energy Commission. At present, he is a senior member of National Academy Engineering of Korea, a member of Committee of Green Growth and the chairman of the energy transition division, the director of Seoul National University Electric Power Research Institute (SEPRI), and a director of Korea Smart Grid Institute. He received his BS from Seoul National University, South Korea, in 1985, and his MS and PhD from The Ohio State University, United States, in 1989 and 1993, respectively, all in electrical engineering. His research interests include power system operation, power quality, flexible AC transmission systems, HVDC control, smart grids, and renewable energy.

Kyeon Hur rejoined Yonsei University, South Korea, in 2010 and leads a smart-grid research group. His current research interests include flexible AC transmission systems/high-voltage direct current (FACTS/HVDC), phasor measurement unit (PMU)-based analysis and control, integration of variable generation and controllable load, and active network management. He is a senior member of IEEE, and is active in IEEE PES and Cigre Working Groups on HVDC and Power Quality. He was an R&D Engineer with Samsung Electronics, Suwon, South Korea, between 1998 and 2003, where he designed control algorithms and user-interface functions, and power-electronic circuits for AC drives. His industrial experience also includes serving in the Electric Reliability Council of Texas (ERCOT), Taylor, Texas, United States, as a Grid Operations Engineer between 2007 and 2008, where he supported real-time grid operations by conducting power flow and stability analyses and managing the state estimator. He was also with the Electric Power Research Institute (EPRI), Palo Alto, California, United States, and conducted and managed research projects in Grid Operations and Planning regarding synchrophasor-based protection and control, reactive power management and control, dynamic load modeling, and FACTS/HVDC, from 2008 to 2010. Dr Hur received his BS and MS in electrical engineering from Yonsei University, South Korea, in 1996 and 1998, respectively, and his PhD in electrical and computer engineering from The University of Texas at Austin , United States, in 2007.

Jang-Mok Kim has been a Professor with the Department of Electrical Engineering, Pusan National University, South Korea, since 2001. He also serves as the Director of the PNU-LG Electronics Smart Control Center (since 2011). From 1997 to 2000, he was a Senior Research Engineer with Korea Electric Power Research Institute, Daejeon, South Korea. He received his BS from Pusan National University (PNU), South Korea, in 1988, and MS and PhD in Electrical Engineering from Seoul National University, South Korea, in 1991 and 1996, respectively. His current interests include control of electric machines, electric propulsions, and high-voltage direct current (HVDC) power transmission.

Gilsoo Jang has been a professor at the School of Electrical Engineering, Korea University, South Korea, since 2000. He served as Associate Dean of the College of Engineering from 2011 to 2103, and as Dean of the School of Electrical Engineering from 2015 to 2017. With the support of the National Research Foundation of Korea since 2017, he has been the director of Grid Connected Power Electronics Research Center. He was a visiting scientist at the Department of Electrical and Computer Engineering, Iowa State University, United States, from 1997 to 1998; a senior researcher at Power Systems Laboratory, Korea Electric Power Research Institute (KEPRI), South Korea, from 1998 to 2000; and a visiting professor at the Department of Electrical Engineering, Cornell University, United States, from 2006 to 2007. He is a senior member of IEEE, a senior member of KIEE, and a member of CIGRE. He is actively participating in the government's establishment of energy-related policies, such as the Electricity Supply and Demand Basic Plan and the National Energy Basic Plan. He was awarded the KIEE Awards in 2006, 2007, and 2013, Yeonam Foundation Fellowship in 2006, and Minister Citations in 2012 and 2017. He has published more than 130 SCI/E-indexed journal papers and holds 15 domestic and international patents.

Contents

Nomenclature

A list of the important symbols that are used in this book can be found in Tables 1–4. A list of acronyms can be found in Table 5

Table 1 Superscripts, subscripts, circumflexes, and prefixes.

*	Complex conjugate
*	Reference
f, F	Filtered value
0	Nominal value
+	Positive sequence
−	Negative sequence
u, l	Upper, lower arm
0	Mean value, zero-sequence component
$a,\ b,\ c$	Phases, a, b, c
$\alpha,\ \beta$	Components of the stationary $\alpha\beta$ reference frame
d, q	Components of the synchronous dq reference frame
$\sum$	Sum
Δ	Difference (Ripple quantity, parasitic quantity, difference, increment)

Table 2 Variables

h	Signed multiple of the fundamental frequency
i	Submodule index
k	Phase number
m	Carrier index
n	Sample index, sideband index
$s = d/dt$	Differential operator (or, where appropriate, complex Laplace variable)
t	Time
z	Forward-shift operator (or, where appropriate, complex z-transform variable)
$\delta = z - 1$	Delta operator
$i_{u,l}$	Arm current
$i_s = i_u - i_l$	Output current
$i_c = \left(i_u + i_l\right)/2$	Circulating current
i_d	DC-bus current
$e = i_s^* - i_s$	Output-current control error
$v^i_{cu,l}$	Capacitor voltage in submodule i

(continued)

Table 2 Variables *(continued)*

$v^{\Sigma}_{cu,l} = \sum_{i=1}^{N} v^{i}_{cu,l}$	Sum capacitor voltage per arm
$v^{\Sigma}_{cu} = v^{\Sigma}_{cu} + v^{\Sigma}_{cl}$	Sum capacitor voltage per phase
$v^{\Delta}_{c} = v^{\Sigma}_{cu} - v^{\Sigma}_{cl}$	Imbalance sum capacitor voltage
$n^{i}_{u,l}$	Submodule insertion index
$n_{u,l} = \left(\sum_{i=1}^{N} n^{i}_{u,l}\right)/N$	Insertion index per arm
$v_{u,l} = \sum_{i=1}^{N} n^{i}_{u,l} v^{i}_{cu,l}$	Inserted arm voltage
$v_s = \left(-v_u + v_l\right)/2$	Output voltage
$v_c = \left(v_u + v_l\right)/2$	Internal voltage
v_a	AC-bus voltage
v_g	Grid voltage
v_{PCC}	PCC voltage
$v_{du,l}$	Pole-to-ground dc-bus voltage
$v_d = v_{du} + v_{dl}$	Pole-to-pole dc-bus voltage
$v^{\Delta}_{d} = v_{du} - v_{dl}$	Imbalance dc-bus voltage
v_R	*R*-part output
$W_{u,l} = \mathrm{C}\left(v^{\Sigma}_{cu,l}\right)^2/(2N)$	Stored energy per arm
$W_{\Sigma} = W_u + W_l$	Stored energy per energy
$W_{\Delta} = W_u - W_l$	Imbalance stored energy
$W_d = C^{''}_{d} v^{2}_{d}/2$	Effective stored dc-bus energy
P	Active output power
P_d	DC-side input power
Q	Reactive output power
$\theta = \int \omega\, dt$	Angle of the control-system *dq* frame

Table 3 Parameters and functions

f_1	Fundamental frequency
$\omega_1 = 2\pi f_1$	Fundamental angular frequency
f_s	Sampling frequency
f_{sw}	Switching frequency
$T_s = 1/f_s$	Sampling period
T_c	Computational time delay
$T_d = T_c + 0.5T_s$	Total time delay
K	Space-vector scaling constant
M	Number of phase
N	Number of submodules per arm
C	Submodule capacitance
C_d	Installed DC-bus capacitance
$C_d^{'} = C_d + 2MC/N$	Effective DC-bus capacitance
L	Arm inductance
R	Parasitic arm resistance
R_I	Insertion resistance
$\hat{V}_s$	Peak value, fundamental component of v_s
$\hat{I}_s$	Peak value, fundamental component of i_s
$\hat{V}_{max}$	Maximum allowed $\hat{V}_s$
$\hat{I}_{max}$	Maximum allowed $\hat{I}_s$
δ_a	AC-bus-voltage phase angle
δ_g	Grid-voltage phase angle
θ_l	Voltage-reference phase angle
ω_c	Carrier angular frequency
θ_c	Carrier phase shift
m_a	Modulation index
Re	Real part
Im	Imaginary part

Table 4 Controller parameters and transfer functions

α_b	PLL low-pass filter bandwidth
α_c	Output-current control-loop bandwidth
α_d	DC-bus-voltage control-loop bandwidth
α_f	Voltage-feedforward-filter bandwidth
α_h	*R*-part bandwidth
α_{id}	DC-bus-voltage integrator bandwidth
α_{ip}	PLL integrator bandwidth
α_l	Power-synchronization control low-pass-filter bandwidth
α_p	PLL bandwidth
α_s	Power-synchronization control-loop bandwidth
K_h	*R*-part gain
K_i	*I*-part gain
K_p	*P*-part gain
K_s	Power-synchronization-control gain
K_v	Voltage droop gain
R_a	"Active resistance" for circulating-current control
R_s	"Active resistance" for power-synchronization control
Φ_h	Compensation angle for resonant filter
F_h	*R* part
G_c	Closed-loop system
G_k	Open-loop system
H_h	Resonant filter
H_p	PLL low-pass filter

Table 5 Acronyms

APOD	Alternative phase opposite disposition
APS	Auxiliary power supply
ATB	Average-voltage tolerance band
BLK	Blocking
BPF	Band-pass filter
CTB	Cell-voltage tolerance band
CTB optimized	Optimized cell-voltage tolerance band
D	Derivative
DBS	Dynamic braking system
DCB	DC circuit breaker
DDSRF	Decoupled double synchronous reference frame

(continued)

Table 5 Acronyms *(continued)*

DFT	Discrete Fourier transform
DPS	Dynamic performance study
DPWM	Digital pulse-width modulation
DSOGI	Double sound-order generalized integrator
DSP	Digital signal processor
EMC	Electromagnetic compatibility
EMF	Electromotive force
EMI	Electromagnetic interference
ESL	Equivalent series inductance
ESR	Equivalent series resistance
FACTS	Flexible alternating-current transmission system
FB	Full bridge
FIT	Failures in time
FPGA	Field-programmable gate array
FRT	Fault ride through
GDU	Gate-drive unit
GTO	Gate turn-off
HB	Half bridge
HMI	Human-machine interface
HVDC	High-voltage direct current
H/W	Hardware
I	Integral
IGBT	Insulated-gate bipolar transistor
IGCT	Integrated gate-commutated thyristor
LVRT	Low-voltage ride through
MMC	Modular multilevel converter
MOV	Metal-oxide varistor
NLC	Nearest-level control
NSCOGI	North sea countries offshore grid initiative
NSI	Negative-sequence injection
PCC	Point of common coupling
PD	Phase disposition
PFC	Power-flow controller
PI	Proportional-integral
PID	Proportional-integral-derivative
PLL	Phase-locked loop
PNSE	Positive/negative sequence extraction
PNSRG	Positive/negative sequence reference generator

(continued)

Table 5 Acronyms *(continued)*

POD	Phase opposite disposition
P	Proportional
PR	Proportional-resonant
PSC	Phase-shifted carrier
PSI	Positive-sequence injection
PWM	Pulse-width modulation
QSG	Quadrature signal generation
RES	Renewable energy source
SCR	Short-circuit ratio
Si	Silicon
SiC	Silicon carbide
SM	Submodule
SOA	Safe operating area
SOGI	Second-order generalized integrator
SSOA	Safe switching operator area
STATCOM	Static synchronous compensator
SVC	Static VAr compensator
THD	Total harmonic distortion
UPS	Uninterruptible power supply
VMM	Voltage-margin method
VSC	Voltage-source converter
VT	Voltage transformer
WTHD	Weighted total harmonic distortion
XLPE	eXtruded cross-bound polyethylene

1

Introduction to High Voltage Direct Current (HVDC) Technology

1.1 Development of High Voltage Direct Current (HVDC) Technologies

In the 1880s, Thomas Edison's first commercially generated electricity and transmission system was low-voltage DC. As low-voltage transmission resulted in high transmission losses, electricity had to be generated near the demand site, Nikola Tesla's AC system, on the other hand, was able to create high voltages with a transformer, thus there are no transmission losses. This made it possible to generate electricity at larger distances away from it end users. This resulted in AC technology being crowned the winner of the "war of currents" in the late 1880s, and the AC system has dominated transmission technology ever since.

A number of high-voltage direct current (HVDC) systems were installed from the 1880s to the 1930s. These system was invented by Rene Thury which converted the voltage by placing generators and loads in series. The Lyon–Mountiers DC transmission project was the most powerful system at that time and operated from about 1906 until 1936 with a final capacity of 20 MW. After that, the mercury arc valve with a vacuum tube was invented. In 1941, Germany tried to install a 60 MW ± 200 kV, 115 km DC transmission system with a mercury arc valve in Berlin but failed with the collapse of the German. The world's first commercial HVDC transmission system was Gotland 1 in 1954, as shown in Figure 1.1. It used mercury arc valves and could transfer 20 MW over a 98 km submarine cable between the Swedish mainland and the island of Gotland [1].

The thyristor was introduced in 1956 by the engineers at General Electric, and starting in the early 1970s, the mercury arc rectifier was replaced with a thyristor valve. Thyristors did not experience the arc-back problem, which is a disadvantage of the mercury arc device. The first HVDC transmission using thyristor was the Eel River project in 1972. The Eel River HVDC scheme was built as a back-to-back converter station to form a 320 MW interconnection between Quebec and New Brunswick [3]. Currently, the thyristor project as shown in Figure 1.2 reaches 8000 MW ± 800 kV at 2210 km (Southern Hami-Zhengzhou UHVDC Transmission Project in 2014), and constructed the Changji-Guquan 1100 kV transmission link to supply 13 GW of power per line. Thyristors can control the conduction of the valve, and a certain minimum strength of AC system is required to operate safe from one valve to the other valve. These operation can be viewed as a DC source that is switched between AC phases by the combined operation of an AC system and a valve control, which is therefore known as a current-sourced converter (CSC) or line-commutated converter (LCC) [4].

Figure 1.1 Gotland Mercury Arc valve hall (from wikipedia)

Figure 1.2 Jeju HVDC thyristor valve converter station

Figure 1.3 ENELFE VSC converter station (courtesy of SIEMENS)

The voltage-sourced converter (VSC) HVDC was developed in the mid-1990s. This type of HVDC uses insulated-gate bipolar transistor (IGBT) valves. The VSC HVDC concept was first demonstrated in 1997 between Hellsjön and Grängesberg in Sweden, with 3 MW and 10 kV. The VSC HVDC has the ability to impose an voltage to an AC system and can thus control active and reactive power injections into the AC system independently. Therefore, it is suitable for a weak system connection such as an island network and has black-start capability. There are two dominant technologies used in VSC HVDC. One uses two- or three-level converters, whereas the other uses modular multi-level converters (MMCs) [8]. The MMC was introduced in 2003 and is the most suitable converter topology for medium/high-voltage applications such as VSC HVDC. MMC HVDCs have some advantages over conventional VSC converters: (1) modularity and scalability, (2) high efficiency, (3) superior harmonic performance, and (4) absence of DC link capacitors as shown in Figure 1.3 [3]. After research for the commercialization of MMC had been completed, the Transbay Cable (TBC) project (which is the first project using MMC HVDCs) started operating in 2010. After the TBC project, the MMC has played a dominant role in the global HVDC transmission market.

1.2 What is HVDC?

In an AC network, the power flowing in each transmission line is determined by the impedance of the line as well as the voltage magnitude and phase angle at the two ends. The system operator can control these by controlling generation, system transformer tap changers, and shunt or series reactive power elements.

In an HVDC system, power electronics (e.g., a converter) are used to convert AC power to DC power at one terminal (rectifier) and vice versa at the other terminal (inverter). A simple representation of an HVDC interconnection is shown in Figure 1.4. AC power is fed to a converter operating as a rectifier. The output of this rectifier is DC power that is independent of the AC supply frequency and phase. The DC power is transmitted through some conduction medium, be it an overhead line, a cable, or a short length of bus bar, and applied to the DC terminals of a second converter. This second converter is operated as an inverter and allows the power to flow into the receiving AC network.

The converter transformers in Figure 1.4 provide galvanic isolation and suitable adjustment of the voltage magnitude by means of their tap changers. Conventional HVDC transmission is therefore termed "line-commutated thyristor technology". Consider a simplified representation of a thyristor as shown in Figure 1.5. When a gate pulse is applied while positive forward voltage is imposed between the anode and the cathode (VTHY), the thyristor will conduct current (I_d). Conduction continues without further gate pulses as long as

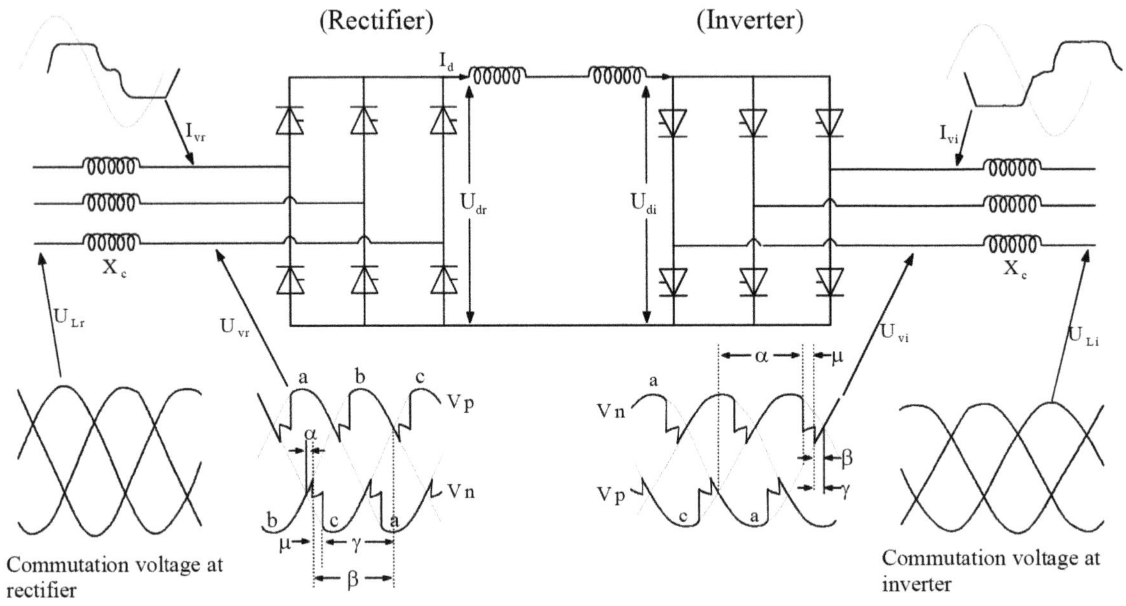

Figure 1.4 Basic HVDC transmission

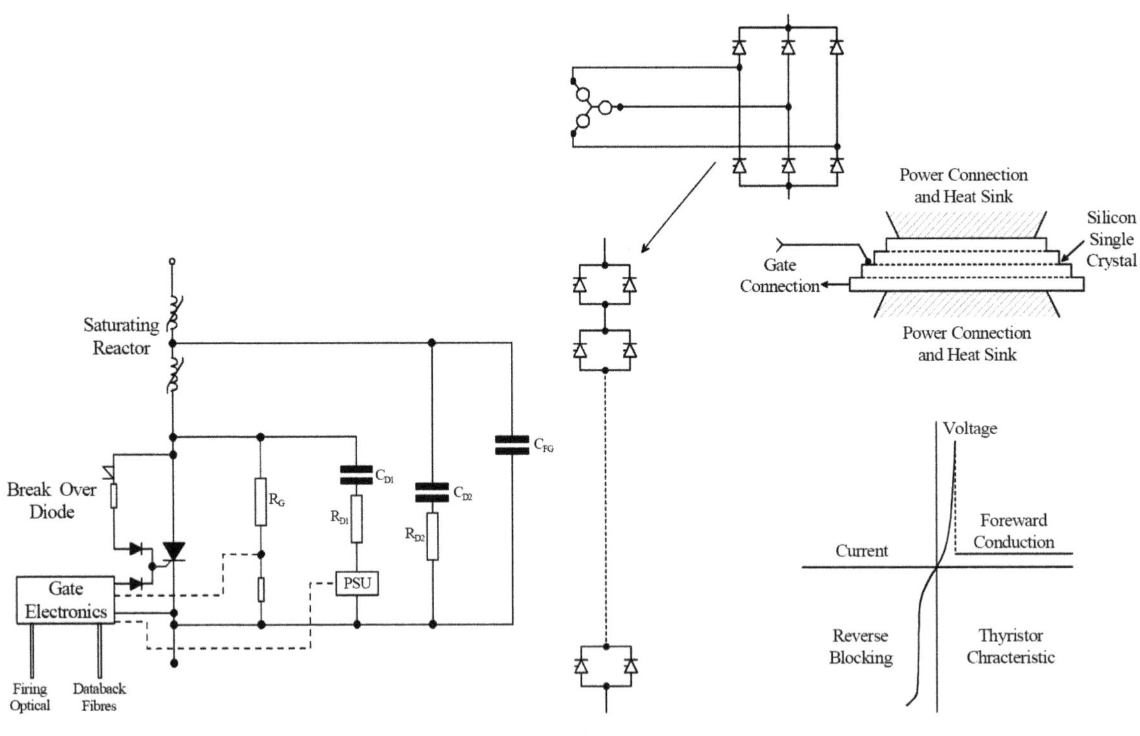

Figure 1.5 Thyristor valve system

the current flows in the forward direction. Thyristor turn-off takes place only when the current tries to reverse. Therefore, a converter comprising of thyristors requires an existing alternating AC voltage (Vac) in order to operate as an inverter. Figure 1.5 shows the thyristor valve system.

The basic building block of an HVDC converter is the Graetz (or 6-pulse) bridge. The switching action of the bridge gives rise to harmonic voltages on the DC side and harmonic currents on the AC side. As shown in Figure 1.6, by connecting two Graetz bridges in series and energizing one from a star-connected transformer winding and the other from a delta-connected transformer winding, the harmonics of the resultant waveform are reduced. The harmonics produced by the converters may still need to be suppressed in order to avoid interference with other electrical equipment, particularly on the AC side. This is achieved by means of harmonic filters, which are shunt connected capacitor banks with auxiliary R, L, and C components to achieve the appropriate harmonic impedance characteristics.

The conversion of power from AC to DC and vice versa is done through the commutation process by the valves. In an LCC, the AC current will always lag the AC voltage such that reactive power is absorbed both at the rectifier and the inverter. Typically, the reactive power is 50% of the real power. The AC harmonic filters are capacitive at fundamental frequency and provide compensation for the reactive power that is absorbed by the converters. The filters are switched as the load varies to achieve satisfactory overall power factor. Normally, the filters are subdivided such that the AC voltage change on filter switching is acceptable.

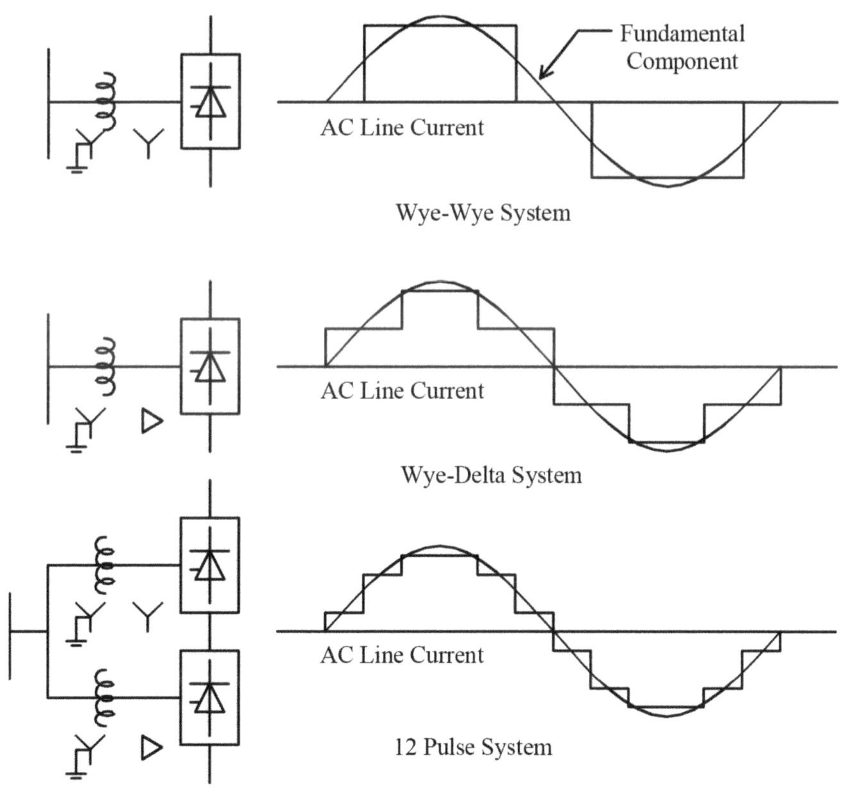

Figure 1.6 (a) The idealized AC current waveforms produced by two 6-pulse thyristor bridges, one connected to a star/star transformer winding and the other connected to a delta/star transformer winding and (b) the resultant idealized AC current waveform when the two 6-pulse AC waveforms are added together at a common connection point (i.e., the converter station AC bus bar)

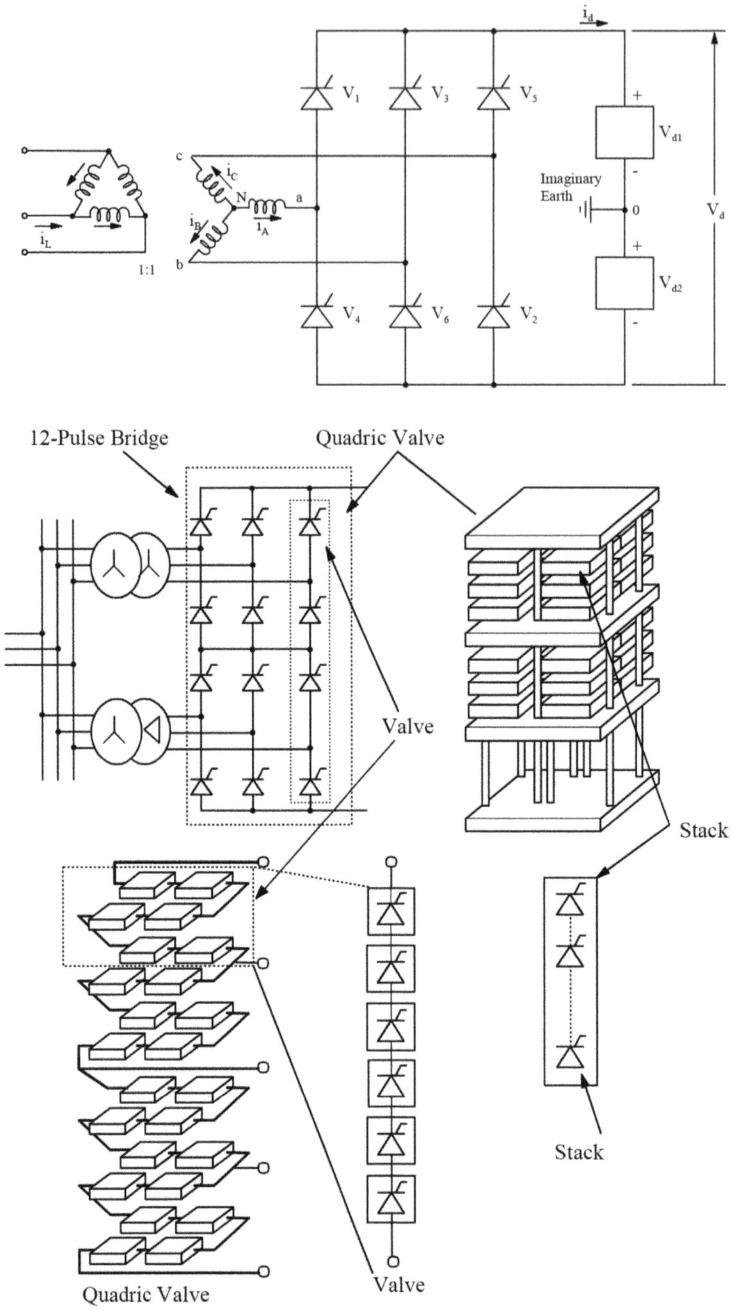

Figure 1.7 A 12-pulse converter bridge

Figure 1.7 shows two 6-pulse bridges combined to form a single 12-pulse converter bridge. All modern HVDC schemes use this 12-pulse bridge arrangement. The practical implementation is as three "quadri-valves," the four valves in a phase being physically mounted on top of each other. Each "valve" within the bridge consists of many devices connected in series in order to obtain the necessary DC voltage for economic power transmission.

Figure 1.8 shows an LS Electric "Tier Assembly" consisting of 14 series-connected 100 mm diameter thyristors, giving it a maximum power rating of over 100 MW. Considering that the highest voltage rated scheme in service to date operates at 600 kV DC, and that the maximum peak voltage-withstanding capability

Figure 1.8 A single "Semi-Tier" containing 14 series-connected thyristors

of thyristors used in HVDC transmission is about 8.5 kV, it is clear that many thyristors need to be connected in series to withstanding the applied voltage.

The converters are controlled by a multi-layered control system. At the lowest level, a very fast phase-locked oscillator controls enable the DC current to be controlled within each power frequency cycle. Intermediate-level controls enable the power output to be varied automatically to benefit the AC network (e.g., power swing damping controls, reactive power and voltage controls, frequency control and emergency power control routines). At the highest levels, the operator is provided with the tools required to change the power and other settings, monitor the performance, and safely execute start/stop and other operating sequences.

1.3 For What Purpose Is HVDC Used?

To date, HVDC power transmission has provided economic benefits in a number of areas. There are five advantages of a power network using HVDC transmission systems; (1) bulk power transmission on overhead lines, (2) power transmission via cables (maybe through submarine cables), (3) power transmission between unsynchronized AC systems, (4) minimizing short-circuit capacity increase, and (5) upgrading existing AC lines to DC system.

1.3.1 Bulk Power Transmission on Overhead Lines

HVDC links use power electronic equipment to convert AC to DC (and vice versa) at each terminal. This requires higher capital outlay than would be required for a substation for an AC interconnection. However, HVDC towers are simpler than AC towers because only two conductors are required for a monopolar HVDC link, which is equivalent to a single AC circuit. In some locations, electrodes can be inserted into the earth (or sea) to provide a low-voltage return path, and in these circumstances only one metallic conductor is necessary. Also, the ROW (Right-of-Way) occupied by DC transmission towers is less than that required for the equivalent AC power transmission. The ability to directly and instantaneously control the power transferred over an HVDC link can be used to benefit the AC system, providing further direct and indirect cost benefits in favor of DC system.

Therefore, even though the cost of an HVDC terminal station is greater than that of an equivalent AC transmission substation, a "break-even" distance can be defined, beyond which it is more economical to transmit bulk power as DC transmission instead of AC transmission. This break-even distance varies depending on the economic factors influencing each case but is typically in the range of 800–1000 km.

1.3.2 Power Transmission via Cable

The construction of a cable (conductor surrounded by insulation enclosed by an earth plane) means that it has a much higher capacitance than that of an overhead line. When an AC voltage is imposed on a cable, not only does the load current flow through the core conductor, but also a capacitive charging current. This charging current changes direction every half cycle, charging, discharging, and then re-charging the cable. Even for a cable of only moderate length (50 km), this reactive charging current can utilize a major part of the total current-carrying capability of the cable, severely limiting the amount of useful real power that can be transmitted. Because of this limitation, the AC capacitive charging current drawn by the cable approaches its total current-carrying capacity, rendering the cable useless as a power transmission medium. Electrodes can also be used for cable schemes as the low-voltage return path.

In the case of HVDC transmission, apart from an initial charging current, the cable draws no capacitive charging current. Therefore, power can be transmitted through any length of cable, being limited only by the cost of the cable and the economics of the I^2R losses incurred. It is a major reason to use an HVDC transmission system for connecting offshore wind farms to the main grid.

1.3.3 Transmission between Unsynchronized AC Systems

When two adjacent AC systems operate at different frequencies, such as 50 and 60 Hz, the only practical way to obtain the advantages of an interconnection between them is by means of a DC connection.

DC power is independent of the frequency and relative phase of the power systems. Therefore, power can be transmitted between two independent AC systems without applying any operational restrictions to either system. An asynchronous HVDC interconnection will not suffer from power swings or risk of tripping arising from overload, which may afflict an AC connection between two AC systems in the event of a fault in one of them. Therefore, even when AC networks operate at the same nominal frequency, an interconnection by HVDC may be preferable to an AC interconnection. There are many examples of this in North America and in India, often using back-to-back HVDC schemes, in which both terminals of the scheme are located at the same site.

1.3.4 Minimizing Short-Circuit Capacity Increases

As power demand increases, additional AC infeeds are usually added. However, there may come a point where any further infeeds will increase the short-circuit level (SCL) in some part of the system beyond the rated capability of the existing AC switchgear. In order to avoid having to upgrade the existing AC switchgear while still delivering more power, HVDC could be used to import the additional power. During a fault in the AC system, the infeed from the HVDC converter is very swiftly limited to a value no larger than the rated current so that the HVDC infeed does not contribute significantly to the fault current.

1.3.5 Upgrading Existing AC Transmission Lines to DC System

In many countries, along with the increasing demand for more power, there are also increasing environmental objections against constructing new overhead lines. One potential solution to this is to modify one or more existing AC lines to be used for HVDC power transmission. The DC power that can be transmitted down a given ROW (Right-of-Way) without affecting the visual impact is significantly greater than that which can be achieved with AC transmission and, as discussed above, the addition of a parallel DC connection can often increase the maximum safe power transfer through existing parallel AC lines.

1.4 VSC HVDC Transmission System

1.4.1 The Development of New Power Semiconductor Devices

The main emphasis of power semiconductor development over recent years has been toward more sophisticated "bimodal" devices such as the IGBT, which provide "turn-off" as well as "turn-on" capability and do not have to rely on current zero. Converters using bimodal semiconductor devices do not require an externally applied AC voltage for successful commutation, so they can be operated at any desired power factor. One application of these devices is in the VSC, which has a relatively large capacitance on the DC side. The semiconductors in a VSC scheme are used to connect the three AC phases to the DC capacitor at appropriate times. With bimodal devices comes the ability to use pulse width modulation (PWM), which can be used to reduce the magnitude of the lower-order harmonics produced by the converter, compared with those produced by a conventional 12-pulse thyristor converter.

Much effort has also been invested in the development of Silicon Carbide (SiC) for semi-conductor devices. Compared with silicon, SiC has the advantage of a higher voltage-withstanding capability with a thinner wafer, which results in lower losses and the ability to operate at higher temperatures. The first SiC devices have been diodes that, when used in conjunction with some new bimodal devices, can enhance the overall performance of the converter. The future development of Sic Thyristor/IGBT will enable significant cost and power loss reductions to be realized in thyristor/IGBT-based HVDC schemes.

Figure 1.9 shows a comparison among the power-carrying capabilities of various semiconductor devices based on presently available ratings, and indicates that increasing sophistication is achieved at the expense of rating. Of the bimodal devices, two types have been particularly taken up by industry; the GTO and the IGBT. Both of these power semiconductor types have advantages and disadvantages depending upon the proposed application, but for industrial applications, the IGBT has won the largest market share. By far, the biggest

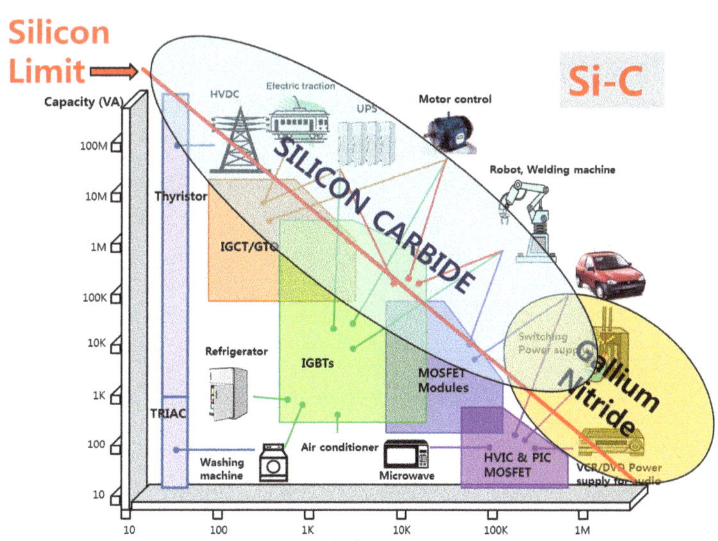

Figure 1.9 Comparative ratings of semiconductor devices as a function of switching frequency

application of the IGBT has been in AC motor drives. The use of bimodal semiconductors allows PWM to be used with the following advantages for motor drives:

- A better approximation of a sine wave than is achievable with the thyristor bridge and hence lower losses in the motor.
- Excellent dynamic response
- Smooth torque/speed control
- Rides through dips in supply voltage
- Good AC power factor over full-speed range

1.4.2 VSC Transmission System

An outline circuit diagram of a VSC Transmission system would look very similar to that shown in Figure 1.4. In its simplest form, one terminal of a two-level converter is used, as shown in Figure 1.10. This 6-pulse bridge uses IGBTs and anti-parallel connected freewheeling diodes.

The AC voltage at its terminals is created by controlled switching of the DC output to the AC terminals by means of the IGBTs. The phase angle and amplitude of the resulting waveform can be fully controlled. By using PWM switching, it is possible to reduce the amplitude of low-order harmonics. However, the more frequently the IGBTs are switched the higher the power losses. Thus, a compromise needs to be made between waveform quality and power losses.

The AC switchyard is much simpler than for a conventional HVDC scheme. The converter transformers are normally simpler than a conventional HVDC converter transformer and will not normally see a DC stress or high harmonics.

An alternative configuration of VSC scheme is shown in Figure 1.11(a). Here, the VSC's semiconductor switches are configured in pairs of "commutation cells," with "floating capacitors" (C_f) connected between pairs of series devices on opposite arms of each leg of the converter. The switches of each cell are always in complementary on/off states. As can be seen from Figure 1.11, the output voltage has several steps. The voltages across the floating capacitors as well as the DC capacitor (C_{dc}) are maintained to a constant value by switching actions. In the case of the four-level VSC shown in Figure 1.11, the voltages across the capacitors will be: $C_{dc} = V_{dc}$, Cf_1 = 2/3Vdc, and Cf_2 = 1/3Vdc. The floating capacitors act as pseudo voltage sources so, by forcing the load current to flow through these capacitors in a particular direction, dictated by the switching states of the commutation cells, the voltage across the capacitor either subtracts or adds to the DC supply voltage. This gives

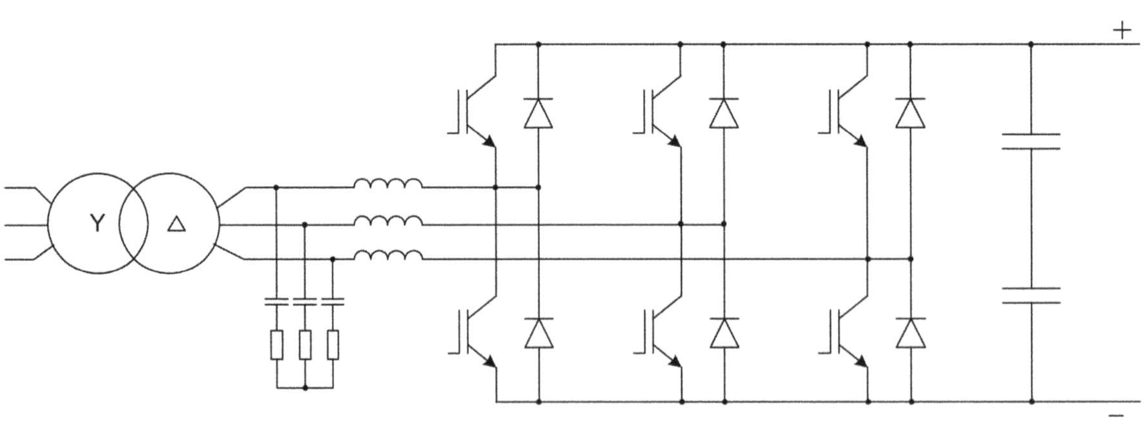

Figure 1.10 Two-level voltage-sourced converter transmission scheme

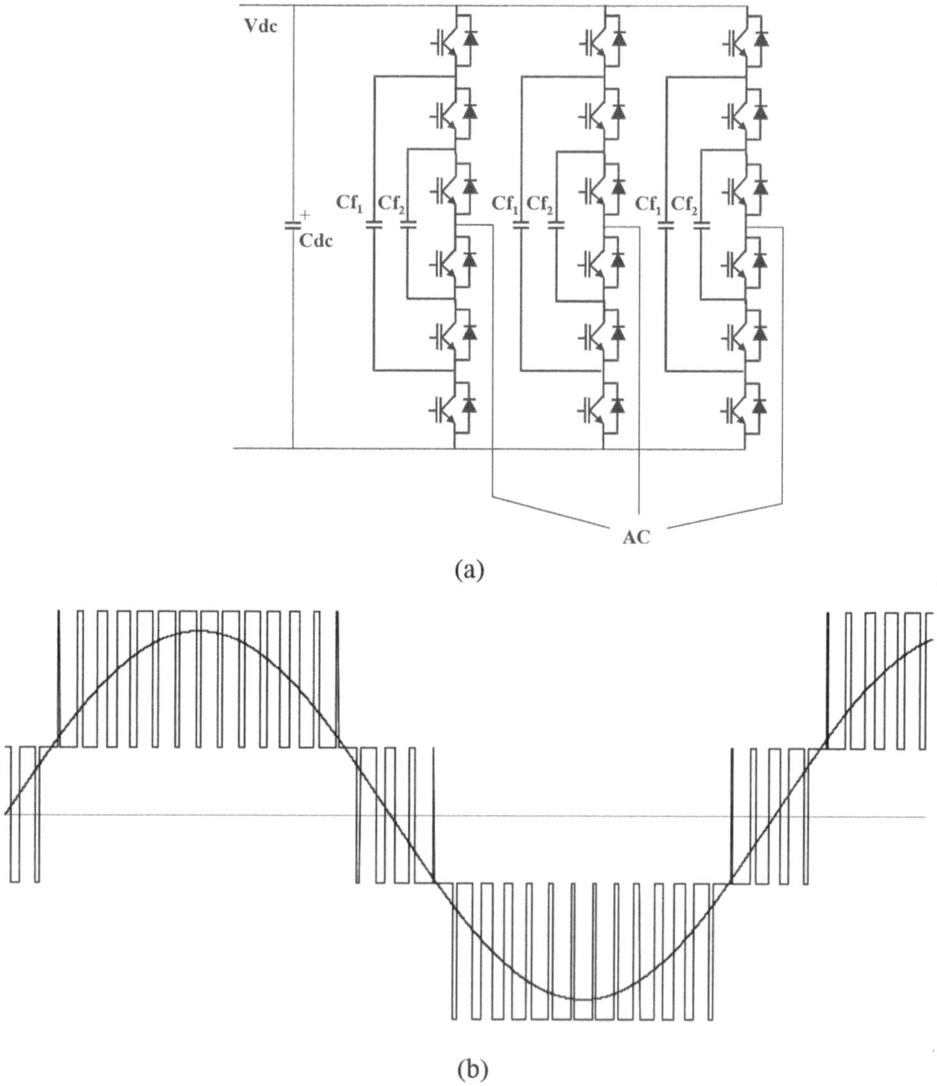

Figure 1.11 (a) A four-level capacitor-clamped voltage-sourced converter and (b) output-voltage waveform of the converter

four possible line-to-ground voltage states, namely +½Vdc, +1/6Vdc, -1/6Vdc, and -½Vdc. The phase voltage is shown in Figure 1.11(b).

1.4.3 The Possible Applications of VSCs in HVDC

Many HVDC schemes have been installed around the world. The following section describes some challenges that have been overcome using conventional technology and what changes may be brought about with the introduction of VSCs.

Rotating Machinery

Normally, the AC voltage required to turn thyristors off by natural commutation is provided by generators that already supply power to the AC network. This is the most typical application in which HVDC transmission provides the interconnection between two distant or non-synchronized power networks (e.g., the 2000 MW

Cross Channel HVDC link between England and France). However, existing generation may not always be guaranteed, for example, where one end of the HVDC link is an island load.

Small island AC networks frequently rely on diesel generators to supply AC power. HVDC may be used to connect the island to a larger AC network whose unit cost of electrical power is much less, thus making cheaper electrical power available to the island system. To avoid having to keep diesel generators operating just so that the HVDC converters operate correctly, a synchronous compensator can be used to provide the necessary rotating AC voltage vector. The synchronous compensator must be designed to have sufficient inertia such that the power flow on the island is not disrupted in the event of a short disturbance to the sending end AC network.

This technique of using an HVDC scheme in conjunction with a synchronous compensator has been used on several schemes, most recently on the 2 x 150 MW KEPCO HVDC project, which was commissioned in 1997, interconnecting the mainland of South Korea to Jeju island. The island can be supported by two synchronous compensators and the entire island can be supplied with power from the HVDC link.

The synchronous compensators are fitted with small gas turbines and clutches enabling their running costs to be minimized. In the event of a complete island blackout, the synchronous compensators can be started up via the gas turbines. Once the compensator is spinning, the HVDC converter can pick up the island load and the turbines can be de-clutched.

Figure 1.12 shows a 300 MW HVDC station in South Korea, with gas-insulated switchgear for filter and station switching on the left, AC filters in the foreground, and the converter transformer with its valve winding bushings protruding into the building that houses the converter, the control, and the auxiliary plant.

By using VSCs to supply islanded AC loads, the requirement for a synchronous compensator to supply a rotating voltage vector would be removed. The island AC frequency would be dictated wholly by the VSC, so the cost, losses, and maintenance requirements of the synchronous compensator would no longer be incurred.

Reactive Power Control

Using conventional LCCs, both the sending end and the receiving end absorb reactive power. Shunt capacitors are usually used to compensate for some or all of the reactive power absorbed by the converters and to give the desired power factor for the complete system. By subdividing the capacitor banks, reactive power can be controlled in "steps." The reactive power absorbed by the converters is dependent on the DC-operating conditions. Therefore, by manipulating these conditions, the converter absorption can be controlled which along with the switching of capacitor banks can be used to meet tight reactive power limits for the converter

Figure 1.12 300-MW converter station in Korea

station. Figure 1.13 shows a graph of reactive vs real power for the Chandrapur 1000 MW back-to-back HVDC interconnector. Note that the line representing converter absorption would, normally, be smooth but converter reactive power absorption has been manipulated to meet the specified reactive power limits.

VSCs using PWM to carry out the AC/DC and DC/AC conversions allow an approximation to a sine wave to be produced by the converter alone, having almost any desired phase relationship with an existing AC system waveform. The relative phase relationship between the two waveforms dictates the direction and magnitude of the reactive power. Therefore, switchable capacitor banks are not required, but at least one fixed high-frequency harmonic filter is required as the PWM technique generates high-order harmonics.

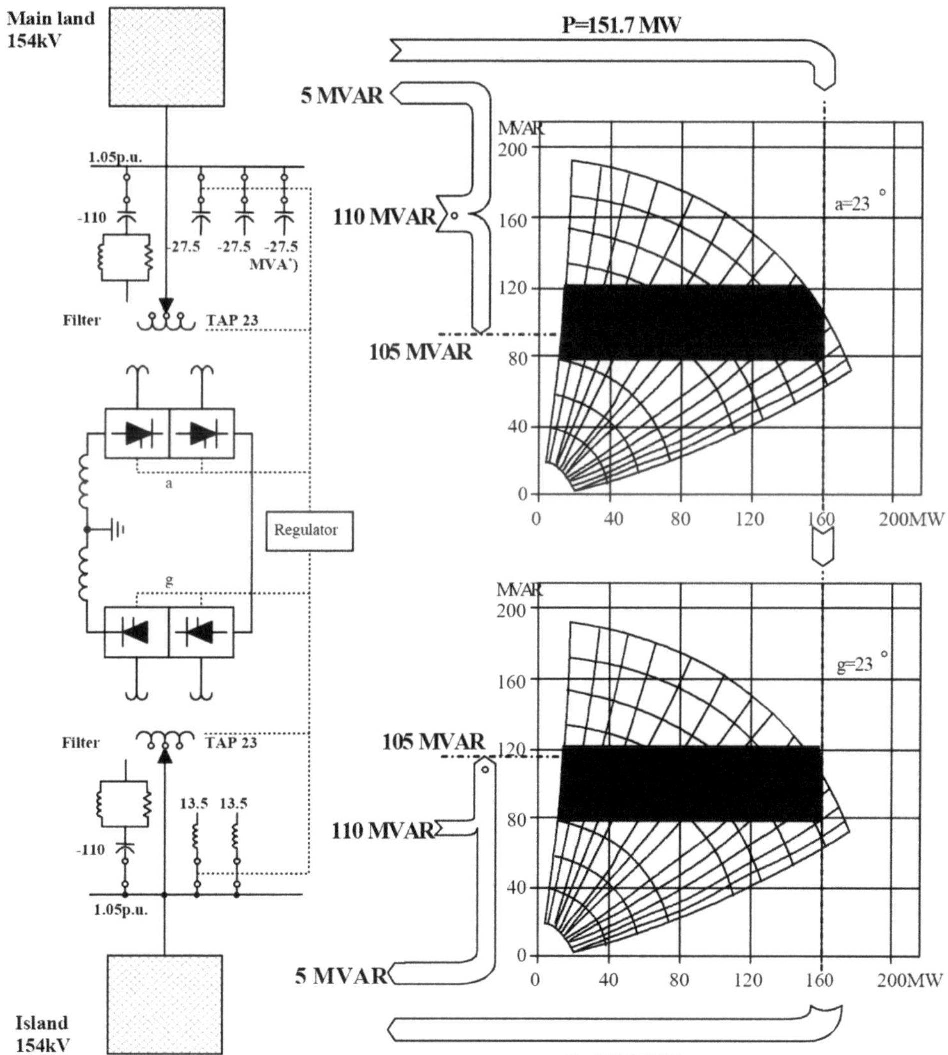

Figure 1.13 Control of the reactive power exchange with the AC system utilizing the converters to meet the working limits

Parallel Technologies

At the same time as the development of the VSC, there have also been developments in cable-manufacturing technologies. Extruded plastic-cable-manufacturing techniques are being developed, and claims are being made that such cables are cheaper than the more commonly used mass-impregnated cables. However, this new cable type is sensitive to rapid voltage reversal. VSCs incorporate reverse-biased diodes, preventing voltage reversal between their DC terminals, thereby potentially allowing this cable technology to be used. Voltage reversals do not normally represent a problem for mass-impregnated cables, so it has not usually been necessary to take special steps to avoid voltage reversals for conventional HVDC schemes. However, a technique developed by ALSTOM in the 1970s utilizes a diode and surge arrester combination at the cable termination to control cable voltage reversals. This technique would enable the new types of extruded cable to be used with conventional LCCs.

VSCs can reduce the amount of electrical plants associated with an HVDC converter station, in particular the amount of switched reactive power/filter banks. In the case of power transmission to an islanded AC system, VSCs can be used without any additional means of generating AC voltage, thereby allowing all local generation to be shut down, and without having to add a synchronous compensator to the island AC system. For practical reasons the maximum voltage of a VSC converter is, at present, limited to about ±320 kVdc.

1.5 Comparison of LCC and VSC HVDCs

HVDC transmission systems involve the use of AC/DC converters. There are two dominant converter technologies used in converting AC to DC or vice versa. These methods are the LCC and the VSC. The success of these two converters became possible with the development of power electronics technology, and both have been commercialized and are standard technological solutions across today's global networks.

As described in previous sections, LCC and VSC HVDCs have different characteristics. The major difference is that a thyristor is utilized for LCC HVDC, whereas IGBT switches are utilized for VSC HVDC. As IGBT switches can be operated actively a VSC HVDC synthesizes AC voltage waveform regardless of connected grid status. Therefore, the active and reactive power of a VSC HVDC can be regulated independently so that more applications for grid service are potentially enabled. On the other hand, a thyristor-based LCC HVDC absorbs certain reactive power, which is generally assumed to be proportional to the transmitting active power. In this section, detailed comparisons of LCC and VSC HVDCs are provided. The contents of comparison include power limit characteristics, overload characteristics, system interactions, commutation failure (CF) characteristics, and fault-blocking capabilities. More detailed explanations and additional supplements on the comparison are presented in the following sections.

1.5.1 Power Limit Characteristics

The power limit has a similar aspect to the overload characteristic, but the power limit is a function that limits the capability of the HVDC system according to AC system conditions. The power limit of the CSC HVDC system determines the output of the HVDC system in consideration of the reactive power condition, such as the On or Off condition of the synchronous compensator and the filter, whereas the power limit of the VSC HVDC system is dependent on AC line conditions according to AC system faults such as a one-phase ground fault or two-phase ground fault.

On the other hand, in the case of VSC HVDC systems, the power limit has no effect on the capability of the AC system and is determined only by the IGBT, which is the switching device. Figure 1.14(a) shows the capability curve of the VSC HVDC system. Figure 1.14(b) shows the power limit in the case of a one-phase fault, Figure 1.14 (c) in the case of a two-phase fault, and Figure 1.14 (d) in the case of a one-phase fault considering a third harmonic injection.

1.5.2 Overload Characteristics

The overload characteristic is related to the power persisting time and stability from the viewpoint of the HVDC's power capability. In the LCC HVDC system, the overload characteristic is determined by the capacitance of the switching element, the thyristor, and the stability of the AC system. Figure 1.15 shows the overload characteristics of LCC and VSC HVDC systems according to SCR (Short Circuit Ratio). The LCC HVDC system in bipole mode operating as a monopole because of a fault in one pole and the overload characteristics of the mono pole LCC HVDC are shown in Figure 1.16. In general, the stability of the LCC HVDC system is based on the maximum available power (MAP) method based on voltage stability, and the AC system voltage is suppressed to within 1.05 [p.u].

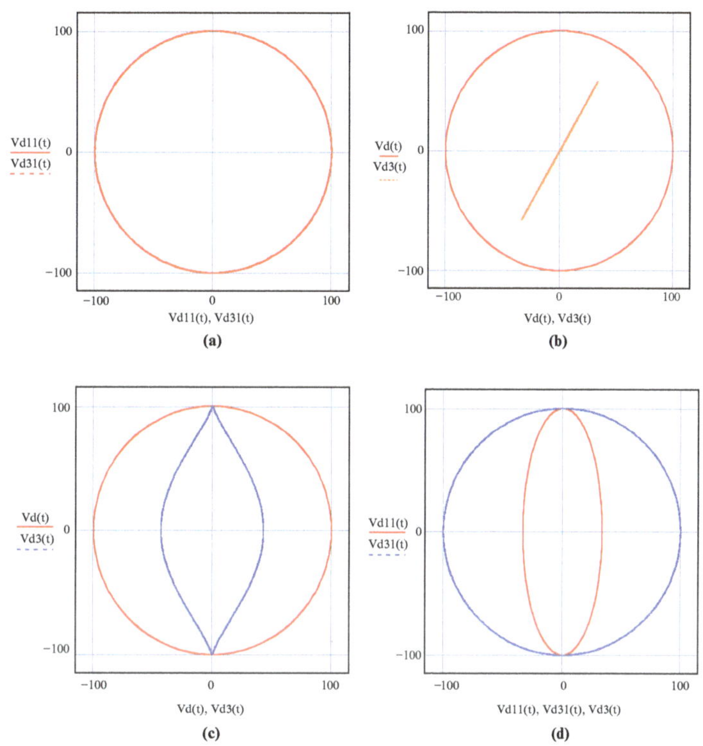

Figure 1.14 Voltage-sourced converter HVDC power limit

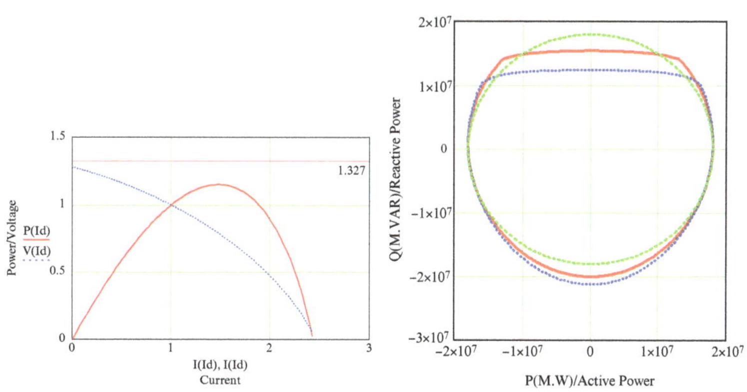

Figure 1.15 Overload characteristics: line-commutated converter (left) and voltage-sourced converter (right)

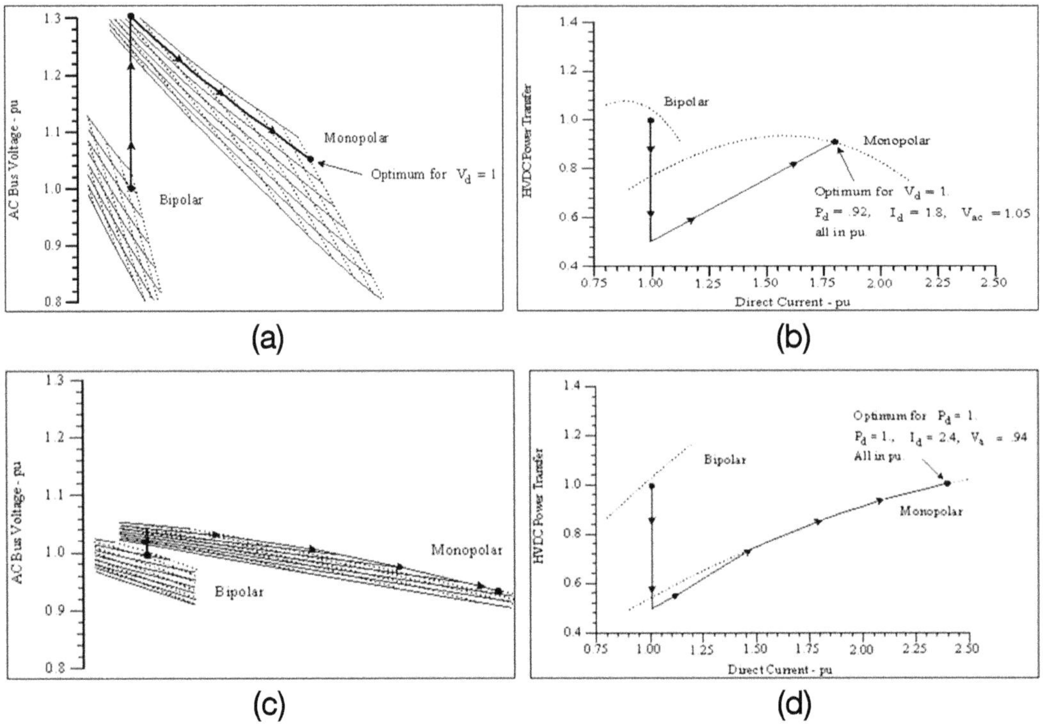

Figure 1.16 Line-commutated converter HVDC overload characteristics; (a) AC Voltage variation (Case 1); (b) Map (Maximum available power) curve (Case 1); (c) AC voltage variation (Case 2); (d) Map (Maximum available power) curve (Case 2)

- **Case 1** shows the overload characteristics of an LCC HVDC system with SCR of 1.8 while Case 2 shows a system with SCR of 6.0. The initial conditions of Case 1 and Case 2 are as follows: AC voltage is 1.05 p.u., operating angle is 18°, reactive power consumption is 0.71 p.u. and DC voltage is 1.0 p.u. Case 1 shows that the overload of the LCC HVDC increases the transient over voltage (TOV) of the AC system to 1.37 p.u. and the firing angle increases by 43° to ensure the stability of the HVDC after the fault (when transferred from bipole mode to monopole). The maximum DC current of the MAP curve is determined to be 1.8 p.u.
- **Case 2** shows that after the failure, the overvoltage of the AC system increases to 1.07 p.u. and the operating angle increases to 20°, increasing the maximum DC current to 2 p.u.

1.5.3 Viewpoint of MIIF

The multi-infeed interaction factor (MIIF) is an indicator of the interaction between HVDCs in cases where multiple HVDC systems are connected to one AC system. It is important to consider MIIF because the CSC HVDC system is highly dependent on the AC system (because the converter consumes reactive power). On the other hand, as VSC HVDC systems independently control the reactive power and the active power, MIIF is less important there, but as the voltage stability is a local problem, MIIF can support the reactive power problem of the nearby CSC HVDC or VSC HVDC system and, therefore, it is necessary to consider MIIF. The following is an example of calculating MIIF in various forms. Figure 1.17 shows the MIIF and SCR calculation method in the case of a combination of several VSC HVDCs. Figure 1.18(a) and (b) shows the combination of CSC HVDCs and VSC HVDCs and Figure 1.18(c) and (d) the case of several CSC HVDCs in combination. In Figures 1.17 and 1.18, the important thing is the reactive power supplying capability. As a VSC HVDC can supply the reactive power, the effective SCR is increased. On the other hand, as CSC HVDCs consumes the reactive power, the effective SCR is reduced. More detailed explanations on the AC/DC interactions of VSC HVDC are included in Chapter 10.

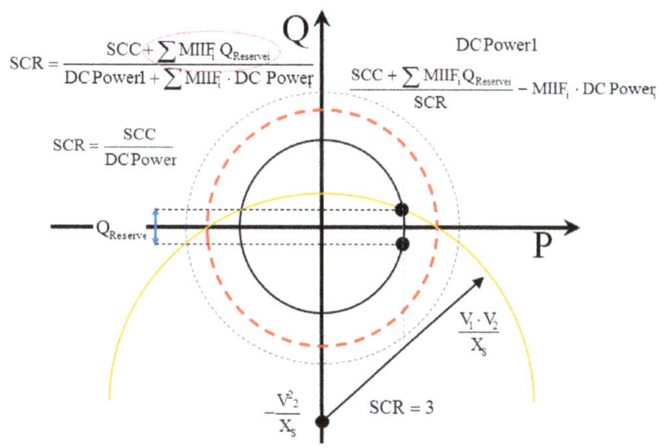

Figure 1.17 Combination of several voltage-sourced converter HVDCs

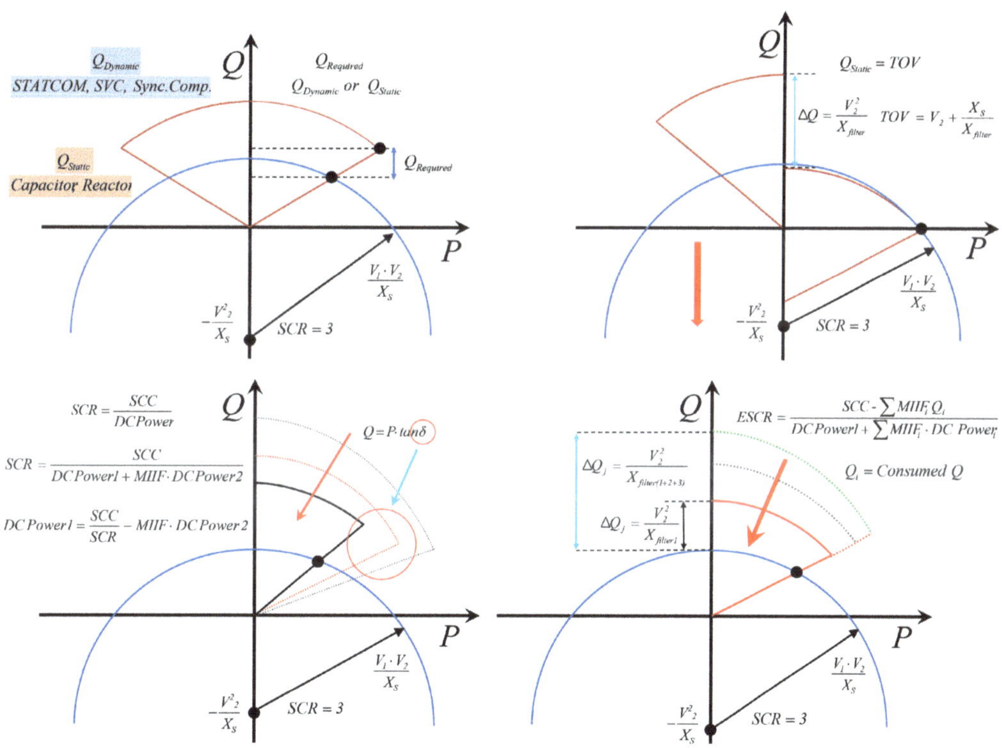

Figure 1.18 Combination of current-sourced converter HVDC and voltage-sourced converter HVDC; (a) CSC HVDC and AC network; (b) CSC HVDC with filter and AC Volitage variation; (c) combination with several CSC HVDCs; (d) combination with several CSC HVDCs and VSC HVDC

1.5.4 CF Characteristics

CSC HVDC, which controls the phase of three phases, causes an instantaneous power interruption in the event of a line voltage drop, line accident, or sudden current increase. This phenomenon is called "Commutation Failure" (CF). Because CFs cause power interruption, TOV, resonance, etc., the impact of these phenomena should be examined from the design stage. To reduce these CFs, the HVDC system employs a gamma control scheme. In a gamma controller, there are two γ measurement methods. The first method is to select the smallest

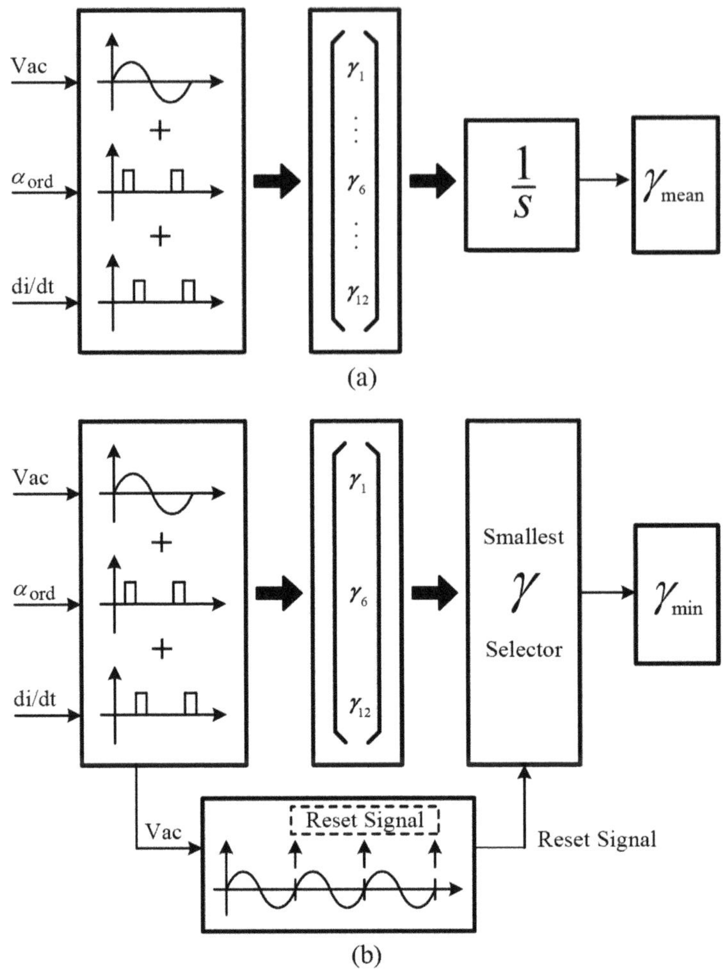

Figure 1.19 γ Measurement methods: (a) Average γ measurement method and (b) Smallest γ measurement method

value among 12 firing pulses per cycle and the second method is to select the average γ value; the measured γ is averaged by a low-pass filter.

These methods have both advantages and disadvantages over each other: The mean gamma measurement ~ method (as shown in Figure 1.19 (a)) in which the measured γ signals are averaged by a low-pass filter, is used as a direct input to the gamma regulator. This method offers the best stability in theory, because there is no delay in the measurement quantity. Conversely, as shown in Figure 1.19's γ measurement methods, the smallest γ value shown in Figure 1.19(b) goes to a feedback signal of the gamma regulator. This method is robust for repeated CFs because of the updated smallest gamma value in one cycle, but it has an inherent delay on increasing the γ of at least one cycle. On the other hand, the VSC HVDC system does not have this phenomenon, so the system design process is simple.

1.5.5 Fault-Blocking Capability

One of the differences between the VSC HVDC and the CSC HVDC systems is their respective fault blocking capability. Although the CSC HVDC system can block the fault current using the thyristor, the VSC HVDC cannot block the fault current because the fault current flows through the diode connected in parallel with the IGBT. The protection schemes of a VSC HVDC system and CSC HVDC systems are different. Figure

1.20(a) shows the protection scheme for the CSC HVDC, and Figures 1.20(b) and (c) for the VSC HVDC. Figure 1.20(b) shows a case where two AC breakers exist on the AC side, and Figure 1.20(c) shows a case where DC breakers exist on the DC side. By comparing the protection scheme of LCC HVDC and that of VSC HVDC, it can be noticed that the protection schemes of both systems are very different. For VSC HVDC systems including Figures 1.20(b) and (c), the fault protection system cannot be implemented through a power electronics approach, which means that the fault-clearing procedure takes much longer for the VSC HVDC system than the LCC HVDC system.

A VSC HVDC systems can be divided into the asymmetry type and the symmetry type according to whether the return cable is grounded or not. The asymmetry type can be configured as a monopole or bipole scheme because the return line is grounded.

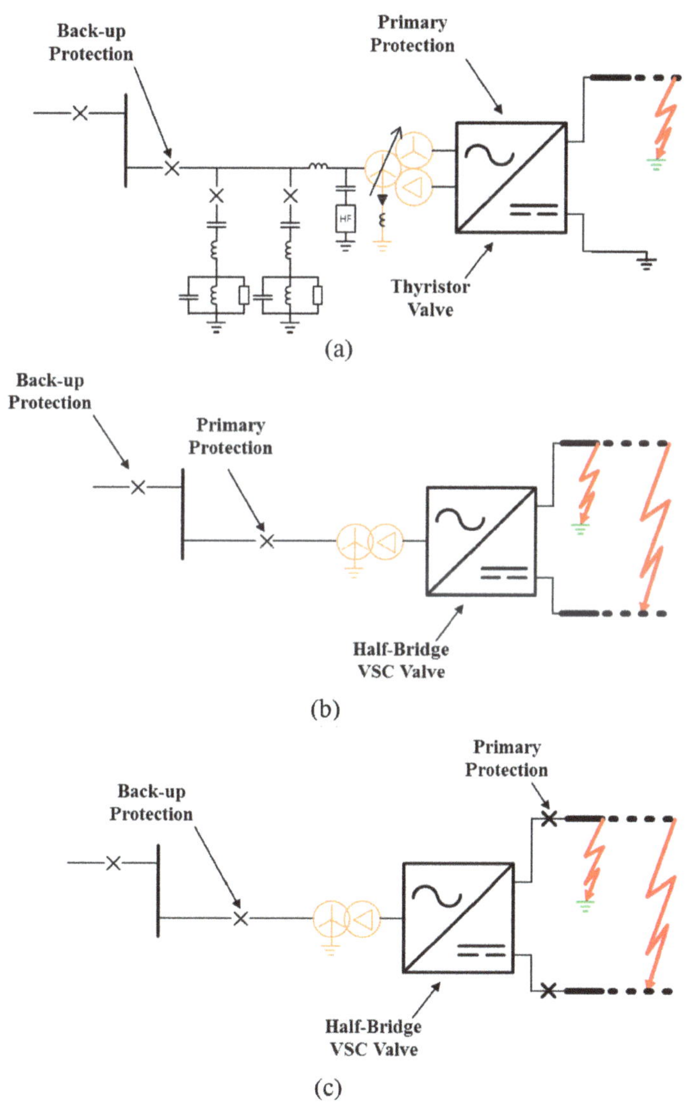

Figure 1.20 HVDC protection scheme; (a) Line-commutated converters use the power electronics as the primary protection; (b) voltage-source converters use the mechanical AC breaker as the primary protection; (c) half-bridge voltage-source converters can use a hybrid DC breaker as the primary protection

In addition, the VSC HVDC system can use a Half-Bridge (HB) type that does not block the fault current and an Full-Bridge (FB) type that has the blocking capability for a fault current. Figure 1.21 shows the protection schemes in case of a DC fault of an HVDC system with asymmetric HVDC and symmetric HVDC systems (a line-to-ground fault or line-to-line fault) when the HVDC system has an HB valve. On the other hand, Figure 1.21 shows the protection scheme when the HVDC system has an FB valve. In Figure 1.21, the main difference between HB HVDC and FB HVDC is protection coordination, that is, the relation between primary protection and back-up protection.

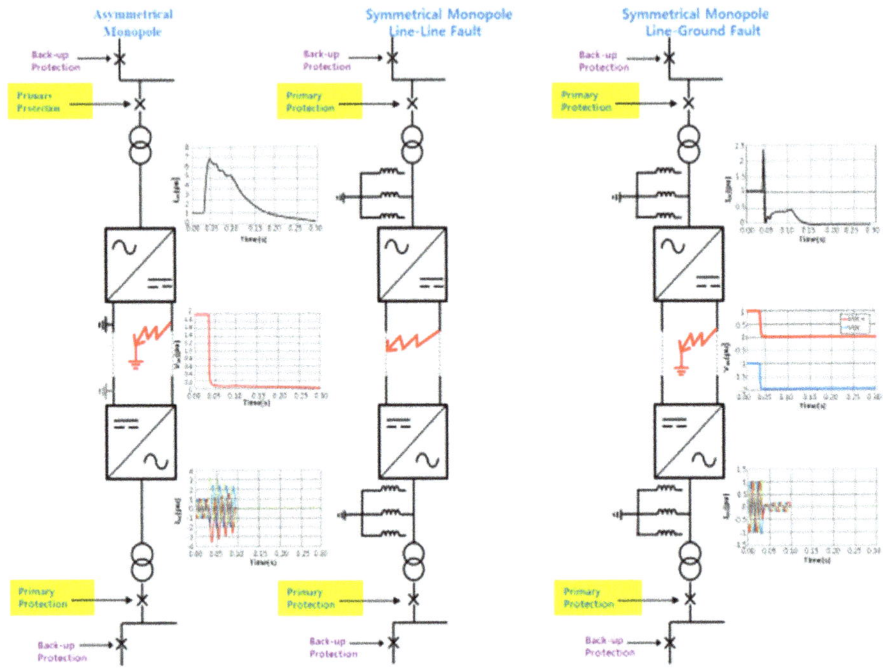

(a)

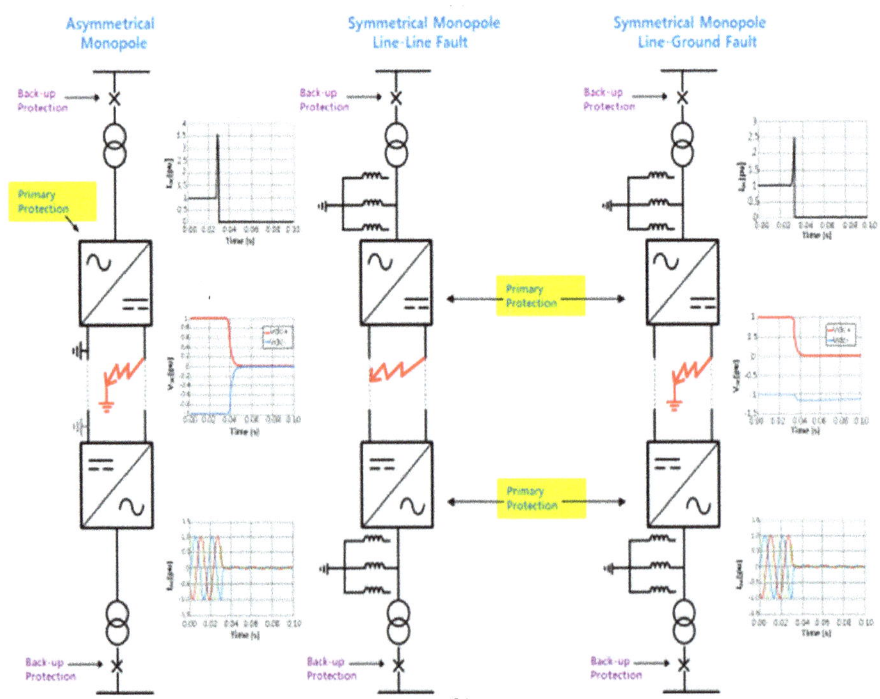

(b)

Figure 1.21 Blocking capability of voltage-sourced converter HVDC system; (a) VSC HVDC (Half Bridge); (b) VSC HVDC (Full Bridge)

1.6 Multi-terminal HVDC (MTDC) Grids and Supergrids

1.6.1 HVDC Development in German System

At the time of writing, HVDCs have become widely used for transmission systems. There are multiple instances where HVDCs provide greater flexibility or functionality, can connect systems that AC cannot, or can transmit more power in a given space than AC can. HVDC is preferred over AC in several cases.

First, if two unsynchronized AC networks or AC networks of differing frequencies (as, e.g. in Japan) need to be connected, HVDC can act as a frequency and phase conversion stage. Examples of this are the HVDC connections between the United Kingdom and Europe, where the two systems operate at a 50 Hz nominal frequency but are not synchronized.

Second, for very long distances, HVDC may be more economical. This is because HVDC lines are cheaper per kilometer than AC, and unlike AC, HVDC lines do not consume reactive power, and are therefore not limited by length or the requirement for periodic reactive power compensation. Moreover, the losses of a DC line are smaller than the losses of an AC line because of high voltages and therefore lower currents. HVDC stations are more expensive than AC stations, so the breakeven point is typically 600–800 km for overhead lines or 50–100 km for cables (which have higher reactive power exchange per kilometer) depending on location, project power, and voltage. Examples of such HVDC connections are many long-distance lines in China, Brazil, Canada, and the United States.

Third, two AC systems may need to be connected without increasing the AC's fault level. The ability of the HVDC system to block AC power flow quickly (in AC timescales) means that power can be fed through the HVDC link into a system, but minimal extra fault current is added to the AC network. Consequently, the AC switchgear doesn't need to be upgraded. Often, this HVDC connection is through back-to-back AC:DC:AC stations on one site. This was one of the advantages of the VSC HVDC for ABB's Mackinac Converter project in Michigan.

Figure 1.22 German planned north–south corridors connections

Fourth, HVDC can transmit more power per given transmission corridor size than AC. Where space is constrained, this may mean that, in the future, HVAC lines may be replaced by HVDC ones. This has been the case in Germany for the Ultranet project. Other examples for this type of HVDC connection are DC links directly into the downtown areas of large cities like New York (Hudson Project) and San Francisco (TBC). Additionally, long and high power HVDC underground or undersea cables are technically and economically feasible, which is not the case for AC cables where the charging current dominates the behavior.

Lastly, VSC HVDCs can provide a variety of power quality support functions. Reactive power support, AC voltage control, and black-start functionality can be provided, all of which again were key factors in ABB's Mackinac project in Michigan. Other functions include: firewalling one AC system so that disturbances do not spread to an adjacent system; providing frequency stabilizing functions and artificial fast frequency response (also called artificial inertia, as implemented in the Caprivi Link project); providing power oscillation damping (such as that which was implemented in the Pacific DC Intertie, INELFE, Barnes *et al.* [12]; stabilizing the AC system (in New Zealand: fault recovery modulation, frequency keeping control, frequency stabilization control, spinning reserve sharing, constant frequency control, Wellington over-frequency brake, and automatic governor control).

Opportunities exist for the integration of the highly flexible VSC HVDC directly as part of smart grid architecture, especially in unbundled markets with a high penetration of renewables, where operators can value new ancillary services to manage variability in power flows. As current and voltage capabilities of VSCs have increased significantly during recent years, additional functionality could be carefully coordinated for most power markets to match the converter's real-power import and export capability.

According to a recent industry report [25], the market is split fairly equally between CSC and VSC technologies. CSC HVDC is presently preferred for very large bulk power transfers. The more compact VSC is most often used when space is at a premium, or when additional services are required at the grid connection point.

Renewables will form an increasing fraction of generation, and transmitting such power to the point of use will be a major driver in, for example, Germany. HVDC connections in Germany bring wind power to shore, and further connections transmit this offshore wind from Northern Germany to industrial centers in Southern Germany (see Figure 1.22). Similarly, LCC HVDC connections, such as the Three Gorges Project in China transmit hydropower to the country's growing mega-cities.

The increase in demand in urban and industrial centers as a result of demographics, lifestyle, and industrial changes will require extra power. As urban space is often constrained and expensive, and utilities do not wish to uprate other utility infrastructure, VSC HVDC and its medium-voltage variant MVDC may become a replacement solution for AC to increase supplied power. They may also be used to connect AC bulk-supply points, reinforcing the network without increasing fault levels.

"Supergrids"—large meshed HVDC networks—have been proposed, and radial networks are starting to be constructed. These would potentially allow transcontinental (or even intercontinental) sharing of resources. For example, in Europe, offshore wind power from the United Kingdom could be shared with Norwegian hydropower and storage, Icelandic thermal energy generation, and solar power from Spain (as well as from other countries and conventional generation). The different demand profiles in the continent's countries could be smoothed out over time, and energy traded, optimizing generation investment and utilization. A number of technical problems remain before this vision can be realized, though, and a common ICT framework would be required to fully integrate HVDC and other smart grid services.

1.6.2 MTDC Grids

HVDC with LCC has largely been a point-to-point solution. Historically, multi-terminal installations have been few and far between, though more recently they have attracted considerable attention as VSC HVDC has developed.

The Hydro-Quebec System designed in the 1980s is often cited as the original HVDC multi-terminal system, based on initial studies for a five-terminal system. The initial point-to-point system was meant to be expanded to this in two stages; in practice, a separate three-terminal link was constructed for phase two, in part because of the consideration that a different vendor might be used. In practice, this three-terminal link has predominantly spent its time operating as a unidirectional system, either transferring hydropower from Radisson to Sandy Pond or Radisson to Nicolet stations.

In 1967, a 200 MW monopolar LCC 200 kV DC link between Italy and Sardinia was established. In 1986–1987, a 50 MW tap was added. This is known as the Sardinia–Corsica–Italy (SACOI) link. However, fast-reversing switches were required to allow rectifier/inverter operation of the Corsican station.

The 2016 Northeast Agra LCC HVDC link is a ±800 kV 6000 MW four-terminal three-converter station. It was designed to supply hydropower from Northeast India to the city of Agra [25].

A collaborative, government-sponsored project to build a back-to-back multi-terminal VSC HVDC station at the Shin-Shinano substation in Tokyo was undertaken by Toshiba, Hitachi, and Mitsubishi Electric in the 1990s. The 300 MW back-to-back station used gate turn-off thyristors (GTOs).

The first VSC HVDC multi-terminal network systems are the Chinese Nan'ao Island (2013) and Zhoushan (2014) VSC HVDC systems (Figure 1.23). The Nan'ao Island project is a ±160 kV three-terminal (200, 150, and 50 MW) collaboration between Rongxin Power Electronic, NR-Electric, and XiDian. Zhoushan is a five-terminal (400, 300, and three times 100 MW) ±200 kV system built by C-EPRI and NR Electric. Both systems are radial networks; as of yet no meshed HVDC grids have been constructed.

1.6.3 Supergrids

In Europe, the European Union (EU) began to link the European region in 1951 with the Union of the Coordination for the Transmission of Electricity (UCTE). Currently, the European Network of Transmission System Operators for Electricity (ENTSO-E) constitutes the European system. ENTSO-E is a member of 42 transmission system operators in 35 countries in Europe. ENTSO-E has a power generation capacity of 1023 GW, and TSOs within the jurisdiction have a capacity of 423 TWh.

The operation of European electricity markets varies by country and region. Most electric power deals are made through bilateral contracts rather than mandatory pools. In addition, the principle of ensuring non-discriminatory transmission network connections and a free bidirectional power trading market is in line with

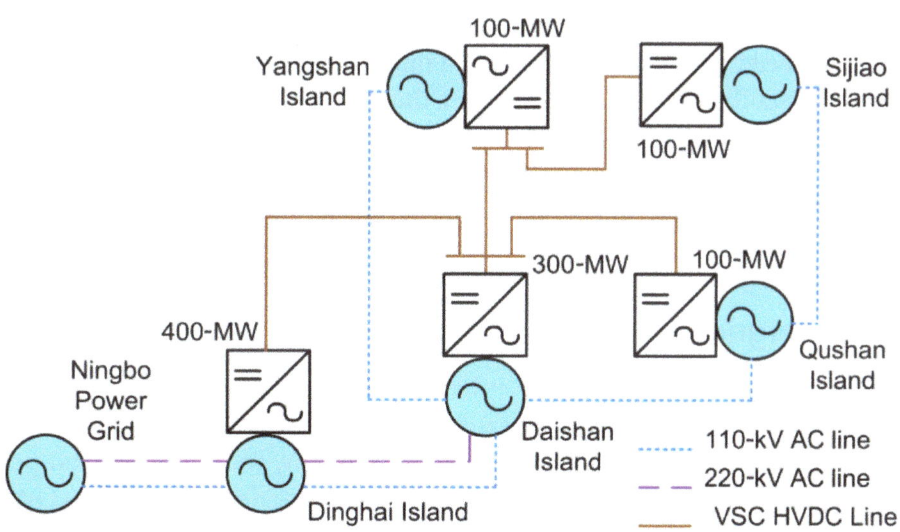

Figure 1.23 Zhoushan five-terminal voltage-sourced converter HVDC network

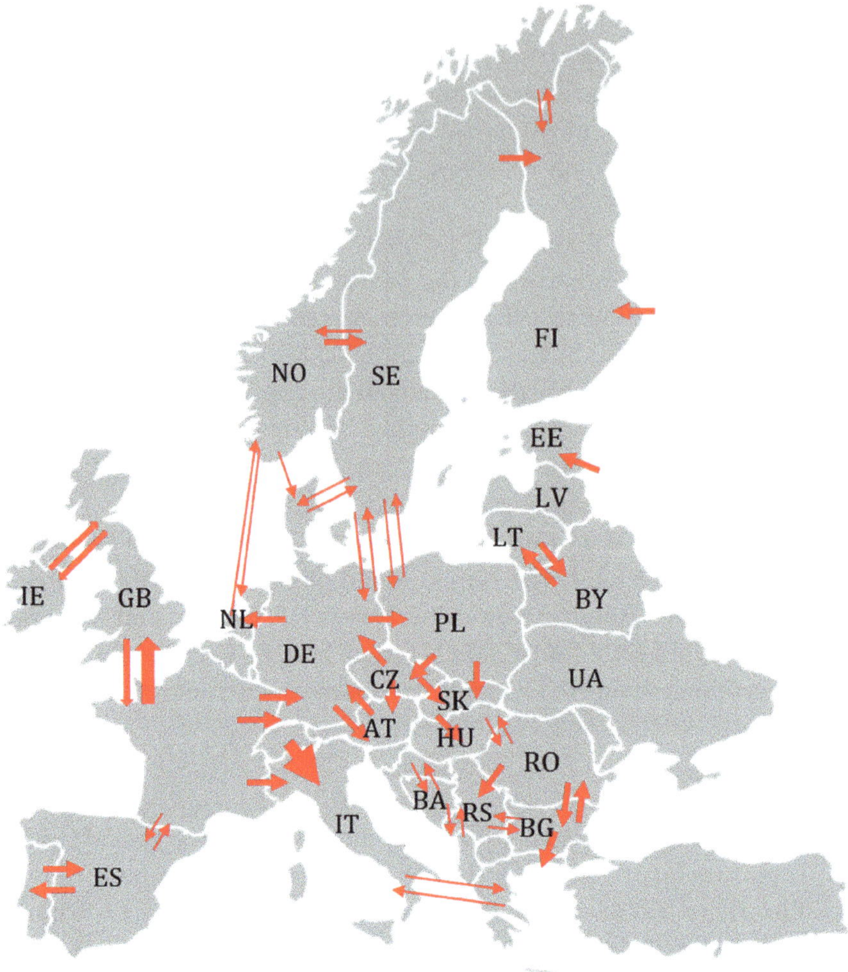

Figure 1.24 Plan of European supergrids

the principle, and it is striving to create conditions for grid operators to freely cross country borders to conduct electricity business. The current status of grid linkage in Europe is shown in Figure 1.24.

Europe and Russia are mainly engaged in bilateral contracts, and 60% of the total electricity is exported to Finland. Russia's export price to China is lower than that of domestic consumers.

Since 1951, Europe has been promoting the European integration network in order to use energy resources reasonably and jointly. Currently, there are 34 countries in the network, including Britain, France, and Germany. In order to maximize the use of renewable energy such as wind power and solar power, and to secure the reliability of the power supply and strengthen cooperation between countries in the network, intercontinental cooperation projects utilizing HVDC technology are actively underway. The proposed supergrid in Europe is as follows.

1.6.4 North Sea Supergrid

The North Sea SuperGrid, as shown in Figure 1.25, is being promoted by the North Sea National Offshore Grid Initiative (NSCOGI). In order to supply European renewable power generation facilities such as offshore wind power in England and Germany, hydropower in Northern Germany, and onshore wind power in Northern Germany, to customers around Europe, the first step is to promote the adoption of the 30 GW by more parts of the European power market before ultimately raising it, 500 GW.

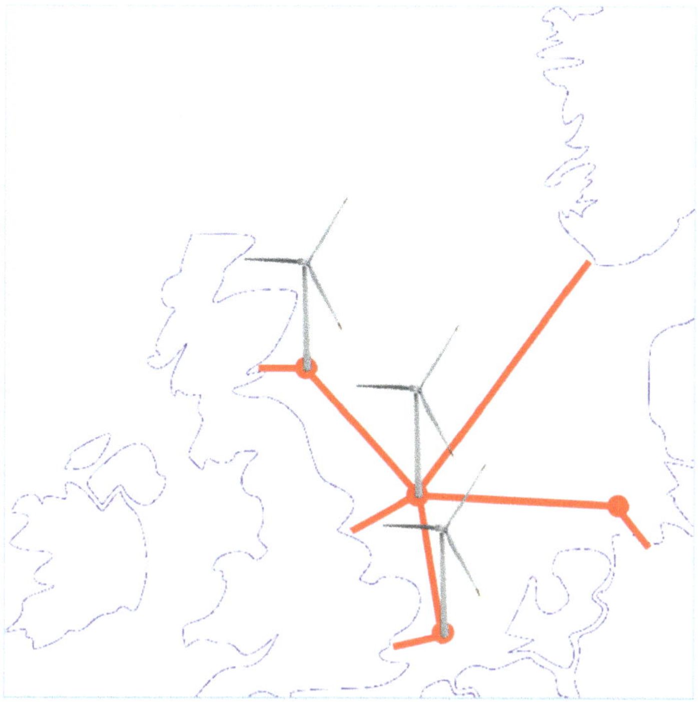

Figure 1.25 Concept of North Sea European supergrids

NSCOGI is a collaborative organization for an integrated marine energy grid that links wind energy and other renewable energy sources to the European North Sea between EU member states and Norway. NSCOGI is divided into three working groups. Working Group 1 studies the composition of the entire grid. Working Group 2 studies markets and regulations. Working Group 3 studies planning and licensing. These working groups are operated under a committee called the Program Board. The structure of this NSCOGI is illustrated in Figure 1.26.

The Nordic Supergrid as shown in Figure 1.28 was proposed by the European Commission in a second strategic energy review published in November 2008. This proposal is considered one of the six priority energy infrastructure projects of the EU. According to the European Commission, the Nordic Supergrid will be one of the key elements of the future European Supergrid. The Declaration of Northern Europe's Supergrid policy of the North Sea National Maritime Grid Planning Organization was signed by the EU Energy Commission in December 2009. This declaration was proclaimed by Germany, England, France, Denmark, Sweden, the Netherlands, Belgium, Ireland, and Luxembourg. The European Commission published the Nordic Supergrid Blueprint in 2010. The Nordic Supergrid transmits electricity using HVDC cables. This allows electricity to be sold and exchanged in all applicable countries. This will optimize energy production and make the entire system less sensitive to climate change. Norwegian hydropower plants will act as "giant batteries" to store the produced power and to discharge the day's power when peak time power or wind power is weak. The proposed HVDC transmission connection between Norway and England is considered to be an integral part of this project. The current location of the offshore wind power in the North Sea is shown in Figure 1.27.

The Nordic system and other countries are interconnected by several DC connections. Jutland in Denmark has several DC connections from Sweden and Norway, and HVDC connections from Sweden to Germany.

The Finnish grid is interconnected with the Swedish grid via two 400 kV AC connections, linked to a 220 kV AC from Norway. The Rauma in Finland and Dannebo in Sweden were completed in December 2011 with a transmission capacity of 800 MW connected to the HVDC Fenno-Skan No. 1 and 2. Currently, Finland and Svenska Kraftnät are conducting a second submarine HVDC link between Rauma and Finnböle in Sweden.

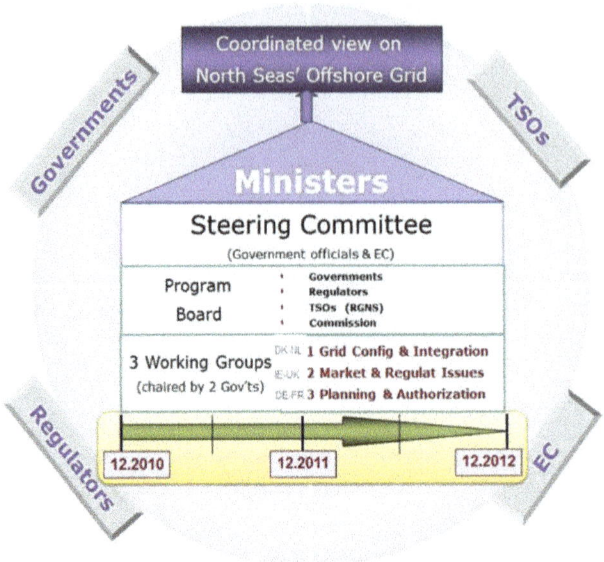

Figure 1.26 Structure of North Sea National Offshore Grid Initiative (NSCOGI)

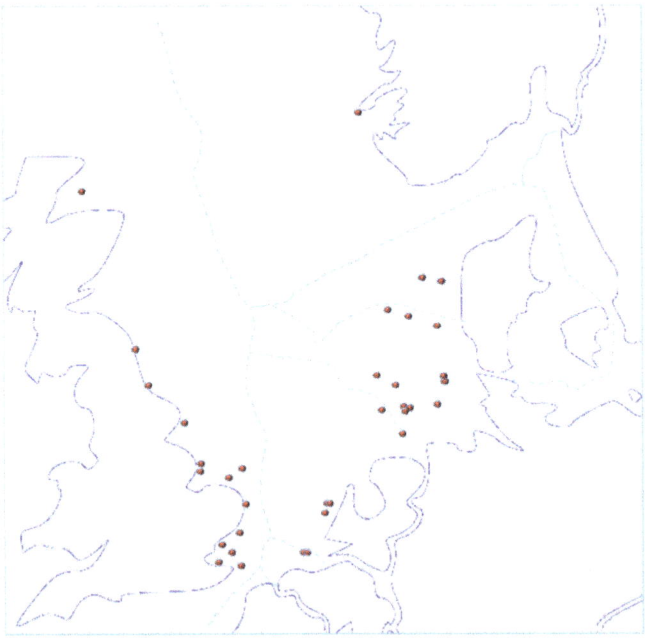

Figure 1.27 Offshore wind farm at North Sea

The Russian and Estonian grids have an Estlink No. 2 HVDC link with a DC link with the Nordic grid as shown in figure 1.28, not with the AC link. The Estlink No. 2 HVDC link has a capacity of 650 MW and was completed in early 2014. In addition, it has three 400 kV transmission links between Finland and Russia's Vyborg, linking Ivalo and Imatra in Finland to 100 kV of power, and through these interconnected lines, Russia's hydropower generation is supplied to the Finnish grid in a stable manner.

The station network is connected to a national transmission grid and also typically to a local distribution grid of one or more 110 kV lines, which are either directly connected to the national grid or are using grid services

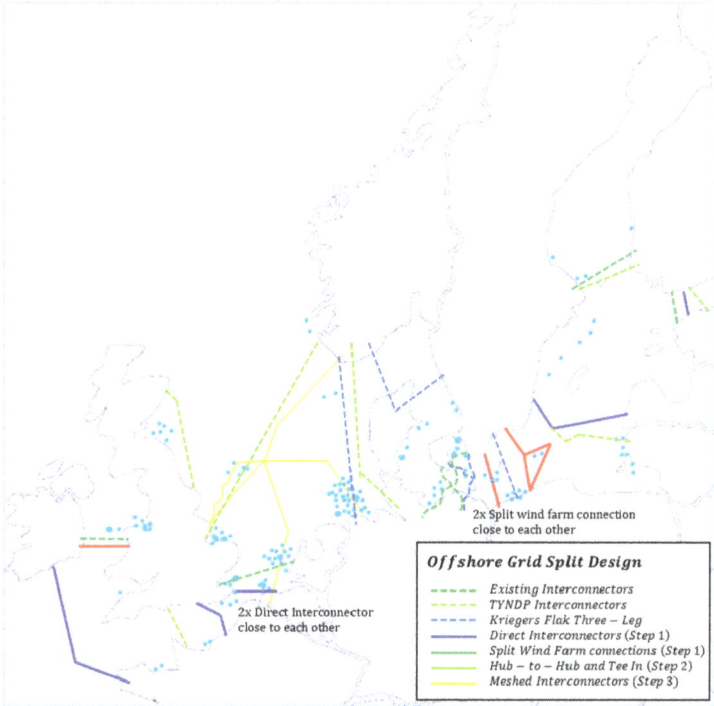

Figure 1.28 Connection of the Nordic Supergrid

through a local network. Distribution networks operate at voltage levels between 0.4 and 110 kV, and household, industrial, trade, service, and other loads (e.g., agriculture) are sometimes connected directly to distribution or local networks, or to the national grid. The plant is connected to the distribution or local network, or to the national grid in each case.

Currently, there is a link between Russia and Finland. The Russian–Finnish linkage as shown in Figure 1.29 is structured to allow only the transmission of electricity from Russia to Finland because of the low electricity prices of Russia, which is the structure in which Finland imports Russian electricity. However, as the gas prices in Russia rose, the price of Russia's electricity grew to higher than the Nord Pool market price. Electricity Consumption of about 1300 MW from Russia by Norway continued to decrease from 2012; and on March 6, 2013, Russia earned no electricity income from Norway for the first time since 1981.

According to Finland's 10-year development plan announced in 2012, Finland is expected to export 350 MW from Russia. These exports are because of Russia's regulatory changes and changes in law due to the power back-up. The tariffs on Russia's capacity payment will also be obstacles to bilateral contracts. As mentioned above, 100 MW of the combined capacity is utilized for frequency control of the Finnish power system.

Southern European Supergrid

In Southern Europe, a Southern European Supergrid, known as the Desertec project, is planned. In order to utilize the solar power sources of the Sahara Desert, it is necessary to connect Southern Europe to the Maglev system. When constructing solar and photovoltaic power in the Sahara Desert, a 1% supply area is expected to become possible. In the Sahara Desert, the Desertec project is intended to supply 470 GW by 2050 to the Mediterranean. The total investment is planned to reach $772.7 billion.

The Desertec concept was developed by a network of international politicians, scholars, and economists, called TREC. Research institutes for renewable energy in Morocco, Algeria, Libya, Egypt, Jordan, Yemen, and the German Aerospace Center (DLR) participated in this supergrid study. The basic research on the

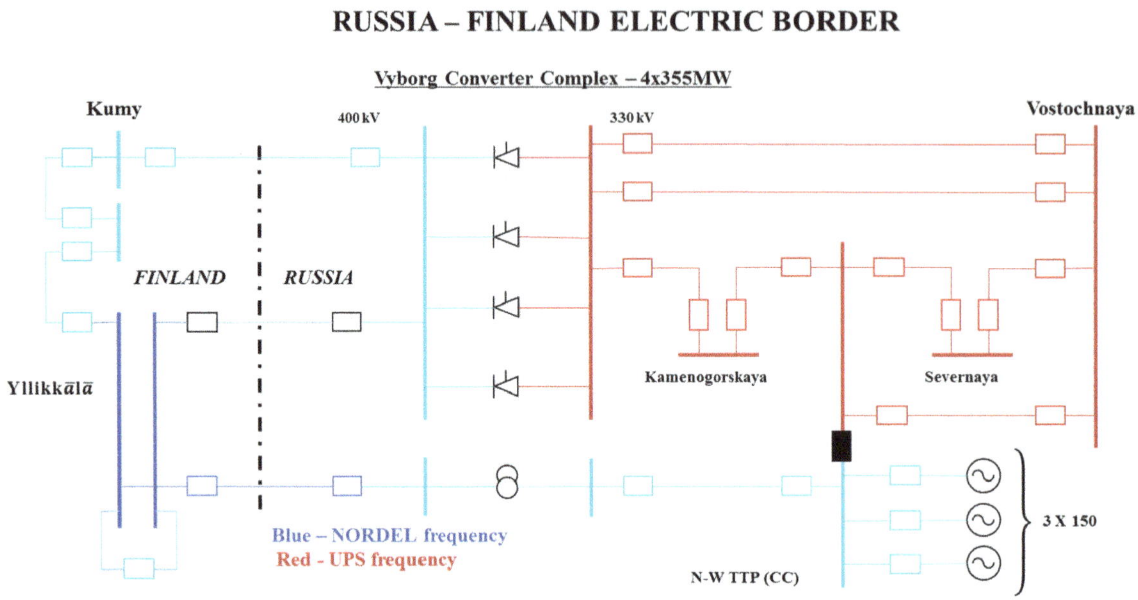

Figure 1.29 Russia–Finland electric border

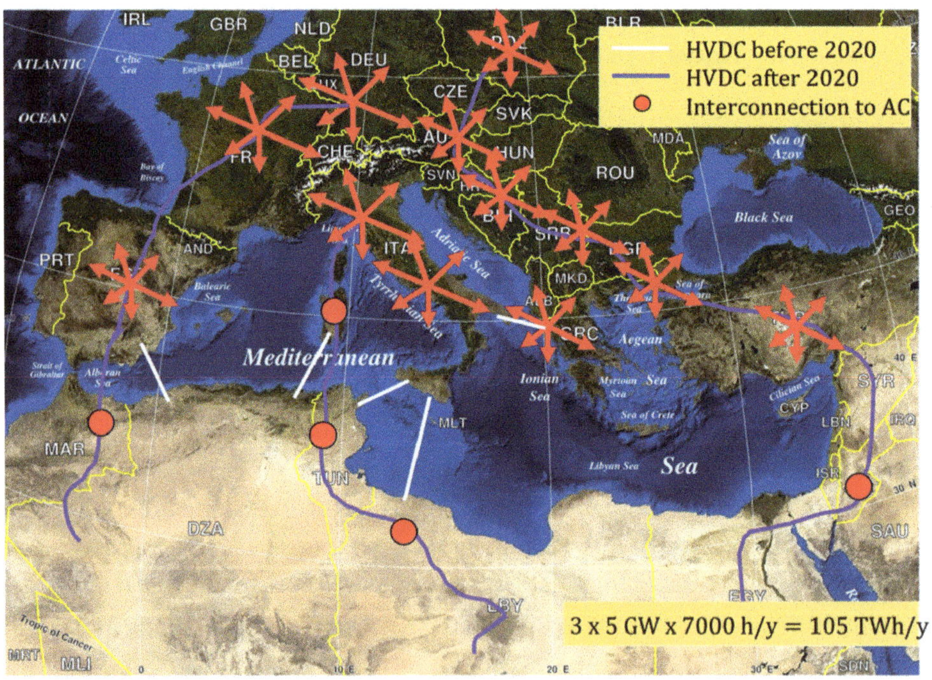

Figure 1.30 Concept of Desertec project

Desertec Supergrid was led by Dr Franz Treib of the DLR Institute of Thermodynamics. The results of the study conducted by the German Federal Agency for Environment, Nature Conservation and Safety are shown in the Figure 1.30.

Studies have shown that transmission losses between North Africa and Europe are expected to be from 10% to 15%. In the deserts of North Africa and the Middle East, solar energy is so abundant that, even considering this transmission loss, it is still advantageous to build solar power plants in North Africa and the Middle East rather than in Europe. As a result of the study, the Sahara Desert was chosen as the best place for solar power generation. In the best case, 80–97% of daylight hours were exposed to bright sunlight.

The annual average solar radiation, which represents the total solar radiation received over a period of time in the region, is 2500 kWh per square meter per year and has risen to 300 kWh in the best case. If the entire desert is covered with solar panels, the annual electricity production reaches up to 130,000 TWh. The size of the desert is very expansive with an area of 9,000,000 km^2 and can support large solar power plants without negatively affecting the local population. Silicon, which is essential for solar panel production, can be obtained in sandy desert areas. However, extreme heat or rough sand winds can have a negative impact on solar power generation.

However, because of political obstacles between Europe, North Africa, and the Middle East, by early 2015, 47 out of 50 consortium shareholders had left the consortium, and the success of the project is uncertain.

Supergrid of the United States

North American power grid connections as shown in Figure 1.31 have been built in Vancouver, Canada, the Atlantic coast of Canada, New York, Texas, and Mexico, and this grid linkage generates revenue using an average of four times the power price gap across the region, as well as secures a stable supply of electric power by avoiding the costly and time-consuming construction of power facilities.

HVDC and AC methods were used as the linking methods. The Quebec and New England connection was about 2 GW, and it was promoted as a project to supply the hydropower of the James Bay area through to Montreal and Boston. The US–Mexico connection was a 150 MW HVDC designed to generate profits through the Peak share. In addition, the Inter Mountain Power project in the United States promoted a 2 GW project to connect California to the HVDC system in Utah, and a project to link HVDC to Oregon–California–Nevada through the Pacific Intertie project at approximately 3 GW. As a new business, we are discussing the TRES-AMIGAS project, which links three asynchronous systems between East and West to Texas with HVDC and superconducting cables. Figure 1.31 shows the HVDC-linked power system in the United States.

Supergrid of Africa

Africa is promoting the Grand Inga project, as shown in Figure 1.32~ which is the construction of the Inga dam in the Democratic Republic of Congo (DR Congo), which has been investigated as the world's largest hydropower resource. It will utilize hydropower of about 40 GW to supply eletricity from Cairo in Egypt to Johannesburg in South Africa.

Inga Falls is 25 miles from the mouth of the Congo River. It is the largest waterfall by volume in the world. The 40 GW hydropower plant to be installed at Inga Falls is the largest in the world and more than twice as large as the Three Gorges Dam in China. Currently, the waterfalls are equipped with Inga I and Inga II dams, with a capacity of 351 MW and 1424 MW, respectively. Currently, these two dams are operating at a capacity of 40% of their installed capacity because of maintenance problems.

The Grand Inga project will proceed in three stages. The first step is the reconstruction of the Inga I and Inga II dams. The Inga I and Inga II dams, built in 1972 and 1982, respectively, were costly and useless to the DR Congo. These dams accounted for a significant portion of the debt burden on the DR Congo and failed to perform for several years because of improper maintenance. The World Bank has proposed a project to restore its current capabilities at a cost of about $500 million.

The second step is to build Inga III, with a capacity of 4300–5000 MW, at an estimated cost of $8 billion. The third step is to build the Inga IV, known as the Grand Ying, at a cost of $8 billion. If the project is completed, it will be able to supply power to 500 million Africans who are not currently receiving electricity. Figure 1.32 shows the Inga project.

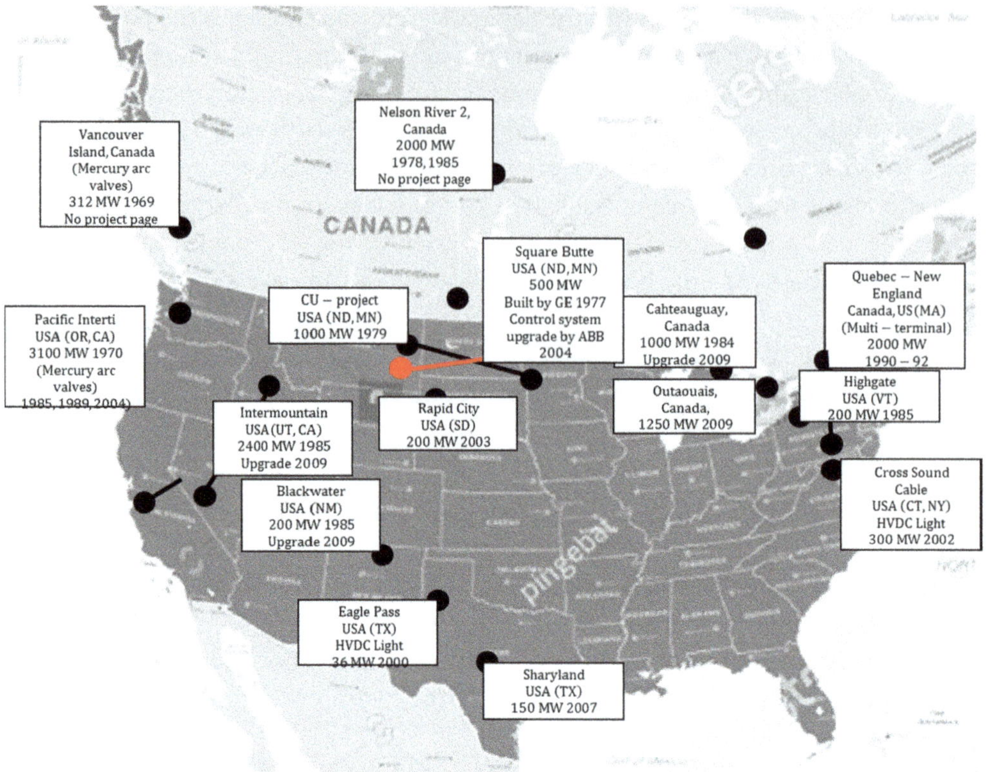

Figure 1.31 Concept of the HVDC-linked power system in the United States [36]

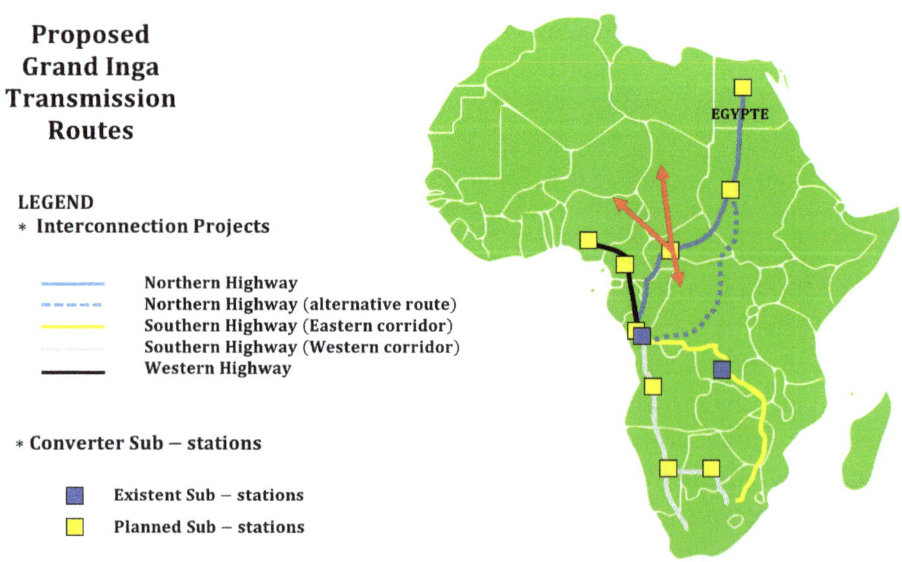

Figure 1.32 Concept of grand Inga transmission routes [36]

References

1. C. Kim, *HVDC Transmission: Power Conversion Applications in Power Systems,* John Wiley & Sons, Hoboken, New Jersey, USA, 2019.
2. P. Kundur, *Power System Stability and Control*, McGraw-Hill, New York, 1996.
3. L. de Andrade and T. P. de Leao, "A brief history of direct current in electrical power systems," in *2012 Thrid IEEE HISTory of Electro-Technology Conference (HISTELCON)*, Pavia, Italy, 2012.
4. M. P. Bahrman, "HVDC transmission overview," in *Transmission and Distribution Conference and Exposition IEEE/PES*, Chicago, IL, USA, 2008.
5. O. E. Oni, I. E. Davidson, and K. N. I. Mbangula, "A review of LCC-HVDC ans VSC-HVDC technologies and applications," in *IEEE 16th International Conference on Environment and Electrical Engineering*, Florence, Italy, 2016.
6. G. Li, et al., "Feasibility and reliability analysis of LCC DC grids and LCC/VSC hybrid DC grids," *IEEE Access*, vol. 7, pp. 22445–22456, Feb. 2019.
7. K. Meah and S. Ula, "Comparative evaluation of HVDC and HVAC transmission systems," in *2007 IEEE Power Engineering Society General Meeting*, Tampa, FL, 2007.
8. C. K. Kim, S. Y. Sim, and J. W. Kang, "Comparison of characteristics between VSC HVDC and LCC HVDC," in *IEEE International Conference on Power Electronics and Drive Systems 2019*, Toulouse, France, Jul. 2019, pp. 9–12.
9. J. Mooney and B. K. Johnson, "HVDC transmission and integration into an AC grid," in *Protection, Automation and Control World*, Sep. 2016.
10. D. Van Hertem, M. Ghandhari, "VSC multi-terminal HVDC for the European supergrid: Obstacles," *Renew Sustai Energy Rev.,* vol. 14, pp. 3156–3163, (2010).
11. Y. Zhang, J. Ravishankar, J. Fletcher, R. Li, and M. Han "Review of modular multilevel converter based multi-terminal HVDC systems for offshore wind power transmission," *Renew. Sustain. Energy Rev.,* vol. 61, pp. 572–586, 2016 .
12. M. Barnes, et al., "HVDC systems in smart grids," *Proceedings of the IEEE.*, vol. 105, no. 11, pp. 2082–2098, Nov. 2017.
13. N. Flourentzou, V. G. Agelidis, and D. G. Demetriades, "VSC-Based HVDC power transmission systems: An overview," *IEEE Trans. Power Electron.*, vol. 24, no. 3, pp. 592–602, March 2009.
14. R. Shah, et al., "Stability and control of mixed AC–DC systems with VSC-HVDC: A review," IET Gener. *Transm. Dis.*, vol. 12, no. 10, pp. 2207–2219, May 2018.
15. J. Setreus and L. Bertling, "Introduction to HVDC technology for reliable electrical power systems," in *Proceedings of the 10th International Conference on Probablistic Methods Applied to Power Systems*, Rincon, 2008.
16. F. Beg and R. Moberly, "Supergrid modeling and operation and control of multi-terminal DC grids for the deployment of a meshed HVDC grid in South Asia," *Int. J. Elect Comput Eng.*, vol. 8, no. 4, pp. 703–712, 2014.
17. D. V. Hertem, O. Gomis-Bellmunt, and J. Liang, *HVDC Grids: For Offshore and Supergrid of the Future,* Wiley-IEEE Press, 2016.
18. N. Flourentzou, V. G. Agelidis, and G. D.Demetriades, "VSC-Based HVDC power transmission systems: An overview," IEEE Trans. Power Electron., vol. 24, no. 3, pp. 592–602, March 2009.
19. ABB, "The birth of commercial HVDC," ABB, 9 7 2014. [Online]. Available at http://www.abb.com/caw p/seitp202/D378AE4401B1E1C8C1257CBA0047ED60.aspx.
20. SIEMENS, "High Voltage Direct Current Transmission–Proven Technology for Power Exchange." [Online]. Available at: www.siemens.com/energy/hvdc.
21. W. Lars, "Bulk power transmission at extra high voltages, a comparison between transmission lines for HVDC at voltages above 600 kV DC and 800kV, AC Indian J. power River Valley Dev., pp. 1–10, 2011.
22. D. Lauria, G. Mazzanti, and S. Quaia, "Comparison between single-circuit overhead transmission lines based online costs and loadability," in *International Symposium on Power Electronics Power Electronics, Electrical Drives, Automation and Motion*, Sorrento, 2012.
23. R. T. Pinto and P. Bauer, "The role of modularity inside the North Sea transnational grid project: Modular concepts for the construction and operation of large offshore grids," in *Proceedings of the Renewable*, Milan, Italy, June 2011.
24. G. Schmidt, B. Fiegl, and S. Kolbeck, "HVDC transmission and the environment," *Power Eng. J.*, vol. 10, no. 5, pp. 204–210, Oct. 1996.
25. R. Leelaruji, et al., "Availability assessment of the HVDC converter transformer system," in *Proceedings of the 10th International Conference on Probablistic Methods Applied to Power Systems*, Rincon, 2008.
26. D. Westerman, et al., *Voltage Source Converter (VSC) HVDC for Power Transmission–Economic Aspects and Comparison with other AC and DC Technologies*, Cigré, Paris, 2012.

27. J. Vobecký, et al., “Silicon thyristors for ultrahigh power (GW) applications,” *IEEE Trans. Electron Devices.*, vol. 64, no. 3, pp. 1–9, Mar. 2017.
28. A. Nami, et al., “Modular multilevel converters for HVDC applications: Review on converter cells and functionalities,” *IEEE Trans. Power Electron.*, vol. 30, no. 1, pp. 18–36, Jan. 2015.
29. R. Marquardt, “Modular multilevel converter impact on future applications and semiconductors,” in Bad Nauheim (Ed.), *Bauelemente der Leistungselektronik*, Bad Nauheim, 2017.
30. H. K. Müller, et al., “The need for a common standard for voltage levels of HVDC VSC technology,” *Energy Policy*, vol. 63, pp. 244–251, 2013.

2

IGBT Device

2.1 IGBT History [1]–[4]

Figure 2.1 shows the applications of the system according to switching and the capacity of semiconductor devices in power electronic systems. As shown in Figure 2.1, thyristors are used in current source converter HVDC systems (CSC HVDC systems), MOSFET in VCR and switching power supply, etc., as well as IGBTs in industrial motor control or UPS. The general versatility of these IGBTs is also applied to voltage source converter HVDCs (VSC HVDCs), and all currently commercialized VSC HVDC systems use IGBTs.

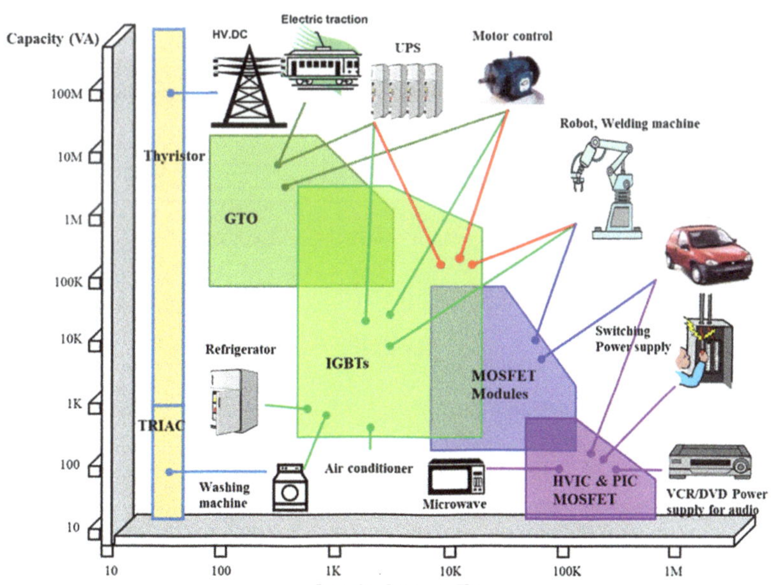

Figure 2.1 Classification according to switching and capacity of semiconductors with industrial appliances

2.1.1. Advent of IGBT

In the 1960s, the Silicon Controlled Rectifier (SCR) became a superior alternative to the mercury arc rectifier and was applied to many power electronics apparatuses, such as the AC drives and home appliances.

Additionally, the improvement of the SCR led to the successful development of the electric power converter. In the 1970s with the development of **P**ulse **W**idth **M**odulation (PWM) control topology, "self-turn-off" devices, which featured fast switching and a high withstanding capability, were developed. That is, **G**ate **T**urn-**O**ff (GTO) thyristors (which can easily handle high voltage and high currents) and the planar-type Darlington bipolar power transistor with multi-diffusion layers were developed. These devices can not only handle high currents by a small base current control but also have a low forward voltage drop.

Because of the relatively wide reverse-biased safe operation area (RBSOA) and high short-circuit withstanding capability of the bipolar power transistors, they are very effective in enabling the electric power converter to reduce power dissipation, even while utilizing high-speed switching. Alongside these device developments , the evoloution of package technology, which enables several transistor devices to be implemented and connected to each other in the H-bridge or 3-phase bridge configuration in one module, was developed. It also coincided with the successful development of bipolar transistor power modules.

In the early 1980s, GTO thyristors and bipolar power transistors were still the main devices used for medium and high-power electronics applications. However, in the case of low-voltage applications, MOS-gated devices with fast-switching abilities were required for PWM control because of further increase in carrier (or sampling) frequency. Although the MOSFET offered the advantage of low control power and fast switching, the sharp increase in its On-Resistance and increase of the breakdown voltage resulted in poor power handling capability when compared with the bipolar power transistor. In order to apply higher voltage systems, a new class of power devices based on a combination of bipolar and MOS physics began to emerge.

It was this reason that led to the development of the IGBT. Yamagami *et al.* introduced the basic idea of a wide base P–N–P transistor driven by n-MOSFET that then became the IGBT from the end of the 1970s to the early 1980s as shown in Figure 2.2. Baliga measured the results of a vertical MOS-gated thyristor structure, which exhibited a current saturation under low-gate-bias conditions. This means that the MOS-gated thyristor

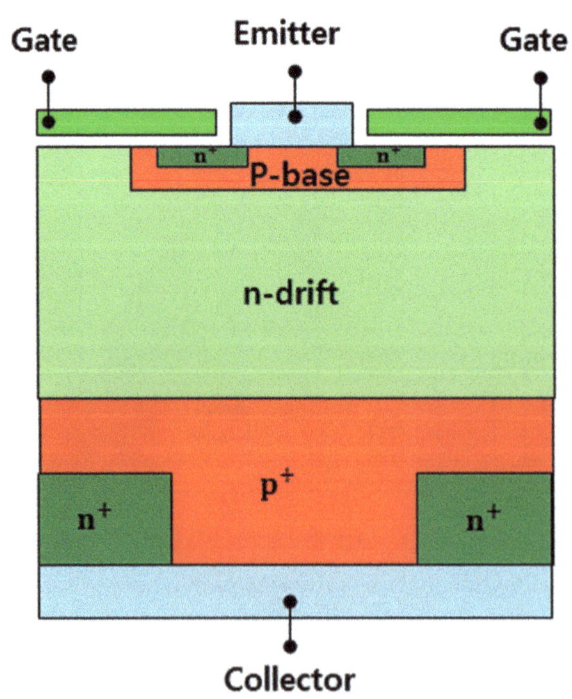

Figure 2.2 Cross section of wide base P–N–P transistor structure driven by N-MOSFET presented in 1972

operates as a wide base P–N–P transistor under this condition. Plummer also exhibited a similar MOS-gated thyristor structure with almost the same operation mechanism. Becke et al. introduced the idea of a new power MOSFET structure with a P-anode region. In this patent, Becke et al. announced that this device does not experience latch-up under any operation conditions. This means that the device can operate as a P–N–P transistor anytime. Based on this characteristic, Russell et al. published the results of this new power MOSFET as a "conductivity-modulated FET" in 1983. Baliga et al. also presented results, though they were fabricated, of the Insulated-Gate Rectifier (IGR), as shown in Figure 2.3. In this paper, Baliga et al. said that a parasitic thyristor action should be suppressed during operation; so, although this device was introduced as a rectifier, it was clear that the device was designed as an MOS-gated bipolar transistor.

Unfortunately, there was still one fatal fault for practical use of these devices: their poor ability to withstand latch-up. Nakagawa successfully overcame this crucial barrier with the so-called "Non-Latch-Up" IGBT structure, which has a surface structure of cell-type N+ sources and high impurity shallow P+ regions, as shown in Figure 2.4. This was the true birth of the IGBT as an applicable power device for practical use. As a result, this new type of MOS-gated bipolar device was successfully delivered to the market for the first time in 1985.

2.1.2. Evolution of the IGBT

The IGR was the first successfully fabricated MOS-gated bipolar transistor; however, its switching speed was too slow, and its RBSOA and short circuit SOA were too small to be an alternative to the power bipolar transistor. Nakagawa successfully solved the SOA problems with the Non-Latch-Up structure development described earlier. From the perspective of improving switching speed, Goodmann et al. proposed a new type of IGBT with an N+ buffer layer in 1983. By applying the N+ buffer layer, the IGBT could attain faster-switching speeds through lifetime control processes with little sacrifice of the forward voltage drop. Technological innovations were required for the IGBT to have breakthroughs in the tradeoff relations of low forward voltage drop,

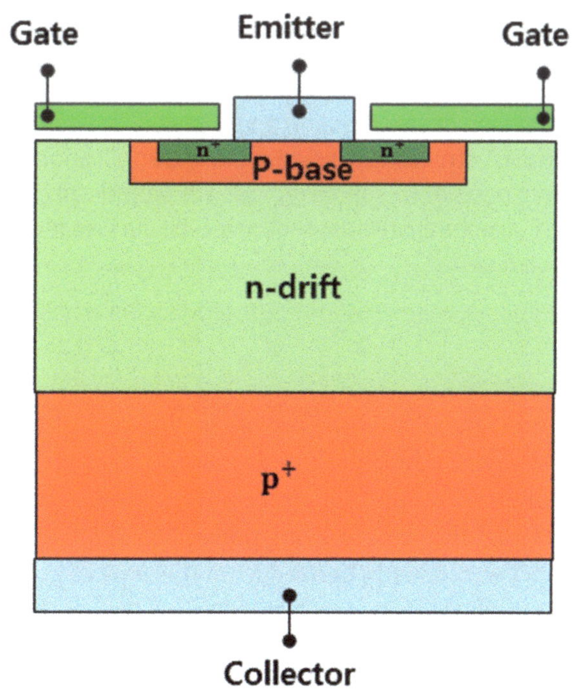

Figure 2.3 Cross section of basic insulated-gate rectifier structure

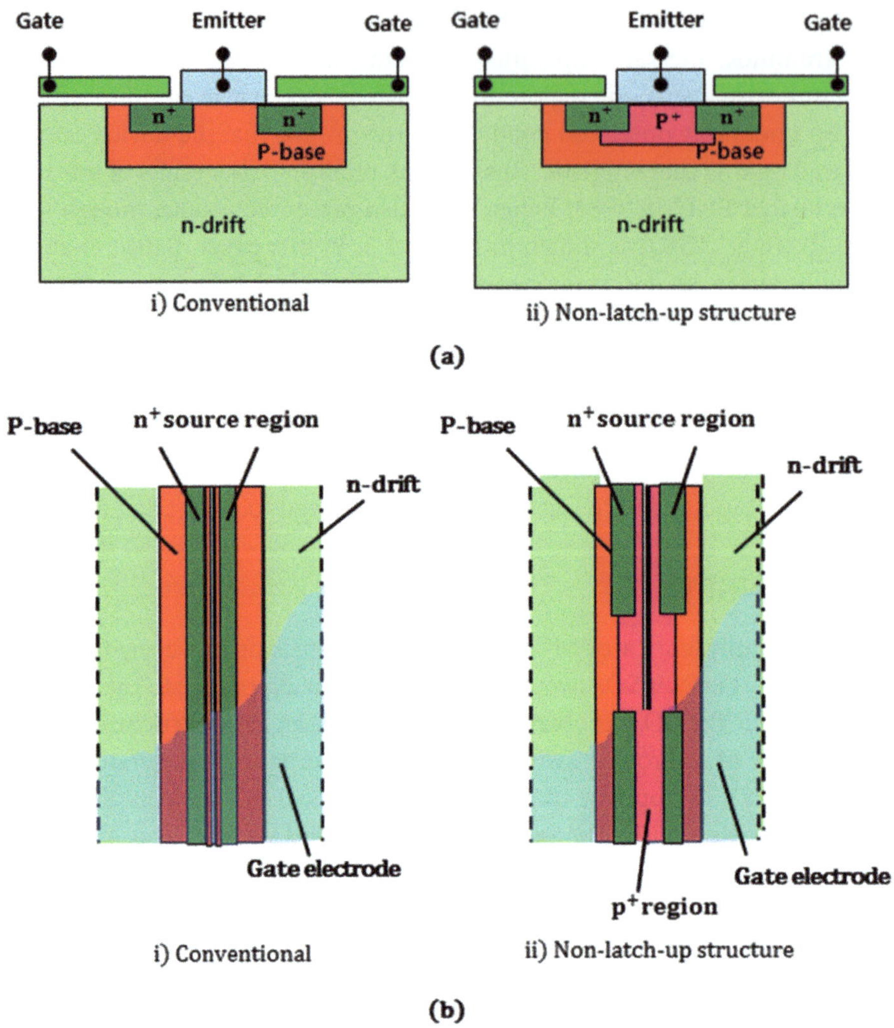

Figure 2.4 Comparison of cross section and top view between a conventional and non-latch-up IGBT structure. In addition to the application of the high impurity P+ region, parts of N- source regions are periodically eliminated in the non-latch-up IGBT structure: (a) cross section of the surface regions and (b) top view of the surface regions

fast-switching, and high-withstand capability; therefore, as in the case of the Ultra-Large-Scale Integration (U-LSI) field, a variety of the most advanced technological innovations were applied to the IGBT with the aim of overcoming these tradeoffs. IGBT cell design and process technologies contributed to the improvement of IGBT characteristics. As the IGBT had the structure of a power MOSFET with a P+ backside layer, designing and fabrication process technologies were mostly derived from this type of power MOSFET. Therefore, in order to improve the forward voltage drop, the trench-gate technologies were applied and the trench-gate IGBT was eventually realized through applying cutting edge ULSI designing and process technologies.

After that, there were several important breakthroughs in IGBT improvement. Firstly, there was much effort to implement trench technology into the IGBT structure. One common way to implement such a structure was the close arrangement of trench cells near each other; the result, however, was a chip with an enormous channel width, completely unable to handle a short-circuit condition without the additional function of an internal or external current limitation circuit for some planar cell IGBTs. Also, the carrier distribution was not suitable for the lower forward voltage drop. One important technology that was used to overcome these issues was the Injection Enhancement (IE) effect. One device, which combined the IE effect with the IGBT structure, was named the IEGT, as shown in Figure 2.5. This device attained a much lower forward voltage drop value by improving the surface gate structure, which increased the resistance of the hole current flow into the emitter electrode, resulting in carrier enhancement around the device's surface. This technology came into wide use for IGBT development and is still being applied to even current state-of-the-art IGBT devices. A modified version of the IEGT, including an additional n-doped layer below the p body of the trench cell, was presented by Takahashi et al as the **C**arrier **S**tored **T**rench-gated **B**ipolar **T**ransistor (CSTBT).

Laska et al. presented 1200 V trench gate IGBT devices with the optimized cell design in terms of shape, size, and distance, which exhibited lower forward voltage drop and excellent short-circuit capability simultaneously. An important moment for IGBT improvement was the development of power device simulation technology which can calculate higher voltage and current condition simultaneously.

Two-dimensional carrier distribution results as the IGR's on-state characteristic analysis were exhibited in 1982. Additionally, Nakagawa showed and analyzed an optimum carrier distribution for the IGBT to achieve lower forward voltage drop and high-speed switching at the same time, and Iwamuro et al. demonstrated a theoretical and accurate analysis of the IGBT's destructive failure mechanism between P-channel and N-channel IGBTs, both by their originally developed 2-D device simulator. As the IGBTs are connected with many different elements, such as resistance, inductances, and capacitances for various fields of application, their switching characteristics have to be solved by both the semiconductor equations and the circuit ones at the same time. Omura and Nakagawa successfully developed their new 2-D device simulator by taking complete external circuits, including snubber circuits and all parasites, into account, as well as analyzing the device failure mechanism more precisely. Since this achievement, power device simulation technology has continued to improve greatly. In addition to its contributions to IGBT improvement, this technology still works as one of the key tools for state-of-the-art IGBT development.

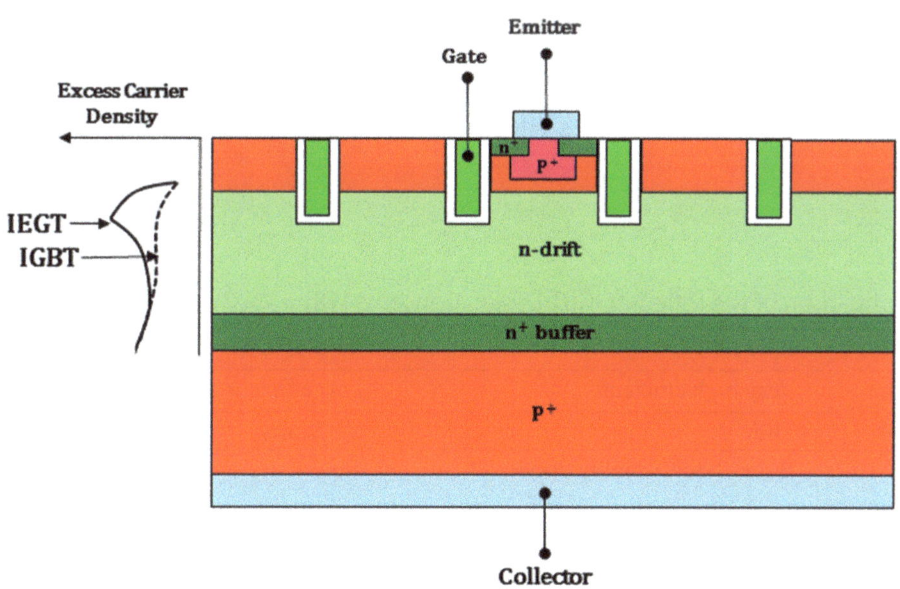

Figure 2.5 Cross section of IEGT structure and comparison of schematic carrier distribution between IEGT and IGBT

2.1.3. IGBT with Thin Wafer Technologies

IGBTs rated below 1200 V were fabricated by a design concept using N-epitaxial drift layers with high-doped P+ substrates. This design made the hole injection from the collector very strong and, consequently, high lifetime control was required, which resulted in poor short-circuit capability and unsuitability for parallel operation. Miller and Sack proposed a new type of IGBT bulk structure called non-punch through (NPT) IGBT. Features of the NPT IGBT are as follows.

1. Applicable to inexpensive bulk wafers;
2. Thinner wafer thickness required;
3. Low carrier injection from the collector achieved;
4. No lifetime control required.

Because of low carrier injection and no lifetime control, the NPT IGBT is able to achieve a flat carrier distribution in the IGBT chip in order to get both lower forward voltage drop and faster switching speed. The low carrier injection and no lifetime control technology led to another advantage, that is, a positive temperature coefficient in the output I–V characteristic. This means the NPT IGBT is suitable for parallel operation and is easy to apply to large capacity applications. Furthermore, it should be noted that this IGBT exhibits higher withstand capability because of the thicker n-drift region, which results in electric field reduction. As a result, the development of the NPT IGBT replaced the conventional power bipolar transistor quickly with the IGBT.

This is possible by implementing a field stop layer; one with a low enough dose to not influence the low dose P collector of the NPT IGBT, but high enough to stop the electrical field under blocking conditions. It is possible to shrink the drift layer thickness of the NPT structure by approximately one third. Figure 2.6 and Table 2.1 show a comparison of the cross section of unit cell and design concepts between the PT and NPT IGBTs and the FS IGBT. The FS IGBT was first published by Laska et al., and it was designed based on the NPT IGBT design concept of low carrier injection and no lifetime control with inexpensive bulk wafers. However, the FS IGBT does not need quite as thick of an N-drift layer as the NPT IGBT because it has a field stop layer that stops the depletion layer IGBTs.

As a result, flat carrier distribution can be successfully achieved in its forward current conduction mode and its forward voltage drop can be lowered for the FS IGBT. Additionally, the FS IGBT can reduce its switching loss and less excess carriers and its thinner drift layer. By adopting a trench gate structure, channel density

Table 2.1 Comparison of the Different IGBT Design Concepts [18]

	PT-IGBT	NPT-IGBT	FS-IGBT
P-emitter	Very high efficient	Low efficient	Low efficient
N-drift	Thin	Medium	Thin
Additional N-layer	Buffer layer = highly doped • to reduce the very high emitter efficiency • to stop the electrical field	No	Field stop layer = weakly doped • to stop only the electrical field
Carrier lifetime	Low (lifetime killing)	High	High

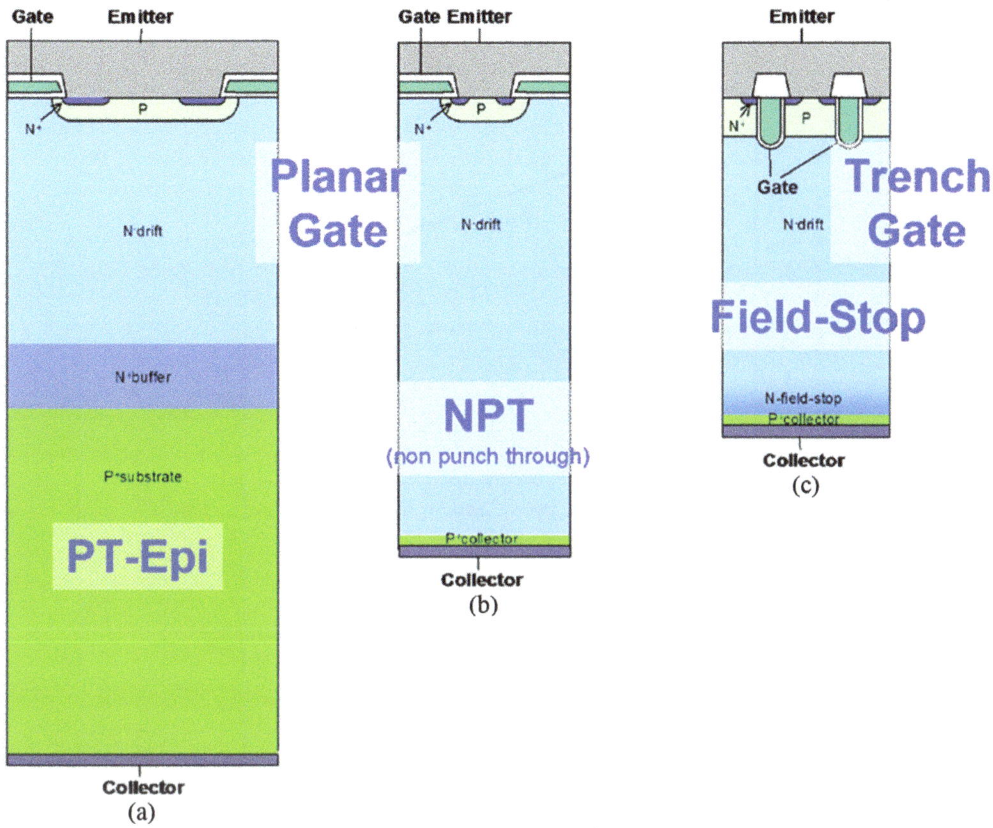

Figure 2.6 Structural change of IGBT cell design: (a) PT-IGBT; (b) NPT-IGBT; and (c) FS-IGBT

can be increased, and the forward voltage drop can be significantly reduced. This is the most advanced type of IGBT. Figure 2.7 shows the chronology of IGBT wafer thickness improvement for the voltage classes 600, 1200, and 1700 V. It should be noted that the thin wafer technology for the IGBT is required not only to be wafer thin, but also to implant and diffuse the impurity and evaporate the metal on the backside of the thin wafer with absolutely no wafer breaking. Therefore, the thin wafer technology for the IGBT can be seen as one of the most advanced technologies in the whole semiconductor industry. This design concept (the thin wafer based upon the field stop vertical cell profile optimization) contributed immensely in expanding the limit of IGBT ratings, especially its voltage-handling capability, which today has reached 6.5 kV. Figure 2.6 shows the structural change of IGBT cell design:

2.1.4. IGBT Module

The aim of the IGBT technological innovation mentioned earlier was a breakthrough in not only in power loss, but also in cost reduction of the IGBT through improving chip characteristics, which results in the shrinkage of die size. IGBT modules consist of multiple IGBT chips mounted on an isolated substrate, which is mounted on a copper base plate. Free-wheeling diodes (FWDs) are also mounted in the module so that components can be assembled easily and to allow minimum lead inductance both inside and outside the module.

The technological innovation was also a breakthrough, not only for power loss and cost reduction, but also for elimination of current/voltage oscillation under unfavorable conditions. The oscillation tendency of FWD increases during turn-off of a lower current in the region of 1/10 of the nominal current. This softness (so-called) can be improved with a higher injection of charge carrier in the die. However, this leads the FWD to

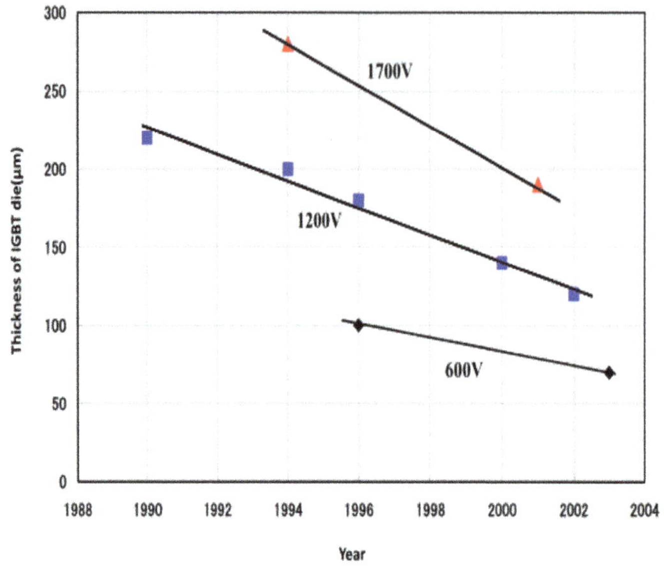

Figure 2.7 Chronology of IGBT wafer thickness improvement

have higher recovery losses. The FWD device was developed to achieve a very good softness using a minimum of injected charge carriers.

As a result, the latest FWD also has softness behavior, as well as the advantages of reduced recovery losses with correct switching in applications with high output power. In the heat-dissipating metal-base-equipped structure, the Direct Copper Bond (DCB) substrate (substrate with ceramic isolation) is soldered to a heat-dissipating metal base, and a cooling fin is attached to the base. In order to miniaturize the IGBT module, it is necessary to shrink the IGBT and FWD chip. However, the miniaturization of the chip results in increased power density, which causes degraded reliability due to a rise in chip temperature.

The new module packages have increased reliability and improved electrical characteristics by making use of the newly developed DCB substrate, whose alumina ceramic layers have low thermal resistance and high strength, leading to an increase in the usable power unit area of the module while achieving lighter weight, smaller size, and lower cost.

Both the improvement of the packaging technology and the shrinkage of chip die size lead to the reduction of the footprint of the IGBT module. It is successfully reduced not only because of the development of trench FS IGBT and its transistor cell improvement but also because of the DCB substrates and the packaging technologies mentioned earlier. Consequently, the smaller size of power electronic apparatus to which the IGBT modules are applied is also achieved. For power electronics systems, both higher energy efficiency and power output with downsizing are required, so the IGBT modules are widely used. A new concept integrating dedicated drive and protection circuits with IGBTs to overcome the IGBT's inherent parasitic limitations and maximize its performance was introduced as an intelligent power module (IPM) in 1984. One of the features of the IPM is the PCB which its gate drive and control circuit are contained on is mounted directly above the power devices. Highly effective over-current and short-circuit protections were achieved through the use of advanced current sense IGBT chips that allow the power device current to be continuously monitored. System reliability was further improved by the IPM's integrated over-temperature protection and so on. Overall, it can be concluded that the IPM has provided an improvement in system performance and reliability over the conventional IGBT module.

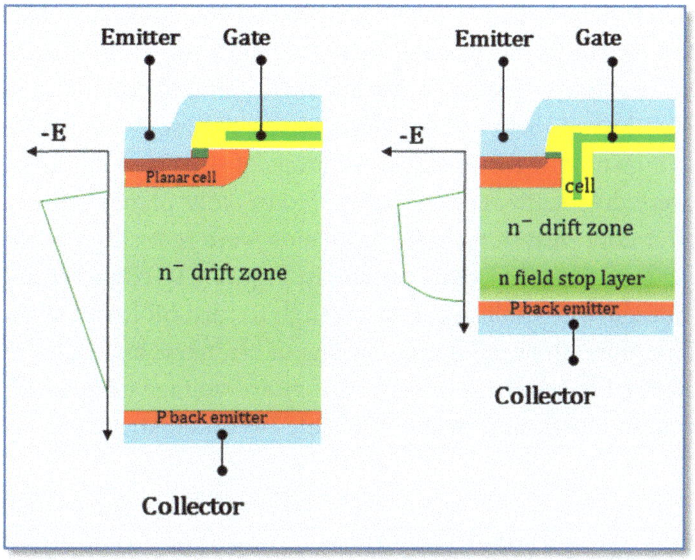

Figure 2.8 Example of an elder IGBT with planar cell and vertical NPT structures (left) and a state-of-the-art IGBT with a trench cell and a vertical field stop concept (right)

2.1.5. Competition between IGBT Structure and MOS-Gated Thyristor Ones

On account of the technological innovations described earlier, the IGBT had been successfully delivered and spread in the market. However, from the mid-1980s to early 1990s, the IGBT was challenged many times by MOS-gated thyristors. The IGBT featured simple gate control and current saturation characteristics, and many studies had been done on its high voltage applications. As an alternative, MOS-gated thyristors, such as the MOS-Controlled Thyristor (MCT) and Base Resistance-Controlled Thyristor (BRCT), had been studied for high-voltage applications because of their simple gate-drive capabilities and low forward voltage drops. However, as these devices did not have the current saturation feature, they showed much poorer short-circuit SOA characteristics when compared with the IGBT, so the market did not accept the MCT and the BRCT devices as suitable alternatives. Dual channel Emitter-Switched Thyristors (EST) and MOS-Controlled Cascade Thyristors (MCCT) are the only MCT structures that were reported to have the capability of the thyristor's gate controlled current saturation. However, as the EST and MCCT had a structure of dual channel MOSFETs for connecting the cathode to the thyristor's N+ floating emitter, these thyristor devices exhibited the shortcoming of higher forward voltage drop. Therefore, because of the invention of IEGT, trench IGBT, and CSTBT (which exhibited much lower forward voltage drop because of carrier enhancement around the device surface in the 1990s), improvement of the forward voltage drop of the EST and MCCT simply wasn't enough for them to become superior alternatives to IGBT devices.

2.1.6. State-of-the-Art IGBTs

Transistor Cell Concepts

When trench cell technology was first introduced, it started in the low voltage range, then stayed in the high-voltage range for a long time. Planar cell concepts with IE effects still existed alongside this technology. Over the whole available voltage-class range (from 400 V up to 6.5 kV), trench cell concepts dominated, whether it was the trench cell-based IEGT, the trench cell structure with defined quite wide distances between the cells, or a structure with more narrow and differing gate and/or source trenches, with or without an additional n-doped charge storage layer below the P body. They all have the basic principle of enhancing the amount of carriers (both holes and electrons) in the top region of the chip, during the on-state condition by a hole confinement or IE effect, in common. As a result, the forward voltage drops of such trench transistor cell-based IGBTs with enhanced carrier amount were drastically reduced compared to older planar cell-based IGBTs. Current trends to further increase the carrier concentration on the front side were reducing cell size and cell pitch to micro- or even sub-micrometer sizes, as shown in one example in Figure 2.9. However, if the increase of carriers in the top region is overemphasized, the result could be too high to turn off losses as the excess of carriers cannot be removed fast enough. Therefore, the carrier engineering of electrons and holes in the top part of the IGBT chip had to be carefully balanced to result in overall low forward voltage drop with reasonable turn-off losses. Figure 2.9 shows the schematic illustration of the ON-state plasma distribution in IGBTs with different cell concepts.

Concepts for Vertical Structures

The second decisive device element of a bipolar power device like the IGBT is the vertical structure. State-of-the-art vertical concept is a punch-through construction (often called a field stop structure or soft punch-through approach) with a lowly doped field stop layer between the drift zone and low efficient back emitter without any carrier lifetime reduction needs. This vertical concept somehow developed self-evidently out of the former NPT and PT concepts, serving the needs of both cost and performance optimized industrialized IGBTs. It was first used with a voltage class of 1200 V, and then up to 6.5 kV, but was also established in the lower voltage arena of 600 V and even as low as 400 V with chip thicknesses of about 45 μm (400 V) to about 700 μm (6.5 kV).

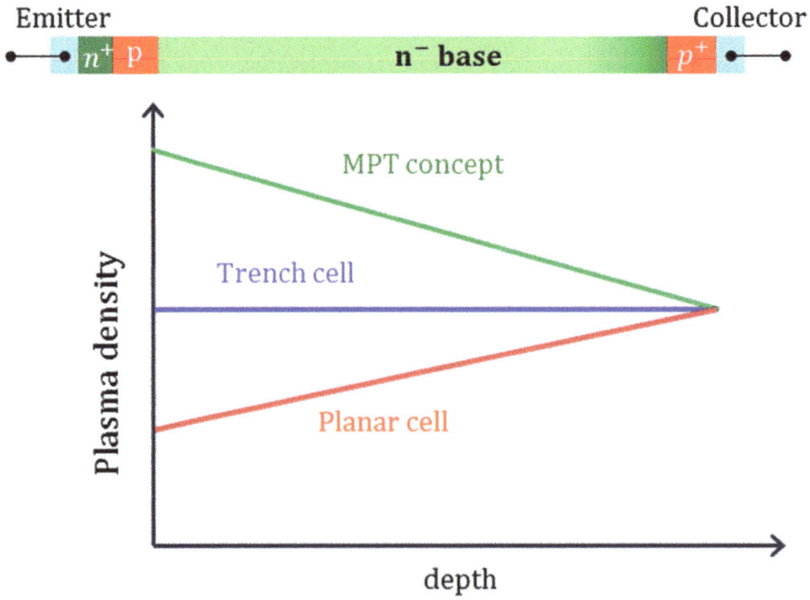

Figure 2.9 Schematic illustration of the ON-state plasma distribution in IGBTs with different cell concepts (planar, trench, and micro-pattern trench)

Reverse Blocking and Reverse Conducting Capability of IGBTs

Further aspects within state-of-the-art IGBTs include integration of additional functions within the chip. One example is the **r**everse **b**locking (RB) IGBT, which has some advantages for special applications like matrix converters, but also has some drawbacks in terms of process complexity and vertical optimization restrictions compared with a standard FS IGBT. Another promising integration topic is the implementation of the FWD function within the IGBT, called the reverse conducting (RC) IGBT. First approaches that fit well only for special applications, such as induction heating with its need for diode hard commutation, could be optimized in terms of the integrated diode function for typical consumer and general purpose drive applications in the 600–1200 V-class range. The challenge is mainly to integrate very differently optimized technological features (lifetime killing of diodes, high P dose in IGBT cell) within one single die. Figure 2.10 shows the reverse conducting IGBT.

Increasing Maximum Junction Temperature

Evolution from very first IGBTs to actual state-of-the-art ones could be seen in the maximum specified junction temperatures (Tj) of the dies within the package. Starting from Tj 125°C today, 150°C as a maximum junction temperature is standard for IGBTs in conventional modules even up to 175°C for discrete IGBTs; meanwhile, first IGBT modules specify a Tj of 175°C, whether as a continuous condition or for overload cases.

Measures taken to be successful with these increased junction temperatures come both from the chip side (reduced hot leakage currents, high intrinsic overcurrent, short-circuit robustness) and from the module side and the interconnected technologies of the dies in the package, where an even higher Tj is discussed.

Increasing Power Density

The motivation for an increased junction temperature is not as clear at first glance as things like improvements in the cell and vertical structure with its reduced forward voltage and switching losses, but combining all these efforts can result in great shrinkage of chip sizes for a given nominal current, to some amount enabled by the increased Tj. This increased current rating per chip area can be used to reduce the size of a module for a given power. Both result in a higher power density and enable customers to significantly reduce the size and weight of the inverters or increase the power of a given inverter size. In Figure 2.11, the example of a standard industrial IGBT module (62-mm dual) is shown with its increase in nominal current and output power over several chip generations enabled by chip shrinkage without additional measures in the module technology. So, power density enhancement has been an important trend and remains a driver for IGBT innovation, enabled by improved transistor cells and vertical technology as well as junction temperature increase.

Figure 2.12 shows the history of IGBT stated earlier

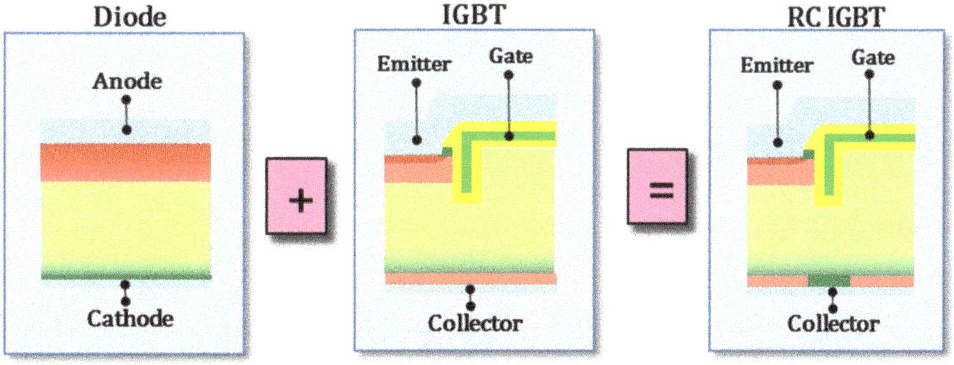

Figure 2.10 Reverse conducting (RC) IGBT as an integration of IGBT and freewheeling diode within one die

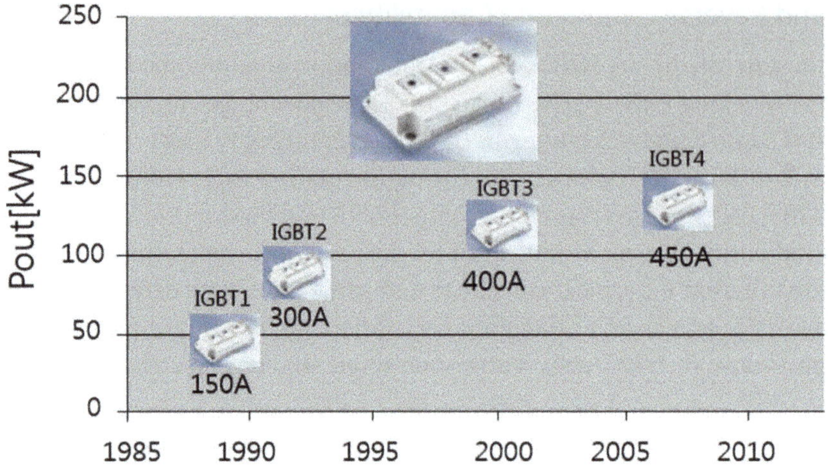

Figure 2.11 Output power increase for a given module footprint (62-mm dual) by improving cell, vertical structure, and increasing junction temperature over four IGBT generations of vendor B

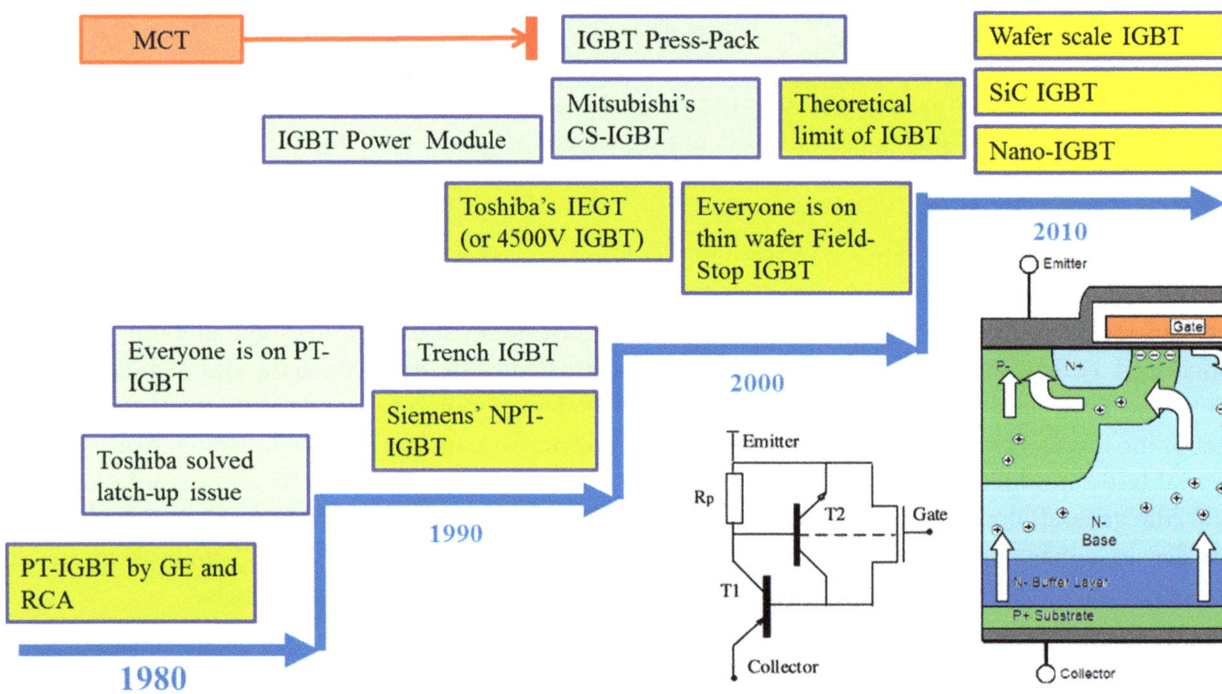

Figure 2.12 Evolution history of IGBT

2.2 IGBT Characteristics [5]

2.2.1. Maximum Ratings

V_{CES}—Collector–Emitter Voltage

This is the maximum voltage that can be supported between the collector and emitter terminals with the gate and emitter terminals shorted together. Depending on the temperature, the maximum permissible collector-emitter voltage could actually be less than the V_{CES} rating.

V_{GE}—Gate–Emitter Voltage

V_{GE} is the rating of the maximum continuous voltage between the gate and emitter terminals. The purposes of this rating are to prevent breakdown of the gate oxide and to limit short-circuit current. The actual gate oxide breakdown voltage is significantly higher than this, but staying within this rating at all times ensures application reliability.

V_{GEM}—Gate–Emitter Voltage Transient

V_{GEM} is the maximum pulsed voltage between the gate and emitter terminals. The purpose of this rating is to prevent breakdown of the gate oxide.

Transients on the gate can be induced not only by the applied gate drive signal, but often more significantly by stray inductance in the gate-drive circuit, as well as feedback through the gate–collector capacitance. If there is more ringing on the gate than V_{GEM}, stray circuit inductances likely need to be reduced and/or the gate resistance should be increased to slow down the switching speed. In addition to the power circuit layout, gate drive circuit layout is critical in minimizing the effective gate-drive loop area and determining the stray inductances of gate drive circuit.

If a clamping zener is used, it is recommended that it be connected between the gate driver and the gate resistor rather than directly to the gate terminal. Negative gate drive, while not necessary, may be used to achieve great switching speed while avoiding dv/dt induced turn-on.

I_{C1}, I_{C2} – Continuous Collector Current

I_{C1} and I_{C2} are ratings of the maximum continuous DC current with the die at its maximum rated junction temperature. They are based on the junction to case thermal resistance rating $R_{\theta JC}$ and the case temperature as follows:

$$P_D = \frac{T_{J(\max)} - T_C}{R_{\theta JC}} = V_{CE(on)} \cdot I_C \tag{2.1}$$

This equation simply states that the maximum heat that can be dissipated, $\frac{T_{J(\max)} - T_C}{R_{\theta JC}}$, equals the maximum allowable heat generated by conduction loss, $V_{CE(on)} \cdot I_C$. There are no switching losses involved in I_{C1} and I_{C2}. Solving for I_C:

$$I_C = \frac{T_{J(\max)} - T_C}{R_{\theta JC} \cdot V_{CE(on)}} \tag{2.2}$$

$V_{CE(on)}$ depends upon I_C (as well as junction temperature). Except at relatively low current, the relationship between I_C and $V_{CE(on)}$ is fairly linear, as shown in Figure 2.13. Thus, a linear approximation can be used to relate I_C to $V_{CE(on)}$.

The curve of $V_{CE(on)}$ is equal to the slope of the die at elevated temperatures. To calculate datasheet values, the maximum $V_{CE(on)}$, which is higher than the typical $V_{CE(on)}$ to account for normal variations between parts, is used. The equation relating $V_{CE(on)}$ to I_C is

$$V_{CE(on)} = I_C \cdot R_{CE(on)} + V_{CE(zero)} \tag{2.3}$$

This equation is substituted in Equation (2.2) for $V_{CE(on)}$ to solve for I_C:

$$\begin{aligned} I_C &= \frac{T_{J(\max)} - T_C}{R_{\theta JC} \cdot V_{CE(on)}} = \frac{T_{J(\max)} - T_C}{R_{\theta JC}\left(I_C \cdot R_{CE(on)} \cdot V_{CE(zero)}\right)} \Rightarrow \\ &= I_C^2 \cdot R_{CE(on)} + I_C \cdot V_{CE(zero)} \frac{T_{J(\max)} - T_C}{R_{\theta JC}} \end{aligned} \tag{2.4}$$

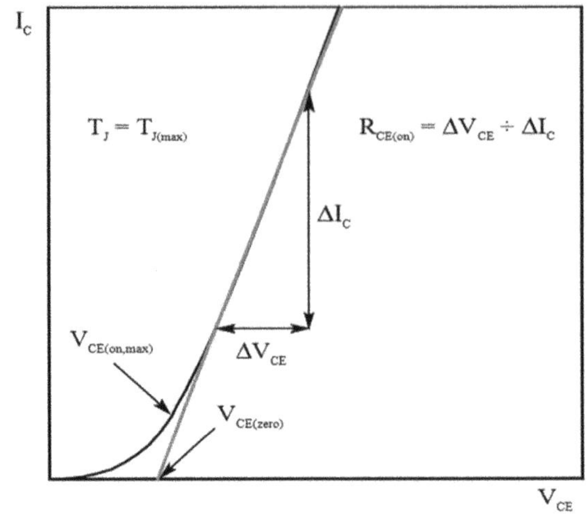

Figure 2.13 Linear approximation of I_C versus V_{CE} (on)

This is in the form of the familiar quadratic equation:

$$x = \frac{-b \pm \sqrt{b^2 - 4 \cdot a \cdot c}}{2 \cdot a}$$

with

$$a = R_{CE(on)},\ b = V_{CE(zero)},\ \text{and}\ c = \frac{-\left(T_{J(\max)} - T_C\right)}{R_{\theta JC}}$$

The solution is

$$I_C = \frac{-V_{CE(zero)} \pm \sqrt{V_{CE(zero)^2} - 4 \cdot R_{CE(on)} \cdot \left(\frac{-\left(T_{J(\max)} - T_C\right)}{R_{\theta JC}}\right)}}{2 \cdot R_{CE(on)}} \tag{2.5}$$

I_C in Equation (2.5) represents the continuous DC current (with the device fully on) that causes the die to heat up to its maximum rated junction temperature. I_{C1} is the solution to Equation (2.5) with T_C equal to 25 °C. I_{C2} is the solution to Equation (2.5) with T_C at an elevated temperature. This is a more useful rating than the traditional I_{C1} rating because operating at a case temperature of only 25 °C is rarely feasible.

Graph of I_C versus T_C

This graph shows the solution to Equation (2.5) over a range of case temperatures. Figure 2.14 shows an example: in this case, the package leads limit the current to 100 A at low temperature.

Using the I_{C1} and I_{C2} Ratings

The I_{C1} and I_{C2} ratings and the graph of maximum collector current versus case temperature indicate the *maximum theoretical continuous DC current* that the device can carry, based on the maximum junction to case thermal resistance. They are mainly considered figures of merit for comparing devices. For a soft switching application, I_{C2} is a good starting point for selecting a device. In a hard or soft switching application, the device might safely carry more or less current depending on:

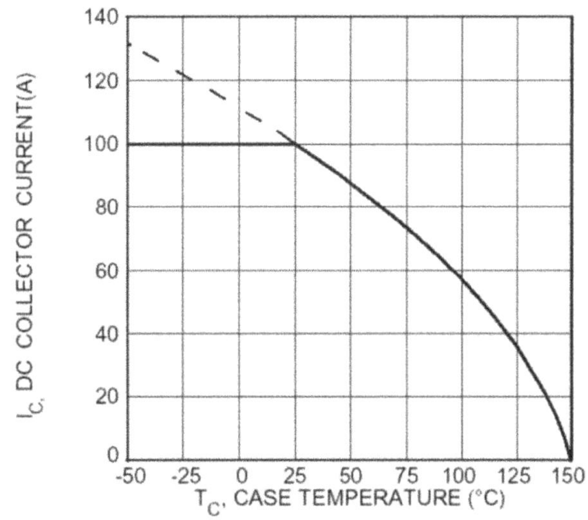

Figure 2.14 Maximum collector current versus case temperature

- Switching losses
- Duty cycle
- Switching frequency
- Switching speed
- Heat sinking capacity
- Thermal impedances and transients

I_{CM}—Pulsed Collector Current

This rating indicates how much pulsed current the device can handle, which is significantly higher than the rated continuous DC current. The purposes of the I_{CM} rating are:

- To keep the IGBT operating in the "linear" region of its transfer characteristic. An IGBT has a maximum collector current for a corresponding gate–emitter voltage that it will conduct. If the operating point at a given gate–emitter voltage goes above the linear region "knee" in Figure 2.15, any further increase in collector current results in a significant rise in collector–emitter voltage, and a consequent rise in conduction loss as well as possible device destruction. The I_{CM} rating is set below the "knee" for typical gate–drive voltages.
- To prevent burnout or latch-up. Even if the pulse width is theoretically too short to overheat the die, significantly exceeding the I_{CM} rating can cause enough localized die feature heating to result in a burnout site or latch-up.
- To prevent overheating of the die. The footnote "*Repetitive rating: Pulse width limited by maximum junction temperature*" implies that I_{CM} is based on thermal limitations depending on pulse width. This is always true for two reasons: (1) there is a small margin in the I_{CM} rating before the risk of damage gets too high other than by the die exceeding its maximum junction temperature and (2) no matter what the failure mechanism really is, overheating is almost always the observed end result.

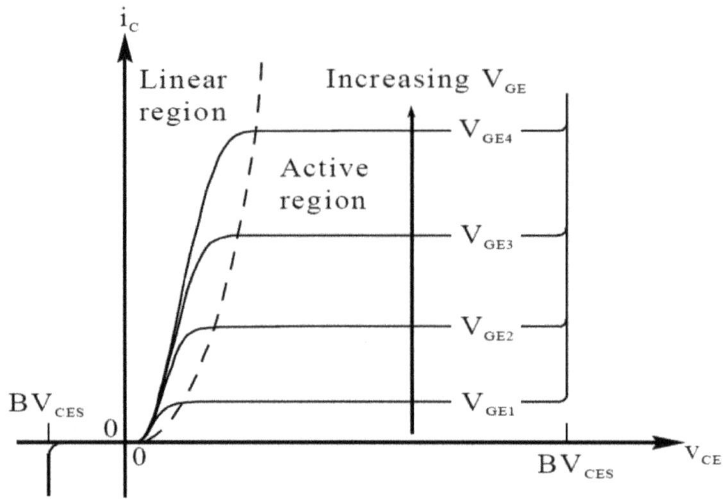

Figure 2.15 IGBT transfer characteristic

- To avoid problems with excessive current through the bond wires (although problems related to excessive current through the die are likely to arise first).

Regarding the thermal limitation on I_{CM}, temperature rise depends upon the pulse width, time between pulses, heat dissipation, and $V_{CE(on)}$, as well as the shape and magnitude of the current pulse. Simply staying within the I_{CM} limit does not ensure that the maximum junction temperature will not be exceeded. See the discussion on the Maximum Effective Transient Thermal Impedance curve for a procedure to estimate junction temperature during a current pulse.

I_{LM}, RBSOA, and SSOA—Safe Operating Areas

These ratings are all related. I_{LM} is the amount of clamped inductive load current the device can safely switch in a snubberless hard switching application. Test circuit conditions such as case temperature, gate resistance, clamp voltage, etc. are specified. The I_{LM} rating is limited by the turn-off transient, where the gate which was positive-biased switches to zero or negative bias. Hence, the I_{LM} rating RBSOA.

Switching safe operating area (SSOA) is simply the RBSOA at full V_{CES} voltage rating. Forward-biased safe operating area (FBSOA), which covers the Turn-On transient, is typically much higher than the RBSOA.

E_{AS}—Single Pulse Avalanche Energy

All devices that are avalanche energy rated have an E_{AS} rating. Being avalanche energy rated is synonymous with being Unclamped Inductive Switching (UIS) rated. E_{AS} is both thermally limited and defect limited and indicates how much reverse avalanche energy the device can safely absorb with the case at 25°C and the die at or below the maximum rated junction temperature. A defect in a closed cell structure can cause the cell to latch-up under avalanche condition.

The E_{AS} rating is equal to $\frac{L \cdot i_C^2}{2}$, where L is the value of an inductor carrying a peak current i_C, which is suddenly diverted into the collector of the device being under test. It is the inductor's voltage exceeding the breakdown voltage of the IGBT that causes the avalanche condition. An avalanche condition allows the inductor current to flow through the IGBT even though it is in the off state. Energy stored in the inductor is analogous to energy stored in leakage and/or stray inductances and is dissipated in the device being under test. In an application, if ringing because of leakage and stray inductances does not exceed the breakdown voltage, then the device will not avalanche and thus does not need to dissipate avalanche energy. Avalanche energy-rated devices offer a safety net depending on the margin between the voltage rating of the device and system voltages, including transients.

P_D—Total Power Dissipation

This is a rating of the maximum power that the device can dissipate. It is based on the maximum junction temperature and thermal resistance $R_{\theta JC}$ at a case temperature of 25°C.

$$P_D = \frac{T_{J(\max)} - T_C}{R_{\theta JC}} \tag{2.6}$$

At case temperatures above 25 °C, the linear derating factor is simply the inverse of $R_{\theta JC}$.

T_J, T_{STG}—Operating and Storage Junction Temperature Range

This is the range of permissible storage and operating junction temperatures. The limits of this range are set to ensure a minimum acceptable device service life, though operating outside these limits can significantly enhance this service life. Although the range is actually an exponential function relating device life to junction temperature, as a "rule of thumb", for thermally induced effects only, every 10°C reduction in junction temperature doubles the device life.

2.2.2. Static Electrical Characteristics

BV_{CES}—Collector–Emitter Breakdown Voltage

Measuring the actual collector–emitter breakdown voltage is practically impossible without destroying the device. BV_{CES} is the collector–emitter voltage at which no more than the specified collector current will flow at the specified temperature and can be used to track the actual breakdown voltage without destroying the device.

As shown in Figure 2.16, BV_{CES} has a positive temperature coefficient. At a fixed leakage current, an IGBT can block more voltage when hot than when cold. In fact, when cold, the BV_{CES} specification is less than the V_{CES} rating.

RBV_{CES}—Reverse Collector–Emitter Breakdown Voltage

This is the reverse collector–emitter breakdown voltage specification, i.e., when the emitter voltage is positive with respect to the collector. As with BV_{CES}, RBV_{CES} is the emitter–collector voltage at which no more than the specified emitter current will flow at the specified temperature. A typical value is about 15 V; however, RBV_{CES} is often not specified because an IGBT is not designed for reverse voltage blocking. Even though in theory NPT IGBTs can block as much reverse voltage as forward voltage, in general they cannot because of the manufacturing process, whereas PT IGBTs cannot block very much reverse voltage because of the n+ buffer layer.

$V_{GE(th)}$—Gate Threshold Voltage

This is the gate-sourced voltage at which the collector current begins to flow. Test conditions such as collector current, collector-emitter voltage, and junction temperature are also specified. All MOS-gated devices exhibit variation in $V_{GE(th)}$ between devices, which is normal. Therefore, a range of $V_{GE(th)}$ is specified, with the minimum and maximum representing the edges of the $V_{GE(th)}$ distribution. $V_{GE(th)}$ has a negative temperature

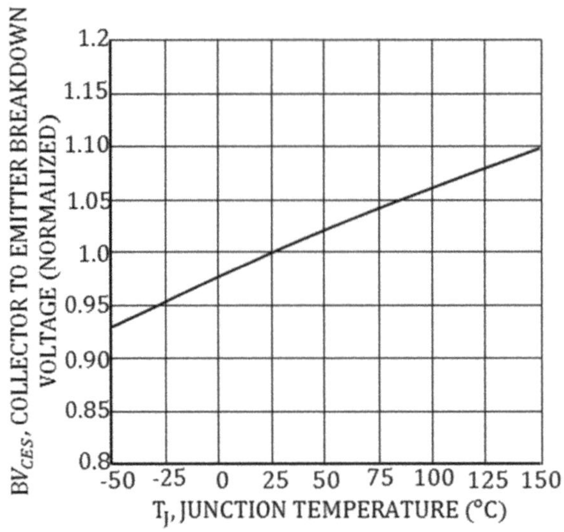

Figure 2.16 Normalized breakdown voltage versus junction temperature

coefficient, meaning that as the die heats up, the IGBT will turn on at a lower gate-emitter voltage. This temperature coefficient is typically about -12mV/°C, the same as that of a power MOSFET.

$V_{CE(on)}$—Collector–Emitter On-Voltage

This is the collector–emitter voltage that falls across the IGBT at a specified collector current, gate–emitter voltage, and junction temperature. As $V_{CE(on)}$ is temperature dependent, it is specified at both room temperature and high temperatures.

I_{CES}—Collector Cutoff Current

This is the leakage current that flows from collector to emitter when the device is off, at a specified collector–emitter and gate–emitter voltage. As leakage current increases with temperature, I_{CES} is specified both at room temperature and high temperatures. Leakage power loss is equal to I_{CES} times the collector–emitter voltage.

I_{GES}—Gate–Emitter Leakage Current

This is the leakage current that flows through the gate terminal at a specified gate-emitter voltage.

2.2.3. Dynamic Characteristics of IGBT

Figure 2.17 shows an equivalent IGBT model that includes the capacitances between the terminals. Input, output, and reverse transfer capacitances are combinations of these.

C_{ies}—Input Capacitance

This is the input capacitance measured between the gate and emitter terminals with the collector shorted to the emitter for AC signals. C_{ies} is made up of the gate-to-collector capacitance (C_{GC}) in parallel with the gate-to-emitter capacitance (C_{GE}), or

$$C_{ies} = C_{GE} + C_{GC} \tag{2.7}$$

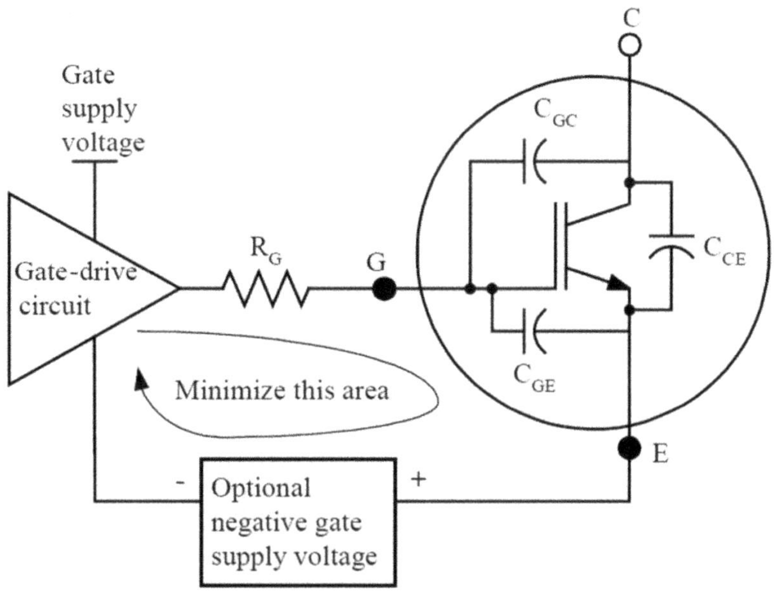

Figure 2.17 IGBT capacitances

The input capacitance must be charged to the threshold voltage before the device begins to turn on and must be discharged to the plateau voltage before the device begins to turn off. The impedance of the drive circuitry and C_{ies} have a direct relationship to turn-on and turn-off delays.

C_{oes}—Output Capacitance

This is the output capacitance measured between the collector and emitter terminals with the gate shorted to the emitter for AC voltages. C_{oes} is made up of the collector to emitter capacitance (C_{CE}) in parallel with the gate to-collector capacitance (C_{GC}), or

$$C_{oes} = C_{CE} + C_{GC} \tag{2.8}$$

For soft switching applications, C_{oes} is important because it can affect the resonance of the circuit.

C_{res}—Reverse Transfer Capacitance

This is the reverse transfer capacitance measured between the collector and gate terminals with the emitter connected to ground. The reverse transfer capacitance is equal to the gate-to-collector capacitance.

$$C_{res} = C_{GC}$$

The reverse transfer capacitance, often referred to as the Miller capacitance, is one of the major parameters affecting the rising and falling times of voltage during switching.

Figure 2.18 shows an example graph of typical capacitance values versus collector–emitter voltage. The capacitances decrease over an increasing collector–emitter voltage range, especially the output and reverse transfer capacitances.

V_{GEP}—Plateau Voltage

Figure 2.19 shows the gate-emitter voltage as a function of gate charge. The turn-on sequence traverses this curve from left to right, whereas turn-off traverses it from right to left. The gate plateau voltage V_{GEP} is defined

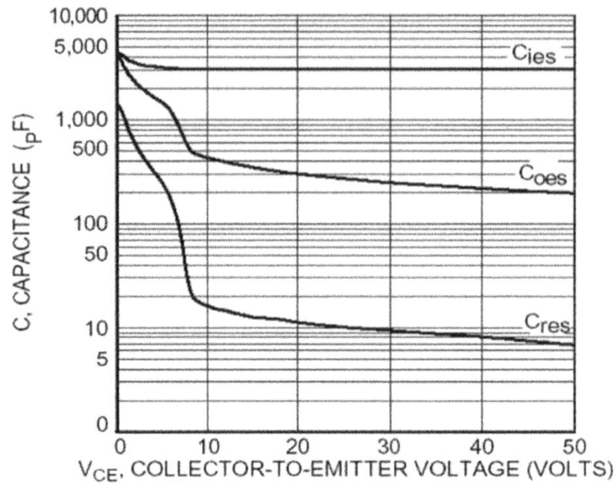

Figure 2.18 Capacitance versus collector–emitter voltage

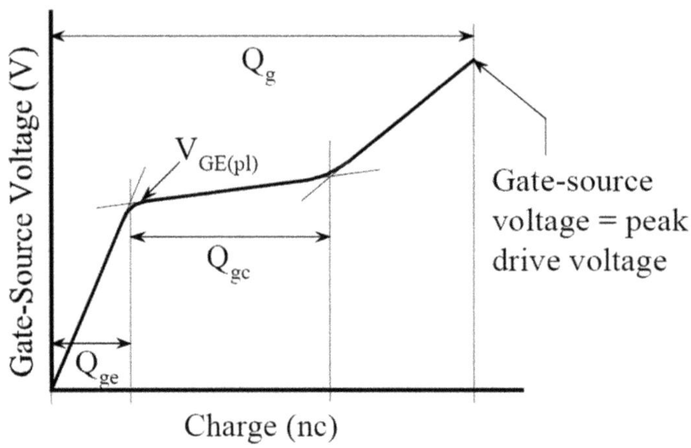

Figure 2.19 V_{GE} as a function of gate charge

as the gate-emitter voltage, when the slope of the gate–emitter voltage first reaches a minimum during the turn-on switching transition for a constant gate current–drive condition. In other words, it is the gate–emitter voltage where the gate charge curve first straightens out after the first inflection in the curve. Alternatively, V_{GEP} is the gate–emitter voltage at the last minimum slope during turn-off. The plateau voltage increases with current but not with temperature.

Q_{ge}, Q_{gc}, and Q_g—Gate Charge

As shown in Figure 2.19, Q_{ge} is the charge from the origin to the first inflection in the curve, whereas Q_{gc} is the charge from the first to second inflection in the curve (also known as the "Miller" charge). Q_g is the charge from the origin to the point on the curve at which v_{GE} equals the peak-drive voltage.

Gate charge values vary with collector current and collector–emitter voltage but not with temperature. The gate charge values reflect charge stored in the internal capacitances. Gate charge is often used for designing gate drive circuitry because it accounts for the changes in capacitance with changes in voltage during a switching transient.

2.2.4. Switching Times and Energies

The switching times and energies for an IGBT are not always easy to predict and will often vary with temperature (except E_{on1}). In general, turn-on speed and energy are either relatively independent of temperature or increase in speed (decrease in energy) very slightly with increasing temperature. Diode reverse recovery current increases with temperature, also resulting in the increase of E_{on2} with temperature. E_{on1} and E_{on2} are defined in the following. Turn-off speed decreases with increasing temperature, corresponding to an increase in turn-off energy. Both Turn-On and Turn-Off switching speeds decrease with increasing gate resistance, corresponding to an increase in switching energies. Switching times and energies also vary strongly with stray inductances in the circuit, including the gate-drive circuit. In particular, a stray inductance in series with the emitter will significantly affect switching times and energies.

$t_{d(on)}$,—Turn-on Delay Time

Turn-on delay time is the time from when the gate–emitter voltage rises past 10% of the drive voltage to when the collector current rises past 10% of the specified inductor current.

$t_{d(off)}$,—Turn-off Delay Time

Turn-off delay time is the time from when the gate–emitter voltage drops below 90% of the drive voltage to when the collector current drops below 90% of the specified inductor current. This gives an indication of the delay before the current begins to transition in the load.

t_r—Current Rise Time

Current rise time is the time during which the collector current rises from 10% to 90% of the specified inductor current.

t_f—Current Fall Time

Current fall time is the time during which the collector current dropping from 90% to 10% of the specified inductor current.

E_{on2}—Turn-on Switching Energy with Diode

This is the clamped inductive turn-on energy that includes a commutating diode reverse recovery current in the IGBT turn-on switching loss.

Turn-on switching energy is an integral part of the product of the collector current and collector–emitter voltage over when the collector current rises past 5% of the test current to when the voltage falls below 5% of the test voltage.

E_{off}— Turn-off Switching Energy

This is the clamped inductive turn-off energy. E_{off} is an integral part of the product of the collector current and collector–emitter voltage over when the gate–emitter voltage drops below 90% to when the collector current reaches zero.

E_{on1}— Turn-on Switching Energy

This is the clamped inductive turn-on energy of the IGBT only, without the effect of a commutating diode reverse recovery current adding to turn-on loss.

g_{fe}—Forward Transconductance

Forward trans-conductance relates collector current to gate–emitter voltage. Forward trans-conductance varies with collector current, collector–emitter voltage, and temperature. High transconductance corresponds with low plateau voltage and fast current rise and fall times.

Usable Frequency versus Current

In an inductive hard switching application, switching frequency is limited by minimum and maximum pulse widths as well as conduction and switching losses. The pulse width limitation is because of transient thermal responses in the die. Back-to-back switching transients do not allow the die time to cool between the large hard switching power loss spikes. Also, repetitively not allowing the switching transient to complete before switching the other way can overheat the die. Depending on operating temperatures and transient thermal impedance, the die junction may become overheated even if the duty cycle is very small.

The minimum duty cycle limitation is a challenge for motor drives, such as those in electric vehicles, where at very low power an exceptionally small duty cycle is required (unless the switching frequency is dropped into the audible range or some types of pulse skipping scheme are implemented).

The total switching time can be estimated by adding the turn-on and turn-off current delay times and current rise and fall times. Examination of the switching waveforms reveals that this gives a good approximation of total switching time. Only the voltage fall time during turn-on is not accounted for, but this is relatively short. The 5% switching period limitation on total switching time provides a large enough error margin for this approximation anyway, so the maximum frequency based on minimum pulse width is

$$f_{max1} = \frac{1}{T_s} = \frac{0.05}{t_{d(on)} + t_{d(off)} + t_r + t_f} \tag{2.9}$$

The thermal limit to frequency is derived as follows:

$$P_{diss} = \frac{T_J - T_C}{R_{\theta JC}} = P_{cond} + \frac{E_{on2} + E_{off}}{t_{diss}} \tag{2.10}$$

P_{cond} is the conduction power loss (collector current times the $V_{CE(on)}$ at that collector current times duty cycle) and t_{diss} is the minimum time in which E_{on2} and E_{off} can be dissipated to maintain the specified junction temperature. As the conduction loss is based on a fixed 50% duty cycle assumption, it is independent of frequency. However, the higher the conduction loss, the more time is required to dissipate the switching losses. Thus, the inverse of t_{diss} is the maximum frequency:

$$f_{max2} = \frac{1}{t_{diss}} = \frac{\frac{T_J - T_C}{R_{\theta JC}} - P_{cond}}{E_{on2} + E_{off}} \tag{2.11}$$

Finally, the maximum switching frequency at a given collector current is simply the minimum of f_{max1} and f_{max2}. Frequency is typically limited thermally except at a very low current.

Figure 2.20 shows an example curve of frequency versus current for a 600 V PT IGBT. It should be noted again that this curve is for inductive hard switching at one particular set of conditions for both the Device Under Test (DUT) and the clamping diode.

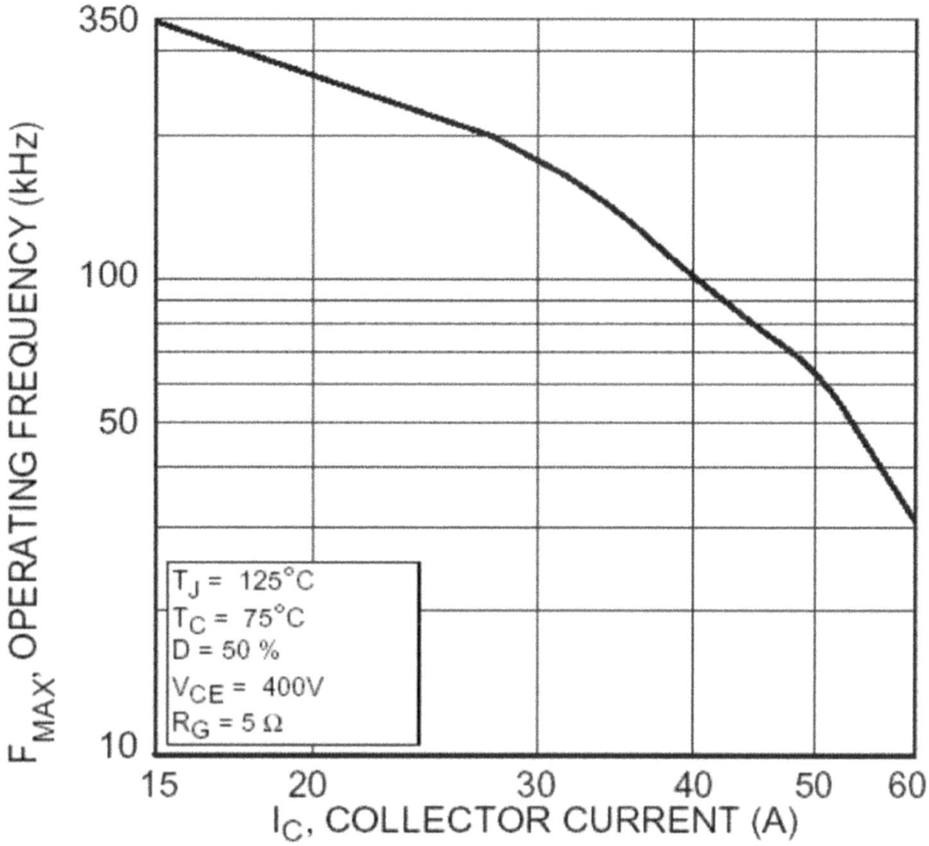

Figure 2.20 Maximum frequency versus current

2.2.5. Evaluation of Switching Losses for IGBT

Diode Turn-On Loss Formula:

In modern fast recovery diodes used with IGBTs, the turn-on loss is negligible (less than 1%) compared to the turn-off loss. Therefore, formulae to model diode turn-on losses are eliminated in this approach.

Diode Turn-Off Loss Formula:

An idealized approximation of diode turn-off waveforms is shown in Figure 2.21.

The waveforms can be characterized using the parameters I_{rrm} (peak reverse recovery current) and t_{rr} (reverse recovery time). Knowing the initial dI_d/dt (which depends on the IGBT turn-on rise time), the parameters t_{rra} and t_{rrb} are estimated as

$$t_{rra} = \frac{I_{rrm}}{\left(\left.\frac{dI_d}{dt}\right|_{t_{1(rr)}}\right)} \tag{2.12}$$

$$t_{rrb} = 1.11(t_{rr} - t_{rra})$$

It is also assumed that the diode voltage drops to 90% of its reverse blocking voltage during the situation this equation represents, $[t_{2(rr)}, t_{3(rr)}] = k_{rr}t_{rrb}$. Energy loss occurs in the diode during the reverse recovery,

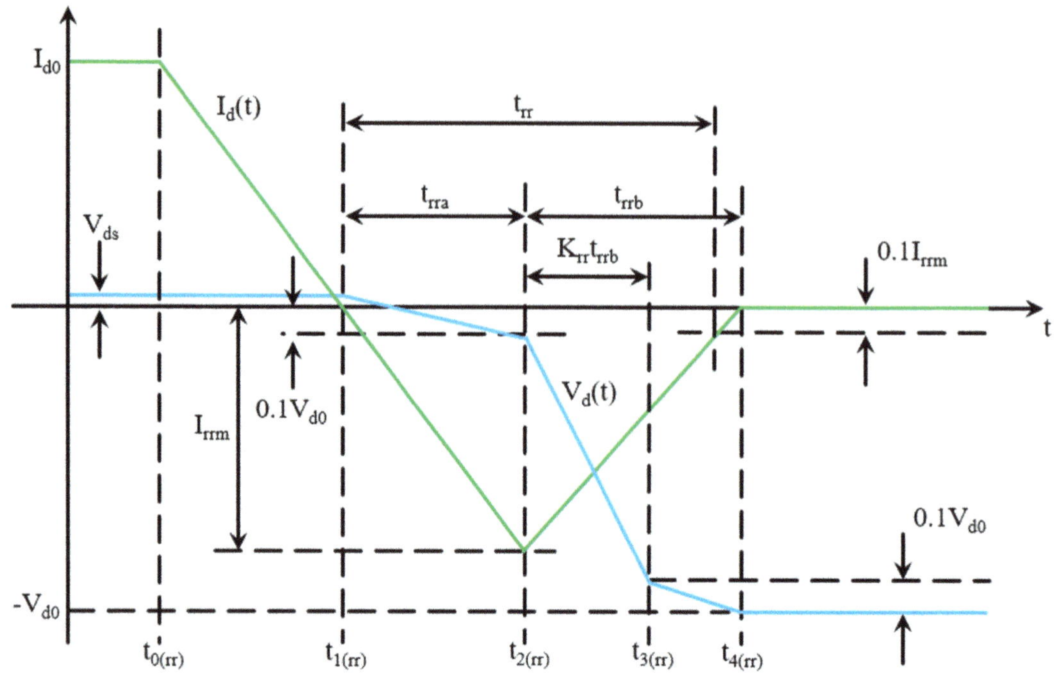

Figure 2.21 Approximated waveforms of diode turn- "OFF" transient

particularly during the building up of reverse voltage. This can now be computed as shown in Equation (2.13):

$$\begin{aligned} &= \int_{t_{0(rr)}}^{t_{4(rr)}} V_d(t) I_d(t) dt \\ W_{rec} &= 0.5 V_{ds} I_{d0} \frac{I_{d0}}{\left(\left. \frac{dI_d}{dt} \right|_{t_1(rr)} \right)} + 0.33 V_{d0} I_{rrm} t_{rra} \\ &+ V_{d0} I_{rrm} (0.467 - 0.433 k_{rr} + 0.15 k_{rr}^2) t_{rrb} \end{aligned} \tag{2.13}$$

IGBT Turn-On Loss Formula:

Figure 2.22 shows approximate waveforms for the hard turn-"on" transient of an IGBT.

Turn-on behavior is characterized by the turn-on delay time $t_{d(on)}$, the rise time t_r, and the turn-on energy W_{on}. The turn-on gate pulse applied at $t_{0(on)}$ raises the gate voltage V_{ge} gradually. When V_{ge} reaches a threshold voltage V_{th}, the collector current I_{ce} starts to rise rapidly and the load current in the FWD of the opposite leg gradually transfers to the IGBT. The rate of rise of the collector current can be determined as shown in Equation (2.14):

$$\frac{dI_{ce}}{dt} = \frac{0.8 I_{ce0}}{t_r} \tag{2.14}$$

During this current rise, the device (collector-emitter) voltage $V_{ce}(t)$ experiences a drop because of parasitic inductance L_p. It is assumed that the voltage $V_{ce}(t)$ drops linearly over the period $\lfloor t_{1(on)}, t_{2(on)} \rfloor$ and reaches a plateau in the voltage waveform. This plateau voltage V'_{ce0} can be found as follows:

$$V'_{ce0} = V_{ce0} - L_P \frac{dI_{ce}}{dt} = V_{ce0} - \frac{0.8 I_{ce0}}{t_r} L_p \tag{2.15}$$

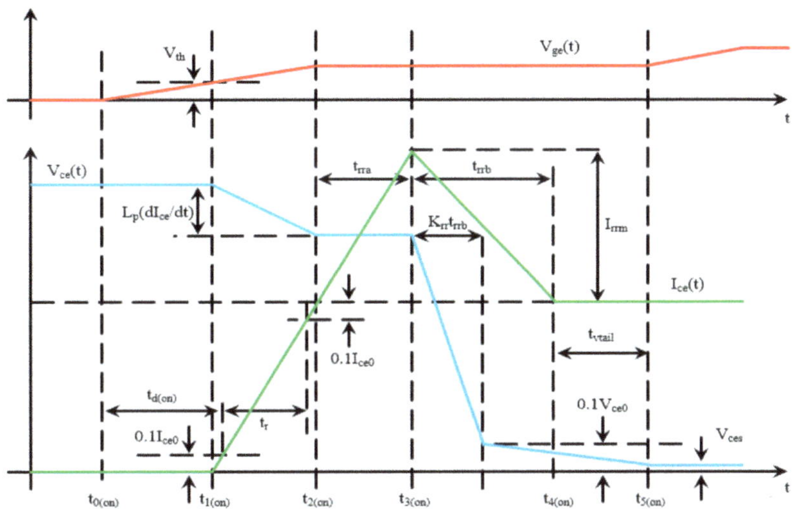

Figure 2.22 Hard switching turn-on transient waveforms for loss calculation

When the load current is fully transferred to the IGBT at $t_{2(on)}$, the outgoing FWD starts to turn-off, and it forces its reverse recovery current through the IGBT. The duration, during which the reverse recovery current increases, can be computed by substituting Equation (2.14) into Equation (2.12):

$$t_{rra} = \frac{I_{rrm}}{0.8 I_{ce}} t_r \tag{2.16}$$

The turn-on energy loss can be computed as follows:

$$\begin{aligned} W_{on} &= \int_{t_{o(on)}}^{t_{4(on)}} V_{ce}(t) I_{ce}(t)\, dt \\ &= 0.05 V_{ceo} I_{ceo} t_{d(on)} + 0.225 V_{ce0} I_{ce0} t_r + V'_{ceo} I_{ceo} \left(0.394 t_r + t_{rra} + 0.55 k_{rr} t_{rrb}\right) \\ &\quad + 0.5 V'_{ceo} I_{rrm} t_{rra} + 0.5 V_{ces} I_{ce0} t_{vtail} + V'_{ceo} I_{rrm} \left(0.033 + 0.483 k_{rr} - 0.167 k_{rr}^2\right) t_{rrb} \\ &\quad + V'_{ceo} I_{ceo} k_{vtail} \left[0.5 \left(1 - k_{rr}\right) t_{rrb} + 0.05 t_{vtail}\right] + 0.167 V'_{ceo} I_{rrm} k_{vtail} \left(1 - k_{rr}\right) t_{rrb} \end{aligned} \tag{2.17}$$

where

$$k_{vtail} = \frac{t_{vtail}}{t_{vtail} + (1 - k_{rr}) t_{rrb}}$$

IGBT Turn-off Loss Formula:

The IGBT's turn-off behavior is characterized by the turn-off delay time $t_{d(off)}$, fail time t_f, and turn-off energy W_{off}. The turn-off process starts on the application of negative gate voltage at time $t_{0(off)}$.

The initial period $\left[t_{0(off)}, t_{1(off)}\right]$ (during which gate–emitter voltage V_{ge} reduces but the collector-emitter voltage V_{ce} remains essentially unchanged) is assumed to be equal to $k_{off} t_{d(off)}$ as indicated in Figure 2.23. The collector–emitter voltage subsequently rises rapidly at the rate shown in Equation (2.18):

$$\frac{dV_{ce}}{dt} = \frac{0.9 V_{ce0}}{(1 - k_{off}) t_{d(off)}} \tag{2.18}$$

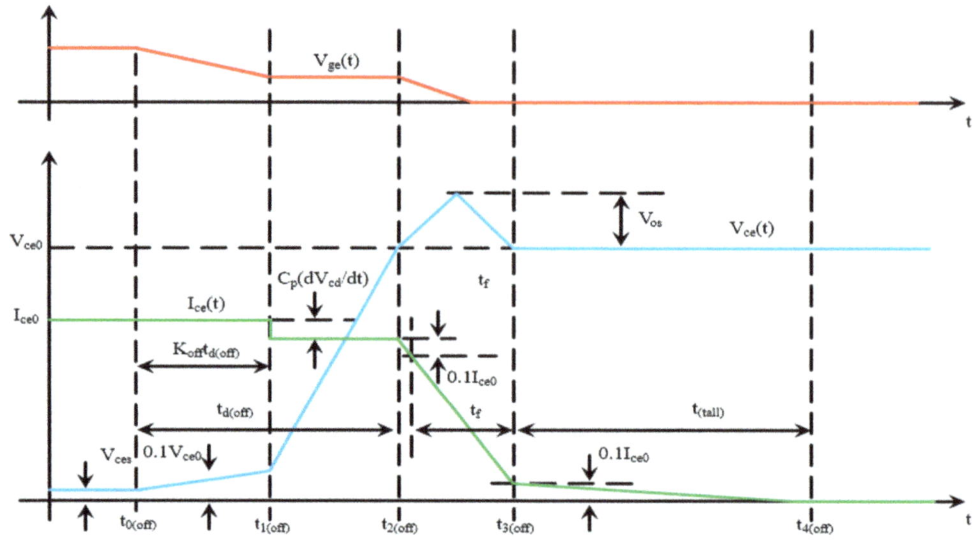

Figure 2.23 Approximated hard switching turn- "OFF" transient waveforms

Considering the idealized waveforms, the resulting plateau in the collector current I'_{ce0} can be approximately determined as

$$I'_{ce0} = I_{ce0} - C_p \frac{dV_{ce}}{dt} = I_{ce0} - \frac{0.9V_{ce0}}{\left(1 - k_{off}\right) t_{d(off)}} C_p \tag{2.19}$$

In Equation (2.19), C_p is the parasitic capacitance. The parasitic inductance produces an overshoot in the voltage V_{ce}:

$$V_{os} = L_p \frac{dI_{ce}}{dt} = \frac{0.8I'_{ce0}}{t_f} L_p \tag{2.20}$$

Based on the idealized waveform, an approximate expression for the turn-off energy can be developed:

$$\begin{aligned} W_{off} &= \int_{t_4(off)}^{t_0(off)} V_{ce}(t) I_{ce}(t) dt \\ &= 0.5V_{ces}I_{ce0}k_{off}t_{d(off)} + 0.05V_{ce0}I_{ce0}k_{off}t_{d(off)} + 0.275V_{os}I'_{ce0}t_f \\ &\quad + 0.55V_{ces}I'_{ce0}(1 - k_{off})t_{d(off)} + V_{ce0}I'_{ce0}[0.55t_f + 0.05t_{itail}] \end{aligned} \tag{2.21}$$

2.2.6. Thermal Performance of IGBT

The junction temperature of the device is an important parameter, as an excessive value can damage the device. The mounting of the device and the heat sink become important in determining the heat removal performance. Therefore, an accurate model is required for thermal performance analysis of the power electronic circuit.

Thermal Model of the IGBT:

A thermal model of the IGBT switching device can be represented as in Figure 2.24. P_{Lj} is the power loss in the device whereas the junction temperature and case temperature are denoted by T_j and T_c, respectively. R_{thi} and

C_{thi} represent the thermal resistance and capacitance of various layers of the semiconductor device. The heat sink can be modeled as a single lumped thermal capacitance C_m and resistance R_{sa} from the sink to the ambient temperature T_a shown in Figure 2.24.

The information required to obtain thermal network parameters is commonly given in the form of a single-pulse junction to case transient thermal impedance Z_{th}, which can be approximated as follows:

$$Z_{th}(t) = A_0 + \sum_{i=1}^{n} A_i e^{-\alpha_1 t} \tag{2.22}$$

In this equation, $A_0 \ldots A_n$ and $\alpha_1 \ldots \alpha_n$ are constants to be found through an appropriate curve fitting technique. Using Equation (2.22), the transfer function of the network can be obtained for a case of two exponential terms as given in Equation (2.23):

$$Z_{th}(s) = \frac{k_1 + sk_2}{1 + sk_3 + s^2 k_4} \tag{2.23}$$

where

$$k_1 = A_0 \quad k_2 = A_0\left(\frac{1}{\alpha_1} + \frac{1}{\alpha_2}\right) + A_1\left(\frac{1}{\alpha_1}\right) + A_2\left(\frac{1}{\alpha_2}\right)$$
$$k_3 = \frac{1}{\alpha_1} + \frac{1}{\alpha_2} \quad k_4 = \frac{1}{\alpha_1 \alpha_2}$$

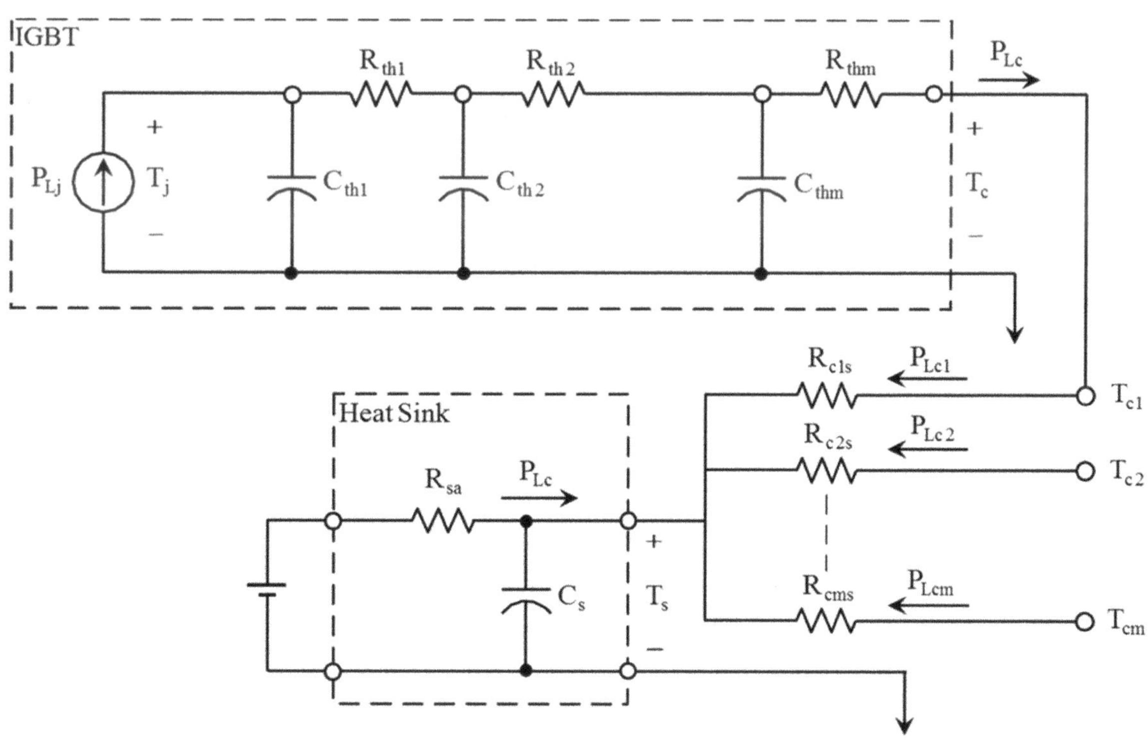

Figure 2.24 Equivalent thermal network of a semiconductor device and heat sink

Alternatively, the transfer function of the thermal equivalent network in Figure 2.24 can be obtained in terms of R_{thi} and C_{thi} as follows:

$$Z_{th}(s) = \frac{(R_1 + R_2) + s\,(R_1 R_2 C_2)}{1 + s\,(R_1 C_1 + R_2 C_1 + R_2 C_2) + s^2\,(R_1 C_1 R_2 C_2)} \tag{2.24}$$

By comparison of the coefficients of Equations (2.23) and (2.24), the expressions for R_{thi} and C_{thi} are obtained as follows:

$$\begin{aligned} R_{th1} &= \frac{k_2^2}{(k_2k_3 - k_1k_4)} \\ R_{th2} &= \frac{k_1 - k_2^2}{(k_2k_3 - k_1k_4)} \\ R_{th3} &= \frac{k_4}{k_2} \\ R_{th4} &= \frac{(k_2k_3 - k_1k_4)^2}{k_2(k_1k_2k_3 - k_1^2k_4 - k_2^2)} \end{aligned} \tag{2.25}$$

2.3 Series and Parallel Connection of IGBT

For high voltage applications, the series-connected two level topologies are attractive because they cost less and are less complex in circuit design and control compared to multi-level circuit topologies. The main concern with series connection is unequal voltage distribution between devices in both transient and steady-state operations. The operating voltage of the series-connected power semiconductor device module is higher than the allowable individual device operating voltage, and voltages ideally must be shared equally between series-connected devices. The unequal voltage sharing is mainly a result of the spread of device dynamic and static parameters and external circuit conditions. This voltage unbalance may exceed the individual device voltage rating and subsequent failure of this device, causing final failure of the entire series string of devices.

Also, the paralleling of power semiconductor devices should be done with care, so as to avoid possible current unbalance in parallel modules. The current sharing has to be balanced in all the operating conditions. The main reasons for asymmetric current sharing are tolerance in device parameters, propagation delay time through the gate unit, and different stray inductances in parallel branches. This current unbalance causes some devices to carry higher currents than their individual ratings, which will lead to device breakdown owing to high-power dissipation.

2.3.1. Series Connection of IGBT [6]–[9]

Voltage Unbalancing of Series connection of IGBT

With the development of power electronic technology, high- voltage and high-power equipments puts forward higher requests to the IGBT. However, because of the structure and material of semiconductor devices, the voltage level of the IGBT is the highest in 6.5kV, which cannot reach the voltage level on many occasions in the power system, so the IGBT series is an inevitable choice for IGBT high-voltage applications. The press pack IGBT has double-side heat dissipation, and the current density is higher under the same voltage level, especially for use in series. As the switching speed of an IGBT is very fast and the device parameters cannot be completely consistent, voltage balancing is key to the application of series-connected IGBTs.

The main reasons for series-connected IGBTs' voltage unbalancing are the gate electrode driver signal being asynchronous or the parameter being different. As shown in Figure 2.25, with two series-connected IGBTs (Q1, Q2) as an example, Un is the average steady-state voltage of each series-connected IGBT.

Voltage balancing can be achieved if each series-connected IGBT's collector–emitter voltage changes at the same time and the slope is the same (i.e., to minimize Δt_{on}, Δt_{off} and the slope of $\Delta\frac{dv}{dt}_{Q1-Q2}$ between collector and emitter). The dynamic voltage unbalancing of series devices can be expressed as $\alpha_{off} = \frac{\Delta U_n}{U_n} \times 100\%$.

Since the IGBT series connection is mainly focused on the dynamic process of turn-off of IGBTs, but not the whole process. The control strategies are the same as Figure 2.26. From Figure 2.26, these methods can be divided into active control directly and passive control indirectly. Active control is done through a closed-loop feedback control circuit of collector–emitter and gate electrode-drive signal, which is complicated and expensive. Passive control is done through the introduction of RCD snubber circuits in the peripheral IGBT, but cost is high and efficiency is low during high frequency.

With the development of digital technology and high-speed optical transmission technology, the difference in gate pulse between the synchronous signals is further reduced, and IGBT parameters are the main factors in IGBT series voltage balancing.

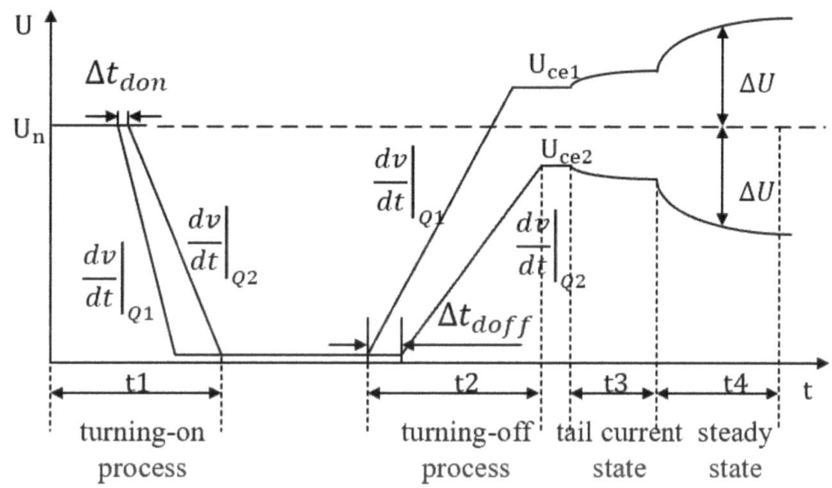

Figure 2.25 Diagram of voltage unbalance during the whole turning on and off period of an IGBT

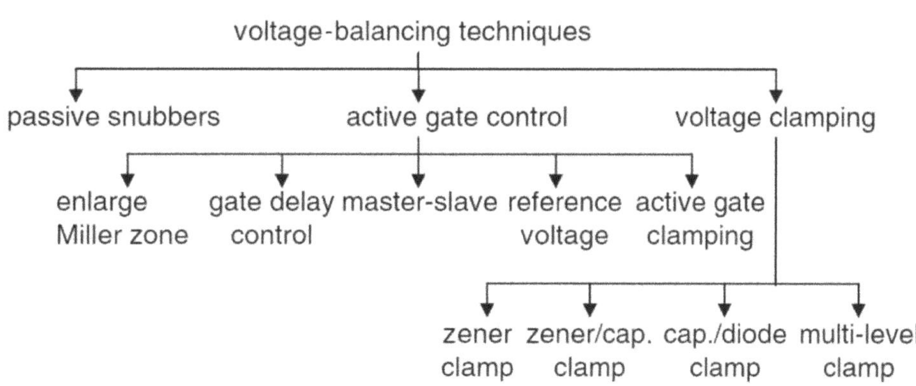

Figure 2.26 Overview of voltage-balancing techniques for series-connected IGBTs

Voltage Unbalancing of Dynamic Turn-On Process

The main reason for voltage unbalancing is that t_{don} (turn-on time) and the down slope of $\frac{dv}{dt}$ are not the same in the dynamic turning-on process. Input capacitance of the gate electrode is charged by the driving power supply in the turning-on process; the turn-on time is

$$t_{don} = R_{gon} \times \left(C_{ge} + C_{gc}\right) \times \ln \frac{U_{cc}}{U_{cc} - U_{ge(th)}} \tag{2.26}$$

To the IGBT and the driver selected, the threshold voltage $U_{ge(th)}$ and the amplitude of the driving power supply U_{cc} can be fixed, so t_{don} is affected by the gate electrode-drive resistance R_{gon}, Gate–Emitter capacitance C_{ge}, Gate-Collector capacitance C_{gc} and so on. Reducing the difference between C_{ge} and C_{gc} of an IGBT as soon as possible can ensure t_{don} is basically synchronous.

When the U_{ce} of the IGBT rapidly dropped, C_{gc} changed with U_{ce} can be approximatly equivalent as shown in Figure 2.27.

At the same time, the approximate expression of the down slope of U_{ce} is as follows:

$$\frac{dU_{ce}}{dt} \approx \begin{cases} -\frac{U_{cc}-U_{mill\_on(I_c)}}{R_{gon}\times C_{gc2}} & U_{ce} \geq U_{ce1} \\ -\frac{U_{cc}-U_{mill\_on(I_c)}}{R_{gon}\times C_{gc1}} & U_{ce} < U_{ce1} \end{cases} \tag{2.27}$$

As shown above, $U_{mill\_on(I_c)}$ is the amplitude of the gate Miller level in the turn-on process, which is related with I_c. For the series of IGBT, I_c is equal, so it can be considered that $U_{mill\_on(I_c)}$ is also equal. Voltage decline rate is mainly affected by R_{gon} and C_{gc} in the dynamic turn-on process.

Voltage Unbalancing of Dynamic Turn-off Process

Similar to the turn-on process, total input capacitance of the gate electrode is charged by the driving power supply in the turning-off process; the turn-off time is

$$t_{doff} \approx R_{goff} \times (C_{ge} + C_{gc}) \times \ln\left(\frac{U_{cc} - U_{gg-}}{U_{mill_{off}(I_c)} - U_{gg-}}\right) \tag{2.28}$$

Where Ugg- is the converter's turn-off voltage.

Therefore, t_{doff} is affected by the gate electrode drive resistance R_{goff}, Gate-Emitter capacitance C_{ge} and Gate-Collector capacitance C_{gc}, and so on.

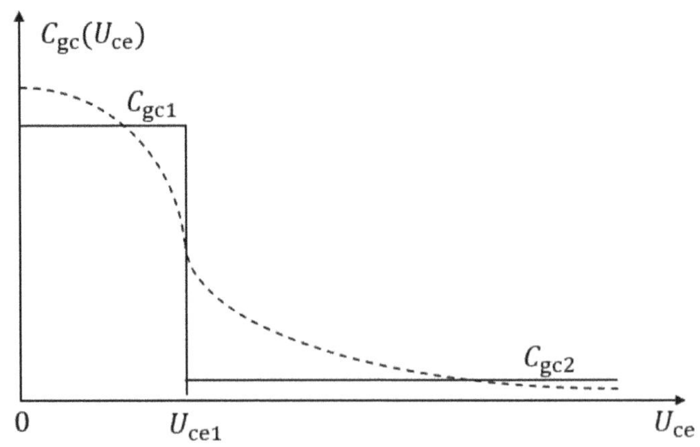

Figure 2.27 Principle of the C_{ge} with U_{ce}.

At the same time, the approximate expression of the rising slope of U_{ce} is as follows:

$$\frac{\mathrm{d}U_{ce}}{\mathrm{dt}} \approx \begin{cases} \dfrac{U_{mill\_off(I_c)} - U_{gg-}}{R_{goff} \times C_{gc1}} & U_{ce} \leq U_{ce1} \\ \dfrac{U_{mill\_off(I_c)} - U_{gg-}}{R_{goff} \times C_{gc2}} & U_{ce} > U_{ce1} \end{cases} \quad (2.29)$$

Then voltage-rising rate is mainly affected by R_{goff} and C_{gc} in the dynamic turn-off process.
In summary, the main factors that affect the dynamic voltage balancing of IGBT are shown in Table 2.2.

According to the analysis above, the dynamic parameters that affect the IGBT in the turn-off process can be divided into two types of external and internal parameters. The external parameters include I_c, drive voltage U_{cc} and U_{gg-}, and gate electrode resistance R_g. The internal parameters include C_{ge} and C_{gc}. Further, I_c ,U_{cc}, and U_{gg-} can be considered equal. Both R_{gon} and R_{goff} are adjustable. Then the key factor affecting the dynamic voltage unbalancing is the difference of C_{ge} and C_{gc} of the IGBT.

A gate compensation capacitor network for dynamic process was then introduced, which is a parallel-connected voltage-balancing capacitor between the gate electrode and the emitter to weaken the inconsistency of the C_{ge} of the IGBT, and a parallel-connected voltage-balancing capacitor between the gate electrode and the collector to weaken the inconsistency of the C_{gc} of the IGBT. Because the parallel capacitor causes IGBT turn-on and turn-off time to be longer, the need for dynamic characteristic optimization is reduced, driving the resistance of R_{gon} and R_{goff} in order to make t_{on} and t_{off} unchanged.

Voltage Unbalancing of the Tail Current State

When the gate voltage of U_{ge} is reduced to below the threshold voltage, the collector current I_c goes into the tail current stage. As the IGBT parameters may not be completely consistent, the current stage of tail attenuation also cannot be exactly the same, which results in voltage unbalancing of the IGBT series. Using two series-connected IGBTs as an example, the voltage unbalancing of the tail current state is shown in Figure 2.28.

At this time, the gate driver lost control of the IGBT, and only through RC voltage-balancing circuit designed for tail current state can it improve voltage unbalancing.

Assume that the decay rate of Q_1 is greater than Q_2, the voltage U_{ce1} of Q_1 rises, U_{ce2} of Q_2 declines, the decline of the tail current decay is approximate to an exponential function, and the tail current time constant is τ_1 and τ_2 of the two series-connected IGBTs, respectively.

Assume that the current I_0 flows through the IGBT when the time is t_0, and voltage unbalancing is absorbed by capacitance C caused by the tail current. Ignoring the limiting current resistor R_s, the relationship between

Table 2.2 Factors Affecting the Dynamic Voltage-Balancing Process of IGBT

Variable		Influence factor
Turning-on	t_{don}	R_{gon}, C_{ge}, C_{gc}
	$\frac{dv}{dt}$	R_{gon}, C_{gc}
		R_{gon}, C_{gc}
Turning-off	t_{doff}	R_{goff}, C_{ge}, C_{gc}
	$\frac{dv}{dt}$	R_{goff}, C_{gc}
		R_{goff}, C_{gc}

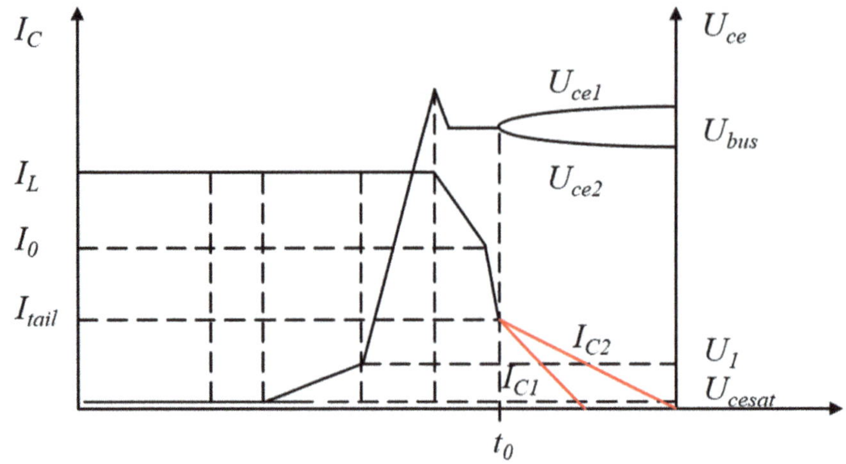

Figure 2.28 Diagram of voltage unbalance during tail current process of series-connected IGBTs

the voltage and the current of the series IGBT is expressed as follows after $\triangle$t:

$$C \times [U_1(t_0+\Delta t) - U_1(t_0)] = I_0 \times \left(e^{-\frac{t_0}{\tau_2}} - e^{-\frac{t_0}{\tau_1}}\right) \times \Delta t + C \times [U_2(t_0+\Delta t) - U_2(t_0)]$$

which is simplified as

$$C \times U_1' = I_0 \times \left(e^{-\frac{t_0}{\tau_2}} - e^{-\frac{t_0}{\tau_1}}\right) + C \times U_2'$$

and

$$U_1 + U_2 = U_{dc} \tag{2.30}$$

The voltage difference in the tail state of the series IGBT is

$$\Delta U = U_1 - U_2 = \frac{I_0}{C} \times \left(\int e^{-\frac{t_0}{\tau_2}} dt - \int e^{-\frac{t_0}{\tau_1}} dt\right) \tag{2.31}$$

Assuming that the current decays to zero after (4~5) τ, absorption capacity is expressed as follows:

$$C = \frac{I_0}{\Delta U} \times (\tau_2 - \tau_1) \tag{2.32}$$

Voltage Unbalancing in the Steady State

Parallel-connected voltage-balancing resistance for the steady state in order to solve the static voltage of series IGBT imbalance when the IGBT is completely turned off. Resistance is selected according to the following formula:

$$R = \frac{1}{10} \times R_{off} \tag{2.33}$$

2.3.2. Parameter Design of IGBT subcircuit

Static State Voltage-Balancing Subcircuit

In the off-state of series-connected IGBTs, the voltage unbalance is mainly caused by the variance of collector–emitter cut-off currents (I_{CES}) of IGBTs. In other words, the difference between off-state equivalent resistances, which can be defined as $R_{off} = V_{CES}/I_{CES}$, leads to the off-state voltage unbalance. Therefore, the general method for static state voltage balancing is connecting a resistor, whose resistance is much lower than R_{off}, in parallel with each IGBT. This resistor is named static state voltage-balancing resistor R_{static}. IGBT with higher I_{CES} has lower R_{off}. In order to achieve a better static state voltage-balancing effect, the highest I_{CES} should be applied to the calculation of R_{static}.

In the condition of a two IGBT (T1 and T2)-series connection, assuming

$$I_{CES1} < I_{CES2} \tag{2.34}$$

that is,

$$R_{off1} > R_{off2} \tag{2.35}$$

T1 withstands higher off-state voltage than T2. In order to ensure that the off-state voltage unbalance ratio is less than 10%, it is required that

$$\frac{R_{static} \cdot V_{CES}/I_{CES1}}{R_{static} + V_{CES}/I_{CES1}} < \frac{R_{static} \cdot V_{CES}/I_{CES2}}{R_{static} + V_{CES}/I_{CES2}} \cdot 110\% \tag{2.36}$$

Moreover, the voltage on T1 mustn't exceed V_{CES}, that is,

$$V_{dc} \cdot \frac{\frac{R_{static} \cdot V_{CES}/I_{CES1}}{R_{static} + V_{CES}/I_{CES1}}}{\frac{R_{static} \cdot V_{CES}/I_{CES1}}{R_{static} + V_{CES}/I_{CES1}} + \frac{R_{static} \cdot V_{CES}/I_{CES2}}{R_{static} + V_{CES}/I_{CES2}}} < V_{CES} \tag{2.37}$$

where V_{dc} represents DC-link voltage. Taking the most adverse situation into account, namely $I_{CES1} = 0$, $I_{CES2} = I_{CESmax}$, where I_{CESmax} is the maximum value of I_{CES}, in order to satisfy (2.35) and (2.36), it is required that

$$R_{static} < \frac{0.1V_{CES}}{I_{CESmax}} \tag{2.38}$$

$$R_{static} < \frac{V_{CES}\,(2V_{CES} - V_{dc})}{I_{CESmax}\,(V_{dc} - V_{CES})} \tag{2.39}$$

The resistance value of R_{static} can be determined through (2.37) and (2.38), while its rated power should satisfy

$$P_{Rstatic} > \frac{V_{dc}^2}{4R_{static}} \tag{2.40}$$

Dynamic State Voltage-Balancing Subcircuit

In an IGBT series connection circuit, many factors might influence the voltage balance during the transient switching process. Therefore, it is more difficult to implement voltage balancing in a dynamic state. The simplest dynamic state voltage-balancing circuit is connecting a capacitor $C_{dynamic}$ in parallel with each IGBT. $C_{dynamic}$ limits the variation rate of collector–emitter voltage V_{CE}, and thus could suppress the voltage unbalance in a dynamic state. However, $C_{dynamic}$ would increase the peak current and loss of IGBT in its switching-on process. In order to solve this problem, a resistor $R_{dynamic}$ is brought in and connected in series with $C_{dynamic}$. But $R_{dynamic}$ would weaken the effect of voltage balancing in the dynamic state and would introduce loss to the voltage-balancing circuit.

Connecting a fast recovery diode in parallel with $R_{dynamic}$ could prevent these drawbacks of the RC snubber circuit to some extent. However, the fast recovery diode must have the same voltage rating as the IGBT. The selection of the required diode is extremely difficult. Therefore, the RC snubber, which is composed of $C_{dynamic}$ and $R_{dynamic}$, is adopted as the dynamic state voltage-balancing sub-circuit for the IGBT series connection. There are two ways to design the parameter of $C_{dynamic}$. One way is to consider the mismatch of IGBT characteristics, whereas the other way focuses on the asynchronies of the drive circuits.

From the view of device characteristics, the dynamic state voltage unbalance is caused by the mismatch of reverse recovered charge Q_{rr} of the anti-parallel diodes in the series -connected IGBT modules. Assuming

$$Q_{rr1} < Q_{rr2} \tag{2.41}$$

during the switching-off process, the load current charges the junction capacitor of IGBT and $C_{dynamic}$; the unbalanced voltage can be expressed as

$$\Delta V < \frac{\Delta Q}{C_{dynamic}} = \frac{Q_{rr2} - Q_{rr1}}{C_{dynamic}} \tag{2.42}$$

In the dynamic state, it must be guaranteed that the voltage on T1 is lower than V_{CES}, that is

$$\frac{V_{dc} + \Delta V}{2} < V_{CES} \tag{2.43}$$

Considering the most adverse situation, which is $Q_{rr1} = 0$ while $Q_{rr2} = Q_{rrmax}$, where Q_{rrmax} is the maximum value of Q_{rr}, combining (2.42) and (2.43), it is obtained that

$$C_{dynamic} = \frac{Q_{rrmax}}{2V_{CES} - V_{dc}} \tag{2.44}$$

From the view of drive synchronicity, the dynamic state voltage unbalance is caused by the delay between switching times of the series-connected IGBTs. Assuming the maximum delay time is Δt_{max} and the current of IGBT is I_C, then the maximum unbalanced voltage during the dynamic process is deduced as

$$\Delta V = \frac{\Delta t_{max} \cdot I_C}{C_{dynamic}} \tag{2.45}$$

In order to guarantee the voltages on IGBTs are lower than V_{CES}, it is required that

$$C_{dynamic} > \frac{\Delta t_{max} \cdot I_C}{2V_{CES} - V_{dc}} \tag{2.46}$$

During the on-state of IGBT, $C_{dynamic}$ should be totally discharged through $R_{dynamic}$, and therefore, it should be satisfied that

$$R_{dynamic} < \frac{T_{Onmin}}{3C_{dynamic}} \tag{2.47}$$

where T_{ONmin} is minimum pulse width of IGBT. The rated power of $R_{dynamin}$ is required as

$$P_{Rdynamic} > \frac{C_{dynamic} V_{CES}^2}{2} \cdot f_s \tag{2.48}$$

where f_s refers to the switching frequency of IGBT.

In an IGBT series connection circuit, because of the high -power density, $P_{Rdynamic}$ calculated through Equations (2.44), (2.46), and (2.48) is usually several kilowatts, which would seriously reduce the efficiency of the system and thus is unacceptable. In this case, the parameter design flow should be reversed, determining a proper $P_{Rdynamic}$ first before calculating $C_{dynamic}$ by Equation (2.48), lastly using Equation (2.47) to obtain $R_{dynamic}$. In this way, the calculated $C_{dynamic}$ may not satisfy Equation (2.44) or (2.46), but the insufficient dynamic state voltage-balancing effect would be compensated by the active clamping sub-circuit at the gate side of IGBT.

Active Clamping Subcircuit

Because of the limit of loss, the collector–emitter-side circuits usually cannot provide a sufficient voltage-balancing effect to series-connected IGBTs in a dynamic state. Furthermore, under some special conditions such as overcurrent and DC-link overvoltage, the collector–emitter-side circuits cannot ensure that voltages on IGBTs are all lower than V_{CES}. Therefore, gate side voltage-balancing circuits are indispensable to an IGBT series connection.

Defining the total breakdown voltage of the series-connected Zener diodes Z_1 to Z_n is V_{Z1}, the total breakdown voltage of Z_{11} to Z_{1m} is V_{Z2}, considering the collector–emitter voltage V_{CE} is approximately equal to the collector–gate voltage of IGBT, the switching-off process of an IGBT with an active clamping sub-circuit can be divided into three phases, as shown in Figure 2.29.

In phase 1, where V_{CE} is lower than V_{Z1}, the active clamping sub-circuit doesn't affect the IGBT, and V_{CE} increases rapidly.

In phase 2, where V_{CE} is higher than V_{Z1}, Zener diodes Z_1 to Z_n are broken down, the collector of IGBT charges C_1 through R_2, Z_1 to Z_n, and D, and therefore the rising rate of V_{CE} is reduced and the voltage unbalance is restrained dynamically. In this phase, the effect of the active clamping subcircuit is equivalent to a capacitor that is connected across the collector–gate, enhancing the Miller effect of IGBT.

In phase 3, where V_{CE} reaches $V_{Z1}+V_{Z2}$, Zener diodes Z_1 to Z_n and Z_{11} to Z_{1m} are all broken down. Considering the voltage drop on R_2, V_{CE} is clamped at a voltage that is a little higher than $V_{Z1}+V_{Z2}$ until the energy stored in stray inductance of the commutation circuit is totally consumed, and then the switching-off process is finished.

During phase 2 and phase 3, the active clamping sub-circuit injects current into the gate of the IGBT and makes it operate in its active region, and thus the voltage spike of V_{CE} is suppressed. However, the loss of IGBT increases rapidly and the switching speed is slowed down in these two phases. Consequently, two requirements should be satisfied simultaneously through the parameter design of the active clamping subcircuit. First, the subcircuit must limit the voltage on an IGBT to lower than its V_{CES} and restrain the overvoltage of device. Second, frequent actions of the active clamping sub-circuit must be avoided in order to reduce the loss and switching time of the IGBT.

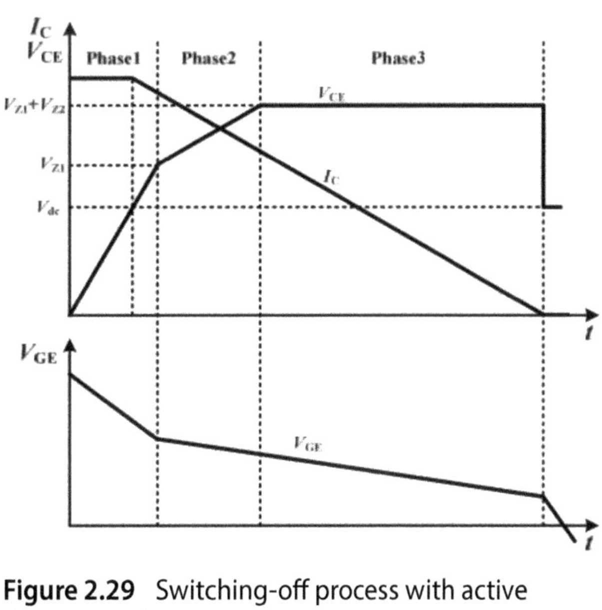

Figure 2.29 Switching-off process with active clamping subcircuit

In the case of a two IGBT-series connection, according to the analyses above, the two threshold values (V_{Z1} and V_{Z1}+V_{Z2}) of the active clamping subcircuit are designed as

$$0.6V_{dc} < V_{Z1} < 0.65V_{dc} \tag{2.49}$$

$$0.65V_{dc} < V_{Z1} + V_{Z2} < 0.75V_{dc} \tag{2.50}$$

Meanwhile, in order to avoid the overvoltage of IGBT,

$$V_{Z1} + V_{Z2} < V_{CES} \tag{2.51}$$

Furthermore, in order to suppress the frequent actions of active clamping, V_{Z1} is required to be higher than the peak voltage of the normal switching-off process, that is,

$$V_{Z1} > \frac{1}{2}\left(V_{dc} + 0.8L_s I_C/t_f\right) \tag{2.52}$$

where L_s refers to the stray inductance of a commutating loop and t_f represents the fall time of the HV-IGBT.

The function of resistor R_2 is limiting the current in the active clamping subcircuit, thus it is required that

$$R_2 > \frac{V_{dc} - V_{Z1}}{I_Z} \tag{2.53}$$

where I_Z is the maximum current of the Zener diodes.

After confirming R_2 and the parameters of the dynamic state voltage-balancing subcircuit, the rising rate of V_{CE} in phase 2 mainly depends on the parameter C_1. The larger C_1 is, the more slowly V_{CE} rises, and the longer the duration of phase 2 is. Parameter C_1should be designed through the simulation considering the characteristics of the IGBT.

After the end of the switching-off process, C_1 is discharged through the resistor R_1. This discharging must finish before the next switching-off process, thus the parameter R_1 should satisfy

$$R_1 < \frac{T_s}{3C_1} = \frac{1}{3C_1 \cdot f_s} \tag{2.54}$$

It can be concluded from the design procedure above that under normal conditions, the active clamping subcircuit would not enter phase 2 and phase 3 during switching-off processes, and voltage balancing of series-connected IGBTs is implemented by the collector–emitter-side circuits. Normally, the active clamping sub-circuit does not have an effect on V_{CE}, which ensures the switching speed and avoids high switching losses of IGBTs. Under special conditions when collector–emitter-side circuits cannot reliably suppress high V_{CE} or high dV_{CE} dt, the active clamping sub-circuit goes into phase 2 and phase 3, and limits V_{CE} to the voltage lower than V_{CES}.

2.3.3. Parallel Connection of IGBT [10]–[13]

The technical issues with paralleling IGBTs can be classified into four main categories, as follows:

1. Steady-state (static) current sharing
2. Dynamic current sharing during switching
3. Thermal stability and maximum junction temperature
4. External influence on sharing and parasitic oscillations

In practice, some of these issues do not cause significant problems, but it is sensible to be aware of them and take precautions to minimize their effects.

The static sharing arises from the spread of on-state resistance (associated with V_{ce} sat) of the IGBT and the wiring impedance of the parallel arm. The dynamic unbalance is mainly because of the spread of transfer characteristics and the external gate-drive circuit. Thus, circuit layout and gate drive play an important part in the parallel operation of IGBTs. The three main techniques used are de-rating the paralleled devices, close-loop control of the active devices, and impedance balancing. The available current-balancing techniques can be overviewed as shown in Figure 2.30.

The equivalent circuit of an IGBT can be depicted by a PNP-transistor, where the base current is controlled by a MOSFET. This equivalent circuit of the IGBT is shown in Figure 2.31, in which the inherent stray capacitors (C_{ge}, C_{gc}, and C_{ce}) are considered as well.

1. Switching phase 0 ~ t_1 (turn-on delay time)

From Figure 2.32, turn-on delay time $t_{d(on)}$ is the time from when the gate–emitter voltage V_{GE} rises past 10% of the drive voltage to when the collector current rises past 10% of the specified inductor current, which will greatly affect the transient current balancing of parallel-connected IGBTs. The time interval of this stage can be roughly equivalent to the turn-on delay time $t_{d(on)}$.

The turn-on delay time $t_{d(on)}$ is

$$t_{d(on)} = -\tau \cdot \ln\left(1 - \frac{V_{GE(th)}}{V_{GE}}\right) \tag{2.55}$$

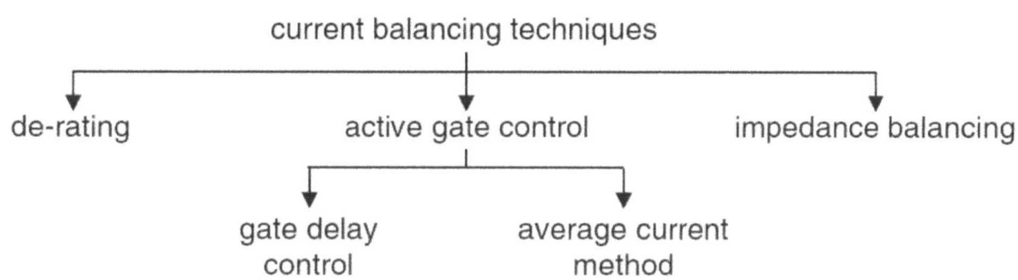

Figure 2.30 Overview of current-balancing techniques for parallel-connected IGBTs

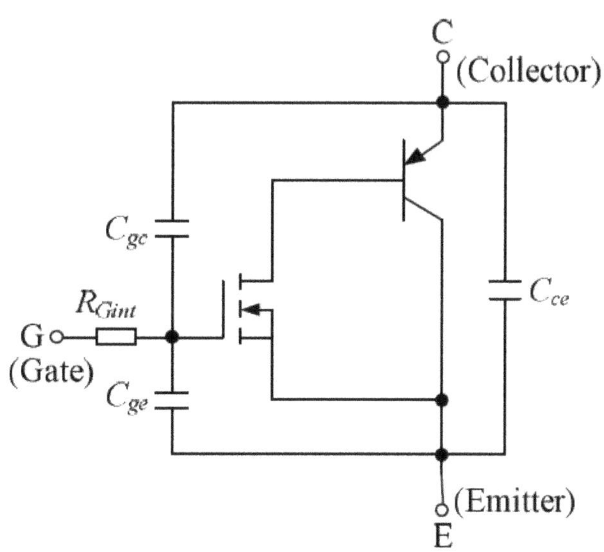

Figure 2.31 Equivalent circuit of IGBT

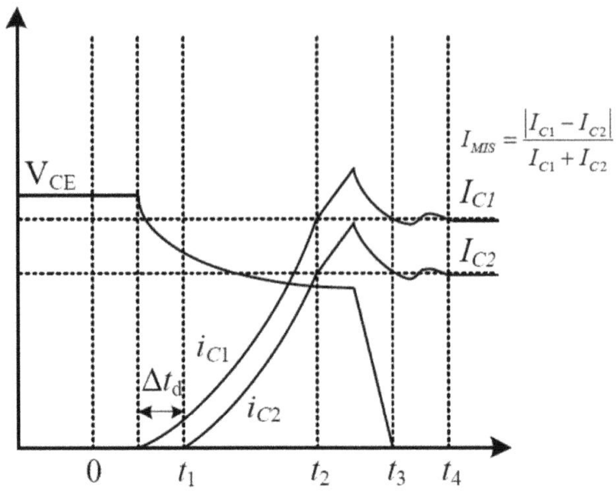

Figure 2.32 Schematic diagram of current-sharing model during turn-on

where the time constant τ is

$$\tau = R_G\left(C_{GE} + C_{GC}\right) \tag{2.56}$$

Equations (2.55) and (2.56) show that the threshold voltage $V_{GE(th)}$, the gate resistance R_G, and the equivalent input capacitance $(C_{GE} + C_{GC})$ have a direct effect on the turn-on delay time $t_{d(on)}$. Therefore, the differences of these parameters between the two IGBT modules will lead to different turn-on delay times. If the two devices are not switched on at the same time, the IGBT module T_1, which is turning on faster, must carry more load current at that moment.

The IGBT module T_2 reaches its threshold voltage $V_{GE(th)2}(t_1)$ at this stage, and the IGBT module T_1 has achieved its threshold voltage $V_{GE(th)1}$ ahead of module T_2 by the time Δt_d:

$$\begin{aligned}\Delta t_d &= t_{d(on)2} - t_{d(on)1}\\ &= -\tau_2 \cdot \ln\left(1 - \frac{V_{GE(th)2}}{V_{GE}}\right) + \tau_1 \cdot \ln\left(1 + \frac{V_{GE(th)1}}{V_{GE}}\right)\end{aligned} \tag{2.57}$$

where the time constants τ_1 and τ_2 respectively, are

$$\tau_1 = R_{G1}\left(C_{GE1} + C_{GC1}\right) \tag{2.58}$$

$$\tau_2 = R_{G2}\left(C_{GE2} + C_{GC2}\right) \tag{2.59}$$

2. **Switching phase $t_1 \sim t_2$ (collector current rise)**
The collector current i_{C2} of IGBT module T_2 starts to rise (t_1), although the collector current i_{C1} of IGBT module T_1 has risen for the time Δt_d. Taking into account the effect of L_e (the parasitic inductance of emitter side), the values of di_{C1}/dt and di_{C2}/dt respectively, are as follows:

$$\frac{di_{C1}}{dt} \approx \frac{V_{GE} - V_{GE(th)1}}{R_{G1}C_{i1}/g_{fs1} + L_{e1}} \tag{2.60}$$

$$\frac{di_{C2}}{dt} \approx \frac{V_{GE} - V_{GE(th)2}}{R_{G2}C_{i2}/g_{fs2} + L_{e2}} \tag{2.61}$$

where C_i is the equivalent input capacitance of this stage and g_{fs} is the forward trans-conductance.

This indicates that the spread of device parameters has less influence on the di_C/dt value than the parasitic inductance of the emitter–ground connection. Thus, the difference of the rising rate of collector current between the two IGBT modules will be unnoticeable ($di_{C1}/dt \approx di_{C2}/dt$) when the parameters ($R_G$, C_i, and g_{fs}) of the two devices differ slightly in a certain manner and the value of the threshold voltage $V_{GE(th)}$ is almost equal.

The entire load current I_{Ctot} will commutate to the parallel-connected IGBTs at the end of this interval, and the FWD will start to block. The relationship between the collector current of the two IGBT modules is expressed as follows:

$$i_{c1}(t = t_2) + i_{c2}(t = t_2) = I_{Ctot} \tag{2.62}$$

$$\begin{aligned} & i_{c1}(t = t_2) - i_{c2}(t = t_2) \\ & = \frac{\left(\frac{di_{C1}}{dt} - \frac{di_{C2}}{dt}\right) \cdot I_{Ctot} + 2 \times \frac{di_{C1}}{dt} \cdot \frac{di_{C2}}{dt} \cdot \Delta t_d}{\frac{di_{C1}}{dt} + \frac{di_{C2}}{dt}} \approx \frac{di_{C1}}{dt} \cdot \Delta t_d \end{aligned} \tag{2.63}$$

3. Switching phase $t_2 \sim t_3$ (transistor fully switched on)

The overshoot in the IGBT collector current i_C during this phase is because of the reverse recovery current of the FWD. A certain share of the reverse recovery current will continue to flow through one of the IGBT modules, whereas the rest flow through the other one.

4. Switching phase $t_3 \sim t_4$ (saturation area)

The two parallel-connected IGBTs are now turned on at t_3. The difference between the collector currents of the two IGBT modules is described as below:

$$\Delta I = I_{C1} - I_{C2} = i_{c1}(t = t_2) - i_{c2}(t = t_2) \approx \frac{di_{C1}}{dt} \cdot \Delta t_d \tag{2.64}$$

This model can be helpful for module selection because the current mismatch can be predicted. Furthermore, the compensation of delay time can be achieved directly when the active gate control method is applied.

2.4 Cosmic Radiation of Influence to IGBT [14]–[17]

Issues with cosmic ray were observed in locomotive converters at high altitudes. At first, the failures of the power semiconductors could not be explained, but after intense study, the origin of the problem was found. Silicon diodes can run into thermal destruction because of cosmic ray. High-power semiconductor devices are designed for operational lifetimes greater than 25 years. Therefore, high-voltage semiconductors must be operated at a voltage level that is significantly lower than the breakdown voltage in order to have a low FIT rate with respect to cosmic-ray-induced failure. The mechanism and cause for these failures are initialized by a neutron from outer space. The effects of cosmic ray are as follows:

Radioactivity from the AI—metalization: It is known from extensive investigations on MOS structures that the aluminum material used for metalization of devices can be a source of alpha particles leading to device failure, so, a careful analysis of the activity of the metallization was required of the metalization. It was found here too that the rates determined were by far too low for the failure rates observed.

Furthermore, the anode side of bare elements without metalization was exposed to the radiation of a high-energy alpha source (Th^{232}, E_{max} = 11,65 MeV) without any effect on the failure rate. These experiments were done on devices with very shallow junctions ($x_j \approx$ 12 pm) with the alpha source put directly on top of the bare silicon.

Radioactivity from the capsule material: Radioactivity from the capsule material was ruled out by operating the devices without the capsule as disc elements where the same failure rates were obtained.

Modeling Cosmic-Ray-Induced Failures: The electrical dipole charge in a fully depolarized filament (zero internal field) of length L and radius R in a field E is approximately (for L >> R)

$$Q_{fil} = \pi \cdot \varepsilon \cdot \varepsilon_0 \cdot E \cdot \frac{L^2}{4 \cdot Log} \cdot \left(1 + \frac{1}{Log}\right)$$

$$\text{Log} = \ln\left(\frac{L}{R}\right) - 1 \tag{2.65}$$

For E = 100 kV/cm, L = 0.3 mm, and R = lμm, we obtain Q_{fil} = 20 pClb. This is very small compared to the stored capacitive charge in the device which is in order of 100 nClb/cm^2 for devices of 1cm^2 or more in area. If we set the energy required to heat up a filament of radius R and length L = 0.5 mm to the melting temperature equal to the stored capacitive energy of a 1cm^2 device at 1000 V, then we find R = 3 μm. This is the order of magnitude of the observed damage structures in failed devices. Two conclusions can be drawn from this:

1. The charge in the nascent breakdown filament is too small for destruction. Destruction requires subsequent charge injection from the cathode and/or anode emitters.
2. There is a minimum capacitive energy below which the filamentary discharge is fully reversible. This explains why this failure mode has never been observed in, for example, low-voltage Zener diodes.

In diodes, GTOs, and thyristors, the electrical field varies slowly on the length scale L. This allows the introduction of the concept of a local failure probability $P_{loc}(E_{loc}, n_{flux})$. E_{loc} is the local electrical field and n_{flux} the flux of cosmic particles. If the particles are neutrons, then the flux will not depend on distance from the surface in the silicon and will only depend on shielding. In the following we will omit the parameter n_{flux} and assume that we have a constant neutron flux characteristic for sea-level and moderate latitudes. The resulting total failure probability P_{tot} is obtained by integrating P_{loc} over the volume. The functional form of the local failure probability absorbs all difficulties in modeling the transition from a nuclear collision to a filamentary discharge. At this point we will not specify the functional form of P_{loc}. The total failure probability becomes

$$P_{tot} = \int_{Vol} P_{loc}(E)\, dV \tag{2.66}$$

In many devices (GTOs, thyristors, diodes) the field distribution in the n-base is, to a good approximation, trapezoidal. This approximation allows us to write Equation (2.66) as

$$\text{P}_{tot} = \text{A} \cdot \frac{dx}{dE} \cdot \int_0^{E_{max}} P_{loc}(E) \cdot dE \tag{2.67}$$

$$\frac{dE}{dx} = \frac{n \cdot e}{\varepsilon \cdot \varepsilon_0} \tag{2.68}$$

where A is the device area, n is the n-base doping and ε the dielectric constant of silicon. By introducing

$$I_{master}(E_{max}) = \int_0^{E_{max}} P_{loc}(E) \cdot dE \tag{2.69}$$

we obtain for an NPT device

$$P_{tot}(V,n) = A \cdot \frac{\varepsilon \cdot \varepsilon_0}{n \cdot e} \cdot I_{master}\left(\sqrt{\frac{2 \cdot n \cdot V}{\varepsilon \cdot \varepsilon_0}}\right) \tag{2.70}$$

A similar expression can be derived for a PT device. This means that the total failure probability is proportional to the area, inversely proportional to the n-base doping n, and a function of the square root of the product of n·V. Introducing the n-base resistivity ρ, Equation (2.68) is described as

$$P_{tot}(S,\rho) = A \cdot \rho \cdot I(S) \tag{2.71}$$

where S is proportional to the maximum electrical field and is given by $S = \sqrt{\frac{V(Volt)}{\rho(\Omega cm)}}$ for an NPT element and $S = 0.2786 \cdot \frac{V(Volt)}{t(\mu m)} + 0.8972 \cdot \frac{t}{\rho(\Omega cm)}$ for PT elements. This means we have succeeded in deriving the scaling parameters for the failure rate without ever specifying the local failure probability. The universal failure integral I(S) can be found experimentally by plotting $\frac{P_{exp}}{A \cdot \rho}$ vs. the field parameter S.

Because the integral P_{tot} is basically exponential, the derivative P_{loc} is also basically exponential. We conclude that the following conditions must be satisfied: (1) "Exponential" field dependence and (2) Saturation at high fields (each collision leads to failure). Condition 1 requires an exponent and condition 2 requires that the field is in the denominator. Therefore P_{loc} can be rewritten as

$$P_{loc}(x) = \nu_0 \cdot \exp\left(-\frac{E_b}{E(x)}\right) \tag{2.72}$$

Note that any coincidence with the functional form of the ionization rate is purely accidental. When comparing this to burst events, the value of ν_0 is too high to represent the total collision rate. This can be accounted for by introducing an additional parameter into the functional form of P_{loc}.

$$P_{loc}(x) = \nu_0 \cdot \frac{1}{\alpha + \exp\left(-\frac{E_b}{E(x)}\right)} \tag{2.73}$$

For $\exp\left(-\frac{E_b}{E(x)}\right) \ll \alpha$ both forms are identical, but saturation occurs at $E(x) > E_b/\text{In}(\alpha)$.

2.5 Catastrophic Failures and fault-tolerance of IGBT

2.5.1. Catastrophic Failures [18]–[21]

Classification of IGBT Catastrophic Failures

IGBT catastrophic failure behaviors can be classified into open-circuit failures and short-circuit failures. Normally, open-circuit failure is considered as not fatal to converters, because the converter can operate with a lower quality of output. On the contrary, short-circuit failure is almost fatal to converters, as the uncontrolled short-circuit current may destroy the failed IGBT and/or other components in the circuit. IGBT catastrophic failure can be classified as shown in Figure 2.33.

Open-circuit Failures

IGBT open-circuit failure can happen after external disconnection because of vibration, as well as bond wire lift-off or rupture because ofhigh short-circuit current. It may lead to pulsating current, output current/voltage distortion, and result in secondary failure of other components after some time.

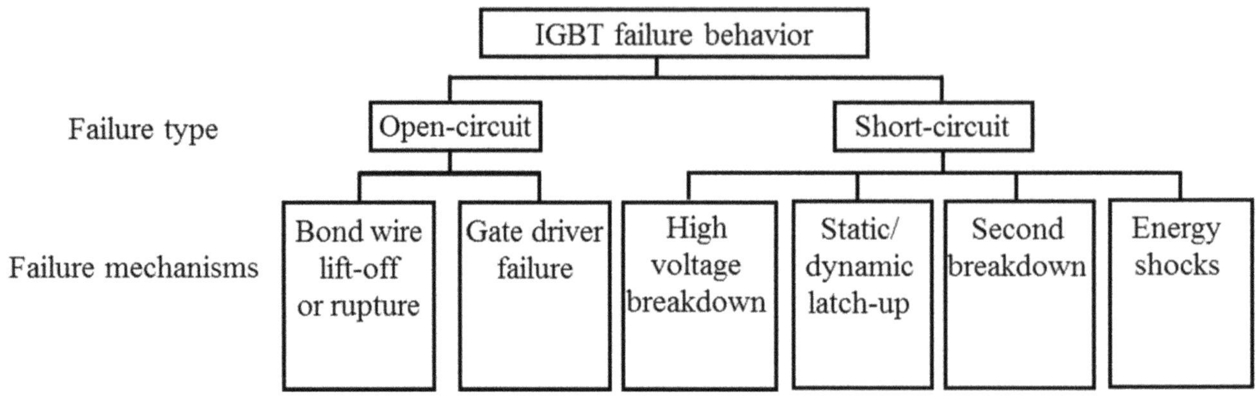

Figure 2.33 Overview of IGBT catastrophic failures

Open-circuit failure can also be because of the absence of gate drive signal. Some common reasons are damage of components in drivers and the disconnection between the driver board and the IGBT.

1. **Bond wire lift-off or rupture**
   Bond wire lift-off failure can happen after short-circuit failure. It is generally because ofmechanical reasons. The main mechanisms are related to mismatch of coefficients of thermal expansion (CTEs) between silicon and aluminum, together with high temperature gradients. Cracking initiates at the periphery of the bonding interface, and the bond wire finally liftsoff when cracking propagates to the weaker central bond area. Central emitter bond wires normally fail first, and the survivor bond wires follow. Another failure mechanism is bond wire rupture, which is slower than lift-off and usually observed after long power cycling tests.

2. **Gate driver failure**
   There are various reasons that cause gate driver failure, such as the damage of power stage devices (e.g. BJTs or MOSFETs) or wires between drive board and IGBT being disconnected. The driver failure may result in an IGBT's intermittent misfiring, degraded output voltage, and overstress of other IGBTs and capacitors.

Short-circuit Failure

IGBT short-circuit failure modes can be classified with respect to timesequence as shown in Figure 2.34. IGBT short-circuit during turn-on can be caused by high gate voltage and external failure. Failure during on-state may be caused by static latch-up or the rapid increase of intrinsic temperature caused by second breakdown, as well as by energy shocks. Failure during turn-off can be caused by dynamic latch-up and high voltage breakdown.

1. **High voltage breakdown**
   High voltage spikes induced by high falling rate of collector current (I_C) and stray inductance can destroy an IGBT during turn-off, especially under repetitive spikes. Because ofthe high turn-off voltage spike, the electric field can reach the critical field and break down one or a few IGBT cells first, which leads to high leakage current as well as high local temperature.

   Subsequently, the heat-flux radially diffuses from the overheated region to the neighboring cells. Collector–emitter voltage (V_{CE}) collapses after the voltage spike, and then I_C rises again. The gate terminal may also fail, which results in gate voltage (V_{GE}) rising.

   High value of V_{CE} and V_{GE} can also lead to shortcircuit during turn-on. An abrupt destruction and peak current happen after several microseconds during turn-on. The hole current caused by avalanche generation concentrates on a certain point (usually high-doped p+ region). The destruction point is always located at the edge of the active area close to device's peripheral region. Therefore, it is critical to clamp V_{GE} and V_{CE} during switching transients.

2. **Static/ dynamic latch-up**

Latch-up is a condition where the collector current can no longer be controlled by the gate voltage. With respect to Figure 2.35, latch-up happens when the parasitic NPN transistor is turned on and works together with the main PNP transistor as a thyristor; then, the gate loses control of I_C. IGBT latch-up can be divided into two types, static and dynamic latch-up.

Static latch-up happens at high collector currents, which turn on the parasitic NPN transistor by increasing the voltage drop across the parasitic resistance R_S.

Dynamic latch-up happens during switching transients, usually during turn-off, when the parasitic NPN transistor is biased by the displacement current through junction capacitance C_{cb} between the deep P+ region and the N-base region. There are two distinct conditions that may lead to dynamic latch-up. One is when the gate voltage drops very fast and induces excessive displacement current through the gate oxide that flows through the parasitic resistance. The other one is when the off-state collector–emitter voltage is quite high and induces excessive charging currents within the IGBT during the switching transient, which will flow through the parasitic resistance. Both conditions may trigger the parasitic NPN transistor and

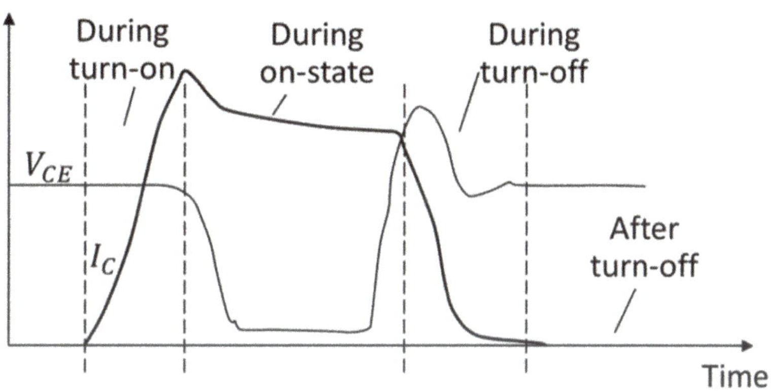

Figure 2.34 Time-sequence classification of IGBT short-circuit failure

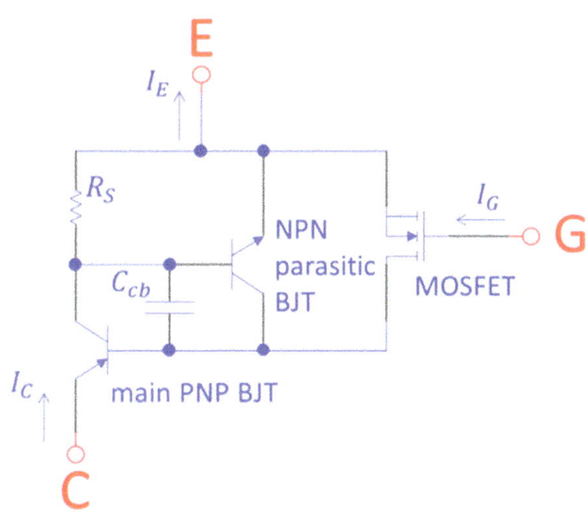

Figure 2.35 The equivalent circuit of IGBT

eventually lead to latch-up. It should be noted that the likelihood of collector current leading to dynamic latch-up is lower than that of static latch-up.

When latch-up happens, the IGBT will be almost inevitably damaged because of the loss of gate control, as confirmed by the 2D finite element simulations. Therefore, several methods are proposed to predict latch-up, especially based on the collector–emitter on-voltage $V_{CE(ON)}$ and turn-off time.

It is worth mentioning, though, that the latest generation of IGBTs with trench-gate structure and heavily doped P-base region under N-emitter has been proven to have good latch-up immunity, and latch-up is not a common failure as it used to be.

1. **Second breakdown**

   Second breakdown is a kind of local thermal breakdown of transistors because of high current stresses, which can also happen to IGBTs during on-state and turn-off.

   The failure mechanism of second breakdown is as follows: with the increase of current, the collector–base junction space-charge density increases, and the breakdown voltage decreases, resulting in a further increase in the current density. This process continues until the area of the high-current-density region reduces down to the minimum area of a stable current filament. Then, the filament temperature increases rapidly because of self-heating and a rapid collapse in voltage across the IGBT occurs. This has been comprehensively simulated and measured. How to improve IGBT ruggedness at high current density and prevent second breakdown is still an interesting research topic.

3. **Energy shocks**

   During short circuit in the on-state, failure may happen because of high-power dissipation. This high-power dissipation within a short time is defined as energy shock. The high short-circuit current will result in energy shock and high temperature. However, the IGBT will not immediately fail even if the junction temperature exceeds the rated temperature. Until reaching the intrinsic temperature (about 250°C for the doped silicon), further rise in junction temperature would lead to an exponential increase in carrier concentration and thermal runaway. With further increase of temperature, the silicon die may become fatally damaged and the contact metal may also migrate into the junctions.

2.5.2. Fault Tolerant

Broadly speaking, it is worth it to note first as a general principle that whatever fault-tolerant circuit or topology adds complexity and cost to the converter may fail by itself. For this reason, the final reliability level is a trade-off between enhancing fault tolerance and increasing weaknesses. With the above in mind, different fault-tolerant designs at circuit level have been proposed, which are classified as shown in Figure 2.36.

Device Redundancy

Contrary to modules, press-pack IGBTs intrinsically short after a catastrophic failure, because of the absence of bond wires and a direct connection between die and metal contacts.

This feature can be profitably used for series redundancy, where several devices work as a single switch. Nowadays it is widely used in traction, high-power drives, and power transmission systems. However, for the sake of completeness, it should be pointed out that failed press-pack IGBTs could be open circuit after some time, because of the interaction of molten aluminum (Al), molybdenum (Mo), and Si leading to various intermetallics, following with poor conductivity as open circuit.

Another way is using bypass switches, as shown in Figure 2.37. The bypass switches are located between the neutral point and the phase output. Failure occurring in switching devices (e.g. S_{ap}) can be cleared by blowing fuses (e.g. F_a) through turning on the bypassing switch (e.g. TR_a). The circuit operates as a four-switch three-phase inverter, with lower quality of the output. This topology can also handle open-phase failure in motor-drive application, which needs to turn on TRn shown in Figure 2.37 and change the phase currents to maintain the motor's torque constant.

This method has been applied in neutral-point-clamped (NPC) converters.

The other method with both series redundancy and bypass switches is cascaded H-bridge multi-level (CHBM) converter or modular multi-level converter (MMC). When a fault occurs to a switch, the faulty H-bridge (or half H-bridge) cell is bypassed. The output voltage magnitude can be maintained with an increase of harmonic distortion.

Phase Redundancy

This concept consists of introducing an additional phase leg to replace a faulty phase leg, as shown in Figure 2.38. The fault-tolerant control scheme is as follows: firstly, the gate-driving signals of the two switches in the faulty leg (e.g. S_1 and S_2) are set to be zero level; secondly, the suited bidirectional switch is triggered (e.g. t_1); finally, the two switches in redundant phase (e.g. S_7 and S_8) are controlled by gate-driving signals to resume the role of the two switches in faulty leg. This method has been applied in three-level converters. Five-leg and six-leg converters also have similar fault redundancy.

Table 2.3 Summay of IGBT Short-Circuit Failure Mechanisms

Failure mechanisms	Failure behavior	Failure location	Thermal behavior
High voltage breakdown	During turn-off: V_{CE} collapses and I_C rises after voltage spike During turn-on: peak I_C results in destruction	At the edge of active area	Overheating on few peripheral cells at first and then spreads to the whole chip
Lath-up	Static latch-up during on-state: high I_C leading to loss of gate control Dynamic latch-up during transients: high dv/dt leading to loss of gate control	Active area	Overheating on a stable subset of cells of the device
Second breakdown	Local thermal breakdown because of high currents	Emitter regions	Very high local temperature spots
Energy shocks	$E > E_C$: thermal runaway after successfully turned off $E < E_C$: degradation of die metallization	Al metalization layer, bond wires lift-off	Local overheating of Al metalization layer after 10^4cycles

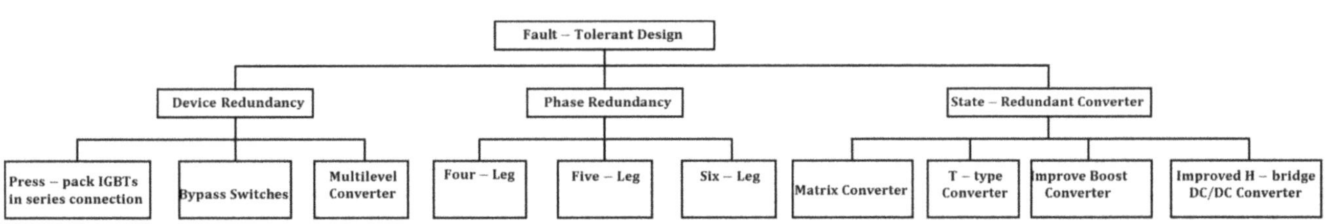

Figure 2.36 Classification of typical fault-tolerant circuits

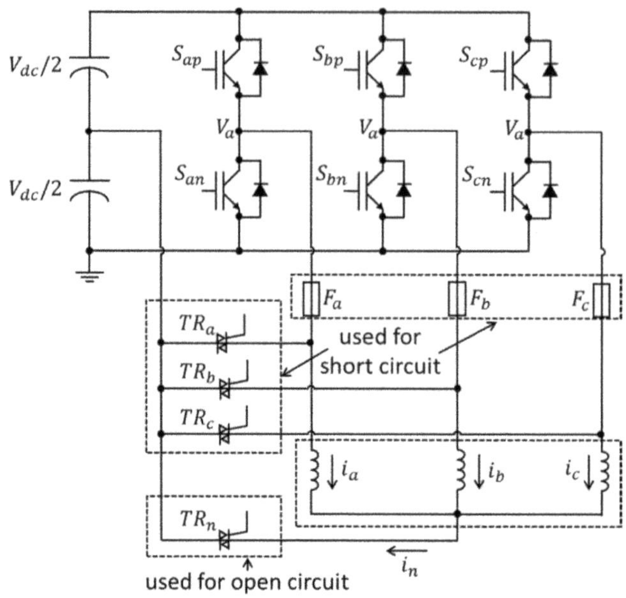

Figure 2.37 Typical bypass switches redundant circuit

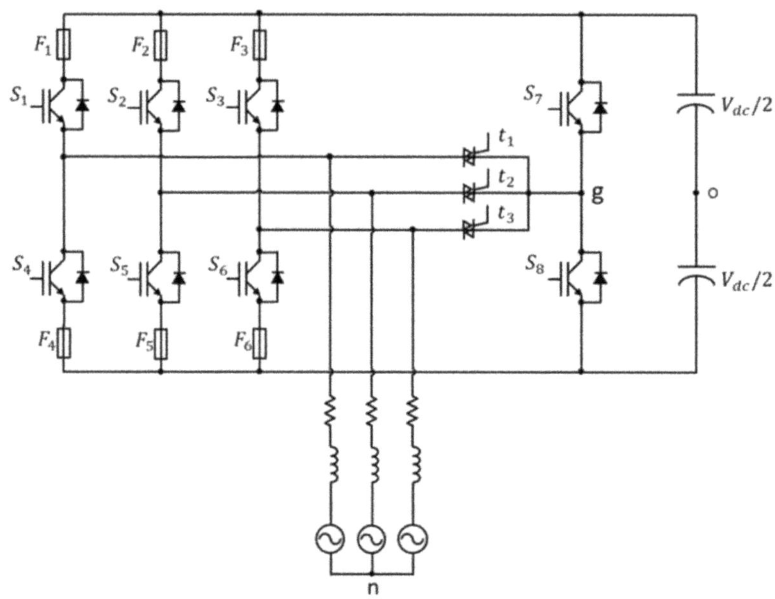

Figure 2.38 Typical phase-redundant circuit

2.6 Protection and Gate Driver of IGBT

2.6.1. Gate Driver [23]–[26]

General Gate-Drive Requirements for High-Power IGBT Modules

The main components of a typical gate-drive circuit are illustrated in the simplified circuit schematic shown in Figure 2.41. The primary function of the gate-drive circuit is to convert logic level control signals into the appropriate voltage and current for efficient and reliable switching of the IGBT module. An output driver stage consisting of small power MOSFETs or bipolar transistors performs the conversion by alternately connecting the IGBT's gate to the appropriate on (V_{ON}) and off (V_{OFF}) voltages. The driver-stage devices and series gate resistance RG must be selected to provide the appropriate peak current for charging and discharging the IGBT's gate. Most gate-drive circuits also provide isolation so that the logic signals are not connected to the dangerous high voltage present in the power circuit. The driver must also be immune to the severe electromagnetic noise produced by the fast-switching, high-voltage, high-current IGBT power circuit. Careful layout and component selection are critical to avoid problems with coupled noise. The following sections will review the critical requirements for a large IGBT module gate drive.

Turn-On Voltage ($\mathbf{V_{on}}$)

In order to establish collector-to-emitter conduction in an IGBT module, a positive voltage must be applied to the gate. The absolute maximum voltage that can safely be applied to the IGBT's gate is usually specified on the device data sheet. Application of voltages greater than 20 V may cause breakdown of the gate oxide, resulting in permanent damage to the device. The 20 V upper limit must be restricted even further if short-circuit survival is required. The short-circuit withstand time (t_w) of a given device is inversely proportional to the product of applied voltage and short-circuit current.

The short-circuit current increases with increasing gate voltage, thus degrading the withstand time. The usable lower limit for the on-state gate voltage is decided by the devices transconductance or gain and acceptable switching losses. For this device it can be seen that a gate voltage of about 10 V is enough to support the devices' peak current rating ($I_{CM} = 2 \times I_{RATED}$). A gate voltage of 10 V would be sufficient to fully turn the device on, but it may not be sufficient to obtain efficient switching.

If 10 V were used for V_{on}, a long dynamic saturation (slow turn-on) will result because the gate voltage takes a long time to reach 10 V as it exponentially charges through the series gate resistance. For optimum performance, using a voltage in this range will ensure that the device stays fully saturated and switches on efficiently while maintaining good short-circuit durability.

Turn-Off Voltage ($\mathbf{V_{off}}$)

A substantial off bias of at least -5 V is recommended for large IGBT modules. Use of an off-bias voltage will reduce turn-off losses and provide additional dv/dt noise immunity. Large IGBT modules generally require a stronger off bias than other power MOS devices for two reasons. First, IGBTs typically operate at higher voltages, resulting in increased dv/dt coupling of switching noise. Secondly, large IGBT modules that are constructed from parallel chips have internal gate resistors in series with each chip. Even if a low-impedance short is applied at the module's external terminals, voltage can develop at the gate of the IGBT chip when Miller effect current flows through the internal resistors.

Gate-Drive Power

When switching, the IGBT consumes power from the gate-drive power supply. The amount of power consumed is a function of operating frequency, on- and off- bias voltages, and total gate charge. The average current that must be supplied by the gate-drive power supply is given by:

$$I_S = Q_G \times f$$

where:

Q_G = is total charge

f = is frequency of operation

Gate Voltage Clamping

During the fault condition, especially the fault under load condition, the gate voltage will be increased and eventually exceed 15 V because of a large *dv*/*dt* across Miller capacitance and gate resistance. As the gate voltage and the collector current increase as the gate resistance increases, measures must be taken to limit the large transient peak current.

The transient short-circuit current is suppressed using the capacitor C_z and the Zener diode D_z, as shown in Figure 2.39. Once a fault is detected, the transistor M_1 is turned on, which causes C_z to charge up to the voltage level of D_z, thus effectively discharging the gate. The value of C_z should be selected carefully because a large C_z value results in unexpected fast turnoff of IGBT modules and a slow ramp up to the clamp current level, whereas a small C_z value results in a high peak fault current because of insufficient gate discharge. The final gate voltage and fault current level are clamped by the Zener diode D_z. The value of the Zener voltage is selected such that the fault current is clamped to a safe current level and the clamped gate voltage is above the threshold voltage to avoid fast turnoff.

Soft Turnoff

An IGBT has to be softly turned off to reduce voltage overshoot because of the effect of power-loop stray inductance and high *di*/*dt* under a short-circuit condition. When the gate voltage has been clamped, the capacitor C_1 is placed in parallel with C_z, and the gate capacitance by turn on of M_2 to turn off the IGBT at a reduced rate of gate voltage change following a delay. Then the buffer input is closed regardless of the input PWM signal.

2.7 Future Prospects IGBT and SiC [27]–[30]

As there is already more than 30 years of history from the very first IGBT prototypes to today's state-of-the-art fully industrialized mass market IGBTs, lot of questions occur about further development steps and perspective, especially with the new wide band successor candidates, mainly SiC (≥1200 V) as well as GaN (<1200 V) in place.

The closer the Si IGBTs are driven to their limits the more important it gets to clarify what properties from an application point of view are really required, and which ones might be "nice to have" or can even be given up altogether. Prominent examples are the required short-circuit robustness in terms of a short-circuit duration time and voltage slopes (dV/dt) restrictions at turn ON and turn OFF.

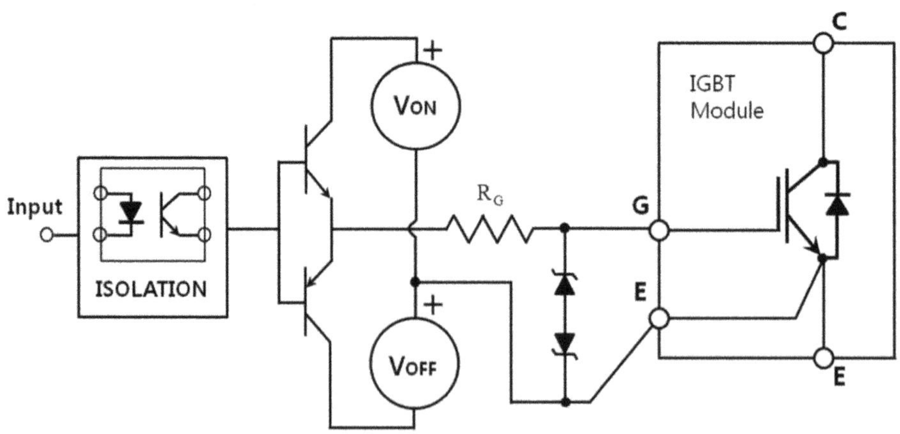

Figure 2.39 Basic gate-drive circuit

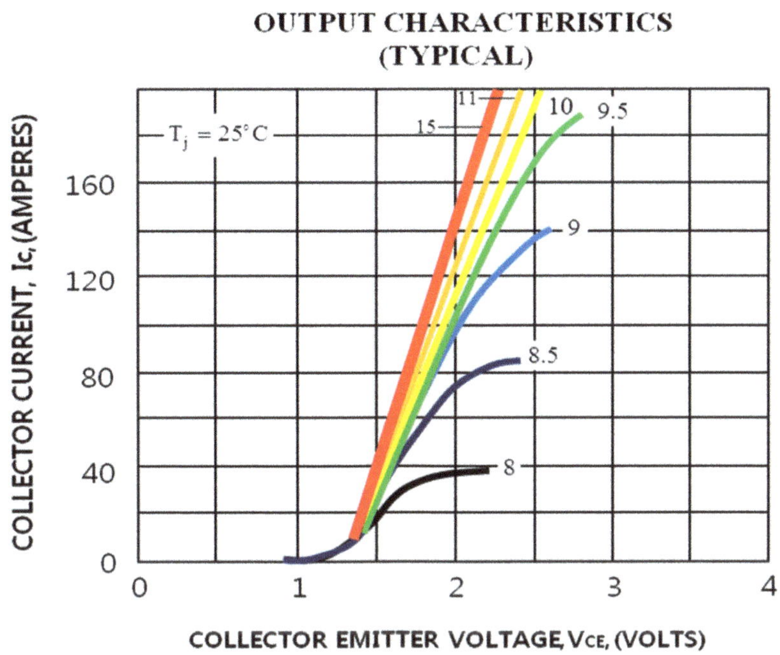

Figure 2.40 IGBT module output characteristic

These aspects do not only define the further path of Si IGBT development, but also must be clarified for wide band gap materials, as especially their advantages of very low switching losses only hold as long as they are allowed to switch very fast, which may be in contradiction with today's requirements from many drives customers in terms of $dV/dt < 10$ kV/μs, coming from both EMI issues and preventing motor deterioration. Allowing higher dV/dt would enable reduction of switching losses also for Si IGBTs. However, especially the correlation of these dV/dt values to real EMI spectrum within the application is so far not always fully clear. As long as these dV/dt requirements remain, a special tradeoff between sufficiently low turn-ON losses under nominal conditions at high temperature and sufficiently low dV/dt at low-current conditions and low temperature has to be carefully addressed by the proper choice of capacities within the trench cell structure using gate trenches, source, or dummy trenches. Reducing the short circuit requirements from today usually 10 to 5μs or even below would open the chance to increase the channel-width to-length ratio, allowing both SiC MOSFETS and Si IGBTs to reduce the on-state losses. Even if today's requirements in short circuit and dV/dt are not changed, there are ways to improve the Si IGBT, but the potential is higher the more of the above-mentioned requirements that can be reduced.

Next Generations Transistor Cell Concepts

With respect to the well-known theoretical limit of IGBT on-state performance, a lot of activities and investigations are running with sub-micrometer mesas between adjacent deep trenches down to mesas below 0.5 μm as shown in Figure 2.41. Out of that, there are indications that the forward voltage drop can indeed be reduced close to 1 V for a 1200 V-blocking-rated device; however, drawbacks come in terms of quite high turn-off switching losses because of a very high amount of carrier plasma, and also short-circuit robustness problems that still remain to be solved. If so, next generation IGBTs may once again make use of significant on-state loss reduction/chip shrink by implementing mesa structures in the deep sub-micrometer region or even below. Figure 2.42 shows the simulated three-dimensional (3-D) tradeoff of the p-emitter variation.

Next Generations Vertical Structure

Within the vertical structure of the IGBTs there is still a room to improve as well. As for every vertical bipolar power device that the thinner the drift zone the lower both on-state and switching losses can be adjusted. Today 1200 V IGBTs with a chip thickness about 100–130 μm are common, as well as 600-750 V IGBTs with around 70 μm. From theoretical limits for 1200 V blocking capability, chip thicknesses at about 80–90 μm should seem feasible, and for 600 V about 45–50 μm. Ultrathin wafer technology today is even able to serve these needs, even in first 300-mm power semiconductor production facilities.

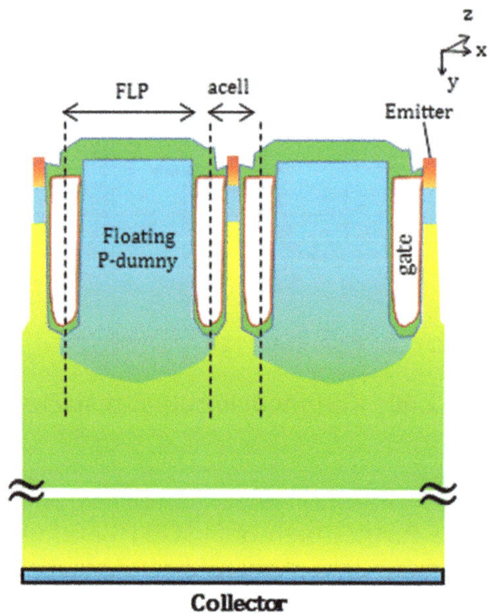

Figure 2.41 IGBT structure with very small mesa

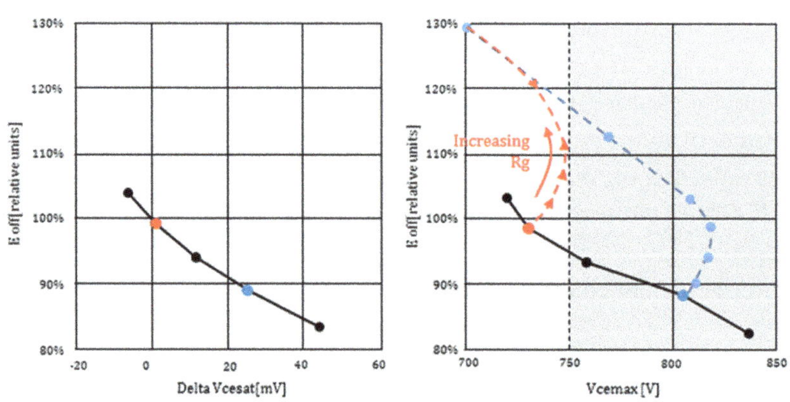

Figure 2.42 Simulated three-dimensional (3-D) tradeoff of the p-emitter variation. Turn-off energy E off versus forward voltage drop V_{cesat} (left) and E off versus maximum collector–emitter voltage VCE_{max} (right): lower p emitter (blue) provides lower E off than the strong p emitter (red) on the cost of much higher surge voltage. If the Lstray x I in the application is too high, the device needs to be slowed down by increasing the R g. In this case, the losses increase dramatically without using the benefit of lower V_{cesat}

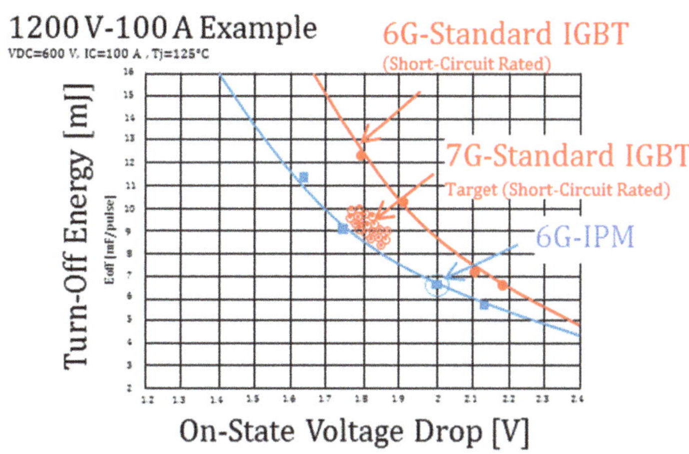

Figure 2.43 Turn-off energy and on-state voltage drop (forward voltage drop) trade-off diagram for standard short-circuit-rated IGBT in comparison with IPM solution with non-short-circuit-rated, but low on-State voltage drop IGBT

However, the thinner the IGBT chips, the more critical switching softness, cosmic ray robustness, and thermal short-circuit capability will become. With respect to the softness topic, low inductive module and inverter setups will play an important role, alongside more sophisticated field stop engineering and the right choice of back emitter efficiency. The latter aspect especially has to be considered and gets clearer through a newly introduced three dimensional (3-D) tradeoff correlation between overvoltage at turn-off, switching, and on-state losses.

It gets obvious that on first look, an unfavorable high-efficient back emitter results in the best compromise in terms of overvoltage and overall losses, at least as long as quite high parasitic stray inductances occur. To keep the cosmic ray robustness on a sufficiently high level, drift zone resistivity can be reduced. A sufficiently high level of thermal short-circuit capability despite a reduced chip thickness can be achieved by implementing a thick metal layer on top of the chip. Figure 2.43 shows the turn-off energy and on-state voltage drop (forward voltage drop) trade-off diagram.

Intelligent Gate Driver

Levers to overcome the tradeoff between short-circuit withstand time and forward voltage drop are using IGBT chips with an increased gate width/length ratio and current sensing and reducing the gate voltage, and thus the short-circuit current reduces to a manageable level within the first few microseconds in short-circuit cases. This method is not that new for IPMs, and as a consequence, such an intelligent gate driving approach may get standard in the future also for standard IGBT modules, as it always enables one additional step in IGBT performance compared with the standard short-circuit-rated IGBT.

A second important improvement out of sophisticated gate driving schemes can arise for the RC IGBTs. Using trench cells for the diode function within the IGBT, a dedicated gate-driving scheme can be used to change the plasma concentration during freewheeling mode from a higher level and shortly before commutation event to a lower level. Doing so, the potential of an RC IGBT with diode control (RCDC) can be expanded to traction and HVDC applications in the high-voltage arena of 3.3 kV up to 6.5 kV. A quite similar approach is demonstrated with the BIGBT, showing the lever in power density enhancement for the 3.3, 4.5, and 6.5 kV power modules, as shown in Figure 2.44. If such more sophisticated gate-driving schemes should become standard, as well as in the lower voltage range (650–1700 V), RCDC IGBT concepts can increase power density compared to the conventional approach of separately optimized IGBTs and diode chips.

Especially in the case of water-cooled systems (traction, hybrid electric vehicles), the integrated reverse conducting concept may give at least 10% more power density compared to every fully optimized IGBT/diode single chip solution. Modulating the plasma concentration with decisive high concentration during on-state conditions and changing to a lower concentration before the switching might become more common also for the IGBT mode in the future.

Advanced Thermal Management

It was already shown that on principle the Si IGBT is able to work at elevated temperature levels up to 200°C, even overload operation (RB SOA, short-circuit SOA) at 200°C seems feasible. Of course, the periphery around the chip (package with its plastic materials, terminals as well as surroundings outside the package) must also be able to somehow sustain increased temperature levels. In detail, the mold compound of a discrete IGBT or the silicone gel within a standard IGBT module will be a challenge at increased temperatures, but there already exist solutions with gels up to 200°C. Another serious issue to be solved is the increased temperature swings, stressing chip to baseplate or lead frame connections as well as wedge bonding on the chip's front side, resulting in reduced numbers of power and temperature cycles until end of life if no countermeasures are addressed.

However, for these challenges, advanced solutions that are already in place may also be feasible up to 200°C, whether it is sintering interconnect technology on the back and front side of the chip or alternatively front side copper wedge bonding, as shown in Figure 2.45. Also, it is clear from Figure 2.46 that if using the .XT technology approach with a thick copper metalization (required on chip front-side for the Cu-bonding process), it also enhances the thermal short-circuit destruction level, which can be paid off in size and vertical chip shrink. It is expected that such combined chip/module solutions will open the path to enhanced maximum junction temperatures up to 200°C in the future, together with innovations in cell and vertical structure as well as gate driving in a higher power density. Of course, all these measures of advanced thermal management will also be of importance for the roll out of wide bandgap power semiconductors with their potential for even higher temperatures above 200°C.

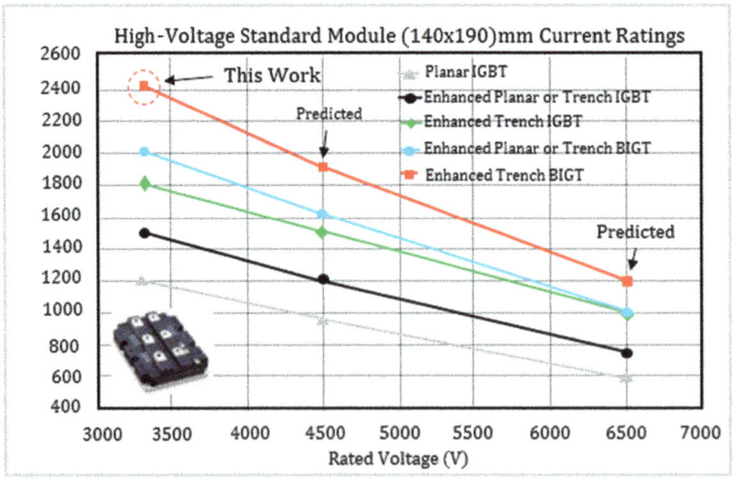

Figure 2.44 High-voltage module current ratings with different generations of IGBT technologies

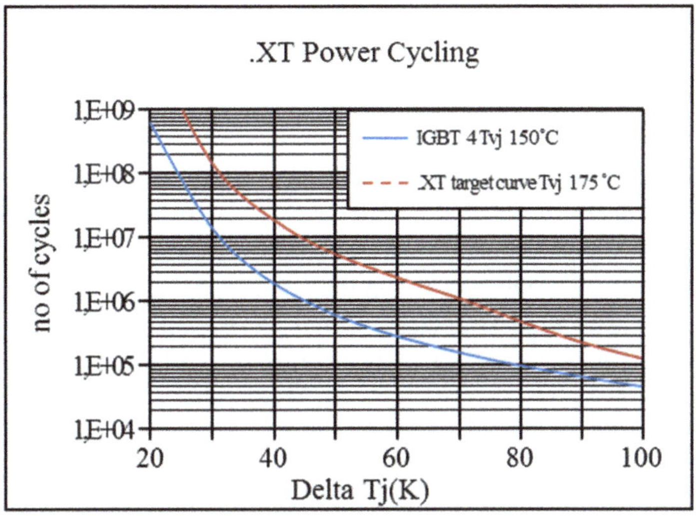

Figure 2.45 Increased power cycling capability of the .XT technology (Cu wedge bonding and improved interconnect chip/DCB and DCB/baseplate) compared with standard interconnect technology. The corresponding lines indicate the amount of cycles that at least can be performed without destruction

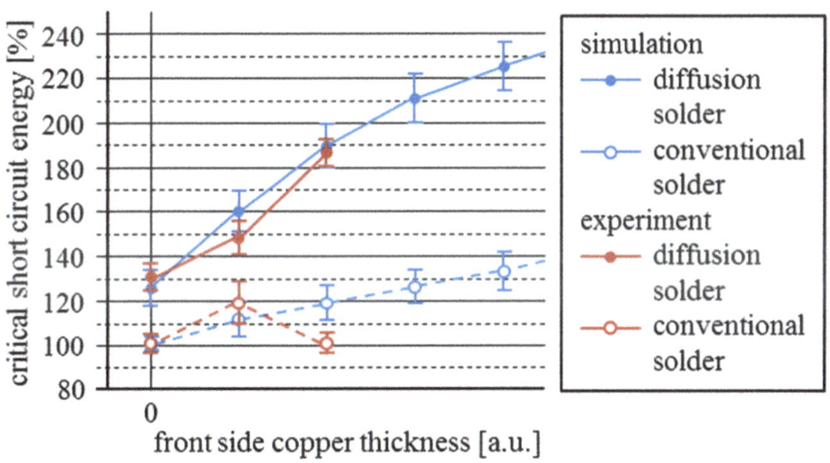

Figure 2.46 Increased critical short-circuit energy of a 1200 V IGBT as a function of different interconnect technologies and Cu chip metal thickness

2.7.1. Challenges and Trends of SiC

Challenges and Trends

Although SiC devices show promising advantages, many challenges exist from both the device and the application perspectives. These guide future research and development trends.

1. The field reliability of SiC devices must be demonstrated for various applications, and a voltage-derating design guideline needs to be established. This is especially important for applications in which reliability is extremely critical, such as aviation systems. On the one hand, a power conversion system with an SiC device potentially provides a higher current rating than one with an Si device, which leads to a larger thermal

ripple. On the other hand, the higher temperature operation of the SiC device indicates more stringent requirements for the packaging materials.

2. Manufacturing processes need to be improved for better yields to make the costs of SiC devices more justifiable for system applications. Most major SiC manufacturers are making a transition to 150-mm SiC epitaxial wafers. In addition, there are discussions about switching to a manufacturing process, in which SiC power devices can be manufactured in Si fabrications. Such a move would lower costs, because the existing facilities and mature manufacturing infrastructure would guarantee the quality. However, the lack of flexibility in processes and SiC tools is challenging.
3. SiC devices switch much faster than Si devices, which poses challenges to the gate driver design. First, the higher *dv*/*dt* of the SiC device injects a higher common mode current into the gate loop through a Miller capacitor, and thus induces a positive spurious gate voltage. An advanced gate driver with active *dv*/*dt* and *di*/*dt* control is a trend. Additionally, the higher *dv*/*dt* injects a higher common mode current to the primary side of the gate driver through the isolation barrier, which sets the limits of the coupling capacitance. Second, SiC devices have a faster current rise during the fault condition, in addition to a smaller die size. Therefore, the short-circuit protection response requirement of the gate driver is higher. Finally, the parallel or series connection of the SiC device is also more sensitive to timing mismatch.
4. The innovative EMI filter design needs to be investigated. A faster switching transient combined with a higher switching frequency contributes to EMI noise that is 10 – 100 times higher. This becomes especially challenging for high-voltage and high-power applications. On one hand, the modeling and prediction methodology of EMI noise for SiC power converters is indispensable. On the other hand, innovative shielding techniques and filter designs are still required.
5. Novel system layouts with a minimized commutation loop are critical, including both the SiC device package and the system busbar layout. Although it is acknowledged that an SiC power converter can deliver higher current, this statement is true only when the device voltage stress during the transition is less than its breakdown voltage. In other words, the current rating of the system may be limited by the voltage overshoot instead of the thermal. For some emerging applications, such as the 1.5 kV DC photovoltaic system, this issue is critically important. Laminated and multi-layer laminated busbar structures are the solution of choice so far. In addition, a low-inductance capacitor needs to be looked into.
6. The smaller die size of SiC chips and the corresponding higher loss density open new design challenges for thermal management strategies. In addition, the smaller thermal capacitance of the SiC chips may result in higher temperature ripple, raising another challenge from the reliability perspective. Some efficient cooling methods such as double-sided cooling, liquid jet impingement cooling, and phase change cooling are potential promising solutions. Furthermore, the emerging 3-D printed technologies can potentially be used to enable creative solutions.

The focus is on the benefits, opportunities, and challenges with SiC-based power electronics. The following are some observations and conclusions:

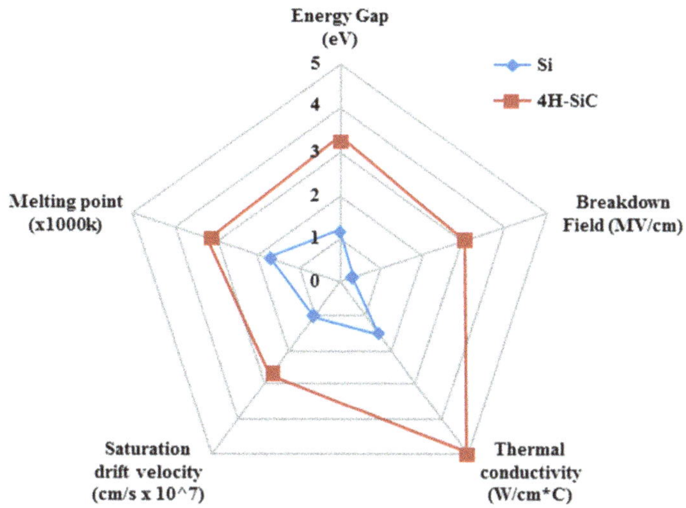

Figure 2.47 Material properties comparison between Si and SiC (ranges 1–5 are marked for all five properties)

1. With smaller size, lower loss, faster switching, and higher temperature capability, SiC devices can improve power conversion systems in several ways: by direct substitution of Si devices in existing circuits for improved efficiency and power density, by simplifying the circuit topology for reduced complexity and further enhanced power density, and by enabling system configuration modification for overall system power density improvement and lower cost.
2. In addition to the converter-level benefits, SiC power electronics can further introduce system-level benefits as a result of fast-switching capability and high available control bandwidth. The examples of system-level benefits include integrated harmonic filtering and stabilizer functions of the SiC-based interface converters in a distribution grid or micro-grid.
3. SiC can also enable new power electronics applications that were previously not feasible with Si technology because ofperformance, cost, efficiency, reliability, or density concerns. More emerging applications can be expected with SiC as a key-enabling technology.
4. With faster switching, higher temperature, and smaller size, SiC also raises new design and application challenges. Gate drive and protection need to be faster and more adaptive. Loads can interfere with converter switching, therefore potentially requiring additional filters. High temperature and fast switching also demand better device packaging. Other challenges include thermal management, EMI filters, sensors, and control. Many solutions have been and are being developed. However, the design and application methodology for SiC power electronics will remain an active research area for the foreseeable future.

Many of the advances presented in this paper are not yet mature and are still under rapid development. It is expected that SiC devices and related technologies will soon become fully developed, resulting in devices with lower cost, higher current ratings, better reliability, better gate-drive and protection technologies, and more robust high-temperature packaging. Consequently, they will be applied in more commercial power electronics converters. One key to overcoming the adoption barrier of SiC and other WBG (Wide-Band Gap) technologies is to train a new generation of engineers in the area of WBG power electronics. Figure 2.47 shows the material properties comparison between Si and SiC.

References

1. N. lwamuro and T. Laska, "IGBT history, state-of-the-art, and future prospects," *IEEE Trans. Electron Devices*, vol. 64, no. 3, 741–752, Mar. 2017.
2. B. J. Baliga, "Enhancement-and depletion-mode vertical-channel m.o.s. gated thyristors," *Electron. Lett.*, vol. 15, no. 20, pp. 645–647, Sep. 1979 .
3. H. W. Becke and C. F. Wheatley, Jr., "Power MOSFET with an anode region," U.S. Patent 4364073, Dec. 14, 1982.
4. N. Iwamuro, et al., "Numerical analysis of short-circuit safe operating area for p-channel and n-channel IGBTs," *IEEE Trans. Electron Devices*, vol. 38, no. 2, pp. 303–309, Feb. 1991.
5. J. Dodge and J. Hess, "*IGBT Tutorial*," Advanced Power Technology, S.W. Columbia, Jul. 2002.
6. L.Miao, et al., "Research of voltage-balancing methods of series-connected IGBTs," *Automat. Control Intell. Syst.* vol. 4, no. 5, pp. 73–79, 2016. doi:http://dx.doi.org/10.11648/j.acis.20160405.1110.11648/j.acis.20160405.11
7. D. Zechun, et al., "Design and characteristic analysis of novel press-contact IGBT module[J]," *Electric Drive for Locomtives*, vol. 1, no. 1, pp. 10–13, 2013 (in Chinese).
8. K. Sasagawa, Y. Abe, and K. Matsuse, "Voltage-balancing method for IGBTs connected in series [J]," *IEEE Trans. Ind. Appl.*, vol. 40, no. 4, pp. 1025–1030, 2004.
9. H. Soonwook, V. Chitta, and D. A. Torrey, "Series connection of IGBT's with active voltage balancing [J]," *IEEE Trans. Ind. Appl.*, no. 35, Vol. 35, pp. 917–923, 1999.
10. Y. Xiao, et al., "Current sharing model of parallel connected IGBTs during turn-on," in *IECON 2014-40th Annual Conference of the IEEE Industrial Electronics Society*, Dallas, TX, USA, Oct. 29–Nov. 1, 2014.
11. N. Y. A. Shammas, R. Withanage, and D. Chamund, "Review of series and parallel connection of IGBTs," *IEEE Proc. Circuits Devices Syst.*, vol. 153, no. 1, pp. 34–39, Feb. 2006.
12. C. Wong and K. Priest, "Current sharing for paralleled IGBTs using statistics method," in *31st IAS Annual Meeting, IAS '96 Conference Record of the 1996 IEEE Industry Applications Conference*, San Diego, CA, USA, vol. 3, pp. 1418–1424, Oct. 1996.
13. D. Yi, Z. Zhengming, and Y. Liqiang, "IGBT modeling for analysis of complicated multi-IGBT Circuits," in *2nd International Conference on Power Electronics and Intelligent Transportation System (PEITS), 2009*, Shenzhen, China, vol. 1, pp. 137–140, Dec. 2009.
14. H. Kabza, et al., "Cosmic radiation as a cause for power device failure and possible countermeasures," in *Proceedings of the 6th International Symposium on Power Semiconductor Devices and Ics*, Davos, Switzerland, Switzerland, May, 31– Jun. 2, 1994.
15. A. E. Waskiewicz, et al., "Burnout of power MOS transistors with heavy ions of Californium-252," *IEEE Trans. Nucl. Sci.*, vol. 33, no. 6, pp. 1710–1713, 1986.
16. J. F. Ziegler and W. A. Lanford, "The effect of sea level cosmic rays on electronic devices," *J. Appl. Phys.* vol. 52, no. 6, pp. 4305–4312, 1981.
17. J. F. Ziegler and W. A. Lanford, "The effect of cosmic rays on computer memories," *Science,* vol. 206, pp. 776–788, 1979.
18. R. Wu, et al., "Catastrophic failure and fault-tolerant design of IGBT power electronic converters–An overview," in *39th Annual Conference of the IEEE Industrial Electronics Society IECON 2013,* Vienna, Austria, Nov. 2013.
19. C. Busca, et al., "An overview of the reliability prediction related aspects of high power IGBTs in wind power applications," *Microelectron. Reliab.*, vol. 51, no. 9–11, pp.1903–1907, Sep./Nov. 2011.
20. M. Ciappa and F. Wolfgang, "Lifetime prediction of IGBT modules for traction applications," in *Proceedings of 38th Annual IEEE International Reliability Physics Symposium*, San Jose, CA, USA, 2000, pp. 210–216.
21. W.S. Loh, et al., "Wire bond reliability for power electronic modules-effect of bonding temperature," in *Proceedings of International Conference on Thermal, Mechanical and Multi-Physics Simulation Experiments in Microelectronics and Micro-Systems*, London, UK Apr. 2007, pp. 1–6.
22. Z. Wang, et al., "A fast overcurrent protection scheme for IGBT modules through dynamic fault current evaluation," in *2013 Twenty-Eighth Annual IEEE Applied Power Electronics Conference and Exposition (APEC)*, Long Beach, CA, USA Mar. 2013.
23. E. R. Motto, "Hybrid circuits simplify IGBT module gate drive," in *PCIM Conference*, Powerex Inc., Youngwood, Pennsylvania, USA 1999.
24. J. Yamashita, et al., "A study on the short circuit destruction of IGBTs," *ISPSD*, Monterey, CA, USA May 1993.

25. M. Harada, et al., "600 V trench IGBT in comparison with planar IGBT," *ISPSD*, Davos, Switzerland, Switzerland May 31, 1994.
26. E. Motto, et al., "Evaluating the dynamic performance of high current IGBT modules," *PCIM/PQ*, Dallas, US, 1994.
27. X. She, "Review of silicon carbide power devices and their applications," *IEEE Trans. Ind. Electron.*, vol. 64, pp. 8193–8205, Oak Ridge, TN, USA 2017.
28. J. G. Kassakian and T. M. Johns, "Evolving and emerging applications of power electronics in systems," *IEEE J. Emerg. Sel. Topics Power Electron.*, vol. 1, no. 2, pp. 47–58, Jun. 2013.
29. *The World Market for Silicon Carbide and Gallium Nitride Power Semiconductors—2013 Edition*, vol. 9790, HIS, Wellingborough, 2013.
30. C. Bodeker, T. Vogt, and N. Kaminski, "Stability of SiC schottky diode against leakage current thermal runaway," in *Proceeding of the International Symposiyum Power Semicond Devices ICs*, Hong Kong, China 2015, pp. 245–248.

3

Background

3.1. Review of Multi-level Converter Topologies [1]–[6]

Multi-level converters are power-conversion systems composed by an array of power semiconductors and capacitive voltage sources that, when properly connected and controlled, can generate a multiple-step voltage waveform with variable and controllable frequency, phase, and amplitude. The stepped waveform is synthesized by selecting different voltage levels generated by the proper connection of the load to the different capacitive voltage sources. This connection is performed by proper switching of the power semiconductors.

The number of levels of a converter can be defined as the number of steps or constant voltage values that can be generated by the converter between the output terminal and any arbitrary internal reference node within it. It is typically a DC-link node, and is usually denoted by N and called neutral. To be called a multi-level converter, each phase of the converter has to generate at least three different voltage levels. This differentiates the classic two-level voltage-source converter (2L-VSC) from the multi-level family. Some single-phase examples of this concept and their respective waveforms are given in Figure 3.1 for different numbers of levels. It is worth mentioning that, generally, the different voltage levels are equidistant from each other in multiples of Vdc.

Multi-level converter technology started in the late 1960s, introducing the multi-level-stepped waveform concept with a series-connected H-bridge, also known as the Cascaded H-Bridge converter (CHB converter). This was closely followed by low power converter development of an flying capacitor (FC) topology in the same year. Finally, in the late 1970s, the diode-clamped converter (DCC) was introduced. The DCC concept evolved into the three-level neutral-point-clamped (3L-NPC) converter we now know and can be considered the first real multi-level power converter for medium-voltage applications. Although the CHB was reintroduced in the late 1980s, it reach more industrial relevance in the mid-1990s. In the same case,the early concept of the FC circuit(introduced for low power converter in the1960s)developed into the medium-voltage multi-level converter topology we know today in the early 1990s.

These three multi-level converter topologies were considered as classic or traditional multi-level topologies that first made it into real industrial products during the last two decades. The power circuits of a single-phase leg of these three topologies are shown in Figure 3.2, featuring the commonly used corresponding semiconductor device. These converters are commercialized by several manufacturers in the field, offering different power ratings, front-end configurations, cooling systems, semiconductor devices, and control schemes, among other technical specifications.

1. The NPC features medium-voltage/high-voltage devices (integrated gate-commutated thyristor (IGCT) and medium-voltage/high-voltage insulated-gate bipolar transistors (IGBTs)), whereas the CHB exclusively uses low-voltage IGBTs (LV-IGBTs).
2. The CHB reaches higher voltages and power levels.

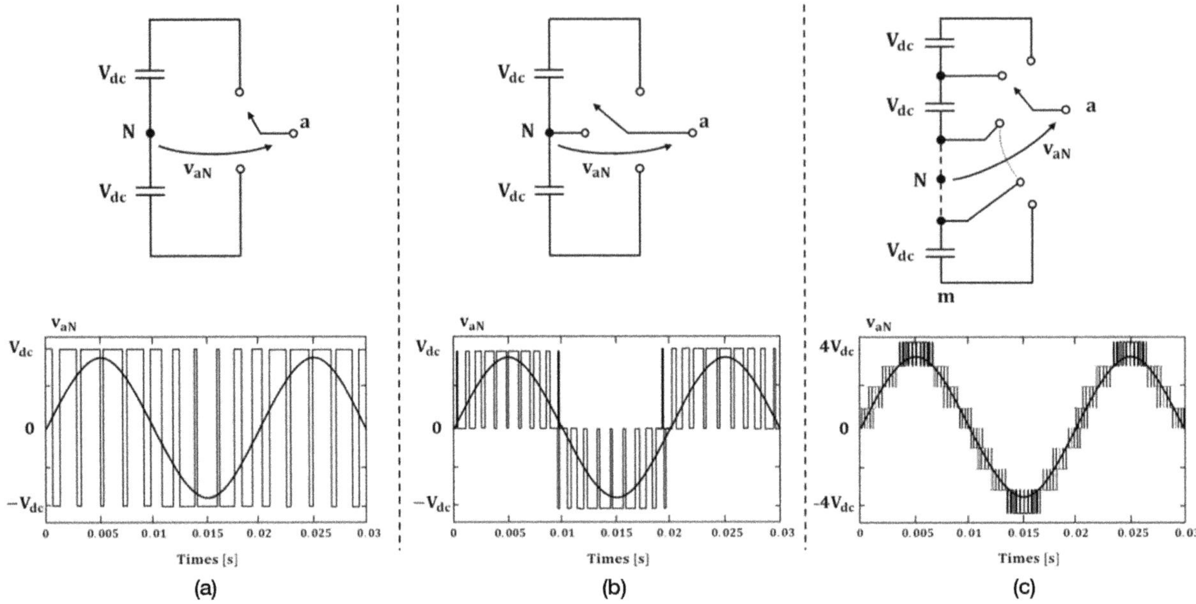

Figure 3.1 Converter output-voltage waveform: (a) Two level; (b) three level; and (c) nine level

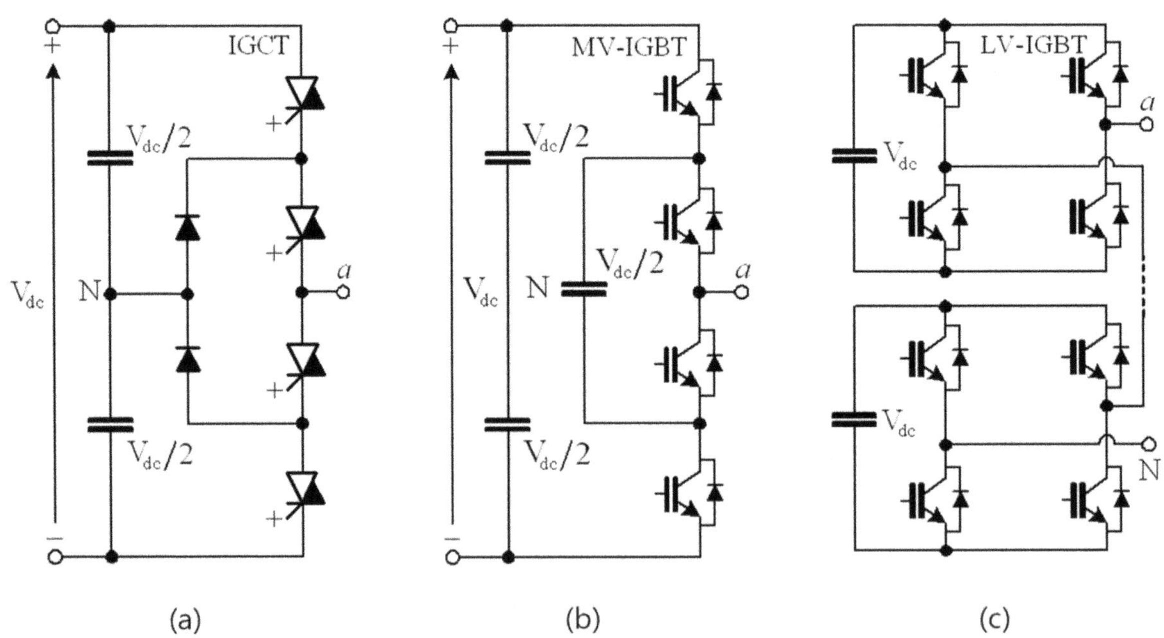

Figure 3.2 Classic multi-level converter topologies (with only one phase shown): (a) Three-level neutral- point- clamped (3L-NPC) featuring integrated gate-commutated thyristors (IGCTs); (b) three-level flying capacitor (3L-FC) featuring MV-IGBTs; and (c) 5L Cascaded H-Bridge (CHB) featuring LV-IGBTs

3. The NPC is definitely more suitable for back-to-back regenerative applications as it needs a substantially higher number of devices to achieve a regenerative option (a three-phase two-level voltage-source inverter (VSI) per cell).
4. The CHB usually needs a phase-shifting transformer to conform a 36-pulse rectifier system. This is more expensive but improves input power quality.
5. The NPC has a simpler circuit structure, leading to a smaller footprint.
6. Although both topologies generate the same amount of levels when using the same number of power switches, commercially available CHBs have more output-voltage levels (up to 17, compared with 3 for the NPC), meaning lower average device switching frequencies are possible for the same output-voltage waveform quality. Therefore, air cooling and higher fundamental output frequency can be achieved without any derating or use of an output filter.

These multi-level VSC topologies belong to the medium-voltage/high-power converter family, the classification of which is shown in Figure 3.3. Note that, generally speaking, the medium-voltage range is considered in the power converter industry to be from 2.3 to 6.6 kV and high power to be in the range of 1–50 MW. This classification also includes direct AC/AC converters and current source converters, which are currently the main competitors of multi-level technology; namely the cyclo-converter and load-commutated inverters

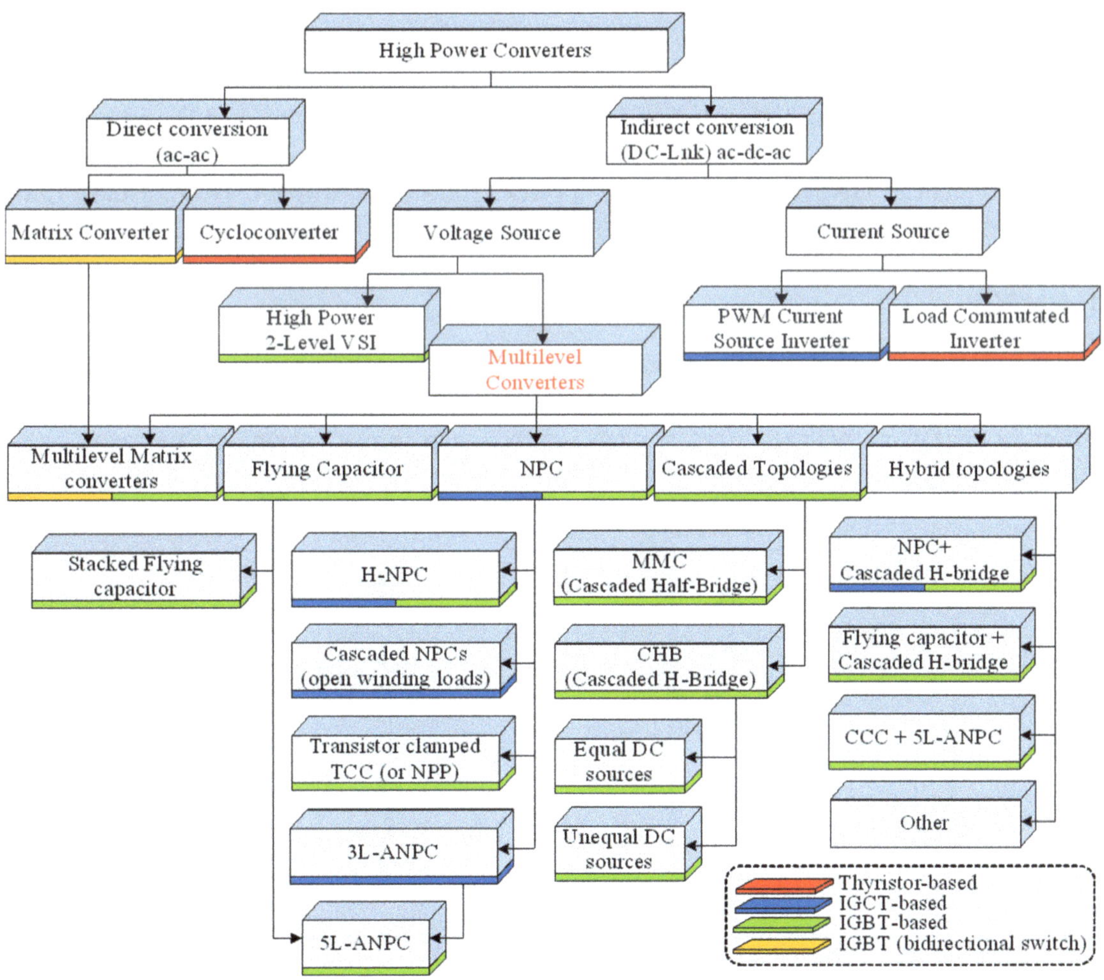

Figure 3.3 Multi-level converter classification

(LCIs) for very high-power, high-torque, and low-speed applications, and the pulse width-modulated current source inverter for high-power variable speed drives.

The 3L-NPC has become quite popular because of its simple transformer rectifier power circuit structure, with a lower device count when considering both the inverter and rectifier and a lower number of capacitors. Although the NPC structure can be extended to a higher number of levels, these are less attractive because of higher losses and uneven distribution of losses in the outer and inner devices. In particular, the clamping diodes, which have to be connected in series to block the higher voltages, introduce more conduction losses and produce reverse recovery currents during commutation that affect the switching losses of the other devices even more. Furthermore, DC-link capacitor voltage balance becomes unattainable in higher level topologies, with a passive front end when using conventional modulation strategies. In this case, the classic multi-level-stepped waveform cannot be retained, and higher *dv/dts* (more-than one-level transitions) is necessary to balance the capacitors for certain modulation indexes.

On the other hand, the CHB is well suited for high-power applications because the modular structure enables higher voltage operation with classic low-voltage semiconductors. The phase shifting of the carrier signals moves the frequency harmonics to the higher frequency side, and this, together with the high number of levels, enables a reduction in the average device switching frequency (≤500 Hz), allowing air cooling and lower losses. However, it requires a large number of isolated DC sources, which have to be fed from phase-shifting isolation transformers. These are more expensive and bulkier compared to the standard transformer used for the NPC. Nevertheless, this has been effectively used to improve the input power factor of this converter, reducing input current harmonics.

Although the FC is modular in structure, like the CHB, it has found less industrial penetration compared to the NPC and CHB, mainly because higher switching frequencies are necessary to keep the capacitors properly balanced whether a self-balancing or a control-assisted balancing modulation method is used (e.g., >1200 Hz). These switching frequencies are not feasible for high-power applications, where they are usually limited to a range of 500–700 Hz. This topology also requires the initialization of FC voltages.

3.1.1. 5L-HNPC

This converter is composed of an H-bridge connection of two classic 3L-NPC phase legs (as shown in Figure 3.4) forming a 5L-HNPC converter, and was first introduced in the 1990s. This topology has been commercialized by two medium-voltage drive manufacturers, and has received increased attention over the years.

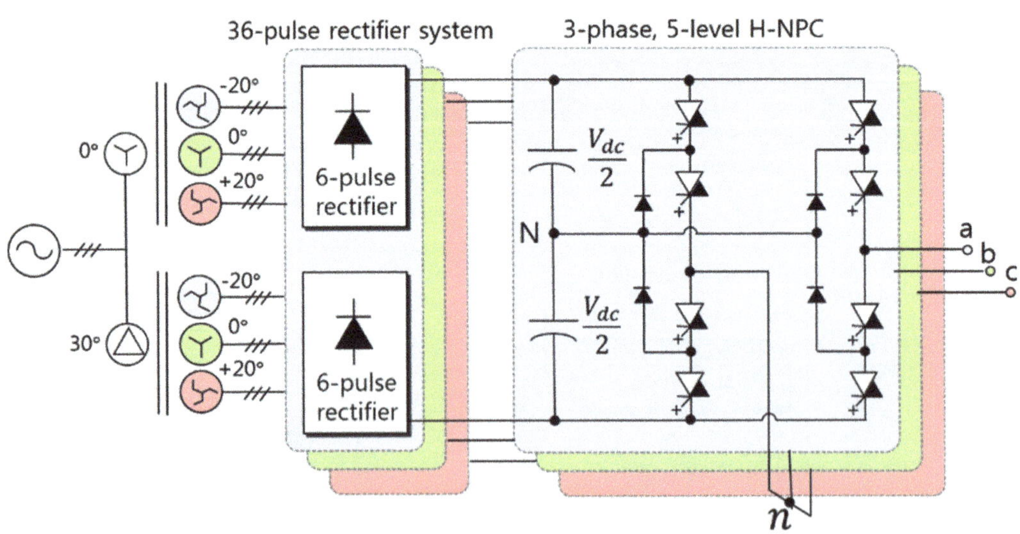

Figure 3.4 Three-phase 5L-HNPC

The combination of the three levels of each leg of the NPC ($V_{dc}/2$, 0, $-V_{dc}/2$) results in five different output levels (V_{dc}, $V_{dc}/2$, $0, -V_{dc}/2, -V_{dc}$). As with the traditional H-bridge, this topology requires an isolated DC source for each H-bridge to avoid a short circuit of the DC links. Therefore, a transformer with three dedicated secondary three-phase windings is necessary to supply the H-bridge of each phase of the converter. Additionally, since the semiconductors of the 3L-NPC leg block half of the total DC-link voltage, higher voltage can be reached with a series connection of two diode bridges. This can lead to a 36-pulse rectifier system, as shown in Figure 3.4.

The disadvantages of a more complex transformer come along with an attractive feature: the enhanced input side power quality obtained with the phase-shifting transformer and multi-pulse rectifier configuration. Low-order harmonics are effectively canceled up to the 25th harmonics (for a 36-pulse rectifier), greatly improving the input current total harmonic distortion (THD) and eliminating the need for filters, just like with the CHB topology. In fact, this topology features a transformer that is identical to the one used for a five-level two-cell CHB, with the same amount of diode rectifier bridges, and number of capacitors and switching semiconductors, but with the addition of 12 clamping diodes and the need to control the neutral point of each H-bridge.

This converter can be found in practice with a 36-pulse rectifier system featuring IGCTs and for a 2- to 7 MW (air cooled) or 5 to 22 MW (water cooled) power range. Other characteristics are given as follows: it is controlled with direct torque control, reaching an output frequency of up to 250 Hz and output voltages of up to 6.9 kV. Alternatively, several configurations of the 5L-HNPC are available: with a 24- or 36-pulse diode rectifier front end with medium-voltage IGBTs (MV-IGBTs), IEGTs, or GCTs; up to 7.86 kV maximum output voltage; up to 120 Hz output frequency; air or water cooling and vector control; and with a power range of up to 120 MVA.

3.1.2. 3L-ANPC

One of the drawbacks of the 3L-NPC topology is the unequal share of losses between the inner and outer switching devices in each converter leg. Since the semiconductors are cooled with separate heat sinks and cooling systems, it results in an asymmetrical semiconductor-junction temperature distribution, which affects the cooling system design and limits the maximum power rate, output current, and switching frequency of the converter for a specific semiconductor technology (which is usually the IGCT).

This issue can be solved by replacing the neutral clamping diodes with clamping switches to provide a controllable path for the neutral current, and thus control the loss distribution among the switches of the converter. In other words, with clamping diodes such as those in the 3L-NPC, the current freewheels through the upper or lower clamping diode depending on the current polarity when the zero-voltage level is generated. However, with clamping switches, the current can be forced to go through the upper or lower clamping path. This can be used to control the power loss distribution and overcome the limitations of the 3L-NPC, enabling substantially higher power rates. These additional devices are called active neutral clamping switches, as shown in Figure 3.5(a), which is what gives this converter its name—active neutral-point-clamped (ANPC).

The 3L-ANPC was introduced as a commercial product with a back-to-back regenerative configuration. It covers a power range of 20–200 MVA and can be connected to a transformer from a 6 kV to a 220 kV grid.

A variation of the ANPC concept has been introduced, that is, a five-level hybrid multi-level converter that combines a 3L-ANPC leg with a three-level FC power cell connected between the internal ANPC switching devices, as shown in Figure 3.5(b). Although it is a hybrid topology, it has been called **5L-ANPC.** It effectively increases the number of levels of the converter with the levels introduced by the FC cell.

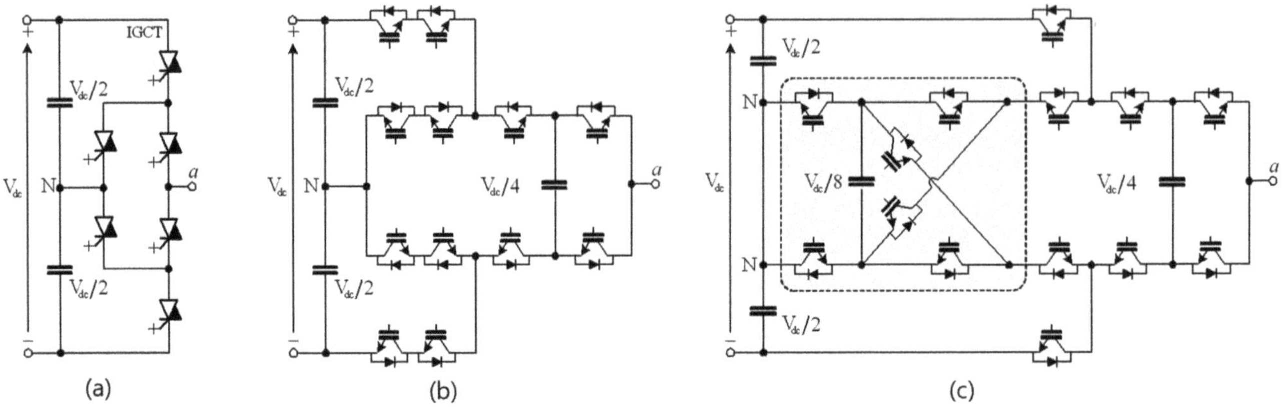

Figure 3.5 Active neutral-point-clamped (NPCs) (with only phase *a* shown): (a) Three-level active neutral point clamped (3L-ANPC) featuring integrated gate-commutated thyristors (IGCTs); (b) 5L-ANPC featuring IGBTs (hybrid between 3L-ANPC and a three-level flying capacitor (3L-FC)); and (c) CCC stage plus 5L-ANPC hybrid nine-level converter

The FC is controlled to $V_{dc}/4$, so that its series connection to the ANPC DC-link capacitors at $V_{dc}/2$ or to the neutral at 0 V, using an appropriate switching state, generate the additional intermediate voltage levels, completing a total of five levels ($\pm V_{dc}/2$, $\pm V_{dc}/4$, and 0). There are several redundancies that can be used to control FC voltage. This hybrid ANPC–FC concept somehow enables the modularity factor that is lacking in the classic NPC converter family just by adding FC cells to reach easily higher-level values, without the need to add series-connected diodes. Moreover, because only a 3L-NPC leg is used, the problems of capacitor voltage balancing when using a passive front end in a higher number of level NPCs are being avoided as well.

These advantages come at the expense of a more complex circuit structure and the need to control the FC voltages and their initialization, aside from the NPC DC-link capacitors voltage unbalance control. In contrast with the CHB topology, this modularity increases not the power rating of the converter, but the number of levels, and the power quality since the FC adds an intermediate voltage level and does not provide active power; thus, the power rating is still limited by the ANPC part. Note that instead of IGCTs, series-connected IGBTs are used in the NPC part of the converter, probably to keep all semiconductors of the same type. This inherently introduces more conduction and switching losses and requires a special gate driver to ensure simultaneous control of both switches. Nevertheless, because of the extra levels and the particular configuration of the power circuit, the outer switches commutate at lower switching frequency, which compensates for the series losses. Other ratios for the FC voltage can be used to increase the number of voltage levels at the output. However, in that case, higher switching frequencies are necessary to properly control the capacitor voltage and make it less attractive.

Numerous variations and even combinations of these topologies have been presented to satisfy particular application requirements or to improve operational features. Some of these variations include the active NPC (ANPC), which produces a more equal semiconductor junction temperature distribution, enabling a substantial increase of the converter output current and power at nominal operation compared to the regular NPC. Another variation is the stacked MC, which can basically be described as a combination of serial- and parallel-connected flying-capacitor power cells.

This converter allows a higher voltage operation, reduction in capacitor volume, and better-quality output voltages than the traditional FC. The CHB, traditionally fed by equal DC voltage sources, has been modified by introducing an asymmetry into the DC voltage sources. This modified version is known as the asymmetric or hybrid CHB. It generates considerably more output-voltage levels than any other topology using less semiconductors and capacitors. Asymmetric or hybrid topologies have also been proposed based on the NPC structure.

3.1.3. Neutral-Point Clamped Converter

In the early 1980s, a new pioneering converter topology was proposed: the three-level NPC voltage-source converter (3L NPC-VSC), also known as the DCC. Since all semiconductors are operated at a commutation voltage of half the DC-link voltage, the topology offered a simple solution to extend voltage and power ranges of the existing 2L-VSC technology, which were severely limited by the blocking voltages of power semiconductors with active turn-on and turn-off capabilities. Consequently, the converter was of particular interest for MV applications (2.3–4.16 kV). The 3L NPC-VSC was soon introduced to the market by leading manufacturers and gained more and more importance. The three-level power circuit is shown in Figure 3.6. NPC converters can be extended to generate more output-voltage levels; however, these topologies have not found industrial acceptance to date.

In a 2L-VSC, a series connection of two switches per switch position is required to enable an operation at the same DC and converter voltage, just like a 3L NPC-VSC. The 3L NPC-VSC features two additional diodes per phase leg, as compared to a 2L-VSC with a direct series connection of two devices per switch position. These so-called NPC diodes link the midpoint of the "indirect series connection" of the main switches to the neutral point of the converter. This allows the connection of the phase output to the converter neutral point N and enables the three-level characteristic of as shown in Figure 3.7. Table 1 shows the switch states of one-phase leg.

Note that two pairs of switches on one leg receive inverted gate signals S_{ak} and S_{ak} (k = 1, 2), both to enable a proper modulation and to avoid forbidden switch states. The overall converter switching state can be defined by $S = (S_a, S_b, S_c)$, where, according to Table 3.1, each phase switching state $S_x(x = a, b, c)$ can be represented by a (+1) when generating Vdc/2, a (−1) when generating $-V_{dc}/2$, or a (0) when connected to the neutral point, thus generating zero-voltage level. Combining the states of all three phases, the 3L NPC-VSC features $n_{ss} = L^3 = 3^3 = 27$ switch states, where L is the number of voltage levels of v_{xN} ($x = a, b, c$).

$$v_s = \frac{2}{3}\left[v_{aN} + av_{bN} + a^2_{cN}\right] = \frac{2}{3}\left[S_a\frac{V_{dc}}{2} + aS_b\frac{V_{dc}}{2} + a^2S_c\frac{V_{dc}}{2}\right] \quad (3.1)$$

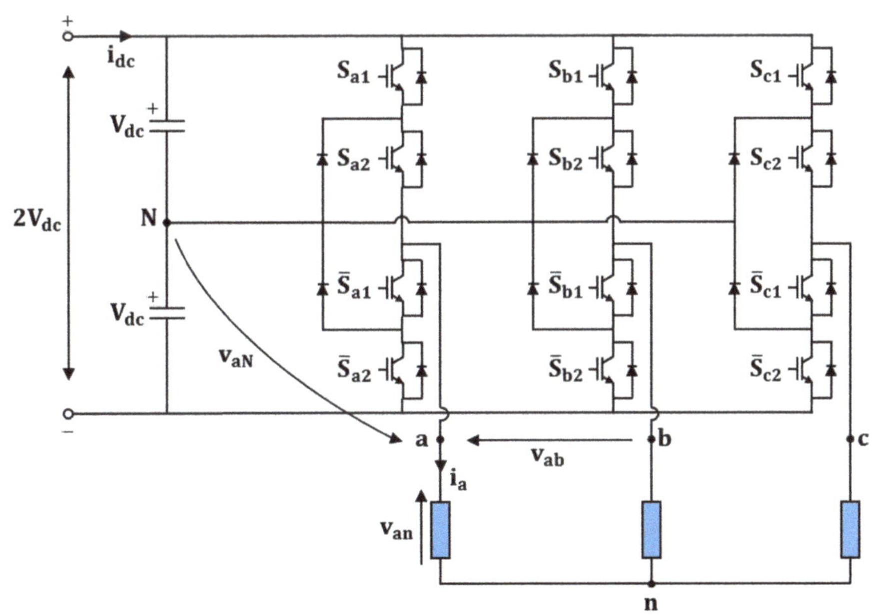

Figure 3.6 Three-level neutral-point-clamped power circuit

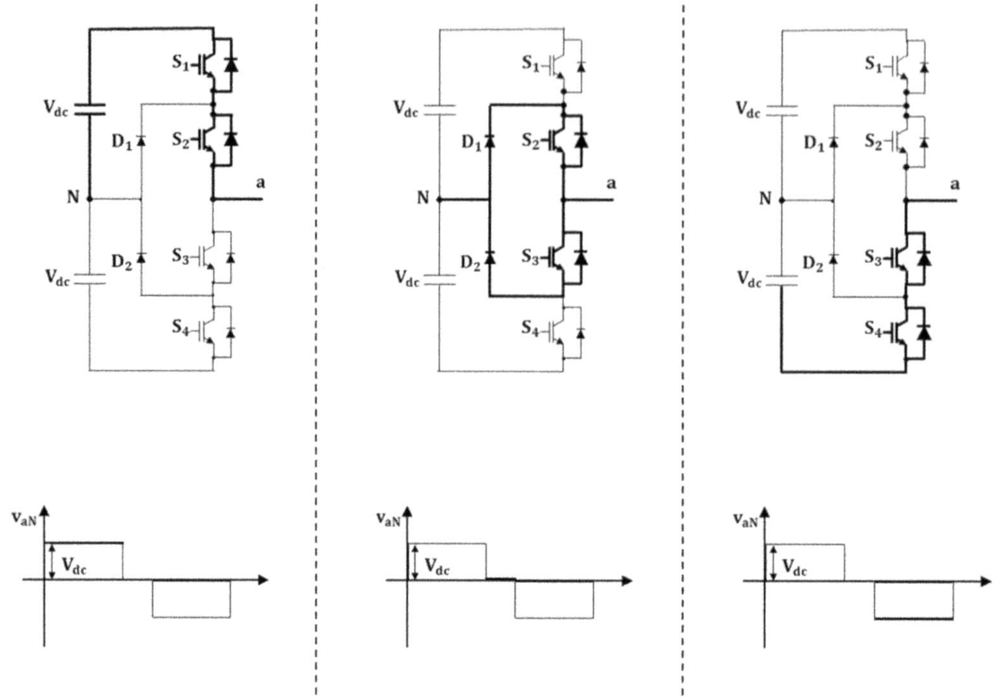

Figure 3.7 Three-level neutral-point-clamped (NPC) switching states and corresponding output-voltage levels

Table 3.1 Three-Level NPC Switch States (PHASE *a*)

Output voltage	Gate signals		Switching state
v_{aN}	S_{a1}	S_{a2}	S_a
$V_{dc}/2$	1	1	(+)
0	0	1	(0)
$-V_{dc}/2$	0	0	(−)

3.1.4. CHB Converter

The CHB is composed of the series connection of H-bridge power cells. For this reason, the CHB is also known as a multi-cell inverter. Each cell includes a single-phase 3L H-bridge inverter, a capacitive DC-link, a rectifier, and an independent or isolated voltage source provided by transformer secondaries or batteries

Consider the inverter side of the power cells presented in Figure 3.8. The switching state of one cell S_k is determined by the logical value of two signals (S_{k1}, S_{k2}), which can be "1" or "0" representing the "ON" and "OFF" state of each switch. This leads to four different binary combinations that generate three different output voltages: $+V_{\mathrm{dc}}$, zero, and $-V_{\mathrm{dc}}$. Since the cells are connected in series, the total inverter output voltage of one-phase x for an inverter with k cells is shown by

$$v_{xN} = \sum_{y=1}^{k} v_{xy} = \sum_{y=1}^{k} v_{dc}\left(S_{y1} - S_{y2}\right), \qquad x = \{a, b, c\} \tag{3.2}$$

where k is the number of power cells per phase and V_{dc} is the DC-link voltage of each cell.

The different output-voltage levels can be determined by replacing all the possible binary combinations of the switching states of each cellin Equation (3.2), obtaining the inverter switching table. An example for one phase of a two-cell five-level CHB inverter is presented in Table 3.2.

More recently, CHB multi-level inverters with unequal DC-source, also known as hybrid or asymmetric CHB inverters, have been introduced. The power circuit is very similar to the regular CHB shown in Figure 3.8; the difference is that the isolated DC sources have different values. When choosing unequal DC

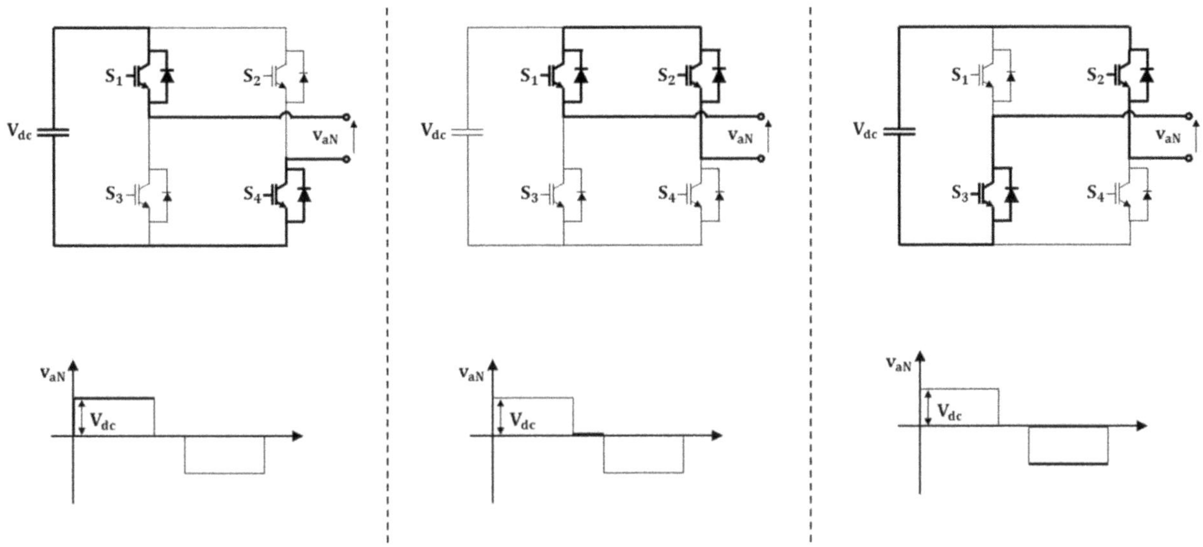

Figure 3.8 Three-level H-bridge switching states (single phase)

Table 3.2 Five-Level Cascaded Inverter Switching States

Output voltage	Cell 1			Cell 2		
v_{aN}	S_{11}	S_{12}	v_{a1}	S_{22}	S_{22}	v_{a2}
$2V_{dc}$	1	0	V_{dc}	1	0	V_{dc}
V_{dc}	1	0	V_{dc}	0	0	0
	1	0	V_{dc}	1	1	0
	0	0	0	1	0	V_{dc}
	1	1	0	1	0	V_{dc}
0	0	0	0	0	0	0
	0	0	0	1	1	0
	1	1	0	0	0	0
	1	1	0	1	1	0
	1	0	V_{dc}	0	1	$-V_{dc}$
	0	1	$-V_{dc}$	1	0	V_{dc}
$-V_{dc}$	0	1	$-V_{dc}$	1	1	0
	0	1	$-V_{dc}$	0	0	0
	0	0	0	0	1	$-V_{dc}$
	1	1	0	0	1	$-V_{dc}$
$-2V_{dc}$	0	1	$-V_{dc}$	0	1	$-V_{dc}$

sources, some switching-state redundancies are avoided, and more different output-voltage levels are generated with the same amount of power cells. This reduces the size and cost of the converter and improves reliability since less semiconductors and capacitors are employed. An additional advantage is that the converter can be controlled appropriately to reduce the switching losses, which is very important in high-power applications. The main drawback is the fact that the modularity of the converter is lost, since the different power ratings of the cells force special design for each power cell (even different power-device families could be needed). In addition, no input-current harmonic cancellation can be achieved, such as in the equally fed case, since the power asymmetry disables the multi-pulse rectifier and transformer function.

The relation between the DC sources that leads to more output levels is to scale them in powers of three

$$[V_1, V_2 \ldots V_k]^T = V_{dc} \cdot \left[3^0\, 3^1 \ldots 3^{(k-1)}\right]^T \tag{3.3}$$

where V_1, V_2, ., V_k are the DC-source voltages of the different power cells and V_{dc} is the smallest DC-source voltage. This asymmetry produces $3k$ different voltage levels: for example, a four-cell converter can generate 81 voltage levels, compared to the nine-voltage levels generated with a four-cell CHB with equal DC sources.

3.1.5. FC Converter

The FC-VSC topology was introduced about 20 years ago. Currently, the four-level FC-VSC (4L FC-VSC) is produced by just one manufacturer of industrial MV drives. The circuit configuration of the topology is depicted in Figure 3.9. This topology can be presented in a more modular way, also known as multi-cell, as shown in Figure 3.10.

One converter phase leg consists of a series connection of three commutation cells. The nominal voltages of the FCs, C_{x1}, and C_{x2}, are $v_{cx1} = 2/3V_{\mathrm{dc}}$ and $v_{cx2} = 1/3V_{\mathrm{dc}}$, respectively. Meynard and Foch describe a simple scheme to charge the FCs to the desired voltage level during the start-up procedure of the converter. The switching states of one phase of the 4L FC-VSC are shown in Table 3.3. The switches of each commutation cell (e.g., S_{x1} for the upper switch position and S_{x1} for the lower switch position) commutate the phase current to the upper and the lower switch positions. The corresponding commutation voltage per cell is equivalent to one third of the DC-voltage ($V_{\mathrm{com}} = V_{\mathrm{dc}}/3$). Redundant switch states with a different polarity of FC currents are used to balance the voltages across the FCs. Combining all three phases, the 4L FC-VSC features $n_{\mathrm{ss}} = L^3 = 4^3 = 64$ different switching states.

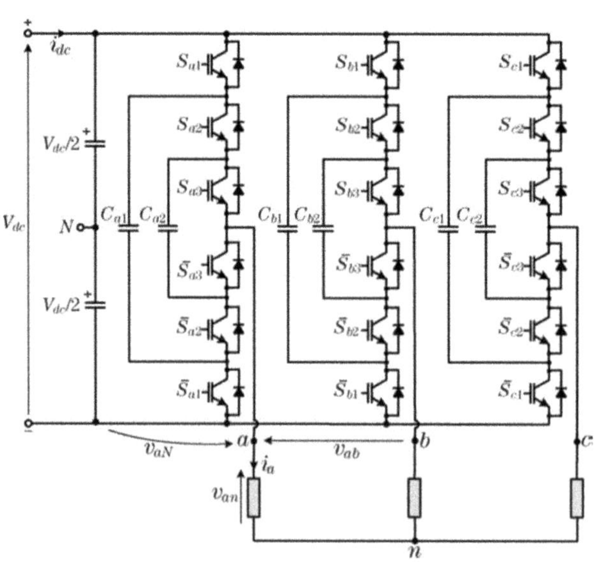

Figure 3.9 Circuit configuration of 4L flying capacitor voltage-source converter (FC-VSC)

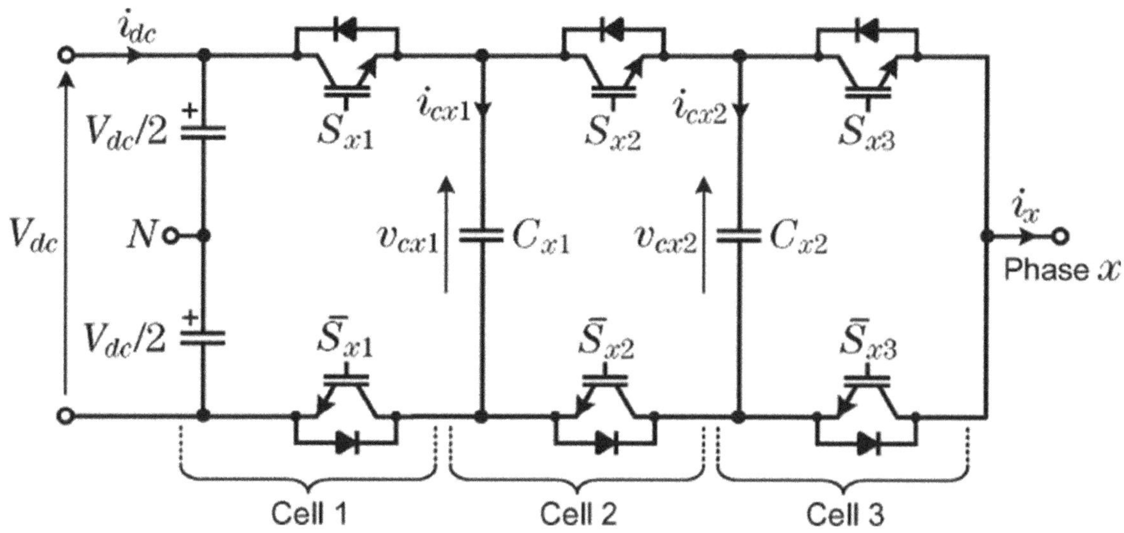

Figure 3.10 Circuit configuration of one-phase leg ($x = a, b, c$) of 4L flying capacitor voltage-source converter (FC-VSC) (multi-cell layout)

Table 3.3 Four-Level FC Switching States (Phase $x = a, b, c$)

Output voltage	Switching state			Capacitor current	
v_{xN}	S_{x1}	S_{x2}	S_{x3}	i_{cx1}	i_{cx2}
$V_{dc}/2$	1	1	1	0	0
$V_{dc}/6$	1	1	0	0	$+i_x$
	1	0	1	$+i_x$	$-i_x$
	0	1	1	$-i_x$	0
$-V_{dc}/6$	1	0	0	$+i_x$	0
	0	1	0	$-i_x$	$+i_x$
	0	0	1	0	$-i_x$
$-V_{dc}/2$	0	0	0	0	0

3.2. Modular Multi-level Converter

3.2.1. Topologies of MMC Systems [7]–[12]

Figure 3.11 shows a schematic diagram of a three-phase MMC. The MMC, as shown in Figure 3.11, consists of two arms per phase leg where each arm comprises N series-connected, nominally identical SMs, as well as a series inductor L_o. Although the SMs in each arm are controlled to generate the required AC phase voltage, the arm inductor suppresses the high-frequency components in the arm current. The upper (lower) arm of three-phase legs are represented by a subscript "p" ("n").

The SMs of the MMC of Figure 3.11 can be realized by the following circuits:

1. The half-bridge circuit: As shown in Figure 3.11(a), the output voltage of a half-bridge SM is either equal to its capacitor voltage v_C (switched-on/inserted state), or zero (switched-off/bypassed state), depending on the switching states of the complimentary switch pairs, that is $S1$ and $S2$.
2. The full-bridge circuit: As shown in Figure 3.11(b), the output voltage of a full-bridge SM is either equal to its capacitor voltage v_C (switched-on/inserted state), or zero (switched-off/bypassed state), depending

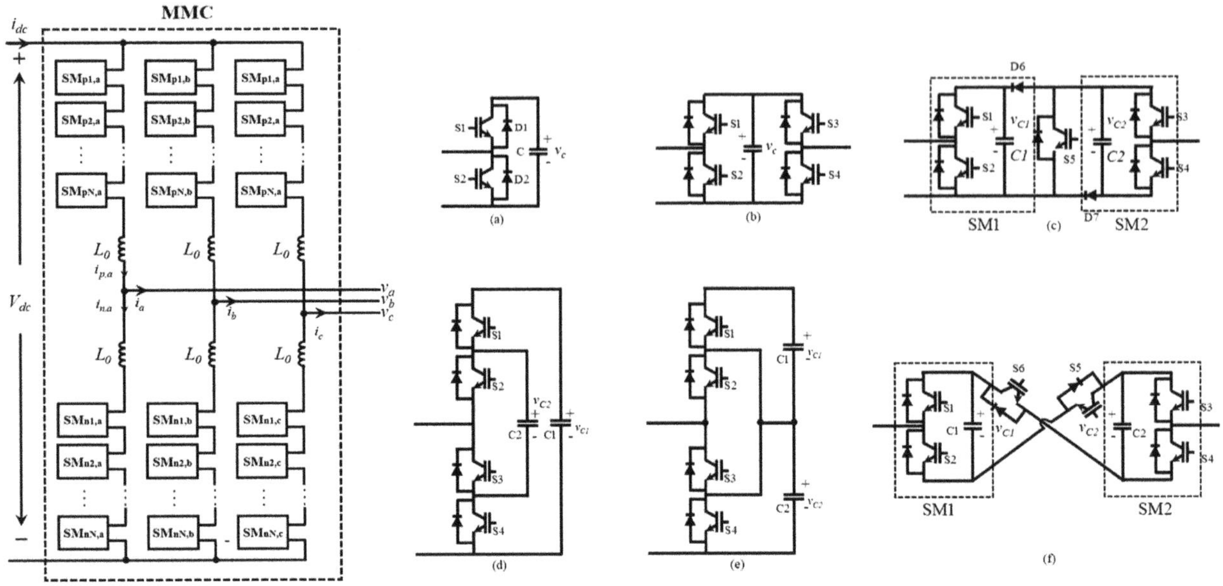

Figure 3.11 Schematic representation of the MMC and various SM topologies: (a) the half-bridge; (b) the full-bridge; (c) the clamp-double; (d) the three-level FC; (e) the three-level NPC; and (f) the five-level cross-connected SM

on the switching states of the four switches $S1$ to $S4$. Since the number of semiconductor devices of a full-bridge SM is twice that of a half-bridge SM, the power losses as well as the cost of an MMC based on the full-bridge SMs are significantly higher than that of an MMC based on the half-bridge SMs.

3. The clamp-double circuit: As shown in Figure 3.11(c), a clamp-double SM consists of two half-bridge SMs, two additional diodes, and one extra IGBT with its anti-parallel diode. During normal operation, the switch $S5$ is always switched ON and the clamp-double SM acts equivalent to two series-connected half-bridge SMs. Compared to the half- and full-bridge MMCs with the same number of voltage levels, the clamp-double MMC has higher semiconductor losses than the half-bridge MMC and lower than the full-bridge MMC [4].
4. The three-level converter circuit: As shown in Figure 3.11(d) and (e), a three-level SM is composed of either a three-level NPC or a three-level FC converter . The three-level FC MMC has similar semiconductor losses to the half-bridge MMC; however, the three-level NPC MMC has higher semiconductor losses than the half-bridge MMC and lower than the full-bridge MMC. From a manufacturing perspective and control, this SM circuit is not very attractive.
5. The five-level cross-connected circuit: As shown in Figure 3.11(f), a five-level cross-connected SM also consists of two half-bridge SMs connected back-to-back by two extra IGBTs with their anti-parallel diodes. Its semiconductor losses are the same as the clamp-double SM.

A comparison of various SM circuits in terms of voltage levels, DC-side short-circuit fault-handling capability, and power losses, is provided in Table 3.4.

From their topologies, MMCs can be classified into the following three categories: (1) Double-Star-connection MMC, (2) Single Star-connection MMC, and (3) Single Delta-connection MMC. Double-Star-connection MMCs can be classified into (1) half-bridge-type MMC and (2) full bridge-type MMC.

The Double-Star-connection MMC topology possesses the common DC-link terminals as shown in Figure 3.12, which enable DC-to-AC and AC-to-DC power conversion. However, the single star/delta connection MMC topology has no common DC-link terminals, as shown in Figure 3.12. As a result, it has no capability

Table 3.4 Comparison of Various SM Circuits

SM circuit	Voltage levels	DC-fault handling	Losses
Half-bridge	0, v_C	No	Low
Full-bridge	0, $+v_C$	Yes	High
Clamp-double	0, v_{C1}, v_{C2}, $(v_{C1} + v_{C2})$	Yes	Moderate
Three-level FC	0, v_{C1}, v_{C2}, $(v_{C1} - v_{C2})$	No	Low
Three-level NPC	0, v_{C2}, $(v_{C1} + v_{C2})$	No	Moderate
Five-level cross-connected	0, v_{C1}, v_{C2}, $+(v_{C1} + v_{C2})$	Yes	Moderate

of achieving DC-to-AC and AC-to-DC power conversion, although it can control active power back-and-forth between the three-phase AC terminals and the floating DC capacitors. This means that the single star/delta connection MMC topology is not applicable to industrial motor drives, though it is suitable for STATCOMs and energy storage systems. This is one of the most significant differences in function and application between the double-star-configured MMC topology and the star-/delta-configured MMC topology.

3.2.2. Modulation Methods for MMC [13]–[19]

Modulation Techniques

Various multi-level modulation techniques, based on using a single reference waveform, have been developed for the MMC system:

Level-Shifted PWM: These techniques require N identical triangular carrier waveforms displaced symmetrically with respect to the zero axis. The comparison of the phase voltage reference waveform with the carriers produces the desired switched output-phase voltage level. Voltage transitions corresponding to a triangular carrier are associated with the insertion/bypass of a particular SM. Based on the phase shift among the carrier waveforms, these techniques are further classified into (1) phase shift carrier (PSC), (2) phase disposition (PD), (3) phase opposition disposition (POD), and (4) alternate phase opposition disposition (APOD), as shown in Figure 3.14(a)–(d). The disadvantages of using these techniques include unequal distribution of voltage ripple across the SM capacitors that impact the harmonic distortion of the AC-side voltages, and large magnitude of circulating currents. To improve the harmonic distortion of the AC-side voltages, a simple carrier rotation technique, a modified carrier rotation technique, or a signal rotation technique is used to equalize the voltage distribution across all the SM capacitors. Despite the several introduced SM capacitor voltage balancing techniques, the output voltages have a relatively high THD. To improve the performance of these techniques, a modified PD PWM technique with an SM capacitor voltage balancing technique is proposed in. In this technique, which is based on the PD carrier waveforms, voltage transitions corresponding to a triangular carrier are no longer assigned to a particular SM. In this technique, comparison of the reference waveform with the carrier waveforms produces an $(N + 1)$-level waveform that determines the number of SMs to be inserted in the upper and lower arms, respectively. Depending on the direction of the arm current and the status of the SM capacitor voltages, the determined number of SMs out of the N SMs in the upper (lower) arm are inserted so as to minimize the difference between the SM capacitor voltages. Mei *et al.* proposed a PD PWM technique with a selective loop bias mapping method for balancing the SM capacitors. This method implements carrier rotation using the following feedback: (1) the maximum/minimum SM capacitor voltages and (2) the direction of arm

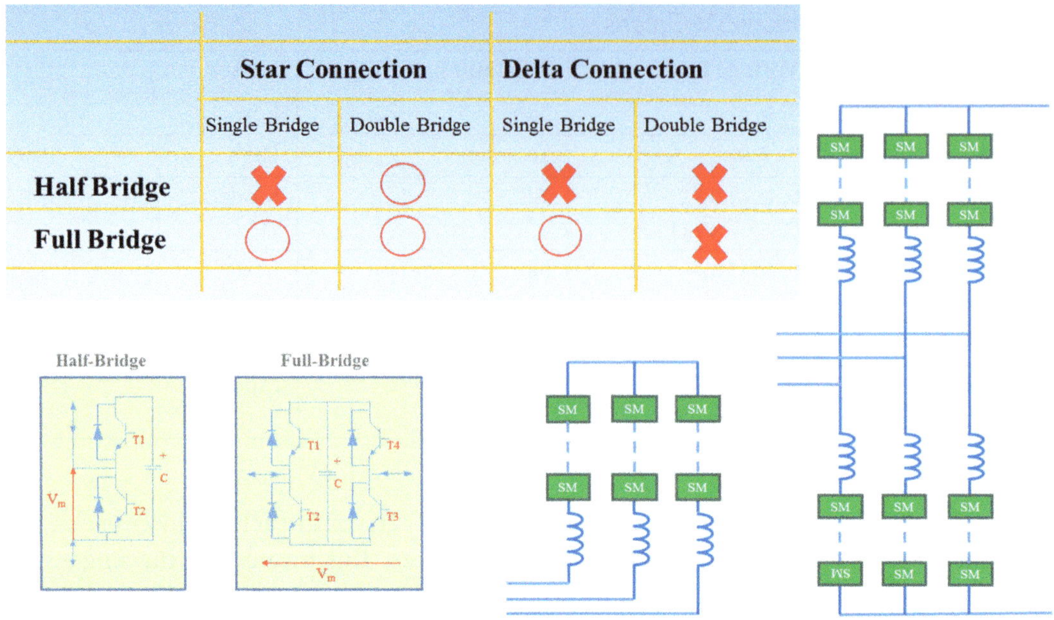

(a) Half Bridge, Full Bridge, Single Bridge and Double Bridge according to connection type

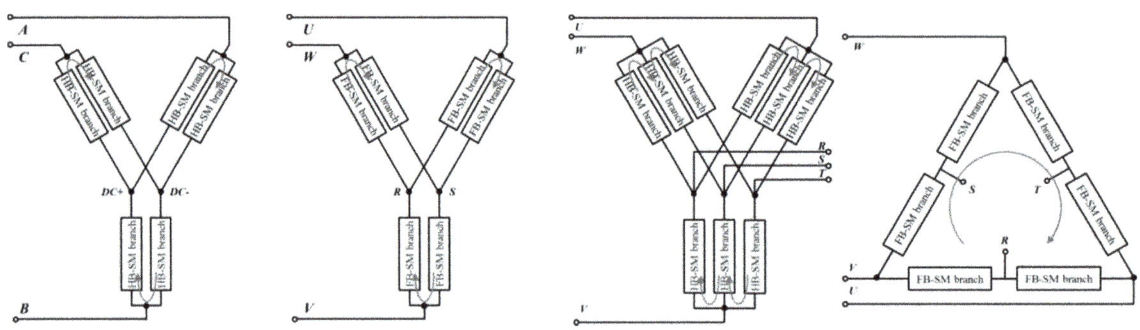

(b) 3-phase to single-phase or 3-phase/DC topology (c) 3-phase to 3-phase topology

Figure 3.12 Circuit configuration of MMCs

current. The advantages of this technique include (1) absence of additional reference signals to control the SM capacitor voltages and (2) ease of implementation in a simple field-implemented gate array (FPGA) even with a large number of SMs.

1. **Phase-shifted carrier-based PWM (PSC-PWM):** PSC-PWM is an extension for multi-level converters of the popular SPWM. The method is based on using $n/2$ carrier signals shifted between them by a phase shift, θ, given by

$$\theta = \frac{360}{n/2} \tag{3.4}$$

Up to this point, two modulation techniques for MMCs based on the number of output-voltage levels have been described. These modulation techniques are called $N + 1$ and $2N + 1$, N being the number of SMs per arm.

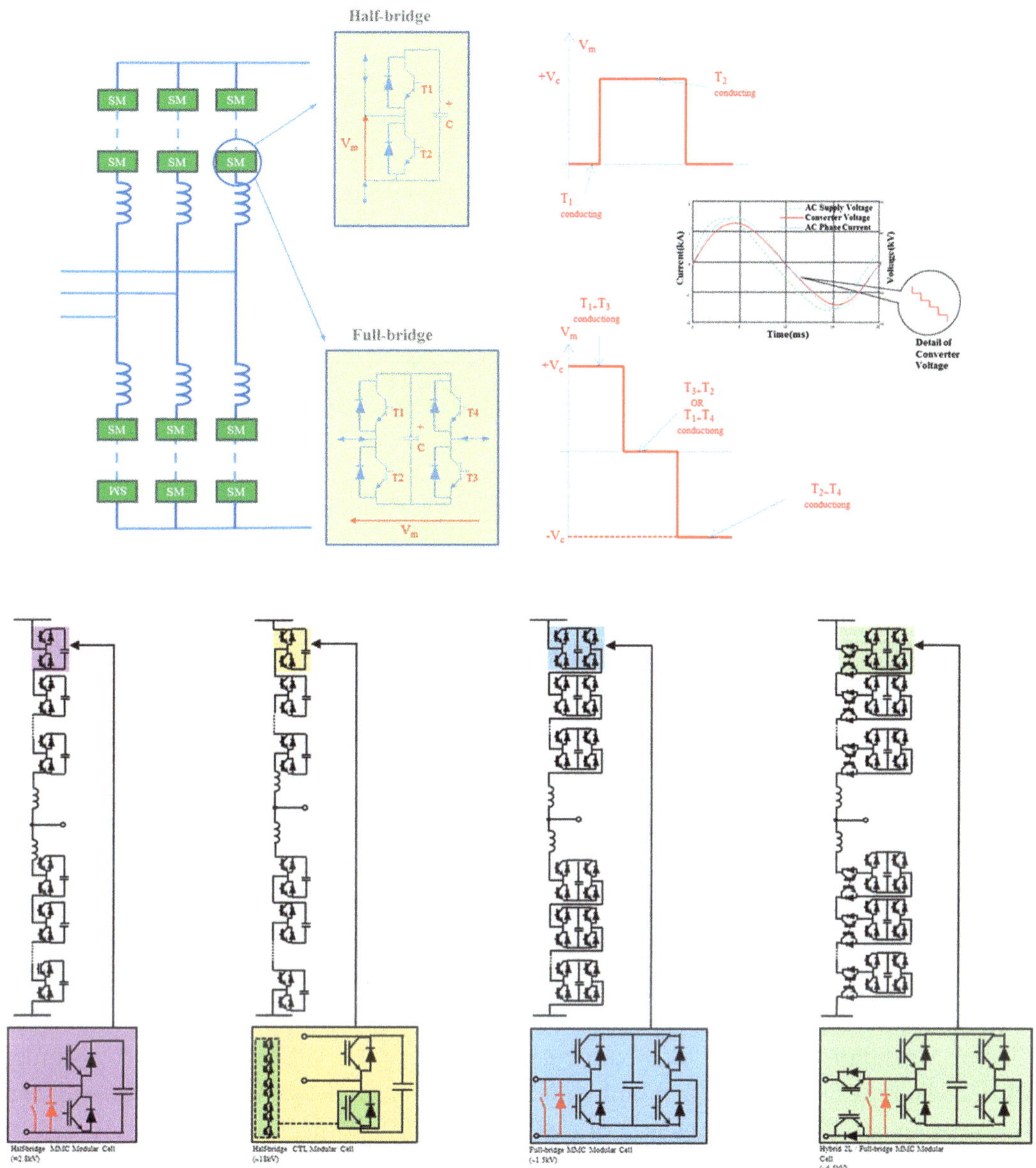

(a) Half-bridge MMC topology – single phase leg, schematic representation

(b) Half-bridge CTL topology – single phase leg, schematic representation

(c) Full-bridge MMC topology – single phase leg, schematic representation

(d) Hybrid MMC topology – single phase leg, schematic representation

Figure 3.13 Classification of MMCs based on single-phase half-bridge or full-bridge converters

In $2N + 1$ modulation, each arm modulates independently. Thus, the output voltage depends on the number of SMs connected in both the upper and lower arm. This approach generates $2N + 1$ output levels. However, because of the number of SMs activated at the same time not being constant, it generates large SM voltage unbalances, especially in the case of MMCs with a low number of SMs

In $N + 1$ modulation, the upper- and lower-phase arms of each phase are inversely commutated. Therefore, $N + 1$ output levels are generated. Despite this technique generating less output levels, it has the advantage of the number of activated SMs always being constant. Consequently, the capacitor voltage oscillations are reduced.

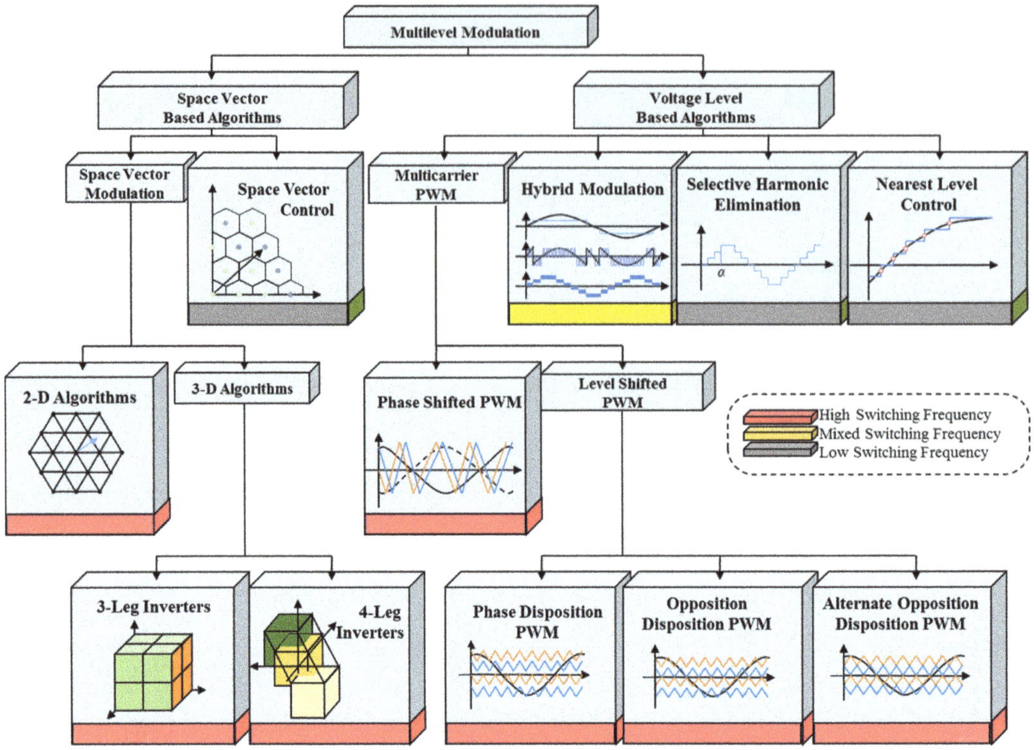

Figure 3.14 Various multi-level modulation technique

Considering a six-level MMC, Figure 3.16 shows the modulation signals, V_x with $x = 1, . . , 5$, for a single phase. The phase shift between the five carriers, VC_x, is $\theta = 72°$. The output waveform is the sum of all the signals.

2. **Space Vector Modulation (SVM):** SVM can conveniently provide more flexibility to optimize switching waveforms, given the existence of redundant switching states and adjustable duty cycles, and is suitable to be digitally implemented. The main drawback of applying SVM to multi-level converters is that the complexity of the algorithm grows rapidly with the number of levels. Consequently, it has not yet been well studied when it is applied to MMCs. In a three-phase n-level converter, there are usually n^3 switching states and $6 \times (n-1)^2$ triangles in the space vector diagram. Figure 3.17 shows the space vector diagram for a six-level converter

   The algorithm used in this article is based on the fast SVM method presented in the 2000s, which avoids the use of coordinate transformations and allows its adaptation for any n-level converter with minimum modifications. The algorithm is based on finding the four closest vectors to the reference voltages, as shown in Figure 3.18, using the normalized line voltages. The vectors are defined by means of the upper- and lower-rounded integer values from the normalized V^n_{ab} and V^n_{bc} voltages according to

$$V_{ul} = \left[\begin{matrix} \lceil V^n_{ab} \rceil \\ \lfloor V^n_{bc} \rfloor \end{matrix} \right], \quad V_{lu} = \left[\begin{matrix} \lfloor V^n_{ab} \rfloor \\ \lceil V^n_{bc} \rceil \end{matrix} \right], \quad V_{uu} = \left[\begin{matrix} \lceil V^n_{ab} \rceil \\ \lceil V^n_{bc} \rceil \end{matrix} \right], \quad V_{ll} = \left[\begin{matrix} \lfloor V^n_{ab} \rfloor \\ \lfloor V^n_{bc} \rfloor \end{matrix} \right] \tag{3.5}$$

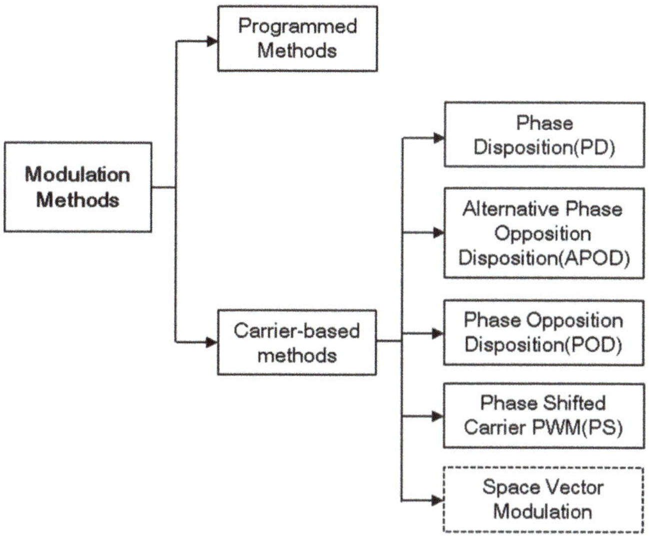

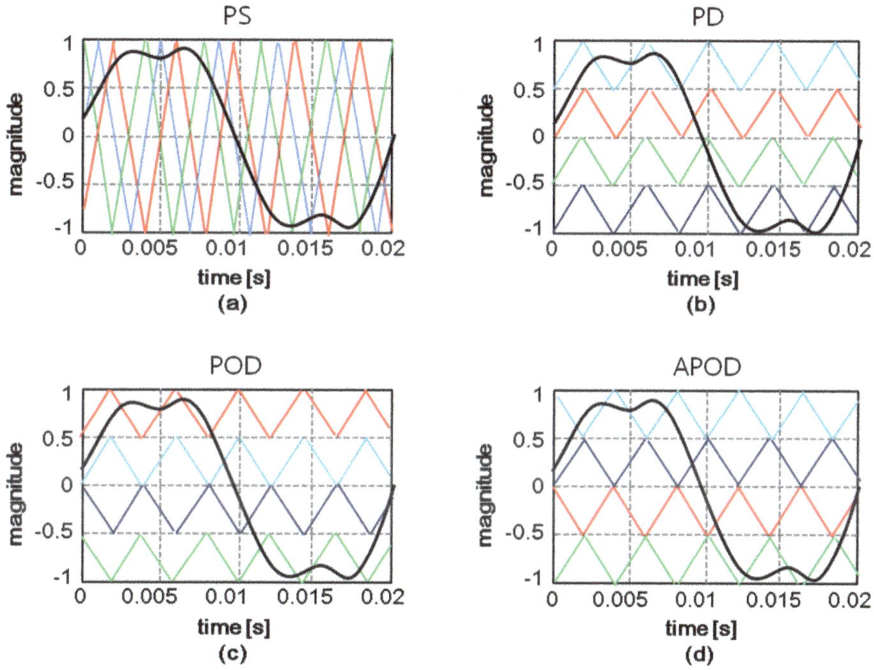

Figure 3.15 Multi-level carriers: (a) phase shift carrier (PSC); (b) phase disposition (PD); (c) phase opposition disposition (POD); and (d) alternate phase opposition disposition (APOD)

V_{ul} and V_{lu} are always two of the three closest vectors. The third vector is chosen according to

$$\begin{cases} V_{uu}, & if\,|V_{ca}^{n} + \left(\lceil V_{ab}^{n} \rceil + \lfloor V_{bc}^{n} \rfloor\right) < 0 \\ V_{ll}, & if\,|V_{ca}^{n} + \left(\lceil V_{ab}^{n} \rceil + \lfloor V_{bc}^{n} \rfloor\right) > 0 \end{cases} \tag{3.6}$$

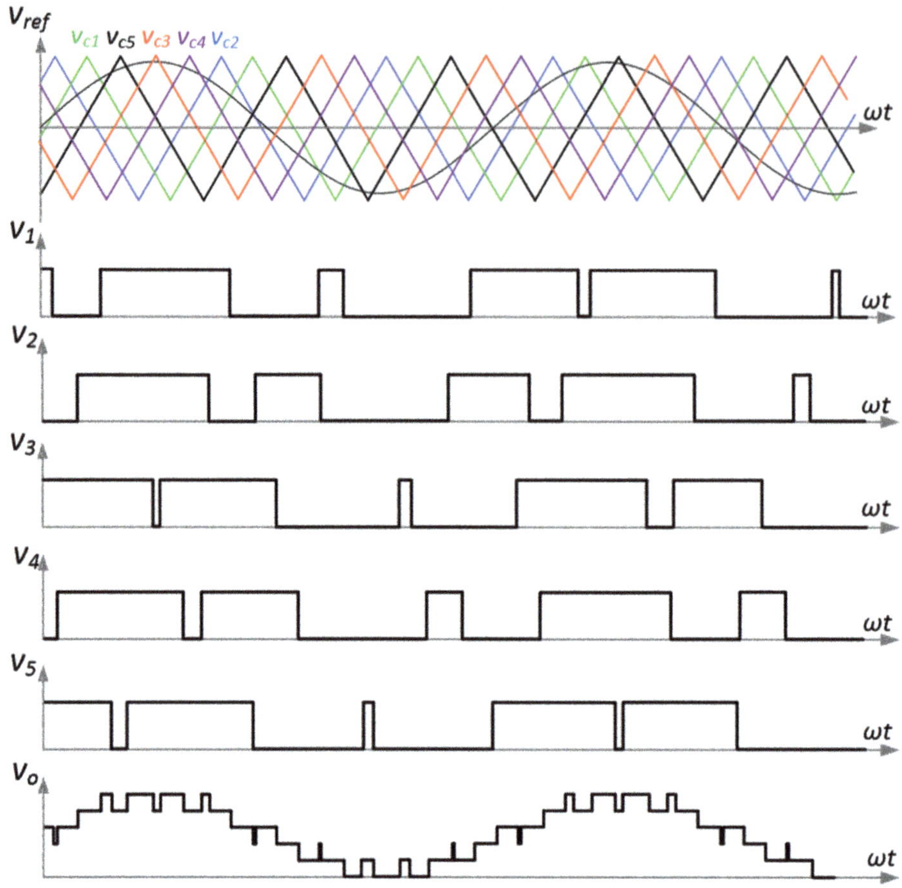

Figure 3.16 Phase shift carrier (PSC)-PWM output waveform generation

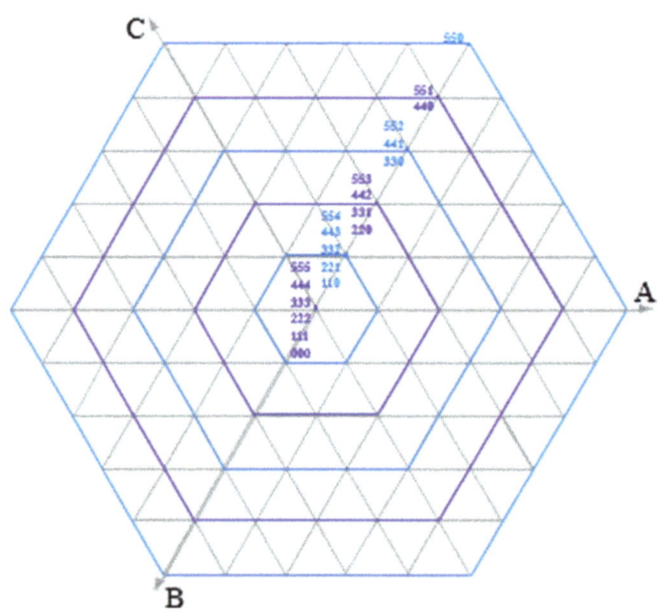

Figure 3.17 Space vector diagram for a six-level converter

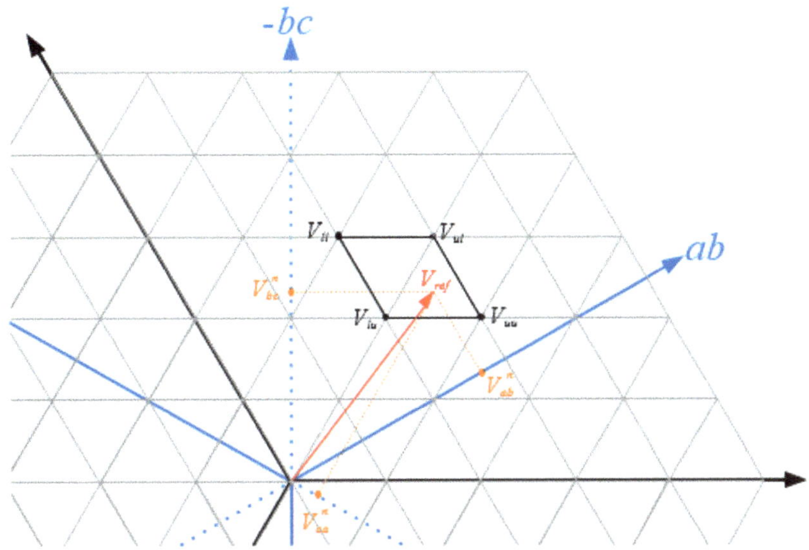

Figure 3.18 Graphic representation of fast space vector modulation (SVM)

The duty cycles are obtained according to the third closest vector. If V_{ll}, the duty cycles are computed as

$$\begin{cases} d_{ul} = \left(V_{bc}^n - \lceil V_{bc}^n \rceil\right) \\ d_{lu} = \left(V_{ab}^n - \lceil V_{ab}^n \rceil\right) \\ d_{uu} = 1 - d_{ul} - d_{lu} \end{cases} \tag{3.7}$$

If the third closest vector is V_{uu}, then

$$\begin{cases} d_{ul} = -\left(V_{bc}^n - \lceil V_{bc}^n \rceil\right) \\ d_{lu} = -\left(V_{ab}^n - \lceil V_{ab}^n \rceil\right) \\ d_{uu} = 1 - d_{ul} - d_{lu} \end{cases} \tag{3.8}$$

The selection of the switching states is done after determining the available switching states for each vector by means of using

$$\begin{bmatrix} k \\ k - V(1) \\ k - V(1) - V(2) \end{bmatrix} \tag{3.9}$$

where $k \in [0, n - 1]$ and $V(1)$ and $V(2)$ are, respectively, the *ab* and the *bc* components of the corresponding vector (V_{ul}, V_{lu}, V_{uu}, or V_{ll}).

A generalized three-dimensional SVM (3D-SVM) algorithm with low computational cost for four-leg multi-level converters was proposed in [68]. This 3D algorithm is a generalization. As the topology changes, the state-space vectors of the converter also change. However, once the subcube is known, almost all the necessary calculations to determine the switching sequence and the duty cycles are exactly the same. This 3D-SVM technique optimizes the switching sequence, minimizing the number of commutations for 4L-4W converters. As for the 3L-3W case, the low computational cost of this method is always the same, and is independent of the number of levels of the converter.

In general, one of the advantages of SVM techniques for multi-level converters is the reduction of computation and implementation complexity compared to carrier-based PWM algorithms, since the number of carriers does not increase as the number of converter levels increases. This advantage makes the digital implementation of the algorithms easier. In addition, the vector redundancies and switching sequences can be used for other control purposes, and can be designed according to specific criteria depending on the application. It must be noted that in order to achieve a proper time average, the modulation period T_s is small, leading to high switching frequencies comparable to carrier-based PWM (above 1 kHz), and is therefore not useful for very high-power applications.

3. **Nearest Level Modulation (NLM):** NLM is a very interesting modulation technique to use when the number of SMs is quite large. Considering that, nowadays, there are MMCs that easily exceed 200 SMs, the advantage of NLM over PSC-PWM or SVM is clear: PSC-PWM will require 100 carrier signals with a 3.6° shift between them, which would need a very precise carrier generator, and the many possible switching states in SVM increases the complexity of the algorithm.

   NLM is based on generating the reference voltage using two different voltage levels. Applying each voltage level during a certain period of time can generate a signal whose mean value is the desired value. Figure 3.20 shows the operating principle of NLM.

   In the MMC, the number of voltage levels in which the output signal can be divided depends on the number of SMs, adding a level for each additional SM. A generalized formula for any level is

$$V_{ref} = V_N \cdot d + V_{N-1}(1-d), \tag{3.10}$$

where V_{ref} is the desired output voltage included between the voltage levels V_N and V_{N-1} and d is the duty cycle. The duty cycle d of each phase is obtained as

$$N = \lceil V_{ref} \rceil \tag{3.11}$$

$$d = V_{ref} - \lfloor V_{ref} \rfloor \tag{3.12}$$

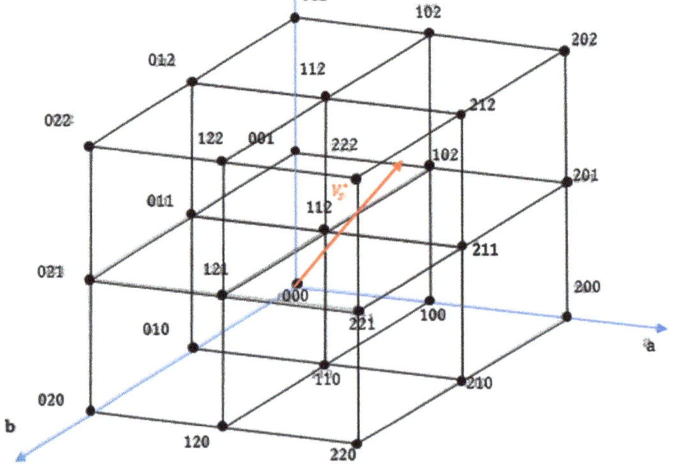

Figure 3.19 State-space vectors of a three-phase three-level converter using the three-dimensional (3D) Cartesian coordinates. This control region is plotted using the normalized phase-to-neutral voltages

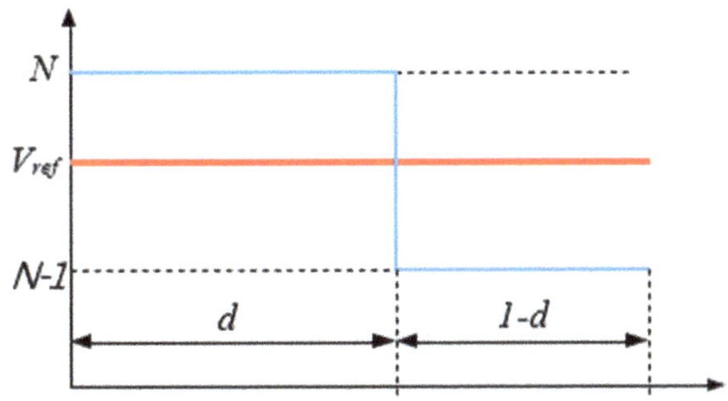

Figure 3.20 Operation principle of the nearest level modulation (NLM)

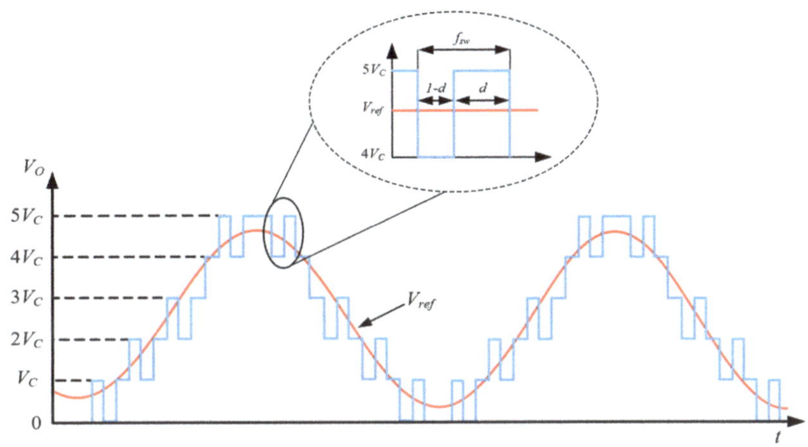

Figure 3.21 Nearest level modulation (NLM) arm waveform ($n = 10$)

Figure 3.21 illustrates the generation of the arm waveform using NLM for an MMC with five SMs per arm.

The main advantage is that the algorithm is greatly simplified in relation to NVC, since it is much easier to find the closest level than the closest vector. In fact, the output-voltage level selection is reduced to a unique simple expression per phase

$$\text{closest voltage level} = v_{aN} = V_{dc} f_{round} \tag{3.13}$$

where V_{dc} is the voltage difference between two levels (usually the DC-link voltage in CHB), which is used to normalize the phase output-voltage reference v^*. The normalized value is then evaluated using the round function (or nearest integer function), which is defined such that $f_{round}\{x\}$ is the integer closest to x. Since this definition is ambiguous for half-integers, the additional convention is that half-integers are always rounded to even numbers; for example, $f_{round}\{1.5\} = 2$. This nearest integer multiplied by V_{dc} corresponds to the closest voltage level to the reference, and is therefore generated by the inverter.

The operating principle is illustrated in Figure 3.23(a) for the first quarter cycle of a sinusoidal reference. Note that the maximum approximation error is $V_{dc}/2$. The implementation of the nearest voltage level generation is presented in Figure 3.23(b). As can be seen in Figure 3.23(a), the round function inherently produces only one commutation between two voltage levels and a maximum dv/dt of V_{dc}, unless the reference presents large step changes.

It is worth mentioning that this method is not a modulation technique, since there is no reference tracking by the time average synthesis between two levels. Instead, the stepped waveform is generated as

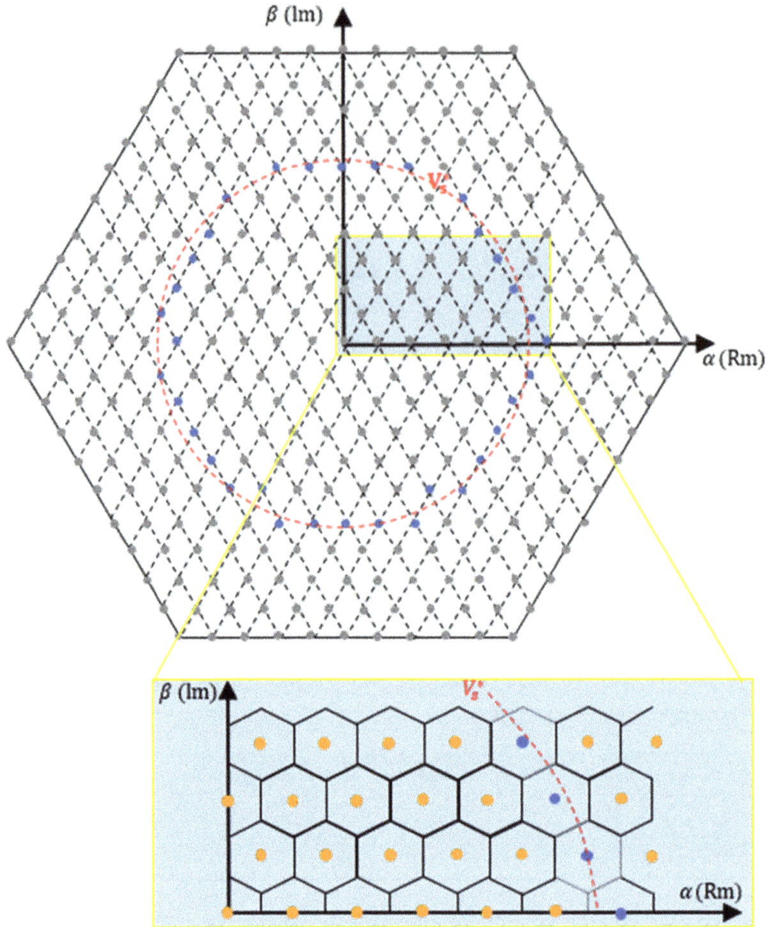

Figure 3.22 Multi-level nearest vector control operating principle

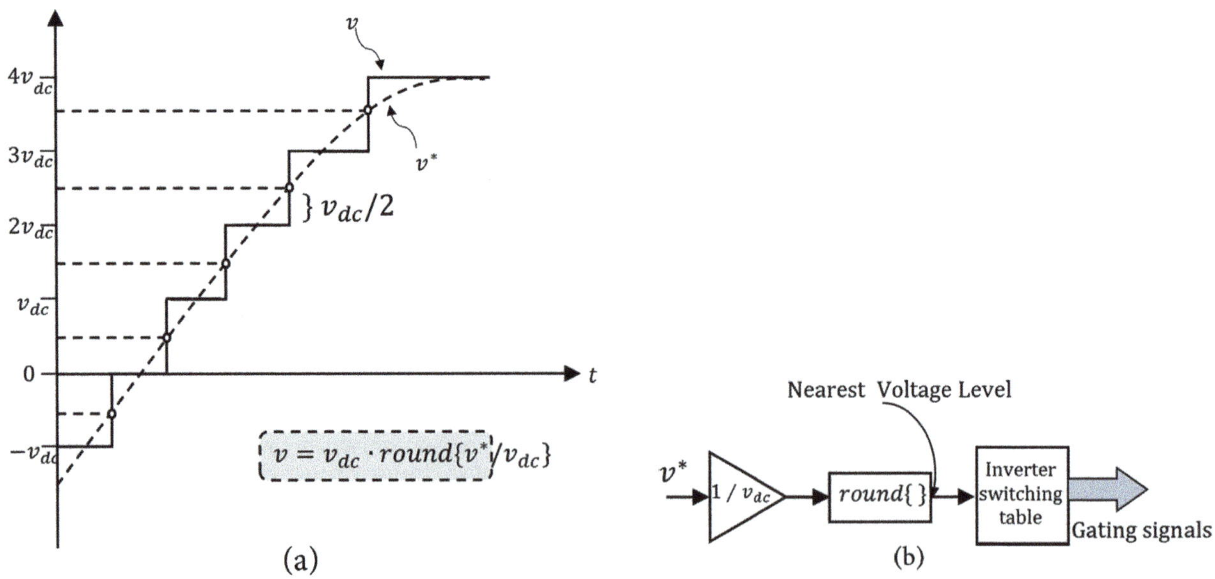

Figure 3.23 Nearest level selection. (a) Waveform synthesis and (b) control diagram

an approximation to the reference. Although the voltage waveform is very similar to the one obtained with staircase modulation, it does not eliminate specific harmonics because it is not really a modulation. Therefore, this control method, like NVC, is not effective for converters with a reduced number of levels, since the approximation error becomes relevant. Thus, it is aimed to be used in converters with a high number of levels to avoid important low-order harmonics at the AC side. Its main advantage is its conceptual and implementation simplicity and the efficiency achieved with this method. An adaptive modulation method that solves this problem by introducing a duty-cycle calculation to NLC has been proposed at the expense of a more advanced control algorithm.

4. **Selected Harmonic Elimination Modulation (SHE PWM):** The main challenge associated with the techniques of SHE PWM is obtaining the analytical solution of the nonlinear transcendental system equations that contain trigonometric terms that, in turn, provide multiple sets of solutions. In the available literature, there are several algorithms for solving the nonlinear transcendental equations that describe selective harmonic elimination. These algorithms make an approach to the iterative Newton-Raphson. In converters with only one cell, two levels, or three levels, the Newton-Raphson method is able to find the solution; however, when increasing the number of cells in the converter, this method is not efficient because it can lead to a local minimum instead of a global minimum of the objective function.

   Furthermore, the DC voltage cells may be unequal. In order to solve the equations that provide switching angles for selective harmonic elimination, an ellipsoidal algorithm of multiple cuts that provided solutions to all regions of operation, which was not presented, was used in this work in an unprecedented way.

   A periodic signal may be represented by

$$V_{output} = \sum_{h=1,3,...}^{\infty} b_h \sin(h\theta), \tag{3.14}$$

where b_h is the coefficient of the Fourier series.

The generalized expression of b_h to any number of pulses is given by

$$b_n = \frac{4V_{DC}}{h\pi}\left[V_1 \sum_{i=1}^{P_1} (-1)^{i+1} \cos(h\alpha_i) \pm V_2 \sum_{i=P_1+1}^{P_2} (-1)^i \cos(h\alpha_i) \pm ... \pm V_M \sum_{i=P_M+1}^{P_M} (-1)^i \cos(h\alpha_i)\right] \tag{3.15}$$

In Equation (3.15), the polarity $\pm$ is negative if P_{M-1} is an odd number, and positive otherwise where M is the number of cells of the DC converter. P_M is the sum of the number of pulses per-quarter cycle of the cells as in

$$P_1 = N_{p1}; P_2 = N_{p1} + N_{p2}; ... \quad P_M = \left[N_{p1} + N_{p2} + ... + N_{pM}\right], \tag{3.16}$$

where N_P is the total number of pulses per quarter cycle of the output waveform converter, represented in

$$N_{P1}, N_{P2}, ..., N_{PM} \quad N_P = [N_{P1} + N_{P2} + ... + N_{PM}] \tag{3.17}$$

and α_i is the switching converter angle. In Figure 3.24, it is possible to observe a general waveform synthesized by a converter that uses selective elimination of harmonics. The switching angles, αi, are obtained by minimizing

$$f(\alpha_1, \alpha_2, \dots, \alpha_{N_P}) = \left(b_1 - \frac{M \cdot \pi \cdot m_i}{4}\right)^2 + b_3^2 + b_5^2 + \dots + b_n^2 \tag{3.18}$$

subject to the constraint

$$\left(0 < \alpha_1 < \alpha_2 < \dots < \alpha_N < \frac{\pi}{2}\right) \tag{3.19}$$

Consequently, the selected harmonics are eliminated. The modulation index is defined in

$$m_i = \frac{b_1}{M \cdot V_{SB}} \tag{3.20}$$

The minimization of Equation (3.18) subject to constraint Equation (3.19) is performed with the multiple cuts ellipsoidal algorithm in order to make the selective harmonic elimination. The fundamental idea of this algorithm is to discard part of the search space at each iteration until it becomes a sufficiently small region that contains an optimal solution under the assumption of quasi-convexity of the minimized function. This strategy has a typically robust performance in practice and deals with constraints seamlessly. Its basic formulation is given by the iterative update of the matrix of an ellipsoid $E_k = \{x \in R^n | (x - x_k)^T Q_k (x - x_k) < 1\}$ with

$$x_{k+1} = x_k + \gamma_k \frac{Q_k^{-1} g_k}{\sqrt{g_k^{-1} Q_k^{-1} g_k}} \tag{3.21}$$

$$Q_{k+1} = \sigma_k \left[Q_k^{-1} - \rho_k \frac{Q_k^{-1} g_k g_k^T Q_k^{-1}}{g_k^T Q_k^{-} g_k} \right] \tag{3.22}$$

where γ_k, σ_k, and ρ_k are values that depend on the cutting plane defined by g_k and a point. Thus, the ellipsoid E_k decreases in size until x_k becomes a good approximation of the solution.

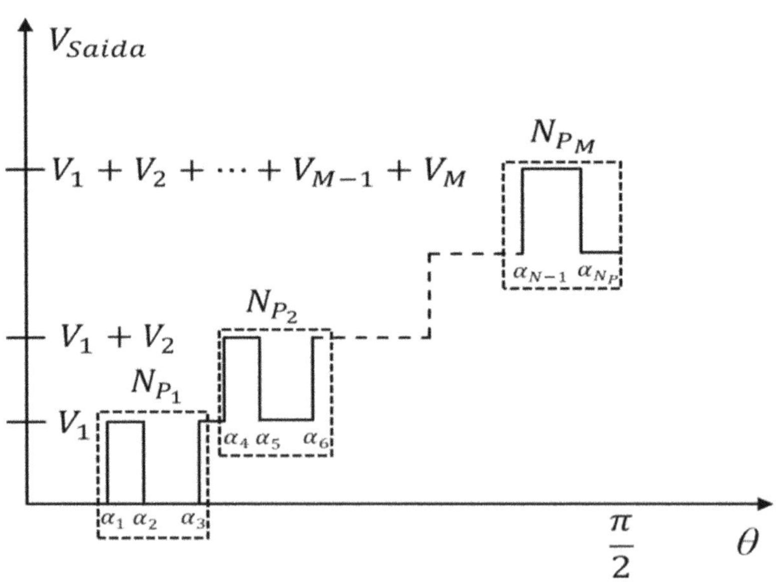

Figure 3.24 Generalized multi-level PWM waveform

3.2.3. Circulating Current and Distortion of Capacitor Voltage

Figure 3.25 shows the equivalent circuit and high-voltage direct current (HVDC) converter switching angle. From Figure 3.25, the equivalent rule for the MMC is where the reference sine wave equals the midpoint threshold level equivalent to having "n" capacitors in circuit in the upper arm and "N-n" capacitors in circuit in the lower arm.

For a pure sinusoidal source, the value of N should be taken to reference the number of switching angles over a quarter cycle. Figure 3.25(b) shows the operation condition for a seven-level converter, which have six submodules in each upper and lower arm, but for which only three angles need to be defined, that is, N = 3.

Defining the series of threshold levels, the following equation can be obtained:

$$\text{Level}_j \frac{2\cdot j+1}{2\cdot N}, \tag{3.23}$$

where "j" is an integer series from 0 to N-1. The angle series α is then given by

$$\frac{V_C}{V_{DC}}\cdot \sin\left(\gamma_j\right) = \text{Level}_j \tag{3.24}$$

or

$$\gamma\,(V_C) := \text{Re}\left(\text{asin}\left(\frac{V_{DC}}{V_C}\cdot \text{Levels}\right)\right) \tag{3.25}$$

Converter Voltage Distortion

To calculate the converter voltage distortion, the angle range is from $\pi/2$ to $5\pi/2$, and the function of arm current is shown in Equation (3.26):

$$i_{CP}\,(\varphi) = \frac{1}{2}\cdot I_{AC}\cdot\left(\sin\,(\varphi+\beta) + \frac{V_{AC}\cdot PF}{2V_{Dc}}\right) \tag{3.26}$$

Current flowing through the capacitors affects the capacitor charge, but as more capacitors are brought into circuit, the rate of change of charge variation increases, which can be represented by a reduction in the capacitor value, so the degree of distortion increases. Thus, the total capacitor voltage variation can be expressed as

$$\Delta V_{CT} = \sum_{2N}\left(\frac{n}{C}\cdot\int_{\gamma_{n1}}^{\gamma_{n+1}} 1_{CP}\,(\phi)\,d\phi\right) \tag{3.27}$$

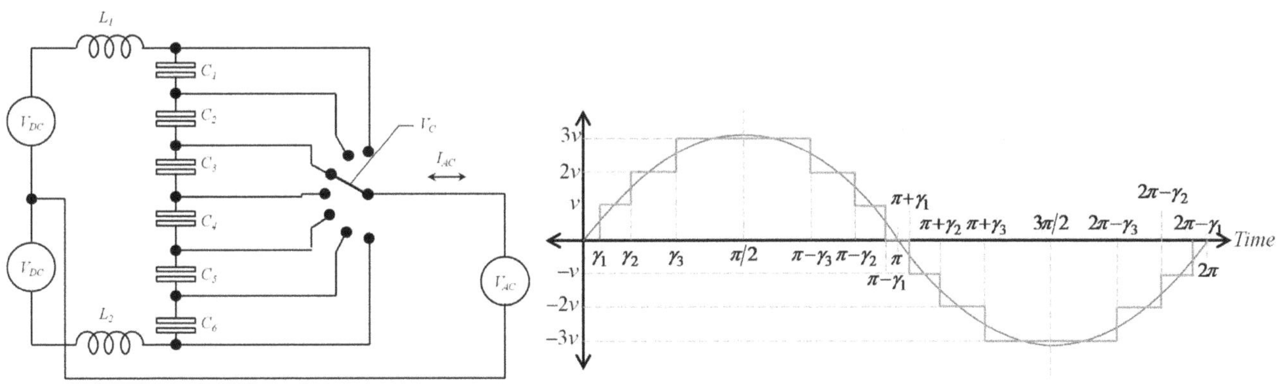

a) Equivalent circuit for MMC b) Converter switching angels

Figure 3.25 Equivalent circuit and switching angle for MMC

At any particular interval between switching angles γ_1 and γ_2 with n modules in circuit the variation in voltage is

$$\Delta v_{CT}(\gamma_1, \gamma_2, n) = \frac{n}{\omega_{50} \cdot C} \cdot \left(\int_{\gamma_1}^{\gamma_2} i_{CP}(\varphi)\, d\varphi \right) \tag{3.28}$$

With the definition of the switching angles only being valid for a single quadrant, the means of defining this function varies according to the quadrant.

Figure 3.26 shows the distorted voltage waveforms. From Figure 3.26(a), the capacitor voltage includes the fundamental frequency and the second harmonic. Figure 3.26(b) and (c) show the reason of voltage distortion.

The current is defined as the difference between the AC voltage from AC networks and the converter voltage; therefore, the voltage distortion has a profound effect on the current of the MMC converter. This effect in turn can cause further effect on the voltage, and so the method for analyzing the full distortion of the voltage and the current is to carry out an iterative process until the variation in current becomes negligible for successive iterations. Through the overall process, also, the DC offset can be predicted.

Current Distortion

Since the impedance of the arm reactor is small, the distortion mentioned previously has a very significant effect on the current. In principle, the variation in current have a further effect on the voltage. Since the effect is too great, the feedback, that is, the iterative process of the MMC system, leads to a runaway effect so that **the capacitor voltage effectively collapses!**

To evaluate the current distortion, the voltage is divided by the AC impedance of the line reactance.

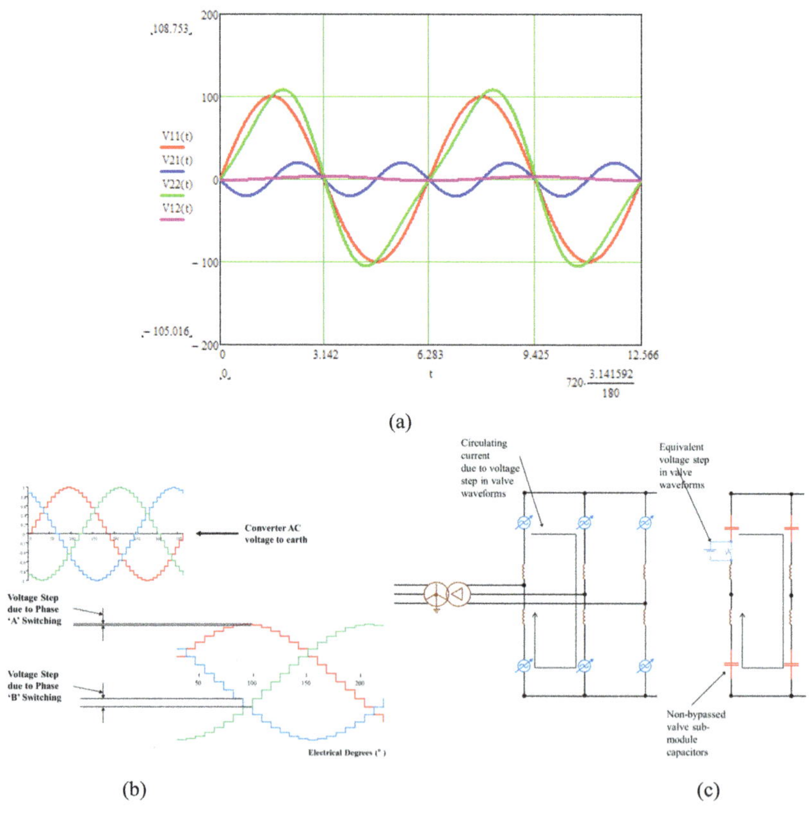

Figure 3.26 Converter voltage distortion in sourcing and sinking current

Since the distorted voltages include harmonics, to calculate the distorted current, the voltage difference is to be taken between the AC voltage and the converter voltage for each harmonic, divided by the AC impedance of the line reactance.

Linear Analysis

To derive an expression to describe the variation of second harmonic magnitude requires a linear approximation to be derived to the stepped waveform.

Combining Equations (3.26) and (3.27), the basic equation becomes

$$\Delta V_C = \frac{n \cdot I_{AC}}{2\omega_{50} \cdot C} \int_{\gamma_1}^{\gamma_2} \left(\sin(\varphi + \beta) + \frac{V_{AC} \cdot PF}{2V_{DC}} \right) \tag{3.29}$$

Equation (3.24) defines the method of deriving the angle set in which the term "level" defines the number of capacitors in circuit, n, referenced to zero voltage. The main variation over the cycle is the number of capacitors in circuit, so the integral given earlier assumes over a given phase range that it is constant. The number of capacitors actually in circuit is N-n, and the equation related to the submodule voltage shows as following:

$$\frac{V_C}{V_{DC}} \cdot \sin(\varphi) = \frac{n}{N} \tag{3.30}$$

The actual number of capacitors in circuit is then given as

$$N - n = N \cdot \left(1 - \frac{V_C}{V_{DC}} \cdot \sin(\varphi) \right) \tag{3.31}$$

Substituting Equation (3.31) into Equation (3.29), the more general linear equation can be obtained as follows:

$$\Delta V_{CL} = \frac{N \cdot I_{AC}}{2 \cdot \omega_{50} \cdot C} \cdot \int_{\frac{\pi}{2}}^{\phi} \left(1 - \frac{V_C}{V_{DC}} \cdot \sin(\varphi) \right) \cdot \left(\sin(\varphi + \beta) + \frac{V_{AC} \cdot PF}{2V_{DC}} \right) d\varphi \tag{3.32}$$

Solving the integral of Equation (3.32), the result is as follows:

$$\begin{aligned} \Delta v_{CL1} = \frac{N \cdot I_{AC}}{2 \cdot \omega_{50} \cdot C} \cdot \Big[& -\cos(\varphi + \beta) + \left(\frac{V_C}{4V_{DC}} - 1 \right) \cdot \sin(\beta) + \left(\frac{V_{AC} \cdot PF}{2V_{DC}} - \frac{V_C}{2V_{DC}} \cdot \cos(\beta) \right) \cdot \left(\varphi - \frac{\pi}{2} \right) \\ & + \frac{V_C}{2V_{DC}} \cdot \left(\frac{1}{2} \cdot \sin(2\varphi + \beta) \right) + \frac{V_{AC} \cdot V_C \cdot PF}{2V_{DC}^2} \cdot \cos(\varphi) \Big] \end{aligned} \tag{3.33}$$

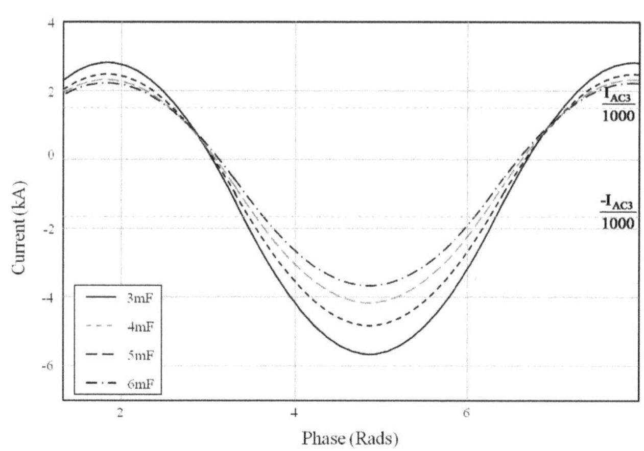

Figure 3.27 Distorted current over a cycle

3.2.4 Condition for Capacitor Charging Balance

If the current and voltage through an MMC valve, which is composed of an inductor and a series of N submodules as shown in Figure 3.28, where the submodules are half-bridge (2 quadrant operation) or full bridge (4 quadrant operation) submodules comprising an array of IGBTs switches and a capacitor as shown in Figure 3.29, the current through the MMC valve can be regarded as the difference between two currents, each related to the voltage on either side of the MMC valve in which there is assumed to be a power balance. For an HVDC system, $V_1 \cdot i_1 = V_2 \cdot i_2$.

Let us assume that there has to be a variation in one or both of the voltages for the submodule capacitances to have a finite value, so it is assumed initially that V_1 has HVDC current and V_2 has AC current at a power frequency. If italicized terms are taken as instantaneous values and capitals are taken as peak or static values, the relation between currents and voltages is as follows.

$$V_1 \cdot i_1 = V_2 \cdot i_2 \cdot \cos(\varphi) \tag{3.34}$$

To force the arm inductor to give the required current, it requires relatively small variations in the number of submodules switched into circuit, so the difference of voltage across the MMC valve is approximately matched

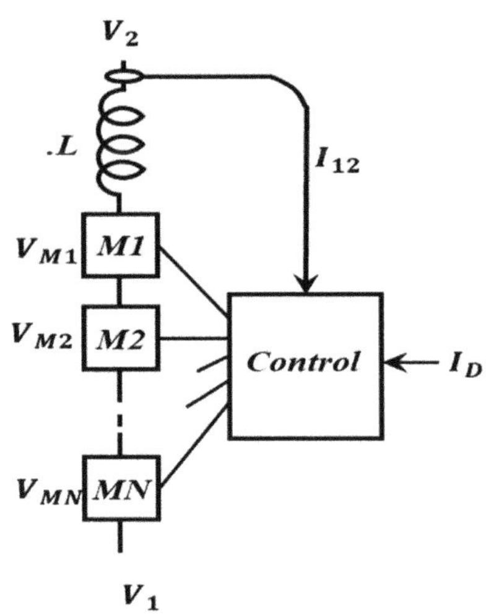

Figure 3.28 Submodule definition according to MMC components

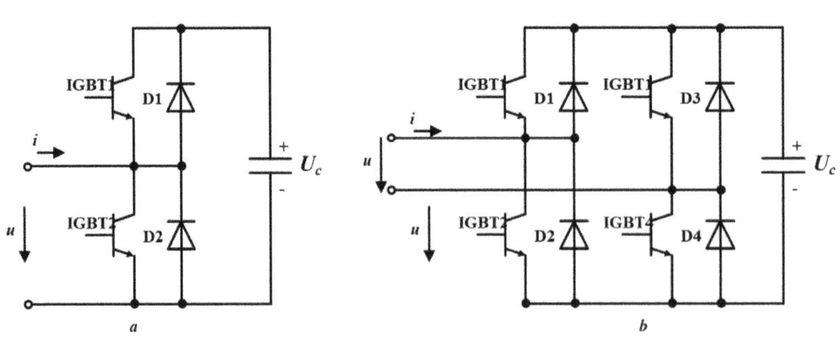

a) Half bridge b) Full Bridge

Figure 3.29 Submodule definition according to operation region by the number of submodules switched into circuit. If a submodule has a capacitance value C and a charge voltage V_M, then the charge variation when it is switched on between phase angle γ_1 and γ_2 is

$$V_{M\theta} = \frac{1}{C} \cdot \int_{\gamma_2}^{\gamma_1} (I_1 - I_2 \cdot \sin(\theta + \varphi))\, d\theta + V_M \tag{3.35}$$

The angles at which the submodules are switched as shown in Figure 3.30 become a sinusoidal function as the number of submodule levels become infinite:

$$V_C \cdot \sin(\gamma_1) = \frac{2 \cdot j + 1}{2 \cdot N} \cdot V_M, \tag{3.36}$$

where, $\boldsymbol{V_C}$ is the peak magnitude of the sine wave being generated, $\boldsymbol{V_M}$ is the maximum voltage from the HVDC converter, and N is the number of switching angles per-quarter cycle in the waveform as j ranges from 0 to N-1. As the number of submodule levels reduces, the ideal values move away from this solution. If the HVDC voltage is greater than the peak of the AC voltage, then submodules are only inserted in a negative direction against the HVDC voltage. When the HVDC voltage is less than the peak of the AC voltage, the submodules are inserted in a positive direction.

Therefore, the relation between the angle and the submodule voltage are as follows:

$$\begin{aligned}
\Delta V_M &= \frac{1}{\omega \cdot C} \cdot [I_1\theta - I_2 \cdot \cos(\theta + \varphi)]_{\gamma_1}^{\gamma_2} \\
\Delta V_M &= \frac{1}{\omega \cdot C} \cdot [I_1(\gamma_2 - \gamma_1) - I_2 \cdot (\cos(\gamma_2 + \varphi) - \cos(\gamma_1 + \varphi))] \\
\Delta V_M &= \frac{1}{\omega \cdot C} \cdot \left[I_1(\gamma_2 - \gamma_1) + 2I_2 \cdot \sin\left(\frac{\gamma_2 + \gamma_1}{2} + \varphi\right) \cdot \sin\left(\frac{\gamma_2 - \gamma_1}{2}\right)\right]
\end{aligned} \tag{3.37}$$

The angle values γ_1 and γ_2 are strongly associated as $\gamma_1 = \pi + \alpha$, and $\gamma_2 = 2\pi + \alpha$ where a ranges from $-\alpha_7 \cdots + \alpha_7$.

$$\Delta V_M = \frac{1}{\omega \cdot C} \cdot \left[I_1(\pi - 2\alpha) + 2I_2 \cdot \sin\left(\frac{3\pi}{2} + \varphi\right) \cdot \sin\left(\frac{\pi - 2\alpha}{2}\right)\right] \tag{3.38}$$

$$\Delta V_M = \frac{1}{\omega \cdot C} \cdot [I_1\,(\pi - 2\alpha) - 2I_2 \cdot \cos(\varphi) \cdot \cos(\alpha)] \tag{3.39}$$

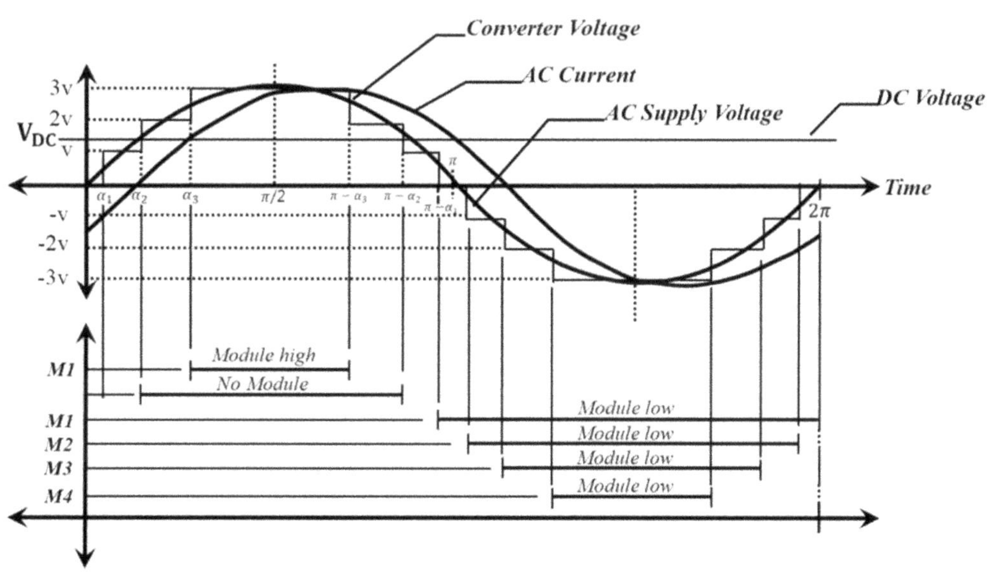

Figure 3.30 Waveforms

The total charging voltage is then

$$\Delta V_{TM} = \frac{1}{\omega \cdot C} \cdot \sum_{-\alpha_7}^{\alpha_7} \left[I_1 \left(\pi - 2\alpha_n\right) \; - 2 I_2 \cdot \cos(\varphi) \; \cdot \cos\left(\alpha_n\right) \right] \tag{3.40}$$

for an even number of angles $\sum_{-\alpha_7}^{\alpha_7} (\pi - 2\alpha_n) = 2N\pi$. Similarly, $\cos(\alpha) + \cos(-\alpha) = 2 \cdot \cos(\alpha)$, so Equation (3.40) can be reduced to:

$$\Delta V_{\mathrm{TM}} = \frac{1}{\omega \cdot C} \cdot \left(2 \cdot I_1 \cdot N \cdot \pi - 4 \cdot I_2 \cdot \cos(\phi) \cdot \sum_{j=0}^{N-1} \left[\cos\left(\alpha_j\right) \right] \right) \tag{3.41}$$

The mathematical description of the harmonic structure of the waveform based on the angles is

$$\alpha_r = \frac{4 \cdot V_M}{r \cdot \pi} \cdot \sum_{i=1}^{n} \cos\left(r \cdot \alpha_i\right) \tag{3.42}$$

Thus, the converter voltage V_c should be approximately given by the condition "$r = 1$"

$$V_C = \frac{4 \cdot V_M}{\pi} \cdot \sum_{j=0}^{N-1} \cos\left(\alpha_j\right) \tag{3.43}$$

Rearranging Equation (3.43):

$$\sum_{N-1}^{j=0} \cos\left(\alpha_j\right) = \frac{V_C \cdot \pi}{4 \cdot V_M} \tag{3.44}$$

Substituting Equation (3.44) into Equation (3.41),

$$\Delta V_{TM} = \frac{2}{\omega \cdot C} \cdot \left(I_1 \cdot N \cdot \pi - I_2 \cdot \cos(\varphi) \cdot \frac{V_C \cdot \pi}{2 \cdot V_M} \right) \tag{3.45}$$

Therefore, the condition for the power balance is as follows:

$$N \cdot V_M \cdot I_1 = \frac{1}{2} \cdot V_C \cdot I_2 \cdot \cos(\varphi) \tag{3.46}$$

V_c and I_C are defined as peak values so that Equation (3.46) defines the power balance condition between AC side and DC side in the case of the constant current. N defines the number of switching angles per quadrant, so the HVDC voltage condition to balance the power the AC side and DC side is shown in Equation (3.47)

$$V_{DC} = N \cdot V_M \tag{3.47}$$

3.2.5 Modeling of MMC System

In a three-phase system, the DC components of the phase leg voltages of the MMC system are equal and the double frequency components are shifted by $2\pi/3$ in the A phase - B phase - C phase sequentially. Including the double frequency component in the arm voltage, the equations can be presented as follows:

$$v_u = \frac{V_{dc}}{2} - \hat{e}_v \cos(\omega t) - \frac{v_{2f}}{2} \sin(2\omega t + \varphi) \tag{3.48}$$

$$v_l = \frac{V_{dc}}{2} + \hat{e}_v \cos(\omega t) - \frac{v_{2f}}{2} \sin(2\omega t + \varphi) \tag{3.49}$$

The double line-frequency voltage excites the leg impedance, creating a double line-frequency component on the difference current also called circulating current $i_{circ;}$

$$i_{circ} = -\frac{v_{2f}}{4Z_{arm}} \sin\left(2\omega t + \varphi - \arg\left(Z_{arm}\right)\right) = \hat{\imath}_{circ} \cos\left(2\omega t + \gamma\right), \tag{3.50}$$

where the arm impedance is calculated from the arm inductance L_{arm} and the equivalent resistance R_{arm}. The arm currents can then be decomposed as

$$i_u = i_{diff} + \frac{i_{ac}}{2} = i_{dc} + \hat{\imath}_{circ} \cos\left(2\omega t + \gamma\right) + \frac{\hat{\imath}_{ac}}{2} \cos\left(\omega t + \varphi\right) \tag{3.51}$$

$$i_l = i_{diff} - \frac{i_{ac}}{2} = i_{dc} + \hat{\imath}_{circ} \cos\left(2\omega t + \gamma\right) - \frac{\hat{\imath}_{ac}}{2} \cos\left(\omega t + \varphi\right) \tag{3.52}$$

Analyzing Equations (3.51) and (3.52), it can be concluded that the circulating current has no effect on the AC current. This current flows through the converter phase leg, increasing the RMS value of the arm currents and resulting in higher converter losses. From the previous equations, it can be noticed that the DC component of the difference current is responsible for the DC/AC power transfer, as shown in Figure 3.31..

From Figure 3.31 and the previous equations, the arm voltage references can be represented as Equations (3.53) and (3.54).

$$v_u^{ref} = \frac{V_{dc}}{2} - e_v = \frac{V_{dc}}{2} - \hat{e}_v \cos\left(\omega t\right) \tag{3.53}$$

$$v_l^{ref} = \frac{V_{dc}}{2} + e_v = \frac{V_{dc}}{2} + \hat{e}_v \cos\left(\omega t\right), \tag{3.54}$$

where $\hat{e}_v$ is the peak value of the generated *emf.*

Considering a pure DC difference current in a one-phase system, also, from Figure 3.31 and the previous equations, the arm current, which can be represented as Equations (3.55) and (3.56) by the DC current i_{dc} and the AC current i_{ac}, is as follows:

$$i_u = i_{dc} + \frac{i_{ac}}{2} = i_{dc} + \frac{\hat{\imath}_{ac}}{2} \cos\left(\omega t + \varphi\right) \tag{3.55}$$

$$i_u = i_{dc} - \frac{i_{ac}}{2} = i_{dc} - \frac{\hat{\imath}_{ac}}{2} \cos\left(\omega t + \varphi\right), \tag{3.56}$$

where $\hat{\imath}_{ac}$ is the peak value of the AC current. Using Equations (3.53)–(3.56), the power in the upper and lower arms is calculated as

$$\begin{aligned} P_u = i_u v_u^{ref} = & \tfrac{i_{dc}V_{dc}}{2} - \tfrac{\hat{\imath}_{ac}\hat{e}_v}{4}\cos(\varphi) - i_{dc}\hat{e}_v\cos(\omega t) - \tfrac{\hat{\imath}_{ac}V_{dc}}{4}\cos(\omega t + \varphi) \\ & -\tfrac{\hat{\imath}_{ac}\hat{e}_v}{4}\cos(2\omega t + \varphi) \end{aligned} \tag{3.57}$$

$$\begin{aligned} P_l = i_l v_l^{ref} = & \tfrac{i_{dc}V_{dc}}{2} - \tfrac{\hat{\imath}_{ac}\hat{e}_v}{4}\cos(\varphi) + i_{dc}\hat{e}_v\cos(\omega t) - \tfrac{\hat{\imath}_{ac}V_{dc}}{4}\cos(\omega t + \varphi) \\ & -\tfrac{\hat{\imath}_{ac}\hat{e}_v}{4}\cos(2\omega t + \varphi) \end{aligned} \tag{3.58}$$

Adding the power in the upper and lower arms to calculate the power in the phase leg results in

$$P_{leg} = P_u + P_l = i_{dc}V_{dc} - \frac{\hat{\imath}_{ac}\hat{e}_v}{2}\cos\left(\varphi\right) - \frac{\hat{\imath}_{ac}\hat{e}_v}{2}\cos\left(2\omega t + \varphi\right), \tag{3.59}$$

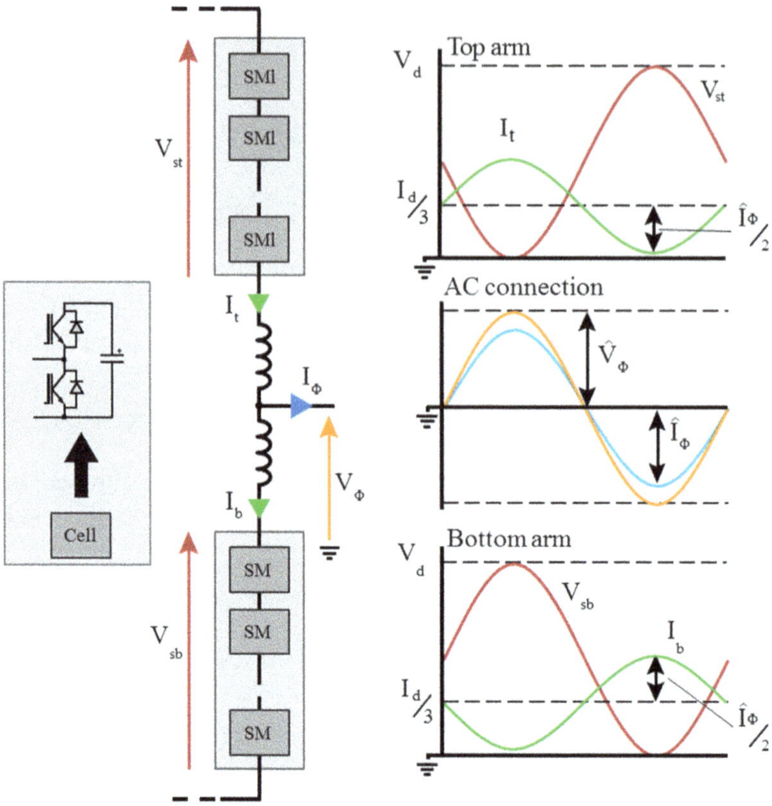

Figure 3.31 Voltage and current waveforms of MMC system

where the first two terms represent the DC power and the third term is a double line-frequency AC component present in the arm. Integrating Equation (3.59), the AC component of the energy stored in the phase leg is given by

$$W_{leg.ac} = \int P_{leg}dt = -\frac{\hat{i}_{ac}\hat{e}_v}{4\omega}\sin(2\omega t + \varphi) \tag{3.60}$$

The energy stored in the phase is located in the submodule capacitors. The energy stored in the capacitors has a quadratic relation with the capacitor voltage; thus, from Equation (3.60) it can be derived that the voltage across the phase leg must contain a double line-frequency AC component and a DC component.

References

1. S. Kouro, et al., "Recent advances and industrial applications of multilevel converters," *IEEE Trans. Ind. Electron.*, vol. 57, no. 8, pp. 2553–2580, Aug. 2010.
2. M. S. Adler, et al., "The evolution of power device technology," *IEEE Trans. Electron Devices*, vol. 31, pp. 1570–1591, Nov. 1984.
3. N. A. Moguilnaia, et al., "Innovation in power semiconductor industry: Past and future," *IEEE Trans. Eng. Manage.*, vol. 52, pp. 429–439, Nov. 2005.
4. L. G. Franquelo, et al., "The age of multilevel converters arrives," *IEEE Ind. Electron. Mag.*, vol. 2, pp. 28–39, Jun. 2008.
5. L. M. Tolbert, F. Z. Peng, and T. G. Habetler, "Multilevel converters for large electric drives," *IEEE Trans. Ind. Appl.*, vol. 35, pp. 36–44, Jan/Feb. 1999.
6. J. Rodriguez, J.-S. Lai, and F.Z. Peng, "Multilevel inverters: A survey of topologies, controls, and applications," *IEEE Trans. Ind. Electron.*, vol. 49, pp. 724–738, Aug. 2002.

7. M. A. Perez, et al., "Circuit topologies, modeling, control schemes, and applications of modular multilevel converters," *IEEE Trans. Power Electron.*, vol. 30, no. 1, pp. 4–17, Jan. 2015.
8. F. Martinez-Rodrigo, etc., "Modular Multilevel Converters: control and applications", energies 10(11), 1709, Oct. 2017.
9. T. C. Lim, et al., "Series-connected IGBTs using active voltage control technique," *IEEE Trans. Power Electron.*, vol. 28, no. 8, pp. 4083–4103, Aug. 2013.
10. A. Lesnicar and R. Marquardt, "An innovative modular multilevel converter topology suitable for a wide power range," in *Proceeding of the IEEE Bologna Power Tech Conference*, Bologna, Italy, vol. 3, Jun. 2003, p. 6.
11. J. Wang, R. Burgos, and D. Boroyevich, "A survey on the modular multilevel converters-Modeling, modulation and controls," in *IEEE Energy Conversion Congress and Exposition,* Denver, 2013, pp. 3894–3991.
12. H. P. Mohammadi and M. T. Bina, "A transformerless medium-voltage STATCOM topology based on extended modular multilevel converters," *IEEE Trans. Power Electron.*, vol. 26, no. 5, pp. 1534–1545, May. 2011.
13. S. Debnath, et al., "Operation, control, and applications of the modular multilevel converter: A review," *IEEE Trans. Power Electron.*, vol. 30, no. 1, pp. 37–53, Jan. 2015.
14. G. Ding, et al., "New technologies of voltage source converter (VSC) for HVDC transmission system based on VSC," in *Proceeding of the IEEE Power Energy Society General Meeting*, Pittsburgh, PA, USA, 2008, pp. 1–8.
15. S. Allebrod, R. Hamerski, and R. Marquardt, "New transformerless, scalable modular multilevel converters for HVDC-transmission," in *Proceeding of the IEEE Power Electronics Specialists Conference,* Rhodes, Greece, pp. 174–179.
16. R. Marquardt, "Modular multilevel converter: A universal concept for HVDC-networks and extended DC-bus-applications," in *Proceeding of the International Power Electronics Conference*, Sapporo, Japan, Jun. 2010, pp. 502–507.
17. G. Carrara, et al., "A new multilevel PWM method: A theoretical analysis," in *Proceeding of the IEEE Power Electronics Specialists Conference,* San Antonio, TX, USA, 1990, pp. 363–371.
18. D. Siemaszko, et al., "Evaluation of control and modulation methods for modular multilevel converters," in *Proceedings of the International Power Electronics Conference*, Sapporo, Japan, Jun. 2010, pp. 746–753.
19. A. Hassanpoor, et al., "Evaluation of different carrier-based PWM methods for modular multilevel converters for HVDC application," in *Proceedings of the Conference in IEEE Industrial Electronics Society*, Montreal, QC, Canada, 2012, pp. 388–393.

4

Sorting and Balancing Control Algorithms of MMC System

Having a series connection of submodules with floating capacitor voltages brings the advantage of a staircase output waveform with low switching frequencies. However, it also increases the level of complexity of the system from the control point of view. To ensure accurate and stable operation of the converter, the submodules must be charged to the same voltage level. Moreover, in order to maintain the output voltage level of the converter, the capacitor voltages must be kept at a reference value.

The balance between the submodules can be maintained through two approaches: first, adding a balancing component to the modulation index of each submodule, or second, sorting the submodules to be inserted during the charging and discharging cycle.

The first approach has the disadvantage of a difficult controller design and implementation, since a distributed control structure must be implemented. Also, because each submodule can modify its own modulation index independently, this approach may generate even harmonics in the output voltage.

In the second approach, the voltage in the capacitors is measured several times per fundamental period. The submodules are sorted in order according to their capacitor voltages. When the arm current is positive, the submodules with the lowest voltage are switched first, so their capacitors get charged. The opposite action is done when the current is negative. This procedure must be done periodically; as a result, even if the number of inserted submodules is not changed, unnecessary switching operations will be performed. This increases the

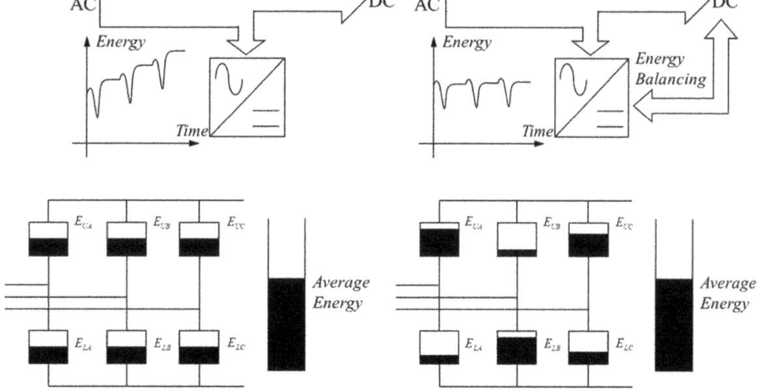

Figure 4.1 Energy balancing and available energy of modular multi-level converter (MMC) converter

switching frequency and therefore the switching losses. Even though these two approaches ensure equal voltage share among the arm submodules, the voltage share between arms must also be taken into consideration.

The difference current contains a strong double line-frequency component that circulates through the converter legs. If not controlled, this current component will increase the losses in the system, but most importantly, it may create unbalances and disturbances during transients. Figure 4.1 shows why the energy of the submodules of the MMC system need to be balanced.

Many strategies have been proposed in order to keep the DC capacitor's voltage to a desired value and eliminate the AC component present in the difference current. In this section, three of the most popular control strategies are discussed in conjunction with their modulation and balancing algorithm.

4.1 Direct Control Algorithm

A first approach of submodule balancing can be an ideal open-loop modulation, where a sinusoidal reference is applied to each arm. Naming ω_N as the angular frequency of the output voltage, the reference signal is

$$\mathrm{m}\,(\mathrm{t}) = \hat{m}\cos(\omega_N t) \tag{4.1}$$

The ideal terminal voltage is given by Equation (4.2).

$$u_V = \frac{u_D}{2} m(t) \tag{4.2}$$

A stiff alternating current source $i_V(t)$ is assumed as the output load. To model different kinds of loads, the phase angle ϕ of the load current can be chosen arbitrarily. Accordingly,

$$i_V(t) = \hat{i}_V \cos(\omega_N t + \varphi) \tag{4.3}$$

The reference signal is translated by the modulator into insertion indices for each converter arm. Thus, the insertion indices for the upper and lower arm are given by Equation (4.4), respectively.

$$\begin{aligned} n_U(t) &= \frac{1-m(t)}{2} \\ n_L(t) &= \frac{1+m(t)}{2} \end{aligned} \tag{4.4}$$

In this control technique, the upper and lower arm voltages of phase-**j** are controlled by two complementary sinusoidal reference waveforms, where $v_{\mathbf{j,ref}}$ represents the reference output voltage and $\mathbf{n_{p,j,ref}}$ and $\mathbf{n_{n,j,ref}}$ are the reference waveforms for the number of inserted SMs in the upper and lower arms, respectively.

$$n_{p,j,\mathrm{ref}} = N\text{-}\frac{\frac{V_{\mathrm{dc}}}{2} - v_{j,\mathrm{ref}}}{V_{\mathrm{dc}}} \tag{4.5a}$$

$$n_{n,j,\mathrm{ref}} = N\text{-}\frac{\frac{V_{\mathrm{dc}}}{2} + v_{j,\mathrm{ref}}}{V_{\mathrm{dc}}} \tag{4.5b}$$

The method shows that the circulating current contains a high second harmonic component that increases the RMS value of the current, thus increasing the losses. Apart from increasing the rating of the components used in the modular multi-level converter (MMC), this current, if left uncontrolled, may create unbalance and disturbances at transients.

Therefore, the circulating current suppressing controller (CCSC) is implemented as vector current control, as shown in Figure 4.2.

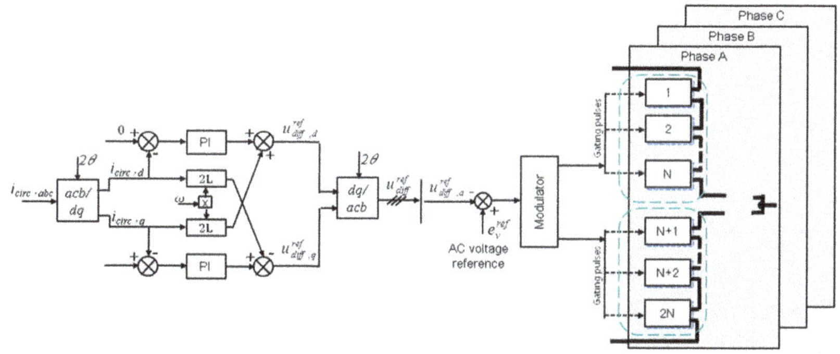

Figure 4.2 Block diagram of circulating current suppression control

From Chapter 3, the circulating current in the MMC may contain double line-frequency components except for the DC one, as shown in Equation (4.6).

$$\begin{aligned}
i_{diffa} &= \frac{i_{dc}}{3} + I_{2f}\sin(2\omega_0 + \varphi_0) \\
i_{diffb} &= \frac{i_{dc}}{3} + I_{2f}\sin\left(2\omega_0 + \frac{2\pi}{3} + \varphi_0\right) \\
i_{diffc} &= \frac{i_{dc}}{3} + I_{2f}\sin\left(2\omega_0 - \frac{2\pi}{3} + \varphi_0\right)
\end{aligned} \tag{4.6}$$

The three-phase difference currents are calculated and transformed to a double line-frequency negative-sequence rotational reference frame. The current references for the control loops are set to zero since the objective is to eliminate the circulating current. The output signal is transformed back to three phase signals, which are subtracted from the upper and lower arm voltage commands of the corresponding phase. In this manner, the arm currents can be influenced without affecting the output voltage.

The major drawback of the direct modulation technique is the presence of circulating currents, which increase converter power losses and rating values of the components.

4.1.1 Indirect Control Algorithm–Energy Control

This method was first proposed by Antonopoulos in [2], where the arm capacitor voltages are kept to a reference through the control of the total stored energy W_C^Σ in the phase leg and the difference between the energy stored in the upper and lower arms W_c^Δ.

In this technique, the upper and lower arm reference waveforms of phase-j are given, where $v_{cx,i,j}$ represents the capacitor voltage of SM-i in arm-x of phase-j, and $v_{\mathrm{reg},j}^{\Sigma}$ and $v_{\mathrm{reg},j}^{\mathrm{circ}}$ are used to control the total energy in the phase-j leg and balance the energy between the arms as shown in Equation (4.7). This technique can be further classified into

$$n_{p,j,\mathrm{ref}} = N\frac{\frac{V_{\mathrm{dc}}}{2} - v_{j,\mathrm{ref}} - v_{\mathrm{reg},j}^{\Sigma} - v_{\mathrm{reg},j}^{\mathrm{circ}}}{\sum_{i=0}^{N} v_{cp,i,j}} \tag{4.7a}$$

$$n_{n,j,\mathrm{ref}} = N\frac{\frac{V_{\mathrm{dc}}}{2} + v_{j,\mathrm{ref}} - v_{\mathrm{reg},j}^{\Sigma} - v_{\mathrm{reg},j}^{\mathrm{circ}}}{\sum_{i=0}^{N} v_{cn,i,j}} \tag{4.7b}$$

* **Closed-loop Control**: In closed-loop control, the term $\sum_{i=0}^{N} v_{cx,i,j}$ in Equation (4.7) is calculated based on the actual measured capacitor voltages. Furthermore, $v_{\mathrm{reg},j}^{\Sigma}$ and $v_{\mathrm{reg},j}^{\mathrm{circ}}$ are obtained from the closed-loop control of the total energy stored in phase-j leg capacitors and the energy balance between the arms, respectively. The balance between the energy stored in each arm relies on temporarily driving a fundamental frequency sinusoidal circulating current. The advantages of this technique lie in (1) control of the average submodule (SM) capacitor voltage, which allows operation under low output voltage with a high number of voltage levels and (2) control of the energy imbalance between the upper and lower arms.

***Open-loop Control:** In open-loop control, the term $\sum_{i=0}^{N} v_{cx,i,j}$ in Equation (4.7) is calculated based on the estimated capacitor voltages. Additionally, $v_{\mathrm{reg},j}^{\Sigma} = 0$ and $v_{\mathrm{reg},j}^{\mathrm{circ}}$ are estimated so as to eliminate the harmonics in the circulating currents and guarantee stable operation of the converter. The estimations are obtained by solving the equations describing the dynamics of the converter, using the measured output currents and DC-link voltage. The advantages of this technique lie in the absence of voltage sensors, as well as simple and fast control. Nevertheless, accurate estimation of the real parameters necessary to describe the dynamics of the system is its main drawback.

A block diagram of this control strategy is presented in Figure 4.3. The nearest-level modulation is used and the submodule balance is done by means of sorting algorithm.

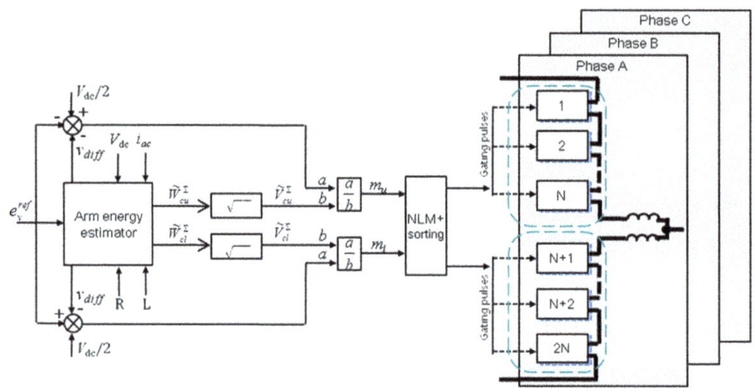

Figure 4.3 Open-loop energy control, block diagram

4.1.2 Averaging Control

In this method, proposed by Akagi, the cell capacitor voltages are controlled independently. The control is implemented in the following two parts:

- Averaging part, implemented per phase-leg
- Balancing part, implemented in each submodule

Figure 4.4 shows the averaging control method; in it, the block diagram of the distributed control method is presented. As it can be observed, the averaging control is implemented in two loops, outer voltage loop and inner current loop. The voltage loop is responsible of controlling the mean value of the capacitor voltages in the leg by influencing each cell individually. The error signal is processed in the controller, resulting in the reference signal for the difference current loop. Under the balanced conditions, the DC component of the difference current is

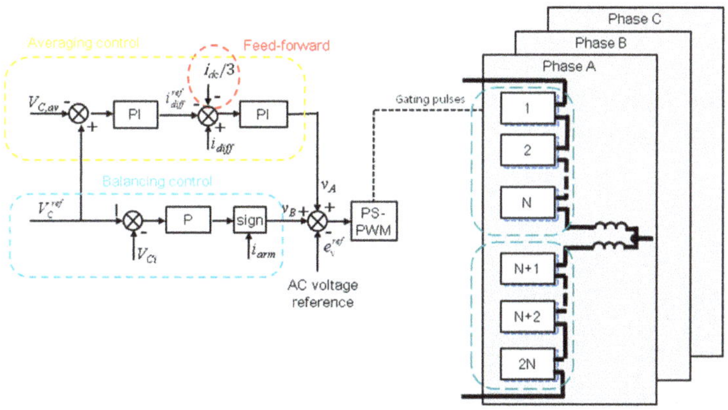

Figure 4.4 Distributed control, block diagram

equal to one third of the DC-link current, therefore a feed-forward term is added to increase the response of the controller as highlighted in Figure 4.4.

If the average voltage $V_{c,av}$ is lower than the desired value, V_C^{ref}, a positive current reference is obtained. The current reference is subtracted from the measured value, reducing the control command. By this means, the DC component of the difference current is increased, raising the charge in the capacitors.

The average charge of the capacitors depends on the DC component of the difference current. If only an integral compensator is used, the DC value of the difference current is controlled, making in no effect on the circulating current. The compensator in the current loop acts on the AC component of the difference current. The balancing control is implemented in each submodule individually. The control signal is generated based on the capacitor voltage and the direction of the corresponding arm current.

The final submodule voltage reference is obtained by adding both averaging and balancing control signals to the voltage reference.

4.2 Submodule-Capacitor Energy Control Strategy [5]–[6]

The method described in this chapter is one of the indirect control methods, which was proposed by Professor Sul of Korea. It has been added to this textbook because the control performance and implementation method are new. According to this method, the value of the upper and lower arm currents of phase x is defined as leg current and is denoted as i_{xo} as shown in Equation (4.8).

$$i_{xo} = (i_{xu} + i_{xl})/2 \tag{4.8}$$

In conventional modeling, the DC bus of the MMC is modeled as a stiff voltage source, whereas the current transient is analyzed independently for each phase [5, 6]. According to Kirchhoff's law, Equations (4.9) and (4.10) can be deduced.

$$-\left(\frac{v_{xu} - v_{xl}}{2}\right) - v_{xg} = \left(\frac{R_o}{2} + \frac{d}{dt}\frac{L_o}{2}\right) i_{xs.} \tag{4.9}$$

$$V_{dc} - (v_{xu} + v_{xl}) = 2\left(R_o + \frac{d}{dt}L_o\right) i_{xo} \tag{4.10}$$

Output AC voltage and leg internal voltage are defined as Equations (4.11) to (4.14), respectively.

$$v_{xs} = -\frac{v_{xu} - v_{xl}}{2} \tag{4.11}$$

$$v_{xo} = \frac{V_{dc} - (v_{xu} + v_{xl})}{2} \tag{4.12}$$

Substituting Equations (4.11) and (4.12) into Equations (4.9) and (4.10), dynamics of output current and leg current can be deduced as follows:

$$v_{xs} - v_{xg} = \left(\frac{R_o}{2} + \frac{d}{dt}\frac{L_o}{2}\right) i_{xs} \tag{4.13}$$

$$v_{xo} = \left(R_o + \frac{d}{dt}L_o\right) i_{xo} \tag{4.14}$$

From Equations (4.13) and (4.14), it can be concluded that output voltage v_{xs} and leg internal voltage v_{xo} can drive output current and leg current independently. To generate the desired v_{xs} and v_{xo} to regulate i_{xs} and i_{xo} independently, references of the upper and lower arm output voltages should be set as Equations (4.15) and (4.16).

$$v_{xu}^* = \frac{v_{dc}^*}{2} - v_{xs}^* - v_{xo}^* \tag{4.15}$$

$$v_{x1}^* = \frac{v_{dc}^*}{2} + v_{xs}^* - v_{xo}^* \tag{4.16}$$

In Equations (4.15) and (4.16), as a feed-forwarding variable, v_{dc}^* should be synthesized as in Equation (4.17).

$$v_{dc}^* = V_{dc} \tag{4.17}$$

To control the arm capacitor energy, the power that flows into each arm in the same leg should be considered. Power that flows into the upper and lower arms in the same "x" phase can be deduced as Equations (4.18) and (4.19), neglecting the conversion losses.

$$P_{xu} = \frac{dE_{xu}}{dt} = v_{xu}i_{xu} = v_{xu}^*i_{xu} = \left(\frac{v_{dc}^*}{2} - v_{xs}^* - v_{xo}^*\right)\left(i_{xo} + \frac{i_{xs}}{2}\right) \tag{4.18}$$

$$P_{xl} = \frac{dE_{xl}}{dt} = v_{xl}i_{xl} = v_{xl}^*i_{xl} = \left(\frac{v_{dc}^*}{2} + v_{xs}^* - v_{xo}^*\right)\left(i_{xo} - \frac{i_{xs}}{2}\right) \tag{4.19}$$

The sum and difference of the power that flows into upper and lower arms in the same leg of "x" phase are calculated as Equations (4.20) and (4.21), respectively.

$$P_x^\Sigma = P_{xu} + P_{xl} = \frac{dE_x^\Sigma}{dt} = \frac{d(E_{xu} + E_{xl})}{dt} = v_{dc}^*i_{xo} - v_{xs}^*i_{xs} - 2v_{xo}^*i_{xo} \tag{4.20}$$

$$P_x^\Delta = P_{xu} - P_{xl} = \frac{dE_x^\Delta}{dt} = \frac{d(E_{xu} - E_{xl})}{dt} = \frac{1}{2}v_{dc}^*i_{xs} - 2v_{xs}^*i_{xo} - v_{xo}^*i_{xs} \tag{4.21}$$

For VSC-HVDC application, the third terms on the right sides of both Equations (4.20) and (4.21) can be neglected. In the conventional arm capacitor energy control strategy, the sum of upper and lower arm capacitor energy E_x^Σ, namely the leg capacitor energy of "x" phase, was controlled by regulating the first term on the right side of Equation (4.20), and the difference of upper and lower arm capacitor energy E_x^Δ was controlled by regulating the second term on the right side of Equation (4.21).

Thus, in the conventional control methods, a DC component of leg current i_{xo} was drawn from the stiff DC bus voltage source to regulate E_x^Σ, and a line frequency component of leg current i_{xo} was injected to regulate

E_x^Δ. A cascade-structured controller consisting of an arm capacitor energy controller as the outer loop and a leg current regulator as the inner loop is shown in Figure 4.5.

The magnitude of the reference of the output EMF is denoted as V_{ms}. A proportional-integral-resonant (PIR) controller was employed for leg current regulation and the resonant regulator resonates at line frequency ω_o. It should be noted that since a considerable twice line frequency component and a line frequency component existed in E_x^Σ and E_x^Δ, respectively, notch filters were employed to cut off the alternating power components. In the steady state, the line frequency component of the leg current vanishes after the upper and lower arm capacitor energy are balanced, and only the DC component of the leg current flows in accordance with the active power of the corresponding phase.

Modification of MMC Modeling: The conventional model and control strategy of the MMC are based on a stiff voltage source in the DC bus. Since actual DC bus in high-voltage direct current (HVDC) transmission application reveals rather current source nature because of the inductances of the smoothing reactor and transmission line, the conventional model is not valid for the HVDC application, and the conventional arm capacitor energy control strategy by drawing leg current from the stiff DC bus would lead to poor dynamics of arm capacitor energy control and even make the system unstable. In this section, a universal model of the MMC is first derived that is valid regardless of the nature of the DC bus. In Figure 4.6, the circuit of the MMC

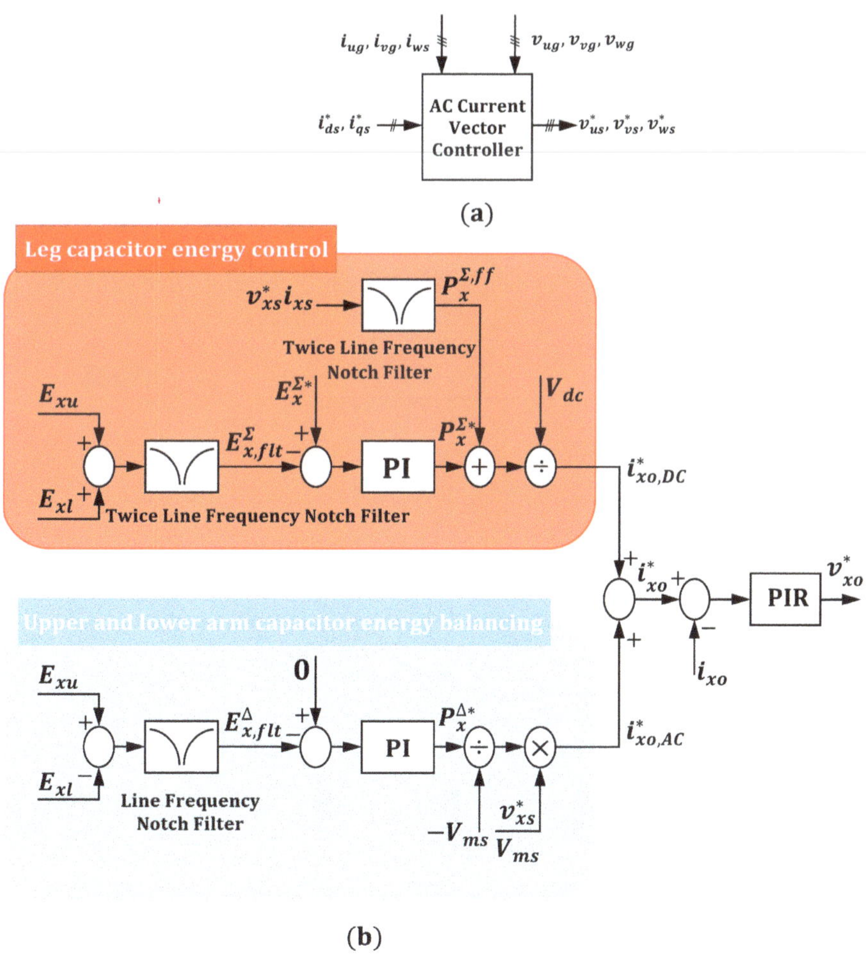

Figure 4.5 Block diagram of the conventional cascade structured arm capacitor energy controller (a) AC grid current vector controller to generate reference of output EMF (b) arm capacitor energy controller by regulating leg current

is shown if the conventional control strategy is employed without any assumption of the nature of the DC bus. For arbitrary "x" and "y" phases, which are two phases among the a, b, and c phases, according to Kirchhoff's law, dynamics of upper and lower arm currents can be derived as Equations (4.22) and (4.23).

$$\begin{cases} \left(-v_{xs}^* + \frac{v_{dc}^*}{2} - v_{xo}^*\right) + \left(L_o \frac{d}{dt} + R_o\right) i_{xu} + v_{xg} \\ -\left(-v_{ys}^* + \frac{v_{dc}^*}{2} - v_{yo}^*\right) - \left(L_o \frac{d}{dt} + R_o\right) i_{yu} - v_{yg} = 0 \end{cases} \tag{4.22}$$

$$\begin{cases} \left(v_{xs}^* + \frac{v_{dc}^*}{2} - v_{xo}^*\right) + \left(L_o \frac{d}{dt} + R_o\right) i_{xl} - v_{xg} \\ -\left(v_{ys}^* + \frac{v_{dc}^*}{2} - v_{yo}^*\right) - \left(L_o \frac{d}{dt} + R_o\right) i_{yl} + v_{yg} = 0 \end{cases} \tag{4.23}$$

And then, Equations (4.22) and (4.23) can be rewritten as Equations (4.24) and (4.25).

$$-v_{xs}^* + \left(L_o \frac{d}{dt} + R_o\right) \frac{i_{xs}}{2} + v_{xg} - (-v_{ys}^*) - \left(L_o \frac{d}{dt} + R_o\right) \frac{i_{ys}}{2} - v_{yg} = 0 \tag{4.24}$$

$$v_{xs}^* + \left(L_o \frac{d}{dt} + R_o\right) \left(-\frac{i_{xs}}{2}\right) - v_{xg} - (v_{ys}^*) - \left(L_o \frac{d}{dt} + R_o\right) \left(-\frac{i_{ys}}{2}\right) + v_{yg} = 0 \tag{4.25}$$

According to Equations (4.24) and (4.25), an equivalent circuit describing the AC grid current can be extracted from Figure 4.6 and Figure 4.7(a). Subtracting Equation (4.24) from Equation (4.25), it can be deduced that as in Equation (4.26)

$$v_{xs}^* - \left(\frac{L_o}{2}\frac{d}{dt} + \frac{R_o}{2}\right) i_{xs} - v_{xg} - v_{ys}^* + \left(\frac{L_o}{2}\frac{d}{dt} + \frac{R_o}{2}\right) i_{ys} + v_{yg} = 0 \tag{4.26}$$

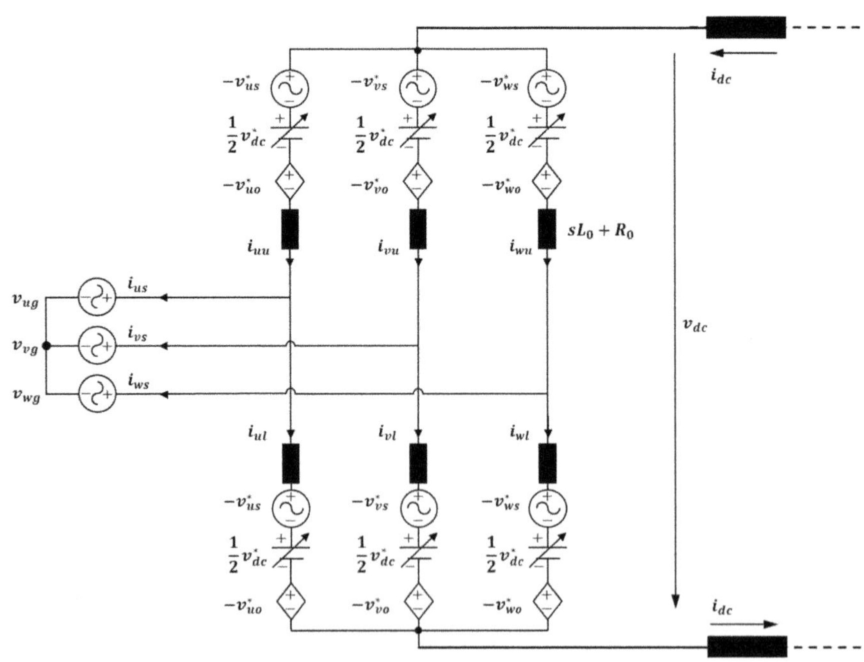

Figure 4.6 Modular multi-level converter (MMC) circuit for generalized DC bus if the conventional control strategy is employed

Then, according to Equation (4.26), the equivalent circuit in Figure 4.7(a) can be simplified as Figure 4.8(a). Similarly, from Figure 4.6, the instantaneous value of the DC bus voltage v_{dc} can be calculated as Equations (4.27)–(4.29).

$$\left(-v_{us}^* + \frac{v_{dc}^*}{2} - v_{uo}^*\right) + \left(L_o\frac{d}{dt} + R_o\right)(i_{uu} + i_{ul}) + \left(v_{us}^* + \frac{v_{dc}^*}{2} - v_{uo}^*\right) \tag{4.27}$$

$$\left(-v_{vs}^* + v_{dc}^*/2 - v_{vo}^*\right) + \left(L_o\frac{d}{dt} + R_o\right)(i_{vu} + i_{vl}) + \left(v_{vs}^* + v_{dc}^*/2 - v_{vo}^*\right) = v_{dc} \tag{4.28}$$

$$\left(-v_{ws}^* + v_{dc}^*/2 - v_{wo}^*\right) + \left(L_o\frac{d}{dt} + R_o\right)(i_{wu} + i_{wl}) + \left(v_{ws}^* + v_{dc}^*/2 - v_{wo}^*\right) = v_{dc} \tag{4.29}$$

Adding Equations (4.27) and (4.28), the instantaneous DC bus voltage can be described as Equation (4.30).

$$v_{dc} = v_{dc}^* + 2\left(L_o\frac{d}{dt} + R_o\right)\left(\frac{i_{dc}}{3}\right) - \frac{2}{3}(v_{uo}^* + v_{vo}^* + v_{wo}^*) \tag{4.30}$$

In accordance with Kirchhoff's current law and Equation (4.30), an equivalent circuit for DC bus current can be extracted from Figure 4.6, as shown in Figure 4.7(b). The equivalent circuit for DC bus current in Figure 4.7(b) can be simplified as shown in Figure 4.8(b). Since the DC bus voltage is instantaneously described as Equation (4.30), dynamics of the leg current of "x" phase, namely i_{xo}, can be deduced as Equation (4.31). It can be seen in Equation (4.31) that, for the case where the DC bus is not a stiff voltage source, the transients of the leg current are not only determined by the corresponding phase leg internal voltage, but also by the mean value of the three-phase leg internal voltages. The leg current will not be regulated independently for each phase if the DC bus is not a stiff voltage source.

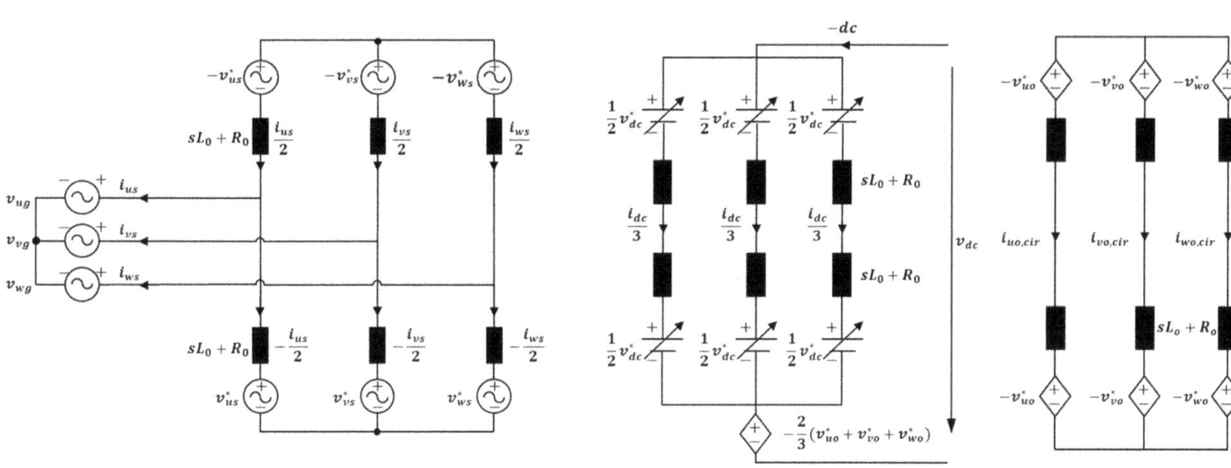

Figure 4.7 Three equivalent circuits for the proposed universal modular multilevel converter (MMC) model under generalized nature of DC bus: (a) equivalent circuit for AC grid current; (b) equivalent circuit for DC bus current; and (c) equivalent circuit for circulating current

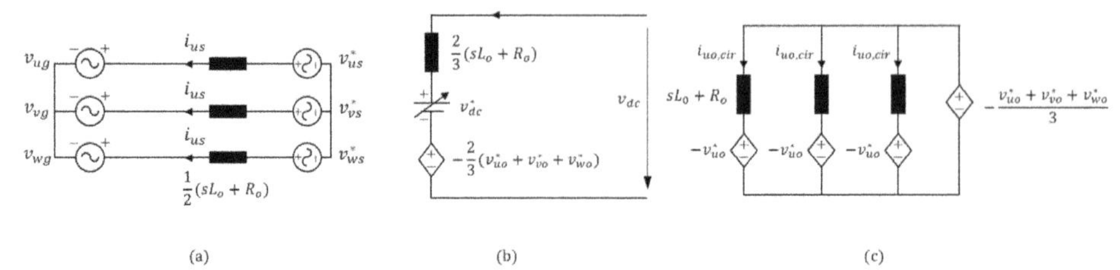

Figure 4.8 Simplified equivalent circuits of the proposed universal modular multilevel converter (MMC) model: (a) simplified circuit of AC grid current; (b) simplified circuit of DC bus current; and (c) simplified circuit of circulating current

$$2\left(L_o\frac{d}{dt}+R_o\right)\left(i_{xo}-\frac{i_{dc}}{3}\right)=2\left(v_{xo}^{*}-\frac{v_{uo}^{*}+v_{vo}^{*}+v_{wo}^{*}}{3}\right) \tag{4.31}$$

Here, a circulating current in "x" phase, $i_{xo,cir}$, is defined as the difference between the corresponding phase leg current i_{xo} and the mean current of the DC bus current that equally flows into each phase, as in Equation (4.32).

$$i_{xo,cir}=i_{xo}-\frac{i_{dc}}{3} \tag{4.32}$$

Substituting Equations (4.32) into (4.33), transients of the circulating current of "x" phase can be described as Equation (4.33).

$$2\left(L_o\frac{d}{dt}+R_o\right)i_{xo,cir}=2\left(v_{xo}^{*}-\frac{v_{uo}^{*}+v_{vo}^{*}+v_{wo}^{*}}{3}\right) \tag{4.33}$$

In accordance with Equation (4.33), an equivalent circuit for three-phase circulating currents can be extracted from Figure 4.6, as shown in Figure 4.7(c). It can be further simplified as Figure 4.8(c). In summary, under the generalized nature of the DC bus, the MMC circuit in Figure 4.6 can be divided into three equivalent circuits in Figure 4.7, which independently describe the dynamics of the AC grid current, the DC bus current, and the circulating current.

In the conventional arm capacitor energy control strategy, the arm capacitor energy was controlled independently for each phase, which means that leg current of the corresponding phase was regulated independently as well. In this case, a common mode voltage component defined as Equation (4.34) can be generated in three-phase leg internal voltages if arm capacitor energy control and leg current regulation are implemented independently for each phase.

$$v_{o,com}^{*}=\frac{v_{uo}^{*}+v_{vo}^{*}+v_{wo}^{*}}{3} \tag{4.34}$$

It is seen in Figure 4.8(b) that the common mode leg internal voltage $v_{o,com}^{*}$ would affect DC bus voltage as seen from Equation (4.30) if the DC bus is not connected to a stiff voltage source. Moreover, it would also affect the regulation of leg current as shown by Equation (4.31), and it may even lead to failure of leg current regulation and make the system unstable. Contrasting with the motor-drive application where the DC bus is connected to a stiff voltage source that is fed by rectifiers, the DC bus in the HVDC application is not a stiff DC source and has rather current source nature, and the common mode leg internal voltage $v_{o,com}^{*}$ should be controlled as zero actively. If an arm capacitor energy control strategy can inherently nullify $v_{o,com}^{*}$, then AC current regulation, DC current regulation, and arm capacitor energy control can be completely decoupled as shown in Figure 4.8, and the MMC for HVDC application may have better dynamic performance compared to the conventional arm capacitor energy control strategy.

Arm Capacitor Energy Control Algorithm: Capacitor energy of the cells inside the same arm can be balanced by proper modulation methods without affecting current regulation. The main target of cell capacitor energy control is controlling the capacitor energy of six arms as rated value E_{rated}, particularly as in Equation (4.35), which is mathematically equivalent to Equations (4.36)–(4.40).

$$E_{uu} = E_{ul} = E_{vu} = E_{vl} = E_{wu} = E_{wl} = E_{rated} \tag{4.35}$$

$$E_{total} = \sum_{x=u,v,w} (E_{xu} + E_{xl}) = 6E_{rated} \tag{4.36}$$

$$E_u^{\Sigma} = E_v^{\Sigma} = E_w^{\Sigma} \tag{4.37}$$

$$E_u^{\Delta} = E_{uu} - E_{ul} = 0 \tag{4.38}$$

$$E_v^{\Delta} = E_{vu} - E_{vl} = 0 \tag{4.39}$$

$$E_w^{\Delta} = E_{wu} - E_{wl} = 0 \tag{4.40}$$

It should be mentioned that Equation (4.36) stands for the regulation of total energy stored in all total MMC cell capacitors. Equation (4.37) means inter-leg capacitor energy balancing, and Equations (4.38)–(4.40) mean upper and lower arm capacitor energy balancing at each phase.

Control of energy stored in the whole MMC cell capacitors: In the conventional leg capacitor energy control method, capacitor energy of each leg was controlled independently, and the energy balancing power had been transferred between each leg and the stiff DC bus voltage source. Contrasting with the conventional control method, here where the HVDC application where the DC bus of the MMC is not a stiff DC voltage source, the whole capacitor energy of an MMC operates in Voltage Regulator (VR) mode. Regulation of the DC bus voltage is done by controlling AC grid side active power as shown in Figure 4.9(a). In this case, the power drawn by the HVDC transmission line is employed as a feed-forward term to improve dynamic performance. Here, it is assumed that the grid voltage vector is oriented to the q-axis of a synchronously rotating reference

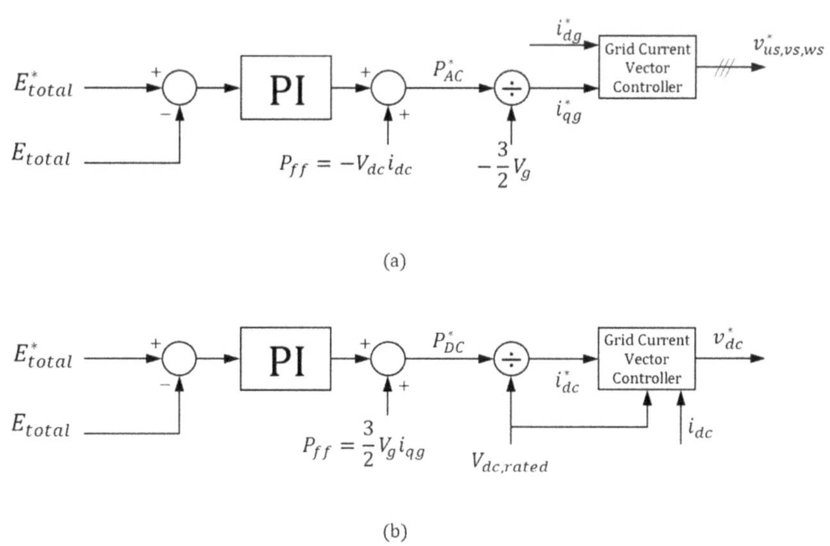

Figure 4.9 Block diagram of the proposed converter total capacitor energy controller: (a) converter total capacitor energy controller for voltage regulator (VR) mode operation and (b) converter total capacitor energy controller for PD mode operation

frame. Otherwise, the whole capacitor energy of an MMC operates in power dispatcher (PD), and controlling the power that flows through the HVDC transmission line is regulated by controlling the DC bus side power, namely by regulating the DC transmission line current actively as shown in Figure 4.9(b). In this case, the active power that flows from the MMC into the AC grid is employed as a feed-forward term to improve dynamic performance. Thus, by the proposed total capacitor energy control strategy, the total capacitor energy of an MMC can be successfully regulated. As shown in Figure 4.8, regulating AC grid current or DC bus current would neither violate the circulating current regulation nor generate the common mode component of leg internal voltages.

Balancing of three-phase leg capacitor energy: Besides the converter total capacitor energy control, the arm capacitor energy of six different arms should be also balanced simultaneously to regulate the capacitor energy of each arm as its rated value. Since the arm capacitor energy balancing is an internal event of the MMC, a balancing strategy should affect neither the AC grid side nor the DC bus side, and it should be generally valid regardless of the nature of the DC bus. The circulating current flows only inside the converter and can be employed to transfer energy among six different arms affecting neither the AC grid side nor the DC bus side. In this, a comprehensive arm capacitor energy balancing strategy is proposed through only injection of the circulating current that flows inside the converter. It is universally valid under the generalized DC bus.

The block diagram of the proposed leg capacitor energy balancing strategy is shown in Figure 4.10(a). When contrasted with the conventional leg capacitor energy control strategy, the most important point of the proposed strategy is that the leg capacitor energy reference is updated every sampling period as the mean capacitor energy of three legs instead of a constant rated reference value. The principle of the conventional leg capacitor energy control method was transferring power between the stiff DC bus and each leg separately, whereas the principle of the proposed method is not transferring power among the AC grid side, the DC bus side, and each phase of the MMC, but instead transferring power among three legs internally through injection of the DC circulating current as shown in Figure 4.10(b). Since the leg capacitor energy reference is updated online as shown in Figure 4.10(a), it can be deduced using Equation (4.41).

$$\sum_{x=u,v,w} (E^*_{leg} - E^{\Sigma}_{x,flt}) = 0 \Rightarrow \sum_{x=u,v,w} i^*_{xo,cirDC} = 0 \Rightarrow \sum_{x=u,v,w} v^*_{xo,DC} = 0 \tag{4.41}$$

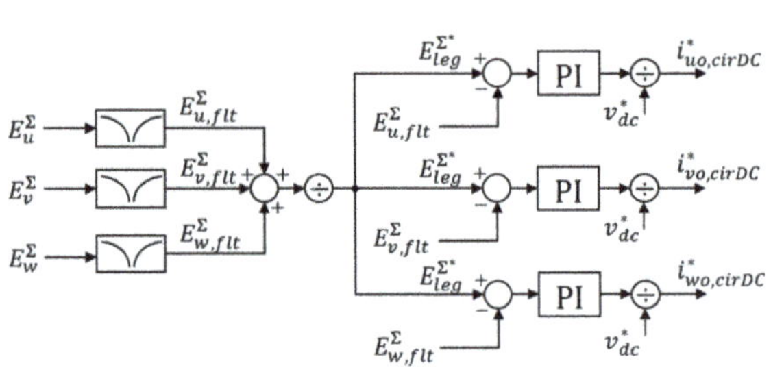

(a)

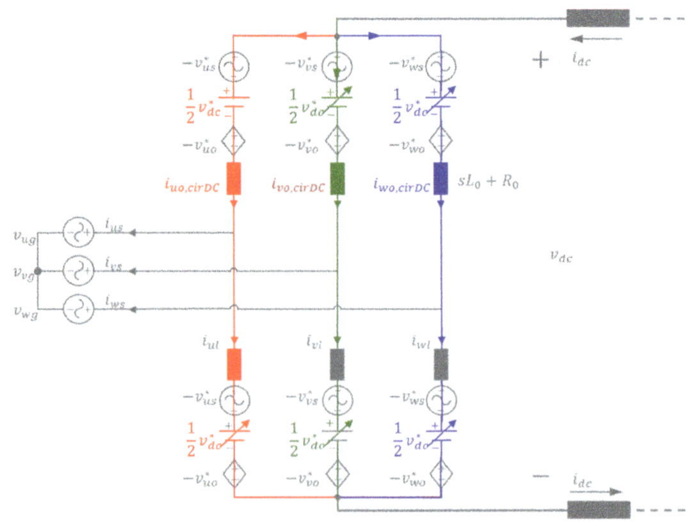

(b)

Figure 4.10 Proposed leg capacitor energy balancing control strategy: (a) block diagram of the proposed leg capacitor energy balancing control strategy and (b) principle of the proposed leg capacitor energy balancing control strategy

Thus, in the proposed leg capacitor energy balancing strategy, the common mode component of leg internal voltages $v^*_{o,com}$ is inherently null. The issues caused by $v^*_{o,com}$ such as poor dynamics of arm capacitor energy control and possible instability of leg capacitor energy balancing can also be inherently prevented. Since there is a considerable twice line frequency fluctuation in the leg capacitor energy, notch filters with a central frequency at twice line frequency are employed in the control loop in Figure 4.10(a). By the proposed control strategy, the leg energy balancing defined by Equation (4.37) can be successfully achieved.

Balancing of upper and lower arm capacitor energy: With the proposed control strategy mentioned in earlier sections, the total capacitor energy in the MMC can be kept by regulating either the AC grid side power or the DC bus side power, and the three leg capacitor energy can be balanced by injecting the DC component of the circulating current. Furthermore, the difference between the capacitor energy of the upper and lower arm of each leg defined by Equations (4.38)–(4.40) should be also balanced to achieve the balancing of all arms of MMC defined as in Equation (4.35).

If the common mode voltage of the leg internal voltages $v^*_{o,com}$ is kept as null by the upper and lower arm capacitor energy balancing control strategy, the common mode component of references of the AC circulating current at line frequency should be zero as in Equation (4.42).

$$\sum_{x=u,v,w} v^*_{xo,AC} = 0 \Rightarrow \sum_{x=u,v,w} i^*_{xo,cirAC} = 0 \tag{4.42}$$

Since the upper and lower arm capacitor energy should be balanced for three phases, three degrees of freedom (DOFs) are necessary to regulate P^{Δ}_u, P^{Δ}_v, and P^{Δ}_w independently. Here, a positive sequence circulating current and a negative sequence circulating current are employed. If the three-phase output voltage can be described by Equation (4.43), then its corresponding space vector can be denoted by Equation (4.44).

$$\begin{cases} v^*_{us} = V_{ms}\cos(\omega_o t + \varphi) \\ v^*_{vs} = V_{ms}\cos\left(\omega_o t + \varphi - \frac{2\pi}{3}\right) \\ v^*_{ws} = V_{ms}\cos\left(\omega_o t + \varphi + \frac{2\pi}{3}\right) \end{cases} \tag{4.43}$$

$$\mathrm{V_s} = V_{ms} e^{j(\omega_o t + \varphi)} \tag{4.44}$$

If a positive sequence circulating current and a negative sequence circulating current are injected as in Equations (4.45) and (4.46), then according to Equation (4.22), the differences in the power that flows into the upper arm and the lower arm caused by the positive and negative sequence circulating currents can be calculated as in Equations (4.47) and (4.48), respectively.

$$\begin{cases} i^*_{uo,cirAC,pos} = I_{cirAC,pos}\cos(\omega_o t + \varphi_{pos}) \\ i^*_{vo,cirAC,pos} = I_{cirAC,pos}\cos\left(\omega_o t + \varphi_{pos} - \frac{2\pi}{3}\right) \\ i^*_{wo,cirAC,pos} = I_{cirAC,pos}\cos\left(\omega_o t + \varphi_{pos} + \frac{2\pi}{3}\right) \end{cases} \tag{4.45}$$

$$\begin{cases} i^*_{uo,cirAC,neg} = I_{cirAC,neg}\cos(\omega_o t + \varphi_{neg}) \\ i^*_{vo,cirAC,neg} = I_{cirAC,neg}\cos\left(\omega_o t + \varphi_{neg} + \frac{2\pi}{3}\right) \\ i^*_{wo,cirAC,neg} = I_{cirAC,neg}\cos\left(\omega_o t + \varphi_{neg} - \frac{2\pi}{3}\right) \end{cases} \tag{4.46}$$

$$\begin{cases} P_{u,pos}^{\Delta} = \frac{dE_u^{\Delta}}{dt}|_{cir,pos} = -V_{ms}I_{cirAC,pos}\cos\left(\varphi - \varphi_{pos}\right) \\ P_{v,pos}^{\Delta} = \frac{dE_v^{\Delta}}{dt}|_{cir,pos} = -V_{ms}I_{cirAC,pos}\cos\left(\varphi - \varphi_{pos}\right) \\ P_{w,pos}^{\Delta} = \frac{dE_w^{\Delta}}{dt}|_{cir,pos} = -V_{ms}I_{cirAC,pos}\cos\left(\varphi - \varphi_{pos}\right) \end{cases} \tag{4.47}$$

$$\begin{cases} P_{u,neg}^{\Delta} = \frac{dE_u^{\Delta}}{dt}|_{cir,neg} = -V_{ms}I_{cirAC,neg}\cos\left(\varphi_{neg} - \varphi\right) \\ P_{v,neg}^{\Delta} = \frac{dE_v^{\Delta}}{dt}|_{cir,neg} = -V_{ms}I_{cirAC,neg}\cos\left(\varphi_{neg} - \varphi - \frac{2\pi}{3}\right) \\ P_{w,neg}^{\Delta} = \frac{dE_w^{\Delta}}{dt}|_{cir,neg} = -V_{ms}I_{cirAC,neg}\cos\left(\varphi_{neg} - \varphi + \frac{2\pi}{3}\right) \end{cases} \tag{4.48}$$

In Equations (4.47) and (4.48), if the positive sequence circulating current is in phase with the output voltage vector $\mathbf{V_s}$, namely $\varphi_{pos} = \varphi$, then $I_{cirAC,pos}$, $I_{cirAC,neg}$, and φ_{neg} provide three DOFs for upper and lower arm capacitor energy balancing under the restriction of Equation (4.42). In accordance with Equations (4.47) and (4.48), the positive sequence circulating current can be injected to eliminate the common component E_{com}^{Δ} of three-phase upper and lower arm capacitor energy differences, and the negative sequence circulating current can be injected to eliminate the differential components E_d^{Δ} and E_q^{Δ} (represented in the stationary d-q reference frame) of three-phase upper and lower arm capacitor energy differences as in Equation (4.49). The principle of the proposed upper and lower arm capacitor energy balancing strategy is shown in Figure 4.11.

$$\begin{cases} P_{com}^{\Delta} = \frac{dE_{com}^{\Delta}}{dt} = \frac{d}{dt}\left(\frac{E_u^{\Delta}+E_v^{\Delta}+E_w^{\Delta}}{3}\right) = -V_{ms}I_{cirAC,pos} \\ P_d^{\Delta} = \frac{dE_d^{\Delta}}{dt} = \frac{d}{dt}\left(\frac{2}{3}E_u^{\Delta} - \frac{1}{3}E_v^{\Delta} - \frac{1}{3}E_w^{\Delta}\right) = -V_{ms}I_{cirAC,neg}\cos\left(\varphi_{neg} - \varphi\right) \\ P_q^{\Delta} = \frac{dE_q^{\Delta}}{dt} = \frac{d}{dt}\left(\frac{\sqrt{3}}{3}E_v^{\Delta} - \frac{\sqrt{3}}{3}E_w^{\Delta}\right) = -V_{ms}I_{cirAC,neg}\sin\left(\varphi_{neg} - \varphi\right) \end{cases} \tag{4.49}$$

Since for upper and lower arm capacitor energy balancing the circulating current reference only includes positive and negative sequence components in accordance with Equation (4.41), Equation (4.50) can be concluded. By the proposed arm capacitor energy balancing strategy, which includes both the leg capacitor energy balancing control and the upper and lower arm capacitor energy balancing control, common mode component of the leg internal voltages v_{xo}^* is inherently nullified and the successful regulation of the circulating current is guaranteed.

$$\sum_{x=u,v,w} i_{xo,cir}^* = \sum_{x=u,v,w} \left(i_{xo,cirDC}^* + i_{xo,cirAC}^*\right) \sum_{x=u,v,w} v_{xo}^* = 0 \tag{4.50}$$

The control block diagram of the proposed upper and lower arm capacitor energy balancing controller is depicted in Figure 4.12. For simplicity, the line frequency notch filters, which should be employed to eliminate fluctuating components, are omitted in the Figure. To make the reference value of the negative sequence circulating current, Equations (4.51) and (4.52) can be derived.

$$i_{do,cirAC,neg}^* + ji_{qo,cirAC,neg}^* = \frac{P_d^{\Delta*} + jP_q^{\Delta*}}{-V_{ms}}e^{j(\omega_o t+\varphi)} = \frac{P_d^{\Delta*} + jP_q^{\Delta*}}{-V_{ms}^2}\mathbf{V_s} \tag{4.51}$$

$$\begin{bmatrix} i^*_{uo,cir AC,neg} \\ i^*_{vo,cir AC,neg} \\ i^*_{wo,cir AC,neg} \end{bmatrix} = \begin{bmatrix} 1 & 0 \\ -\frac{1}{2} & -\frac{\sqrt{3}}{2} \\ -\frac{1}{2} & \frac{\sqrt{3}}{2} \end{bmatrix} \begin{bmatrix} i^*_{do,cir AC,neg} \\ i^*_{qo,cir AC,neg} \end{bmatrix} \quad (4.52)$$

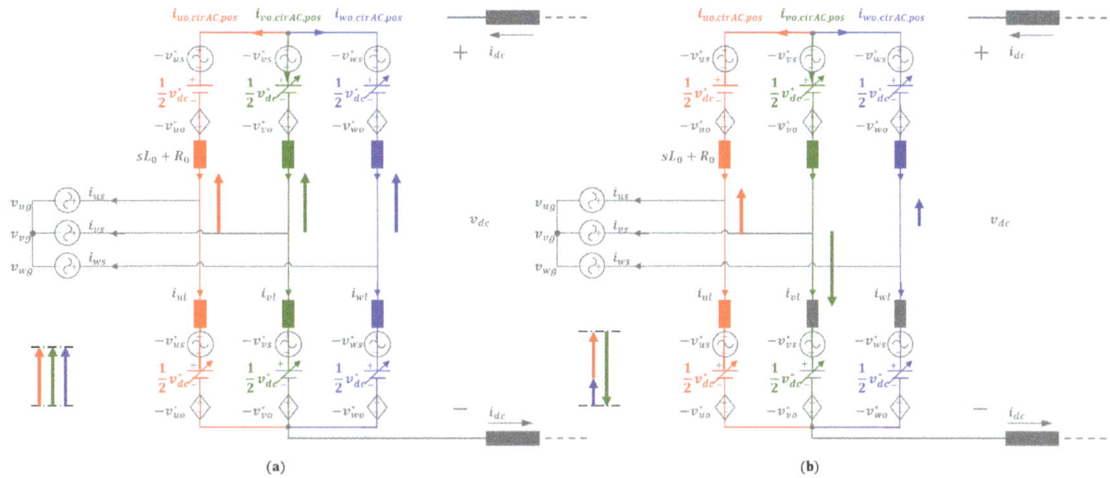

Figure 4.11 Principle of the proposed upper and lower arm capacitor energy balancing strategy (a) elimination of the common component of the upper and lower arm capacitor energy differences (b) Principle of the proposed upper and lower arm capacitor energy balancing strategy

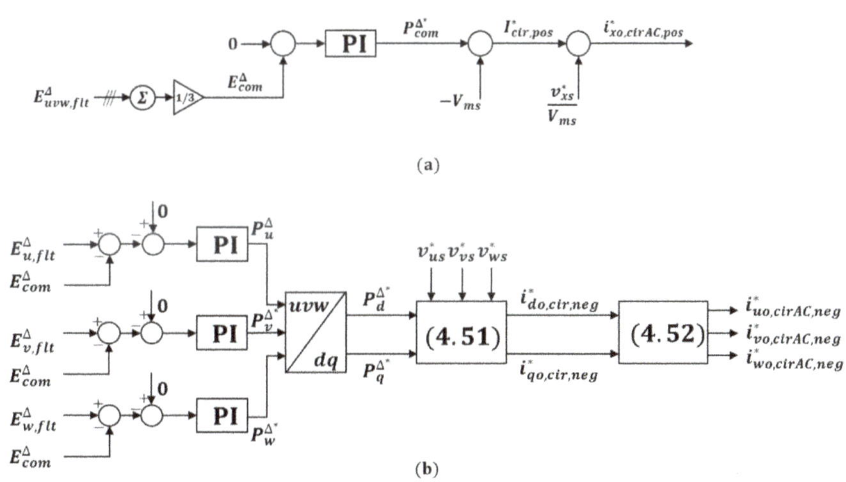

Figure 4.12 Block diagram of the proposed upper and lower arm capacitor energy balancing controller: (a) controller of the common component of the upper and lower arm capacitor energy differences and (b) block diagram of controller of the differential component of the upper and lower arm capacitor energy differences

Experimental Verification of Control Algorithm

Performance at the No Load Condition: First, validity of the proposed control strategy is investigated without DC load. The R-L load is separated from the MMC DC bus by a circuit breaker. To verify the validity of the proposed control strategy, controllers proposed in Section IV are activated successively. Experimental results of the arm capacitor energy by the proposed control strategy are shown in Figure 4.14. In Figure 4.14(a), the converter total capacitor energy is boosted up to its rated value after the converter total capacitor energy controller is activated, and the leg capacitor energy of three phases converges to the rated value as soon as the leg capacitor energy balancing controller is activated. In Figure 4.14(b), the differential component of the upper and lower arm capacitor energy differences is eliminated after the activation of the differential error controller, and the upper and lower arm capacitor energy differences of three phases become equal. Then, the common component of the upper and lower arm capacitor energy *differences is eliminated after the activation of the common error controller.*

Current waveforms during arm capacitor energy control procedures by the proposed control strategy are recorded as shown in Figure 4.15. In Figure 4.15(a), the MMC starts to draw active current from the AC grid to boost up the converter total capacitor energy to the rated value after the converter total capacitor energy controller is activated. In Figure 4.15(b) and 4.15(c), a negative sequence circulating current is injected inside the converter to eliminate the differential component of upper and lower arm capacitor energy differences,

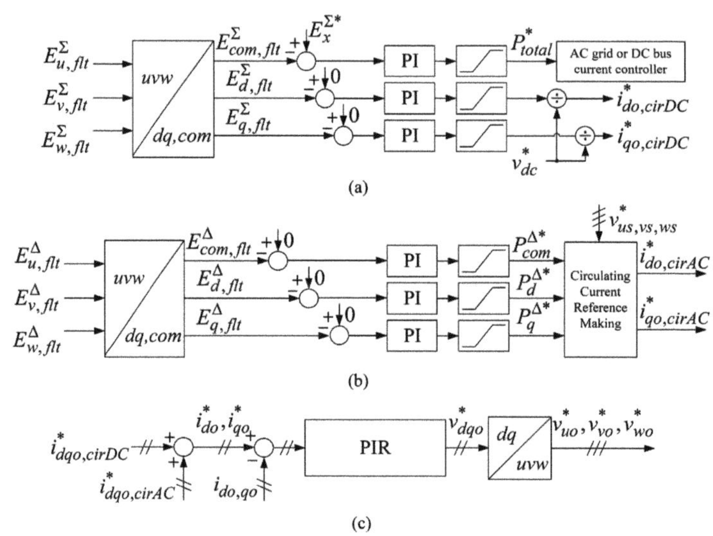

Figure 4.13 Block diagram of the proposed arm capacitor energy balancing controllers constructed in the stationary d-q reference frame: (a) leg capacitor energy balancing controller; (b) upper and lower arm capacitor energy balancing controller; and (c) circulating current regulator

Table 4.1 Parameters of Prototype modular multi-level converter (MMC) System Under Test

Quantity	Values
Number of cells per arm	6
Rated DC bus voltage	300V
Rated cell capacitor voltage	50V
Cell capacitor capacitance	5.4mF
Grid voltage	110V
Arm inductor inductance	4.0 mH
DC bus R-L load inductance	8.0 mH
DC bus R-L load resistance	30 Ω
Digital controller sampling frequency	5.0 kHz

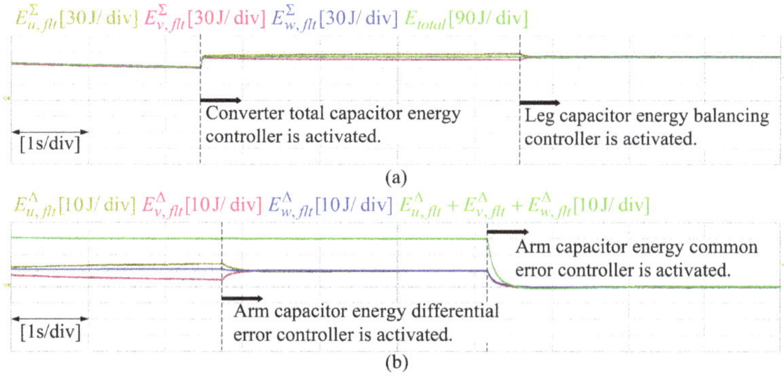

Figure 4.14 Experimental result of arm capacitor energy control by the proposed control strategy: (a) leg capacitor energy and (b) difference of upper and lower arm capacitor energy

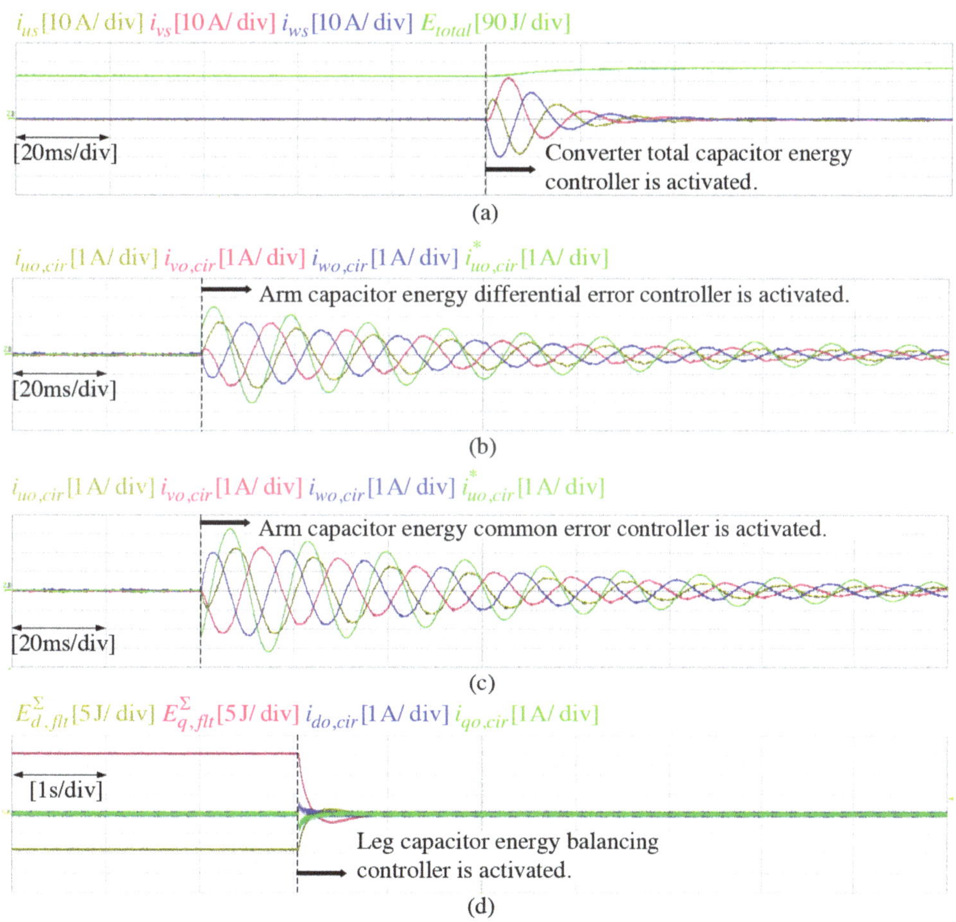

Figure 4.15 Experimental result of the currents during arm capacitor energy control by the proposed control strategy: (a) AC grid side current and total capacitor energy; (b) circulating current and its reference during elimination of differential component of upper and lower arm capacitor energy differences; (c) circulating current and its reference during elimination of common component of upper and lower arm capacitor energy differences; and (d) circulating current and leg capacitor energy in the stationary reference frame

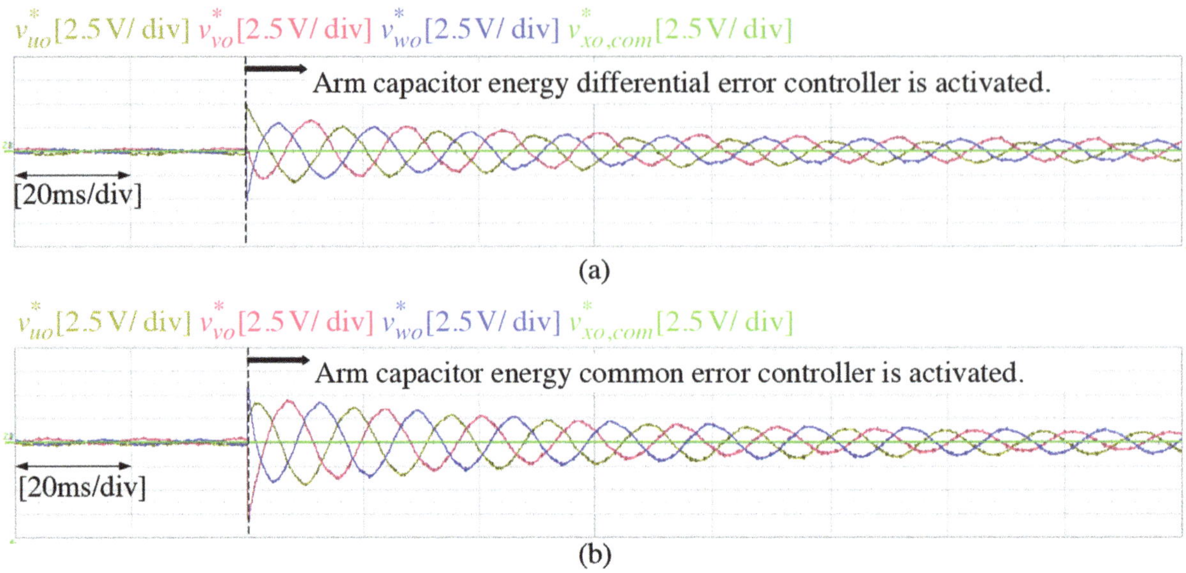

Figure 4.16 Experimental result of references of leg internal voltages (a) leg internal voltages during elimination of differential component of upper and lower arm capacitor energy differences (b) Experimental result of references of leg internal voltages

whereas a positive sequence circulating current is injected to eliminate the common component of the differences. The circulating current tracks its reference with reasonably good dynamics by the proposed control strategy. In Figure 4.15(d), a DC circulating current is injected to balance the leg capacitor energy of three phases. It should be emphasized that the circulating current vanishes as long as the arm capacitor energy is balanced by the proposed control strategy.

The references of the leg internal voltages during upper and lower arm capacitor energy balancing control are presented in Figure 4.16. By the proposed arm capacitor energy balancing control strategy, the common mode component is nullified, and it guarantees successful regulation of the circulating current.

Performance at the Loaded Condition: For the second, performance of the proposed control strategy is studied at the loaded condition. The MMC is connected to the AC grid, and all of the proposed controllers are activated before the converter is connected to the R-L load. Then the circuit breaker between the MMC DC bus and the R-L load is switched on.

Performance of the proposed arm capacitor energy control strategy is investigated in the loaded condition. After the DC bus of the MMC is suddenly connected to the R-L load, the three-phase leg capacitor energy is kept as the rated value as shown in Figure 4.17(a), and the upper and lower arm capacitor energies are well balanced with good dynamics as shown in Figure 4.17(b). Even though the transient process of the R-L load is very fast and within less than 1 ms, the cell capacitor voltages are controlled at the rated voltage with enough dynamic performance using the proposed method as shown in Figures 4.17(c) and 4.17(d).

The current waveforms at the loaded condition are recorded as shown in Figure 4.18. The MMC starts to draw active current from the AC grid to regulate the total capacitor energy as soon as the DC bus is loaded as shown in Figure 4.18(a). The circulating current is injected during the loading transient process to balance the arm capacitor energy and vanishes after the arm capacitor energy is balanced as shown in Figure 4.18(b).

The injected circulating current is dominated by the negative sequence component, since the power that flows into the upper three arms and the lower three arms is kept almost equal instantaneously, even during rapid transient, as indicated by Equations (4.18) and (4.19). It means that the disturbance term of E^{Δ}_{com} during the transient is negligible. It can be observed from Figure 4.18(c) to 4.18(e) that, although the circulating current vanishes after the balancing transient process of arm capacitor energy, the leg current is almost pure DC, and both the upper arm current and the lower arm current include a DC component and a line frequency AC component.

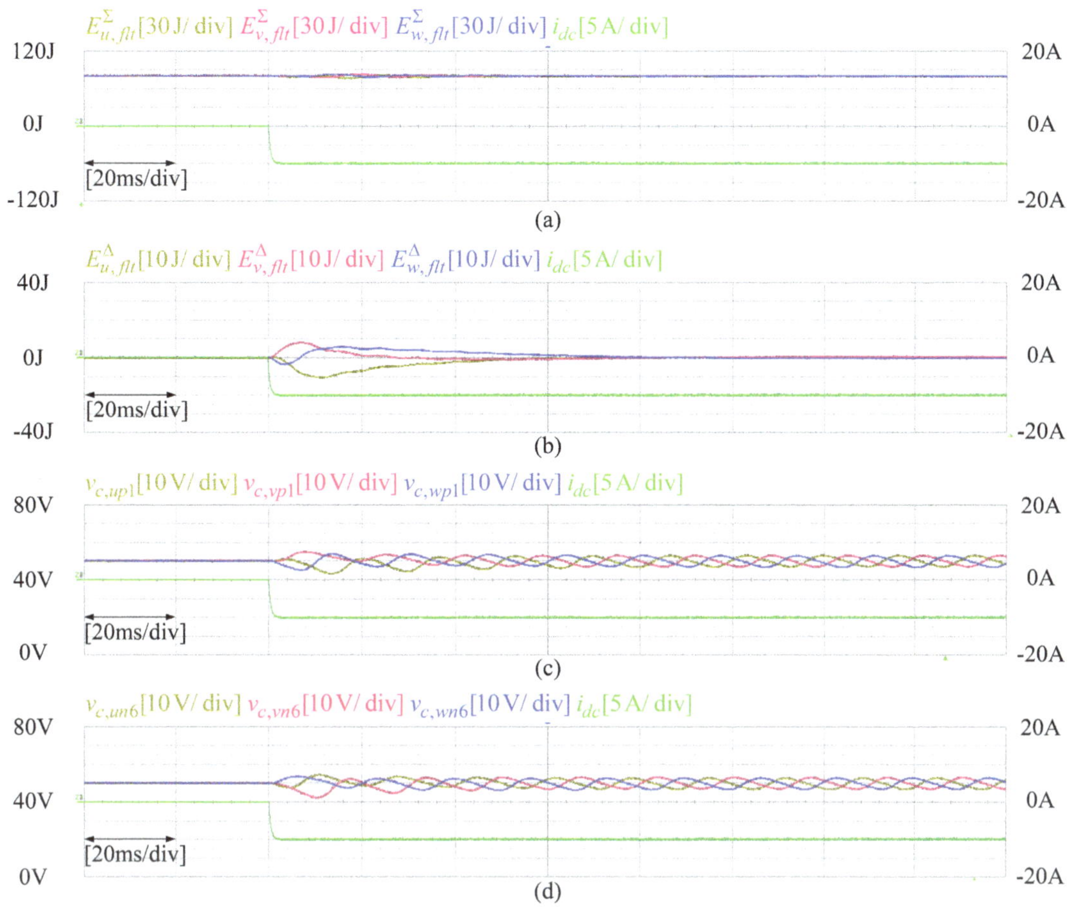

Figure 4.17 Performance of the proposed arm capacitor energy control strategy at loaded condition: (a) leg capacitor energy; (b) upper and lower arm capacitor energy difference; (c) selected three phase cell capacitor voltages of upper arms; and (d) selected three phase cell capacitor voltages of lower arms

Investigation of the DC bus: The control performance of the DC bus by the proposed control strategy is investigated here. The DC bus current waveform is recorded, whereas the DC bus of the MMC is connected to the R-L load directly as shown in Figure 4.19. Since the $v^*_{o,com}$ is nullified by the proposed method, the DC bus of the MMC is equivalent to a high bandwidth-controlled DC voltage source behind a DC inductor as shown in Figure 4.8(b). Then, the transient of the DC bus current is supposed to be that of a first-order R-L circuit as theoretically calculated by Equation (4.53). It can be observed clearly in Figure 4.19 that the experimental result of the DC bus current coincides with the calculated transient with reasonable accuracy.

The transient performance of the DC bus by the proposed control strategy is also investigated by changing the DC bus voltage command v^*_{dc}. As stated in the previous section, v^*_{dc} is fully decoupled from the capacitor energy, and can be updated every sampling period. Also, the DC bus voltage can be controlled instantaneously using the proposed control method. This is clearly demonstrated in Figure 4.20, where the DC bus voltage is changed abruptly by the proposed method, and the DC bus current varies in accordance with a first-order R-L transient process while v^*_{dc} is updated.

$$i_{dc,cal} = \frac{-v^*_{dc}}{R_{load}}\left\{1 - exp\left(-(t - t_{load})/\left\{\left(\frac{2}{3}L_o + L_{load}\right)/R_{load}\right\}\right)\right\} \tag{4.53}$$

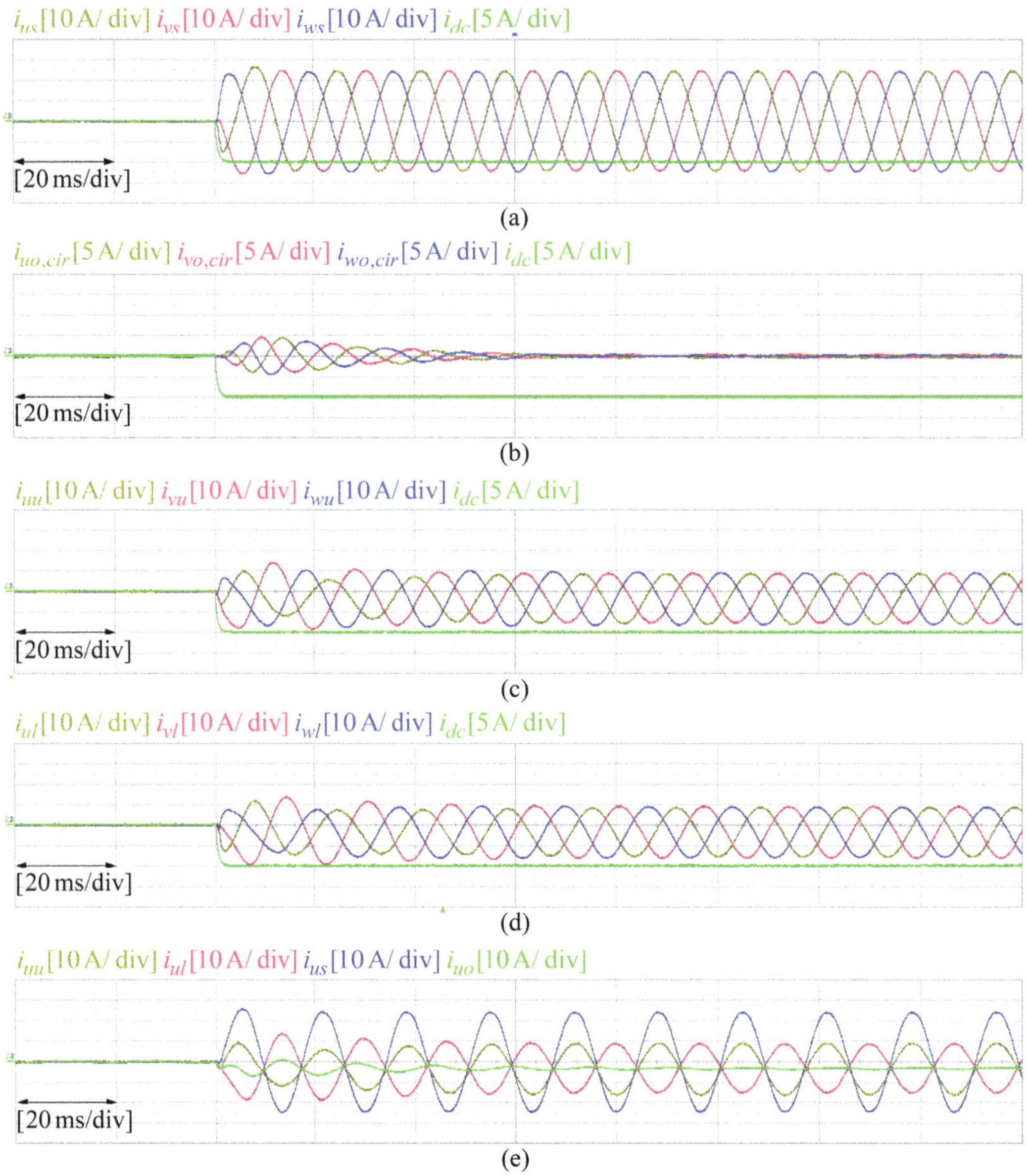

Figure 4.18 Experimental result of the currents at the loaded condition: (a) AC grid side current; (b) circulating current; (c) upper arm current; (d) lower arm current; and (e) upper arm current, lower arm current, AC grid side current, and leg current of phase U

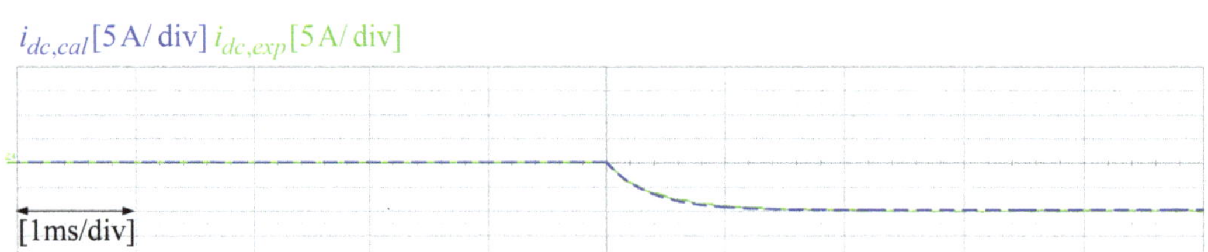

Figure 4.19 Transient of the DC bus current after the R-L load is connected

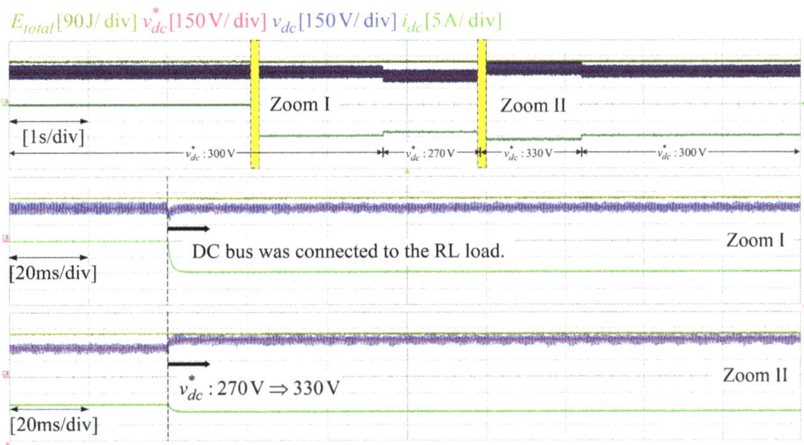

Figure 4.20 DC bus voltage and current at loaded condition

4.3 Singularity Instability [6]–[9]

4.3.1 Control Strategies Under Unbalanced Grid Conditions

Under unbalanced grid conditions, the interaction between the symmetrical components gives rise to uncontrolled active and reactive power oscillations between the converter and the grid. The instantaneous powers in a four-wire connection under unbalanced grid conditions can be written as follows:

$$\begin{bmatrix} P \\ Q \\ P_0 \end{bmatrix} = \begin{bmatrix} \overline{P} + P_{c2}\cos(2\omega t) + P_{s2}\sin(2\omega t) \\ \overline{Q} + Q_{c2}\cos(2\omega t) + Q_{s2}\sin(2\omega t) \\ \overline{P}_0 + P_{0c2}\cos(2\omega t) \end{bmatrix} \tag{4.54}$$

where $\overline{P}$, $\overline{Q}$, $\overline{P}_0$ are the average powers and P_{c2}, P_{s2}, Q_{c2}, Q_{s2}, P_{0c2} represent the magnitude of the oscillating terms.

The subindex 0 is used to denote the active power delivered by the zero sequence. If the zero sequence current is eliminated by appropriate control, P_0 term can be neglected. Transforming the three-phase currents and voltages into synchronous reference frame and splitting into positive and negative, the power components can be represented as follows:

$$\begin{bmatrix} \overline{P} \\ P_{c2} \\ P_{s2} \\ \overline{Q} \\ Q_{c2} \\ Q_{s2} \end{bmatrix} = \frac{3}{2} \begin{bmatrix} V_{sd}^{+} & V_{sq}^{+} & V_{sd}^{-} & V_{sq}^{-} \\ V_{sd}^{-} & V_{sq}^{-} & V_{sd}^{+} & V_{sq}^{+} \\ V_{sq}^{-} & -V_{sd}^{-} & -V_{sq}^{+} & V_{sd}^{+} \\ V_{sq}^{+} & -V_{sd}^{+} & V_{sq}^{-} & -V_{sd}^{-} \\ V_{sq}^{-} & -V_{sd}^{-} & V_{sq}^{+} & -V_{sd}^{+} \\ -V_{sd}^{-} & -V_{sq}^{-} & V_{sd}^{+} & V_{sq}^{+} \end{bmatrix} \begin{bmatrix} i_d^{+} \\ i_q^{+} \\ i_d^{-} \\ i_q^{-} \end{bmatrix} \tag{4.55}$$

The given relation of sequence components and instantaneous power can be used for calculating the current references depending on the grid conditions and application requirements. It can be noticed in Equation (4.55) that the converter has four degrees of freedom to control the AC currents. Selecting the average active and reactive powers to follow the reference, two DOFs are left, which means that two control targets can be selected. Three simulation cases are presented with the following objectives:

$$\begin{aligned} \overline{P} &= P_s^{ref} \\ \overline{Q} &= Q_s^{ref} \end{aligned} \tag{4.56}$$

- Elimination of negative sequence currents
- Suppression of active power ripples
- Suppression of reactive power ripples

Case 1. Elimination of negative sequence currents: The elimination of the negative sequence currents results in balanced converter AC currents under unbalanced grid voltages. This feature is important for grid-connected inverters because of the restrictions provided by grid operators regarding the injection of unbalanced currents into the grid. The control objectives are defined as follows:

$$\begin{aligned} i_d^- &= 0 \\ i_q^- &= 0 \end{aligned} \tag{4.57}$$

As a result, having $V_{sq}^+ = V_{sq}^- = 0$ ensured by positive and negative sequence PLL, the references for the positive sequence current are the following:

$$\begin{aligned} i_d^{+ref} &= \frac{2}{3V_{sd}^+} P_s^{ref} \\ i_q^{+ref} &= -\frac{2}{3V_{sd}^+} Q_s^{ref} \end{aligned} \tag{4.58}$$

Case 2. Elimination of active power oscillations: In an HVDC system rated in hundreds of MW, the active power oscillations between the converter station and the grid might affect the AC system. This could trip the grid relay protections, worsening the fault conditions. To compensate for the active power oscillations, the control objectives are defined as follows:

$$\begin{aligned} P_{c2} &= 0 \\ P_{s2} &= 0 \end{aligned} \tag{4.59}$$

Using Equation (4.55), the following positive and negative sequence current references are obtained:

$$\begin{aligned} i_d^{+ref} &= \frac{2V_{sd}^+ P_s^{ref}}{3\left[\left(V_{sd}^+\right)^2 - \left(V_{sd}^-\right)^2\right]} \\ i_q^{+ref} &= -\frac{2V_{sd}^+ Q_s^{ref}}{3\left[\left(V_{sd}^+\right)^2 + \left(V_{sd}^-\right)^2\right]} \\ i_d^{-ref} &= \frac{2V_{sd}^- P_s^{ref}}{3\left[\left(V_{sd}^+\right)^2 - \left(V_{sd}^-\right)^2\right]} \\ i_q^{-ref} &= -\frac{2V_{sd}^- Q_s^{ref}}{3\left[\left(V_{sd}^+\right)^2 + \left(V_{sd}^-\right)^2\right]} \end{aligned} \tag{4.60}$$

Case 3. Elimination of reactive power oscillations: Reactive power oscillations between the converter station and the grid have negative impact on the AC voltage profile, which could worsen the fault conditions. To compensate for the mentioned oscillations, the control objectives are defined as follows:

$$\begin{aligned} Q_{c2} &= 0 \\ Q_{s2} &= 0 \end{aligned} \tag{4.61}$$

Operating Equation (4.55), the following current references are obtained:

$$\begin{aligned} i_d^{+ref} &= \frac{2V_{sd}^{+}P_s^{ref}}{3\left[\left(V_{sd}^{+}\right)^2-\left(V_{sd}^{-}\right)^2\right]} \\ i_q^{+ref} &= -\frac{2V_{sd}^{+}Q_s^{ref}}{3\left[\left(V_{sd}^{+}\right)^2-\left(V_{sd}^{-}\right)^2\right]} \\ i_d^{-ref} &= \frac{2V_{sd}^{-}P_s^{ref}}{3\left[\left(V_{sd}^{+}\right)^2-\left(V_{sd}^{-}\right)^2\right]} \\ i_q^{-ref} &= -\frac{2V_{sd}^{-}Q_s^{ref}}{3\left[\left(V_{sd}^{+}\right)^2-\left(V_{sd}^{-}\right)^2\right]} \end{aligned} \tag{4.62}$$

4.3.2 Reference Calculation Problem

The reference values of all components of the current are calculated using the power references and given the instantaneous-measured AC and DC grid voltages. A variation of the AC grid voltage causes the AC components to change, whereas a variation of the DC grid voltage causes the DC components to change.

The choice of AC grid current reference during an unbalanced voltage sag has been discussed in the past and the most common approach is to set its negative sequence to zero and export active and reactive power using positive sequence current only.

In such a scenario, the active power exchanged between each leg of the converter and the AC grid is different, resulting in a sustained drift of their energy that must be compensated for using additive DC current.

On the other hand, the additive AC current is normally used to control the differences between the energy of the upper arms and that of the lower arms, which may suffer from deviations during transients. The output voltage of the converter is close to the grid voltage, so the power exchanged between upper and lower arms to eliminate the oscillatory terms is expressed as Equations (4.58)–(4.62).

The solution runs into discontinuity when $V_g+ = V_g-$ (singular voltage condition), resulting in infinite current being requested when positive and negative sequence voltages are very close in magnitude. This singularity causes the additive current references to saturate, which compromises the performance of energy balancing controllers.

This reference calculation problem could also appear in other control strategies different than the control scheme, provided that particular control method uses AC additive currents to balance the energy stored in the upper and lower arms.

Countermeasure for the problem: Three different methods that can be used to overcome the problem presented in the previous part follow as

Method 1—Temporarily Disabling the Upper to Lower Arms Energy Balance Control:* Unbalanced AC voltage sags do not cause sustained drift between the energy of the upper and lower arms of the converter. However, stepwise deviations may be caused by transients. The normal approach to controlling these deviations would require a very large amount of additive AC current under voltage imbalances close to the singular voltage condition. Therefore, one possible solution is to disable the energy controller that balances the differences between the energy of the upper and lower arms during the voltage sag. The main drawback of this method is that the energy deviation may build up during very long fault clearance times, and will not be reduced until the fault is cleared and the energy regulator is activated again.

Method 2—Kernel-Based Approach: The second approach is based on the idea of avoiding the use of current that does not contribute to power exchange. Such current can be found by studying the kernel of matrix **X** in Equation (4.55), which loses rank under singular voltage conditions.

Method 3—Anti-Windup Implementation for the Reference Current: The third method is able to maintain the additive current references bounded during singular voltage conditions. However, when the fault is cleared and normal operating conditions are restored, integral windup problems may appear in the energy regulators that control the difference between the energy of the upper and the lower arms. These problems are caused by the mismatch between the power set points of the energy controllers and the actual power exchange.

To overcome this problem, an anti-windup structure controlling the integral terms of the energy regulators must be implemented. This structure calculates the difference between the power reference given by the energy regulators and the actual achievable power exchange recalculated from the chosen current references. The error is then subtracted from the input of the integrators to keep them from increasing indefinitely.

Method 4—Singular Voltage Condition and Application of Alternative Reference Calculation Methods: The AC additive reference calculation uses the grid voltage V_g. The positive V_g+ and negative V_g- sequences of the overall grid voltage V_g can be obtained through the phase-locked loop (PLL). Whenever the difference in magnitude between the positive and negative sequence voltages is below a defined threshold, one of the three proposed methodologies can be applied in order to avoid a large AC additive current reference value. The aforementioned method assumes that AC grid voltage is similar to the voltage applied by the converter, considering that the impedances of the filters are small. Alternatively, the voltages applied by the converter could be estimated using the AC grid voltage, the filter impedance, and the AC grid current references.

4.4 Submodule Sorting Algorithms [12]–[17]

This chapter investigates a crucial engineering trade-off in designing capacitor voltage balancing schemes for an MMC: voltage regulation performance and switching loss. The MMC has numerous stacks of identical converters called submodules.

For the half bridge configuration, a submodule has two switching devices implemented by anti-parallel connections between a fully controllable insulated-gate bipolar transistor (IGBT) and a capacitor to generate the output voltage (see Figure 4.21(b)). As the level of the MMC or the number of submodules increases to enable higher power transmission, the capacitor voltage balancing task becomes the primary technical challenge because the balancing scheme may adversely impact system loss, device ratings, and control performance if not properly designed.

As conceptually illustrated in Figure 4.21(a), there is a trade-off between voltage regulation performance and loss. The higher the switching frequency, the lower the voltage ripple (ΔV), leading to better voltage regulation for a given capacitance despite the higher switching loss. Large capacitance should allow for small voltage ripple and reduced loss because of reduced switching needs among submodules for the purpose of voltage balancing.

Note an ideal point in Figure 4.21(a) where extremely large capacitance provides perfect voltage regulation and minimum loss comprising inherent conduction loss dominantly. This point is, however, prohibitively uneconomical for high-voltage applications, and the trade-off between voltage regulation and loss is balanced on the dotted line in practice.

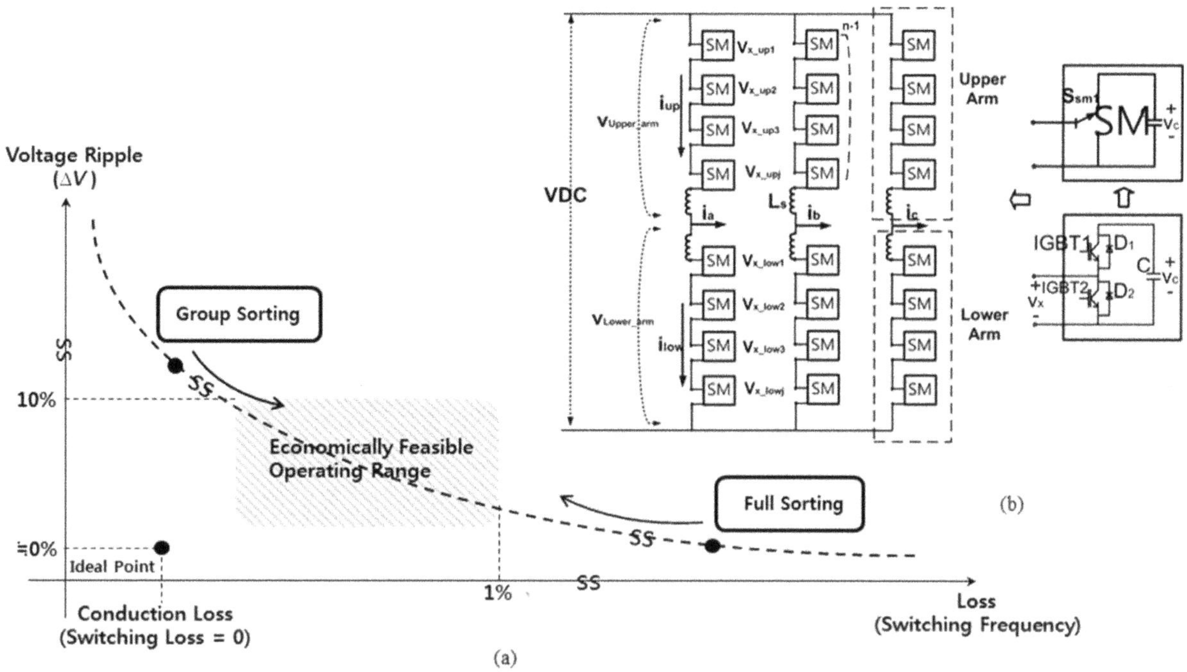

Figure 4.21 (a) Engineering trade-off in designing modular multi-level converter (MMC) system. (b) MMC topology

Ranking and switching all submodules by their capacitor voltages (often called "Full Sorting") at every switching instant should provide the highest voltage regulation for the designed number of submodules and its capacitance at the cost of highest switching loss. Thus, first a technically and economically feasible operation range and then a point should be determined based on grid conditions and desired MMC performance.

Here, let's discuss three well-reported methods for balancing submodule capacitor voltage-balanced control: Full Sorting, Advanced Full Sorting, and Group Sorting. It then presents an effective balancing based on Group Sorting with a supplementary ad hoc switching signal incorporating the design trade-off of voltage regulation and loss for each balancing method.

4.4.1 Full Sorting Method

Full Sorting may be the most basic but intuitive method for ranking all the measured voltages of submodule capacitors at every switching instant and sequencing an order for turning on or off their switches. The governing law states that on-state submodules charge the corresponding capacitors for positive arm current and discharge the capacitors for negative arm current. The off-state submodules are bypassed regardless of the current direction. If the arm current is in the positive direction, then the on-state submodules in an upper or lower arm with the lowest voltages are selected to be turned on. The corresponding capacitor voltages then increase because of the charging process. If the arm current is negative, then the on-state submodules with the highest voltages are selected to be turned off, and the corresponding capacitor voltages decrease. The balancing control using this Full Sorting scheme is shown in Figure 4.22. To reduce sensing cost, directional information can be derived from the voltage difference between two subsequently measured voltage samples.

Since it determines the on-state by checking the voltage of each submodule capacitor at each sampling time, on/off switching of the submodule occurs frequently. Therefore, voltage regulation of the capacitor performs best, and harmonics of the generated voltage waveforms could be minimal; a method focusing on

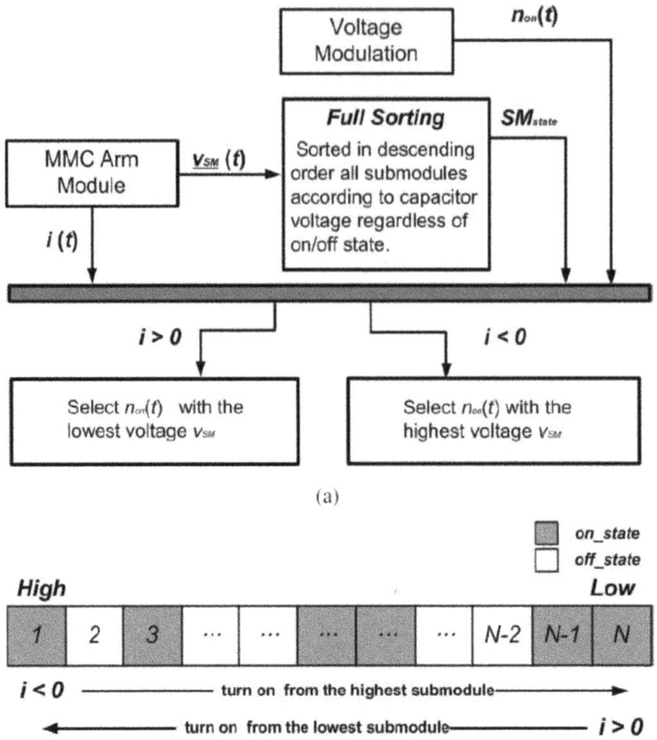

Figure 4.22 Full sorting: (a) schematic of full sorting and (b) arrangement and operation of each submodule

the improvement of voltage regulation performance is represented on the y-axis in Figure 4.21(a). However, undesired high loss because of frequent switching is a critical constraint for high-voltage applications.

4.4.2 Advanced Full Sorting

Advanced Full Sorting optimizes by adjusting the already measured submodule capacitor voltages before the next sorting process. At each sampling time, if the off-state (on-state) submodule capacitor voltage is smaller (greater) than the on-state (off-state) submodule capacitor voltage with the current direction, the submodule of the on/off state exchanges. Multiplying the actual capacitor voltage of the on-state submodule by a maintaining factor (>1) increases the likelihood of maintaining the present state, and thus avoids unnecessary switching as illustrated in Figure 4.4. The Advanced Full Sorting method may be understood as a way to move toward less loss on the x-axis of Figure 4.21(a) while not compromising voltage regulation too much; balancing a trade-off between voltage regulation and switching loss.

4.4.3 Group Sorting

To avoid successive switching of the same submodule, the previous gate signal of the submodule must be considered. The first step of the Group Sorting scheme is classification of groups by the previous state (on-state and off-state) of the submodules at the switching instant. Each group is sorted independently by its submodule capacitor voltages. Also, the $n_{on}(t)$, which is the number of submodules, must be turned on at the current time step (see Figure 4.23), and the arm current is then used to determine the individual gate signals at the current time step. As shown in Figure 4.24, when a difference (*Diff*) between the number of on-state submodules in an arm at the previous state and that of the current time step is positive, one of the off-group submodules

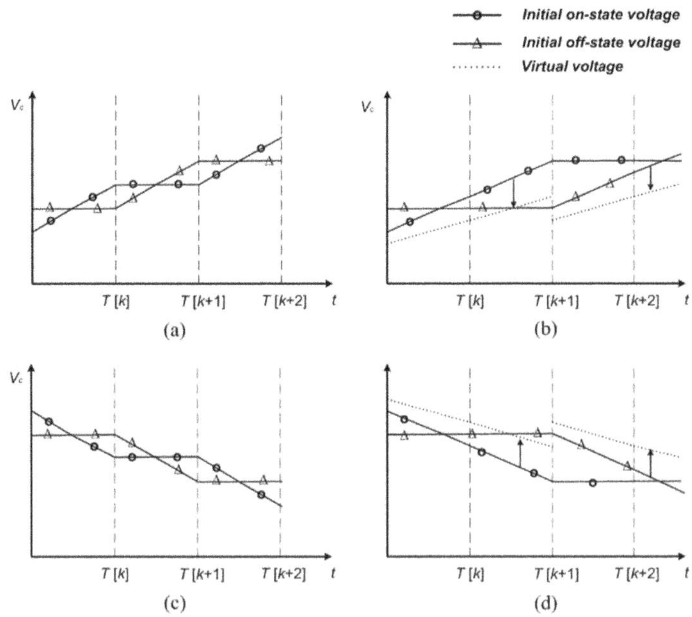

Figure 4.23 Diagram of the switching state changes: (a) positive arm current without virtual voltage; (b) positive arm current with virtual voltage; (c) negative arm current without virtual voltage; and (d) negative arm current with virtual voltage

is commanded to be turned on. The submodule with the highest (lowest) voltage value in the off-group is selected to charge (discharge) the submodule when the arm current is flowing negative (positive). When the sign of the *Diff* is negative, one of the on-state group submodules has to be turned off. Depending on the arm current direction, either the highest or lowest submodule is selected. To deal with numerous *Diffs* in a single time step, the previous process is repeated for each *Diff*. This scheme produces equally distributed capacitor voltage and switching commutation, which increases the mechanical reliability of power devices and reduces power losses for the installed device.

Group Sorting focuses on reducing switching loss. By switching the highest or lowest submodule depending on the current direction at each sampling time, successive on/off switching of the same submodule rarely occurs. Therefore, switching frequency of the submodule device, along with switching loss, are the lowest when compared to the previous algorithms.

4.4.4 Ad-hoc Switching Method for Voltage Balancing

The aforementioned Group Sorting method complements the disadvantages of the Full Sorting method and improves converter efficiency remarkably. By dividing into two groups from the gate signal at the previous time step ($t - \Delta t$), and operating only the submodules in one group depending on the *Diffs*, all submodules can have one or less than one time state change denoted as N during a half period of the output waveform. Since the output waveform is synthesized by switching in or out submodule capacitors, the output voltage step transition is mostly one submodule capacitor voltage in steady-state operation. The submodule that maintains each on or off state at the current time step does not participate in the switching until state variation of the total switching

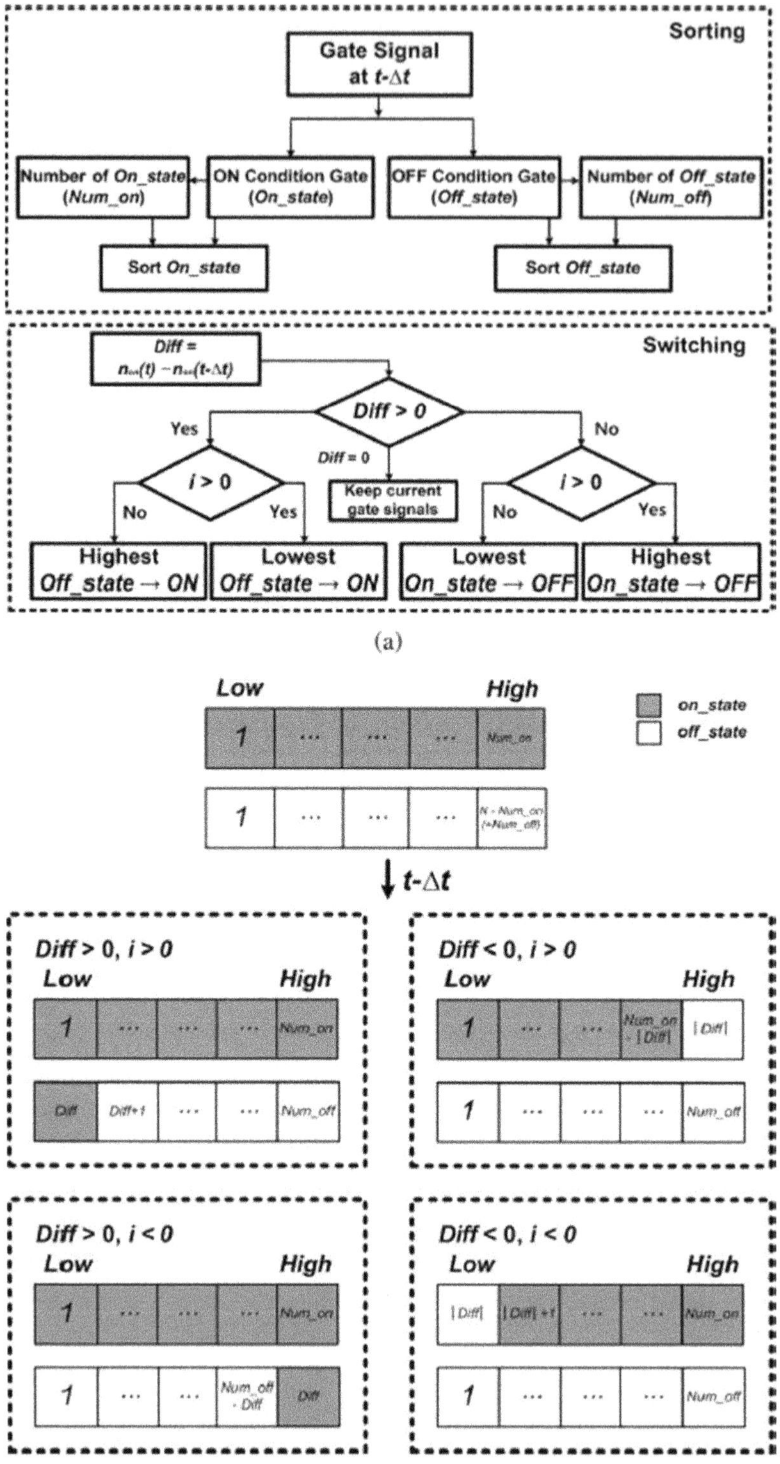

(b)

Figure 4.24 The group sorting: (a) schematic of group sorting and (b) arrangement and operation of each submodule

number occurs, which may adversely impact capacitor voltage balancing. The proposed method is designed to overcome these weaknesses.

Balancing is improved by augmenting the switching of the submodules determined from the Group Sorting. Additional switching is fired by actuating a signal whose frequency is determined according to the external variable as being distinct from firing because of N change. The frequency of the actuating signal in this case is calculated as (4.63).

$$f_{as} = 2k \cdot N \cdot f_s \tag{4.63}$$

where f_{as} is the actuating signal frequency, k is the number of exchange submodules for balancing, N is the number of submodules in an arm, and f_s is the system frequency.

The basic operating strategy is illustrated in Figure 4.6 and is described as follows:

1. Under positive *Diff* condition, the second lowest (second highest) module in the off-state is turned on if the current direction is positive (negative), since the highest (lowest) module is already turned on. Then, in order to synchronize the number of on-state submodules with N change, the highest (lowest) submodule in the on-state is turned off.
2. Under negative *Diff* condition, the second highest (second lowest) module in the on-state is turned on if the current direction is positive (negative), since the highest (lowest) module is already turned on. Then, in order to synchronize the number of off-state submodules with state variation, the lowest (highest) submodule in the off-state is turned.
3. In all cases, the number of on/off submodules for improving voltage balance can be flexibly set or adjusted in either direction by an external command as needed.

Compared with the Group Sorting method, the proposed balancing algorithm may remarkably reduce THD by reducing the voltage ripple. This method brings the desired computational efficiency for high-voltage applications, as well as flexibility in providing additional switching signals adaptive to the changing grid conditions.

4.4.5 Tolerance Band Modulation [24]

Average Submodule Capacitor Voltage Tolerance Band (ATB) Method: Each submodule capacitor voltage varies based on the capacitor conducting current, and the best-case scenario consists of evenly distributing the total voltage ripple among all submodules. Accordingly, each cell should contribute by managing the average cell capacitor voltage (v_{cap}^{av}), which is defined as

$$v_{cap}^{av}(t) = \frac{\sum_{n=1}^{N} v_{cap_n}(t)}{N} \tag{4.64}$$

Over time, $v_{cap}^{av}(t)$ indicates the optimum set point for the submodule capacitor voltage at each time step. Departure of any submodule capacitor voltages from this set point can be the basis for a balancing action. For the ATB method, constant monitoring of the error between the actual submodule capacitor voltages and the average submodule capacitor voltage ($v_{cap}^{av}(t)$) is required. Additionally, sorting action is a process to rank the submodules based on their capacitor voltages and create a sorted list L, such as

$$L = \begin{cases} \left[v_{cap}(\text{lowest}) \cdots v_{cap}(\text{highest})\right] & \text{ascending order,} \\ \left[v_{cap}(\text{highest}) \cdots v_{cap}(\text{lowest})\right] & \text{descending order.} \end{cases} \tag{4.65}$$

The following condition determines the sorting instants for ATB:

$$\forall \{t \in \mathbb{R} | 0 \le t < \infty\} \& \{n \in \mathbb{N} | 1 \le n < N\}, |v_{cap_n}(t) - v_{cap}^{av}(t)| \le \delta_r \tag{4.66}$$

where δ_r defines the acceptable tolerance band for submodule capacitor voltage deviation from its set point.

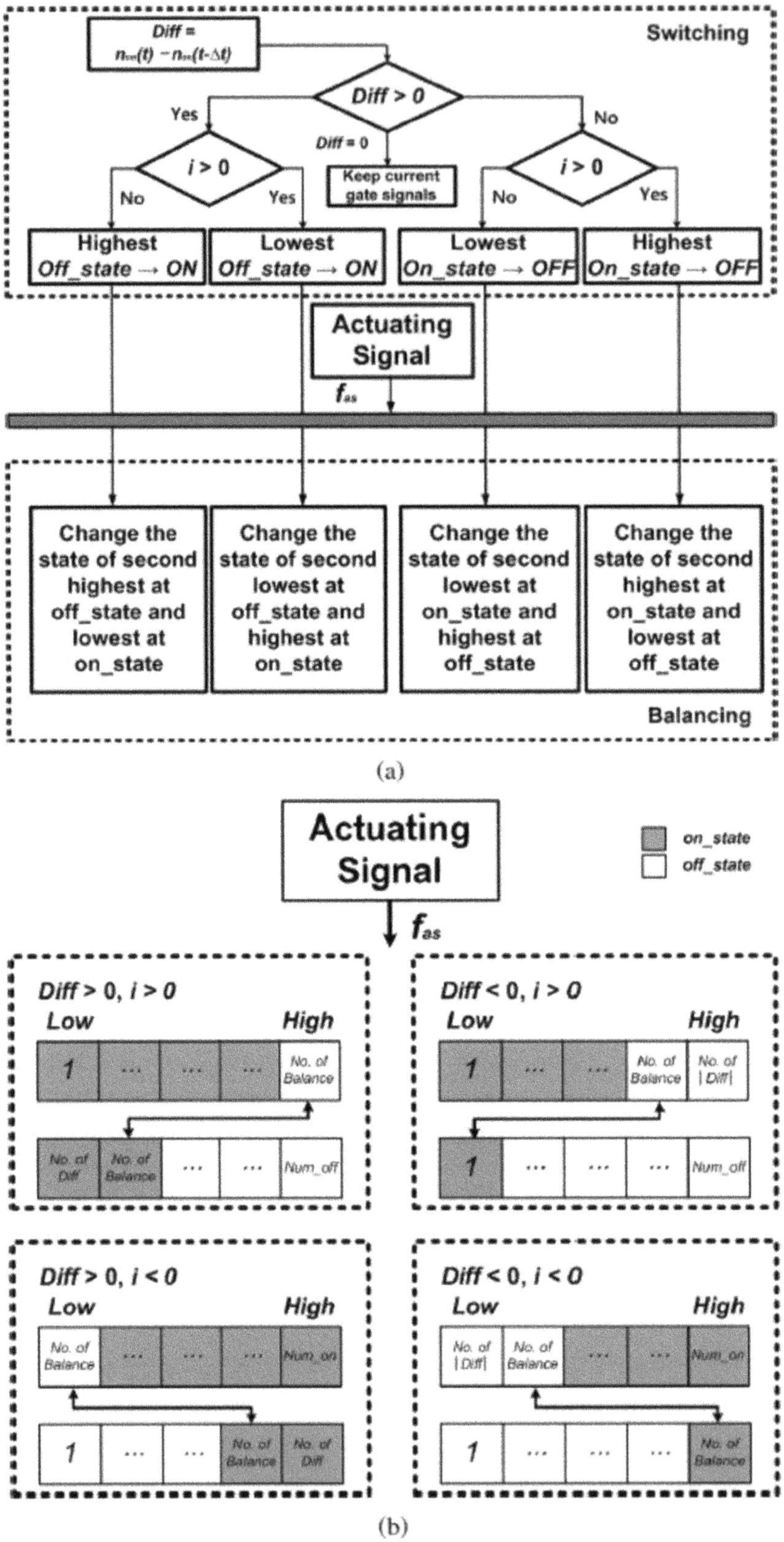

Figure 4.25 Modified balancing method (a) Schematic of additional switching method, (b) A Modified balancing method

The principle of the ATB method is illustrated in Figure 4.26a. The switching pulse pattern (S) can be generated by following as

Algorithm 1: δ_r determines how much deviation is acceptable for cell capacitor voltages from the average, which is defined by (4.64). Increasing δ_r results in larger cell voltage imbalances but lower switching frequency, while decreasing δ_r provides better voltage balancing but higher switching frequency. Generally, a larger deviation is acceptable between the submodule capacitor voltages during steady-state mode when compared to transient mode, so varying the tolerance band (δ_r) in different operation modes can achieve a proper compromise between voltage balancing and switching frequency.

Individual Submodule Capacitor Voltage Tolerance Band (CTB) Method: In practice, submodule capacitors are designed for rated transferred power and for handling certain fault cases. The capacitor voltages are allowed to deviate from the nominal submodule capacitor voltage (v_{nom}^{cap}) to certain maximum and minimum voltages. These maximum and minimum allowable capacitor voltages (V^{max} and V^{min}) are design factors related to IGBT blocking voltage, submodule electronics, submodule size, and so on.

To utilize the total available voltage range of each capacitor, V^{max} and V^{min} can be set as reference levels for conducting balancing actions. According to the CTB method, it is required to constantly monitor each individual submodule capacitor voltage and perform sorting actions, according to (4.65), when any capacitor voltages hit the boundary level of V^{max} or V^{min}. This concept is depicted in Figure 4.26b.

Algorithm 2 shows the steps toward pulse pattern generation when employing the CTB method. This method utilizes the maximum permitted deviation for each submodule, and avoids any additional switching action for balancing purposes since the capacitor voltage level is in the acceptable range.

Optimized Submodule Capacitor Voltage Tolerance Band (CTB-optimized) Method: IGBT switching losses are proportional to the arm current ($i_{p,n}$) at the switching instant. However, neither of the introduced methods consider the current level at switching.

The CTB method can be further optimized in respect to switching power losses. A heuristic method is introduced based on the introduced optimization problem, called CTB-optimized. It first inserts the submodules with high capacitor voltages during low currents, while keeping the submodules with low capacitor voltages for high current intervals.

Consequently, the number of switching events is minimized during high current intervals, which leads to less switching losses. This method utilizes zero-current crossings to do most of the switching events for balancing purposes. Similarly to the CTB method, voltage tolerance bands are applied for each individual submodule in CTB-optimized; however, if any submodule capacitor voltage hits the boundaries, that specific submodule will be replaced by another submodule.

Algorithm 3 gives the pseudocode for the CTB-optimized method. The features of this method can be summarized as follows:

1. It utilizes zero-current crossing to perform balancing switching events at lossless conditions.

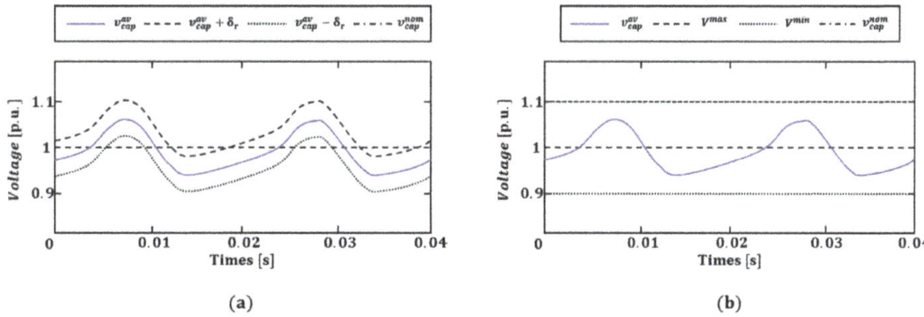

Figure 4.26 Tolerance band: (a) over average cell capacitor voltage (ATB) (b) over individual cell capacitor voltage (CTB)

2. It minimizes switching events at high current intervals by reserving the submodules with low capacitor voltages.
3. The submodule capacitor voltages are directly controlled by fixed tolerance bands.
4. Sorted lists need to be created only twice in each cycle.
5. Fundamental switching frequency is achievable for some operating points.

Tolerance band methods are based on the NLC modulation technique. NLC is a simple method for practical implementation since it does not require any carrier, but the resulting spectrum is non-characteristic and its harmonic content depends on the number of levels (N), the modulation index (m), and the sampling frequency (f_s) of the reference signal. To utilize all available levels, a minimum sampling frequency is required, which is related to the number of levels as

$$f_s^{min} = \pi N f_1 \tag{4.67}$$

where f_1 is the AC side voltage fundamental frequency. Consequently, control system requirements become more and more complex when increasing the number of submodules.

Dynamic Adjustment of Tolerance Band Methods: The MMC topology inherently generates ripple in the arm average capacitor voltages. This ripple is independent of the switching scheme, but related to the converter transferred power (the operating point) and the energy storage density (C_{arm}^{eq}). Therefore, the capacitors are not fully charged to the maximum allowable voltage (V^{max}) in case of low power demand.

Consequently, there is a risk of some submodules not switching for a few cycles, which is not desired. To avoid this risk, a dynamic boundary setting method is introduced.

Generally, HVDC converter stations are working in steady-state mode unless a new operating point is ordered or a fault case occurs. Thus, it is reasonable to design the modulation and submodule selection scheme with a focus on minimum power losses at steady-state.

However, such a focus may result in a switching scheme not suitable for fault cases. For instance, the ATB and CTB methods can handle all transient cases, but CTB-optimized is vulnerable in fault cases.

Dynamic Boundary Setting: In grid-connected MMCs, energy is exchanged between the AC side and the DC side through submodule capacitors, and causes capacitor voltage ripple. The amount of exchanged energy is related to the converter operating point, that is, the active and reactive power. The average submodule capacitor voltage maxima decreases while the exchanged power demand is reduced. So, in order to ensure proper capacitor voltage balancing, the CTB method boundaries should also be adjusted in accordance with the converter operating point. The following two formulas were expressed to dynamically scale the upper and lower boundaries for CTB and CTB-optimized methods in each fundamental cycle:

$$V_{cycle}^{max} = \frac{V_{dc}}{N}\left(1 + k_B \frac{V_{cap,av}^{max} - \frac{V_{dc}}{N}}{V_{av,rated}^{max} - \frac{V_{dc}}{N}}\right) \tag{4.68}$$

$$V_{cycle}^{min} = \frac{V_{dc}}{N}\left(1 - k_B \frac{\frac{V_{dc}}{N} - V_{cap,av}^{min}}{\frac{V_{dc}}{N} - V_{av,rated}^{min}}\right) \tag{4.69}$$

k_B is the percentage of the maximum allowable capacitor voltage based on nominal submodule capacitor voltage, and $V_{av,rated}^{max}$ and $V_{av,rated}^{mian}$ are maxima and minima of the average submodule capacitor voltages at rated power. A control block diagram for practical implementation of Equations (4.68) and (4.69) is given in Figure 4.27. To have a robust controller, submodule capacitor voltage feedbacks are used to calculate extrema of average submodule capacitor voltages at current zero crossing.

Switching Scheme Scheduling: The converter transient mode can be distinguished from the steady-state mode by monitoring certain system variables. This makes it possible to switch between different switching schemes in different converter modes. Thus, not only methods that can handle transient cases are feasible, but also any method with acceptable steady-state performance can be considered. The switching scheme scheduling controller is illustrated in Figure 4.28. The system can benefit from an efficient optimized method for steady-state operation, such as CTB-optimized, whereas it switches to a method more suitable for transient conditions, such as ATB, during fault cases.

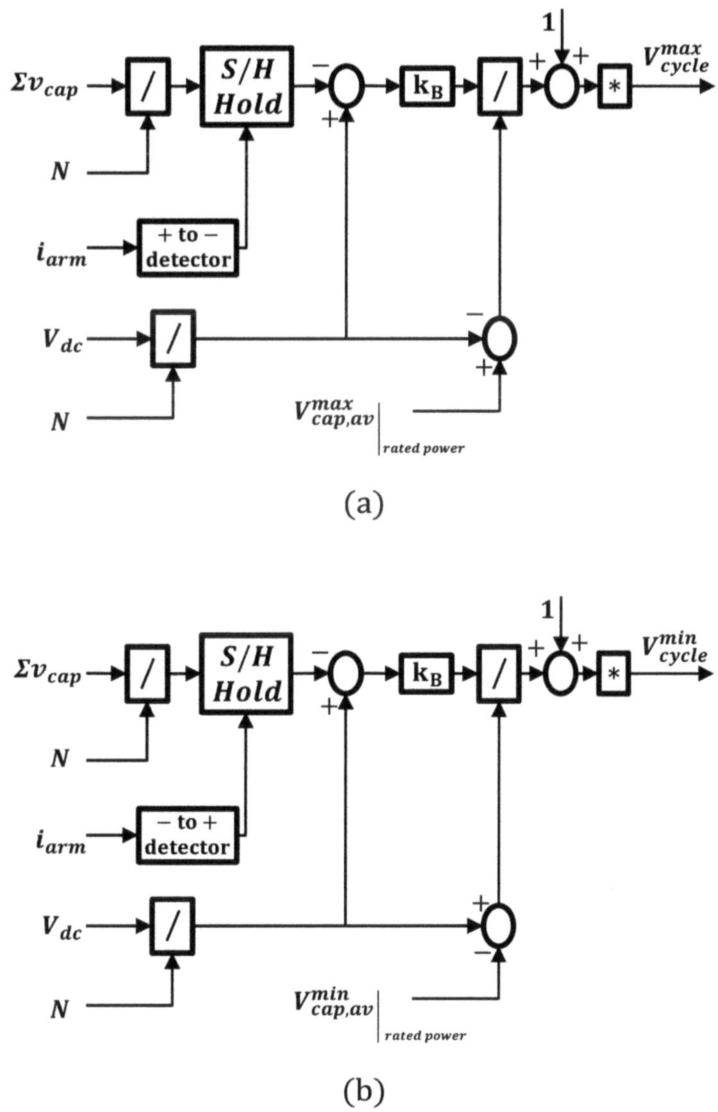

Figure 4.27 Control block diagram for dynamic boundary setting (Σv_{cap}: sum of capacitor voltages, N: number of cells, i_{arm}: arm current) (a) upper limit calculation, (b) lower limit calculation.

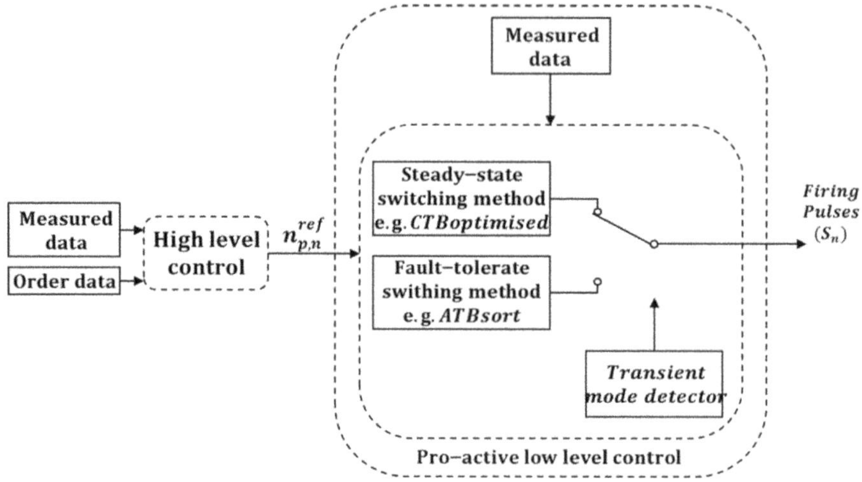

Figure 4.28 Low-level controller adopting switching scheme scheduling

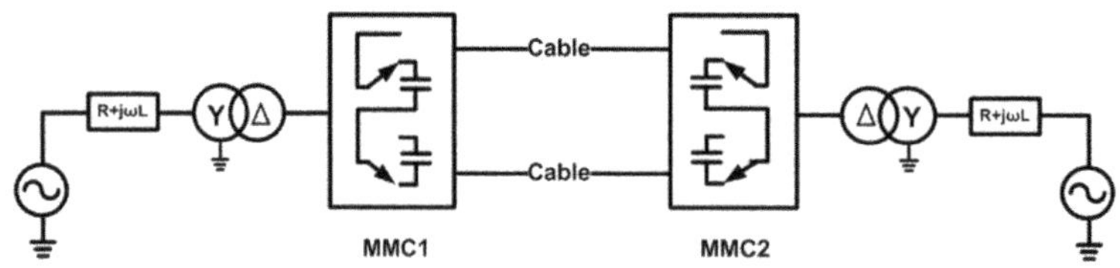

Figure 4.29 Schematic diagram of MMC-HVDC system

4.4.6 Simulation Results

The performance of the proposed capacitor voltage balancing method is evaluated for an MMC-HVDC using PSACD/EMTDC. As depicted in Figure 4.29, the HVDC transmission system used in the simulation benchmarked the Trans Bay Cable project designed as a mono-polar structure with DC voltage of ±200 kV, a submodule capacitance of 4.5 mF, and number of submodules per arm of 200. The system is designed to transmit 400 MW from MMC 1 to MMC 2. The main circuit parameters are listed in Table 4.2. The NLM and desired system control functionalities (*P*control and V_{dc}control, etc.) are commonly implemented in the simulation system. To evaluate the performance of the proposed capacitor voltage balancing scheme, the capacitor voltages of 20 submodules among the 200 submodules in the upper arm of phase A are selected and compared with those of three conventional voltage balance methods.

For the convenience of presenting the results, Full Sorting, Advanced Full Sorting, Group Sorting, and finally the ad hoc switching method are referred to as Methods I–IV in the following section:

Figures 4.30 and 4.31 illustrate the capacitor voltage waveforms and corresponding gate signals to the switching submodules. Figure 4.8 compares the submodule capacitor voltages of the upper arm in phase-A during six cycles of a period for four different balancing methods (Full Sorting, Advanced Full Sorting, Group Sorting, and proposed ad-hoc switching method are referred to as Methods I–IV in this section). Figure 4.30(a) shows that the capacitor voltages become perfectly balanced with Method I; the voltage performance should be the best out of the four different methods.

However, the switching loss is the highest when compared with the others because of a higher average switching frequency. Figure 4.30(b) presents the results of Method II, with a switching frequency of 169 Hz. Figure 4.30(b) shows that the capacitor voltages are well balanced, but the capacitor voltage ripple is larger than Method I because the average switching frequency is lower than Method I. Figure 4.30(c) is the result employing Method III. The switching frequency is calculated to be 50 Hz, which is the lowest among the four methods. However, the submodule capacitor voltages are not well balanced, and the peak-to-peak voltages are remarkable. In comparison with Figure 4.30(a), this figure shows that the capacitor voltages become unstable and the magnitude of the capacitor ripple is very large because of the low switching frequency. Figure 4.30(d) illustrates the results of Method IV, which are similar to Figure 4.30(b). This resulted in the same 169 Hz switching frequency as Method II, and the proposed ad hoc switching method successfully balances the capacitor voltages. Peak-to-peak capacitor voltage ripples are compared to Figure 4.30(c). Figure 4.31(a)–(d) compare the gate signals of one phase of the upper arm submodules under the four different modulations during three cycles to illustrate the switching operation or switching frequency mentioned earlier.

Table 4.3 summarizes the study results in terms of the switching frequency, total harmonic distortion (THD), switching, conduction, and total losses for the mentioned voltage balancing methods. The estimated switching and conduction losses are calculated as the percentage of transferred active power based on the datasheet of 5SNA1200G330100 IGBTs. For accurate comparison, we set the same switching frequency for the proposed ad hoc switching method (Method IV) and the Advanced Full Sorting (Method II). As presented in Table 4.3, the same switching frequency leads to the same loss: compare Methods II and IV. However, the total harmonic distortion (THD) of Method II is higher than that of the Method IV, since the use of virtual voltage does not allow for uniformly distributed switching. Method IV is computationally efficient.

Table 4.2 Main Circuit Parameters of the Simulation System

Items	Values	Comments
Active power P	400 MW	1.0 pu
AC system voltage at rectifier U_r	230 kV	L-L, rms
AC system voltage at inverter U_i	115 kV	L-L, rms
Short circuit ratio (SCR) at the PCC	3.5	
Transformer ratio at rectifier U_r/U_v	230 kV/180.5 kV	Y_0/Δ
Transformer ratio at inverter U_i/U_v	115 kV/180.5 kV	Y_0/Δ
DC bus voltage $U_{dc}/2$	200 kV	$U_{dc} = \pm 200$ kV
Number of submodules per arm N	200	
Submodule capacitance C_o	4.5 mF	
Arm inductance L_o	15 mH	
Submodule capacitor voltage U_c	2 kV	
Modulation index k	0.75	
Sampling frequency f_s	10 kHz	
DC cables	88 km	

Table 4.3 Characteristic of each balancing method

Balancing method	f_{sw} (HZ)	THD % of v_{ac}	P_{Loss}^{sw}%	P_{Loss}^{cond}%	P_{Loss}^{total}%
Method I	9140	0.25	8.11	0.45	8.56
Method II	169	1.03	0.15	0.45	0.60
Method III	50	1.59	0.04	0.45	0.49
Method IV	169	0.61	0.15	0.45	0.60

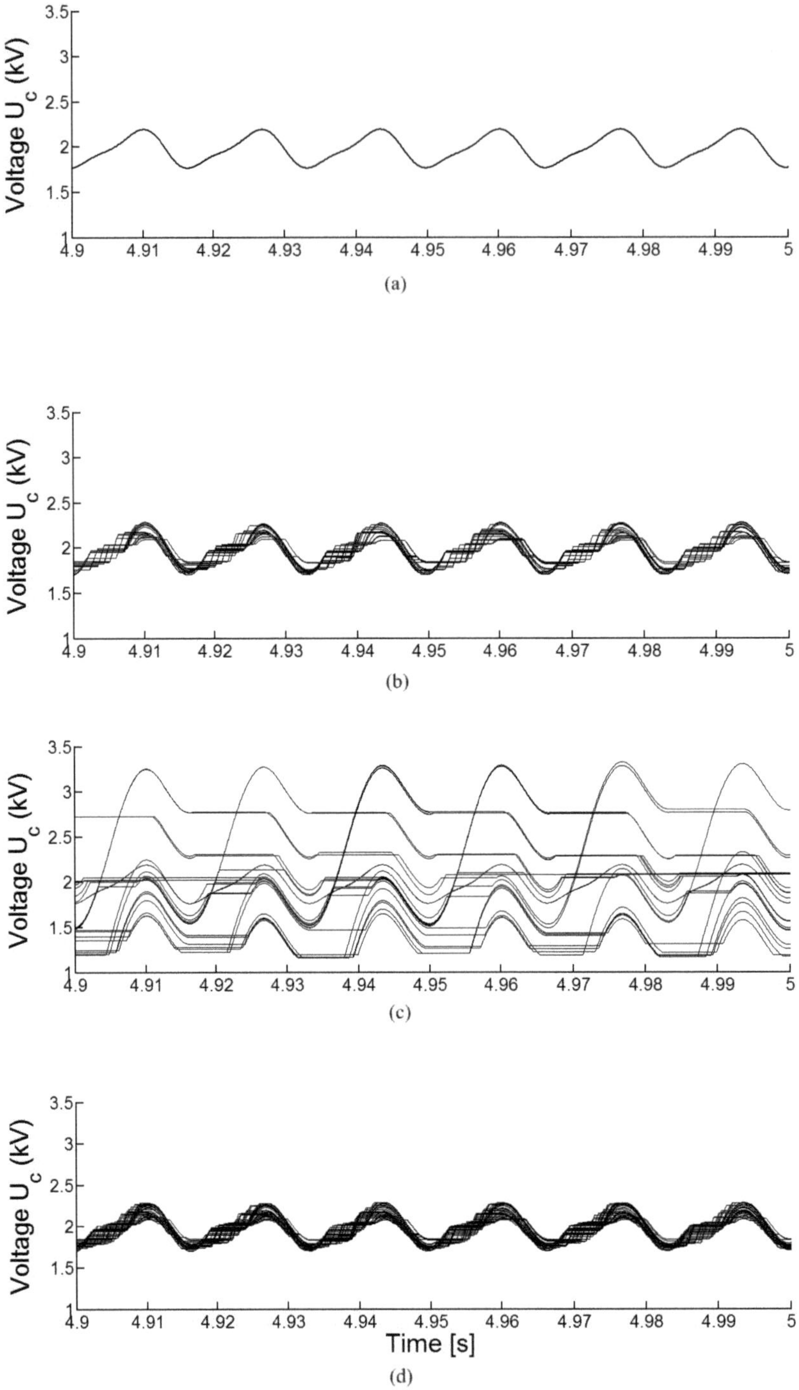

Figure 4.30 Comparison of submodule capacitor voltage with each balancing method: (a) Method I; (b) Method II; (c) Method III; and (d) Method IV

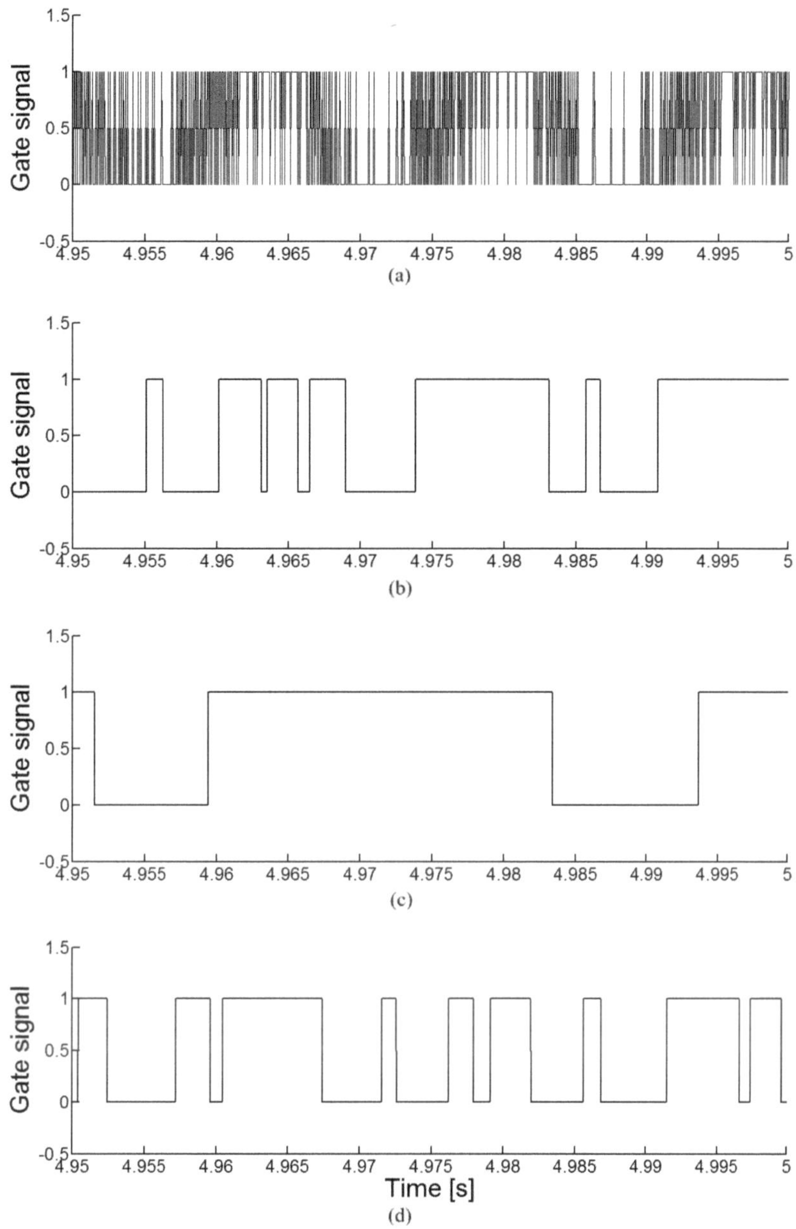

Figure 4.31 Gating signal of one switching device in submodule with each balancing method: (a) Method I; (b) Method II; (c) Method III; and (d) Method IV

References

1. MARQUARDT, Rainer. "Modular multilevel converter: An universal concept for HVDC-networks and extended DC-bus-applications." In: *The 2010 International Power Electronics Conference-ECCE ASIA-. IEEE*, 2010. p. 502–507.
2. Lennart Angquist, Antonios Antonopoulos, Daniel Siemaszko et al., "Open-Loop Control of Modular Multilevel Converters Using Estimation of Stored Energy," *IEEE Transactions on Industry Applications*, vol. 47, no. 6, pp. 2516–2524, Nov-Dec. 2011
3. Makoto Hagiwara, Hirofumi Akagi, "Control and Experiment of Pulse width-Modulated Modular Multilevel Converters," *IEEE Transactions on Power Electronics.*, vol. 24, no. 7, pp. 1737–1746, Jul. 2009

4. A. Antonopoulos, L. Angquist, and H.-P. Nee, "On dynamics and voltage control of the modular multilevel converter," in *Proc. European Power Electronics Conf.*, Barcelona, Spain, Sep. 2009 pp. 1–10
5. Shenghui Cui, Sungmin Kim, Jae-Jung Jung, Seung-Ki Sul, "A Comprehensive Cell Capacitor Energy Control Strategy of a Modular Multilevel Converter (MMC) without a Stiff DC Bus Voltage Source," *2014 IEEE Applied Power Electronics Conference and Exposition - APEC 2014*, Mar. 2014
6. Jae-Jung Jung, Hak-Jun Lee, and Seung-Ki Sul, "Control strategy for improved dynamic performance of variable-speed drives with the Modular Multilevel Converter," *in IEEE Energy Conversion Congress and Exposition*, pp. 1481–1488, 2013
7. Moon, J.W., Park, J.W., Kang, D.W., & Kim, J.M. (2014) "A Control Method of HVDC-modular multilevel converter based on arm current under the unbalanced voltage condition" *IEEE Transactions on Power Delivery*, 30(2),pp. 529–536.
8. Q. Tu, et al., "Suppressing DC voltage ripples of MMC-HVDC under unbalanced grid conditions," *IEEE Trans. Power Del.*, vol. 27, no. 3, pp. 1332–1338, Jul. 2012.
9. Q. Tu, et al., "Parameter design principle of the arm inductor in modular multilevel converter based HVDC," in *Proceedings of the International Conference on Power System Technology*, Hangzhou, China, 2010, pp. 1–6.
10. S. Li, T. Haskew, and L. Xu, "Control of HVDC light system using conventional and direct current vector control approaches," *IEEE Trans. Power Electron.*, vol. 25, no. 12, pp. 3106–3118, Dec. 2010.
11. U. Gnanarathna, A. Gole, and R. Jayasinghe, "Efficient modeling of modular multilevel HVDC converters (MMC) on electromagnetic transient simulation programs," *IEEE Trans. Power Del.*, vol. 26, no. 1, pp. 316–324, Jan. 2011.
12. T. Nam, et al., "Trade-off strategies in designing capacitor voltage balancing schemes for modular multilevel converters," *J. Electr. Eng. Technol.*, vol. 11, no. 4, pp. 829–838.
13. J. Qin and M. Saeedifard, "Reduced switching-frequency voltage-balancing strategies for modular multilevel HVDC converters," *IEEE Trans. Power Del.*, vol. 28, no. 4, pp. 2403–2410, Oct. 2013.
14. F. Deng and Z. Chen, "A control method for voltage balancing in modular multilevel converters," *IEEE Trans. Power Electron.*, vol. 29, no. 1, pp. 66–76, Jan. 2014.
15. Q. Tu, Z. Xu, and L. Xu, "Reduced switching-frequency modulation and circulating current suppression for modular multilevel converters," *IEEE Trans. Power Del.*, vol. 26, no. 3, pp. 2009–2017, Jul. 2011.
16. M. Guan and Z. Xu, "Control and modulation strategies for modular multilevel converter based HVDC system," in *Proceedings of the 37th Annual Conference IEEE Industrial Electronics Society*, pp. 849–854, Nov. 2011.
17. K. Wang, et al., "Voltage balancing and fluctuation suppression methods of floating capacitors in a new modular multilevel converter," *IEEE Trans. Ind. Electron.*, vol. 60, no. 5, pp. 1943–1954, May 2013.
18. G. Ding, et al., "New technologies of voltage source converter (VSC) for HVDC transmission system based on VSC," in *Proceeding of the IEEE Power Energy Society General Meeting*, 2008, pp. 1–8.
19. C.-K. Kim, et al., "Virtual group control algorithm for modular multilevel converter," in *2018 International Conference on Smart Grid (icSmartGrid)*, Nagasaki, 2018.
20. Q. Tu, and Z. Xu, "Impact of sampling frequency on harmonic distortion for modular multilevel converter," *IEEE Trans. Power Del.*, vol. 26, no. 1, pp. 298–306, Jan. 2011.
21. D. Zhong, et al., "Reduced switching-frequency active harmonic elimination for multilevel converters," *IEEE Trans. Ind. Electron.*, vol. 55, no. 4, pp. 1761–1770, Apr. 2008.
22. D. Zhong, L. M. Tolbert, and J. N. Chiasson, "Active harmonic elimination for multilevel converters," *IEEE Trans. Power Electron.*, vol. 21, no. 2, pp. 459–469, Mar. 2006.
23. J. N. Chiasson, et al., "A unified approach to solving the harmonic elimination equations in multilevel converters," *IEEE Trans. Power Electron.*, vol. 19, no. 2, pp. 478–490, Mar. 2004.
24. Lil, R., Fletcher J.E., & Williams, B.W. "Influence of third harmonic injection on MMC-based HVDC Transmission Systems, IET Generation, Transmission and Distribution, 10(11)2764–2770.
25. U. N. Gnanarathna, A. M. Gole, and R. P. Jayasinghe, "Efficient modeling of modular multilevel HVDC converters (MMC) on electromagnetic transient simulation programs," *IEEE Trans. Power Del.*, vol. 26, pp. 316–324, 2011.

5

Subcontrol Algorithms of MMC

5.1 Nearest Level Control Method and Cluster Stream Buffer

In the voltage-source converter (VSC) high-voltage direct current (HVDC) system, Modular Multi-Level Converter (MMC) and nearest level control (NLC) are used as basic standard topology because they are excellent in all respects. The MMC topology proposed by R. Marquardt has many advantages, such as high-power capacity possibility and low harmonics and so on.

In addition, as the MMC method is that insulated-gate bipolar transistor (IGBT) is connected in series, it requires an electrical bypass switch (thyristor) and mechanical bypass switch (permanent magnetic actuator [PMA] type or pyromatic type).

The NLC method has been used to reduce switching loss to within 1%. It is simple and suitable for the NLC method to implement high levels in the MMC HVDC system.

On the other hand, some companies use method-grouping submodules in hardware. This method controls group modules known as SMs, which are connected in series to compose a group module. As this method causes high-frequency harmonics and unbalance between the IGBT devices, it requires a snubber circuit to reduce the $\frac{di}{dt}$ and $\frac{dv}{dt}$.

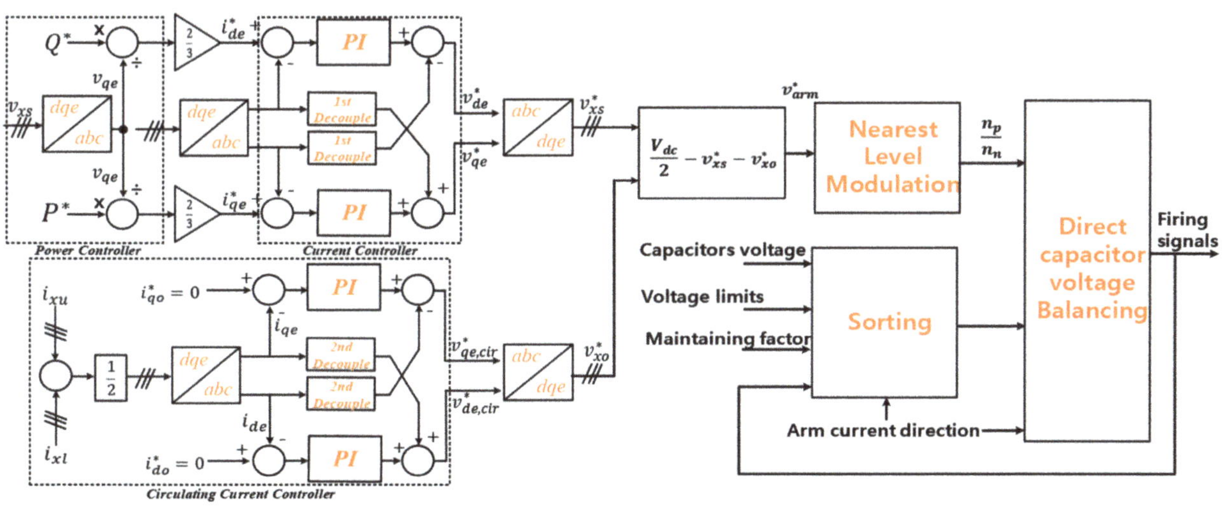

Figure 5.1 Modular multi-level converter control diagram with nearest level control ~ modulation method.

One of the main issues in the MMC HVDC system is increasing the capacity to 1–2 GW with the currently developed IGBT (Max. 3.3 kV, 1500 A), as the DC voltage is 500 kV and 500 IGBTs of SM are connected in series. It requires high performance of the controller hardware and high speed of sampling time and high-speed communication to turn-on/off IGBTs in real time. Another additional method [1] is the cluster stream buffer (CSB) method, which consists of 32 levels as a cluster that can overlap the clusters in series. The CSB method can be increased over 1 GW without changing the sampling time.

5.1.1 NLC Method

As shown in Figure 3.14 in Chapter 3, low-switching frequency modulation algorithms used in multi-level converter systems are as follows: selective harmonic elimination (SHE); nearest vector control (NVC); and nearest level control (NLC) in order to reduce the switching loss.

Among these, the SHE method requires a lot of calculated switching angles, and needs to be computed offline and stored in lookup tables. Therefore, the SHE method is too complex and needs too many calculations.

On the other hand, the NLC modulation is used widely in HVDC systems because it is easier to implement and simple. If the HVDC system has a sufficiently large number of SMs, a relatively low harmonic amplitude and low total harmonic distortion (THD) can be achieved. Figure 3.23 shows the control diagram of the NLC modulation method.

It is important to mention that in digital-signal-processing (DSP) systems, the uniform sampling frequency f_s of the reference voltage v_{ref} must be considered. When the uniform sampling time is applied to control systems with large numbers of SMs, a high enough sampling frequency is needed to ensure that the resulting voltage level utilizes all SMs (Figure 5.2(a); every SM generates a one-step voltage level, so each voltage step is v_c).

Using a smaller sampling frequency reduces the level of the output voltage that can be generated (see Figure 5.2(b); the first and third voltage steps are $2v_c$). The voltage reference shown in Figure 5.2(a) and (b), v_{ref}, is a staircase waveform with the sampling interval $Ts = \frac{1}{fs}$ instead of a sinusoidal waveform (the blue line in Figure 5.2(a) and (b)). The closest voltage level is determined by dividing the sampled reference voltage v_{ref} by the value of the capacitor voltage (here, v_c is considered to be a constant).

Finally, the closest voltage level is used to determine the number of SMs to be turned on. The round (x) function is used to determine the number of turn-on SMs. This function returns the nearest integer to the input number (e.g., round (3.4) = 3, round (3.6) = 4). In each arm, the total number of SMs to be switched on n_{on} is calculated.

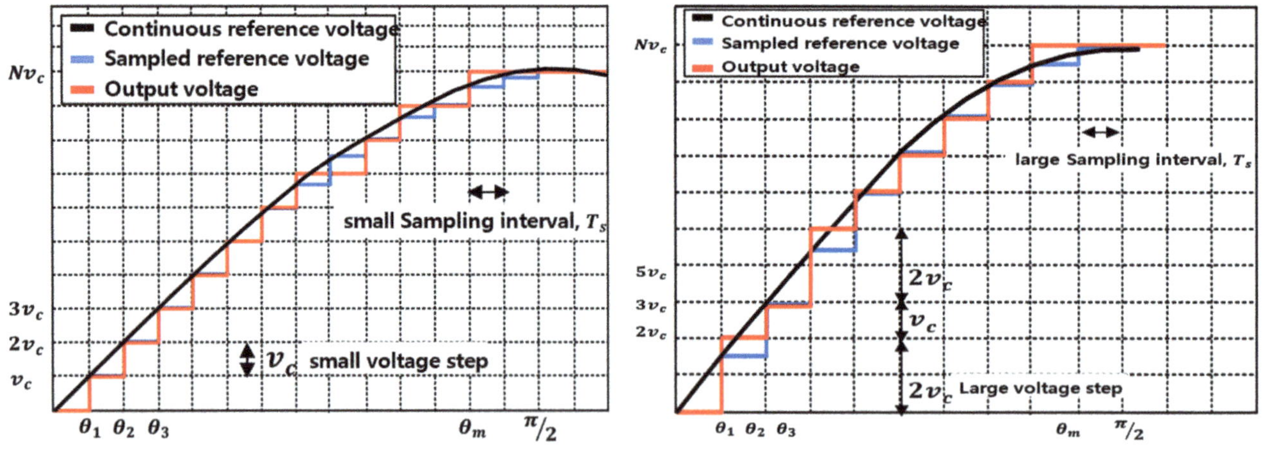

(a) Small sampling interval (b) Large sampling interval

Figure 5.2 Modular multi-level converter output characteristics according to sampling interval variation

5.1.2 Cluster Stream Buffer (CSB) Method [1]

The main issue in HVDC systems adopting MMC and NLC methods is keeping the sampling time MMC output constant as the number of SMs increases.

When the interval of sampling time is large as shown in Figure 5.2(b), it shows the possibility that the switching of the SM can implement two to three SMs or five SMs at the same time.

If multiple SMs switch simultaneously, it is necessary to add a snubber circuit to the SMs for each SM's voltage and current distribution equally. This increases the loss of the HVDC system. Therefore, although keeping the sampling time constant, the requirement of a system with a uniform switching number of SMs (no snubber circuit required) is necessary for a large capacity HVDC system.

Figure 5.3 shows the actual implementation of the CSB method, which shows that the number of switching levels is commanded from the phase controller to CSB's controller. Figure 5.4(a) and (b) shows the algorithm of the CSB method, which is a combination of clusters each having 32 SMs, and the number of clusters increases as the SM level (or voltage level) increases.

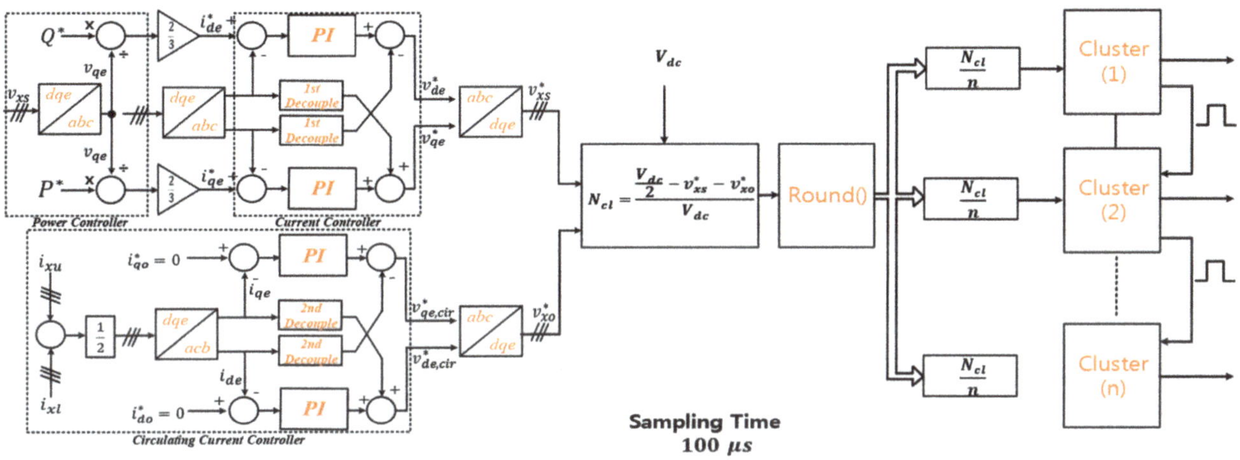

Figure 5.3 Actual implementation of cluster stream buffer method

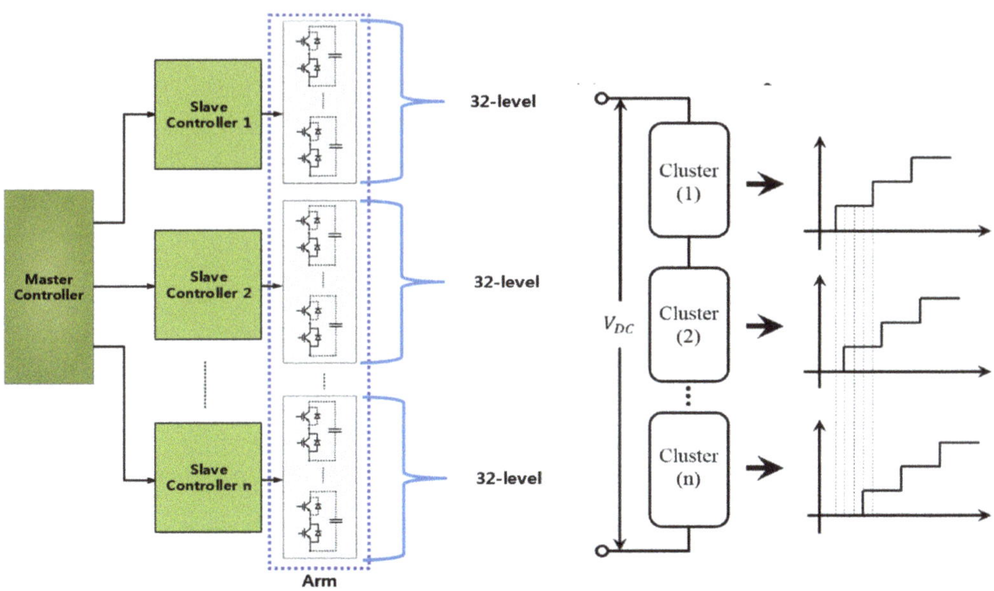

(a) Virtual group of submodules (b) Operation characteristic

Figure 5.4 Cluster stream buffer method

The advantage of this method is that the expansion of the scalability is unrestricted (60 MW→600 MW) while keeping the fixed sampling time of the system and without degradation of the system performance and complexity. Unlike the module-grouping method shown in the approach in Figure 5.3, which switches the IGBT devices simultaneously, the CSB algorithm guarantees switching the IGBT devices one by one.

The CSB algorithm would be described as follows: the total number of SMs that need to be turn-on in the MMC system N_{cl} is divided by the total number of clusters n and the result is transmitted to each cluster. The total number of SMs divided into several clusters based on the system frequency as shown in Figure 5.4 and the time delay of the SMs are switched in the same order as Cluster 1, Cluster 2, . . . , Cluster n.

The output of the MMC-based VSC HVDC system using the proposed algorithm shown in Figure 5.5(b) has a much higher level than the one without the algorithm shown in Figure 5.5(a).

The cluster shown in Figure 5.4(b) can be made of 32 levels or 16 levels. The number of IGBTs in a cluster can be chosen to be 16, 32, or 64; however, selecting 16 IGBTs in a cluster means many cluster controllers are required while expanding the MMC system, and 64 IGBTs in a cluster cause a reduction in the reliability of the system. Therefore, each cluster is designed to have 32 IGBTs.

In order to verify the performance of the algorithm, the simulation is confirmed through MATLAB/Simulink. The CSB modeled in MATLAB consists of 11 clusters and 1 master/phase controller. Each cluster consists of 32 SMs. The target system in this paper is a system with ±300 kV, a Symmetry Monopole HVDC System, and 325 SMs per arm (without redundancy), while the total number of SMs including both rectifier and inverter is 3900. The sampling time of the phase controller of the HVDC system is 100 μs, where the VBE (Valve Based Electronics) part and sorting part are skipped.

Figure 5.6 shows the performance characteristics of the proposed system. Figure 5.6(a) shows that the SM is not switched to the group but increases by one. Figure 5.6(b) shows the operation waveform of the 32-level cluster.

Figure 5.6(c) and (d) shows the operating waveform when the third harmonic is added to the HVDC-voltage command value. In this case, it does not operate up to the 32 levels, only up to 27. The reason for injecting the third harmonic in the converter system is to reduce the loss by reducing the magnitude of the current, and by increasing the HVDC voltage to 32 levels by injecting the third harmonic (i.e., 15% of the fundamental current), the current is reduced to 15% of its original magnitude. The loss is also reduced by about 15% in Figure 5.6(d).

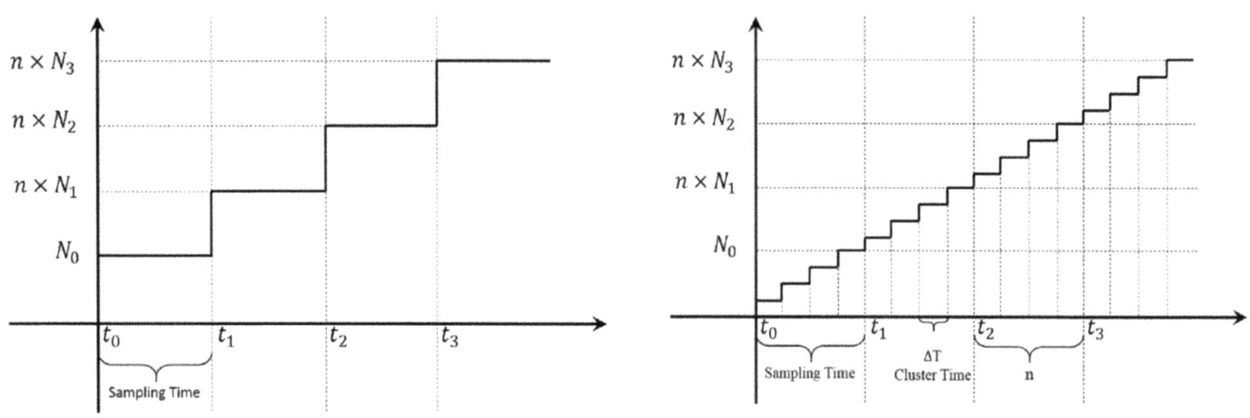

(a) Without cluster stream buffer method (b) With cluster stream buffer method

Figure 5.5 Waveform characteristics of modular multi-level converter system

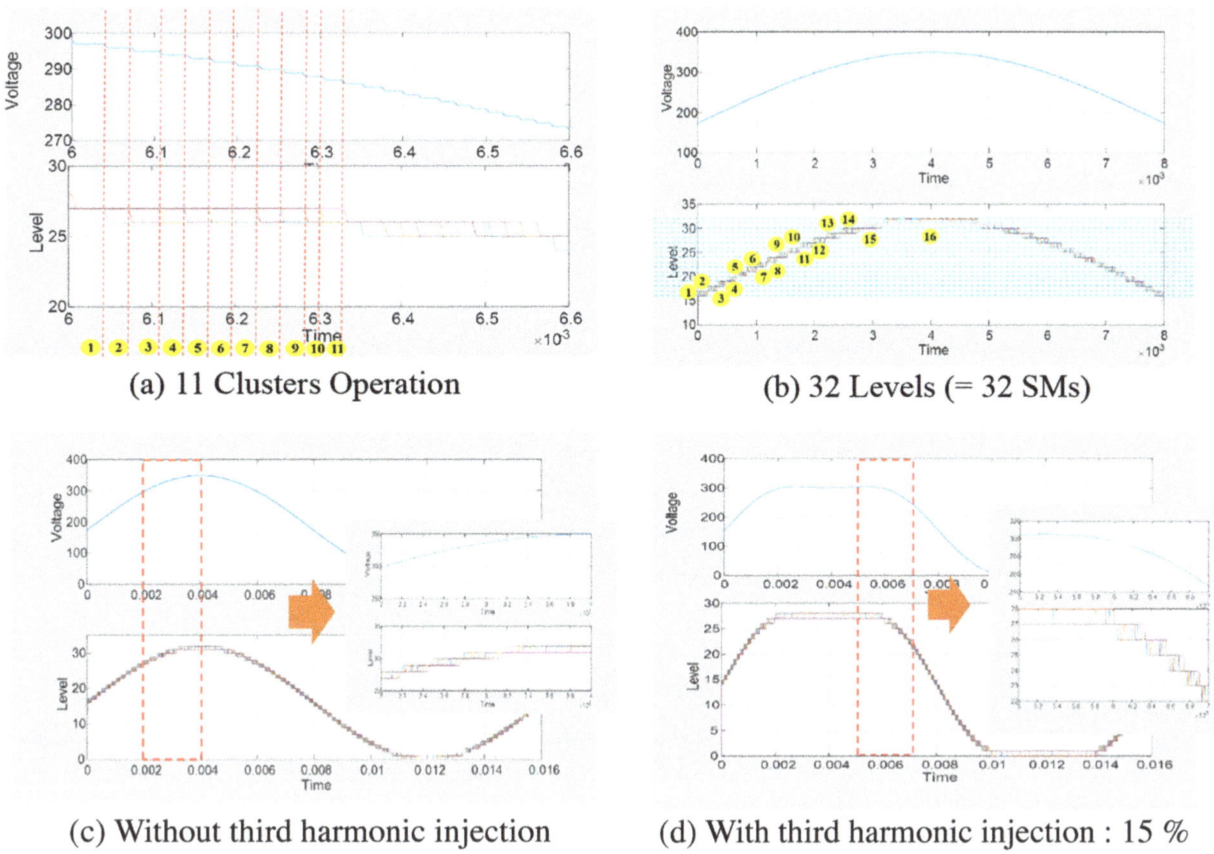

Figure 5.6 Simulation result of the proposed algorithm: cluster stream buffer method

5.2 Impact of Sampling Frequency on Harmonic Distortion [3]

The generation of switching pulses is based on fast monitoring the voltages of each SM capacitor and the direction of arm current. The principles of the balancing algorithm of SMs are as follows:

- When the arm current charges the capacitors, the submodules with the lowest storage capacitor voltages will be switched on to form the desired arm voltage.
- When the arm current discharges the capacitors, the submodules with the highest storage capacitor voltages will be switched on to form the desired arm voltage.

To choose the SM with the highest or lowest voltage, this balancing algorithm needs to sort the SM voltages in one arm. To avoid unnecessary switching operations, a predefined SM voltage-deviation band ΔV_{max_ref} is used. If the actual maximum SM voltage difference ΔV_{max} exceeds this predefined value

$$\Delta V_{max} = V_{c\_max} - V_{c\_min} > \Delta V_{max\_ref} \tag{5.1}$$

where V_{c_max} and V_{c_mim} denote the maximum and minimum SM voltages at a certain instant in one arm, respectively, then the aforementioned balancing algorithm will be performed.

Otherwise, if Equation (5.1) is not satisfied, the switching principles are as follows:

- If extra SMs need to be switched on during the following control cycle (i.e., Δn_{on} is positive), no switching is applied to those SMs currently in on-state. The conventional balancing algorithm mentioned before will only be applied to those SMs currently in off-state.

- If some SMs that are currently in on-state need to be switched off during the following control cycle (i.e., Δn_{on} is negative), no additional SMs that are currently in off-state will be switched on. The conventional balancing algorithm mentioned before will only be applied to the SMs currently in on-state.

Based on the voltage-balancing algorithm and the switching pulse generation principles as noted before, the number of switching operations and the total losses of the converter can be reduced significantly.

5.2.1 Harmonic Analysis of MMC

To calculate the harmonics of the staircase output voltage, assuming that the voltage of each SM capacitor Vc is kept constant as shown in Equation (5.2). Then, the Fourier series expansion is used. This theoretical method to analyze the harmonics of the staircase waveform has been reported, but the result is based on an infinite sampling frequency for the reference voltage. In other words, each SM corresponds to one voltage level.

$$v_{oVtpVt} = \frac{4V_C}{\pi} \sum_{h=1,3,5}^{\infty} \left[\frac{1}{h} \cdot \sum_{i=1}^{m} \cos(h\theta_1) \cdot \sin(h\omega_0 t) \right] \tag{5.2}$$

where $\theta_i (1 \leq i \leq m)$ is the ith switching angle shown in Figure 5.2, m is the total number of voltage levels in a quarter period of time, and h is the harmonic order.

But, as shown in Figures 5.2, if the voltage levels are unequal, Equation (5.2) is not suitable. In this case, the sampling frequency f_s must be considered because the actual sampling frequency in the digital control system cannot be infinite. As a result, a single output voltage level may contain more than one SM capacitor voltage V_c especially when n_{SM} is large and f_s is not high enough.

To analyze the waveforms in Figure 5.2, the following assumptions should be made:

- The waveform are odd and have half-wave symmetry, so there are not any even-order harmonics
- The amplitude modulation index is $k \leq 1$

Neglecting the DC component $V_{dc}/2$, the sampled reference voltage can be expressed as

$$\begin{gathered} u_{ref}(t) = \frac{k}{2} V_{dc} \sin(\omega_0 t) = V_{ref} \sin(\omega_0 n T_s) \\ \left(0 \leq n \leq \frac{\frac{\pi}{2}}{\omega_0 T_s} = \frac{f_s}{4 f_0}, n = 0, 1, 2, 3, \ldots \right) \end{gathered} \tag{5.3}$$

where V_{ref} is the peak value of $V_{ref,}$ n is a discrete variable, ω_0 is the radian fundamental frequency, and $T_s = 1/f_s$ is the sampling interval as shown in Figure 5.2.

At a given sampling interval $T_{s,}$ in order to find all of the switching angles θ_i in Figure 5.7, define

$$\Delta L = \text{round}\{V_{ref} \sin(\omega_0 n T_s)\} - \text{round}\{V_{ref} \sin[\omega_0 (n-1) T_s]\} \tag{5.4}$$

which means the voltage step V_{step} contains a number of ΔL submodule voltages V_c ($V_{step} = \Delta L \cdot V_c$) at the sampling interval. Then, for different values of n, the corresponding switching angle θ_i can be acquired

$$\begin{cases} \theta_i \text{ does not exist (if } \Delta L = 0) \\ \theta_i = \omega_0 n T_s \text{ (if } \Delta L \geq 1). \end{cases} \tag{5.5}$$

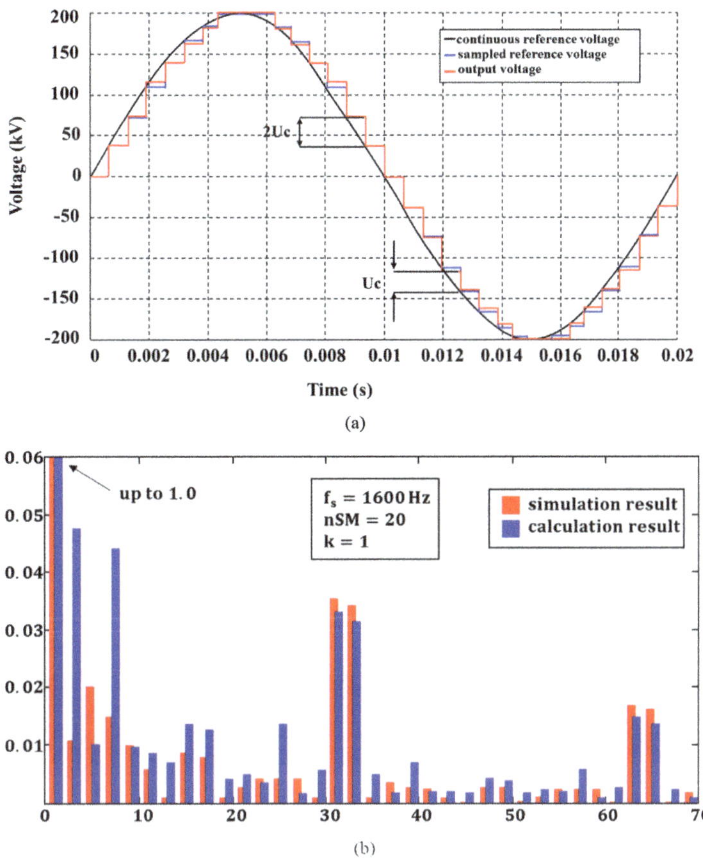

Figure 5.7 Output voltage with the nearest level control method and voltage spectrum: (a) voltage waveform with uniform sampling and (b) simulated and calculated spectrum

So, the Fourier series expansion of the output voltage is given by

$$v_{output} = \frac{4V_c}{\pi} \sum_{h=1,\,3,\,5}^{\infty} \frac{1}{h} \cdot \left[\sum_{i=1}^{m} (L_i - L_{i-1}) \cos(h\theta_i) \right] \cdot \sin(h\omega_0 t)\,(L_0 = 0) \tag{5.6}$$

where θ_i $(1 \le i \le m)$ is the ith switching angle in a quarter fundamental period, which is calculated by Equation (5.5); L_i $(1 \le i \le m)$ is the corresponding number of on-state SMs at θ_i, so $L_i V_c$ is the corresponding voltage level shown in Figure 5.7; and h is the harmonic order.

Figure 5.8 is the staircase output voltage and the corresponding spectrum simulated and calculated. The simulation result is based on numerical fast Fourier transform (FFT) analysis, whereas the calculation result is based on Equation (5.6).

To evaluate voltage quality two concepts must be distinguished:

n_{SM}: the number of submodules per arm; it is determined by the DC-bus voltage V_{dc} and the steady-state capacitor voltage V_c as $n_{\text{SM}} = V_{dc}/V_c$.

n_{level}: the number of output voltage levels, which directly affects the waveform quality.

If n_{SM} is small (5 or 7, etc.), the n_{level} can be directly calculated as

$$n_{\text{level}} = n_{\text{SM}} + 1 \tag{5.7}$$

But when MMC is extended to a very high–voltage range, the number of cascaded SMs is considerably large. So, the effect of sampling frequency should be considered, and Equation (5.7) is no longer correct.

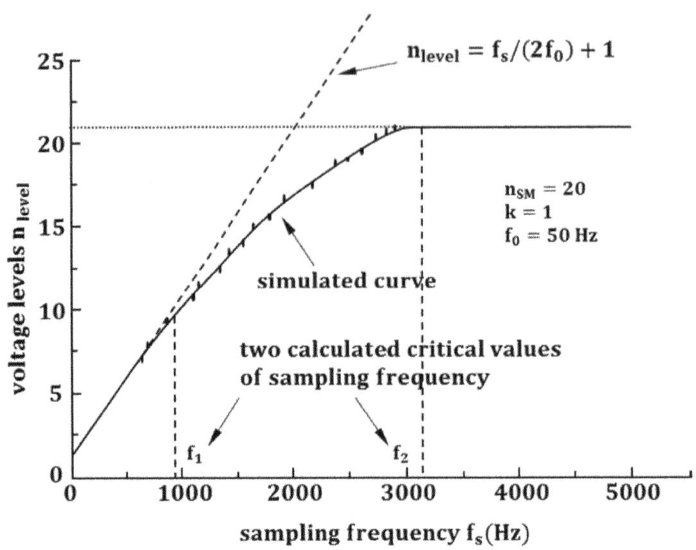

Figure 5.8 Voltage-level number as a function of the uniform sampling frequency

In this case, n_{level} is determined by three major factors:

- the sampling frequency f_s (as shown in Figure 5.7)
- the number of submodules n_{SM}
- the amplitude modulation index k

5.2.2 Impact of Sampling Frequency f_s

Figure 5.8 shows the relationship between the number of output voltage levels n_{level} and the sampling frequency f_s given $n_{SM} = 20$ and $k = 1$. It is concluded that the voltage levels n_{level} and the THDs are determined by the sampling frequency f_s if n_{SM} and k are supposed to be constant. From Figure 5.8, there are two critical values of the sampling frequency:

- f_1: the lower limit of f_s, when $f_s < f_1$, n_{level} decreases significantly and be linear with f_s
- f_2: the upper limit of f_s, when $f_s > f_2$, n_{level} keeps constant and every SM composes a voltage level.

Consequently, the relationship between n_{level} and f_s is

$$n_{level} = \begin{cases} \frac{f_s}{2 \cdot f_0} + 1 \ (f_s < f_1) \\ n_{SM} + 1 \ (f_s < f_2) \end{cases} \tag{5.8}$$

where f_0 is the fundamental frequency. It is important that when $f_s < f_1$, n_{level} just completely depends on f_s and is not influenced by n_{SM}.

These two critical values f_1 and f_2 serve as useful indicators when selecting the sampling frequency of the controller f_s. It is shown that the amplitudes of harmonics will increase rapidly with the decrease of f_s, in the condition of $f_s < f_1$.

Certainly, it is unnecessary to choose $f_s > f_2$ from the viewpoint of reducing voltage harmonics, because in this frequency range, all of the SMs are fully utilized and raising sampling frequency will not lead to a remarkable reduction of the voltage THD.

To calculate the two critical values of the sampling frequency f_1, f_2 from Equation (5.8), a sinusoidal voltage output waveform described in Equation (5.3) is considered. In one sampling interval T_s, the change of the reference voltage is

$$\begin{aligned} du_{ref} &= \frac{k}{2} V_{dc} \cdot \omega_0 \cdot \cos(\omega_0 t) \cdot dt \\ &= \frac{k}{2} V_{dc} \cdot \omega_0 \cdot \cos(\omega_0 t) \cdot T_s \\ &= k V_{dc} \cdot \pi \frac{f_0}{f_s} \cdot \cos(\omega_0 t) \end{aligned} \tag{5.9}$$

The lower critical value f_1 means that the minimum voltage change during one sampling interval T_s is just equal to a single capacitor voltage V_c. This occurs at the flattest point of a sine wave ($\omega_0 t = \pi/2$; here, we use $\omega_0 t = \pi/2 - \omega_0 T_s$ as an approximation for the convenience of calculation)

$$du_{ref}|_{min} = k V_{dc} \cdot \pi \frac{f_0}{f_s} \cdot \cos(\omega_0 t)\,|_{\omega_0 t = \pi/2 - \omega_0 T_s} = V_c \tag{5.10}$$

$$\Rightarrow k \cdot n_{SM} \cdot \pi \cdot \frac{f_0}{f_s} \sin \frac{2\pi \cdot f_0}{f_s} = 1. \tag{5.11}$$

If we suppose $f_s \gg 2\pi \cdot f_0$ the earlier equation can be simplified as

$$2k \cdot n_{SM} \cdot \frac{\pi^2 \cdot f_0^2}{f_s^2} = 1 \Rightarrow f_1 = \pi \cdot f_0 \cdot \sqrt{2k \cdot n_{SM}}. \tag{5.12}$$

Similarly, if we set $\omega_0 t = 0$, the upper critical value of the sampling frequency f_2, which is reached when every SM is utilized to form a voltage level, can be calculated as

$$du_{ref}|_{max} = k U_{dc} \cdot \pi \frac{f_0}{f_s} \cdot \cos(\omega_0 t)\,|_{\omega_0 t = 0} = U_c \tag{5.13}$$

$$\Rightarrow f_2 = \pi \cdot f_0 \cdot k \cdot n_{SM}. \tag{5.14}$$

5.3 Influence of Third Harmonic Injection [7]

The third harmonic injection that is used in MMC control has a significant advantages over sinusoidal modulation, from the viewpoints of the system power losses, submodule capacitance, circulating current (CC), and fault current.

By injecting the triplen harmonics, the DC-voltage utilization ratio is increased from $\frac{1}{2}$ to $\frac{1}{\sqrt{3}}$. Compared with sinusoidal modulation, the maximum output voltage peak with reference to the midpoint of the DC link is reduced from $\frac{1}{2}V_{DC}$ to $\frac{1}{4}\sqrt{3}V_{DC}$ as demonstrated in Figure 5.9 and 5.10, whereas the amplitude of the output fundamental voltage remains $\frac{1}{2}V_{DC}$, where V_{DC} is the DC voltage. For simplicity, it is assumed that $\frac{1}{2}\sqrt{3}N$ is an integer, where N is the SM number per arm. Only $\frac{1}{2}\sqrt{3}N$ of the SMs are utilized when third harmonic injection is used rather than sinusoidal modulation. In real application, the potentially used SMs are ceil $\left(\frac{1}{2}\sqrt{3}N\right)$, which is close to $\frac{1}{2}\sqrt{3}N$ given that HVDC systems typically use hundreds of SMs per arm.

In HVDC systems, by increasing the transformer ratio between the converter side and the grid side by 15.5%, higher voltage is induced on the converter side of the transformer when the voltage of the grid connected to the station remains unchanged. Therefore, the higher voltage output capability that is required for the MMC and the redundant SMs, introduced by the higher DC-voltage utilization capability with third harmonic injection, can be effectively utilized.

5.3.1 Characteristics of Third Harmonic Injection with Higher Transformer Ratio

Third harmonics are injected into the three-phase sinusoidal voltages v_a, v_b, and v_c by

$$\begin{cases} v_{aSV} = v_a - v_{hSV} \\ v_{bSV} = v_b - v_{hSV} \\ v_{cSV} = v_c - v_{HSV} \end{cases} \tag{5.15}$$

where v_{aSV}, v_{bSV}, and v_{cSV} are the three-phase voltages with third harmonics, referenced to the mid-point of the DC link, and

$$v_{hSV} = \frac{1}{2}\left[\max\left(v_a, v_b, v_c\right) + \min\left(v_a, v_b, v_c\right)\right] \tag{5.16}$$

Besides the third harmonic, the voltage v_{hSV} contains other triplen harmonics. However, the third harmonic is the dominant frequency in the triplen harmonics, and the influence of other triplen harmonics ($h = 9^{th}$, 15th, ...) is ignored (purely third harmonic injection is a valid concept for increasing the DC-voltage utilization ratio by 15.5%). The adopted third harmonic injection approach described by Equations (5.15) and (5.16) has the same effect as space vector(SV) modulation. Moreover, it is easy to implement and does not cause extra computational burden for the controller, even for MMCs with several hundred SMs per arm.

Using the third harmonic injection with a higher transformer ratio, the maximum output voltage peak is increased to $\frac{1}{2}V_{DC}$, so the modulation index is increased from $\frac{1}{2}\sqrt{3}$ to 1 and all the SMs are effectively utilized, as shown in Figure 5.9(b). As the transferred power is fixed, the AC current i_{ACSV} is reduced by 13.4%:

$$i_{ACSV} = \frac{1}{2}\sqrt{3}I_m \sin\left(\omega t - \varphi\right) \tag{5.17}$$

where I_m is the AC current amplitude in sinusoidal modulation control, ω is the grid angular velocity, and φ is the phase angle between the phase voltage and current. The AC current amplitude I_{m0} can be expressed in terms of the DC current i_{DC}, then Equation (5.17) can be rewritten as

$$i_{ACSV} = \frac{\sqrt{3}}{2}\frac{4i_{DC}}{3m\cos\varphi}\sin\left(\omega t - \varphi\right) = \frac{2\sqrt{3}i_{DC}}{3m\cos\varphi}\sin\left(\omega t - \varphi\right) \tag{5.18}$$

where m_0 is the modulation index with sinusoidal modulation.

5.3.2 Influence of Third Harmonic Injection on MMC

The influence of a third harmonic injection with a resultant higher transformer ratio are discussed in normal operation as well as during pole-to-pole DC fault.

Power Losses

The conduction losses are higher than the switching losses in MMCs for HVDC application and are the dominant part of station converter losses. Therefore, only the conduction losses will be discussed. Considering the DC and fundamental frequency components, the upper and lower arm currents are described as

$$i_{uSV} = \frac{1}{3}i_{DC} + \frac{1}{2}i_{ACSV} = \frac{1}{3}i_{DC}\left[1 + \frac{\sqrt{3}\sin\left(\omega t - \varphi\right)}{m_0\cos\varphi}\right] \tag{5.19}$$

$$i_{lSV} = \frac{1}{3}i_{DC} - \frac{1}{2}i_{ACSV} = \frac{1}{3}i_{DC}\left[1 - \frac{\sqrt{3}\sin\left(\omega t - \varphi\right)}{m_0\cos\varphi}\right] \tag{5.20}$$

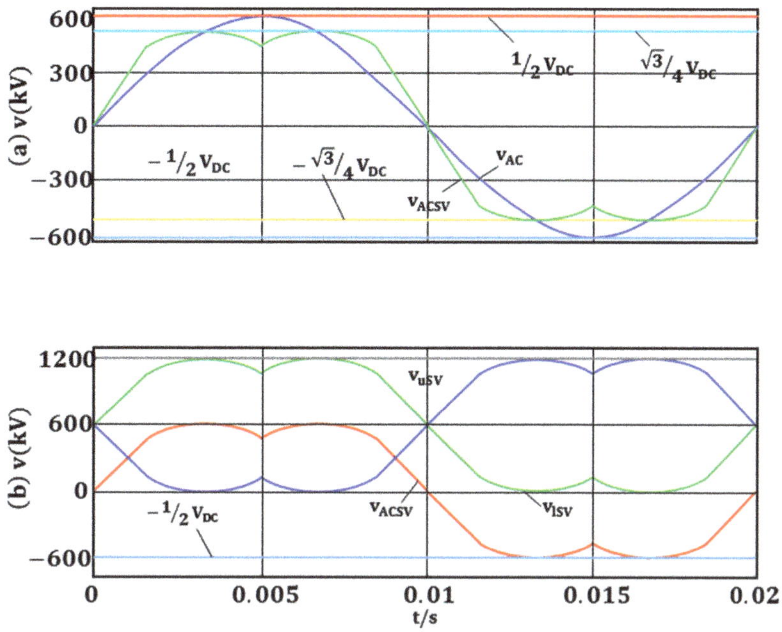

Figure 5.9 Reference voltages: (a) comparison between sinusoidal modulation and third harmonic injection and (b) third harmonic injection with higher transformer ratio

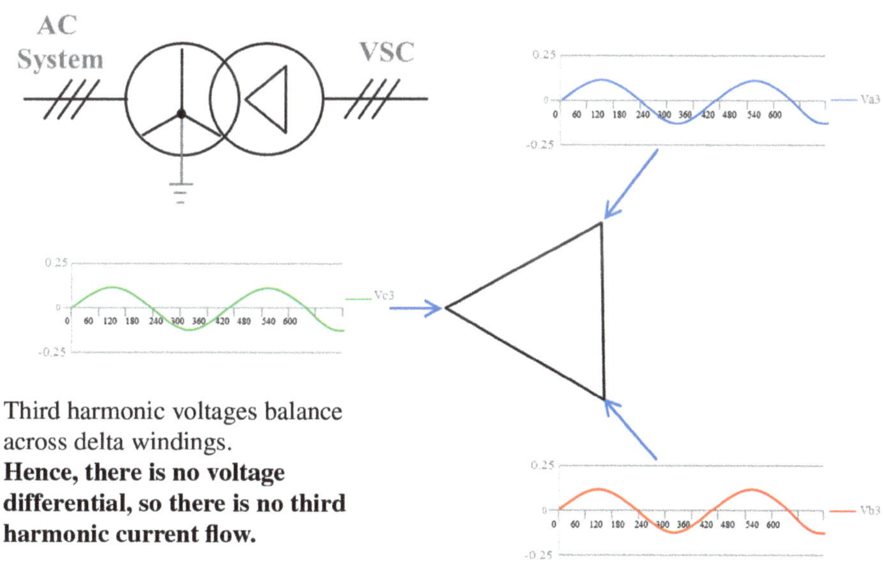

Figure 5.10 Third harmonic voltage balance across delta windings of transformer secondary side

For simplicity, the forward voltages of the IGBT and freewheel diode are assumed to be identical and independent to the current flowing through the semiconductors. This simplification significantly reduces the calculation complexity compared to the alternative approach and is adopted in this study to evaluate the influence of the third harmonic injection on the station converter conduction losses. As the semiconductor number in the current path per arm is N, according to Equation (5.19), the station converter conduction loss can be calculated as

$$P_{lossSV} = \frac{6}{2\pi}\int_0^{2\pi} NV_{fd}|i_{uSV}|d(\omega t) = \frac{4NV_{fd}i_{DC}}{\pi}\left(\frac{\sqrt{3 - m_0^2\cos^2\varphi}}{m_0\cos\varphi}\right) + \arccos\sqrt{1 - 13m_0^2\cos^2\varphi} \tag{5.21}$$

where V_{fd} is the simplified forward voltages of the IGBT and freewheel diode. Similarly, the conduction loss of the station converter with the sinusoidal modulation is

$$P_{lossSV} = \frac{4NV_{fd}i_{DC}}{\pi}\left(\frac{\sqrt{4 - m_0^2\cos^2\varphi}}{m_0\cos\varphi}\right) + \arccos\sqrt{1 - 14m_0^2\cos^2\varphi} \tag{5.22}$$

SM Capacitance Requirements

The specified maximum capacitor voltage ripple typically determines the SM capacitance requirement. This section presents the process of sizing the capacitances for the MMC with a third harmonic injection. Because of the third harmonic, the converter can generate higher fundamental voltage than with sinusoidal modulation:

$$V_{ACSV} = \frac{V_{DC}}{\sqrt{3}}\left[m\sin\omega t + m_{hSV}\sin 3\omega t\right] \tag{5.23}$$

where m_{hSV} is the modulation index of third harmonic voltage. From Equation (5.23), the upper and lower arm voltages are expressed as

$$v_{uSV} = \frac{1}{2}V_{DC}\left[1 - \frac{2}{\sqrt{3}}m_0\sin\omega t - \frac{2}{\sqrt{3}}m_{hSV}\sin 3\omega t\right] \tag{5.24}$$

$$v_{lSV} = \frac{1}{2}V_{DC}\left[1 + \frac{2}{\sqrt{3}}m_0\sin\omega t + \frac{2}{\sqrt{3}}m_{hSV}\sin 3\omega t\right] \tag{5.25}$$

From Equations (5.19), (5.20), (5.24), and (5.25), integrating the arm voltage multiplied by the arm current, the upper and lower arm energy variations are

$$\begin{aligned}\Delta E_{uSV}(\omega t) &= \frac{1}{\omega}\int_0^{\omega t} v_{uSV}i_{uSV}d(\omega t)\\ &= \frac{V_{DC}i_{DC}}{6\omega}\left\{-\frac{\sqrt{3}}{m\cos\varphi}\left[\cos(\omega t - \varphi) - \cos\varphi\right] + \frac{2}{\sqrt{3}}m_0\left[\cos(\omega t) - 1\right]\right.\\ &\left.+ \frac{1}{2\cos\varphi}\left[\sin(2\omega t - \varphi) + \sin\varphi\right]\right\}\\ &- \frac{m_{hSV}V_{DC}i_{DC}}{3\sqrt{3}\omega}\left\{-\frac{\cos(3\omega t) - 1}{2}\right.\\ &\left.+ \frac{\sqrt{3}}{2m\cos\varphi}\left[\frac{1}{2}\sin(2\omega t + \varphi) - \frac{1}{4}\sin(4\omega t - \varphi) - \frac{3}{4}\sin\varphi\right]\right\}\end{aligned} \tag{5.26}$$

$$
\begin{aligned}
\Delta E_{lSV} &= \frac{1}{\omega}\int_0^t v_{lSV} i_{lSV} d(t) \\
&= \frac{V_{DC} i_{DC}}{6\omega}\left\{\frac{\sqrt{3}}{m\cos\varphi}\left[\cos(\omega t-\varphi)-\cos\varphi\right]-\frac{2}{\sqrt{3}}m_0\left[\cos(\omega t)-1\right]\right. \\
&\left.+\frac{1}{2\cos\varphi}\left[\sin(2\omega t-\varphi)+\sin\varphi\right]\right\} \\
&+\frac{m_{hSV} V_{DC} i_{DC}}{3\sqrt{3}\omega}\left\{-\frac{\cos(3\omega t)-1}{3}-\frac{\sqrt{3}}{2m_0\cos\varphi}\left[\cos(\omega t-\varphi)-\cos\varphi\right]\right. \\
&\left.+m\left[\cos(\omega t)-1\right]+\frac{1}{2\cos\varphi}\left[\sin(2\omega t-\varphi)+\sin\varphi\right]\right\}
\end{aligned} \tag{5.27}
$$

Circulating Voltages

Assuming that SM capacitor voltages in each arm remain balanced, the stored energy in each arm is

$$E_{arm} = \frac{1}{2}NC_{SM}\left(\frac{V_{DC}}{N}+\Delta v_{SM}\right)^2 \tag{5.28}$$

where, CSM is the SM capacitance and Δv_{SM} is the SM capacitor voltage ripple. Based on Equation (5.28), the arm energy variation can be approximated as

$$\Delta E_{arm} = NC_{SM}\left(\frac{V_{DC}}{N}+\Delta v_{SM}\right)\Delta V_{SM} \approx C_{SM}V_{DC}\,\Delta v_{SM} \tag{5.29}$$

Then the SM capacitor voltage ripple can be expressed by the arm energy variation:

$$\Delta v_{SM} = \frac{\Delta E_{arm}}{C_{SM}V_{DC}} \tag{5.30}$$

According to Equations (5.24), (5.25), and (5.30), and considering SM capacitor voltage ripple, the upper and lower arm voltages are

$$v_{uSV} = \frac{1}{2}\left(1-\frac{2}{\sqrt{3}}m_0\sin\omega t-\frac{2}{\sqrt{3}}m_{hSV}\sin 3\omega t\right)\left(V_{DC}+\frac{N\Delta E_{uSV}}{C_{SM}V_{DC}}\right) \tag{5.31}$$

$$v_{lSV} = \frac{1}{2}\left(1+\frac{2}{\sqrt{3}}m_0\sin\omega t+\frac{2}{\sqrt{3}}m_{hSV}\sin 3\omega t\right)\left(V_{DC}+\frac{N\Delta E_{lSV}}{C_{SM}V_{DC}}\right) \tag{5.32}$$

Thus the sum of the upper and lower arm voltages is

$$
\begin{aligned}
v_{uSV}+v_{lSV} &= V_{DC}+v_{cirSV} \\
&+\frac{Ni_{DC}(m_0\sin\omega t+m_{hSV}\sin 3\omega t)}{3\omega C_{SM}}\left\{\frac{\cos(\omega t-\varphi)-\cos\varphi}{m_0\cos\varphi}-\frac{2}{3}m_0\left[\cos(\omega t)-1\right]\right\} \\
&-\left\{\frac{2m_{hSV}}{9}\left[\cos(3\omega t)-1\right]\right\}
\end{aligned} \tag{5.33}
$$

where v_{cirSV} is the dominant circulating voltage, specifically

$$v_{sirSV} = \frac{N}{2C_{SM}V_{DC}}(\Delta E_{lSV}+\Delta E_{uSV}) = \frac{N}{2C_{SM}V_{DC}}\Delta E_{pSV} \tag{5.34}$$

According to Equations (5.26) and (5.27), the phase energy variation with third harmonic injection is

$$\begin{aligned}\Delta E_{pSV} &= \Delta E_{uSV} + \Delta E_{lSV} \\ &= \frac{V_{DC} i_{DC}}{6\omega \cos\varphi}\left[\sin(2\omega t - \varphi) + \sin\varphi\right] \\ &- \frac{m_{hSV} V_{DC} i_{DC}}{3m_0 \omega \cos\varphi}\left[\frac{1}{2}\sin(2\omega t - \varphi) - \frac{1}{4}\sin(4\omega t - \varphi) - 34\sin\varphi\right]\end{aligned} \tag{5.35}$$

The phase energy variation with sinusoidal modulation can be similarly obtained as

$$\Delta E_{p0} = \frac{V_{DC} i_{DC}}{6\omega \cos\varphi}\left[\sin(2\omega t - \varphi) + \sin\varphi\right] \tag{5.36}$$

Pole-to-Pole DC Fault Currents

A significant challenge with MMC VSC HVDC transmission systems is DC fault protection. In the event of a DC short circuit, high current flows through the freewheel diodes in the half-bridge MMC, which is currently the preferred HVDC converter configuration, from the AC grid into the fault on the DC side, even if the station converter is blocked immediately. The low impedance of the short-circuit path leads to a steep rise in fault current, which may cause serious damage to the converters and may result in shutdown of the entire HVDC network. To limit CCs and DC fault currents, the arm inductance L_{arm} is required and can be calculated as:

$$L_{arm} = \alpha \frac{V_{LRMS}^2}{\omega P} = \frac{3m^2\eta^2 V_{DC}^2}{40\omega P} \tag{5.37}$$

where V_{LRMS} is the root mean square (RMS) line voltage, P is the rated power, m is the modulation index, α is the pu value of arm inductance, and η is the DC-voltage utilization ratio. With the same arm inductance pu value α, the relationship between arm inductances with sinusoidal modulation and third harmonic injection is derived as Equation (5.38), according to Equation (5.39):

$$L_{armSV} = \frac{4}{3} L_{arm0} \tag{5.38}$$

Similarly, the converter-side transformer leakage inductance L_{TcSV} increases by a third, whereas the grid-side leakage inductance L_{TgSV} remains unchanged:

$$L_{TcSV} = \frac{4}{3} L_{Tc0}, \quad L_{TgSV} = L_{Tg0} \tag{5.39}$$

Transforming the converter-side inductances to the grid-side, the total AC –side inductance L_{ACSV} with third harmonic injection is the same as with sinusoidal modulation:

$$\begin{aligned}L_{ACSV} &= \frac{L_{armSV} + L_{TcSV}}{n_0^2} + L_{TgSV} + L_g = \frac{43\left(L_{arm0} + L_{Tc0}\right)}{\left(2n_0/\sqrt{3}\right)^2} + L_{Tg0} + L_g \\ &= \frac{L_{arm0} + L_{Tc0}}{n_0^2} + L_{Tg0} + L_g = L_{AC0}\end{aligned} \tag{5.40}$$

$$i_{fcSV} = \frac{i_{fgSV}}{n_{SV}} = \frac{i_{fg0}}{2n_0/\sqrt{3}} = \frac{1}{2}\sqrt{3}\, i_{fc0} = 0.866 i_{fc0} \tag{5.41}$$

where L_g is the grid inductance and n_0 and n_{SV} are the transformer ratio for sinusoidal modulation and third harmonic injection, respectively.

For a pole-to-pole DC fault at the station terminals, which is the most serious fault case for the station, the fault current on the transformer grid-side i_{fgSV} remains unchanged because of the constant grid voltage and the unchanged total AC–side inductance L_{ACSV}, as shown by Equation (5.40). However, because of the increased transformer ratio n_{SV}, the fault current on the transformer converter side i_{fcSV} is expected to be reduced by

13.4% using third harmonic injection, which reduces the current stress on the freewheel diodes and DC circuit breakers during a DC fault.

Equation (5.41) shows the fault current reduction ratio in the case with third harmonic injection.

5.4 Precharging [12]–[19]

Generally speaking, the starting process is complex since at the start-up of the MMC, all capacitors in each module of the MMC are required to be charged to a certain equal value before they can be transferred to the normal operation. Therefore, the process of starting up the MMC from its de-energized condition is a crucial issue, because the SM capacitors cannot be charged to the rated DC voltage. Early schemes always adopted external voltage sources to precharge the SM capacitors one by one. But this method needs dedicated and auxiliary voltage sources, and is not a cost-effective solution for industrial applications.

To get rid of the auxiliary power supply, other start-up schemes were proposed to directly precharge the capacitors by absorbing power from the DC-side main voltage, or from the AC-side main voltage.

Also, it is hoped that the starting process will be as smooth as possible so that the current stress in the power switching devices and filters is minimized. At the same time the surge current to the grid will be reduced, which is especially required for high-power application. For these reasons, a soft start-up process is necessary.

According to the states of the IGBTs and the directions of current, a submodule has three kinds of operation modes. When all the IGBTs work, submodule capacitor will be shorted. We should avoid this situation when the system is in operation. Table 5.1 and Figure 5.11 illustrate the operation modes and current path.

Table 5.1 Operation Modes of Submodule

The direction of current	T_1	T_2	D_1	D_2	The stage of capacitor
A→B	0	0	1	0	Charge
B→A	0	0	0	1	Bypass
A→B	0	1	0	0	Bypass
B→A	0	1	0	1	Bypass
A→B	1	0	1	0	Charge
B→A	1	0	0	0	Discharge

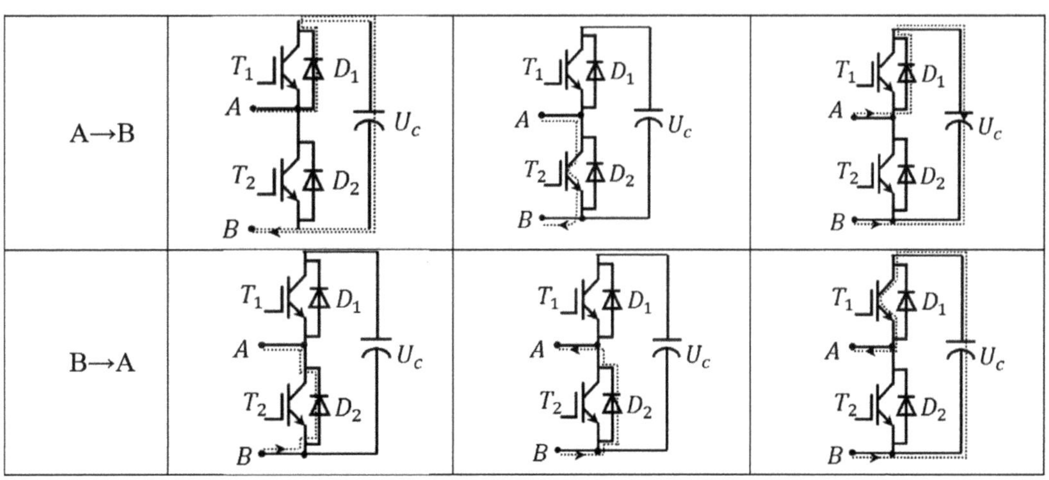

Figure 5.11 Current paths of submodule

Mode 1. When all the IGBTs are turned off, current flows through the diode D_1 and the capacitor or the diode D_2, so the capacitor will be charged or bypassed. Without driving IGBTs, this mode can charge capacitor voltage to a certain value. This mode is called the uncontrolled charging method.

Mode 2. When IGBT T_1 is turned off and T_2 is turned on, current flows through the IGBT T_2 or capacitor and diode D_2, so the capacitor will be bypassed. Capacitor voltage is stable when IGBTs work on this mode.

Mode 3. When IGBT T_1 is turned on and T_2 is turned off, current flows through the diode D_2 or IGBT T_1, so the capacitor will be charged or discharged.

5.4.1 Uncontrolled Charging Method from the AC-Side Main Voltage

During the uncontrolled charging method stage, all IGBTs are turned off. Current flows through the diodes and capacitors. Without driving IGBTs, it is widely used in the system that IGBT-drive voltage is generated by capacitor voltage.

In the early stage of the uncontrolled charging method, capacitor voltages are low, and the path of the current is only concerned with the instantaneous value of line voltage. Figure 5.12 illustrates the equivalent circuit diagram of this situation.

Because of the rotating frame of the MMC system, the equivalent charging loop between phase-a and phase-b is shown in Figure 5.13. C, L_0, R_s, and L_s are the equivalent capacitance, equivalent inductance of the arm, equivalent resistance, and equivalent inductance of the source grid, respectively. By Kirchhoff's voltage law, the characteristics can be described by

$$\begin{aligned} \dot{U}_{ab} &= 2\dot{I}_{ab}R_s + j\dot{I}_{ab}X \\ X &= \omega(2L_s + L_0) - \frac{N}{2\omega C} \end{aligned} \tag{5.42}$$

$$I_{ab(max)} = \frac{U_{ab}}{\sqrt{4R_s^2 + X^2}} \tag{5.43}$$

where $U_{ab} = \sqrt{3}U_m$, U_m is the peak of the phase voltage.

Thus, the current–limiting resistance can be described by

$$R_s = \frac{1}{2}\sqrt{\left(\frac{U_{ab}}{I_{ab(max)}}\right)^2 - X^2} \tag{5.44}$$

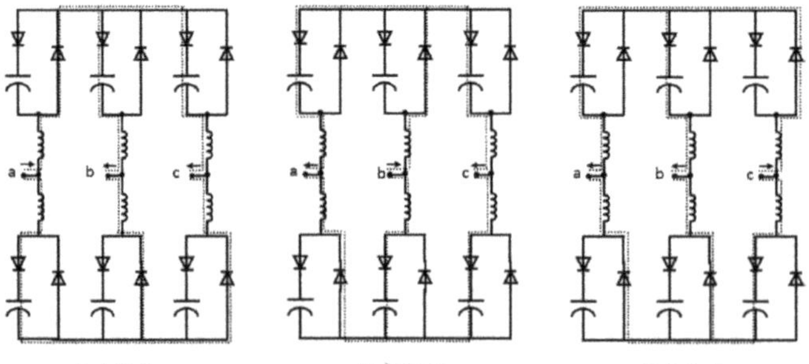

Figure 5.12 Early stage equivalent circuit of uncontrolled charging method

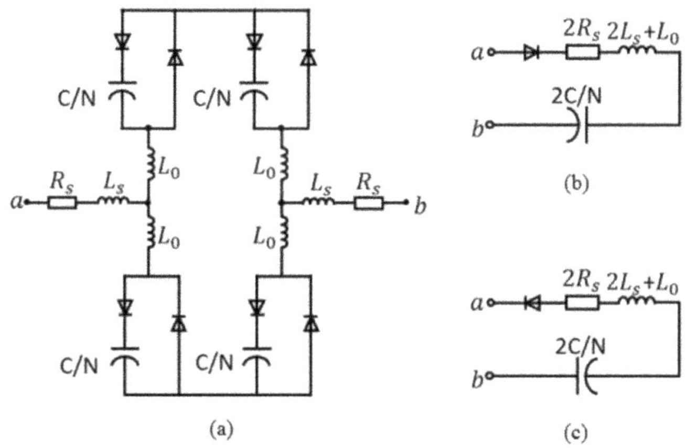

Figure 5.13 Charging circuit of phase-a-to-phase-b: (a) equivalent circuit; (b) simplified circuit when $i_{ab} > 0$; and (c) simplified circuit when $i_{ab} < 0$

From the simplified circuit, and from the following equation, we can see that the time constant is

$$\tau = 2R_s \cdot \frac{2C}{N} \tag{5.45}$$

Because the length of time that an arm is charged takes one third of the total length of time, the time constant should be adjusted to

$$\tau = 2R_s \cdot \frac{2C}{N} \times \frac{3}{1} = 12\frac{R_s C}{N} \tag{5.46}$$

As Figure 5.13 shows, the maximum value of the capacitor voltage can be described by

$$U_{c(max)} = \frac{\sqrt{3}U_m}{N} \tag{5.47}$$

Hence, the submodule capacitor voltage can be described by

$$\begin{aligned} &\tau = 12\frac{R_s C}{N}, \\ &U_c = \frac{\sqrt{3}U_m}{N}\left(1 - e^{-\frac{t}{\tau}}\right) \end{aligned} \tag{5.48}$$

5.4.2 Precharge Process from the DC-Side Main Voltage

The charging process can be divided into two stages. In the first stage, the DC-bus voltage will be charged to its rated value, meanwhile each module capacitor will be charged to half of its rated voltage. In the second stage each module capacitor will be boosted up to its rated voltage, which is the main part of the start-up scheme. A detailed description of the two-charging stage is given as follows:

- Stage 1: In the first stage, AC breaker S1 is closed, and S2 is open (in Figure 5.14). All the modules are inserted. The DC-bus voltage will be charged to the rated value, whereas the capacitor voltage in each module is equally charged to half of its rated value Ud/(2×3). Using resistors in the AC side, the charging current can be limited. Once the DC-bus voltage reaches its rated value, then the second stage starts.
- Stage 2: In the second stage, both S1 and S2 are closed and the DC side is simplified as shown in Figure 5.14. After stage 1, the sum of all the module capacitor voltages in one phase equals the DC-bus voltage; one or more modules have to be bypassed to produce current in the arms to charge the inserted

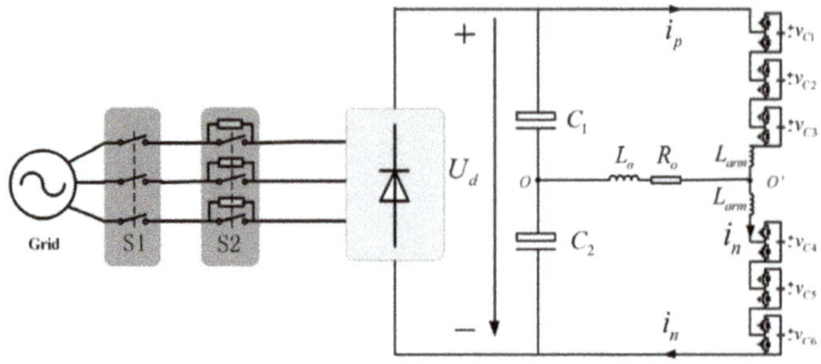

Figure 5.14 Test case for explaining DC-side precharging

module. The number of inserted modules in the upper arm will be equal to that of the lower arm so that $V_{00'} = 0$ is fulfilled throughout the charging process, and there will be no load current. In this way, the module in the upper arm and those in the lower arm are charged simultaneously. And load condition will have no impact on the arm current, which makes this start-up scheme suitable for a three-phase MMC. Also, the charging process of stage 2 can be further divided into two substages as follows:

- Substage 1: Figure 5.15 shows the circuit for substage 1. In this substage, two modules are inserted in the upper arm and the third module is operating in the pulse width modulation (PWM) switching mode with switching period T_s. The state of the PWM-mode module will change from insert to bypass in one switching period. When the PWM-mode module is bypassed, the arm current will rise. Therefore, the module capacitors can be charged up. When the PWM-mode module is inserted, the arm current will decrease, which can effectively suppress the arm current. To make the DC capacitor voltage even for all three modules in upper arm, the PWM switching operating mode has to swap between the upper three modules. A selecting algorithm is used here to determine which module should be chosen to work in the PWM–mode. And the duty of the PWM mode module will decrease from 1 to 0 linearly with a certain slope. In this way, DC capacitor voltage increases as the PWM duty decreases. Similarly, three modules in the lower arm follow the same working mode as that of the upper arm. The substage ends once the PWM duty reaches 0.
- Substage 2: Figure 5.16 shows the circuit for substage 2. In this substage, three modules operate at three different states in the upper arm. One module is inserted, one bypassed and the third module is operating in the PWM switching mode. Again the duty of the PWM–mode module will decrease from 1 to 0.5 linearly. When the duty reaches 0.5, each module will be charged to Ud/3, and the charging process will be over. To make the DC capacitor voltage even for all three modules in one arm, the PWM switching operating mode is also swapped between three modules in one arm.

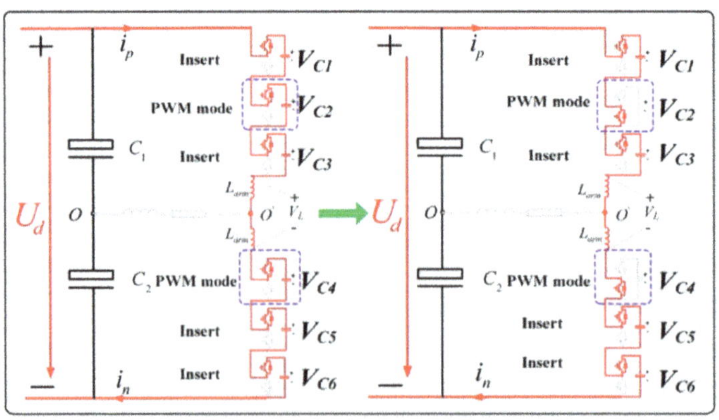

Figure 5.15 Circuit for substage 1

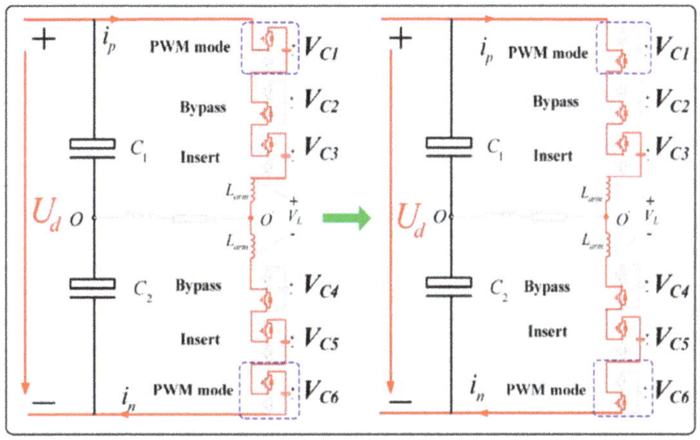

Figure 5.16 Circuit for substage 2

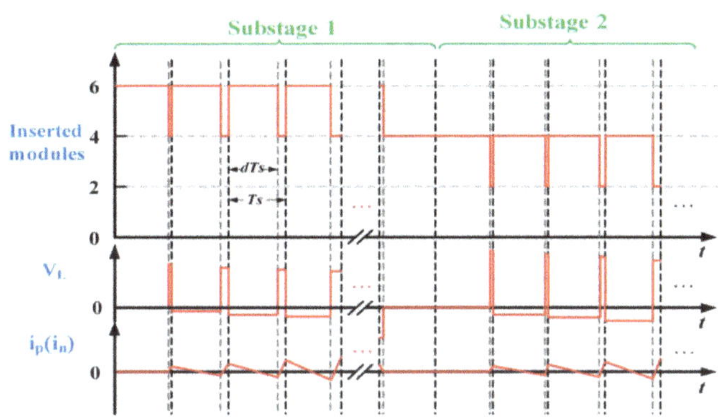

Figure 5.17 Waveforms of the charging process

Figure 5.17 shows the waveforms in the charging process. When the PWM-mode module is in bypass state, a positive voltage will be placed on the arm inductors, and the arm current will rise. When the PWM-mode module switches to insert mode, there will be a negative voltage drop in the arm inductors, then the arm current will fall. Before the circuit begins working in its steady state, the average arm current over one switching period is always positive. So when executing the selecting algorithm, the direction of the current is defaulted as positive. The modulewith the highest voltage will always be selected to work in the bypass state, and the module with the lowest voltage will work in the insert state.

It should be noted that though the DC-bus voltage is derived from a noncontrolled rectifier, it will limit the application of the start-up scheme. Actually, the first stage of the start-up scheme is mainly decided by how the DC bus voltage is obtained. Generally, the soft-start of the DC- bus is done by the end-front converter. For a modular multi–level converter, the main attention is paid to the second stage of the charging process.

5.4.3 Precharging Method Without Inrush Current

Li et al. [19] proposed a novel start-up control scheme for MMC, in which the closed-loop control is adopted to regulate a constant RMS precharge current with a unity power factor. As such, the SM capacitors will be linearly precharged to the rated value without any inrush current and with a much shortened charging time. Moreover, this scheme requires no extra power supply and is very flexible as it can be easily implemented into any MMC application regardless of the power rating and the number of SMs.

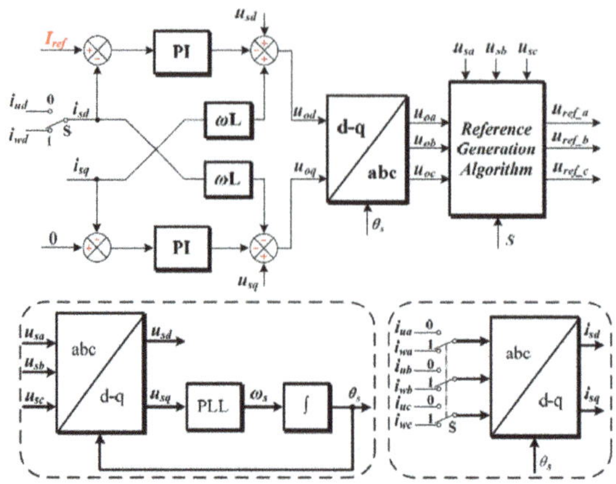

Figure 5.18 Block diagram of the precharge control scheme with constant current

After SM capacitor voltage has increased to $U_{\text{sm(initial)}}$ by aforementioned uncontrolled process, the proposed precharge scheme will be put into action to further charge the SM capacitors to the rated value. The contactors at the AC-side terminals of MMC are then activated to bypass the current-limiting resistor. Figure 5.18 shows the block diagram of the precharge control scheme. It comprises the classical *d-q* decoupling current control scheme. Meanwhile, the reference of the d-axis current is denoted as I_{ref}, which determines the value of the precharge current during the start up process. The q-axis current reference is set to 0 so that the AC current of the MMC during this precharge process will be in phase with grid voltage. Note that the upper-arm SMs and the lower-arm SMs are precharged separately. When the switch S is turned to 0, the upper-arm currents i_{uj} (j = a, b, c) will be selected as the feedback. On the other hand, SMs in the lower arm will get charged by taking the lower arm currents i_{wj} as feedback.

Thus, the dq reference voltage of MMC can be derived as

$$\begin{cases} u_{od} = u_{sd} + \omega L i_{sq} - K_{p1}\left(I_{ref} - i_{sd}\right) - K_{i1}\displaystyle\int\left(I_{ref} - i_{sd}\right)dt \\ u_{oq} = u_{sq} - \omega L i_{sq} - K_{p2}\left(0 - i_{sq}\right) - K_{i2}\displaystyle\int\left(0 - i_{sq}\right)dt \end{cases} \tag{5.49}$$

where K_{p1}, K_{i1}, K_{p2}, and K_{i2} are the control parameters, u_{sd}, u_{sq} and i_{sd}, i_{sq} represent the grid voltage and ac current in the dq-axis, respectively.

5.5 Analysis of MMC under Unbalanced Conditions [20] ~ [43]

As in grid-connected converters, the occurrence of grid-side faults is rather common, leading to imbalances and distortions of the AC-side voltages. Therefore, the operation and control of MMC HVDC systems under grid imbalances is of great significance in order to satisfy the grid codes and reliability requirements of the system.

Jinke Li et al. [20] provide a comprehensive review of operation and control methods applied to MMC HVDC systems with a specific focus on unbalanced AC-grid conditions.

Under unbalanced AC-grid conditions, the control systems of a VSC are sensitive to grid voltage changes, and thus, stricter requirements on VSCs have been imposed by transmission system operators as a VSC can exhibit undesirable performance that includes output current distortion, DC-link voltage oscillations, and output power oscillations. Therefore, a variety of control strategies have been proposed to improve the performance of VSCs, which enable diverse characteristics of fault current and power in different ways.

Compared to conventional two-level and three-level VSCs in HVDC applications, an issue specific to the MMC is the control of its CC, which originates from the SM capacitor voltage variations. In addition to the DC component and double-line frequency negative-sequence component, a double-line frequency positive and zero–sequence component caused by grid imbalances will increase the difficulty of designing and tuning CC controllers also causing DC-side power oscillations.

5.5.1 General Consideration

Classification of MMC HVDC Control Methods Under Grid Imbalances

A general classification of high-level and low-level MMC HVDC control under grid imbalances is given in Figure 5.19, which includes DC-voltage control, power control, and AC-side voltage/frequency control. The DC-voltage control maintains the DC-voltage stability, and the AC-side voltage/frequency control is used to generate a desired voltage supplying a passive network.

VSC HVDC System Operation Under Grid Imbalances

The analysis presented in this chapter assumes a coordinated control of the VSC HVDC based on a single adjustable coefficient k for the smooth regulation of both the active and the reactive power as defined in the following equation:

$$\mathrm{i}_{\mathrm{pq}}^{*} = \frac{S}{\|v^{+}\|^{+} + k\|v^{-}\|^{2}} R(\varphi)\left(v^{+} + kv^{-}\right) \tag{5.50}$$

The range of coefficient k is $[-1, 1]$. The value of k defines the control target of the MMC with $k = 0$ leading to balanced ac output currents, and active power oscillations can be eliminated by setting $k = -1$, whereas reactive power oscillations can be eliminated when $k = 1$.

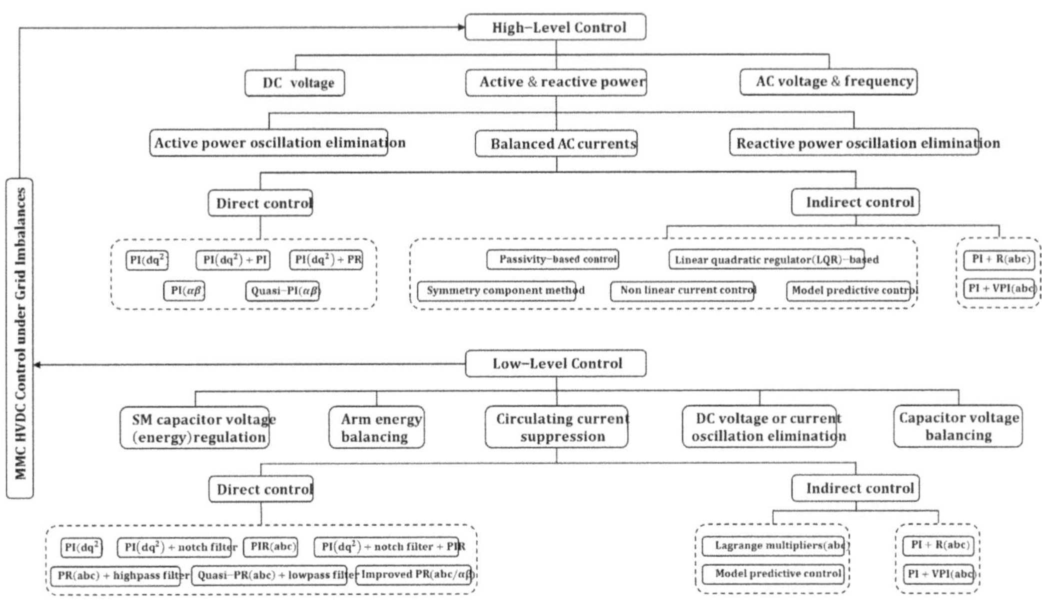

Figure 5.19 Classification of MMC HVDC control methods under grid imbalances

The CC under unbalanced AC-grid conditions can be calculated by an AC/DC decoupled model. Although the AC–side voltage does not contain zero-sequence components, the double-line frequency zero-sequence component is present in the CC. The impact of the control target k in positive, negative, and zero-sequence components of the CC $icir$ between the three phases is analyzed in [38].

When the system control target is selected to eliminate active power oscillations ($k = -1$), the DC components among the three phases are equal and the negative-sequence component is not affected by the voltage imbalance degree or k. Table 5.2 summarizes the fault characteristic of an MMC HVDC system under a single-line-to-ground fault.

MMC HVDC System Structure

From Figure 5.3, the arm currents of phase j ($j = a, b, c$) can be written as $i_{ju} = (1/2)\, i_j + i_{jcir}$ and $i_{jl} = (1/2)\, i_j - i_{jcir}$. The CC of each phase-leg i_{jcir} can be controlled by generating an appropriate differential voltage reference v^*_{jcir}. u_{jl} and u_{ju} are the lower and upper arm voltages; e_j is the voltage generated in phase j and can be expressed as $e_j = \left(\left(u_{jl} - u_{ju}\right)/2\right)$.

Under unbalanced AC-grid conditions, the phase voltage command e_j is generated by an inner loop controller that regulates the AC-side line currents i_j. The references of arm voltages can be defined as following:

$$\begin{cases} v^*_{jq} = \frac{1}{2}v_{dc} - e^*_j - v^*_{jcir} \\ v^*_{jl} = \frac{1}{2}v_{dc} + e^*_j - v^*_{jcir} \end{cases} \tag{5.51}$$

Table 5.2 Comparison of System Control Targets

	CC components	Characteristics
Target 1	Negative- and zero–sequence	Balanced AC output currents Lowest peak arm current and AC output current
Target 2	Positive-, negative-, and zero sequence	Constant active power in AC side Larger SM voltage ripple active power oscillation in DC-side
Target 3	Positive-, negative-, and zero-sequence	Constant reactive power Smallest arm energy variation Lower SM voltage ripple

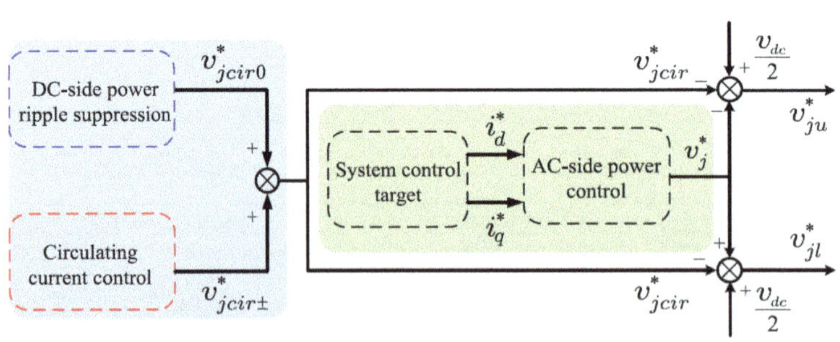

Figure 5.20 Typical control scheme of a modular multi-level converter

Figure 5.20 shows the overall control scheme of the MMC, whereas in the considered system, MMC2 works in the P/Q mode and MMC1 controls the DC-link voltage, V_{dc}, and Q. The sum of the references generated by positive, negative, and zero–sequence current regulation forms the phase voltage command e_j^*. The control reference v_{jcir}^* comprises a DC-side power ripple suppression reference v_{jcir0}^* and a CC control reference $v_{jcir\pm}^*$. As indicated in Equation (5.51), the DC-side power ripple suppression controller does not affect the phase voltage generation e_j. Hence, both the unbalanced current and the DC-side power ripple can be eliminated at the same time, enabling the desirable performance of MMC HVDC under grid imbalances.

5.5.2 AC-Side Power Control Strategies

AC-side power controllers are designed to ride-through grid faults by maintaining normal power flow, eliminating the unbalanced capacitor voltage. It may also be required for the MMC to provide reactive power (up to 100% current capacity) to assist with grid voltage recovery.

Several control strategies have been proposed to enhance the transient performance of two-level VSC under unbalanced AC-grid conditions. Proportional–integral (PI) controllers, PR controllers, PI plus resonant (PIR) controllers, reduced-order generalized integrators, reduced-order resonant controllers, deadbeat control, model predictive controls (MPCs), linear-quadratic regulators (LQRs), fuzzy logic controller (FLCs), and feedback linearizing controls (FBLCs) have been applied to MMCs under unbalanced conditions.

Direct Power Control Methods

To control the AC-side power, a dual current control scheme, including positive- and negative-sequence vector current controllers in the *dq* synchronous rotating frame (SRF), performs well under unbalanced AC-grid conditions. The zero-sequence components, which may lead to over-currents and/or over-voltages when transferred to the DC-side, are unavoidable in the event of asymmetrical faults on the MMC in a transformerless system. Thus, zero-sequence current controllers are based on the PR structures and PI controllers shown in Figure 5.21(b) and (c) in addition to the positive- and negative-sequence current controllers. Although tuning the PI controller in the *dq* SRF is relatively easy, a dual current control scheme has certain disadvantages, one being its relatively complicated overall structure.

A hybrid current vector controller-based PIR was used in the positive SRF to uniformly control the DC current and second-order harmonic components of the AC output currents shown in Figure 5.21 (d). Although this control structure is simple, the controller parameters are hard to design and an additional zero-sequence current controller is needed under a transformerless system. Because of a relatively concise control structure, PR-based control strategies on the $\alpha\beta$-frame have become popular in an MMC HVDC transmission system under grid imbalances shown in Figure 5.21(e).

As PR controllers introduce an infinite gain at the resonant frequency, any small frequency deviation of the reference signal may lead to a large tracking error. Quasi-PR in the $\alpha\beta$ frame was used to control the positive- and negative-sequence currents uniformly shown in Figure 5.21(f).

Indirect Power Control Methods

Existing control methods regulate the AC output and CC by separating each component under grid imbalances. Figure 5.22(a) shows a method that controls each arm current utilizing PIR controllers. It regulates both AC-side and CCs without separating the current components. However, PIR controllers cannot achieve ideal 0-dB amplitude response and 0 degree phase response at the resonant frequency in the closed-loop transfer function. An arm current control strategy based on decoupler and PI controller plus vector PI (VPI) controller is shown in Figure 5.22(b). The two methods [22], [23], have definite physical meaning; however, the complex controller design and the interacting current components lead to complex practical implementations.

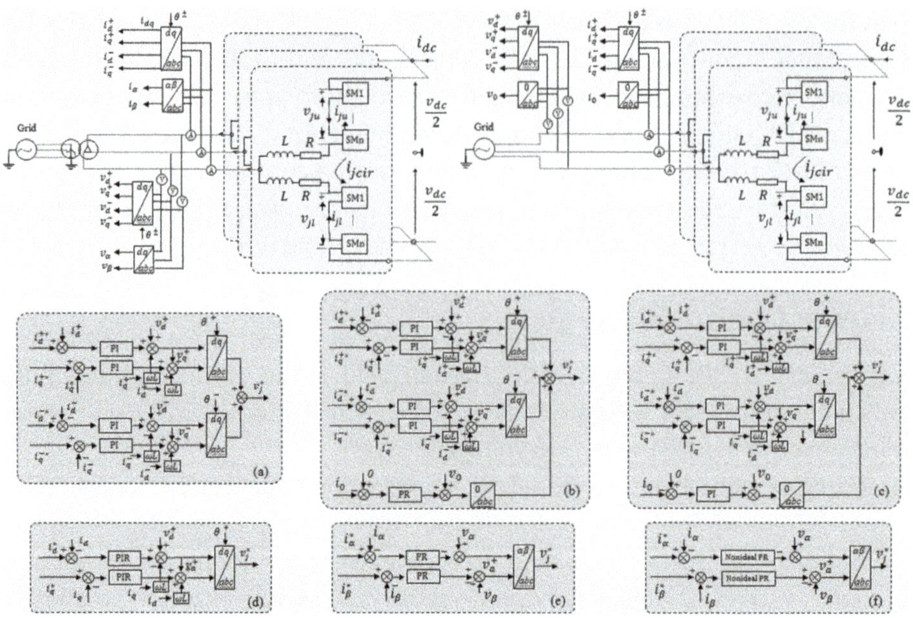

Figure 5.21 AC-side power control strategies of a modular multi-level converter under grid imbalances—direct methods: (a) positive- and negative-sequence vector current control in the *dq* frame; (b) zero-sequence current control using a proportional-resonant (PR) controller under a transformerless system; (c) zero-sequence current control using a proportional–integral (PI) controller under a transformerless system; (d) vector current control-based PI plus resonant (PIR) in the positive synchronous rotating frame; (e) PR control strategy under the $\alpha\beta$ frame; and (f) Quasi-PR control strategy under the $\alpha\beta$

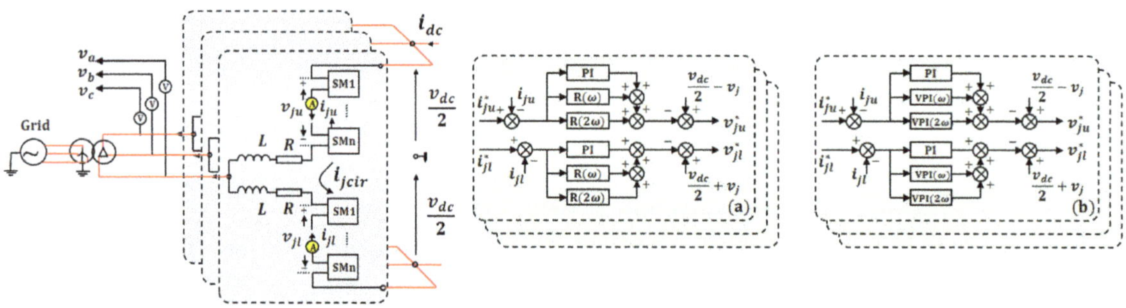

Figure 5.22 AC-side power control strategies of a modular multi-level converter under grid imbalances: indirect methods: (a) the arm current control method using proportional–integral (PI) plus resonant (PIR) controllers and (b) the arm current control method using PI + vector P I (VPI) controllers

A *dq* Euler–Lagrange model of the MMC was derived in [24], and a passivity-based control strategy was proposed with the consideration of a desired equilibrium point, state, and state error. This algorithm eliminates the negative-sequence AC current, but it may have limitations under the presence of zero-sequence current. Passivity-based PI control was also used in [25] with a nonlinear global tracking controller that consists of a linear PI regulating two constructed signals where the incremental model of the MMC becomes passive.

Reference [26] presented a *dq* model of indirect MMC by a generalized state-space representation. Correspondingly, a complete LQR-based state feedback control was applied to regulate the multiple currents. This method shows the multi-input multi-output characteristics and proposes a multi–variable and optimal control strategy to fulfill diversified control requirements. However, it is sensitive to system parameter variations. An alternative control structure based on instantaneously calculated values of negative- and zero-sequence currents was proposed in Ref. [27]. The calculations are performed instantly without a current regulator which results in a fast response; however, the open-loop compensation control is inflexible in the negative-sequence component control. A nonlinear current control technique defined by a hysteresis band was used in Ref. [28] to achieve a fast response.

MPC is a kind of strategy that can enable control of real and reactive power, eliminate the CCs, and carry out the voltage-balancing task [29]. Thus, an MPC with a reduced number of states for ac current control, CC control, and the SM capacitor voltage-balancing algorithm under grid imbalances was used [30]. The method is divided into three stages according to the control goals, and the cost function in each stage is designed separately without weighting factors. A summary of the AC-side power controllers, including a comparison based on their key advantages and disadvantages, is given in Table 5.3.

Table 5.3 Summary and Comparison of AC-Side Power Control Strategies

Name	Advantages	Disadvantages
4PI $(dq^{\pm})$	Easy to design controller	No zero-sequence current controller Relatively complicated structure
4PI $(dq^{\pm})$ + PR (0)	Enhance zero-sequence current control	Relatively complicated structure High digital control resources
4PI $(dq^{\pm})$ + PI (0)	Zero-sequence current control	Relatively complicated structure Control effect of zero-sequence current
2PI + $R_{2\omega}$ (dq^{+})	Simplified control structure	No zero-sequence current controller Complicated controller parameters design
2PR $(\alpha\beta)$	Concise control structure	Sensitive to system frequency variation
2quasi-PR $(\alpha\beta)$	Concise control structure Enhanced frequency adaptive operation	Complicated control parameter design
6PI + $R_{\omega+2\omega}$ (abc)	Simultaneous suppression of CC	Complicated control parameter design High digital control resources Response at the resonant frequency
6PI + $VPI_{\omega+2\omega}$ (abc)	Simultaneous suppression of CC Enhance controller performance	Complicated control structure Complicated control parameter design High digital control resources
Four PI and passivity control $(dq^{\pm})$	*dq* Euler–Lagrange model Enhanced controller performance	No zero-sequence current controller Model is sensitive to system parameters
Passivity PI (abc)	Improved active power control Reduced dc-link voltage ripple	Difficulty of practical implementation High– order harmonics is neglected
LQR-based State feedback control	MIMO(Multi-Input Multi-Output) of indirect MMC *dq* model Multi–variable and optimal control	Estimator is sensitive to system parameters
2PI $(dq^{\pm})$ and Symmetrical components	No additional current regulator Faster control in negative and zero-sequence currents	Accuracy of the open-loop control Inflexible negative-sequence current control
2PI $(dq^{\pm})$ and NLCC(Non-Linear Current Control) (abc)	Faster response during transients Fewer PI controllers	Low-order harmonics in grid currents
MPC	Reduced number of states algorithm Transient characteristics	Computational complexity

5.5.3 Circulating Current(CC) and Energy Balancing Control Strategies

A CC suppressing controller based on the double-line frequency negative-sequence rotating frame has been widely used in the literature. PR and PIR controllers, state observers, and repetitive control have been proposed to suppress the AC components of the CC. In the meantime, control methods based on harmonic injection into the CC were also defined in order to reduce the SM voltage ripples.

Although a proper CC control can maintain a safe system operating, SM capacitor energy control is also one of the main concerns in the MMC. Thus, SM capacitor energy balancing and arm energy balancing methods were proposed to ensure reliability and safety of the MMC system under balanced conditions.

Direct Methods of CC Control

The direct methods of CC control shown in Figure 5.23(a)–(c) eliminate the CC by transforming it into the negative-sequence *dq* rotating frame and then controlling it with two PI controllers. This method is easy to design and implement, and is widely used in the rotating frame. However, it can only be used in three-phase systems and cannot eliminate the positive- and zero-sequence CCs under unbalanced AC-grid conditions.

The single-phase vector control shown in Figure 5.23(d) was used for each phase of the three-phase MMC to eliminate the second-order component in the CC. Although this method achieves a good performance under grid imbalances, it requires six properly tuned PI controllers for a three-phase system. The positive- and negative-sequence components were controlled through a PIR controller in a three-phase stationary frame

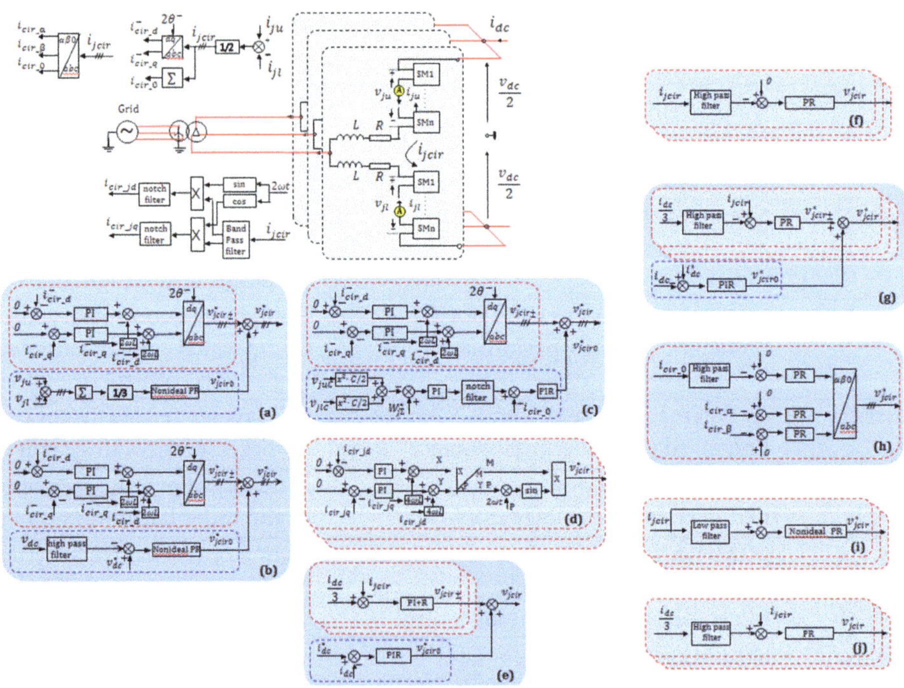

Figure 5.23 Circulating current control strategies of a modular multi-level converter under grid imbalances: direct methods I: (a)–(c) proportional–integral (PI) controllers under the negative-sequence *dq* frame; (d) single-phase vector control; (e) PI plus resonant (PIR) controller under the three-phase stationary frame; (f) proportional-resonant (PR) controller under the *abc* stationary frame; (g) additional zero-sequence controller; (h) high-pass filters (HPFs) under the $\alpha\beta 0$ stationary frame; and (i) and (j) nonideal PR controllers in the *abc* stationary frame

without the need for separating the positive- and negative-sequence components shown in Figure 5.23(e). However, an additional zero-sequence component controller is also needed.

A CC control method in the *abc* stationary frame shown in Figure 5.23(f) was implemented through a PR controller, considering positive-, negative-, and zero-sequence components. However, the transient response during the unbalanced AC-grid conditions is slow compared with other methods. Similar to Ref. [30] a control method based on PR controllers in the *abc* stationary frame had been proposed. The PR controller reference was set as one third of the DC currents under balanced AC-grid conditions $\left(\frac{i_{dc}}{3}\right)$, and an additional zero-sequence controller was also used (red box in Figure 5.23(g)). However, the reference $\frac{i_{dc}}{3}$ limits the system control flexibility and accuracy because of the DC-current variations during AC-grid imbalances.

The control method for CC suppression in the $\alpha\beta$ 0 stationary frame as shown in Figure 5.23(h), uses high-pass filters (HPFs) to extract the double-line frequency component. The filters influence the dynamic performance of the controller, and the response of the ideal PR controller may deteriorate under frequency variations. Alternative CC suppression methods based on nonideal PR controllers in the *abc* stationary frame were used with a low-pass filter (LPF) for the measurement of the CC DC component as shown in Figure 5.23(i) or using a reference value of $i_{dc}3$ as shown in Figure 5.23(j). The two methods improve the dynamic response and are also frequency-adaptive. A control model based on PR controllers was proposed in Ref. [30].

Indirect Methods of CC Control

A method that controls each arm current utilizing PIR controllers was proposed as shown in Figure 5.22(a). It regulates both the AC-side current and the CC without separating each of the current components. An arm current control strategy based on decoupler and PI + VPI controllers was proposed as shown in Figure 5.22(b). The two methods, [22, 23], have definite physical meaning; however, the complex controller design and interacted current components make them hard to implement.

The MPC method also includes CC control as one of the aspects in the formulation of the cost function. The method is divided into three stages according to the control goals, and the cost function in each stage is designed separately without weighting factors. A passivity-based PI controller has been applied to the MMC in the *abc* frame. Two CC control references generated from an analytical optimization problem based on Lagrange multipliers were proposed. However, the estimation was model-based, which is a drawback as there is parameter uncertainty during practical implementations. General methods for calculating the CC references of the MMC based on mathematical optimization by means of Lagrange multipliers in the *abc* frame were shown in Refs. [31–33]. Elimination of DC-side power oscillations can be achieved with either constant capacitive energy sum or with only DC components in CC per phase. Different system control targets do not affect the control performance.

A CC injection method-based instantaneous information was used in Ref. [33]. It can reduce the capacitor voltage without complex, complicated calculation, but DC power ripples occur by this method. The second-order harmonic injection was used in Ref. [34] considering the effects of CC control in capacitive storage reduction as well as DC-side power oscillation elimination. Table 5.4 shows the summary and Comparison of Circulating Current (CC) Control Methods

Energy Balancing Control Methods

The arm energy regulation of SM for the MMC under AC-grid imbalances has also been performed. By exploiting the negative-sequence component in the AC-side voltage for capacitor balancing, the dynamics of SM voltage balancing can be improved. However, the zero-sequence component of AC-side voltage was neglected, and a control performance relies on the detection of the phase and amplitude of the positive- and negative-sequence AC-grid voltages.

A leg average energy control and an arm unbalanced energy control were combined to regulate the SM energy. The method only uses the six arm energies and controls the leg current in the $\alpha\beta 0$ reference frame.

A feedback/feedforward control method to improve the voltage balancing of the MMC under grid imbalances was proposed in Ref. [38]. Based on a detailed analysis of the power terms disturbing the MMC arm energies, the control method is tailored to improve the voltage transient response by adding feedforward terms for arm and phase balancing. The control performance of this method also relies on the detection of the phrase and amplitude of the positive- and negative-sequence AC-grid voltages and CC.

Three alternative methodologies that enable the operation of the converter under all possible faults were proposed in Ref. [39]. Again, six arm energy regulators, six independent current controllers, and two reference calculation stages are used, whereas two different calculation methods that improve energy deviation during the grid imbalances are also derived.

Table 5.4 Summary and Comparison of Circulating Current (CC) Control Methods

Name	Advantages	Disadvantages
2PI(dq^-)	Easy to design	Positive- and zero-sequence CC are neglected Only used in three-phase systems
6PI and 6 notch filters(abc)	Individual control of each phase Dynamic response of DC-voltage regulation	Positive- and zero-sequence CC are neglected Number of controllers
3PI + notch filter + 1PIR	DC-side voltage ripple elimination	Positive-sequence CC is neglected
3PIR(abc)	Individual control of each phase	Additional zero-sequence CC controller CC reference can only achieve Target 2
3PR and 3 high-pass filters (HPFs)	DC-side current ripple elimination	Complex filter parameter design Weak frequency adaptiveness
3PR(abc)	Simplified control structure	No DC-side current ripple elimination Weak frequency adaptiveness
3PR($\alpha\beta 0$) and 1 HPFs	DC-side current ripple elimination	Complex filter parameter design Weak frequency adaptiveness
3quasi-PR(abc) and 3 low-pass filters (LPFs)	No additional DC-side current controller Dynamic response and frequency adaptive	Additional current reference transformation
3quasi-PR(abc) and 3 LPFs	No additional DC-side current controller Dynamic response and frequency adaptive Current reference on the $\alpha\beta$ frame	Filter parameters influence control effect
3 improved PR($abc/\alpha\beta 0$)	Control model with inherent characteristics DC-bus voltage ripple elimination can be implemented in different frames	Additional zero-sequence CC control Accurate filter and estimation are required
6PI + $R_{\omega+2\omega}(abc)$	AC-side power is controlled Simplified model	Complicated control parameter design High digital control resource cost Response at the resonant frequency
6PI + $VPI_{\omega+2\omega}(abc)$	Ac-side power is controlled Enhanced controller performance Simplified model	Complicated control structure Complicated control parameter design High digital control resource cost

(continued)

Table 5.4 Summary and Comparison of Circulating Current (CC) Control Methods *(continued)*

Name	Advantages	Disadvantages
MPC	Reduced number of states Good transient characteristics	Computational complexity Control errors depend on model
Lagrange multipliers(*abc*)	Mathematical optimization DC-side power oscillation elimination Constant energy sum	Complex practical implementation
Instantaneous information(*abc*)3PI(*abc*) and 3PR(*abc*)	Simplified calculation Improved arm stored energy	Additional DC power ripples
6PR(*abc*) and 6 notch and 3 PIR(*abc*)	Constant DC-link active power transfer Improved arm stored energy	Complex control structure Accurate AC voltage parameter is required

5.5.4 DC-Side Voltage or Current Ripples Elimination Control Strategies

The CC consists of zero-sequence components placed alongside the positive- and negative-sequence components at twice the line frequency. The zero-sequence CC flows through the DC bus and converter legs, leading to a ripple in DC-bus power. CC control methods can suppress the DC-side power ripple, but under certain conditions and control methods, additional DC-side power control strategies are also needed. A supplementary DC-voltage ripple suppressing controller that eliminates the second-order harmonic in the DC-bus voltage was expressed in Figure 5.23(a) requiring six additional arm voltage sensors to regulate the DC-bus voltage. A quasi-PR feedback control was proposed in Refs. [42] and [43] to mitigate the DC-bus voltage ripple. Figure 5.23(b) shows that a second HPF is employed to eliminate any DC components. In a practical implementation, such calculations can be realized by subtracting the DC-bus voltage from its low-frequency components through an LPF.

PIR controllers can also be used to reduce the DC-current ripple. The output of the DC-bus current ripple controller is added to the output of the CC suppression controller. The zero-sequence components were also controlled independently through a DC-bus current controller, and a PIR controller was used to regulate the DC current directly. A simple method based on a zero-sequence control loop scheme with a notch filter at twice the grid fundamental frequency and a PIR controller at twice the fundamental frequency was shown in Figure 5.23(c).

5.6 Phase–Locked Loop [42] ~ [47]

Phase-Locked Loop Synchronization Techniques

The most widely accepted synchronization solution to a time-varying signal can be described by the basic structure shown in a block diagram form in Figure 5.24, where the difference between the phase angle of the input and that of the output signal is measured by the phase detection (PD) and is passed through the loop filter (LF). The LF output signal drives the voltage-controlled oscillator (VCO) to generate the output signal, which could follow the input signal.

Synchronous Frame Phase-Locked Loop

Synchronous frame phase-locked loop (PLL) (SF-PLL) is widely used in three-phase systems. The block diagram of SF-PLL is illustrated in Figure 5.25, where the instantaneous phase angle θ is detected by synchronizing

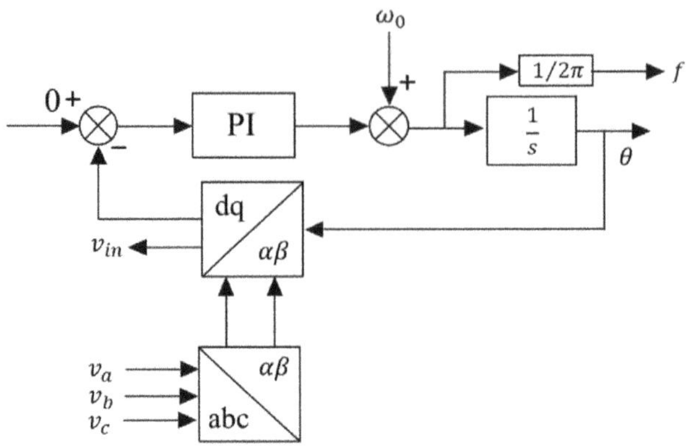

Figure 5.24 Closed-loop synchronization structure

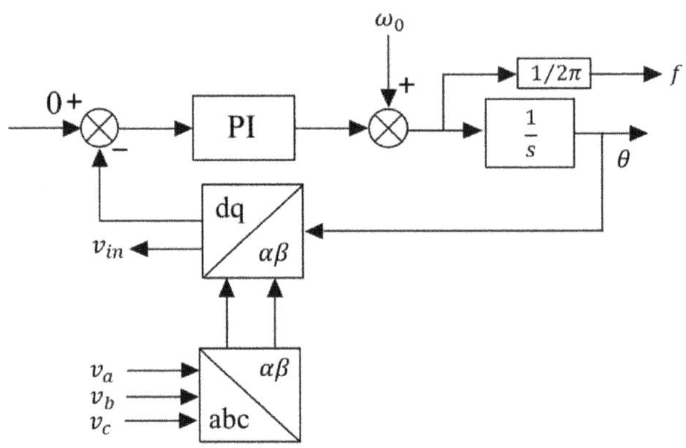

Figure 5.25 Block diagram of synchronous frame phase-locked loop

the PLL rotating reference frame to the utility voltage vector. The PI controller sets the direct or quadrature axis reference voltage $(v_d \text{ or } v_q)$ to zero, which results in the reference being locked to the utility voltage vector phase angle. In addition, the voltage frequency f and amplitude v_m can be obtained as byproducts. Under ideal utility conditions, and without any harmonic distortions or unbalance, SF-PLL with a high bandwidth can yield a fast and precise detection of the phase and amplitude of the utility voltage vector.

In the case of the utility voltage being distorted with high-order harmonics, the SF-PLL can still operate if its bandwidth is reduced at the cost of the PLL response speed reduction in order to reject and cancel out the effect of these harmonics on the output. However, the PLL bandwidth reduction is not an acceptable solution in the presence of unbalanced utility voltage.

Note that the amplitude, phase, and frequency values provided by SRF-PLL are not individual phase, but average information, and SRF-PLL may not be applied to single-phase systems in a straightforward manner. However, it provides a useful structure for single-phase PLLs, as long as the 90-degree-shifted orthogonal component of the single-phase input signal is created.

PQ-PLL

Rolim et al. point out that PLL may fail in tracking the system voltage during start–up under some adverse conditions, and oscillations caused by the presence of subharmonics can pull the stable point of operation synchronized to the subharmonic frequency. To settle these problems, a robust digital synchronization of PLL based on the instantaneous real and imaginary power theory (PQ-PLL) is presented to maintain

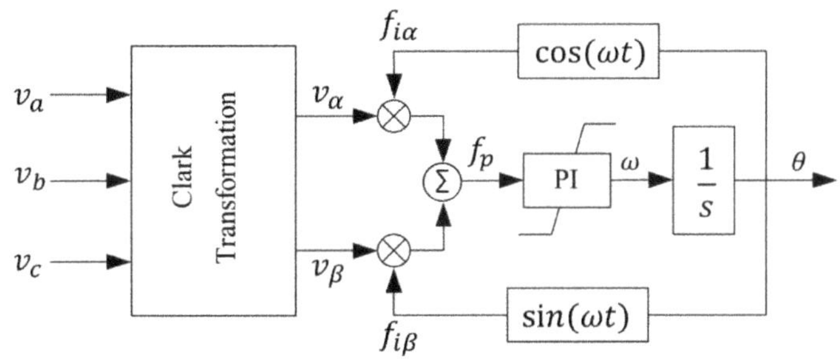

Figure 5.26 Block diagram of PQ phase-locked loop

synchronization in the presence of subharmonics, harmonics, and negative-sequence unbalances. The block diagram of PQ-PLL is shown in Figure 5.26.

It has a similar structure with that of the conventional SF-PLL, but it can be easily understood from the perspective of the instantaneous power theory. Besides, the fundamental positive-sequence components can be obtained as a by–product.

Double Synchronous Frame PLL

The double synchronous frame PLL (DSF-PLL) is based on transforming both the positive–and–negative sequence components of the utility voltage into a double synchronous frame, which can completely eliminate the detection errors of the conventional SF-PLL. The block diagram of the DSF-PLL is shown in Figure 5.27.

Its unique decoupling network can cancel out double frequency oscillations at 2ω in $v^*_{q^+}$, so there is no need to reduce the PLL bandwidth for accurate positive-sequence component estimation compared to the conventional SF-PLL. It is very suitable for the control of grid-interfaced converters operating in severe frequency derivation and unbalanced conditions.

$$\begin{bmatrix} v_{d^+} \\ v_{q^+} \end{bmatrix} = \begin{bmatrix} v^*_{d^+} \\ v^*_{q^+} \end{bmatrix} + \begin{bmatrix} \tilde{v}_{d^+} \\ \tilde{v}_{q^+} \end{bmatrix} \quad \begin{bmatrix} \tilde{v}_{d^+} \\ \tilde{v}_{q^+} \end{bmatrix} = \begin{bmatrix} \cos(2\theta) & \sin(2\theta) \\ -\sin(2\theta) & \cos(2\theta) \end{bmatrix} \begin{bmatrix} \bar{v}_{d^-} \\ \bar{v}_{q^-} \end{bmatrix} \quad \begin{bmatrix} v^*_{d^+} \\ v^*_{q^+} \end{bmatrix} = \begin{bmatrix} v_{d^+} \\ v_{q^+} \end{bmatrix} - \begin{bmatrix} \cos(2\theta) & \sin(2\theta) \\ -\sin(2\theta) & \cos(2\theta) \end{bmatrix} \begin{bmatrix} \bar{v}_{d^-} \\ \bar{v}_{q^-} \end{bmatrix} \tag{5.52}$$

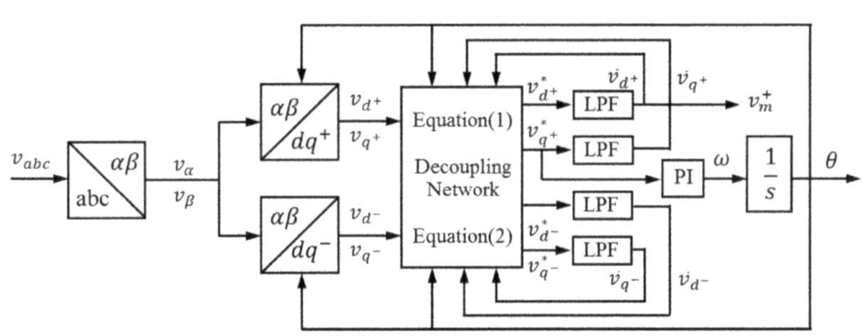

Figure 5.27 Block diagram of double second-order generalized integrator phase-locked loop

$$\begin{bmatrix} v_{d^-} \\ v_{q^-} \end{bmatrix} = \begin{bmatrix} v^*_{d^-} \\ v^*_{q^-} \end{bmatrix} + \begin{bmatrix} \tilde{v}_{d^-} \\ \tilde{v}_{q^-} \end{bmatrix} \quad \begin{bmatrix} \tilde{v}_{d^-} \\ \tilde{v}_{q^-} \end{bmatrix} = \begin{bmatrix} \cos(-2\theta) & \sin(-2\theta) \\ -\sin(-2\theta) & \cos(-2\theta) \end{bmatrix} \begin{bmatrix} \bar{v}_{d^+} \\ \bar{v}_{q^+} \end{bmatrix} \quad \begin{bmatrix} v^*_{d^-} \\ v^*_{q^-} \end{bmatrix} = \begin{bmatrix} v_{d^-} \\ v_{q^-} \end{bmatrix} - \begin{bmatrix} \cos(-2\theta) & \sin(-2\theta) \\ -\sin(-2\theta) & \cos(-2\theta) \end{bmatrix} \begin{bmatrix} \bar{v}_{d^+} \\ \bar{v}_{q^+} \end{bmatrix} \tag{5.53}$$

Sinusoidal Signal Integrator-PLL

The sinusoidal signal integrator PLL (SSI-PLL) tracks the utility voltage by extracting the fundamental positive sequence to the SF-PLL. It operates well under voltage distortions and imbalances. The block diagram of an SSI-PLL is shown in Figure 5.28, where the constant K controls the bandwidth and the response speed of the SSI-PLL. On the other hand, there is another similar structure with a single SSI shown in Figure 5.29, where the positive-sequence component v_α^+ is extracted using a single SSI acting as a filter for the measured utility voltage, and v_β^+ is calculated from v_α^+ by delaying 90 degrees. Constant K controls the filtering response and the PLL bandwidth. The sign is chosen by a sequence detector. If the voltage sequence at the PCC is known in advance, the sequence detector can be eliminated. The main advantage of SSI-PLL is immunity to voltage distortion and unbalance. In addition, it can be extended into single-phase system applications with a few modifications.

Double Second-Order Generalized Integrator PLL

The double second order generalized integrator PLL (DSOGI-PLL) [19-21] is similar to SSI-PLL, which extracts the fundamental positive sequence to the SF-PLL. But it utilizes a DSOGI to implement the quadrature-signal generator.

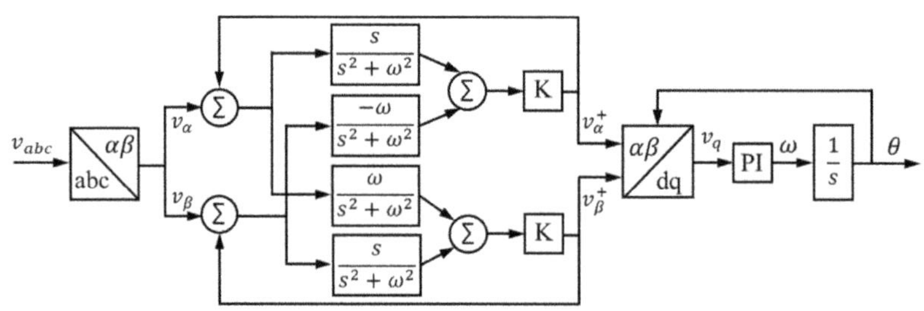

Figure 5.28 Block diagram of phase-locked loop with multiple sinusoidal signal integrators

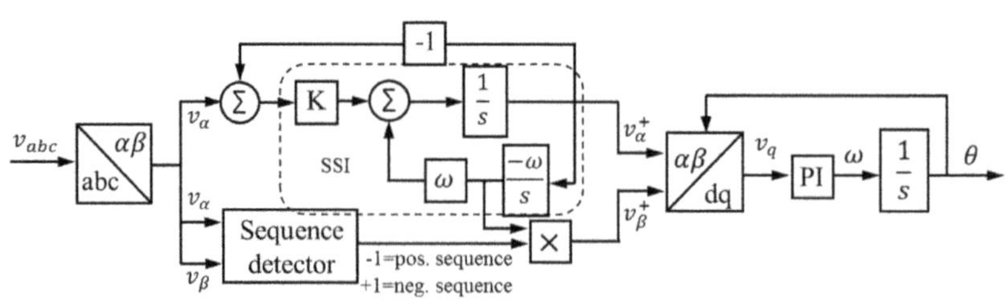

Figure 5.29 Block diagram of phase-locked loop with single sinusoidal signal integrator

The three-phase voltage vector can be expressed as $v_{abc} = [v_a, v_b, v_c]^T$, and its positive-sequence component can be obtained as follows:

$$v_{abc}^{+} = \left[v_a^{+}, v_b^{+}, v_c^{+}\right]^T = [T]\, v_{abc}$$

$$[T] = \frac{1}{3}\begin{bmatrix} 1 & a^2 & a \\ a & 1 & a^2 \\ a^2 & a & 1 \end{bmatrix} a = e^{-j\frac{2\pi}{3}} \tag{5.54}$$

And then the positive-sequence component of the voltage vectors can be obtained as follows:

$$v_{\alpha\beta}^{+} = C_{32} v_{abc}^{+} = C_{32}\,[T]\, v_{abc} = C_{32}\,[T]\, v_{\alpha\beta} = \frac{1}{2}\begin{bmatrix} 1 & -q \\ q & 1 \end{bmatrix}$$

$$C_{32} = \frac{2}{3}\begin{bmatrix} 1 & -\frac{1}{2} & -\frac{1}{2} \\ 0 & \frac{\sqrt{3}}{2} & -\frac{\sqrt{3}}{2} \end{bmatrix} C_{23} = \begin{bmatrix} 1 & -\frac{1}{2} & -\frac{1}{2} \\ 0 & \frac{\sqrt{3}}{2} & -\frac{\sqrt{3}}{2} \end{bmatrix}^T \tag{5.55}$$

In the same way, the negative-sequence component of the voltage vectors can be also obtained as follows:

$$v_{\alpha\beta}^{-} = \frac{1}{2}\begin{bmatrix} 1 & q \\ -q & 1 \end{bmatrix} v_{\alpha\beta} \quad q = e^{-j\pi/2} \tag{5.56}$$

Therefore, the accurate sequence component extraction here requires a 90-degree phase shift for v_α and v_β. There are two solutions for this purpose. They are the transport delay buffer method and the all pass filter method. However, such methods are not frequency-adaptive, which could give rise to errors in positive sequence estimation. Moreover, they can't block harmonics from the input signals. In order to solve these problems, a smart combination of an LPF and a band-pass filter (BPF) is utilized, where the BPF (6) only provides harmonic-filtering functions but the LPF (7) offers both harmonic filtering and the 90-degree phase shift. Besides that, the filter parameter ω is updated from the PLL for frequency-adaptive purposes. The block diagram is shown in Figure 5.30.

$$\mathrm{BPF}(s) = \frac{v'}{v} = \frac{k\omega s}{s^2 + k\omega s + \omega^2} \tag{5.57}$$

$$LPF(s) = \frac{qv'}{v} = \frac{k\omega^2}{s^2 + k\omega s + \omega^2} \tag{5.58}$$

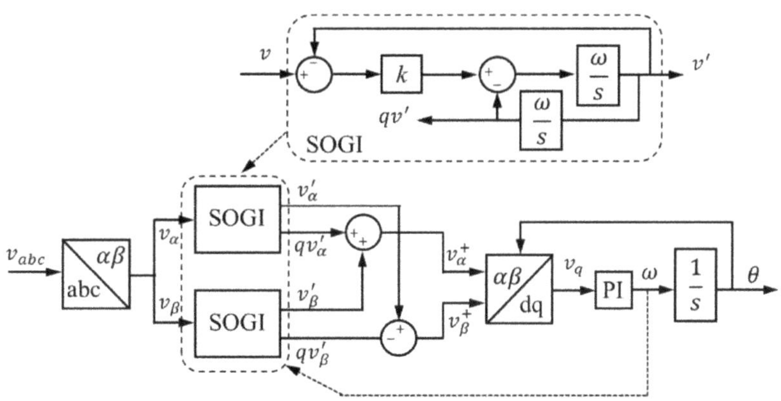

Figure 5.30 Block diagram of double second-order generalized integrator phase-locked loop (DSOGI-PLL)

Having the same principle as SSI-PLL, DSOGI-PLL extracts the fundamental positive sequence v_{α}^{+} and v_{β}^{+} to the conventional SF-PLL. Therefore, it can provide accurate positive–sequence information for grid synchronization even under grid faulty conditions. In addition, it can be also extended into single-phase system applications as SSI–PLL because 90-degree phase shift information can be easily obtained.

Enhanced PLL

Enhanced PLL (EPLL) is a frequency-adaptive nonlinear synchronization approach. The block diagram of EPLL is shown in Figures 5.31 and 5.32.

Its major improvement over the conventional PLL lies in the PD mechanism that allows more flexibility and provides more information such as amplitude and phase angle. There are three independent internal parameters K, K_pK_v and K_iK_v. Parameter K dominantly controls the speed of the amplitude convergence. K_pK_v and K_iK_v control the rates of phase and frequency convergence.

EPLL can provide a higher degree of immunity and insensitivity to noise, harmonics, and unbalance of the input signal. It is an effective method for the synchronization of grid-interfaced converters in polluted and variable-frequency environments. In addition, EPLL can provide the 90–degree shift of the input signal. Therefore, it is an attractive solution in some single-phase system applications.

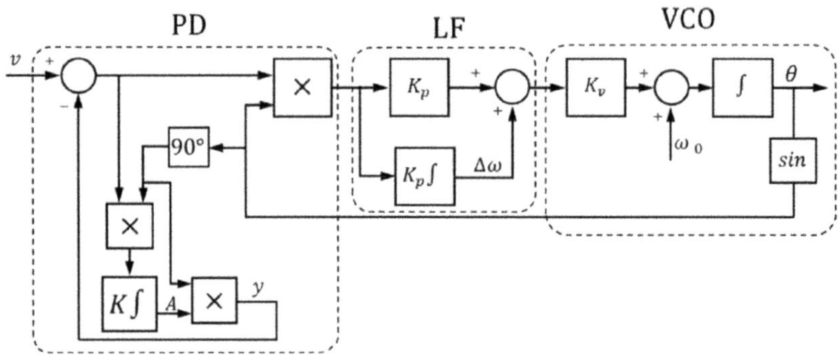

Figure 5.31 Block diagram of single-phase enhanced phase-locked loop

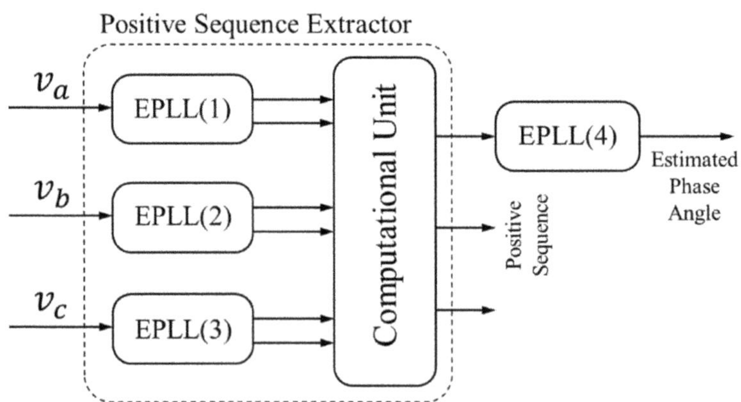

Figure 5.32 Block diagram of three-phase enhanced phase-locked loop (EPLL)

Three-Phase Magnitude PLL

Three-phase magnitude PLL (3MPLL) adaptively tracks and estimates the magnitude, phase angle, and frequency of the input signal. The block diagram is shown in Figure 5.33, which can be mathematically described with the differential Equations (5.59).

$$\begin{cases} \dot{V} = \mu_v \left[e_a \sin\phi + e_b \sin\left(\phi - \frac{2\pi}{3}\right) + e_c \sin\left(\phi + \frac{2\pi}{3}\right) \right] \\ \dot{\omega} = \mu_\omega \left[e_a \cos\phi + e_b \cos\left(\phi - \frac{2\pi}{3}\right) + e_c \cos\left(\phi + \frac{2\pi}{3}\right) \right] \\ \dot{\phi} = \omega + \mu_\delta \dot{\omega} \end{cases} \tag{5.59}$$

The performance of 3MPLL is controlled by three controlling parameters μ_v, μ_ω, and μ_δ. It is an extension of the already presented single-phase PLL. In particular, it can be used as a three-phase nonlinear antialiasing filter with no phase-shift and no amplitude bias. Four unique features of such a filter are frequency adaptation, distortion/unbalance mitigation, structural simplicity, and robustness.

Quadrature PLL

The operation principle of this phase-locked approach is based on estimating in-phase and quadrature-phase amplitudes of the fundamental component of the input signal and is thus called quadrature PLL (QPLL). The block diagram is shown in Figure 5.34, where its output *y(t)* can be formulated as a sum of the in-phase and quadrature-phase components.

$$y(t) = K_s(t)\sin\phi(t) + K_c(t)\cos\phi(t) \tag{5.60}$$

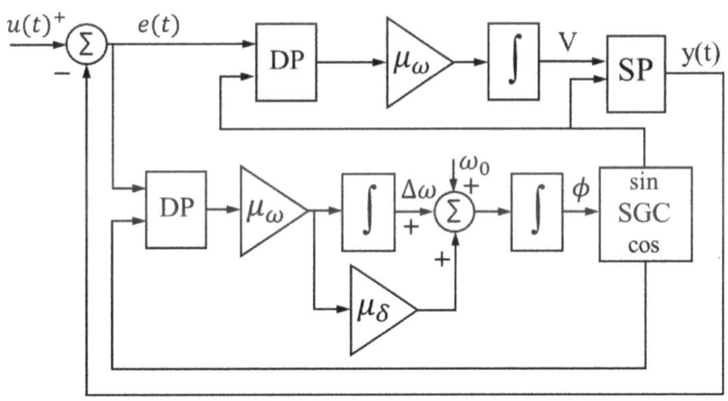

Figure 5.33 Block diagram of three-phase magnitude phase-locked loop

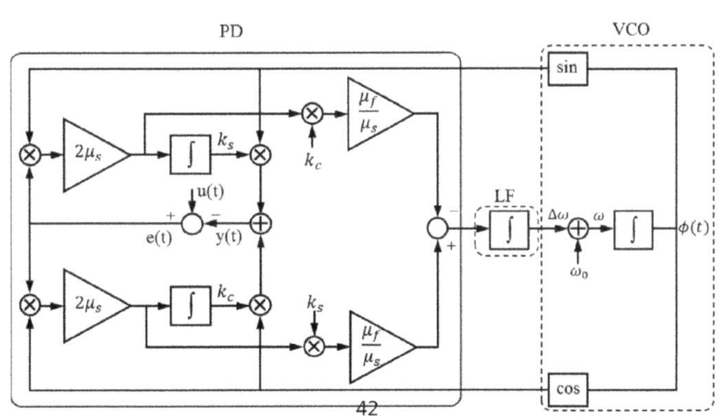

Figure 5.34 Block diagram of quadrature phase-locked loop

Parameters μ_s, μ_c, and μ_f control the behavior of QPLL, each of which corresponds to the in-phase and quadrature-phase amplitudes and frequencies.

QPLL can provide a precise estimation of the in-phase amplitude K_s, quadrature amplitude K_c, and phase ϕ (t) in an adaptive way. It is applicable to both distributed generation and communication system applications.

Robust Single-Phase PLL

Shinnaka has recently proposed a robust single-phase PLL (RPLL) structure [29-30] as shown in Figure 5.35. It consists of a two-phase signal generator, a vector rotator, an LPF, a phase synchronizer, and a multi-rate sample-holder.

Different from other single-phase PLLs, the phase synchronizer is based on the "generalized integral-type PLL method." The unique multi-rate sample-holder keeps the output constant for a specified holding period even for varying input so as to avoid instability phenomena because of instant reuse of the frequency estimate. Besides, an important observation pointed out by Shinnaka is that the differential-type two-phase signal generator is better than the integral-type one in transients. This RPLL can instantly estimate the phase, frequency, and amplitude information of single-phase signals with frequency variation, phase jump, amplitude sag/swell, harmonics distortion, and/or contaminated noise.

Predictive PLL

Predictive PLL (PPLL) is a wide-range synchronization method. The block diagram of PPLL is shown in Figure 5.36. It is composed of a predictor, an oscillator, a phase shifter, and data acquisition blocks. Its phase-locking mechanism is similar to the conventional PLL in the sense that PPLL locks the phase of input signals, but it employs the analytical expressions for computing the frequency, amplitude, and phase difference in a predicted manner. The predictor is the core of PPLL. It acquires three 120° equidistant samples from the input signal: s_0, s_{120}, s_{240}, which are used to characterize one period of a sinusoidal waveform.

Its unique feature lies in its wide-range synchronization capability. The synchronization information can be extracted within two cycles of the input signal period in even the worst cases such as perturbations in frequency, amplitude, and phase angle.

Adaptive Linear Combiner PLL

B. Han proposed a novel PLL that is composed of an adaptive linear combiner (ALC) and proportional integral control. The block diagram is shown in Figure 5.37. The associated formulae are as follows:

$$\begin{cases} V_M \sin(\omega t + \phi) = V_{M\,sin} \sin(\omega t + \theta^*) + V_{M\,cos} \cos(\omega t + \theta^*) \\ \hat{Y} = \begin{bmatrix} W_1 & W_2 \end{bmatrix} \begin{bmatrix} X_1 \\ X_2 \end{bmatrix} = \begin{bmatrix} V_{M\,cos} & V_{M\,sin} \end{bmatrix} \begin{bmatrix} \sin(\omega t) \\ \cos(\omega t) \end{bmatrix} \end{cases} \tag{5.61}$$

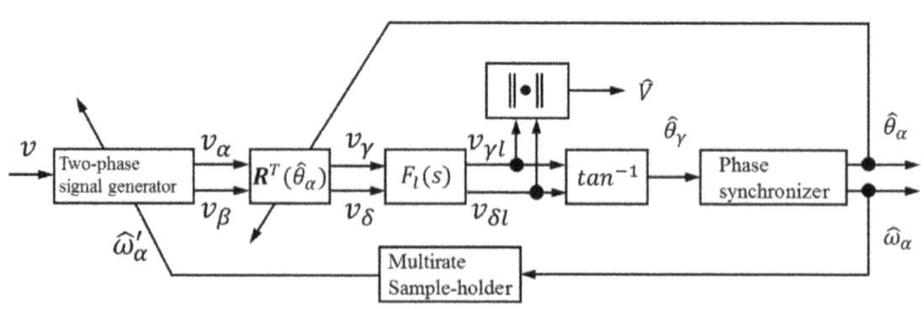

Figure 5.35 Block diagram of robust single-phase phase-locked loop

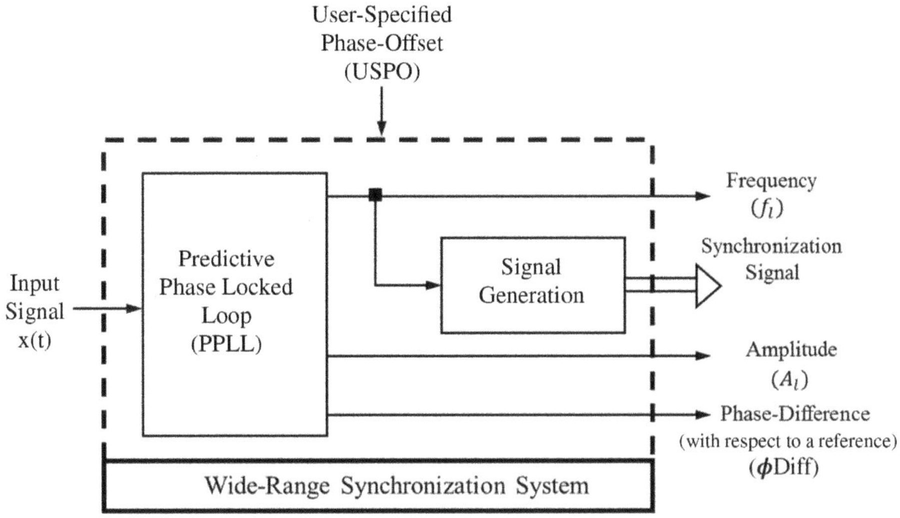

Figure 5.36 Block diagram of predictive phase-locked loop

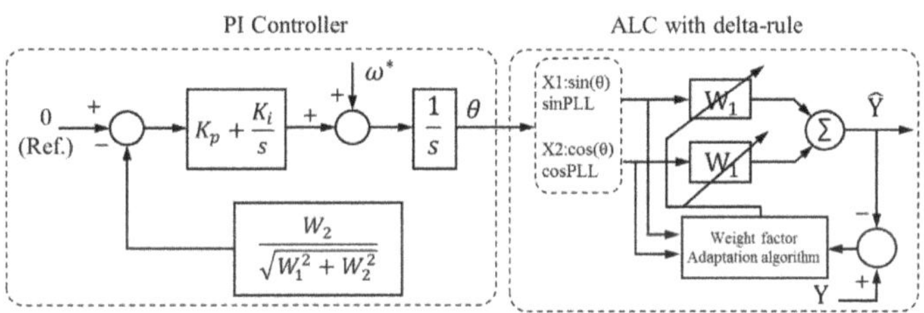

Figure 5.37 Block diagram of adaptive linear combiner PLL (ALC-PLL)

ALC-PLL forces the value of $V_{M\ cos}$ to zero, and thus the phase angle of X_1 will track the fundamental component of the input signal. It can provide accurate performance in phase locking under voltage disturbances, such as sag, harmonics, unbalance, and phase-angle jump.

Multi-rate PLL

The multi–rate PLL (MR-PLL) is a variable sampling-rate system. Compared with the single-rate PLL, it operates at two different sample rates. One sample rate is much higher than the input-signal (carrier) frequency and the other sample rate is equal to the carrier frequency. MR-PLL can provide accurate phase synchronization to severely disturbed signals, and it is an all-digital approach. The disturbance rejection is mainly determined by the frequency response of a high-order band-pass anti-aliasing filter, which provides more disturbance rejection compared to other approaches. The anti-aliasing filter automatically adapts to the input-signal frequency variations through the system's variable sample-rate operation, which is achieved by modifying the processor timer and applying the oversampling technique to the input signal.

Adaptive PLL

The adaptive PLL (APLL) regulates the system gain in an adaptive manner. It consists of three control units that individually control frequency, phase angle, and voltage magnitude. The voltage controller output is used to compensate the reduced gain caused by voltage sag. The output phase angle and its derivative, the frequency signal, are controlled to enable elimination of frequency and phase error without compromising transient responses. The main advantage of the adaptive PLL is that settling time and overshooting are significantly lower in the case of reduced AC voltage magnitude.

Other Synchronization Methods

The simplest synchronization solution is the zero-crossing detection (ZCD) method shown in Figure 5.38, where the zero-crossing point is detected every period and the sin-table pointer is reset from scratch. It can be only used where the input is a stable and sinusoidal signal because of its transient sensitivity and sensitivity to noise.

The SV method is another synchronization solution. The basic idea of the SV method is to detect the angular frequency of the SV ($v = v_\alpha + jv_\beta$). The block diagram is shown in Figure 5.39. The instantaneous angular frequency can be obtained as follows:

$$\omega_v = \frac{d}{dt}\left(a\tan\frac{v_\beta}{v_\alpha}\right) = \frac{v_\alpha \frac{dv_\beta}{dt} - v_\beta \frac{dv_\alpha}{dt}}{v_\alpha^2 + v_\beta^2} \tag{5.62}$$

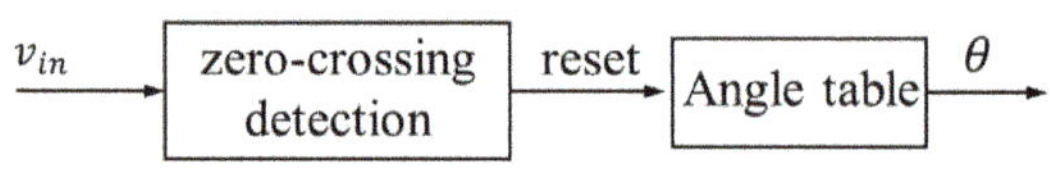

Figure 5.38 Zero-crossing detection method

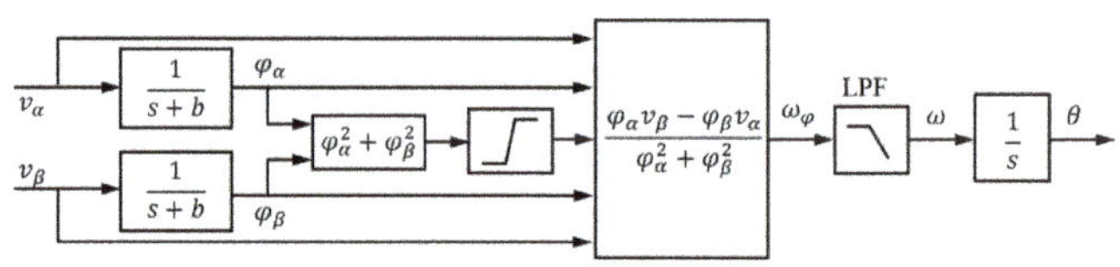

Figure 5.39 Space vector method

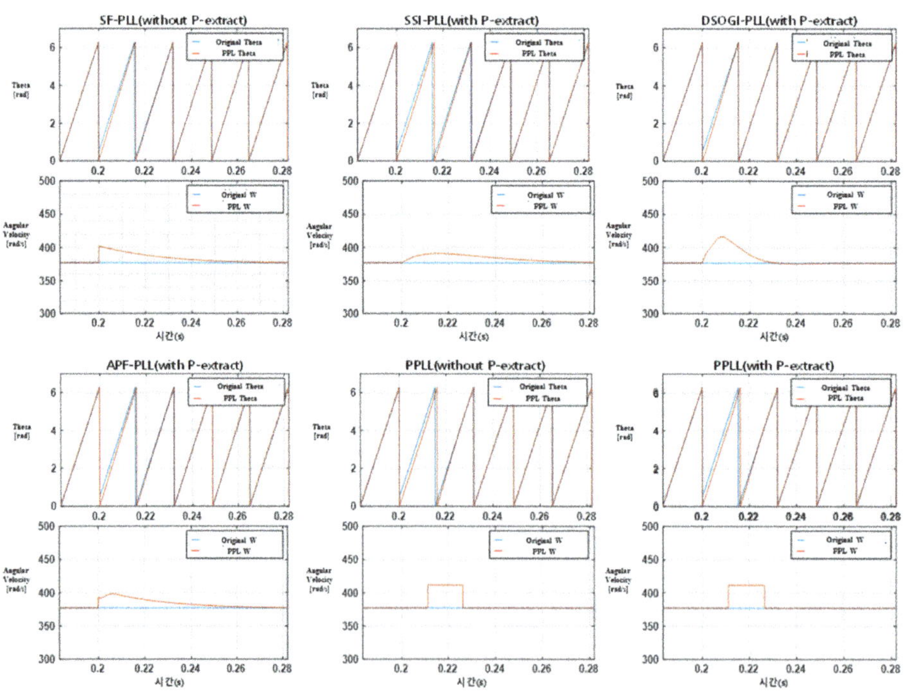

Figure 5.40 Simulation results according to the kinds of PLL(injected shifted phase 30 degree at 0.2s)

To avoid the numerical derivations in Equation (5.62), another approach should be attempted. The following SV ($\varphi = \varphi_\alpha + j\varphi_\beta$) is considered by applying the transfer function

$$\varphi(s) = \frac{1}{s+b} v(s) \tag{5.63}$$

where b is a damping coefficient that ensures the asymptotic stability of Equation (5.63). Note that with b=0, Equation (5.63) would be sensitive to the initial conditions and would accumulate errors.

In steady state, the vector v has the same average frequency as φ, whereas the instantaneous frequencies are different. Equation (5.62) of the angular frequency can be applied to the vector as

$$\omega_\varphi = \frac{\varphi_\alpha \frac{d\varphi_\beta}{dt} - \varphi_\beta \frac{d\varphi_\alpha}{dt}}{\varphi_\alpha^2 + \varphi_\beta^2} = \frac{\varphi_\alpha v_\beta - \varphi_\beta v_\alpha}{\varphi_\alpha^2 + \varphi_\beta^2} \tag{5. 64}$$

The average value of Equation (5.64) delivers the frequency of the vector Equation (5.62), by using an LPF with a cut-off frequency chosen specifically to obtain the desired dynamics. The main advantages of this method are that it is robust and also less sensitive to disturbances.

Another SV method is based on the fact that the $\alpha\beta$ components of the grid voltage are mutually dependent. It is formulated as follows:

$$\begin{cases} X[n+1] = \gamma R(\omega_0 T_s) X[n] + BU[n] \\ Y[n] = \gamma R(\omega_0 T_s) X[n] + DU[n] \end{cases} \tag{5.65}$$

where $U[n] = [v_\alpha[n], v_\beta[n]]^T$, $B = D = (1-\gamma)I_2$ in which $I_2$is the 2×2 identity matrix, and $\gamma \in (0, 1)$ is called the forgetting factor. A closer value of γ to 1 provides better filtering of the inputs, and therefore, results in less distortion in the estimated phase angle. SV method can be tuned to provide a nearly distortion-free estimation. The main drawback of this method is sensitivity to input frequency variations and imbalance.

The last method presented in this paper is the weighted least-squares estimation (WLSE) method, which employs the recursive WLSE algorithm. This method rejects the impact of negative sequence and accommodates variations in the frequency. In addition to threshold and change detection problems, it exhibits other shortcomings such as long transient time intervals in detecting frequency changes and computational problems associated with the (Least-Squares) methods. Figure 5.40~5.43 show the simulation results of according to the kinds of PLL

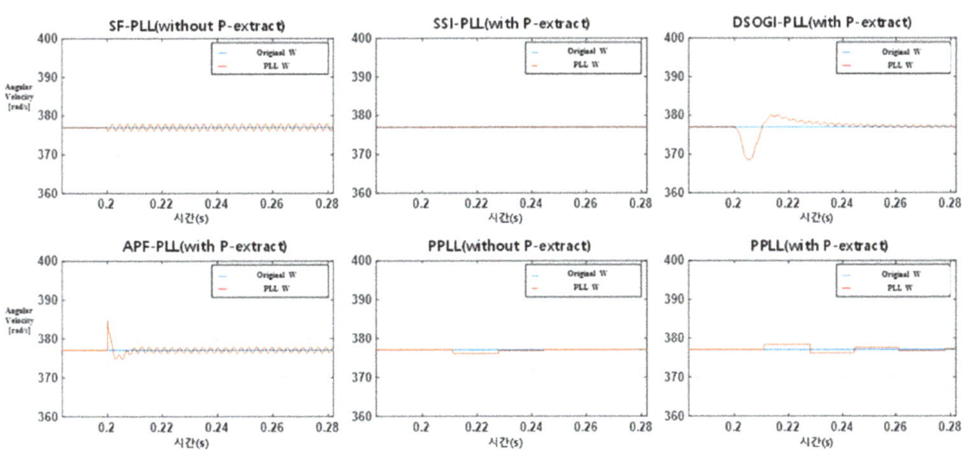

Figure 5.41 Simulation results of angular velocity according to the kinds of PLL(injected shifted phase 30 degree at 0.2s)

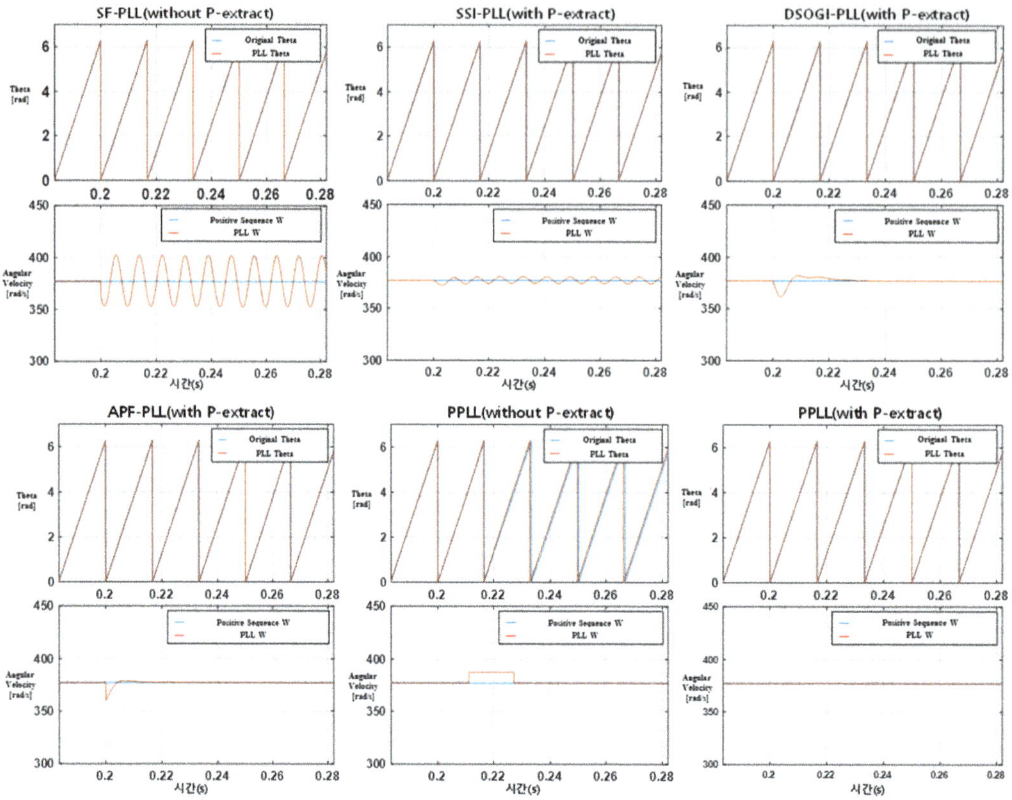

Figure 5.42 Simulation results of angular velocity according to the kinds of PLL(injected 50% negative sequence at 0.2s)

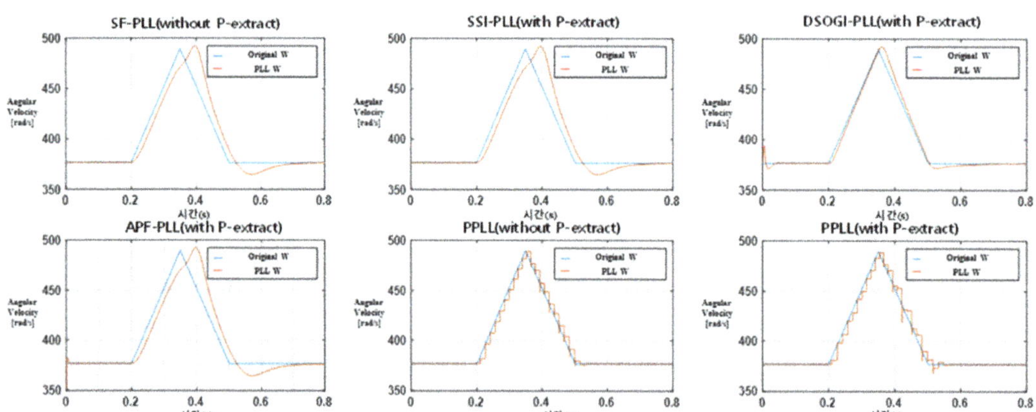

Figure 5.43 Simulation results of angular velocity according to the kinds of PLL(reference change to Ramp up at 0.2s)

5.7 Digitalization of System Functions [48] ~ [53]

The transfer function of a second-order generalized integrator (SOGI) shown in Figure 5.30 for a single sinusoidal signal is

$$G(s) = \frac{2s}{s^2 + \omega_0^2} \tag{5.66}$$

Another important application of the SOGIs is to generate quadrature signals. The new, more detailed, mathematical expression is introduced as follows:

$$G_{SOGI}(s) = \frac{s\omega_0}{s^2 + \omega_0^2} \tag{5.67}$$

The new transfer function has two poles placed in $\pm j\omega_0$ and a zero in origin, exactly equal to Equation (5.66). The only difference between the two expressions is the gain.

Discrete-Time Implementation of SOGIs Based on the Discretization of the Transfer Functions

A method to implement in the discrete time-domain consists of developing continuous time-domain transfer functions and discretizing them with one of the techniques indicated in Table 5.5.

From these and for the SOGI structure shown in Figure 5.44, Table 5.6 shows the different transfer functions in the z-domain. The implementation in digital processors is achieved with the differential equations obtained from these transfer functions. For the SOGI, better results are obtained with the Tustin and ZOH(Zero Order Holder) techniques. The main advantage of this method is a very good resolution independently of the sampling time, whereas the main disadvantage is that the difference equations have to be calculated off-line, especially for ZOH and matched methods, and it is then not possible to have an adaptive response with the grid frequency.

Discrete-Time Implementation of SOGIs using Discrete Integrator.

In this chapter, a method based on discretizing each one of the continuous time-domain integrators of the SOGI of Figure 5.45 is presented. Also, Table 5.7 shows the mathematical expressions in the discrete time-domain and in the difference equations of the discrete integrators, which will be used in the development of the discrete SOGIs.

Table 5.5 Methods for discretizing continuous-time transfer functions

Method	Relation	Method	Relation
Euler Forward	$s = \frac{1-z^{-1}}{T_s z^{-1}}$	ZOH	$H(z) = (1 - z^{-1})\,Z\left[\frac{H(s)}{s}\right]$
Backward Euler	$s = \frac{1-z^{-1}}{T_s}$		
Tustin	$s = \frac{2}{T_s}\frac{1-z^{-1}}{1+z^{-1}}$	Matched Pole-Zero	$z = e^{T_s s}$

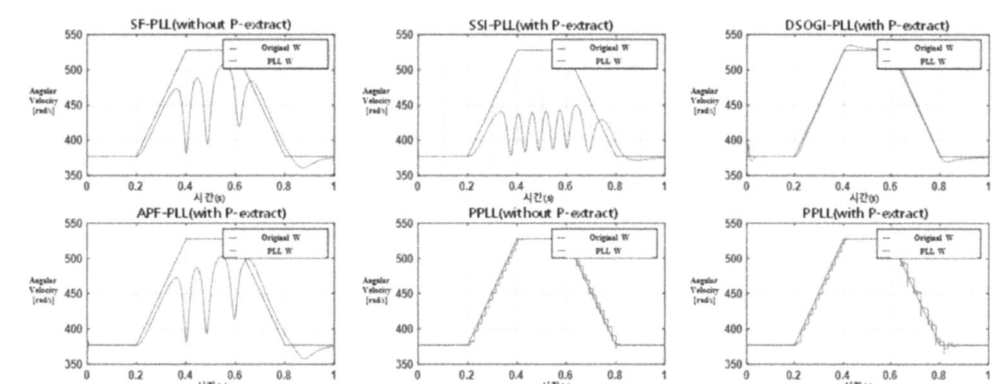

Figure 5.44 Following angular frequency in change with ramp reference (2), respectively, with phase-locked loop (PLL)

Table 5.6 Discrete Integrators

Type of integrator	z-domain equation Difference equation
Euler Forward	$H(z)=\dfrac{\omega_o T_S(z-1)}{z^2-2z+\omega_o^2 T_S^2+1}$
Backward Euler	$H(z)=\dfrac{\omega_o T_S}{1+\omega_o^2 T_S^2}\cdot\dfrac{(z-1)z}{z^2-\dfrac{2}{1+\omega_o^2 T_S^2}z+\dfrac{1}{1+\omega_o^2 T_S^2}}$
Tustin	$H(z)=\dfrac{2\omega_o T_S}{4+\omega_o^2 T_S^2}\cdot\dfrac{z^2-1}{z^2+\dfrac{\omega_o^2 T_S^2-8}{4+\omega_o^2 T_S^2}z+1}$
ZOH	$H(z)=\dfrac{\sin(\omega_o T_S)(z-1)}{z^2-2\cos(\omega_o T_S)z+1}$
Matched Pole-Zero	$H(z)=\dfrac{k'(z-1)}{z^2-2\cos(\omega_o T_S)z+1}$

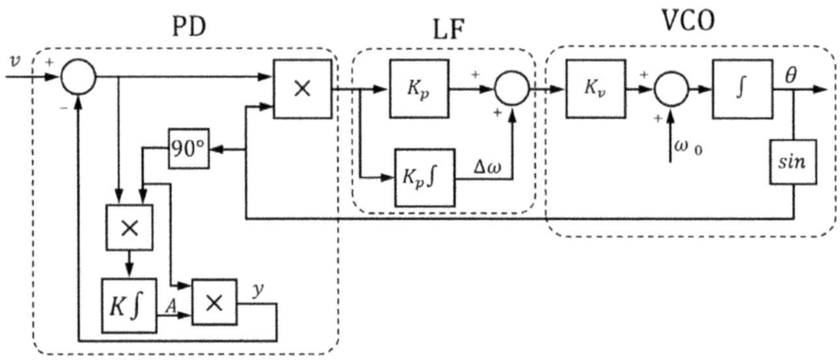

Figure 5.45 Signal flow of second-order generalized integrator in continuous time-domain

Table 5.7 SOGI Possible Implementations (Figure 5.44)

Direct Gain	Feedback Gain	Computat. Delay
Tustin	Tustin	No
Tustin	Tustin	Yes
Euler Backward	Euler Backward	No
Euler Backward	Euler Backward	Yes
Euler Forward	Euler Forward	No
Euler Forward	Euler Forward	Yes
Euler Forward	Euler Backward	No
Euler Forward	Euler Backward	Yes

From the discrete integrators indicated in Table 5.7, different implementations modify the discrete integrators used in direct gain and feedback gain. Moreover, in some realizations, a computational delay has been included in the third column in series with the feedback gain, which models the inherent delay produced in the programming process. Column 4 indicates where is represented in Figure 5.6, for each option, the temporal response of the output $y(z)$ faced up a step with amplitude 1 in the input $x(z)$. The used sampling time is $Ts = 200\ us$.

According to Figure 5.46, only the temporal responses of options a, d, and g coincide with the response described for the continuous time-domain SOGI. These options are represented through the z-domain equations in Figures 5.47, 5.48, and 5.49, respectively. Of these three options, the programming of Figure 5.47 is not easy because there are algebraic loops, and it is rejected. Regarding the structures of Figures 5.46 and 5.47, they are very similar; the difference is only one of them has a delay in the output response. Under these circumstances, the topology of Figure 5.46 is chosen because it has a structure more similar to a classical PI controller.

Thus, a correct operation of the SOGI presented in Figure 5.48, based on Euler Backward Discrete Integrators, is assured as resonant controller.

Tuning of the Proportional Resonant Controller to Grid Current Control of a Grid Converter with L-Filter

In this section, a numerical method to tune the proportional resonant controller based on a discrete time-domain SOGI to control a grid converter with L-filter is presented. The SOGI is discretized by the method explained previously in this chapter. As it was indicated in previous sections, the SOGIs are an ideal solution to current control in stationary reference frames ($\alpha\beta$-frames) of grid converters. In this case, there is no cross

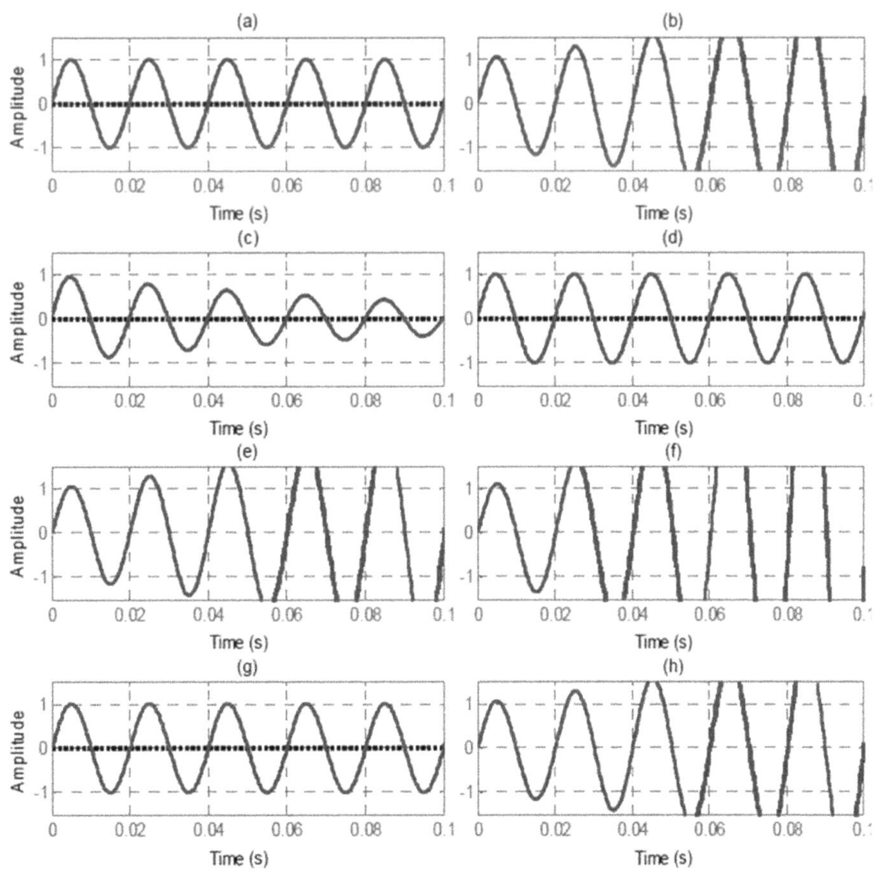

Figure 5.46 Temporal response of $y(z)$ under a step in the input $x(z)$ of the proposals

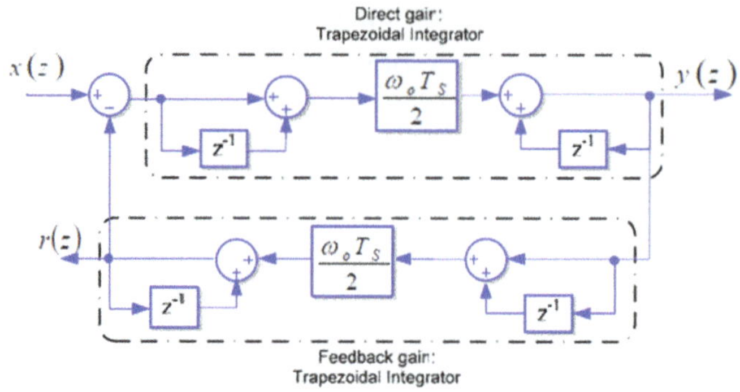

Figure 5.47 Second-order generalized integrator based on Tustin integrators

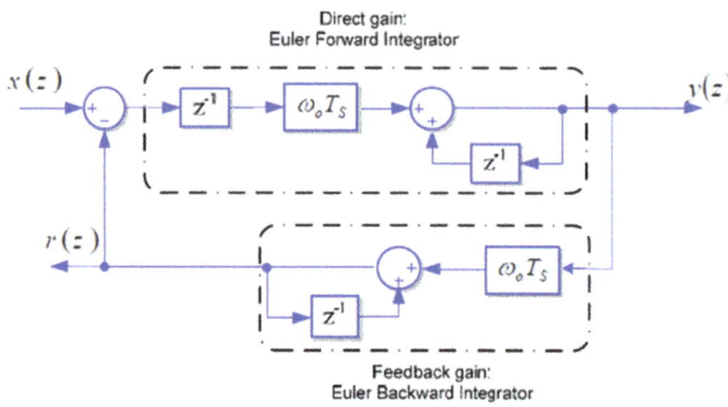

Figure 5.48 Second-order generalized integrator based on Euler backward integrators and computational delay

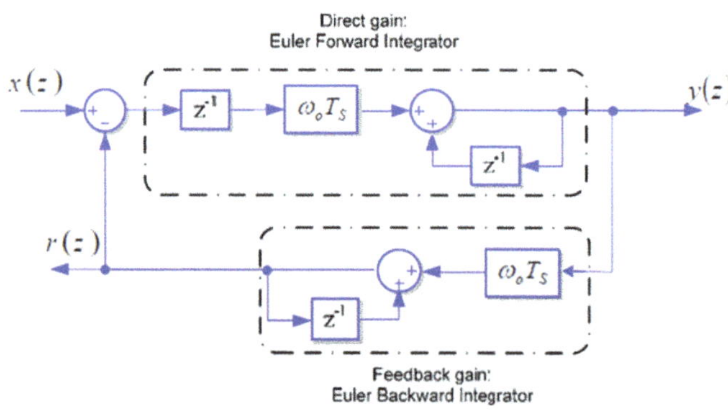

Figure 5.49 Second-order generalized integrator based on an Euler forward integrator and an Euler backward integrator

coupling between axes, and the control problem can be reduced to one axis or β, giving rise to the control loop for each one of the axes represented in Figure 5.50.

The transfer function of the SOGI implemented in the previous section (Figure 5.48) is

$$G_{SOGI}(z) = \frac{K_i \omega_0 T_S z(z-1)}{(z-1)^2 + \omega_0^2 T_s^2 z} \tag{5.68}$$

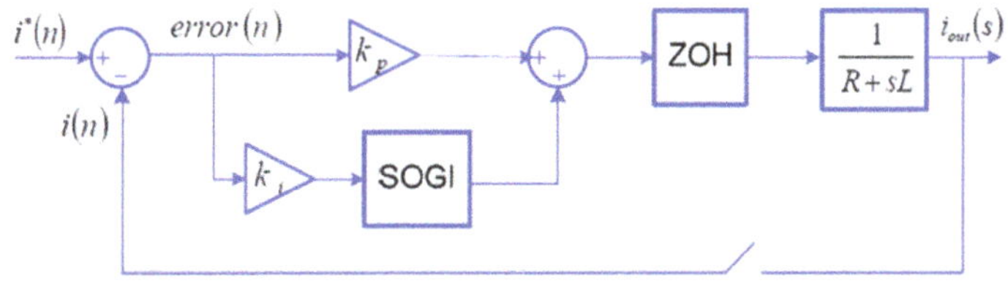

Figure 5.50 Current control loop of α or β axis for a grid converter connected to the grid through an L-filter

If $\omega_0^2 T_s^2 << 2$, which is verified for $T_S < 1.4ms$, if $\omega_0 = 2\pi 50\frac{rad}{s}$ and supposing a factor much less than 10, the previous equation can be approximated as

$$G_{SOGI}(z) \approx \frac{K_i \omega_0 T_S z}{(z-1)} \tag{5.69}$$

So, the proportional resonant controller can be expressed as

$$G_{P+SOGI}(z) = k_p + k_i \frac{\omega_0 T_S z}{(z-1)} = K_p \frac{z-\alpha}{z-1} \tag{5.70}$$

where $K_p = k_p + k_i \omega_0 T_S$ and $\alpha = \frac{k_p}{k_p + k_i \omega_0 T_S}$.

If the proportional resonant controller is used as a current controller of a grid converter connected to the grid through an L-filter, then the transfer function of the plant in the discrete time-domain is

$$G_{RL}(z) = \left(1 - z^{-1}\right) Z\left[\frac{G_{RL}(s)}{s}\right] = \frac{\frac{1}{R_1}\left(1 - e^{-R_1 T_S L_1}\right)}{z - e^{-R_1 T_S L_1}} = \frac{a}{z-b} \tag{5.71}$$

The loop gain will have two poles placed in $p_1 = 1$ and $p_2 = b$, very near to 1, and a zero located in $z = \alpha$.

Depending on the controller gain, the closed-loop system $G_t(z)$ can have two conjugate poles as shown. In this case, the general form of the denominator polynomial of $G_t(z)$ is

$$P(z) = \left(z - \rho e^{j\varsigma}\right)\left(z - \rho e^{-j\varsigma}\right) = z^2 - (2\rho \cos) z + \rho^2 \tag{5.72}$$

where $\rho = e^{-(\xi \omega_n T_S)}$ and $\varsigma = \omega_n T_S \sqrt{\left(1 - \xi^2\right)}$. To verify these conditions, K_p and α have to take the next values:

$$K_p = \frac{1 + a - 2\rho\rho \cos}{b}; \alpha = \frac{a - \rho^2}{bK_p} \tag{5.73}$$

and then $k_p = aK_p$ and $k_i = \frac{K_p - k_p}{\omega_0 T_S}$.

For a concrete application, the next values are established: $T_S = 200us$, $\xi = 0.707$, and a settling time of t_s=2 ms. The last one takes the indicated value so that the computational delay, because of the reference calculated in *k* being applied to the PWM generator in *k+1*, does not affect to the system response. From settling time and damping coefficient values, ω_n can be calculated with the expression $t_s \approx \frac{4}{\xi \omega_n}$ for the second-order system. Figure 5.51 represents the response ($i_{out}(s)$, the reference ($i^*(n)$), and the error ($error(n)$)) of the current control loop shown in Figure 5.50. As can be verified in Figure 5.51, the system responds as it has been configured since, for example, the error takes a null value in the 2 ms that is the settling time.

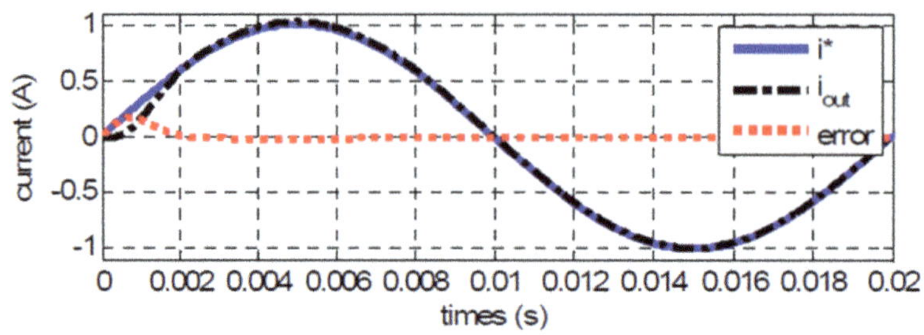

Figure 5.51 Reference ($i^*(n)$), output ($i_{out}(s)$) and error ($error(n)$) currents of the control diagram

References

1. C. K. Kim, "Level selection algorithm of MMC converter," in *IEEE International Conference on Power Electronics and Drive Systems*, 2018.
2. G. Dinget al "New technologies of voltage source converter (VSC) for HVDC transmission system based on VSC," in *Proceedings of IEEE Power & Energy Society General Meeting*, pp. 1–8. 2008.
3. Q. Tu and Z. Xu, "Impact of sampling frequency on harmonic distortion for modular multilevel converter,"*IEEE Trans. Power Del.*, vol. 26, no. 1, pp. 298–306, 2011.
4. D. Zhong et al., "Reduced switching-frequency active harmonic elimination for multilevel converters," *IEEE Trans. Ind. Electron.*, vol. 55, no. 4, pp. 1761–1770, Apr. 2008.
5. D. Zhong, L. M. Tolbert, and J. N. Chiasson, "Active harmonic elimination for multilevel converters," *IEEE Trans. Power Electron.*, vol. 21, no. 2, pp. 459–469, Mar. 2006.
6. J. N. Chiasson et al., "A unified approach to solving the harmonic elimination equations in multilevel converters," *IEEE Trans. Power Electron.*, vol. 19, no. 2, pp. 478–490, Mar. 2004.
7. R. Lil, J. E. Fletcher, and B. W. Williams, "Influence of third harmonic injection on MMC-based HVDC Transmission Systems," IET Gener. *Transm. Dis.*, vol. 10, no. 11, pp. 2764– 2770, 2016.
8. U. N. Gnanarathna, A. M. Gole, and R. P. Jayasinghe, "Efficient modeling of modular multilevel HVDC converters (MMC) on electromagnetic transient simulation programs," *IEEE Trans. Power Del.*, vol. 26, pp. 316–324, 2011.
9. M. A. Perez et al "Circuit topologies, modeling, control schemes, and applications of modular multilevel converters," *IEEE Trans. Power Electron.*, vol. 30, pp. 4–17, 2015.
10. R. Picas, et al., "Improving capacitor voltage ripples and power losses of modular multilevel converters through discontinuous modulation," in *39th Annual Conference of the IEEE Industrial Electronics Society*, 2013, pp. 6233–6238.
11. M. Saeedifard and R. Iravani, "Dynamic performance of a modular multilevel back-to-back HVDC system," *IEEE Trans. Power Del.*, vol. 25, pp. 2903–2912, 2010.
12. L. Zhang, et al., "A generalized precharging strategy for soft startup process of the modular multilevel converter-based HVDC systems," *IEEE Trans. Ind. Appl.*, vol. 99, pp. 1-1.
13. K. Shi, et al., "A novel startup scheme for modular multilevel converter," in *2012 IEEE Energy Conversion Congress and Exposition (ECCE)*, Sept. 2012, pp. 4180–4187.
14. T. Li ,et al., "Start-up scheme for hvdc system based on modular multilevel converter," in *Renewable Power Generation Conference (RPG 2013), 2nd IET.*
15. A. Das, H. Nademi, and L. Norum, "A method for charging and discharging capacitors in modular multilevel converter," in *37th Annual Conference on IEEE Industrial Electronics Society IECON 2011*, Nov. 2011, pp. 1058–1062.
16. P. Wang et al., "Start-up control of an offshore integrated MMC multi-terminal HVDC system with reduced dc voltage," *IEEE Trans. Power Syst.*, vol. 31, no. 4, pp. 2740–2751, 2016.
17. P. Dai, G. Guo, and Z. Gong, "A selection precharge method for modular multilevel converter," *Int. J. Control Automation.*, vol. 9, no. 4, pp. 161–170, 2016.
18. B. Li "Closed-loop precharge control of modular multilevel converters during start-up processes," *IEEE Trans. Power Electron.*, vol. 30, no. 2, pp. 524–531, Feb. 2015.
19. B. Li, Y. Zhang, D. Xu, "Start-up control with constant precharge current for the modular multilevel converter," *Proceedings of the 2014 IEEE 23rd International Symposium on Industrial Electronics*, 2014, pp. 673–676.

20. J. Li, et al., "Operation and control methods of modular multilevel converters in Unbalanced AC Grids: A Review," *IEEE J. Em. Sel. Top. Power Electron.*, vol. 7, no. 2, pp. 1258–1271, 2018.
21. J.-W. Moon, et al., "A control method of HVDC-modular multilevel converter based on arm current under the unbalanced voltage condition," *IEEE Trans. Power Del.*, vol. 30, no. 2, pp. 529–536, Apr. 2015.
22. Y. Liang, et al., "Arm current control strategy for MMC-HVDC under unbalanced conditions," *IEEE Trans. Power Del.*, vol. 32, no. 1, pp. 125–134, Feb. 2017.
23. X. Cai, et al., "Passivity-based control of HVDC transmission system based on modular multilevel converter under unbalanced grid conditions," in *Proceeding of the IET Renewable Power Generation*, Sep. 2013, pp. 1–4.
24. G. Bergna, et al., "Modular multilevel converter passivity-based PI control suited for balanced and unbalanced grid conditions," in *Proceedings of the IEEE International Conference on Industrial Technology*, Mar. 2015, pp. 3072–3078.
25. Y. Wan, S. Liu, and J. Jiang, "Systematic modeling and control of indirect modular multilevel converter (MMC) with grid unbalance estimation," in *Proceedings of the IEEE Industrial Electronics Society Conference*, Nov. 2013, pp. 1349–1354.
26. G. Falahi and A. Huang, "Control of modular multilevel converter based HVDC systems during asymmetrical grid faults," in *Proceeedings of the IEEE Industrial Electronics Society Conference,* Oct. 2014, pp. 4595–4600.
27. D. Ramirez, et al., "Assessment of a non linear current control technique applied to MMC-HVDC during grid disturbances," *Renew. Energy*, vol. 101, pp. 945–963, Feb. 2017.
28. J. Qin and M. Saeedifard, "Predictive control of a modular multilevel converter for a back-to-back HVDC system," *IEEE Trans. Power Del.*, vol. 27, no. 3, pp. 1538–1547, Jul. 2012.
29. J.-W. Moon, et al., "Model predictive control with a reduced number of considered states in a modular multilevel converter for HVDC system," *IEEE Trans. Power Del.*, vol. 30, no. 2, pp. 608–617, Apr. 2015.
30. Y. Ma and L. Fan, "Circulating current and DC current ripple control in MMC under unbalanced grid voltage," in *Proceedings of the North American Power Symposium*, Oct. 2015, pp. 1–6.
31. G. Bergna, et al., "Analysis of MMCs under unbalanced grid conditions with different load current control strategies and Lagrange-based differential current control," in *Proceedings of International Future Energy Electronics Conference*, Nov. 2013, pp. 669–674.
32. G. Bergna, et al., "Generalized ABC frame differential current control ensuring constant DC power for modular multilevel converters under unbalanced operation," in *Proceedings of the IEEE European Conference*, Sep. 2013, pp. 1–10.
33. G. Bergna, et al., "MMC circulating current reference calculation in ABC frame by means of Lagrange multipliers for ensuring constant DC power under unbalanced grid conditions," in *Proceedings of the IEEE European Conference on Power Electronics*, Aug. 2014, pp. 1–10.
34. J. Li, et al., "Impact of circulating current control in capacitor voltage ripples of modular multilevel converters under grid imbalances," *IEEE Trans. Power Del.*, vol. 33, no. 3, pp. 1257–1267, Jun. 2018.
35. M. Vasiladiotis, "Operation of modular multilevel converters under grid asymmetries," in *Proceedings of the IEEE Industrial Electronics Society Conference*, Nov. 2013, pp. 6281–6286.
36. M. Vasiladiotis, N. Cherix, and A. Rufer, "Impact of grid asymmetries on the operation and capacitive energy storage design of modular multilevel converters," *IEEE Trans. Ind. Electron.*, vol. 62, no. 11, pp. 6697–6707, Nov. 2015.
37. J.-J. Jung, et al., "A cell capacitor energy balancing control of modular multilevel converter considering the unbalanced AC grid conditions," in *Proceedings of the IEEE International Power Electronics Conference*, May 2014, pp. 1268–1275.
38. A. E. Leon and S. J. Amodeo, "Energy balancing improvement of modular multilevel converters under unbalanced grid conditions," *IEEE Trans. Power Electron.*, vol. 32, no. 8, pp. 6628–6637, Aug. 2017.
39. E. Prieto-Araujo "Control of modular multilevel converters under singular unbalanced voltage conditions with equal positive and negative sequence components," *IEEE Trans. Power Syst.*, vol. 32, no. 3, pp. 2131–2141, May 2017.
40. X. Shi, et al., "Analysis and control of DC voltage ripple for modular multilevel converters under single line to ground fault," in *Proceedings of the IEEE Energy Conversion Congress and Exposition*, Sep. 2013, pp. 4112–4118.
41. X. Shi, et al., "Characteristic investigation and control of a modular multilevel converter-based HVDC system under single-line-to-ground fault conditions," *IEEE Trans. Power Electron.*, vol. 30, no. 1, pp. 408–421, Jan. 2015.
42. X.-Q. Guo, W.-Y. Wu, and H.-R. Gu, "Phase locked loop and synchronization methods for grid interfaced converters: a review," *Przegld Elektrotechniczn.*, vol. 87, no. 4, pp. 182–187, 2011.
43. E. V. Appleton, "Automatic synchronization of triode oscillators," *Proc. Cambridge Phil. Soc.*, vol. 21, no. 3, p. 231, 1923.

44. S. L. Goldman, "Second-order phase-lock-loop acquisition time in the presence of narrow-band Gaussian noise," *IEEE Trans. Commun.*, vol. 21, no. 4, pp. 297–300, 1973.
45. M. Moeneclaey, "The optimum closed-loop transfer function of a phase-locked loop used for synchronization purposes," *IEEE Trans. Commun.*, vol. 31, no. 4, pp. 549–553, 1983.
46. N. Margaris, V. Petridis, and D. Efthymiatos, "Phaselocked loop control of a nonlinear DC motor," *IEEE Trans. Ind. Electron.*, vol. 29, no. 1, pp. 91–93, 1982.
47. H. Miura, et al., "A synchronous rectification using a digital PLL technique for contactless power supplies," *IEEE Trans. Magn.*, vol. 41, no. 10, pp. 3997–3999, 2005.
48. F. J. Rodriquez, et al., "Discrete-time implementation of second order generalized integrators for grid converters," in *2008 34th Annual Conference of IEEE Industrial Electronics*, Orlando, FL, 2008.
49. P. Rodriguez, et al., "New positive-sequence voltage detector for grid synchronization of power converters under faulty grid conditions," in *Proceedings of the IEEE Power Electronics Specialists Conference*, 2006.
50. L. R. Limongi, et al., "Analysis and comparison of phase locked loop techniques for grid utility applications," in *Power Conversion Conference*, Nagoya, 2007, pp. 674–681.
51. X. Yuan, et al., "Stationary frame generalized integrators for current control of active power filters with zero steady-state error for current harmonics of concern under unbalanced and distorted operating conditions," *IEEE Trans. Ind. Appl.*, vol. 38, pp. 523–532, 2002.
52. D. N. Zmood and D. G. Holmes, "Stationary frame current regulation of PWM inverters with zero steady-state error," *IEEE Trans. on Power Electron.*, vol. 18, no. 3, pp. 814–822, May 2003.
53. M. Ciobotaru, R. Teodorescu, and F. Blaabjerg, "A new single-phase PLL structure based on second order generalized integrator," in *37th IEEE Power Electronics Specialists Conference*, 2006.

6

MMC Topology Extra

6.1 Low-Voltage Ride Through [1]–[5]

Basic Control Principles:

Assuming that a high-voltage direct current (HVDC) converter in inverter mode has positive current and power, and in rectifier mode it has negative. The direction reference is therefore consistently defined for both the AC and the DC side. This leads to regular AC-frequency-droop curves having a negative slope, and DC-voltage-droop curves (e.g., Figure 6.2) having a positive slope.

The control of the DC voltage shows a lot of similarities with frequency control in an AC power system. Whenever a deficit occurs in the current that flows in/out of the DC grid, the voltage at the different nodes will immediately react to this change as a result of the capacitors and cable capacitance discharging or charging. This current deficit can occur, for example, when a converter faces an outage. If the converter was exporting to the AC system, the outage will cause a current surplus leading to a voltage increase. On the contrary, if the converter was importing into the DC grid, the converter outage causes a current shortage, leading to a voltage decrease. DC voltage can therefore be taken as a balance indicator, similar to AC frequency in AC power systems.

In AC systems however, the frequency is the same throughout the entire system. The active power flow on a line is observed through the phase angle. In DC systems, DC voltage is influenced both by the global balance and by the currents on the lines. Therefore, DC voltage is not a true global measure, and it is more challenging to use DC voltage as a reference for balancing control. Despite these difficulties in terms of control, DC voltage still appears to be the best indicator for stable DC-grid operation. Based on the described behavior, it is clear that the current balance has to be restored as soon as possible in order to keep the voltage from falling or rising. All control structures discussed further on in this article rely on the fact that deviation with controlled DC voltage (local measurement at each DC bus) is determined and corrected by the control action, which comes down to restoring the current balance.

Assuming that a system is lossless, all node voltages would be identical, and the current balance can be directly rewritten as a power balance. In a real system, DC line losses will add to the power balance equation and cause a mismatch between imported and exported power. The general phenomena are the same: a change in the power balance will cause the DC voltage to deviate.

There are two basic principles of controlling DC voltage, that is, current-based control and power-based control. The larger the DC voltage deviates from the reference value, the larger the differences between power and current-based controls become. So, it is important to mention that both current-based control and power-based control behave identically at the operating point.

Current-Based Control: In a current-based control scheme, the relationship used to control the DC voltage is an I–V characteristic. The main advantage of a voltage–current control characteristic is that it reflects linear control action from the viewpoint that a voltage deviation will result in an equivalent current deviation. As the voltage is linked with charging or discharging the capacitive DC system, the control is linear, and is the same for all voltage deviations from the reference value. The voltage–power relation is consequently nonlinear (parabolic). Current-based control can be directly linked with DC network dynamics: the charging of the capacitances in the DC network relies on a linear current–voltage relation that is represented as resistance. This makes current-based control somehow intuitive from a physical perspective as it has the same physical units as the DC line impedance.

Power-Based Control: As an alternative to current-based control, DC-voltage control may be expressed in terms of active power. So, the power–voltage relation is linear. Contrary to current-based control, power-based control shows nonlinear (hyperbolic) control action. DC-voltage control, therefore, is nonlinear. Power-based control has the advantage of being more intuitive from a power system perspective, where the focus is often on transmitted power. It is also similar to power frequency control in AC systems. The power–voltage relation would have a physical unit of 1/A or Ω/V, which does not have a direct intuitive physical meaning.

Power-based control can easily be integrated with existing vector control schemes, which is well suited to be used with the standard dq-current controllers used in Voltage-sourced converter (VSC) HVDC systems.

Converter system limits: There are several limits for HVDC converter control. The three most important are briefly described and shown in Figure 6.1.

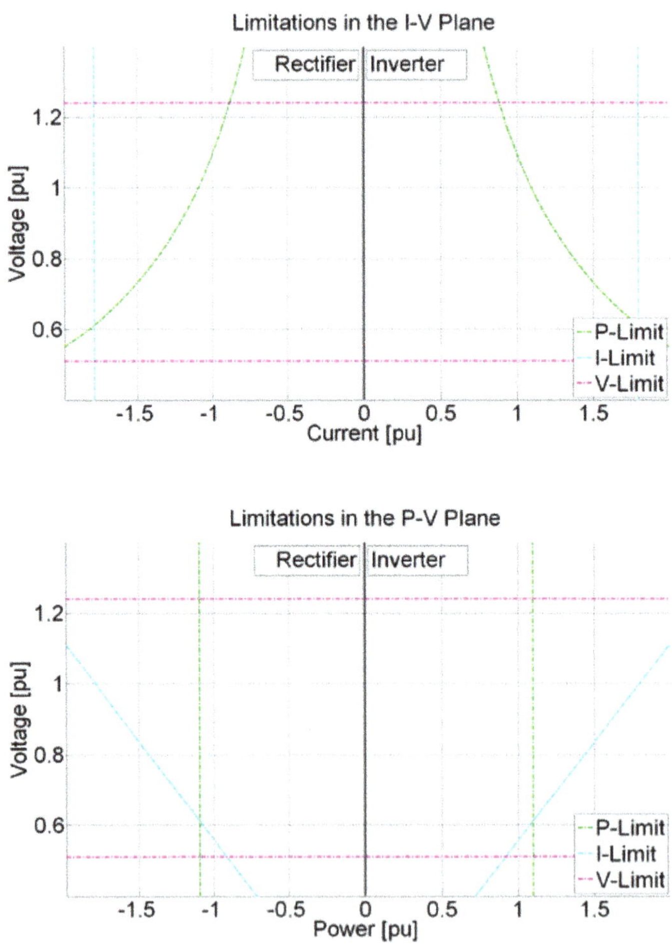

Figure 6.1 Power-based control with limits

- *DC-voltage limits:* DC voltage has an upper and a lower limit. The upper limit is determined by the insulation of the switching components. The lower limit is more complex, as it is based on a limitation of the modulation index, and also depends on the converter topology and the converter control implementation.
- *Power limits:* The active power has an upper limit. This power limit is caused by the semiconductor current limit, which limits the AC current and consequently the power (assuming a constant AC voltage). The power limit appears as a hyperbolic curve in the IV curves, and as a straight vertical line in the PV curves.
- *DC current limits:* The DC current has an upper limit, based on the current rating of the connected DC components. This appears as a straight vertical line in the IV curves and as a straight line toward the axes intersection in the PV curves.

Basic Converter Control Strategies: In the previous section, emphasis was put on how a converter can control the DC–voltage control and the distinction that can be made based on a P–V or I–V relationship. In this section, different control schemes and how they can be approached as a generalized case of a voltage-droop control will be discussed. The discussion holds both for I–V- and P–V-based droops.

The control is complicated because of the fact that the voltage in the DC-grid varies as a result of the DC-grid power flow. To obtain a certain power flow, the voltage set points have to be different for all converters and should reflect the steady-state power flow solution of the DC network.

The explanation provided in this section reflects the mathematical representation of the control objective in terms of droop characteristics. The actual implementation, however, does not necessarily rely on the proportional droop controller, as discussed in the following. The term "basic strategies" refers to consistent control strategies that do not depend on the operating point. The control strategies have only a single parameter, which is the droop value or control gain. The control is therefore linear. The advanced control schemes in Section 5 can be seen as a combination of the three control types discussed in this section.

Voltage-droop control (positive-droop constant): Voltage-droop control creates a proportional relationship between the voltage and the control base (current or power) and is shown in Figure 6.2.

The voltage, current, and power deviations are defined as $\Delta V = V - V_{ref}$, $\Delta I = I - I_{ref}$ and $\Delta P = P - P_{ref}$.

The I–V droop relation can be written as

$$\Delta I_I = \frac{1}{k_I}\Delta V \tag{6.1}$$

The droop constant is called k. The control base is indicated by the subscript "I" or "P." The subscript "ref" indicates the voltage, current, and power set-points, with the natural relationship: Pref = Vref · Iref.

Rewriting Equation (6.1) yields the corresponding power equation, which is parabolic:

$$\Delta P_I = \left(\frac{V_{ref}}{k_I} + \frac{P_{ref}}{V_{ref}}\right)\Delta V + \frac{1}{k_I}\Delta V^2 \tag{6.2}$$

The equation for the P–V-droop relation, shown in Equation (6.3), is similar to Equation (6.2):

$$\Delta P_P = \frac{1}{k_P}\Delta V \tag{6.3}$$

Rewriting Equation (6.3) in terms of current yields the Equation (6.4) that is hyperbolic:

$$\Delta I_P = \left(\frac{1}{k_P} - I_{ref}\right)\frac{\Delta V}{\Delta V + V_{ref}} \tag{6.4}$$

The slope of the I–V or P–V characteristic is determined by the droop value that is the inverse of the proportional controller gain. The droop in constant k in these equations originates from the slope of the characteristics in Figure 6.1 and is the inverse of the proportional controller gain used in actual implementation.

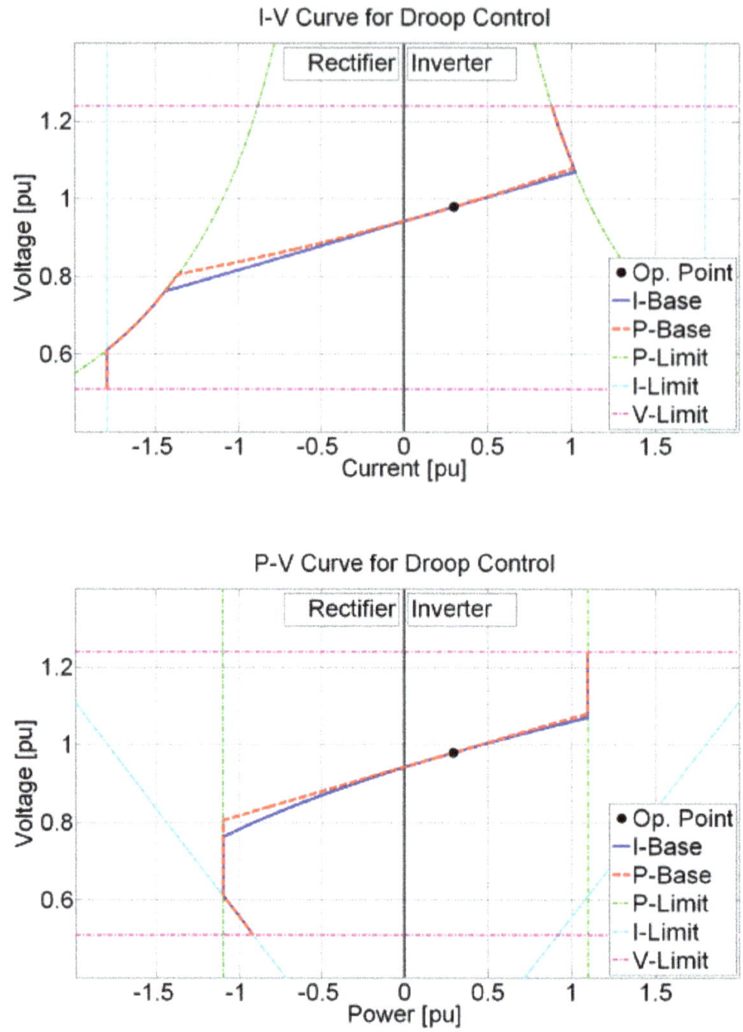

Figure 6.2 Voltage-droop control

The stability of the droop controller depends on the droop value and the actual implementation. If well designed, the voltage-droop control leads to a stable operation. When a contingency occurs at faults, the droop control is determined by a steady-state deviation from the voltage reference as a result of the proportional control action.

Constant flow control (infinite-droop constant): Constant flow control means constant current control and constant power control and can, respectively be represented as a vertical line segment in the I–V and P–V plane. Mathematically, constant flow control can be described as a limited case of the aforementioned voltage droop, with a droop constant equal to infinity, thereby not changing the current/power flow when the DC voltage changes. The converter control will, at all costs, try to maintain the current/power injection constant, irrespective of the value of the DC voltage at its DC bus.

Figure 6.3 shows the characteristic curves for constant flow control modes in both the P–V and I–V plane. Constant current control leads to a linear action in the P–V plane. Constant power control leads to hyperbolic action in the I–V plane.

However, in a real system, constant flow control is implemented using a PI controller to keep the current/power equal to its reference value. This is because of many reasons, but theoretically comes down to the limiting of an infinite-droop value.

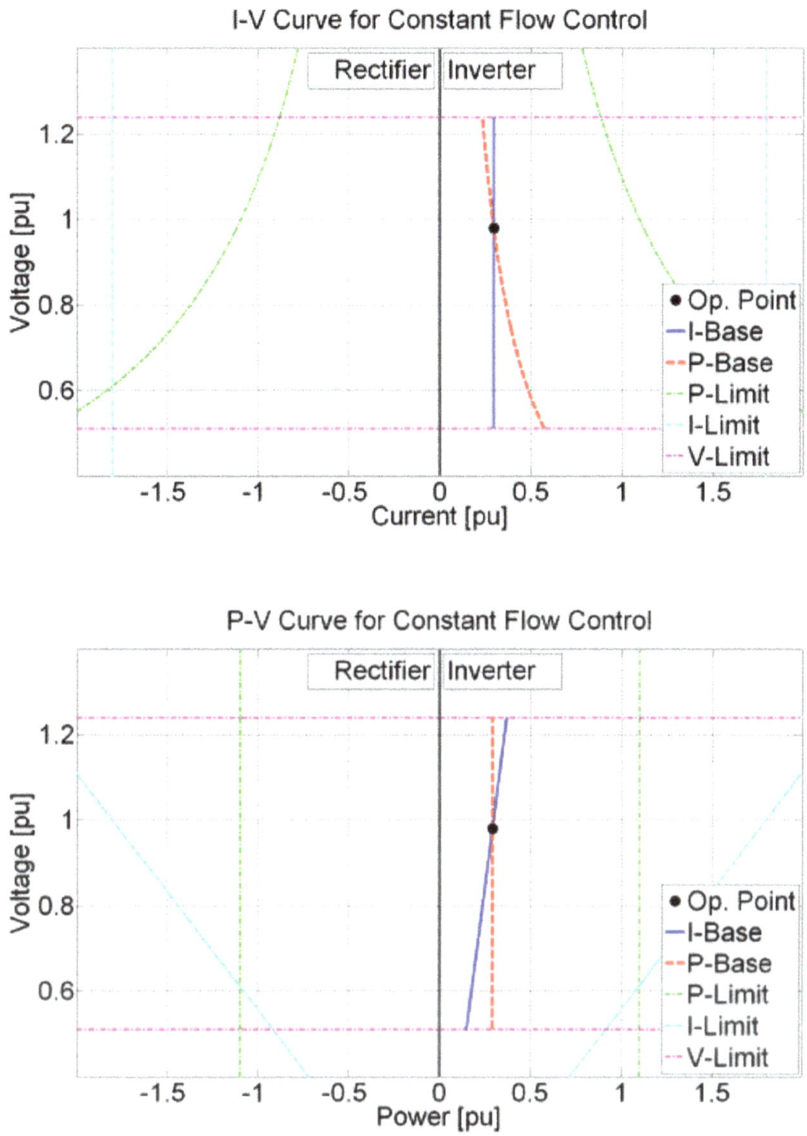

Figure 6.3 Constant flow control

Constant voltage control (zero-droop constant): Constant voltage control can be represented as a horizontal curve in the I–V or P–V plane shown in Figure 6.4. DC-voltage control can be regarded as the limiting case for which the droop constant goes to zero. From Equations (6.1) to (6.3), that the converter in this case controls the voltage limit to its reference value is clear. In Figure 6.4, no distinction between current and power-based control is shown, because the curves are identical. A converter with this method of control is often referred to as a DC slack bus.

This control mode relates to a proportional control gain of infinity. This, of course, is unrealistic and will lead to instability. Constant voltage control is therefore in reality not realized with an infinite gain, but with a PI controller. In theoretical steady state, the outcome is the same, but because of dynamic reasons, the infinite control gain is not viable.

Advanced Converter Control Strategies: There are special control methods that cannot be described by a single control parameter. They are therefore nonlinear, or to be more precise, piecewise-linear. The actual

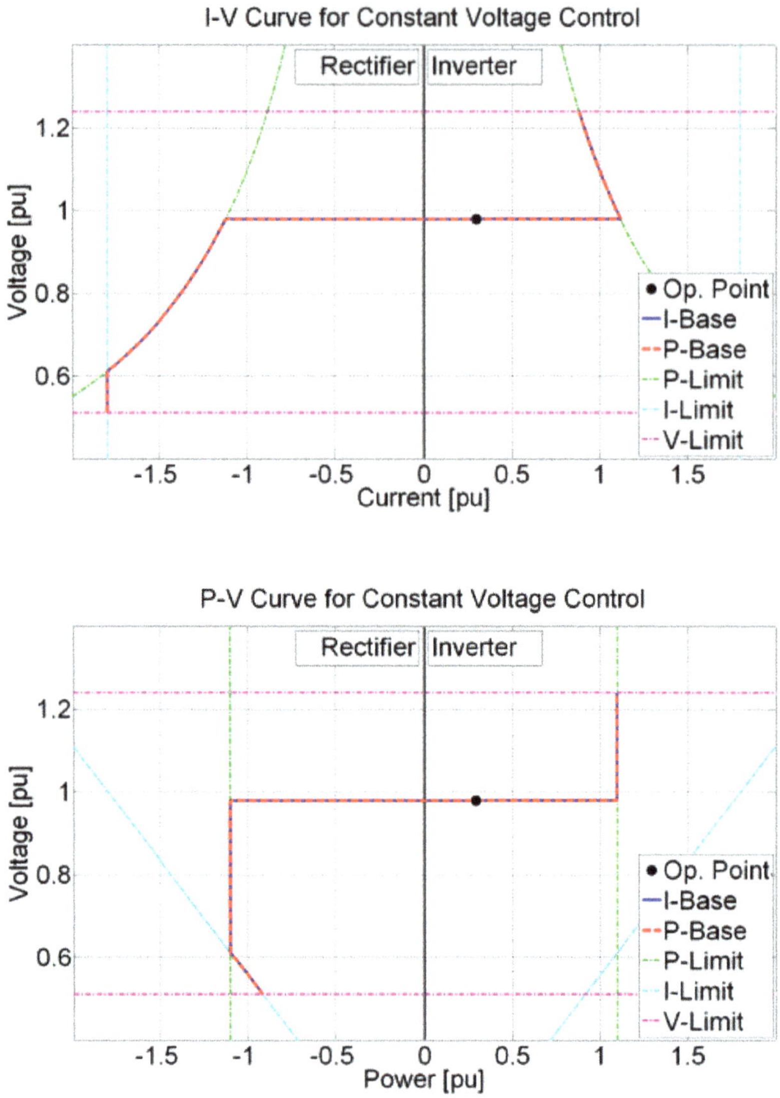

Figure 6.4 Constant voltage control

control behavior depends on the operating point, thereby making a distinction between normal and disturbed operation. The advanced converter control strategy's behavior changes when large disturbances occur.

Voltage Margin Control: Voltage margin control is that it is a combination of constant current/power control and constant voltage control. A converter with voltage margin control normally operates in constant current/power control mode (as long as the voltage is within the normal operation voltage margin). If the voltage deviation reaches the limit of the voltage margin, the converter controller switches to constant voltage control, clamping the voltage at the margin limit to prevent further voltage deviation. Figure 6.5 shows the voltage margin control.

Dead-Band-Droop Control: Another advanced control strategy is droop control with an additional dead-band. Just like with voltage margin control, the converter operates at constant current/power control mode as long as the voltage is within the normal operation band or margin. If the voltage deviates to the limit of the dead-band, the controller switches to droop control. Voltage margin control can be considered as the special case of dead-band-droop control with a zero-droop constant. Figure 6.6 shows the dead-band droop control.

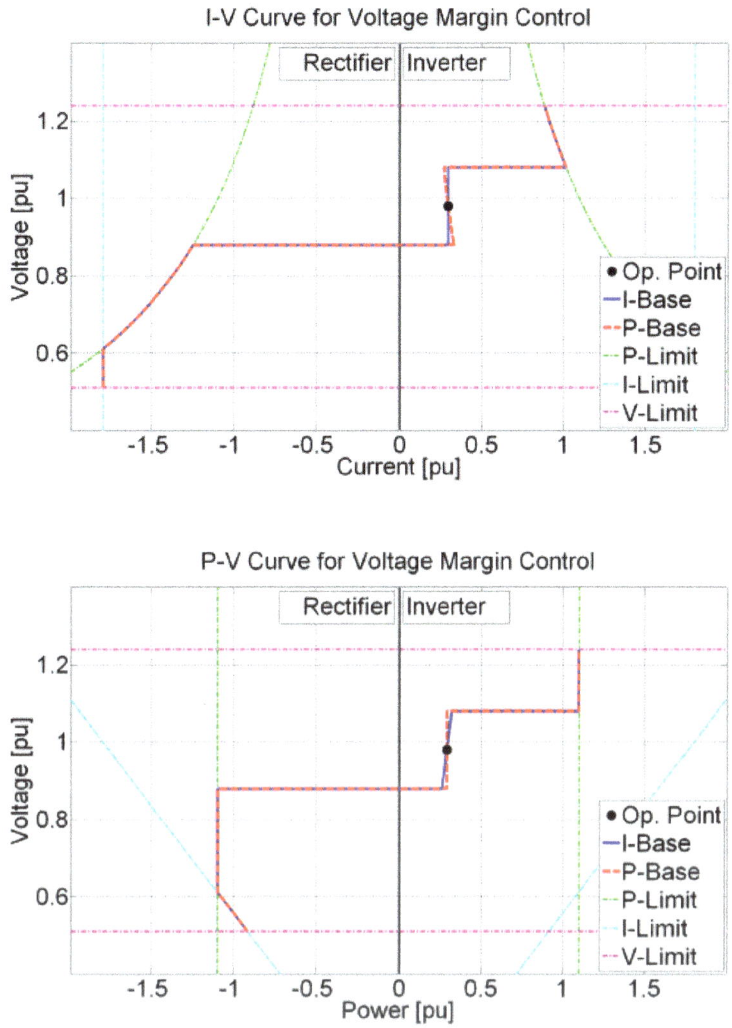

Figure 6.5 Voltage margin control

Undead-Band-Droop Control: The undead-band-droop control is fully based on voltage droop but distinguishes between normal and disturbed operation by defining different droop constants for these two operation conditions (normal and disturbed operation). It offers the possibility to optimize the droop constants and the overall control separately for normal and disturbed operation. Figure 6.7 shows the undead-band-droop control

The undead-band-droop control is the most general control method, and all the other mentioned control methods can, in fact, be considered specific examples of parameter sets of the undead-band-droop control. For example, dead-band-droop control can be considered as a special case of undead-band-droop control with an infinite-droop constant (constant current/power) under normal operation. Similarly, voltage margin control can be considered as a special case of undead-band-droop control with an infinite–droop constant under normal operation (constant current/power) and a zero-droop constant under disturbed operation (constant voltage). Regular droop control can be seen as a limiting case of undead-banddroop control with the same droop constant in both operation modes. As a compliment to dead-band-droop control, a zero-droop constant under normal operation (constant voltage) can be combined with droop control under disturbed operation. Unlike a dead-band, where the converter does not react to voltage changes, and keeps the voltage constant under normal operation. Another possible variation is a controller with droop control under normal operation and with zero droop (constant voltage) under disturbed operation.

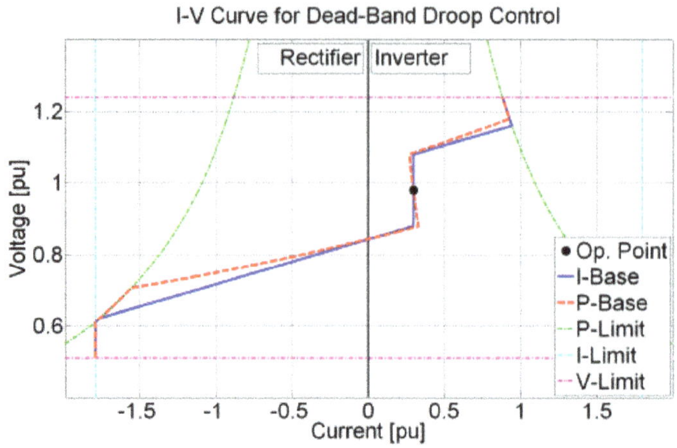

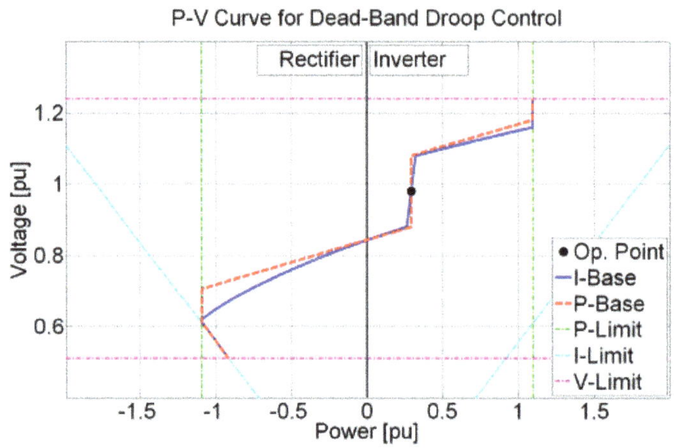

Figure 6.6 Dead-band-droop control

6.2 Grounding of Modular Multi-level Converter (MMC) System [10]–[15]

6.2.1 Grounding Methods of MMC System

A. Y Winding and Delta Winding for Transformer of Conventional VSC System [21]

The Y/Δ arrangement of the three-phase transformer, with the Δ-connection on the VSC side as illustrated in Figure 6.8, excludes zero sequence so that the issue is avoided altogether. However, the system of Figure 6.8 (a) is met with two problems:

Firstly, in the event of a single-phase-to-ground fault on the Δ side, the line-to-line DC voltage reaches a peak over-voltage of about 1.5 p.u. as shown in Figure 6.8 (b), and secondly, in the event of a DC-line-to-ground fault, the upper and lower DC line voltages become unbalanced, and they remain permanently unbalanced even after the fault is cleared as shown in Figure 6.8 (c).

From Figure 6.8, the important conclusion is that the delta connection does not provide a return path for the ground current (zero sequence) to equalize the charges in the capacitors in order to re-establish voltage balance in upper DC voltage and lower DC voltage.

On the other hand, it is found that the aforementioned two problems are solved when the VSC side of the three-phase transformer is Y connected with its neutral point grounded as shown in Figure 6.9.

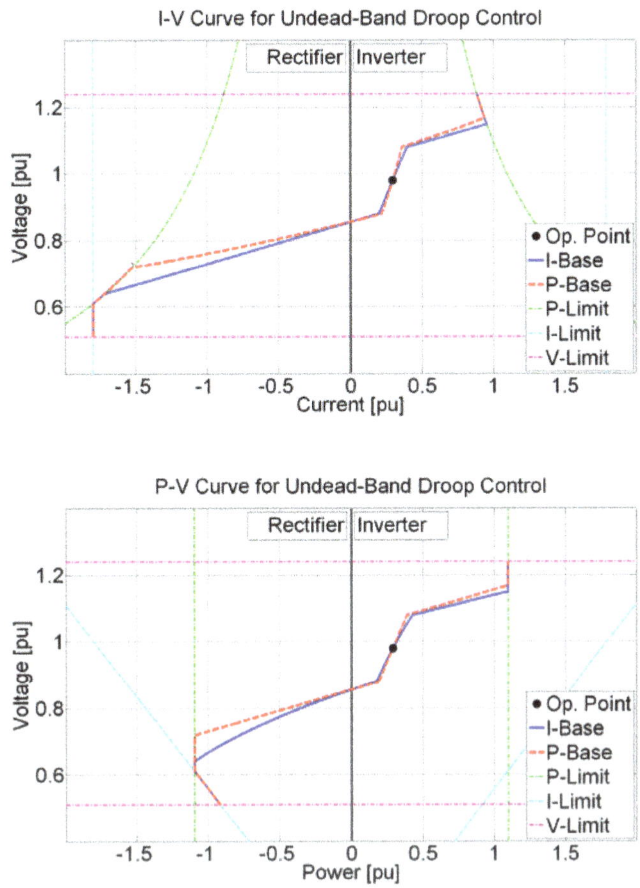

Figure 6.7 Undead-band-droop control

As the zero sequence can flow through the ground from the mid-point of the DC capacitors to the neutral point of the Y connection, it is imperative that the impact of a zero-sequence current on the VSC is fully understood.

B. Equivalent Circuit of Zero Sequence

Although admitting the zero sequence has solved the two aforesaid problems, it is necessary to be assured that the impact of the zero sequence on the VSC is understood and their side effects can be managed. Figure 6.10 shows the rigorous derivation of the properties of the zero sequence in the basic VSC circuit. On the AC-side, the zero-sequence current 3i0=(ia+ib+ic) is given a path to flow at the grounded neutral point of the AC-source voltages ea, eb, and ec. On the DC side, the currents flowing through the upper and the lower DC buses are ip and in, respectively. The grounded midpoints of the DC capacitors and the DC-voltage sources provide two return paths.

Also, the equivalent circuit in Figure 6.11 summarizes the properties of the zero sequence in the VSC system. From Figure 6.11, on the AC side, Raco and Laco are zero-sequence equivalent resistance and inductance. The zero-sequence components of the AC- source voltage and VSC voltage are represented by e0 and uo, respectively.

The equivalent circuit is derived through two transitions: from Figure 6.11(a) to Figure 6.11(b) and then to Figure 6.11(c). Figure 6.11(a) shows separate upper and lower buses and capacitors on the DC side, the values of the circuit elements being identical to those of the DC side in Figure 6.11. The zero-sequence current, which has

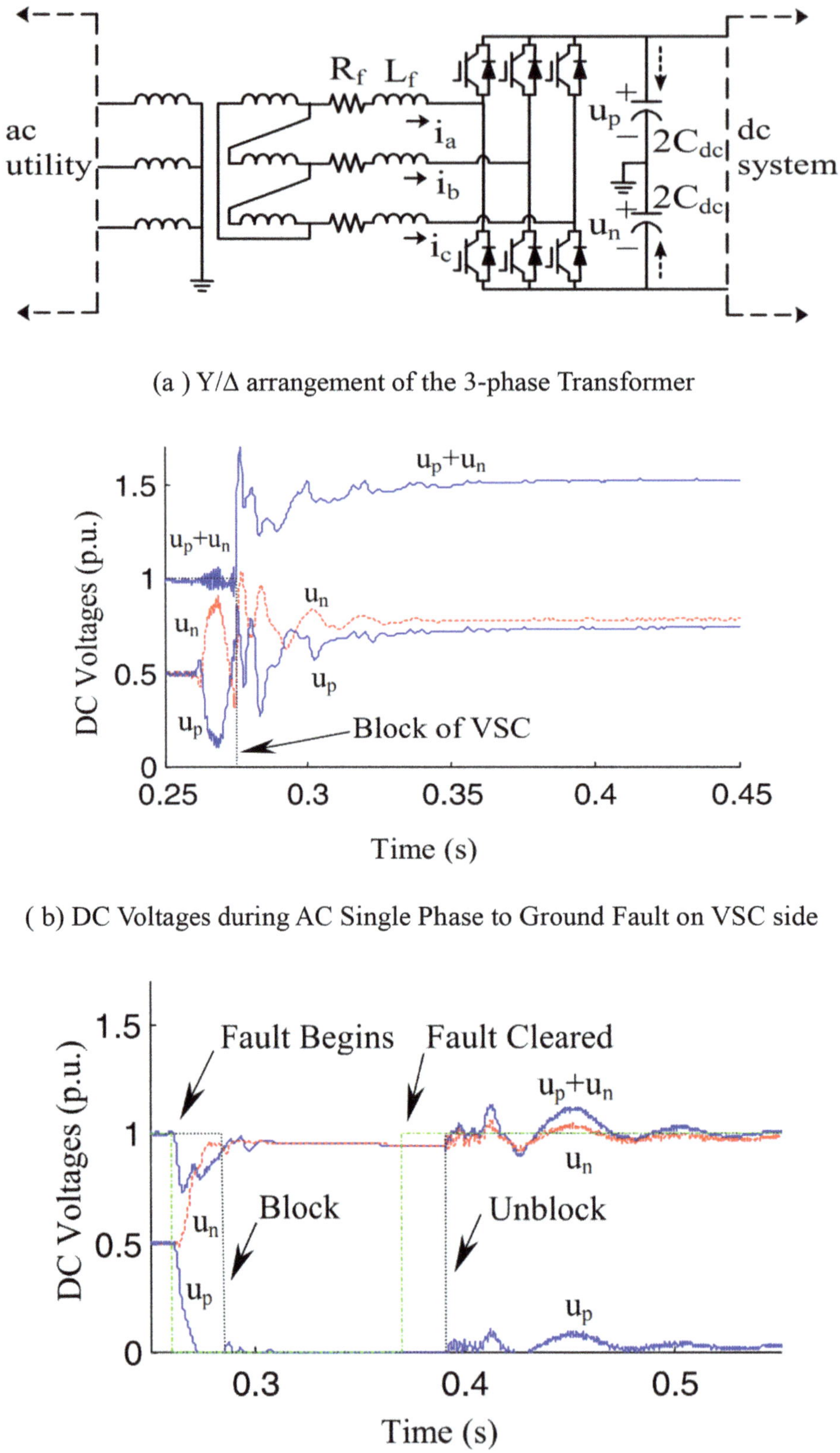

(a) Y/Δ arrangement of the 3-phase Transformer

(b) DC Voltages during AC Single Phase to Ground Fault on VSC side

(c) DC Voltages during DC Line-to-Ground Fault in DC system

Figure 6.8 Voltage-sourced converter (VSC) grounding topology with the Δ-connection on the VSC side

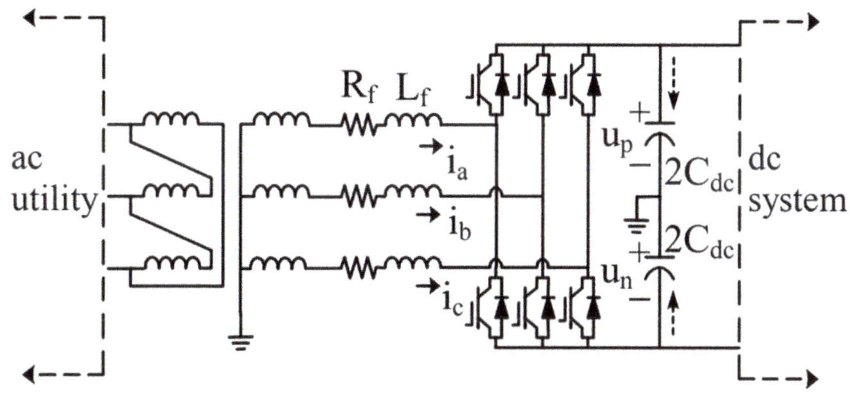

(a) Δ/Y arrangement of the 3-phase transformer

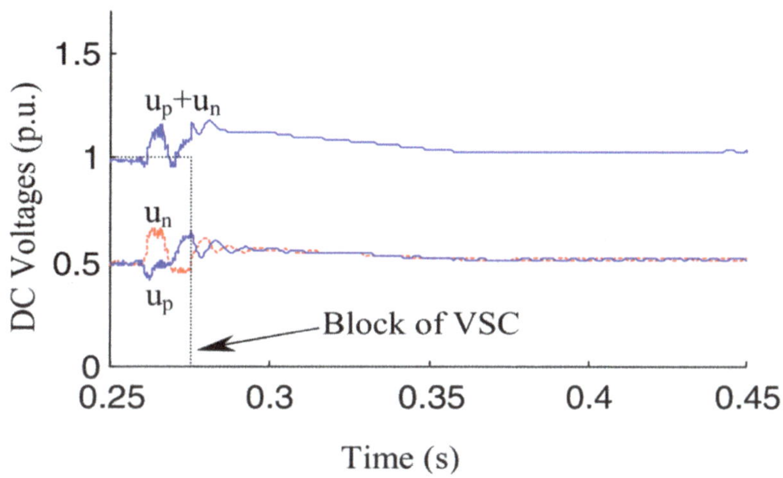

(b) DC Voltages during AC Single Phase to Ground Fault on VSC side

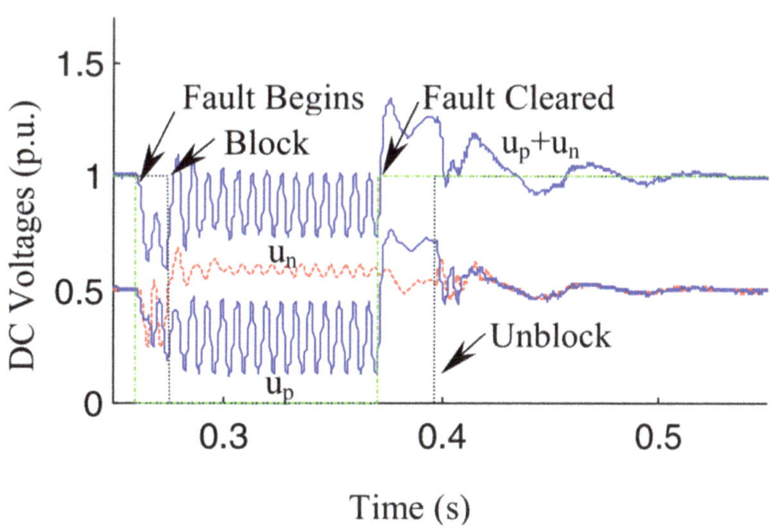

(c) DC Voltages during DC Line-to-Ground Fault in DC system

Figure 6.9 Voltage-sourced converter (VSC) grounding topology with the Y-connection on the VSC side

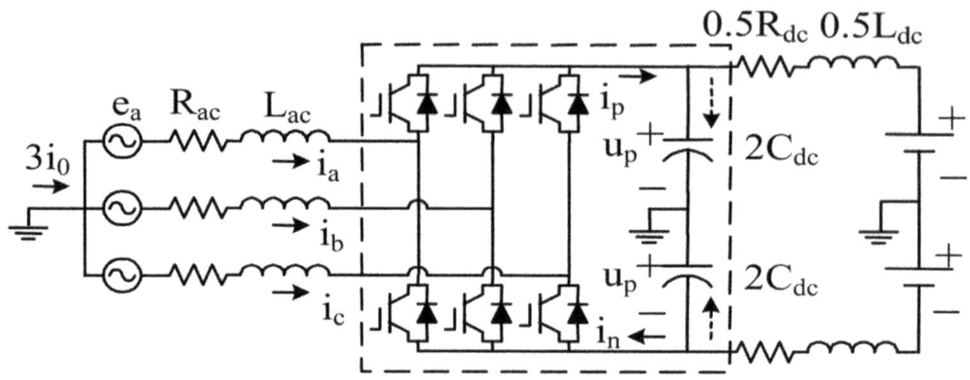

Figure 6.10 Basic voltage-sourced converter system with AC system grounded

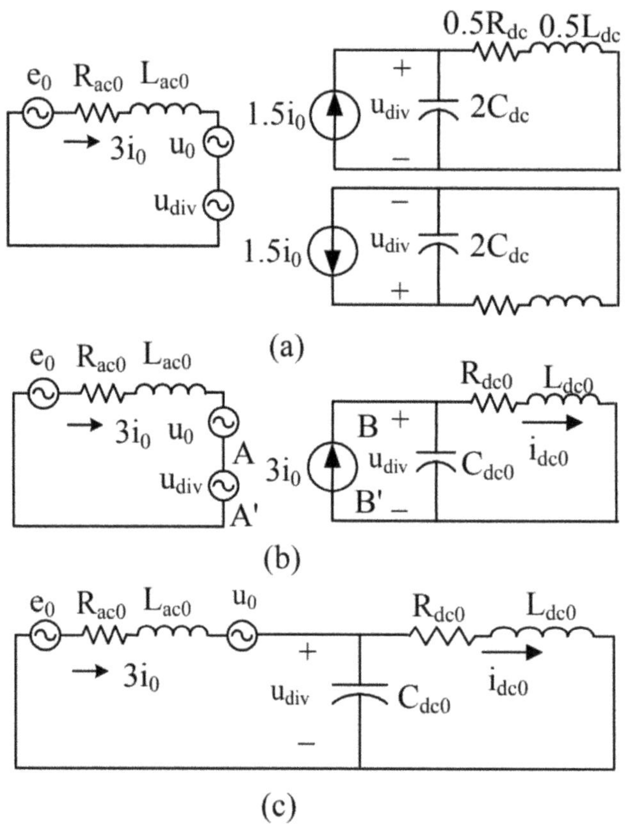

Figure 6.11 Zero-sequence equivalent circuit

divided into two equal halves, flows in the directions indicated by the arrows in the two ideal current sources, As the two halves can be viewed as parallel, they can be integrated into one circuit, as shown in Figure 6.11(b).

6.2.2 Star-Point Reactor [21]–[24]

In the Δ-connected MMC system, grounding equipment has to be installed on the DC side or the AC side to provide a zero potential for the MMC HVDC system. The reasons for grounding in the MMC system are as follows:

- Minimize DC-offset because of parameter error between Cable or Overhead line (OHL)

- Compensate unbalance of converter internal because of MMC characteristics (charging and discharging)
- Make a zero potential of DC converter

For the symmetrical MMC approach, a special grounding device located on the AC side can be used to provide high impedance for the AC-grid frequency and low impedance for the DC current, that is, in case of AC grounding, ***a star-point reactor, zigzag transformer, or grounding transformer*** can be used.

Of course, a grounding arrangement may also be located on the DC side to provide high impedance to the DC current, but in this case, high-resistance grounding equipment is required and can lead to high power loss, low reliability, and high cost in an MMC system. The grounding options available for MMC HVDC are shown in Figure 6.12.

***Star-Point Reactor*:** Figure 6.12(a) shows the star-point reactor, which is a Y-connected inductive coil and non-reactive resistor, and is installed on the AC side. The inductor provides high impedance for the AC frequency and a low impedance path for the DC current. This type has a simple configuration and a low cost. However, it is difficult to apply to a high voltage of 220 kV or more because of insulation problem, and there is a disadvantage: reactive power consumption is large.

Zigzag Grounding Transformer: Figure. 6.12(b) shows the zigzag transformer, a transformer in which A phase, B phase, and C phase are zigzagged in the order of B phase to C phase to A phase. In this case, although the required installing space reactive power consumption is small, the weight of the grounding equipment is large and the cost is high.

In Figure 6.13, the grounding reference point G of the AC network, the grounding GT of an interface transformer, the neutral point N of the DC side, and the common points a, b, and c of the converter should always be the same. However, the common points a, b, and c of the converter can be changed for the following reasons:

- Unbalancing between upper arm and lower arm through charging and discharging of the submodule (SM)
- Unbalancing because of parameter and control error between upper arm and lower arm
- Unbalancing through DC line faults
- Unbalancing because of impedance error of cable line or overhead line located on DC side

Therefore, the purpose of the sta -point (S.R from Figure 6.13) is to make the common mode voltage constant.

***Third Harmonic Instability and Modeling*:** Figure 6.14 shows AC equivalent modeling considering a star point reactor and the injection of the third harmonic offset voltage. The injected third harmonic offset voltage is a zero-sequence voltage and exists in each phase equally. Figure 6.15(a) shows the equivalent circuit that separates the zero sequence and the third harmonic from AC equivalent modeling using the principle of

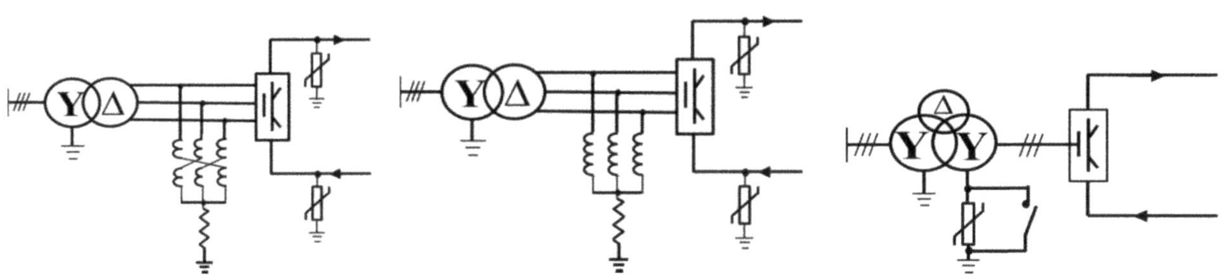

(a) (b) (c)

Figure 6.12 AC-side grounding options

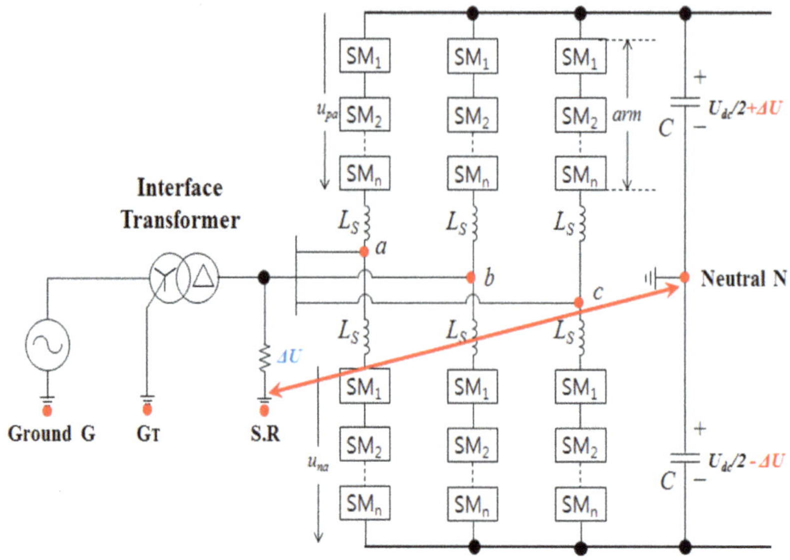

Figure 6.13 Grounding mode in modular multi-level converter system

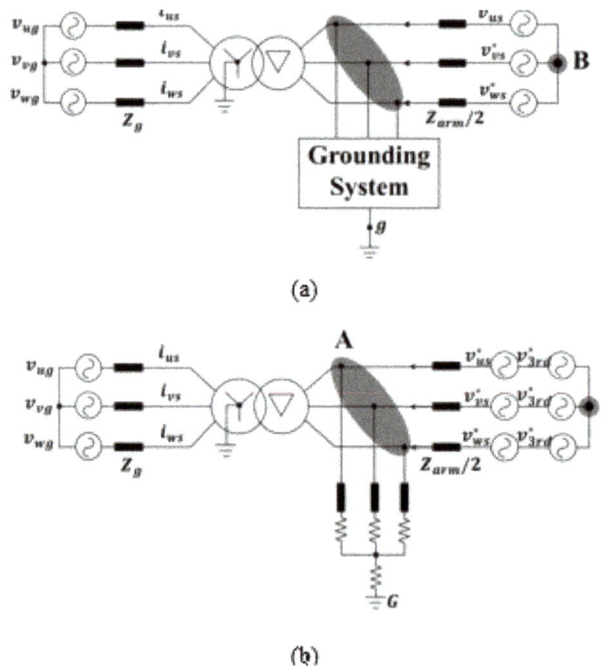

Figure 6.14 AC equivalent modeling considering star-point reactor and injection the third harmonic offset voltage

superposition. As the Δ-connected secondary voltage of the interface transformer is not excited with respect to the zero-sequence voltage, it appears to be open.

Figure 6.15(b) shows the equivalent circuit that simplifies Figure 6.14, as well as the impedances existing in three phases being reduced to 1/3 of each other. In the traditional converter equivalent circuit concept, there is no path through which the current flows in the opened B point. Therefore, the voltage of each node (A and B) with respect to the ground voltage can be obtained by Equation (6.5).

$$v_{AG} = 0, v_{BG} = v^*_{3rd} \tag{6.5}$$

According to Equation (6.5), the injected third harmonic voltage is seen from the pole-to-ground point at the DC side than at the AC side. However, Equation (6.5) leads to different result: There is a path through which a current can flow from the node B. Actually, this assumption is valid for point-to-point (PTP) because of the line-to-ground impedance of cable line or overhead line.

Generally, as the zero-sequence impedance of a star point reactor is very large, the voltage of each node can be expressed as Equations (6.6) and (6.7). From Equation (6.6), C is the line-to-ground capacitor expressed as the capacitor of a cable or overhead line:

$$v_{AG} = \frac{Z_{star}}{Z_{star} + Z_{arm/6} + Z_{DC_{side}}} v^*_{3rd1} \approx v^*_{3rd1} \tag{6.6}$$

$$v_{DC_{side}} = \frac{-Z_{BG}}{Z_{star} + Z_{arm/6} + Z_{DC_{side}}} v^*_{3rd1} \approx 0 \tag{6.7}$$

$$\text{where, } Z_{star} = \frac{Z_{star\_1}}{3} + Z_{star\_2}, Z_{BG} = \frac{1}{j2\pi fC}$$

As in Equations (6.6) and (6.7), seeing the third harmonic voltage on the AC side, the impedance of a star point reactor has to be a sufficient amount larger than the line-to-ground impedance of the DC side. Figure 6.6 shows an alternative choice to determine the DC side impedance.

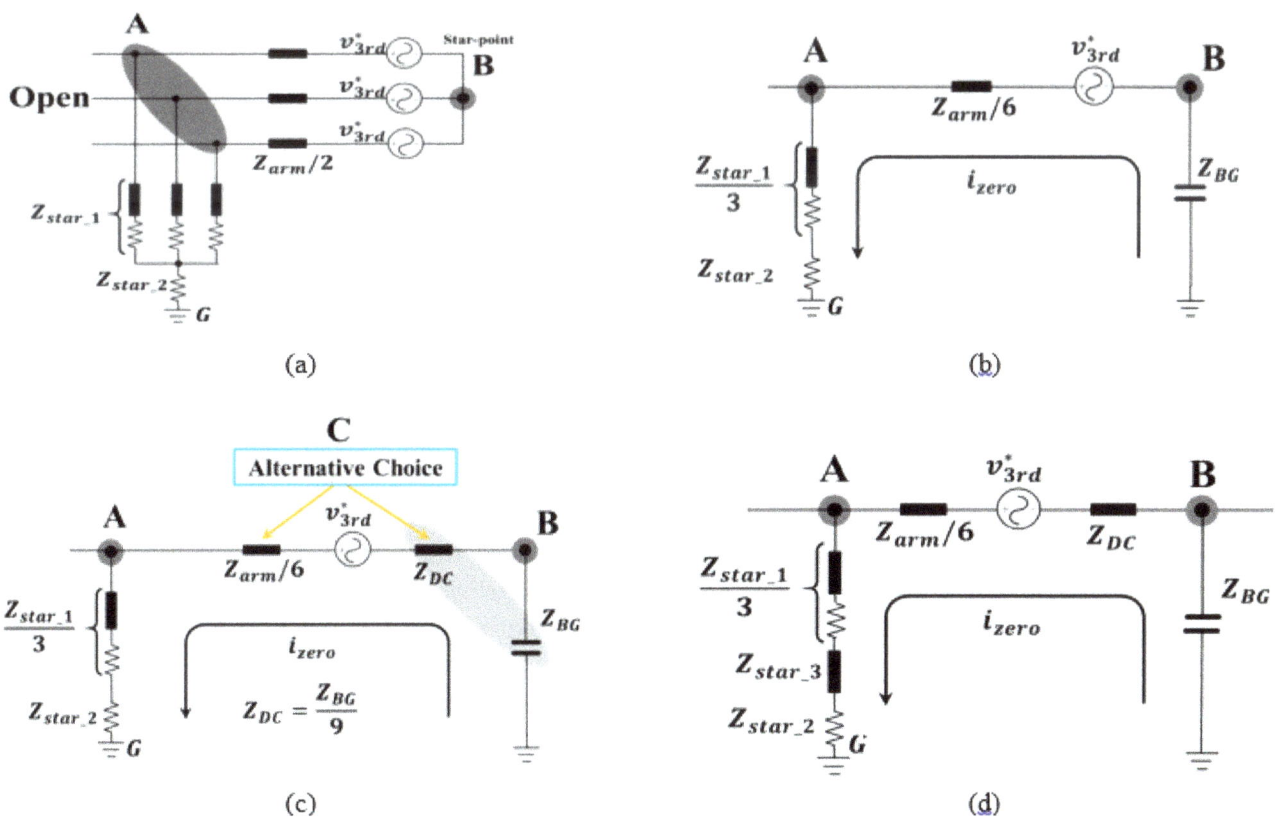

Figure 6.15 Zero-sequence impedance modeling of modular multi-level converter HVDC system considering the star-point reactor

As Figure 6.15(d) shows, in the case of installing an inductor in the neutral line of a star point reactor, when DC-side impedance is resonated at the third harmonic frequency the design of the star-point reactor becomes easy.

Also, the zigzag transformer can be used as AC-side grounding equipment. The zigzag transformer is connected to the Δ wiring at the secondary side of the transformer like a star-point reactor to form a neutral point. As the positive- and negative-sequence impedance of the zigzag transformer is seen as the excitation impedance of the transformer, in general, the rated excitation current of a single-phase transformer has a value of about 1% of the rated current, and the magnitude of the excitation impedance is about 100 times larger than the base impedance. However, only the leakage impedance from the viewpoint of the zero-sequence impedance is seen, because the electromotive forces of each transformer generated by the zero-sequence current act in the direction of canceling each other. In general, the magnitude of the leakage impedance of the transformer is very small.

In conclusion, the zigzag transformer has very large impedance for positive sequence and negative sequence, so it can be operated with a proper grounding system during theoretic operation. However, in case of the injecting the third harmonic voltage, the zigzag transformer itself, other than the neutral grounding resistor (NGR), may have insufficient zero-sequence impedance.

Therefore, it is necessary to add a large neutral grounding impedance to make the third harmonic voltage appear on the AC system side instead of the pole-to-ground voltage of the DC side.

6.2.3 Parameter Design of Grounding Devices [22]

The unbalanced voltage on the DC side of the MMC HVDC transmission system can be kept balanced using a grounding device and can produce a corresponding DC current. The DC current generates DC magnetic flux, and the corresponding magnetic density in the iron depends on the structure of the interface transformer. To avoid interface transformer super-saturation, the DC current should be less than the given value.

To determine the resistance value of the NGR, we should understand the electric and magnetic circuit of the interface transformer has a three-phase three-column structure, a three-phase five-column structure, and three single-phase cores as shown in Figure 6.16(a)–(c).

As for the interface transformer with three-phase three-column cores, there is no closed iron magnetic path for the DC magnetic flux. The magnetic reluctance of the DC magnetic flux is the maximum; the DC magnetic flux is the minimum. Therefore, the allowable DC current I_{dc} of the interface transformer is the maximum among the three interface transformers.

For the interface transformer with three-phase five-column cores, the cross-sectional area of the synthetical yoke sections may be reduced to the same values that are used in the outer legs and in the top and bottom yokes, and the DC magnetic flux is relatively small because of a closed iron magnetic path. Therefore, the allowable DC current I_{dc} of the interface transformer is in the middle among the three interface transformers.

As for the interface transformer consisting of three single-phase cores, there is an independent and closed iron magnetic path for the DC magnetic flux of each phase. Therefore, the allowable DC current I_{dc} of the interface transformer is the least among the three interface transformers.

According to the characteristics of the DC magnetic path of the interface transformer, we have found that an interface transformer with a three-phase three-column structure is the preferred transformer. The transformer, however, is limited in applications of high capacity because of the transport capacity limit. For a given transformer, the allowable DC current is smaller, and the resistance value of the NGR will be higher.

Neutral Grounding Resistor: In the MMC HVDC transmission system with a grounding device, there may be DC current in the winding of the interface transformer. The DC current may result in DC magnet bias for the

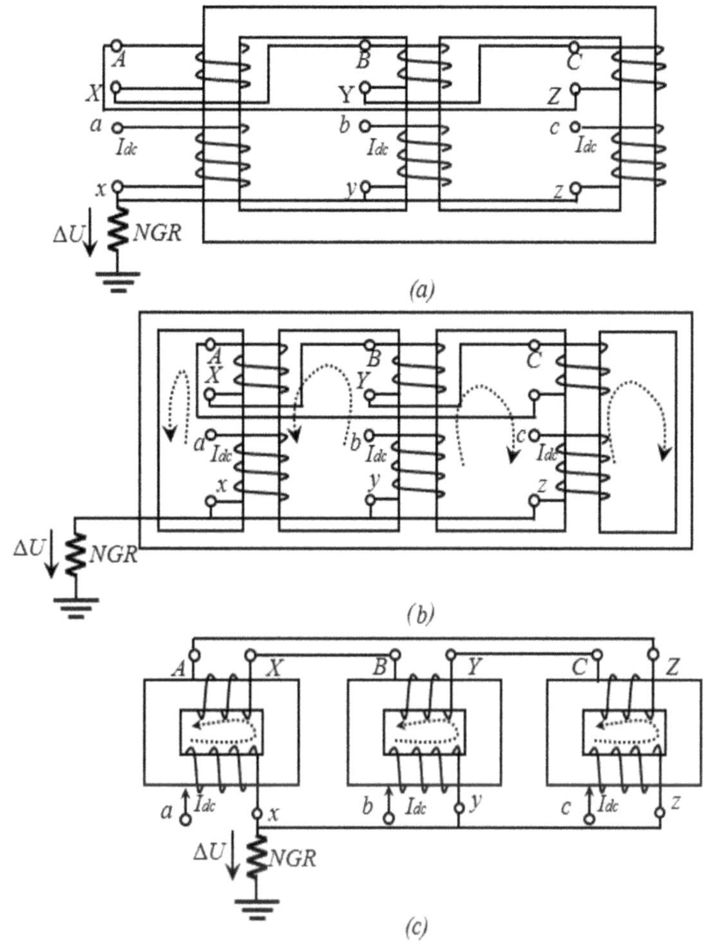

Figure 6.16 Electric and magnetic circuit of the interface transformer: (a) interface transformer with three-phase three-column cores; (b) interface transformer with three-phase five-column cores; and (c) interface transformer consisting of three single-phase cores

interface transformer, and bring adverse influence including a noise increase, non-sinusoidal output voltage, iron loss, increasing reactive power, and so on. To keep the DC current at an acceptable level, the interface transformer with a Y-connection on the MMC side can usually adopt proper grounding resistance.

However, it is difficult to determine the acceptable I_{dc} of the interface transformer accurately. The acceptable I_{dc} of each phase is not to exceed 0.7% of the rated AC phase current or two times the rated exciting current I_0 under the no-load condition, which is confirmed by Teshmont Consultants Inc. Zhong et al. tested the DC current for different transformers and found that: Firstly, for the transformer with the single- or three-phase five-column structure, the acceptable DC current is about 1.1–1.5 times the rated exciting current under the no-load condition; Secondly, for the transformer with a three-phase three-column structure, the acceptable DC current is about two times the rated exciting current under the no-load condition. Thus, the resistance value of the NGR R should satisfy

$$R \geq \Delta U/3I_{dc} = \Delta U/3kI_0 \tag{6.8}$$

where, R is the ratio of I_{dc} to I_0, and K is changing with the structure of different transformers.

When the exciting current of the transformer does not exceed the defined value, the DC current can be accepted while considering the DC magnet bias. From Equation (6.8), we can find that the resistance value of the NGR is decreased by decreasing the unbalanced voltage ΔU. Besides the structure of the interface transformer, the resistance value of the NGR also depends on the rated exciting current of the transformer under the no-load condition and DC voltage U_{dc} of the MMC. For a given MMC HVDC transmission system, the maximal unbalanced voltage ΔU is half of the interpolar voltage U_{dc}. To decrease the resistance value of the NGR, we should choose a transformer with a three-phase three-column structure first.

Zigzag Transformer: In the MMC HVDC transmission system, if the interface transformer is the Y_n/Δ arrangement, and the Δ-connection is on the MMC side, we can use the transformer to construct the neutral point artificially, thus integrating the characteristics of the zigzag transformer.

A zigzag transformer is a special connection of either three single-phase transformer windings or three-phase transformer windings as shown in Figure 6.17(a). Figure 6.17(b) shows the phasor diagram of Figure 6.17(a). From Figure 6.17(b), we can find that the phase voltage U_{PN} of the three-phase system is multiplied by the winding voltage U_{PN} of the zigzag transformer. The zigzag transformer has been used to create a neutral point in medium voltage or power distribution systems.

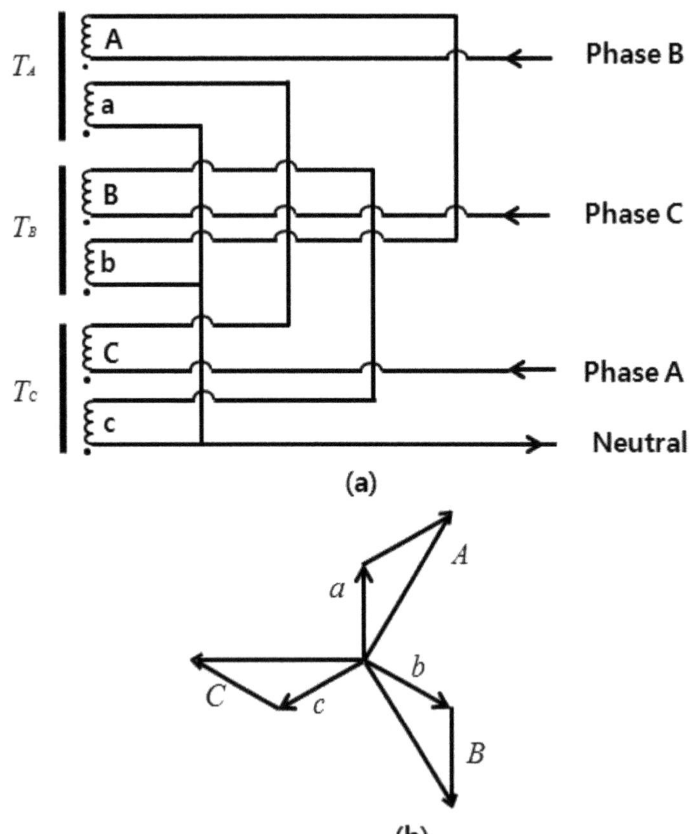

Figure 6.17 Zigzag transformer: (a) Zigzag transformer connection and (b) phasor diagram

The positive- or negative-sequence impedance is the excitation impedance Z_{Tm}, which is usually very large. To determine the different impedances of the zigzag transformer, we still use the usual parameter of a single-phase two-winding transformer, which includes the use of rated capacity S_N, voltage U_{TN}, current I_N, and the exciting current I_0. We take the reference impedance Z_{TN} as U^{2TN}/S_N in the system with the transformer. Then, the positive-and negative-sequence impedances are Z_{Tm}, which is

$$Z_{Tm} = \frac{U_{TN}}{I_0} \times \frac{U^{2TN}}{S_N} = k_m Z_{TN} \tag{6.9}$$

where, k_m is the ratio of I_N to I_0

From Equation (6.9), we find that the positive-and negative-sequence impedances depend on k_m, which is in the range of 50–100 when rated exciting current I_0 is in the range of 1–2% of the rated current I_N. Let us assume that the rated exciting current I_0 is about 1% of the rated current I_N for the usual transformer Z_{Tm} and is about $100Z_{TN}$.

In the zigzag transformer shown in Figure 6.17, all three-phase positive- and negative-sequence currents flowing through the zigzag transformer are equal, and the vector addition of three-phase positive-or negative-sequence currents is zero. Taking T_A as an example, the magnetomotive force (MMF) $\vec{F}_{TA\pm}$ of T_A is

$$\vec{F}_{TA\pm} = N\left(I_{B\pm} - i_{A\pm}\right) \tag{6.10}$$

where, N is the number of turns of coil A and a; $i_{B\pm}$ and $i_{A\pm}$ are the positive- and negative-sequence currents of phase B and phase A, respectively. According to the phase between $i_{B\pm}$ and $i_{A\pm}$, Equation (6.10) can be rewritten as

$$\vec{F}_{TA\pm} = \sqrt{3}Ni_{B\pm e^{\pm 20^\circ}} \tag{6.11}$$

From Equation (6.11), we find the composite magnetic potential of two coils wound on the same leg is $\sqrt{3}$ times the magnetic potential of one coil. This means the equivalent turns of the zigzag transformer's excitation coil is $\sqrt{3}$ times its single-phase transformer's turns. According to Equation (6.11), the ratio of the positive- or negative-sequence impedances $Z_{\pm}(Z)$ of the zigzag transformer and the single-phase two-winding transformer is

$$\frac{Z_{\pm}}{Z_{T\pm}} = \left(\frac{\sqrt{3}N}{N}\right)^2 = 3 \tag{6.12}$$

The positive- or negative-sequence impedances of the zigzag transformer are about 300 Z_{TN} on the condition of each two-winding transformers with 1% current ratio between I_0 and I_N.

As for the impedance of zero-sequence currents, MMFs of two coils wound on the same leg are equal and opposite in direction. The composite magnetic potential of T_A with coil and α is zero. There is only leakage magnetic flux in T_A. Accordingly, the zero-sequence impedance Z_0 is

$$Z_0 = Z_{A0} + Z_{a0} = 2Z_{T0} \tag{6.13}$$

where, the impedances Z_{A0} and Z_{a0} are the leakage impedance of the coils A and α, respectively, and Z_{T0} is the leakage impedance of the coils or α.

Let us assume that the leakage impedance in each winding is about 0.08, and Z_0 is about $0.16Z_{TN}$, which is far less than the $Z_{\pm}$ of the zigzag transformer.

For the DC current, the MMFs of two coils wound on the same leg are also equal and opposite in direction. Accordingly, the zigzag transformer presents only leakage reactance for the DC current. Because of the angular frequency of the DC current being zero, the transformer presents only two times the winding resistance for the DC current under steady state. The impedance of the DC current is

$$Z_{dc} = (Z_{A0} + Z_{A0})|_{\omega=0} = 2r \tag{6.14}$$

where, ω is the angular frequency of the current of the zigzag transformer.

According to Equations (6.12)–(6.14), one can see that $Z_{\pm}$ is the greatest, and much larger than Z_0 and Z_{dc}. Z_{dc} is the least among the three kinds of impedances for the zigzag transformer. In addition, the difference between Z_0 and Z_{dc} is only the leakage reactance.

Figure 6.18 shows the connection of a zigzag transformer with NGR in an MMC-HVDC transmission system. In this system, the voltage of the neutral point may include the zero-sequence voltage and/or DC voltage. To hold off the zero-sequence current and/or DC current through the neutral point, the appropriate resistance can be a series connection between the neutral point and the ground.

Design of the Zigzag Transformer: We assume that the zigzag transformer consists of three single-phase transformer windings with the wiring connection of Figure 6.18.

For the positive- and negative-sequence currents, the zigzag transformer provides a high-impedance path, which is the corresponding excitation impedance. The positive- or negative-sequence currents of two coils on the same leg are equal and 60-phase difference, respectively.

The positive- and negative-sequence impedances are equal and depend on the structure, ferromagnetic material, and wiring turns of the zigzag transformer. The magnetizing impedance Z_T is

$$Z_T = \frac{U^{2TN}}{S_0/3} \tag{6.15}$$

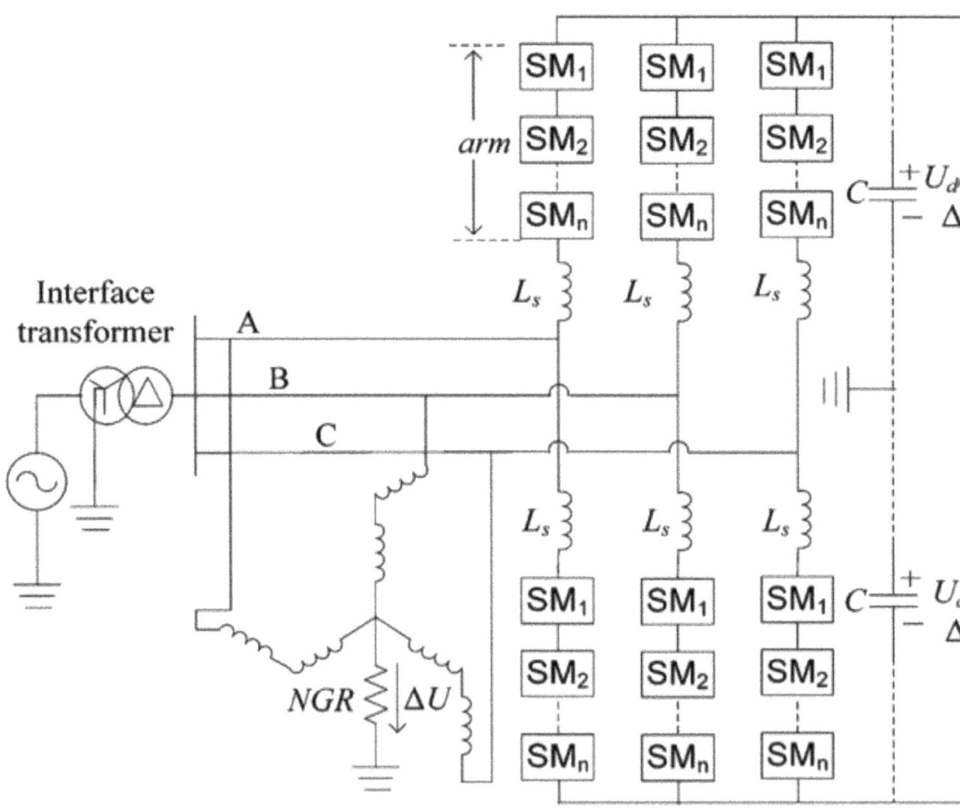

Figure 6.18 Zigzag transformer with the neutral grounding resistor

For a zero-sequence current, the zigzag transformer provides a low-impedance path, which is the corresponding leakage reactance 2 Z_{T0}. When a single phase-to-ground fault takes place on the AC side of the MMC, the zero-sequence current I_{RZ} flowing through the NGR is

$$I_{RZ} = \frac{3U_{PN}}{2Z_{T0} + 3R} \approx \frac{U_{PN}}{R} \tag{6.16}$$

where, U_{PN} denotes the RMS of the phase voltage, which is equal to $\sqrt{3}U_{TN}$.

When a monopole ground takes place on the DC side, the steady-state DC current I_{Rdc} flowing through the NGR R is

$$I_{Rdc} = \frac{U_{dc}/2}{2r + R} \approx \frac{U_{dc}}{R} \tag{6.17}$$

The modulation ratio m is defined as

$$m = \frac{\sqrt{2}U_{PN}}{U_{dc}/2} = \frac{2\sqrt{2}U_{PN}}{U_{dc}} \tag{6.18}$$

According to Equation (6.16)–(6.18), the relation between I_{Rdc} and I_{RZ} is

$$I_{Rdc} = \sqrt{2}I_{RZ}/m \tag{6.19}$$

Generally, the modulation ratio m is less than 1, and $I_{Rdc} \geq \sqrt{2}I_{RZ}$. Thus, the resistance value of the NGR R mainly depends on the allowable DC current of the zigzag transformer.

The allowable DC current of each phase is two times the rating exciting current under the no-load condition for the three-phase three-column transformer, and the DC current mainly generates leakage magnetic flux. The main magnetic potential composite phasor of the DC currents of two coils on the same leg is zero.

Then there is only leakage magnetic flux in the transformer iron, at least in theory. Thus, the zigzag transformer is not easily saturated when compared with the three-phase three-column transformer. The allowable DC current flowing through the wiring of the zigzag transformer is mainly limited by the wiring current carrying the capacity or copper losses of the zigzag transformer. From the standpoint of cost savings, an appropriate grounding resistor should be connected between the neutral point of the zigzag and the earth point.

6.3 MMC Control under HVDC Pole Voltage Imbalances [10]–[23]

6.3.1 Causes and Countermeasures for Pole-Voltage Imbalances

HVDC cables for VSC HVDC transmission are usually single core with a grounded sheath. When a cable is severely damaged, the conductor is likely to touch the sheath. This creates a low impedance path, resulting in a quick discharge of the energy stored in the damaged cable. If no DC circuit breaker is available to quickly isolate the damaged cable during the first few milliseconds after the fault, the stray capacitance of the healthy pole will charge through the HVDC converters, and its voltage will build up until the AC-side circuit breakers isolate the transmission system from the grid. Afterwards, in a system capable of operating as an asymmetric monopole, DC switchgear could be used to disconnect the damaged cable and the system could be re-energized to carry on exporting power at reduced loading using the healthy cable.

On the other hand, slow drifting and small pole imbalance can occur when there is a mismatch of the impedance between the poles and the ground. Such a situation can happen when high-impedance short circuits exist in the cable because of mechanical damage or ageing, or when small loads are tapped in an asymmetric way from one pole to the ground.

The action required in such situations may depend on the source of the imbalance, which can be identified through different means. If the cable is only lightly damaged, it is desirable to keep the system operating to give the transmission system operator (TSO) time to reallocate generation and bring the HVDC link to a safe stop afterwards. In contrast, if the imbalance is caused by an asymmetric load, HVDC power converters will be required to be able to operate in this state indefinitely by controlling the current injected to the ground to return balance to the pole voltages.

System description and modeling: The analysis focuses on a VSC HVDC transmission system based on half-bridge cell MMCs. To study the steady-state operation of the converter, a simplified equivalent circuit of the MMC is used (see Figure 6.19). Such a circuit aggregates each arm of the converter as a voltage source capable of applying a positive voltage obtained as the addition of an AC plus a DC component, done in order to control the power exchange between the converter modules and the AC and DC grids. A simplified model of the AC grid plus the transformer is also included in the model. The connection of the transformers is delta in the converter side; however, a star-equivalent with an infinite impedance, Zgn, connected between the neutral and the ground, is used in the model to simplify the derivation of the equations. The SR is modeled as a three-legged-core star-connected inductor with the neutral grounded through an impedance Z0n. The DC grid is modeled as an ideal DC bus.

System Steady-State Equations: To simplify the equations, it is convenient to introduce a change of variables and define the additive and subtractive voltages and the additive current, V_{sum}, V_{diff}, and I_{sum} respectively, as

$$\begin{cases} V_{diff} \triangleq \frac{1}{2}\left(-\underline{V}_u + \underline{V}_l\right) \\ V_{sum} \triangleq \underline{V}u + \underline{V}_l \\ I_{sum} \triangleq \frac{1}{2}\left(\underline{I}_u + \underline{I}_l\right) \end{cases} \tag{6.20}$$

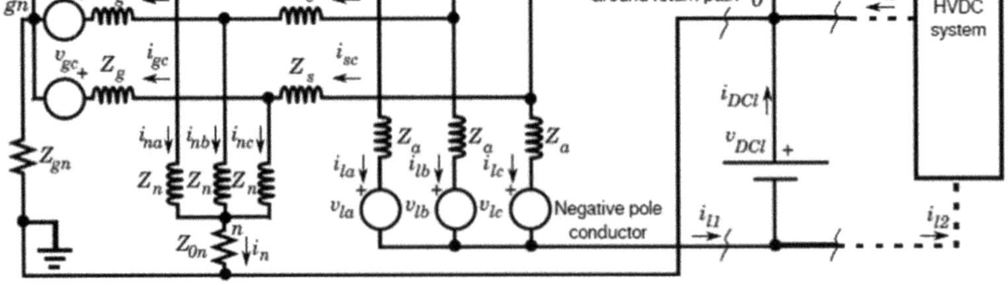

Figure 6.19 Equivalent circuit of the HVDC converter station

which implies

$$\begin{cases} V_u = -\underline{V}_{diff} + \frac{1}{2}\underline{V}_{sum} \\ \underline{V}_l = \underline{V}_{diff} + \frac{1}{2}\underline{V}_{sum} \\ \underline{I}_u = \frac{1}{2}\underline{I}_s + \underline{I}_{sum} \\ \underline{I}_l = -\frac{1}{2}\underline{I}_s + \underline{I}_{sum} \end{cases} \tag{6.21}$$

The power exchange between the converter and the AC grid is controlled through the subtractive voltage. Neglecting the star-reactor AC current, the current injected to the grid becomes

$$\underline{I}_g^{+-} \approx \underline{I}_s^{+-} \approx \frac{1}{\underline{Z}_s + \underline{Z}_g + \frac{1}{2}\underline{Z}_a}\left(\underline{V}_{diff}^{+-} - \underline{V}_g^{+-}\right) \tag{6.22}$$

The actual positive- and negative-sequence current flowing through the SR can be calculated as

$$\underline{I}_n^{+-} = \frac{1}{\underline{Z}_s\underline{Z}_g + \frac{1}{2}\underline{Z}_a\underline{Z}_g + \underline{Z}_n(\underline{Z}_s + \underline{Z}_g + \frac{1}{2}\underline{Z}_a)}\left(\underline{Z}_g\underline{V}_{diff}^{+-} + \left(\underline{Z}_s + \frac{1}{2}\underline{Z}_a\right)\underline{V}_g^{+-}\right) \tag{6.23}$$

The zero-sequence AC current flowing through the SR and returning through the DC grid is controlled to be zero using the zero-sequence subtractive voltage of the converter:

$$\underline{I}_s^0 = \underline{I}_n^0 = \frac{1}{3}\underline{I}_n = \frac{1}{\underline{Z}_s + \frac{1}{2}\underline{Z}_a + R_n + 3\underline{Z}_{0n}}\underline{V}_{diff}^0 \tag{6.24}$$

The internal power exchange between the upper and lower arms of the converter can be achieved using positive and negative sequence additive current, whereas the zero-sequence additive current is controlled to be zero. These variables depend on the additive voltage as

$$\underline{I}_{sum}^{0+-} = -\frac{1}{2\underline{Z}_a}\underline{V}_{sum}^{0+-} \tag{6.25}$$

Under normal operation, the power exchange between the converter and the DC grid is achieved using the zero-sequence additive DC current; this can be controlled using zero-sequence additive DC voltage:

$$I_{sum}^{0DC} = \frac{1}{2R_a}\left(V_{DCu} + V_{DCl} - V_{sum}^{0DC}\right) \tag{6.26}$$

Internal power exchange between the arms of the three phases of the converter is achieved using α- and β-additive DC currents; both are controlled using additive DC voltage:

$$I_{sum}^{\alpha\beta DC} = -\frac{1}{2R_a}V_{sum}^{\alpha\beta DC} \tag{6.27}$$

Positive- and negative-subtractive DC currents flow through the AC of the converter, which can cause saturation of the transformer and the SR; therefore, it is controlled to be zero through positive and negative subtractive DC voltage:

$$I_s^{\alpha\beta DC} = \frac{R_g + R_n}{R_sR_g + \frac{1}{2}R_aR_g + R_n\left(R_s + R_g + \frac{1}{2}R_a\right)}V_{diff}^{\alpha\beta DC} \tag{6.28}$$

Finally, DC-current injection through the SR, which is required for the asymmetric operation described here, can be controlled using the zero-sequence subtractive DC voltage:

$$I_n^{0DC} = I_s^{0DC} \frac{1}{R_s + R_n + 3R_{0n} + \frac{1}{2}R_a)} \left(V_{diff}^{\alpha\beta DC} + \frac{1}{2}(V_{DCu} - V_{DCl}) \right) \tag{6.29}$$

6.3.2 MMC with a Zigzag Transformer for Bipolar HVDC [34]–[40]

Compared to AC transmission, DC transmission has higher efficiency, higher reliability, and higher flexibility. There are two types of DC systems: the monopole configuration and the bipolar configuration as shown in Figure 6.20(a) and 6.20(b), respectively. Compared to the monopole configuration, a bipolar system has significant advantages, for example, providing different voltage levels, similarly to a three-phase AC system, and increased reliability because of the independent operation of the two poles. To establish a bipolar DC system, two AC–DC converters (e.g., two-level converters (TLC) with galvanically isolated AC sides) are required as shown in Figure 6.20(b). It increases the system complexity, and the secondary side of the transformer is subjected to a DC offset voltage with respect to the ground. To address this issue, one general approach is to employ an auxiliary voltage balancer to maintain the DC pole voltage balance in case of a power unbalance as shown in Figure 6.20(c). The main drawback of this approach is that it requires an extra power-electronic converter as the voltage balancer. A three-level converter with two split DC-link capacitors (e.g., a neutral-point-clamped converter or a T-type converter) can establish a bipolar DC-link voltage by controlling the DC-link neutral-point current via manipulating the zero-sequence voltage reference.

However, the unbalanced operation capability of two DC-link poles is very limited because of the dependency of the maximum available neutral-point current on the AC-side current magnitude and phase, and it is not suitable for establishing a bipolar DC distribution system. Thus, it can be concluded that an auxiliary balancing leg has to be added in order to enable full bipolar operation capability. A novel approach of integrating a voltage balancer into a TLC is shown in Figure 6.21(a). A three-phase grounding inductor was connected to the AC terminal of the TLC to provide a current path from the AC side to the DC side neutral so that the positive and negative poles on the DC side can be controlled independently. A three-arm core inductor shown in Figure 6.21(b) was employed as a grounding inductor for a couple of reasons: (a) a three-arm core inductor presents large positive- and negative-sequence impedance that results in low reactive-power consumption and (b) a three-arm core inductor presents small zero-sequence impedance that leads to a fast response of the zero-sequence current and enables a fast dynamic control of the two DC poles.

However, the aforementioned concept has several crucial drawbacks. The grounding inductor is subjected to the transformer's secondary-side voltage and two thirds of the rated DC-link current would flow through each winding of the inductor.

Thus, the power rating of the grounding inductor is 70–80% that of the transformer. Moreover, the inductance has to be very large in order to decrease the reactive power consumption, and it would result in high conductive losses. In case of power unbalance at the DC pole, a zero-sequence DC current flows through the grounding inductor and induces significant zero-sequence MMF.

Because of the lack of a zero-sequence magnetic return path in the three-arm core inductor, the magnetic flux returns outside the core and might result in interference to surrounding electronic components. In addition, such a concept is applicable to two/three-level VSCs and is not suitable for MVDC distribution where the DC voltage is in the range of dozens of kilovolt. An auxiliary zigzag transformer is introduced as a grounding device for symmetric monopole MMC HVDC systems in order to provide a virtual neutral point for grounding purposes and keep balance between the positive and negative pole-to-ground voltages of the HVDC cables.

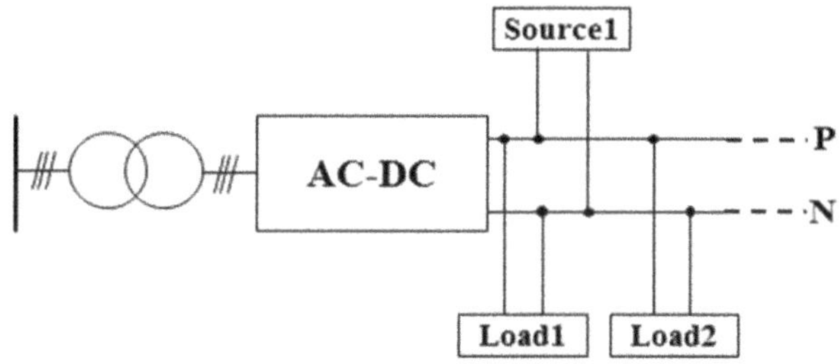

(a)

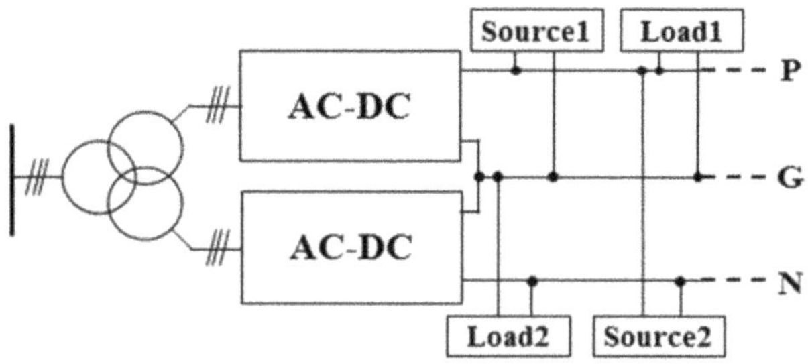

(b)

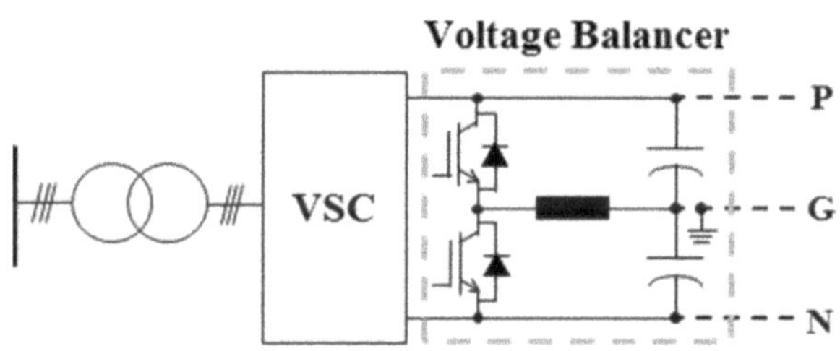

(c)

Figure 6.20 Conceptual configurations of DC systems:(a) a monopole DC system; (b) a bipolar DC system with two AC–DC converters; and (c) a bipolar DC system with a single AC–DC converter and a voltage balancer

Besides the aforementioned functionalities, the auxiliary zigzag grounding transformer can also be used to establish a bipolar DC link for a voltage-source converter (e.g., a TLC or an MMC) as shown in Figure 6.22.

One crucial drawback of the configuration shown in Figure 6.22 is that a significant DC current flows through the zigzag transformer winding in case of unbalanced DC-link pole powers, and its maximum root-mean square (RMS) value is as high as two thirds of the rated DC-link current. Thus, the power rating of the auxiliary zigzag grounding transformer has to be around 60% that of the grid interface transformer.

By employing a delta-zigzag transformer for AC-grid connection with a dedicated operation scheme, an MMC can provide a full bipolar operation capability such that the positive and negative poles on the DC side can be independently controlled. One significant advantage of the zigzag scheme is that it requires no additional component and results in no penalty of increased current rating to the grid interface transformer. In addition, the voltage rating of the MMC can be increased simply by increasing the number of SMs, and the flexibility of the MMC can be exploited.

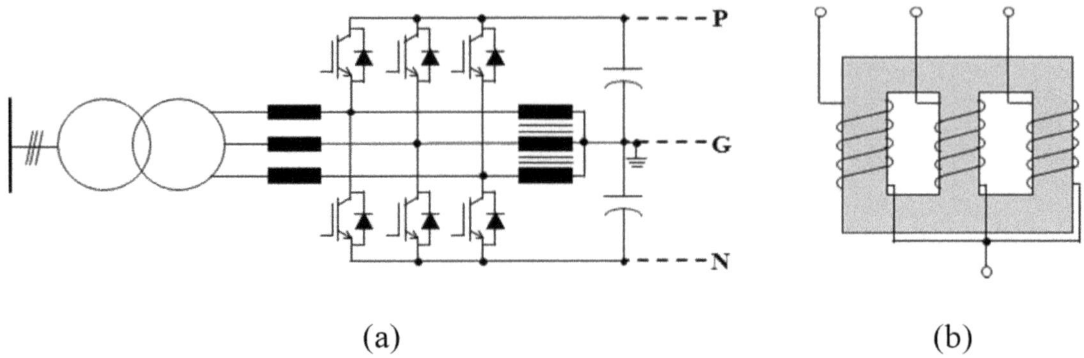

Figure 6.21 A TLC with a grounding inductor for bipolar DC systems: (a) circuit diagram and (b) grounding inductor structure

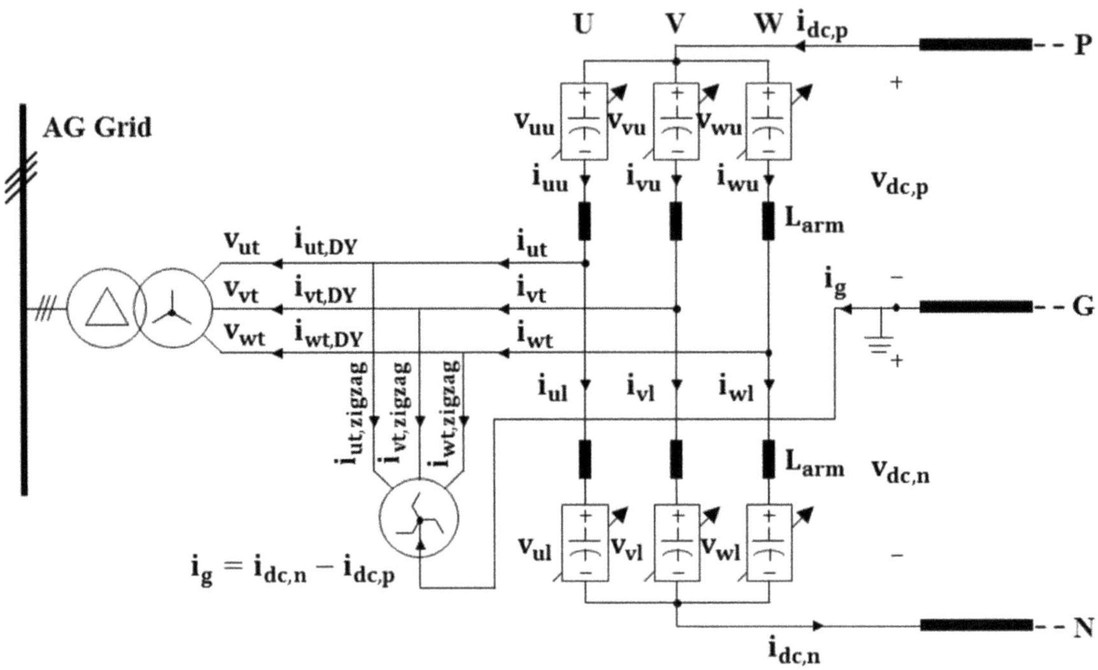

Figure 6.22 An MMC with an auxiliary zigzag grounding transformer for bipolar operation capability

Figure 6.23(a) shows the schematic diagram for an MMC and a zigzag transformer, that is, the MMC is connected to an AC grid via the zigzag transformer. The neutral point of the MMC system is connected to the ground and is also connected to the star point of the zigzag transformer. The primary side of the transformer is delta-connected in order to provide a circulating path for the third-order current caused by the core nonlinearity phenomena.

On the secondary side of the transformer, the turn numbers of the two windings are identical, and they are connected in such a way that the magnetic fluxes induced by the zero-sequence winding current are cancelled out at each core. In case of the DC-pole power unbalance, a DC current i_g flows via the star point of the transformer secondary side into the ground.

If the winding current is controlled such that the DC ground current i_g homogenously distributes in three phases, then it does not result in core flux bias of the transformer. Thus, the zigzag transformer can operate

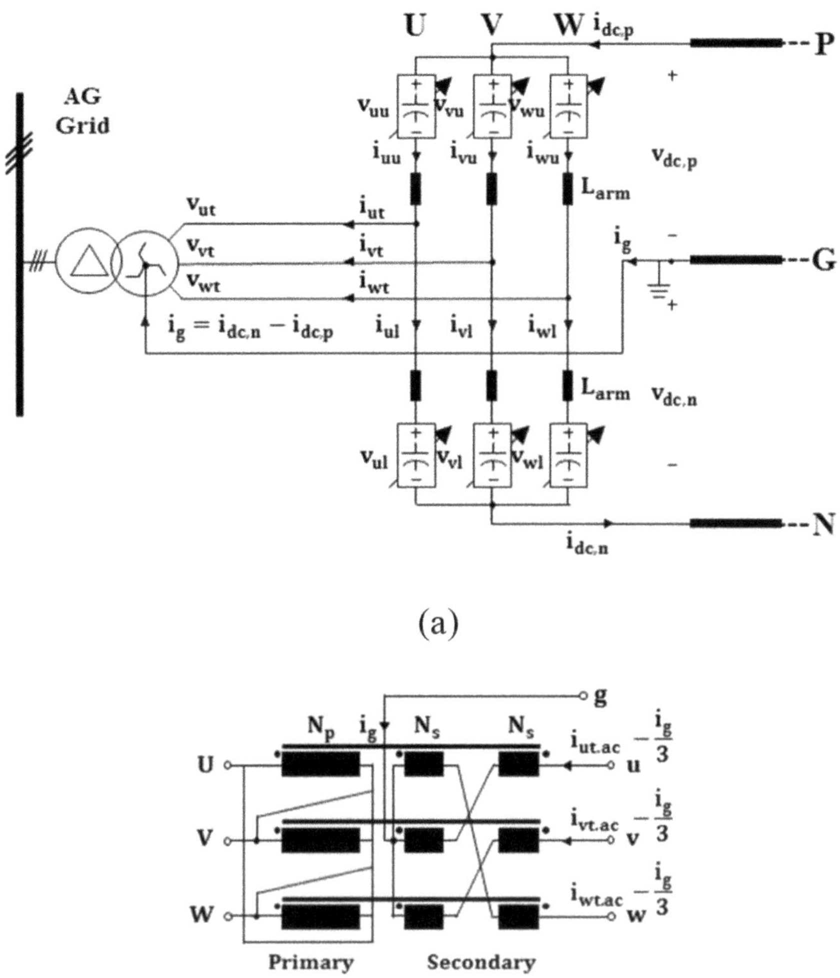

(b)

Figure 6.23 Proposed configuration for bipolar MVDC distribution: (a) schematic of the MMC with the zigzag transformer, and (b) schematic of the winding connection of the zigzag transformer

continuously without saturation even if a DC current flows via the star point. As the DC magnetic flux is cancelled out in the core, it would not result in stray flux outside the core as in a three-arm inductor. It should be also noted that the zero-sequence impedance of the zigzag transformer is in the same range of the leakage inductance, which enables a fast dynamic response of the zero-sequence current that occurs in case of the DC-pole power unbalance.

When the star point is floated, an arm current of an arbitrary x phase of the MMC includes three components: the transformer current i_{xt}, the DC-link current i_{dc}, and the circulating current i_{xcir}, and each component can be controlled independently. When the star point is connected to the DC-side neutral point, an arm current would also include the ground current component. The transformer winding current includes two components, namely the AC current, which is a positive-sequence component, and the DC ground current, which is a zero-sequence component as in Equation (6.30). The upper and lower arm currents of x phase are presented by

Equations (6.31) and (6.32).

$$i_{xt} = i_{xt,ac} - \frac{i_g}{3} = i_{xt,ac} - \frac{i_{dc,n} - i_{dc,p}}{3} \tag{6.30}$$

$$i_{xu} = \frac{i_{xt,ac}}{2} - \frac{1}{2} \cdot \frac{i_g}{3} + \frac{1}{3} \cdot \frac{i_{dc,p} + i_{dc,n}}{2} + i_{xcir} \tag{6.31}$$

$$i_{xl} = -\frac{i_{xt,ac}}{2} + \frac{1}{2} \cdot \frac{i_g}{3} + \frac{1}{3} \cdot \frac{i_{dc,p} + i_{dc,n}}{2} + i_{xcir} \tag{6.32}$$

The block diagram of the transformer current controller is depicted in Figure 6.24, where the term v_{xs} is the voltage component synthesized by the MMC that can regulate the transformer current independently. The transformer current is regulated in the stationary reference frame by a proportional-integral-resonant (PIR) controller. The resonant controller enables the line-frequency AC component of the winding current tracking its reference, which is associated with the reactive and active power control, and the integral controller nullifies the DC differential component in the transformer current. Thus, it is guaranteed that the DC ground current flows homogeneously into three-phase windings, and transformer core saturation is consequently avoided.

In case of a power unbalance on the DC side, it results in SM capacitor energy unbalance among upper and lower arms if no countermeasure is taken. The power disturbance introduced by a DC-side power unbalance can be compensated for by injecting the positive-sequence AC component of the circulating current. To improve dynamic performance of SM capacitor energy regulation, the unbalanced power between the positive and negative poles on the DC side should be utilized for feed-forward control. The block diagram of the feed-forward controller is presented in Figure 6.25.

The power disturbance introduced by the DC-side power unbalance is calculated directly (depicted in the dashed block), so the required magnitude of the positive-sequence circulating current can be calculated in a feed-forward manner for maintaining the SM-capacitor energy balance among upper and lower arms. The power errors caused by something such as power conversion losses are compensated for by the PI regulator in the arm-capacitor energy balancing controller in the steady state.

In the steady state of the DC-side power unbalance, positive sequence circulating current continuously flows inside the MMC. The sum of the powers of two DC-side poles is handled by the transformer-side active power. In accordance with power balance in the steady state, and assuming no power conversion loss, the relationship can be described mathematically as Equation (6.33), where $\hat{V}_t$ and $\hat{I}_{t,ac}$ denote the magnitudes of transformer voltage and AC current. The difference between the powers of two DC-side poles is handled by the injected AC circulating current. In accordance with power balance in the steady state, assuming no power conversion loss, the relationship can be described mathematically as Equation (6.34), where $\hat{I}_{cir}$ denotes the magnitude of the

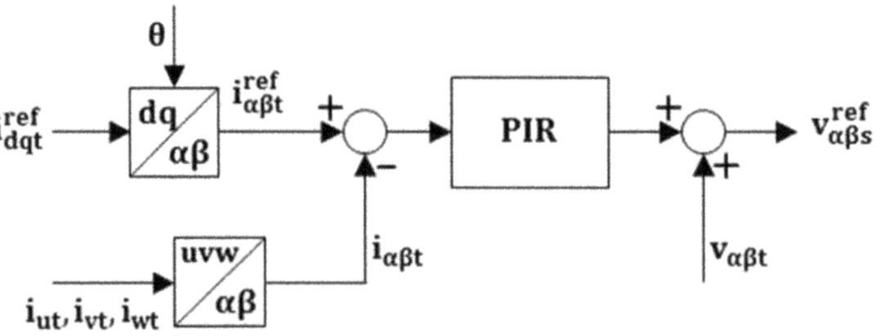

Figure 6.24 Block diagram of the transformer current controller

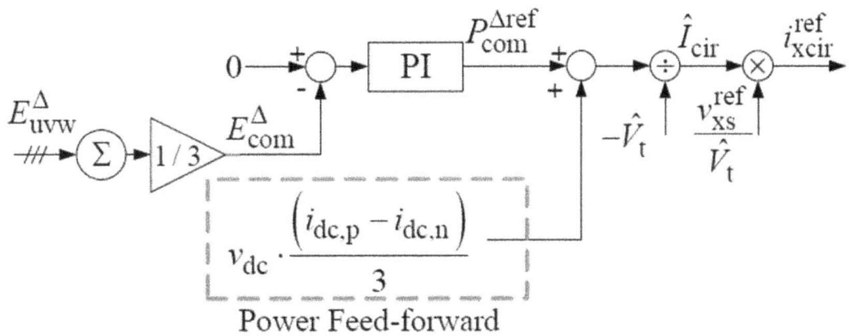

Figure 6.25 Block diagram of the power unbalance feed-forward controller

injected circulating current.

$$-\frac{3}{2}\cdot\hat{V}_t\cdot\hat{I}_{t,ac}=\frac{v_{dc}}{2}\cdot\left(i_{dc,p}+i_{dc,n}\right) \tag{6.33}$$

$$-\frac{3}{2}\cdot 2\cdot\hat{V}_t\cdot\hat{I}_{cir}=\frac{v_{dc}}{2}\cdot\left(i_{dc,p}-i_{dc,n}\right) \tag{6.34}$$

Analysis of Voltage and Current Stress: Regarding the application of the zigzag configuration for bipolar MVDC distribution, one of the main concerns is whether it will result in a penalty of increased voltage or current stress on the components. In this section, it is analyzed in a mathematical way to show that no extra stress would be imposed on the components.

A. Arm Current

As the arm current directly flows through the power-semiconductor devices of SMs, its magnitude and RMS value would be a concern. Assuming that the MMC consumes or provides no reactive power and the voltage across the arm inductor is negligible compared to the transformer voltage, the arm currents of x phase can be represented as Equations (6.35) and (6.36) by substituting Equation (6.33) and (6.34) into Equations (6.31) and (6.32):

$$i_{xu}=i_{xt,ac}\cdot\frac{i_{dc,p}}{i_{dc,p}+i_{dc,n}}+\frac{i_{dc,p}}{3}=-\frac{v_{dc}}{3\hat{V}_t}\cdot i_{dc,p}\cdot\sin(\omega\cdot t)+\frac{i_{dc,p}}{3} \tag{6.35}$$

$$i_{xl}=-i_{xt,ac}\cdot\frac{i_{dc,n}}{i_{dc,p}+i_{dc,n}}+\frac{i_{dc,n}}{3}=-\frac{v_{dc}}{3\hat{V}_t}\cdot i_{dc,n}\cdot\sin(\omega\cdot t)+\frac{i_{dc,n}}{3} \tag{6.36}$$

From Equations (6.35) and (6.36), it can be concluded that the arm current is virtually linear with respect to the power of the corresponding pole, which means that two poles can operate independently. Thus, the MMC can operate in a bipolar manner without resulting in increased arm current stress.

SM-capacitor voltage may be another concern because it is directly imposed on power-semiconductor devices. The dynamic of the SM-capacitor energy of upper and lower arms can be modeled as Equations (6.37) and (6.38), both of which lead to SM-capacitor voltage ripples. As the arm output voltage v_{xu} and v_{xl} are almost identical over different operating points, and the arm current is linear with respect to the corresponding DC-pole power, no increased voltage stress results from a DC-pole power unbalance:

$$\frac{\mathrm{d}E_{xu}}{\mathrm{dt}}=v_{xu}i_{xu} \tag{6.37}$$

$$\frac{dE_{xl}}{dt} = v_{xl} i_{xl} \tag{6.38}$$

Regarding the manufacturing of the transformer, the RMS value of the winding current is one of the key influencing factors on costs. The RMS value of the winding current can be calculated analytically as in Equation (6.39). The winding RMS current is contributed by one third of the DC ground current and the AC active current. The DC ground current is proportional to the power difference between two poles, and the AC active current is proportional to the total power of two poles as shown in Equation (6.39). It is interesting to note that if the powers of the DC poles are unbalanced, the DC ground current increases, whereas the AC active current decreases compared to the balanced situation because of the reduced total DC-link power. From Equation (6.39), it can be concluded mathematically that the maximum of the winding RMS current occurs when $i_{dc,p} = 1\ pu,\ i_{dc,n} = 1\ pu$ or $i_{dc,p} = -1\ pu,\ i_{dc,n} = -1\ pu$, which corresponds to full-power situations when DC-pole powers are balanced. It means that the current rating of the zigzag transformer just needs to be designed to be as large as that of the grid interface transformer used in the case of a monopole DC link:

$$I_{t,rms} = \sqrt{\left(\frac{i_g}{3}\right)^2 + \left(\frac{\hat{I}_{t,ac}}{\sqrt{2}}\right)^2} = \sqrt{\left(\frac{i_{dc,p} - i_{dc,n}}{3}\right)^2 + \left(\frac{v_{dc}(i_{dc,p} + i_{dc,n})}{3 \cdot \sqrt{2} \cdot \hat{V}_t}\right)^2} \tag{6.39}$$

The manufacturing cost of the zigzag transformer would be higher than that of a normal wye- or delta-connected grid interface transformer with the same voltage and current ratings because of its more complicated winding connection. However, it can enable full bipolar operation capability of a single MMC without increasing transformer current rating and can significantly enhance the flexibility of DC grids. Moreover, it does not require an auxiliary electronic or magnetic device with a considerably high power rating.

Consequently, a power unbalance of two poles on the DC side does not result in any extra voltage or current stress on the components. Thus, by connecting the transformer secondary side in a zigzag structure as shown in Figure 6.23(b) and employing the dedicated operation scheme, an MMC can establish a bipolar DC distribution system without any extra cost.

Simulation and Experimental Results

To verify the validity of the zigzag transformer configuration, a computer simulation has been performed. The simulation parameters are listed in Table 6.1. A saturable core model is used in the simulation of the zigzag transformer.

In Figure 6.26(a), simulation results are shown from when the DC-pole power $(p_{dc,p}, p_{dc,n})$ changes from (1 p.u., 1 p.u.) to (1 p.u., 0 p.u.) when t=1.5 s, and when it changes from (1 p.u., 0 p.u.) to (1 p.u., -1 p.u.) when t=1.6 s. In Figure 6.26(b), simulation results are shown from when the DC-pole power $(p_{dc,p}, p_{dc,n})$ changes from (−1 p.u., −1 p.u.) to (−1 p.u., 0 p.u.) when t=2.5 s, and when it changes from (−1 p.u., 0 p.u.) to (−1 p.u., 1 p.u.) when t=2.6 s.

In the simulation results shown in both Figure 6.26(a) and (b), the positive-pole power is kept constant, whereas the negative-pole power varies. The upper-arm current is kept constant, whereas the lower-arm current varies linearly in accordance with the negative-pole power.

It is clearly shown in the simulation results that the arm current and SM-capacitor voltage over the whole operation range do not exceed the values of the symmetric rated power operation. The maximum transformer RMS current occurs while two DC poles operate at the rated power in a symmetric manner. This verifies that by employing a 20-MVA zigzag transformer as a grid interface transformer, it is possible to operate a 20-MW MMC in a bipolar manner such that both DC poles can operate fully independently at a power rating of 10 MW.

Table 6.1 Parameters of the Simulated System

Transformer	
Transformer secondary-side voltage	10 kV
Transformer power rating	20 MVA
Modular multi-level converter	
Rated power	20 MW
Rated DC-link voltage	±10 kV
Rated DC-link current	1 kA
Number of submodules (SMs) per arm	20
Rated SM capacitor voltage	1.0 kV
Capacitance of SM capacitor	9 mF
Inductance of arm inductor	1 mH
Controller sampling frequency	10 kHz

The MMC is well controlled, and the zigzag transformer does not saturate even if a DC current flows through the windings. An identical MMC with a delta-wye transformer as a grid interface transformer and a zigzag transformer as a grounding device as shown in Figure 6.22 is simulated with the same scenarios for comparison. The dedicated operation method proposed in this letter is performed to the MMC, and the simulation results are shown in Figure 6.27. The arm current and SM-capacitor voltage of the MMC are identical to the case of the zigzag configuration shown in Figure 6.26. The maximum RMS current of the grid interface transformer occurs while two DC poles operate at the rated power in a symmetric manner, and the maximum RMS current of the auxiliary zigzag grounding transformer occurs while two DC poles operate at the rated power but with opposite direction.

Thus, in order to enable a bipolar operation capability, a 20-MW MMC requires a 20-MVA grid interface transformer and an 11.5-MVA zigzag grounding transformer. On the other hand, the grid interface transformer and zigzag grounding transformer are fully integrated to a single 20-MVA delta-zigzag grid interface transformer in the zigzag transformer configuration.

6.4 Analysis of AC Coupling Effects on PTP MMC HVDC System [17]

In this chapter, the AC coupling effects on DC transmission lines are modeled in a PTP MMC HVDC system. Then, the influence of the induced current on the MMC is analyzed based on the decoupled control strategy. As with much research about AC coupling effects on HVDC lines, the effects can be simply modeled as an AC-voltage source with fundamental frequency. Considering that the coupling effects can induce voltages on both positive poles and negative poles, the PTP HVDC system with AC coupling can be represented as shown in Figure 6.28, where $\boldsymbol{v}_{ind\_p}$ and $\boldsymbol{v}_{ind\_n}$ stand for the induced voltages on the positive pole and the negative pole, respectively. Also, the DC-side equivalent model of MMC can be represented as a variable DC-voltage source in series with equivalent impedance. Based on this model, the whole system can be simplified as in Figure 6.29. From the simplified model, it can be easily deduced that the induced voltages cause the induced AC current $\boldsymbol{i}_{ind}$ in the loop of DC line, and it actually flows into the legs of the MMC. Because of the symmetry of leg impedances of MMC in three phase, the induced current is equally distributed by one third in three phase legs. When single line ground fault occurs in a parallel AC system, the induced voltages increase very rapidly, three to four times higher than before, and the current also increases.

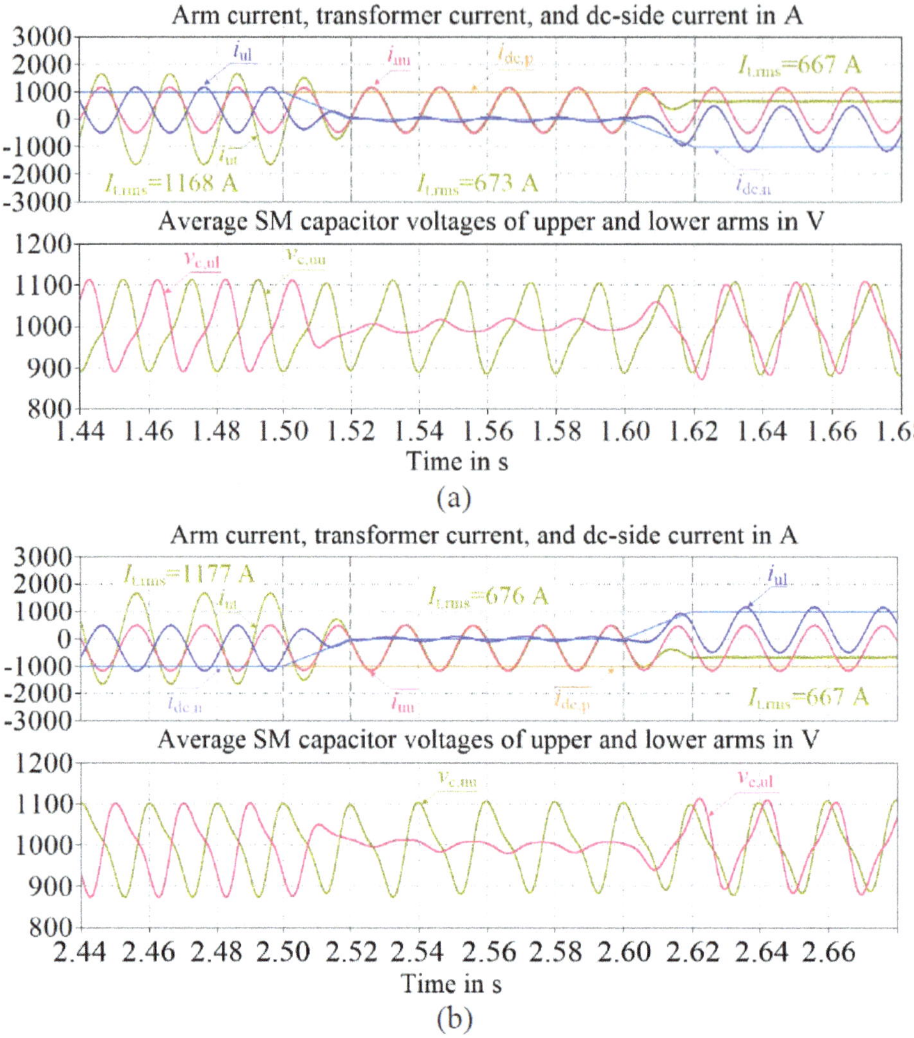

Figure 6.26 Simulation results of the zigzag configuration in different operation conditions: (a) the negative-pole power changes from 1 to −1 p.u. and the positive-pole power keeps 1 p.u. and (b) the negative-pole power changes from −1 to 1 p.u. and the positive-pole power keeps -1 p.u.

To analyze the influence of induced current on the MMC, the average power flowing into each arm should be firstly considered. The average power that flows into the upper and lower arms because of the induced current in the same leg is in Equation (6.40), where x denotes u- or v- or w-phase and upper bar stands for time average value. Then the sum and difference of the power between upper and lower arms can be deduced as in Equations (6.41) and (6.42):

$$\begin{cases} \overline{P}_{xu} = \dfrac{d\overline{E}_{xu}^{\Sigma}}{dt} \left\langle v_{xu}^{*} \dfrac{i_{ind}}{3} \right\rangle \approx -\dfrac{1}{3} v_{xs}^{*} i_{ind} \\ \overline{P}_{xl} = \dfrac{d\overline{E}_{xl}^{\Sigma}}{dt} \left\langle v_{xl}^{*} \dfrac{i_{ind}}{3} \right\rangle \approx -\dfrac{1}{3} v_{xs}^{*} i_{ind} \end{cases} \tag{6.40}$$

$$\begin{cases} \overline{P}_{x}^{\Sigma} = \overline{P}_{xu} + \overline{P}_{xl} = 0 \\ \overline{P}_{x}^{\Delta} = \overline{P}_{xu} - \overline{P}_{xl} = -\dfrac{2}{3} v_{xs}^{*} i_{ind} \end{cases} \tag{6.41}$$

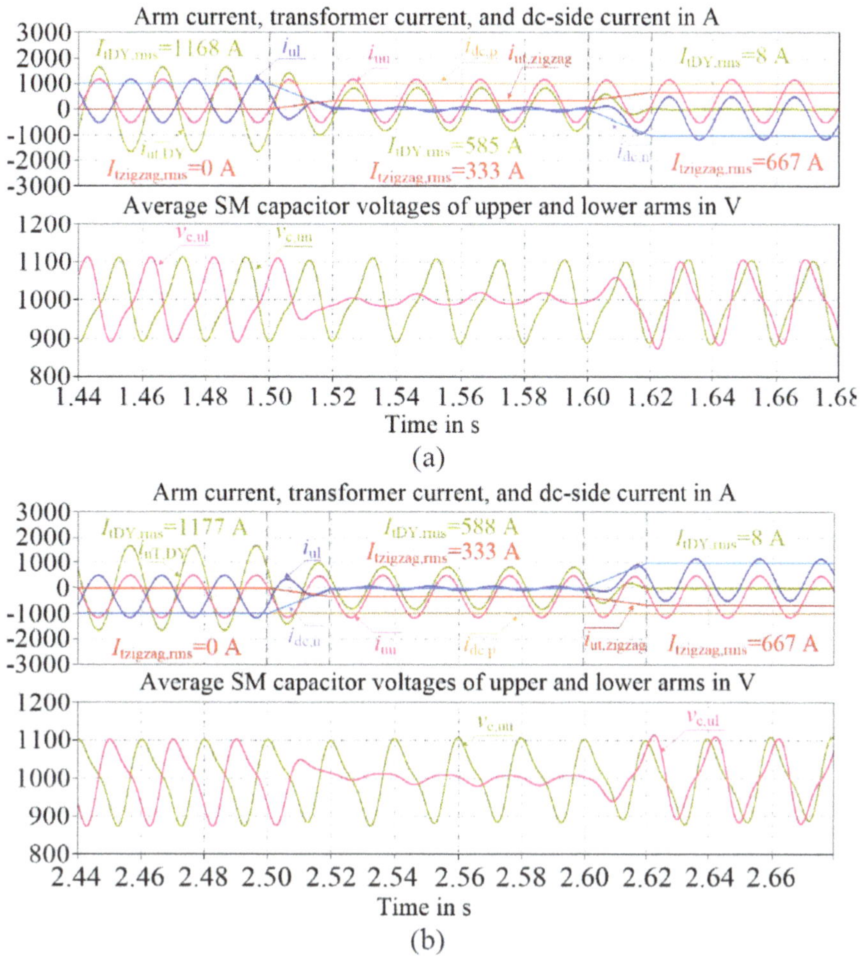

Figure 6.27 Simulation results of the configuration with an auxiliary zigzag grounding transformer in different operation conditions: (a) the negative-pole power changes from 1 to –1 p.u. and the positive-pole power keeps 1 p.u. and (b) the negative-pole power changes from -1 to 1 p.u. and the positive-pole power keeps –1 p.u.

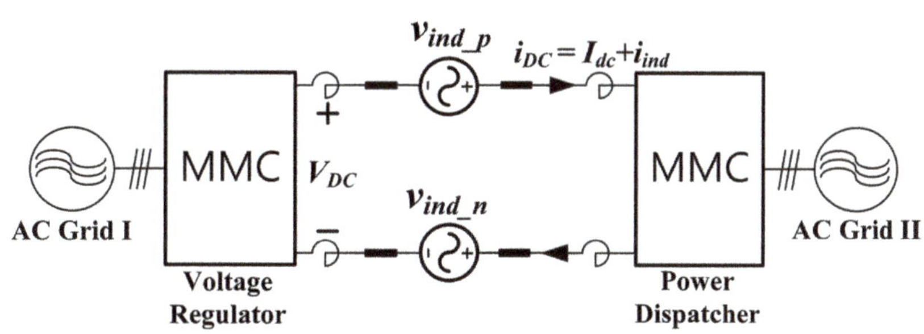

Figure 6.28 Point-to-point modular multi-level converter HVDC system with AC-coupling effects

$$\begin{cases} \overline{P}^{\Delta}_{com} = \frac{1}{3} \sum\limits_{x=u,v,w} \overline{P}^{\Delta}_{x} = -\frac{2}{9}\left(v^*_{us} + v^*_{vs} + v^*_{ws}\right) i_{ind} = 0 \\ \overline{P}^{\Delta}_{x,diff} = \overline{P}^{\Delta}_{x} - \overline{P}^{\Delta}_{com} = -\frac{2}{3} v^*_{xs} i_{ind} \end{cases} \tag{6.42}$$

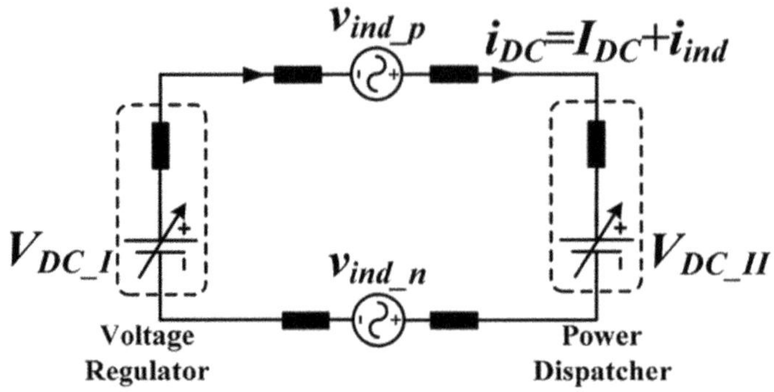

Figure 6.29 Simplified circuit of the DC side of point-to-point modular multi-level converter HVDC system

From Equation (6.42), it can be noted that the induced current in MMC leads to the power component $\overline{P}^{\Delta}_{x,diff}$, which is the time derivative of the differential difference of the arm energies in each leg. Meanwhile, the power $\overline{P}^{\Delta}_{com}$ (the time derivative of the common difference of arm energies) is not affected by the induced current. Like previously mentioned internal balancing control in Section II, the energy unbalance because of $\overline{P}^{\Delta}_{x,diff}$ could be reduced by controlling the negative sequence circulating current. However, the power $\overline{P}^{\Delta}_{x,diff}$ exists even at the steady state, which makes circulating current flow continuously, and thus may add considerable extra losses in the MMC.

Besides the average power, the induced current also makes the total energy fluctuating at the same frequency of induced current like Equations (6.43) and (6.44). As the feedback value $\tilde{E}_{Total}$ in the total energy controller described in a previous chapter is fluctuating, the output of the controller i_q^{+*} is also fluctuating. Through the AC-grid current controller, the fundamental frequency component in synchronous reference frame q-axis is converted to DC current and twice frequency harmonic current in stationary frame. When the DC current is injected into an AC grid, it could make the converter transformer saturated and might cause core saturation instability.

$$\begin{cases} \tilde{P}_{xu} = \dfrac{d\tilde{E}_{xu}}{dt} \approx \dfrac{V_{dc}i_{ind}}{6} \\ \tilde{P}_{xl} = \dfrac{d\tilde{E}_{xl}}{dt} \approx \dfrac{V_{dc}i_{ind}}{6} \end{cases} \tag{6.43}$$

$$\tilde{P}_{Total} = \frac{d\tilde{E}_{Total}}{dt} = \sum_{x=u,v,w} (\tilde{P}_{xu} + \tilde{P}_{xl}) = V_{dc}^{*} i_{ind} \tag{6.44}$$

The induced AC current to DC line should be reduced, because it has detrimental effects on both MMC and transformer on the AC-grid side I, a novel DC-current regulator of a power dispatcher MMC for induced current mitigation is proposed as shown in Figure 6.30. The induced voltage on DC line, v_{ind}, is considered as a disturbance term in this controller, whose magnitude and frequency would vary depending on the situation of the parallel AC system. Therefore, the goals of the AC suppression controller are to eliminate the disturbance perfectly at the steady state and to respond to the rapid variation of the disturbance during the fault state of the parallel AC grid. Conventionally only proportional and integral (PI) controllers were used for regulating DC current. It had poor performance from the perspective of sinusoidal disturbance rejection because there should, in theory, have been infinite loop gain at the same frequency with the disturbance to reject the disturbance. To

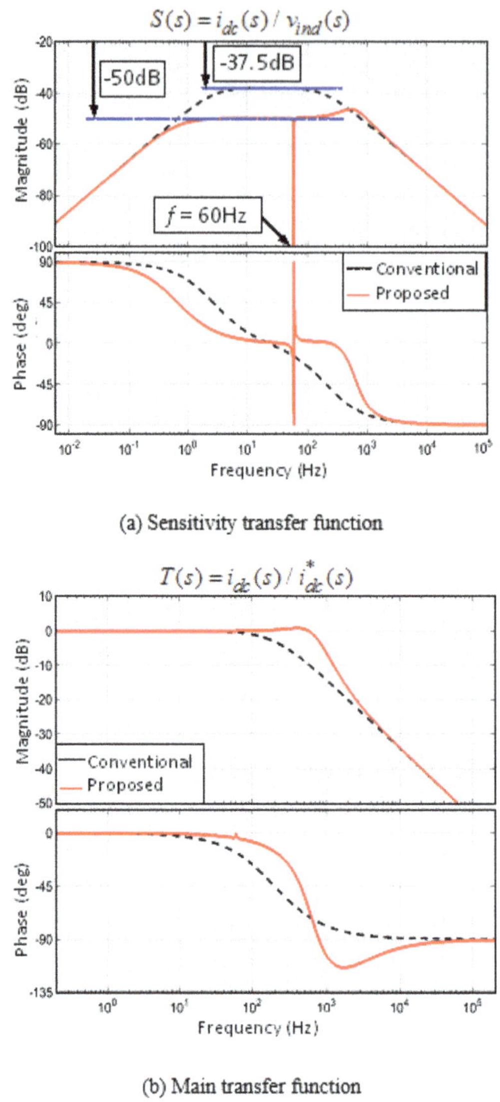

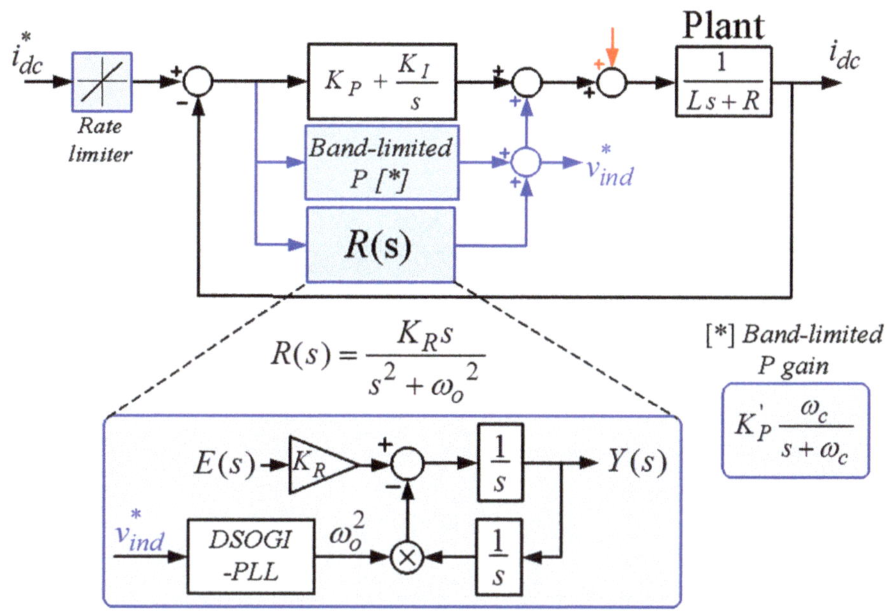

Figure 6.30 Block diagram of the AC suppression control

address this issue, the proposed controller has two additional controllers: the resonant controller *R(s)* with frequency adaptive capability, and the band-limited high gain proportional (BLHGP) controller.

Firstly, the resonant controller provides infinite gain in an open loop at the resonant frequency ω_o without delays in phase. As the command of the controller only has DC value, the resonant controller has little impact on the ability of system to track the command while minimizing the effect of the disturbance. Secondly, the BLHGP controller is utilized for fast responses to disturbance variation. Although the resonant controller guarantees the rejection of induced voltage at steady state, it cannot respond immediately in transient state because of the intrinsic properties of integrators in the resonant controller. By adding the BLHGP controller, the system can react fast enough to limit the AC current caused by the sudden rise of disturbance while filtering the noise amplification caused by high proportional gain. Lastly, the phase-locked loop (PLL) is adopted for frequency adaptive capability of the resonant controller. If the frequency of the induced voltage varies from its nominal value according to that of parallel AC-grid voltage, the resonant frequency of *R(s)* also needs to be adjusted for perfect rejection of AC components. Among the numerous signals taken from the system, $v_{ind,est}$ (the sum of output voltages of the resonant and the BLHP controller) has information about the frequency of disturbance even at steady state. By taking $v_{ind,\,est}$ as the input signal of the PLL, the frequency of the disturbance could be identified by PLL.

For comparison of the control performance between the conventional PI controller and the proposed one, Bode plots of sensitivity transfer function $S(s) = i_{dc}(s)/v_{ind}(s)$ and main transfer function $T(s) = i_{dc}(s)/i^*_{dc}(s)$ are shown in Figure 6.31, based on the parameters in Table 6.1.

From Figure 6.31(a), it can be noted that the AC suppression controller can eliminate the effect of disturbance $\boldsymbol{v_{ind}}$ on the output current, $\boldsymbol{i_{DC}}$, at the steady state because the magnitude response of S(s) is almost zero at the frequency of induced voltage ($\omega_o = 2\pi \cdot 60$ $[rad/s]$) by the resonant controller.

Also, the magnitude response in S(s) of the proposed controller is lower than that of the conventional controller by 12.5 dB in the wide range of frequency by the BLHGP controller. This means that, although $\boldsymbol{v_{ind}}$ surges because of SLG fault in a paralleled AC system, $\boldsymbol{i_{DC}}$ can be significantly suppressed in the transient state. Meanwhile, the dynamics of the DC-current control are determined by the main transfer function, $T(s)$.

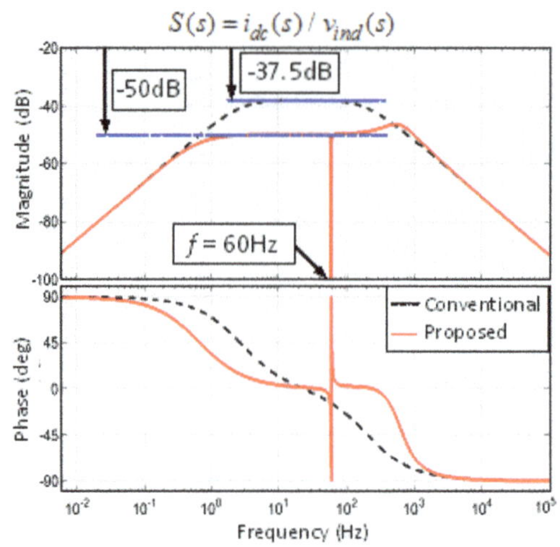

(a) Sensitivity transfer function

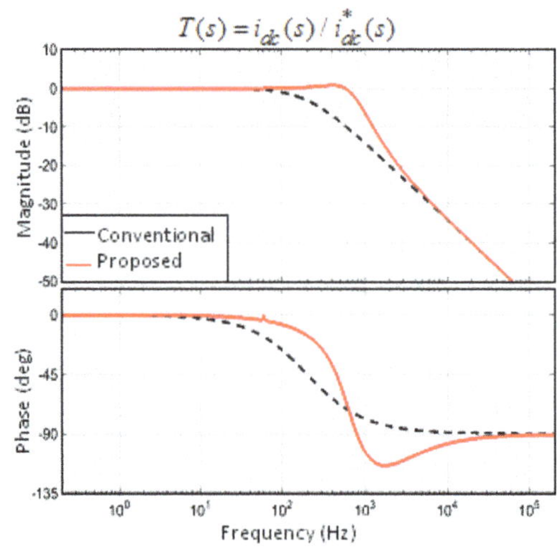

(b) Main transfer function

Figure 6.31 Bode plots of closed loop system with the AC suppression control

As shown in Figure 6.31(b), the AC suppression controller has wider bandwidth than the conventional one, while an undesired minor overshoot occurred at around $f = 440$ Hz. This overshoot may make the system underdamped and cause oscillation at the transient state. Therefore, the rate limiter to suppress a sudden change of command is included in the AC suppression controller to ensure stable operation.

Simulation Results

To verify the validity of the AC suppression controller, the simulations are carried out in a 400 MW full-scaled PTP MMC HVDC system using PSIM software. The schematic of the simulated system is depicted in Figure 6.28. The rated DC voltage is 400 kV and the line-to-line voltages of AC grids both I and II are 180.5 kV/60 Hz. To see the performance of the AC suppression controller, the scenario of the simulation is created, which is analyzing the induced voltage on a VSC HVDC system. The details are shown in Table 6.2. In the following scenario, the magnitudes of the induced voltages are represented as a percentage of the rated DC voltage of the HVDC system. The waveforms of that are shown in the first traces of Figures 6.32 and 6.33.

Table 6.2 Parameters of the Experimental Setup

Modular multi-level converter	
Transformer secondary-side voltage	135 kV
Rated DC-link voltage	±150 kV
Rated DC-link current	10 A
Number of submodules (SMs) per arm	6
Rated SM capacitor voltage	50 V
Capacitance of SM capacitor	5.4 mF
Inductance of arm inductor	4 mH
Controller sampling frequency	10 kHz

The first simulation results are shown in Figure 6.32, when the conventional controller is applied. As the conventional controller cannot mitigate the coupling effects, the induced AC current flows in DC line from t = 3 s as shown in the second trace of Figure 6.32(a). At the same time, in the third trace, DC and second order harmonic currents are injected into the AC grid connected to the voltage regulator MMC, because the induced current makes the total energy, and thus the q-axis current reference, fluctuate. The DC and second-order harmonic current injected into the AC grid are changed to 0.5 and 120.5 Hz AC current from t = 5 s because the frequency of arm-energy-induced voltage changes to 60.5 Hz. In addition, through the balancing controller, the negative-sequence circulating currents start to flow inside the MMC to reduce the differential difference of arm energies as shown in the fourth trace. All of these become worse from t = 7 s when an SLG fault occurs in the paralleled AC system.

The induced AC voltage rises very rapidly, so the induced current also increases up to 498 A, almost 50% of the rated DC current. And, as the frequency of the induced current changed to 58 Hz, 2 Hz and 118 Hz AC currents are injected into the AC grid. The magnitude of the circulating current is almost four times larger than that of normal operation and would cause additional loss in the MMC.

The second simulation results are depicted in Figure 6.32(b), when the AC suppression controller is applied. As represented in the second trace, with the resonant controller, the induced current is eliminated. Even in the case of SLG fault condition, the induced current is significantly suppressed by the BLHGP controller at the transient state. The peak is only 110 A, which is less than a quarter ($\approx$ 12.5dB = 50dB-37.5dB) of the 498 A from the conventional controller. From this point, it can be confirmed that the AC suppression controller is effective for suppressing an AC component induced to DC line. The AC-grid current of the voltage regulator MMC is also well controlled as a balanced three-phase current, because there is no induced AC current on the DC line. The circulating current of the voltage regulator MMC is almost null for the same reason. However, in this case, the circulating current of the power dispatcher MMC flows inevitably, because the DC voltage of the power dispatcher MMC is fluctuating to cancel out the induced AC voltage. Compared to the magnitude of circulating current of voltage regulator MMC in the first simulation, it is reduced to almost one third. As the loss is proportional to the square of current, it can be said that the loss of MMC is also significantly reduced when the AC suppression controller is engaged. Table 6.3 shows the simulation scenario.

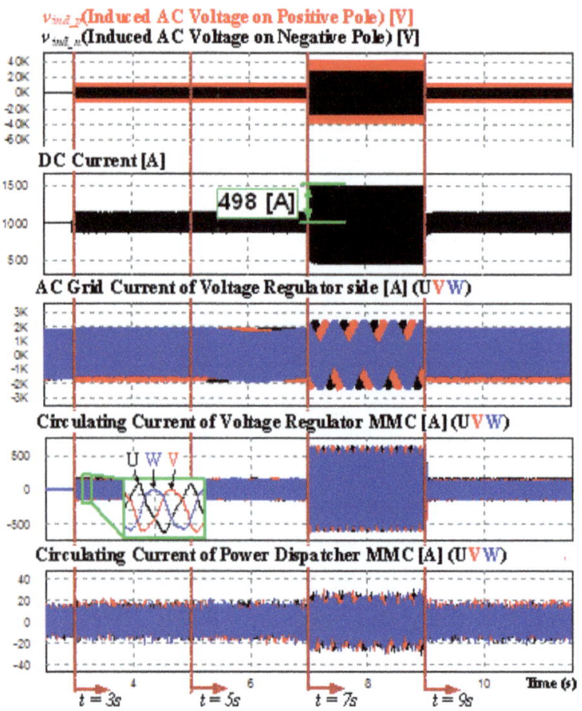

(a) Simulation results when the conventional controller is applied

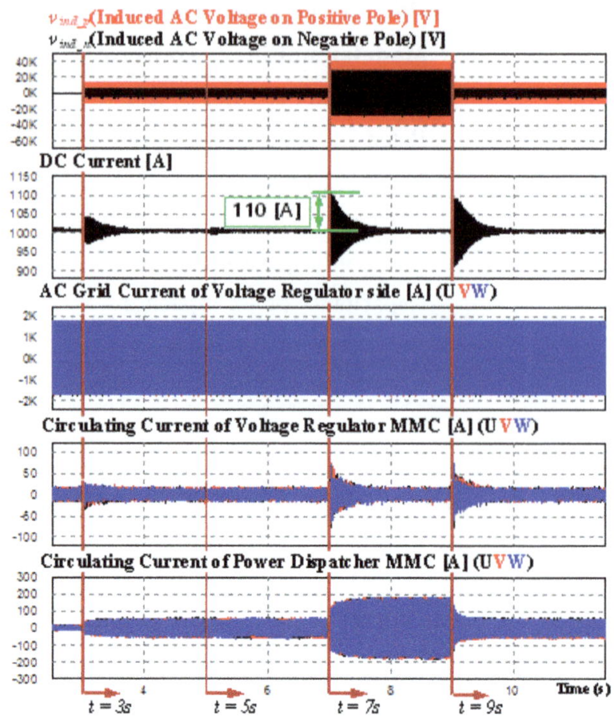

(b) Simulation results when the proposed controller is applied

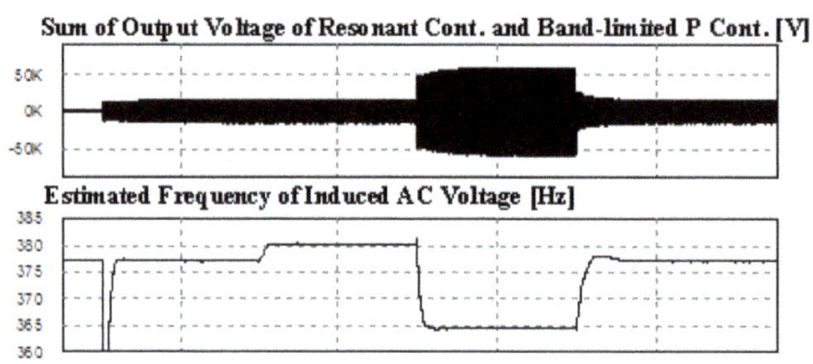

(c) Simulation results of frequency estimation of PLL

Figure 6.32 Simulation results (a), (b) and (c)

Table 6.3 Simulation Scenario

Time	Situations
t = 3s	The parallel AC system starts to operate normally. → $\lvert v_{ind\_p} \rvert$ = 3%, $\lvert v_{ind\_n} \rvert$ = 1.5% and $\boldsymbol{f}$ (of v_{ind}) = 60 Hz
t = 5s	Frequency of the parallel AC system changes. → $\boldsymbol{f}$ (of v_{ind}) = 60.5 Hz
t = 7s	The SLG fault is occurred in the paralleled AC system. → $\lvert v_{ind\_p} \rvert$ = 10%, $\lvert v_{ind\_n} \rvert$ = 7% and $\boldsymbol{f}$ (of v_{ind}) = 58 Hz
t = 9s	The fault is cleared. → Magnitude and frequency of v_{ind} are restored as normal.

References

1. T. K. Vrana, et al., "A classification of DC node voltage control methods for HVDC grids," *Electr. Pow. Syst. Res.*, vol. 103, pp. 137–144, Oct. 2013.
2. W. Wang, M. Barnes, and O. Marjanovic, "Droop control modelling and analysis of multi-terminal VSC-HVDC for offshore wind farms," in *Proceedings of the IET International Conference on AC and DC Power Transmission*, Birmingham, UK, Dec. 4–6, 2012.
3. E. Prieto-Araujo, et al., "Methodology for droop control dynamic analysis of multiterminal VSC-HVDC grids for offshore wind farms," *IEEE Trans. Power Del.*, vol. 26, no. 4, pp. 2476–2485, 2011.
4. B. Berggren, et al., "Method and control device for controlling power flow within a dc power trans- mission network," Publication Number WO 2012/000 549A1, International Filing Date: 30 June 2010.
5. T. K. Vrana, L. Zeni, and O.B. Fosso, "Dynamic active power control with improved undead-band droop for HVDC grids", in *Proceedings of the IET International Conference on AC and DC Power Transmission*, Birmingham, Dec. 4–6, 2012.
6. T. Nakajima and S. Irokawa, "A control system for HVDC transmission by voltage sourced converters," in *Proceedings of the IEEE Power Engineering Society Summer Meeting*, Edmonton, Alta, Canada, 2002.
7. Y. Sekine, et al., "Application of power electronics technologies to future interconnected power system in Japan," in *Proceedings of the 1995* CIGRE Tokyo Symposium, May 1995.
8. H.Suzuki, et al., "Development and testing of prototype models for a high performance 300 MW self-commutated AC/DC converter," *IEEE Trans. Power Del.*, vol. 12, no. 4, Oct. 1997; K. Sakamoto , et al., "Development of a control system for a high-performance self-commutated converter," IEEE Trans. Power Del., vol. 13, no. 1, Jan. 1998.
9. T. Nakajima, et al., "A converter transformer with series-connected line-side windings for a DC link using voltage source converters," in *Proceedings of the IEEE Power Engineering Society Winter Meeting*, New York, USA, Feb. 1999.
10. H. Wang, et al., "Efficient grounding for modular multilevel HVDC converters (MMC) on the AC side," *IEEE Trans. Power Del.*, vol. 29, no. 3, pp. 1262–1272, Jun. 2014.
11. R. A. Walling and A. H.Khan, "Characteristics of transformer excitingcurrent during geomagnetic disturbances," *IEEE Trans. Power Del.*, vol. 6, no. 4, pp. 1707–1714, Oct. 1991.
12. L. Zhong, et al., "The influence of current of DC earthing electrode on directly grounded transformer," *High Voltage Eng.*, vol. 29, no. 8, pp. 12–13, Aug. 2003.
13. DL/T 437-12, Guide rule for HVDC earth electrode (in Chinese), 2012
14. Y. You, et al., "Reactive power demand of transformers with DC bias," *IEEE Ind. Appl. Mag.*, vol. 2, no. 4, pp. 45–52, Jul. 1996.
15. W. Leterme, et al., "Overview of grounding and configuration options for meshed HVDC Grids," *IEEE Trans. Power Del.*, vol. 29, no. 6, pp. 2467–2475, Dec. 2014.
16. K. De Kerf, et al., "Wavelet-based protection strategy for DC faults in multi-terminal VSC HVDC systems," IET Gener. *Transm. Distrib.*, vol. 5, no. 4, pp. 496–503, 2011.
17. O. Nanayakkara, A. Rajapakse, and R. Wachal, "Traveling-wave-based line fault location in star-connected multiterminal HVDC systems," *IEEE Trans. Power Del.*, vol. 27, no. 4, pp. 2286–2294, Oct. 2012.
18. J. Yang, J. Fletcher, and J. O'Reilly, "Short-circuit and ground fault analyses and location in VSC-based DC network cables," *IEEE Trans. Ind. Electron.*, vol. 59, no. 10, pp. 3827–3837, Oct. 2012.
19. ABB, "HVDClight: It's time to connect," ABB, Ludvika, Sweden, Tech. Rep. POW-0038, 2012.
20. J. Candelaria and J.-D. Park, "VSC-HVDC system protection: A review of current methods," in *Presented at the IEEE PSCE*, Phoenix, AZ, Mar. 20–23, 2011.
21. L. Tang and B.-T. Ooi, "Managing zero sequence in voltage source converter," in *37th IAS Annual Meeting Conference Record of the 2002 IEEE Industry Applications Conference*, Pittsburgh, PA, USA, 2002.
22. H. Wang, et al., "Effeccient grounding for modular multilevel HVDC converters (MMC) on the AC side," *IEEE Trans. Power Del.*, vol. 29, no. 3, pp. 1262–1272, Jun. 2014.
23. W. Leterme, et al., "Overview of grounding and configuration options for meshed HVDC grids," *IEEE Trans. Power Del.*, pp. 2467–2475, Jun. 2014.
24. R. Li, J. E. Fletcher, and B.W. Williams, "Influence of third harmonic injection on modular multilevel converter-based high-voltage direct current transmission systems," *IET Gener. Transm. Dis.*, Nov. 2016, pp. 2764–2770.
25. J. Pou, et al., "Circulating current injection methods based on instantaneous information for the modular multilevel converter," *IEEE Trans. Ind. Electron.*, vol. 62, no. 2, pp. 777–788, Feb. 2015.

26. A. Junyent-Ferré, "Operation of HVDC Modular Multilevel Converters under DC pole imbalances," in *16th European Conference on Power Electronics and Applications*, Lappeenranta, Finland, 2014.
27. J. Yang, et al., "Characteristics and recovery performance of vsc-hvdc dc transmission line fault," in *2010 Asia-Pacific Power and Energy Engineering Conference (APPEEC)*, IEEE, Chengdu, China, 2010, pp. 1–4.
28. T. Westerweller, et al., "Trans bay cable worlds first hvdc system using multilevel voltage-sourced converter," in *CIGR´E Paper B4-101*, Paris, 2010.
29. X. Chen, C. Zhao, and C. Cao, "Research on the fault characteristics of hvdc based on modular multilevel converter," in *Electrical Power and Energy Conference (EPEC)*, IEEE, Winnipeg, MB, Canada, 2011, pp. 91–96.
30. D. Kwon, et al., "Modeling of HVDC system to improve estimation of transient DC current and voltages for AC line-to-ground fault an actual case study in Korea," Energies, vol. 10, p. 1543, 2017.
31. A. Bidadfar, et al., "Power system stability analysis using feedback control system modeling including HVDC transmission links," *IEEE Trans. Power Syst.*, vol. 31, pp. 116–124, 2016.
32. C. Karawita and U. D. Annakkage, "Multi-infeed HVDC interaction studies using small-signal stability assessment," *IEEE Trans. Power Deli.*, vol. 24, pp. 910–918, 2009.
33. Z. Wei, et al., "Direct-current predictive control strategy for inhibiting commutation failure in HVDC converter,"*IEEE Trans. Power Syst.*, vol. 29, pp. 2409–2417, 2014.
34. S. Cui, et al., "A modular multilevel converter with a zigzag transformer for bipolar MVDC distribution systems," *IEEE Trans. Power Electron.*, vol. 34, pp. 1038–1043, 2019.
35. R. W. De Doncker, "Power electronic technologies for flexible dc distribution grids," in *IEEE International Power Electronics Conference*, Hiroshima, Japan, 2014, pp. 736–743.
36. H. Kakigano, Y. Miura, and T. Ise, "Low-voltage bipolar-type dc microgrid for super high quality distribution,"*IEEE Trans. Power Electron.*, vol. 25, pp. 3066–3075, 2010.
37. J. Lago, J. Moia, and M. L. Heldwein, "Evaluation of power converters to implement bipolar dc active distribution networks–DC-DC converters," in *IEEE Energy Conversion Congress and Exposition*, Phoenix, AZ, 2011, pp. 985–990.
38. Y. Gu, W. Li, and X. He, "Analysis and control of bipolar LVDC grid with dc symmetrical component method," *IEEE Trans. Power Syst.*, vol. 31, pp. 685–694, 2016.
39. H. Wang, et al., "Efficient grounding for modular multilevel HVDC converters (MMC) on the ac side," *IEEE Trans. Power Del.*, vol. 29, pp. 1262–1272, 2014.
40. S. Cui, et al., "A comprehensive cell capacitor energy control strategy of a modular multilevel converter (MMC) without a stiff dc bus voltage source," in *IEEE Applied Power Electronics Conference and Exposition,* Fort worth, TX, USA, 2014, pp. 602–609.
41. J.-H. Lee, J.-J. Jung, and S.-K. Sul, "Analysis and mitigation of AC coupling effects on overhead line of modular multilevel converter (MMC) based HVDC transmission system,"in *Proc. IEEE Energy Convers. Congr. Expo.*, Oct. 2017, pp. 97–104.

7

Equivalent Modeling and Simulation of MMC System

7.1 Modeling of MMC on Electromagnetic Transient Programs [1–2]

The modular multi-level converter (MMC) topology has become the standard configuration for high-voltage direct current (HVDC) schemes that are based on voltage-sourced converter (VSC) systems. MMC uses a large number of power-electronic switching devices that operate independently. The reason is to minimize the harmonic content of the converter output voltage waveform and reduce the voltage rating of the individual submodule. Therefore, the MMC has a very large number of switch elements. In half-bridge submodule converters, each of the N submodules of the six multi-valves in a three-phase MMC contains two switch elements, giving a total of $12N$ switches per converter. With 100 submodules per multi-valve, this means 1200 switches per converter. This number is doubled when the full bridge submodules are used in the converter.

Modeling such converters with individual switches in an electromagnetic transient (EMT) simulation program would lead to extremely slow runs, and hence seems to be impractical. To overcome this issue, MMC systems can be modeled as a detailed equivalent circuit, which is an efficient model for the converter using the Nested Fast and Simultaneous Solution method. It partitions the solution into two parts, with the external network solution being implemented in the main EMT solver. Each phase of the MMC is interfaced as a specially designed Thévenin equivalent, thereby greatly reducing the number of nodes. The MMC is solved separately in an efficient manner by exploiting its simple topology.

Traditional admittance matrix solution for MMC is using Dommel's algorithm, shown in Figure 7.1. From Figure 7.1 each dynamic element in the electrical system can be modeled as a Norton equivalent current source in parallel with a conductance through the application of trapezoidal integration. The current source is updated each time-step and is a function of the history of the network in earlier time-steps.

The trapezoidal rule is used for integrating the ordinary differential equations of lumped inductors and capacitors. It is simple, numerically stable, and accurate enough for practical purposes. The memory function of the integration process is contained in the current source I, which for the inductor is defined as

$$\mathrm{I}_{\mathrm{km}}(t-\Delta t) = i_{km}(t-\Delta t) + \frac{\Delta t}{2L}\left(e_k(t-\Delta t) - e_m(t-\Delta t)\right) \tag{7.1}$$

and for the capacitor as

$$\mathrm{I}_{\mathrm{km}}(t-\Delta t) = -i_{km}(t-\Delta t) - \frac{2C}{\Delta t}\left(e_k(t-\Delta t) - e_m(t-\Delta t)\right) \tag{7.2}$$

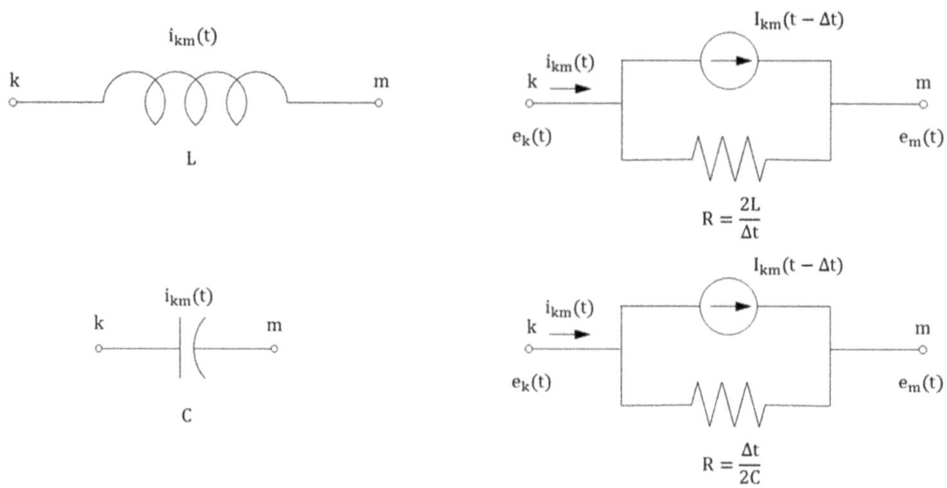

Figure 7.1 Method of equivalent inductors and capacitors into resistors with current sources

where

Δt = time interval between successive calculations (i.e., time-step)
$e_k(t-\Delta t)$ = volts at node k at the past time-step
$e_m(t-\Delta t)$ = volts at node m at the past time-step
$i_{km}(t-\Delta t)$ = current through branch from node k to node m at the past time-step

Thus it follows that at a given time step, the current through an inductor or capacitor branch is defined by

$$i_{km}(t) = \frac{e_k(t) - e_m(t)}{R} + I_{km}(t-\Delta t) \tag{7.3}$$

where

$$R = 2L/\Delta t \text{ for an inductor} \tag{7.4}$$

$$R = \Delta t/2C \text{ for a capacitor} \tag{7.5}$$

7.1.1 Nested Fast and Simultaneous Simulation Approach

The equations necessary for solving the full circuit can be formulated using nodal analysis, from the resulting network of current sources and conductances. The number of rows and columns of the nodal admittance matrix is each equal to the number of nodes in the network.

A recent enhancement to the Dommel's algorithm is the use of Nested Fast and Simultaneous Solution. This multi-step hierarchical approach partitions the network into smaller subnetworks and solves the admittance problem for each separately. Although, this increases the number of steps in the solution, the size of each of the admittance matrices is much smaller than that of the admittance matrix of the full network, which can lead to a reduction in simulation time.

The algorithm is explained using the two-part network shown in Figure 7.2. The equivalent admittance matrix formulation for this network can be written as Equation (7.6)

$$\begin{pmatrix} Y_{11} & Y_{12} \\ Y_{21} & Y_{22} \end{pmatrix} \begin{pmatrix} V_1 \\ V_2 \end{pmatrix} = \begin{pmatrix} J_1 \\ J_2 \end{pmatrix} \tag{7.6}$$

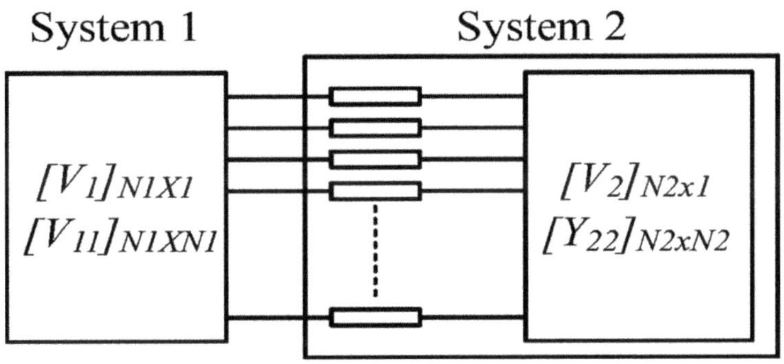

Figure 7.2 Partitioning of the system into smaller subsystems

where

Y_{11}, Y_{22} are admittance matrices for the partitions 1 and 2, respectively
Y_{12}, Y_{21} are admittance matrices for interconnections
V_1, V_2 are unknown node voltage vectors for the partitions
J_1, J_2 are source current vectors (obtained from branch history and other sources)

Assuming the number of nodes in the two partitions to be $N_1$and, N_2, respectively, a straightforward, direct solution for the unknown voltage vector $[V_1,\ V_2]^T$would require an admittance matrix of size $(N_1 + N_2) \times (N_1 + N_2)$ to be re-triangularized.

The second row of Equation (7.6) can be rearranged in the form of Equation (7.7) to express voltage V_2 as a function of V_1 and other sources. Substituting Equation (7.7) into the first row of Equation (7.6) yields Equation (7.8). Note that the quantity in parenthesis in Equation (7.8) is the Norton equivalent of V_1 as seen from the boundary nodes of System 1 in Figure 7.2 and includes a term for the Norton current source $(Y_{12}Y_{22}^{-1}J_2)$ and a term $(Y_{12}Y_{22}^{-1}Y_{21})$ for the Norton admittance.

$$V_2 = -Y_{22}^{-1}Y_{21}V_1 + Y_{22}^{-1}J_2 \tag{7.7}$$

$$J_1 = Y_{11}V_1 + Y_{12}\left(Y_{22}^{-1}J_2 - Y_{22}^{-1}Y_{21}V_1\right) \tag{7.8}$$

$$V_1 = \left(Y_{11} - Y_{12}Y_{22}^{-1}Y_{21}\right)^{-1}(J_1 - Y_{12}Y_{22}^{-1}J_2) \tag{7.9}$$

Equation (7.8) can be further simplified to Equation (7.9), which is a set of linear equations of dimension N_1. It can be solved for V_1. Then, V_1 can be substituted into Equation (7.8) to calculate V_2. Once all voltages are known, all currents can readily be calculated.

Note that this procedure requires the inversion of two matrices Y_{22} of dimension $N_2 \times N_2$ and $(Y_{11} - Y_{12}Y_{22}^{-1}Y_{21})$ of dimension $N_1 \times N_1$. Although described for a partition of the original system into two subsystems, the approach can be generalized to multiple subsystems. In the model developed here, each multi-valve is modeled as its own subsystem, which reduces the maximum subsystem size even further.

However, rather than applying the procedure in a formal mathematical manner using Equations (7.7) and (7.8) as in the original paper on Nested Fast and Simultaneous Solutions, which considers circuits in general, as the Norton equivalent-based equation is used here. This Norton equivalent for the MMC can be obtained in a very straightforward manner directly.

7.1.2 Thévenin or Norton Equivalent Circuit for the Submodule

The insulated-gate bipolar transistor (IGBT) switches can be treated as two-state resistive devices. Since the parallel connection of an IGBT and a diode acts as a bidirectional switch and only one device is conducting at a given instance, the pair is still considered as a single two-state resistance. Using the trapezoidal integration method, the capacitor can be represented as an equivalent voltage source and a resistor as in Equation (7.11).

$$V_c(\mathrm{t}) = \frac{1}{C}\int_0^t I_c(t) \approx V_c(t-\Delta T) + \frac{1}{C}\left(\frac{I_c(t-\Delta T)+I_c(t)}{2}\right)\Delta T \tag{7.10}$$

hence

$$V_c(\mathrm{t}) = R_C \cdot I_C(t) + V_{cEQ}(t-\Delta T) \tag{7.11}$$

where, $R_C = \frac{\Delta T}{2C}$, $V_{cEQ}(t-\Delta T) = \left(\frac{\Delta T}{2C}I_c(t-\Delta T) + V_c(t-\Delta T)\right)$

From Equation (7.11), the equivalent resistance R_C depends on the capacitance of the submodule and simulation time-step only. The equivalent voltage V_{cEQ} is simply calculated using the capacitor current and voltage values in the past time-step (history terms). The aforementioned capacitor representation as shown in Equation (7.11), including switch resistances R_1 and R_2, generates the mathematically equivalent circuit shown in Figure 7.6.

The values of resistor R_1 (similarly, R_2) depend on the switch state and is either the "ON" state resistance R_{ON} or the "OFF" state resistance R_{OFF}. The terminal voltage of an individual submodule can be determined as a function of its current as Equation (7.12).

$$V_{SM}(t) = R_{SMEQ} \cdot I_{SM} + V_{SMEQ}(t-\Delta T) \tag{7.12}$$

where,

$$R_{SMEQ} = R_2\left(1 - \frac{R_2}{R_1+R_2+R_C}\right)$$

$$V_{SMEQ}(\mathrm{t}-\Delta\mathrm{T}) = \left(\frac{R_2}{R_1+R_2+R_C}\right) \times V_{cEQ}(t-\Delta T)$$

7.1.3 Dynamic Model for Multi-valve

By daisy-chaining the individual submodules of Figure 7.3, the equivalent circuit of Figure 7.4 for the multi-valve can be constructed (as shown in Figure 7.4). This structure corresponds to system 2, and because of the daisy-chaining, the resulting matrix Y_{22} is sparse and has a banded diagonal, making its inversion more computationally efficient. However, as it turns out, its Thévenin equivalent can be determined very simply, and so even this mathematical step can be simplified. Since output terminals of the submodules are series connected, the multi-valve voltage is the sum of all the individual submodule voltages as shown.

$$V_{MV}(\mathrm{t}) = \sum_{i=1}^{NP} V_{SM_i}(t) = \left[\sum_{i=1}^{NP} R_{SMEQ_i}\right] \cdot I_{MV}(t) + \left[\sum_{i=1}^{NP} V_{SMEQ_i}(t-\Delta T)\right] = R_{EQ} \cdot I_{MV}(t) + V_{EQ} \tag{7.13}$$

where, NP number of submodules in a multi-valve.

Equation (7.13) can be converted into a Thévenin equivalent circuit consisting of a series-connected equivalent voltage source and a resistor. Since the same multi-valve current passes through all submodules, this equation can be used to develop a dynamic equivalent model for the multi-valve as in Figure 7.5, which is simple, yet exact. The inputs to the equivalent shown in Figure 7.5 are those that are required to determine the value of the Thévenin equivalent's parameters. For example, from Equations (7.12) and (7.13), it is evident that the determination of R_{EQ}, requires knowledge of the ON/OFF state of each submodule.

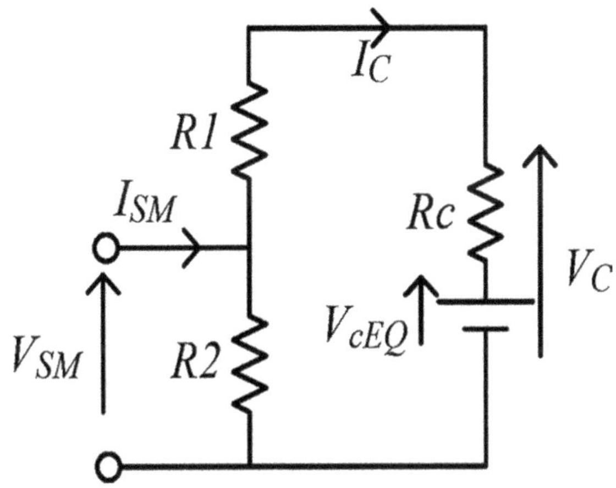

Figure 7.3 Equivalent circuit for a submodule

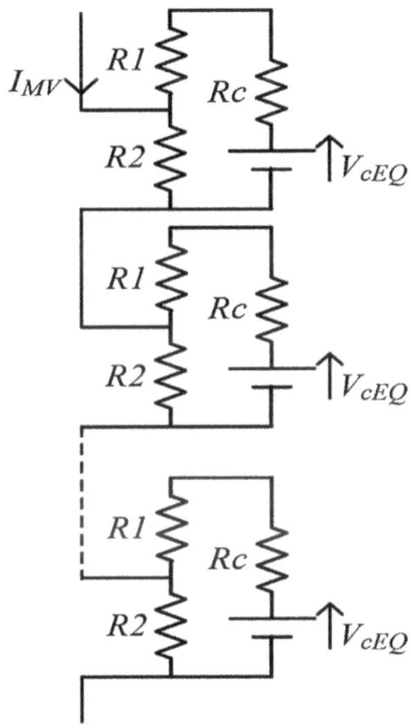

Figure 7.4 Equivalent circuit representation for a multi-valve

This is determined from the firing pulses (input "*FP*" in Figure 7.5) that are generated by the MMC control system. Also, since all of the submodules are now collapsed into a single equivalent, their individual identities are no longer available in the main EMT solver. The Thévenin equivalent solver, however, does consider each submodule separately and so has a record of the individual capacitor voltages. These are outputs from the block and made available for the controller since they are needed by the capacitor voltage balancing controller.

The nodal admittance matrix solver requires a Norton equivalent, for which the conversion from the Thévenin equivalent is straightforward and trivial. Each multivalve in the MMC is thus reduced to a single two-node element in the main EMT solver as shown in Figure 7.6 and, thus, the size of the resulting admittance matrix of the full network is reduced typically by several orders of magnitude.

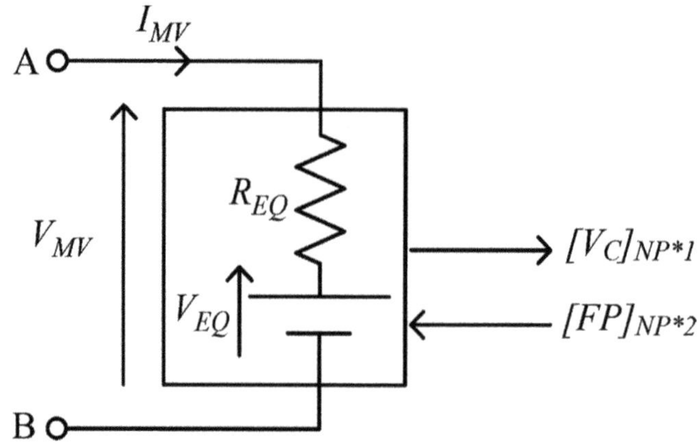

Figure 7.5 Thévenin equivalent of a multi-valve of modular multi-level converter (MMC)

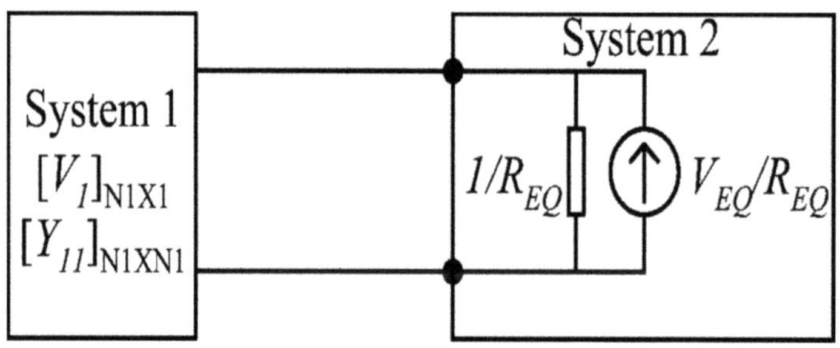

Figure 7.6 Two-part network with a reduced number of nodes in subsystem 2

Another benefit of the model is that the inner loop controls (e.g., IGBT switching, capacitor voltage balancing) are packaged into the block and, hence, the user is relieved from the details of modeling these, and can concentrate on system-level modeling. However, the interface between the inner-loop controls and the power electronic module is accessible, permitting the advanced user to override the built in controls.

7.1.4 Numerical Integration Methods

The trapezoidal rule has been traditionally used in EMT-type programs to construct a time-domain Norton equivalent model of typical power system elements. The form for the submodule equivalent derived in Equation (7.11) uses this method. The main advantages of the trapezoidal rule are that it is A-stable and reasonably accurate. A stability implies that the method does not diverge for any time step, assuming that the actual system is stable. However, recently there have been other A-stable methods used for obtaining the equivalent that claim superior accuracy to trapezoidal integration. The two-stage diagonally implicit Runge-Kutta (2S-DIRK) is one of these methods.

7.1.5 Selection Switch ON/OFF Status

Considering the gate signal, device terminals voltage (*Vce*) and current direction (*Ice*), a decision can be made as to whether the IGBT (*T*) or diode (*D*) of the switching device is "ON" at any moment in time. An IGBT/diode switch can have one of three switching states depending on the device conduction:

T - ON/D - OFF   *T - OFF/D - ON*   *T - OFF/D - OFF*

In each of these states, the switch resistance is *RT_ON-D_OFF, RT_OFF-D_ON* and *RT_OFF-D_OFF*, respectively. The possible transitions when the switch is in each of the previous states are the following and can be summarized by the state transition diagram shown in Figure 7.7.

1. When the switch is in the *T-ON/D-OFF* state, that is, it is carrying forward current (*Ice* > 0) and there is a gate pulse, the switch resistance is *RT_ON-D_OFF*. The switch can make a transition to the other states as follows:
   a. If the current reverses (*Ice* < *0*), then the IGBT turns "OFF," and the diode spontaneously turns "ON." The switch thus attains the *T-OFF/D-ON* state.
   b. If the gate pulse is stopped and the current is in the forward direction (GATE = 0), then the IGBT turns "OFF." The switch then attains the *T-OFF/D-OFF* state.
2. When the switch is in the *T-OFF/D-ON* state, the current is in the reverse direction (*Ice* < 0). The gate pulse may be 0 or 1 (note GATE = 1 will not result in IGBT conduction if the IGBT is reverse biased). The switch resistance is *RT_OFF-D_ON.*
   a. If GATE = 1 and the current suddenly changes direction (*Ice* > 0), the IGBT turns "ON" and the diode turns "OFF." The switch now attains the *T-ON/D-OFF* state.
   b. If GATE = 0 and the current attempts to go positive, the diode instantly turns "OFF" and the system attains the *T-OFF/D-OFF* state.
3. When the switch is in the *D-OFF/T-OFF* state, neither IGBT nor diode is conducting (*Ice* = 0), and the switch resistance is *RT_OFF-D_OFF.* The switch can transition to one of the following conducting states:
   a. If GATE = 1 and the IGBT has a forward-biased voltage (*Vce* > 0), the IGBT turns "ON." The switch now attains the *T-ON/D-OFF* state.
   b. If the voltage across the switch is negative (*Vce* < 0), the diode turns "ON." It is not necessary to apply a gate pulse in this situation.

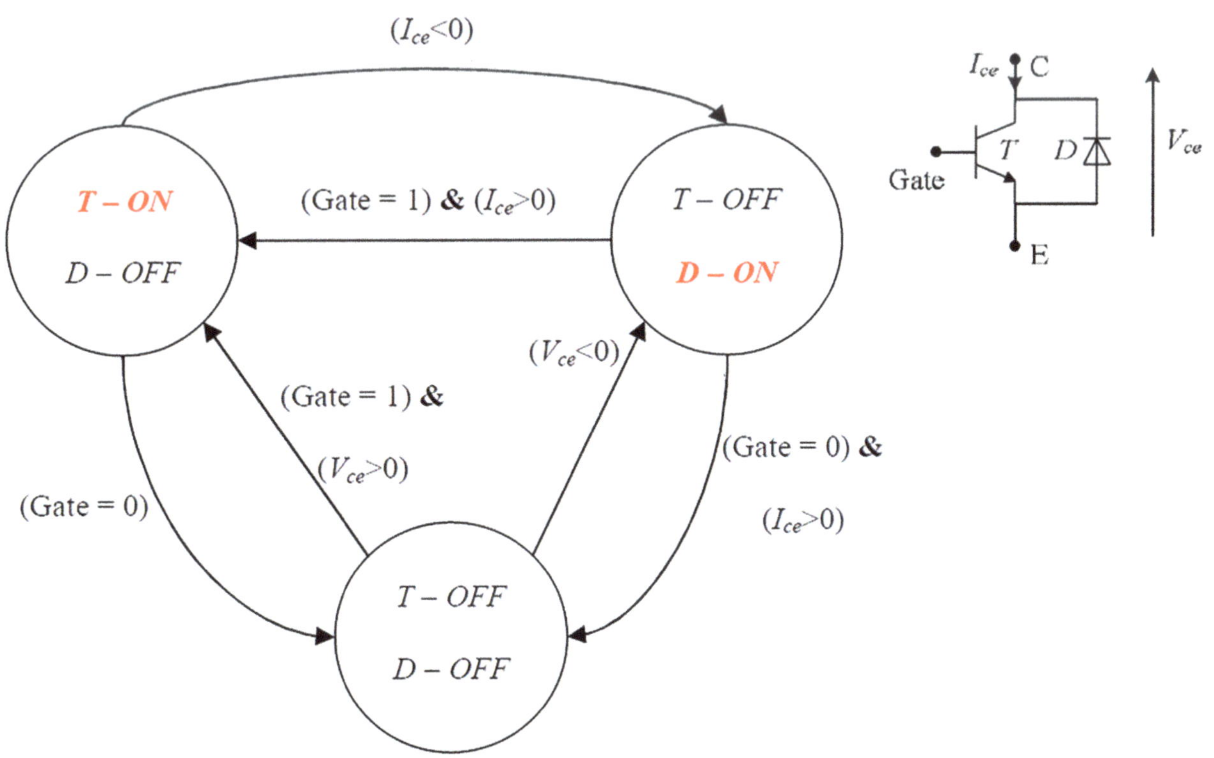

Figure 7.7 State transition diagram for an IGBT/diode switch combination

Note that, instead of having separate "ON"" resistance values for the diode and IGBT, a single "ON" resistance R_{ON} for the "ON" combination would also be possible in this model. Then, states 1 and 2 would have the same resistance R_{ON}.

When testing the proposed model with the previous switch conduction logic, it readily became apparent that a deficit in the model was because of the lack of a "blocking" facility. The converter is said to be in the "blocked" state when applying "OFF" gate signals to all submodule switches. In a typical MMC circuit operation, immediately after blocking on the converter, the anti-parallel diode starts to conduct and the IGBT current does not get chopped to zero instantaneously.

Since the proposed model is composed of a series of discrete calculations, and because of the consequence of the order of these calculations, it was artificially chopping the current for a single time-step, and as the circuit is inductive, it resulted in a large spurious overvoltage spike across the valve reactor. To avoid such a phenomenon, a modification is introduced to the formal diode conduction logic. Using the preceding current direction of the IGBT, the appropriate diode conduction is determined immediately after blocking. This prevents the occurrence of the large spurious voltage across the IGBT that the inductive current chopping causes.

Figure 7.8 shows the possible conduction states of diodes (time = t) by considering the normal operation in the previous time-step prior to blocking. For the proceeding time-step (i.e., time $\geq t + \triangle T$), the conduction states are selected based on the state transition diagram shown in Figure 7.7.

7.2 Validation for Compatible MMC Models Using PSCAD/EMTDC

7.2.1 Configuration of Seven-Level MMC System

A seven-level MMC test systems are designed to validate the detailed equivalent models. The test system parameters are given in Table 7.1; several simulation cases are in the following:

- -Case A: Full-detailed model, which consists of two IGBTs and a capacitor for each SMs. 12 IGBTs and 6 capacitors are used for each converter arm.
- -Case B: Detailed equivalent model developed by Manitoba HVDC Research Centre.
- -Case C: Detailed equivalent model developed by Yonsei University.
- -Case D: Combined detailed equivalent model. The model consists of detailed equivalent model for upper arm of each phase in Case B and detailed equivalent model for lower arm of each phase in Case C.

Table 7.1 Parameters of AC System and the Modular Multi-level Converter (MMC) System

Rated power	50 MW
AC system nominal voltage	138 kV
Short circuit ratio (SCR) at the PCC	10
Nominal frequency	60 Hz
Transformer voltage	138 kV / 30 kV (Y/Δ)
Nominal DC voltage (V_{DC})	60 kV
Submodule capacitor(C_{SM})	2500 μF
Smoothing reactor (L_s)	3 mH
Switching frequency	2.1 kHz
DC line	Overhead 100 km

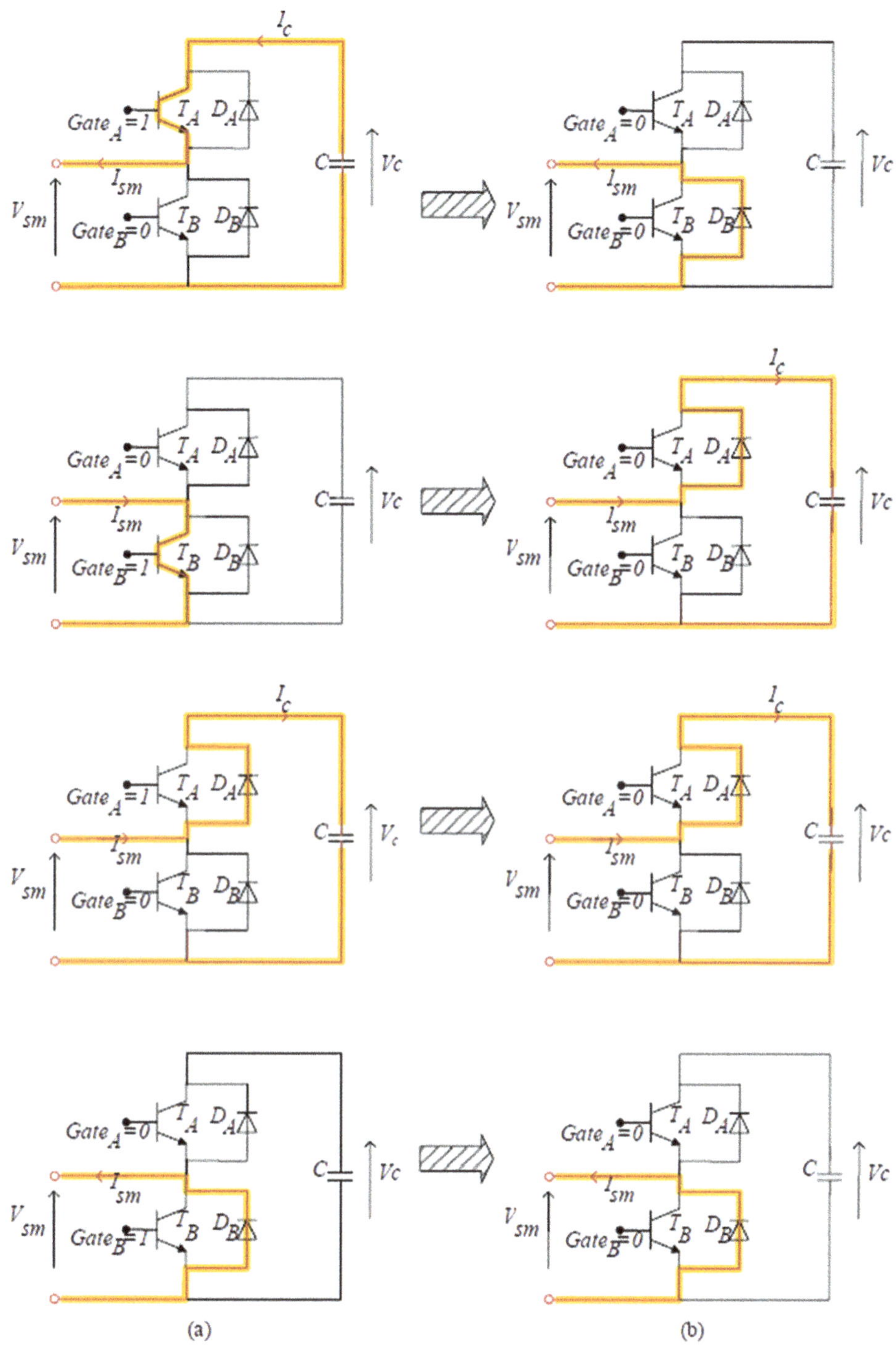

Figure 7.8 Diode selection diagram on half-bridge submodule blocking: (a) prior to blocking (time = t-ΔT) and (b) immediately after blocking (time = t)

Normal Operation

Figure 7.9 shows that the overall performance of seven-level MMC HVDC system in four cases. In this scenario, the HVDC system operates at rated power of 50 MW. The results show that output performance is very similar except DC current. DC current in full-detailed model (Case A) shows an oscillatory that differs from Cases B, C, and D.

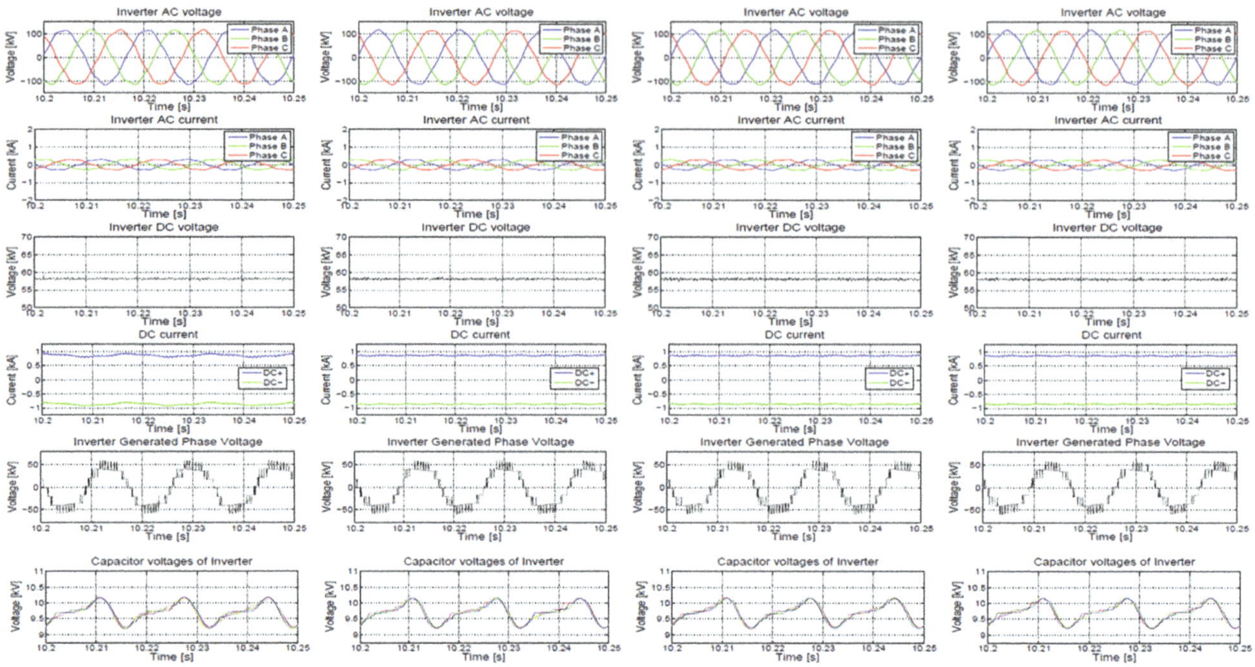

(a) Case A (b) Case B (c) Case C (d) Case D

Figure 7.9 Normal operation

Transients

Remote Three-Phase Fault at Inverter Side: An AC fault occurs near the inverter side. In this scenario, the HVDC system operates at rated power of 50 MW and a remote AC fault is applied for 40 ms at inverter side at 10.3 sec. Figure 7.10 shows rectifier and inverter AC voltage, DC voltage, DC current, and capacitor voltage of phase A upper arm at rectifier and inverter with four cases: Case A through Case D. The results show that the detailed equivalent model represents the MMC well, and Case D shows that the models are compatible.

DC Fault Near Inverter Side: The DC fault is located near the inverter side. In this scenario, the HVDC system operates at rated power of 50 MW and a DC fault is applied for 40 ms near inverter side at 10.3 sec. Figure 7.11 shows rectifier and inverter AC voltage, DC voltage, DC current, and capacitor voltage of phase A in upper arm at rectifier and inverter with four cases: Case A through Case D. The results show that the detailed equivalent model represents the MMC well, and Case D shows that the models are compatible.

7.2.2 Model Validation with High-Level MMC System

This section shows that the operation of further advanced 201-level model compare to 7-level MMC from last section to verify the model can be adopted in testing large system. The AC system, DC voltage, system rating, and the number of level are brought from the Trans Bay Cable project. The model uses the conventional circulating current suppression control (CCSC). Detailed system data are given in Table 7.2. The test system is divided into three cases, like the simulation conducted in the previous section: normal operation, a remote three-phase line fault at inverter side, and DC fault near inverter side.

Normal Operation

Figure 7.12(a) shows the dynamic performance of modeled 201-level MMC HVDC system in normal operation. Sending 230-kV grid transmits the rated 400 MW power to the 115-kV grid while keeping DC voltage ±200 kV. The system response, including AC voltage and current, DC voltage and current, and capacitor voltages, are presented in Figure 7.12(a). The results show that detailed equivalent model represents MMC as well, and the models are compatible.

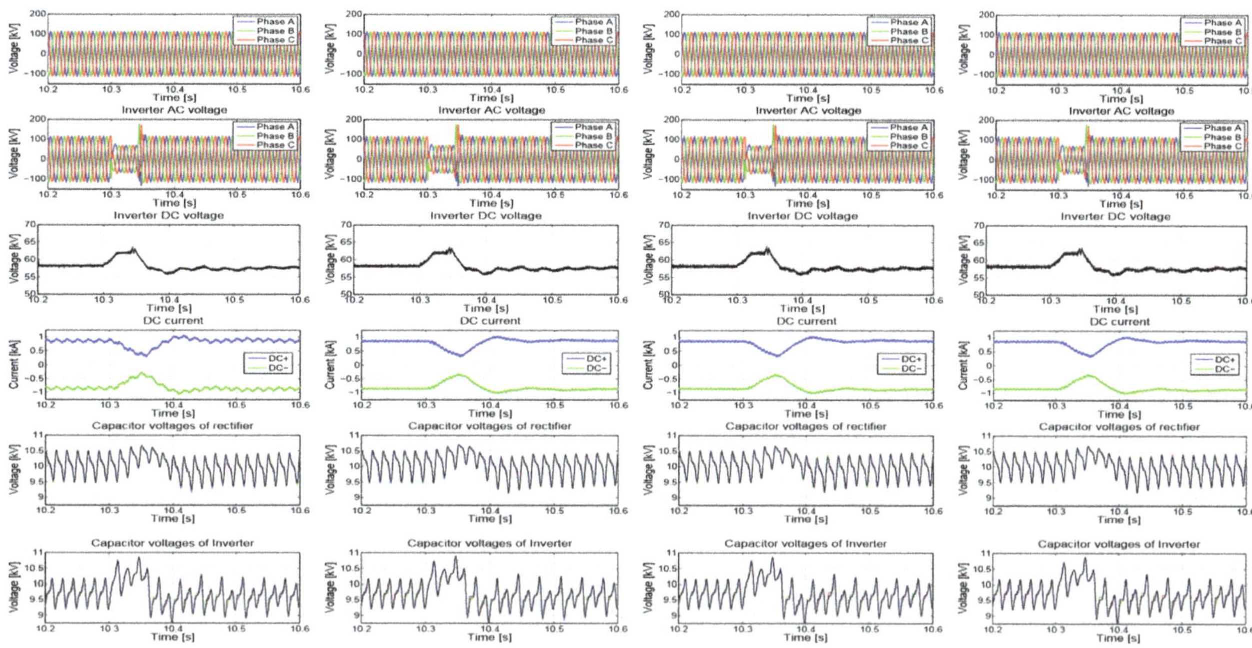

(a) Case A (b) Case B (c) Case C (d) Case D

Figure 7.10 AC fault

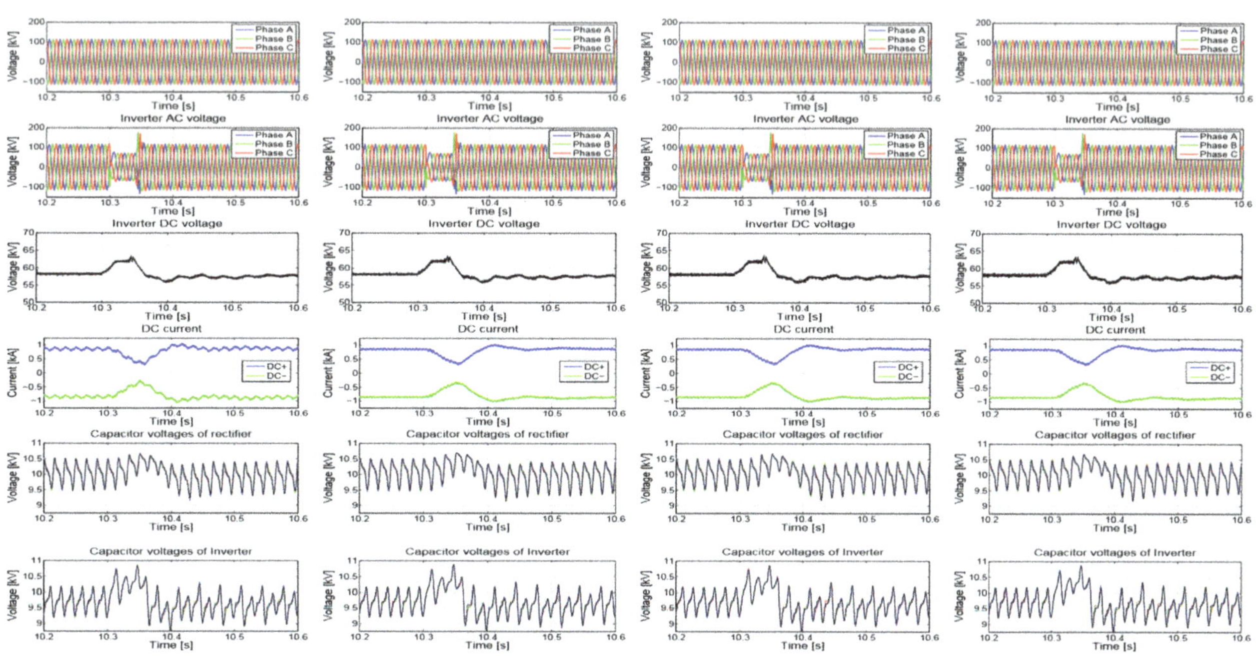

(a) Case A (b) Case B (c) Case C (d) Case D

Figure 7.11 DC fault

Table 7.2 Parameters of the High-Level Modular Multi-level Converter MMC System

Rated power	400 MW
AC system nominal voltage at rectifier	230 kV
AC system nominal voltage at inverter	115 kV
Short circuit ratio (SCR) at the PCC	3
Nominal frequency	60 Hz
Transformer voltage at rectifier	230 kV / 180 kV (Y/Δ)
Transformer voltage at inverter	115 kV / 180 kV (Y/Δ)
Nominal DC voltage (V_{DC}) Number of submodule	±200 kV 2400 (200 per each Arm)
Submodule capacitor(C_{SM})	6000 μF
Smoothing reactor (L_s)	20 mH
Switching frequency	60 kHz
DC line	Overhead 88 km

A Remote Three-Phase Fault at Inverter Side

Figure 7.12(b) shows dynamic response of AC fault. A remote three-phase line fault is applied at 10.3 sec for 40 ms on high-voltage side of the 230 kV/180 kV transformer. It is observed that MMC experiences a voltage sag and other oscillatory transients as same as previous seven-level case. The rectifier side is slightly affected by fault occurred at inverter side. The results show that detailed equivalent model represents MMC as well, and the models are compatible.

DC Fault

Figure 7.12(c) shows dynamic performance of MMC HVDC system when DC fault is applied at 10.3 sec for 40 ms on inverter side. Even though DC fault are less frequent in HVDC system, it is an important concern for MMC HVDC system. When the fault occurs, converter experiences severe voltage drop and both AC and DC inrush currents are observed since the energy stored in capacitor is discharged in a short time (15 ms). The results show that detailed equivalent model represents MMC as well, and the models are compatible.

7.3 Transmission Line Modeling

Generally in EMT simulations, there are two main ways to represent transmission lines. The most familiar method is using PI line sections to build up the transmission line. Single- or multi-phase lines can be modeled. The second method is to use a distributed transmission line, which is most suited for transient line response modeling using a digital computer. A distributed model operates on the principle of traveling waves. A voltage disturbance travel along a conductor at its propagation velocity (near the speed of light), until it is reflected at the line's end. In a sense, a transmission line or cable is a delay function. Whatever is fed into one end will appear at the other end, perhaps slightly distorted, after some delay. However, there are some other considerations that must be dealt with which include mutual coupling with other conductors, and wave-shape attenuation as it travels along the line.

There are three basic transmission line or cable modeling techniques in EMT model: PI sections, the Bergeron Model, and the Frequency-Dependent Line Model.

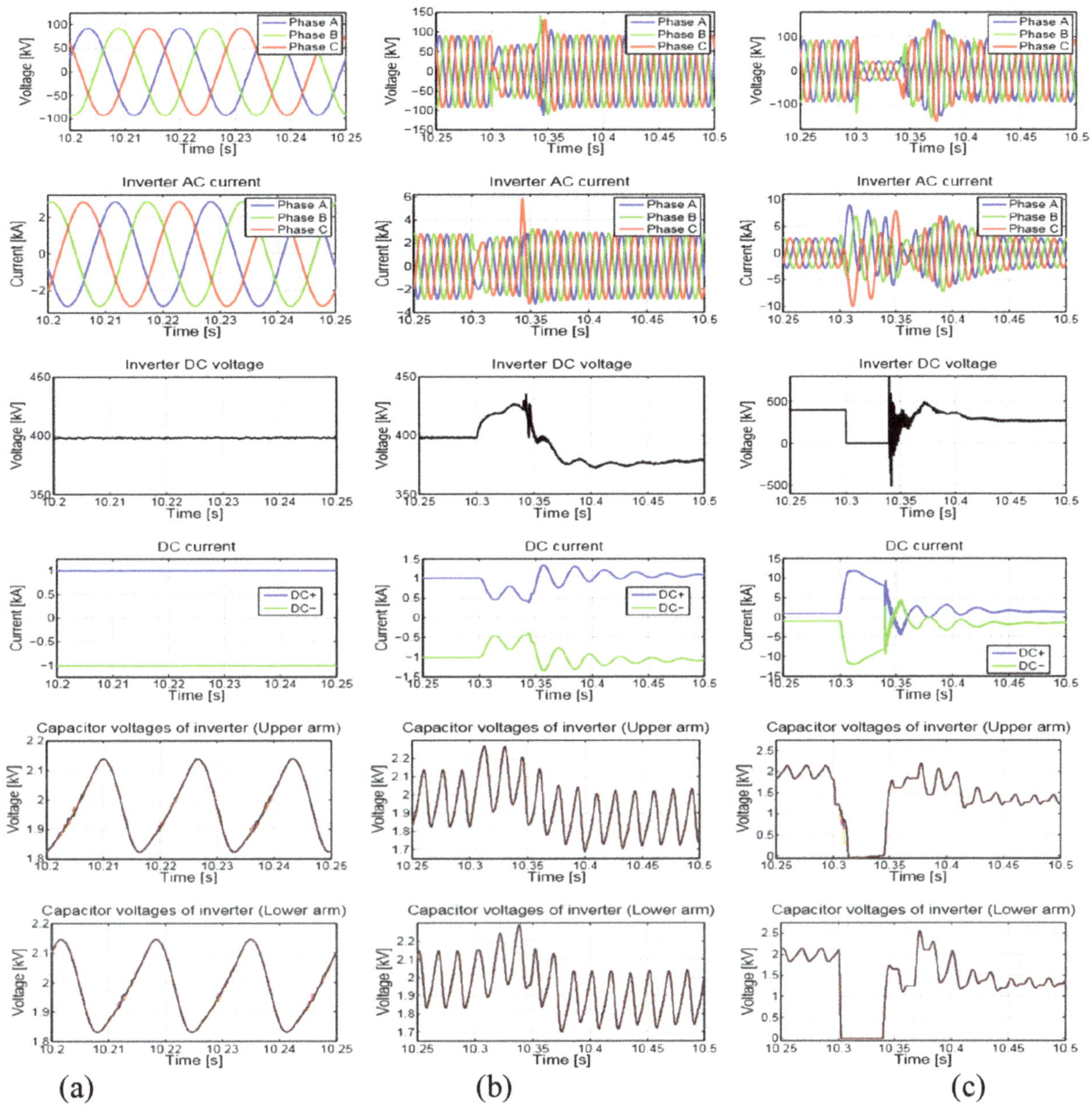

Figure 7.12 Dynamic performance of modular multi-level converter (MMC) HVDC using Combined Detailed Equivalent model

PI section is very simple and one limitation of the Bergeron model is that the line parameters are not frequency dependent; thus, the skin effect cannot be included into the Bergeron model. The frequency-dependent model parameters are solved in frequency domain, and are then convolved into their equivalent time-domain characteristics. It represents R, L, and C as distributed parameters.

7.3.1 PI Section Model

A simple PI section model shown in Figure 7.13 gives the correct fundamental impedance, but cannot accurately represent other frequencies unless many sections are used (which is inefficient). It also cannot accurately represent the frequency-dependent parameters of a line. It is suitable for very short lines where

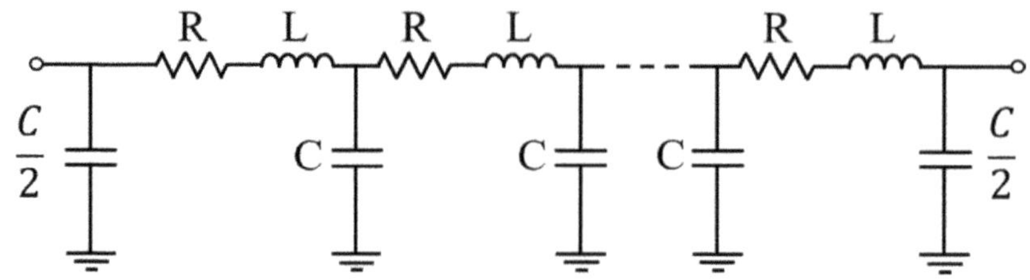

Figure 7.13 Configuration of PI model

the traveling wave models cannot be used. For EMT simulation, PI sections are not considered a very elegant means of transmission line modeling for the following reasons:

- Greater computational time and increased matrix sizes
- Not practical with a large number of mutually coupled conductors
- Frequency-dependent attenuation of the traveling waves is not easily or accurately incorporated

The required number of sections for the distributed parameter model is a function of frequency of interest and cable length.

An equation for calculating is Equation (7.14).

$$f_{max} = \frac{N \cdot v}{8l_{total}} \tag{7.14}$$

where f_{max} is the approximation of the maximum frequency range represented by the PI line model, N is the number of PI sections, v is the propagation speed, and l_{total} is the length of the line.

The result of this relationship means that as the cable length increases, for a given frequency of interest, the required number of sections makes the model cumbersome. Additionally, the insertion of discrete lumped components introduces extra resonant combinations between capacitors and inductors such that false resonances may be observed when using this model. The time domain implementation for the PI model is shown in Figure 7.14.

7.3.2 Bergeron Model

The Bergeron model represents the L and C elements of a PI section in a distributed manner (not using lumped parameters like PI sections). It is roughly equivalent to using an infinite number of PI sections except that the resistance is lumped (1/2 in the middle of the line, 1/4 at each end). Like PI sections, it also accurately represents the fundamental frequency. It also represents impedances at other frequencies, except that the losses do not change. This model is suitable for studies where the fundamental frequency load—flow is most important (i.e., relay studies).

To appreciate the concept of a distributed transmission line, a single conductor overhead transmission line is considered. Multi-phase and mutually coupled transmission lines follow similar concepts. A distributed transmission line interfaces to the EMT electric network as shown in Figure 7.15.

The "line constants" routing will have generated:

Z_0 = The characteristic or surge impedance (ohms)

R = The total line resistance (ohms)

τ = The travel time of the line (seconds)

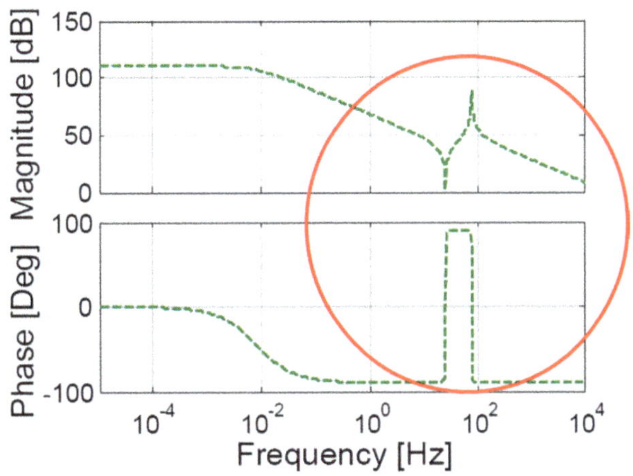

Figure 7.14 Frequency response result of PI model

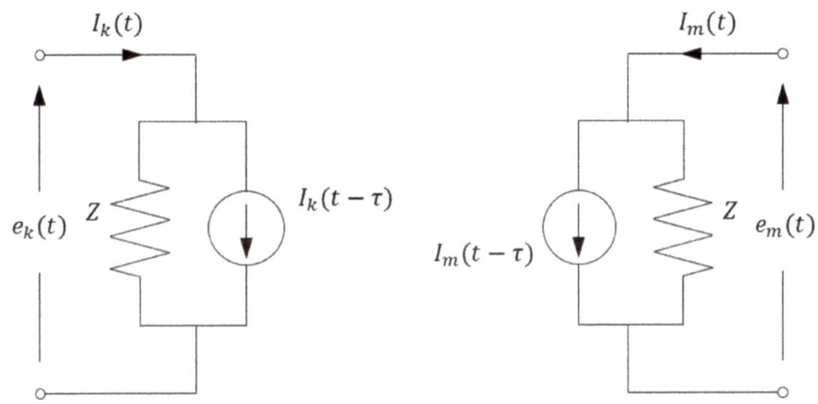

Figure 7.15 Interfacing the distributed line to the circuit

The Norton interface impedance at each end of the line in Figure 7.15 is

$$Z = Z_0 + \frac{R}{4} \tag{7.15}$$

The Norton current injected at each end of the line has been determined by Dommel to be

$$I_k(t-\tau) = \left[\frac{e_m(t-\tau)}{z} + H \cdot i_m(t-\tau)\right] \tag{7.16}$$

$$I_m(t-\tau) = \left[\frac{e_k(t-\tau)}{z} + H \cdot i_k(t-\tau)\right] \tag{7.17}$$

where, $H = \frac{Z_0 - R/4}{Z_0 + R/4}$

These equations are derived by breaking a lossless line into two sections, inserting R/2 in the middle and R/4 at each end of the line.

The EMT model actually has two frequency paths, a lower and an upper frequency path. This enables the transmission losses to be accurately represented at the fundamental frequency, and to have increased losses at higher frequencies. The line is not exactly modeled at the higher frequency, however, because the surge impedance, travel time, and transformation matrix are still the steady state values.

The low frequency should be the fundamental frequency of the studied system and would generally be then either 50 or 60Hz for an AC transmission line. For a DC line, a low frequency such as 5 Hz should be used. The surge impedance and travel time are calculated at this frequency.

The second higher frequency is used for calculating the high-frequency attenuation of the line. T-LINE and CABLE will create a split at the log average of the two frequencies specified to determine high- and low-frequency attenuation. A general high frequency to enter would be between 100 and 2 kHz. This high frequency can, if desired, be chosen to accommodate specific resonance conditions the user might be interested in.

A block diagram of a single-phase or single-mode-distributed transmission line used in the EMT model is shown in Figure 7.16.

Just before the current is injected into the line terminations, it is processed through a real—pole with a time constant of Tc (seconds).

Selection of wave-shaping constant *Tc* can be made in order to shape the steep front of a traveling wave to a known attenuation and slope. If one is concerned about matching the steep front from the EMT model line model to a known response, adjustment by trial and error of *Ts* will enable this to be done. (By default, the shaping time constants are set to 0.05 msec from the line constants. This essentially provides no wave shaping). Care should be taken when using wave shaping time constant, because although the real pole will attenuate the high-frequency components of a line, it also introduces an additional lag or time delay that affects the effective overall travel time, and thus the line impedance.

The Bergeron model of a transmission line would generally be chosen over a PI equivalent type circuit for any lines for which the travel time of the wave (assuming a speed close to that of light) is in the order of the simulation time-step. For a general simulation time-step of 50 microseconds, lines over 15 Ian should at least be represented by the Bergeron model.

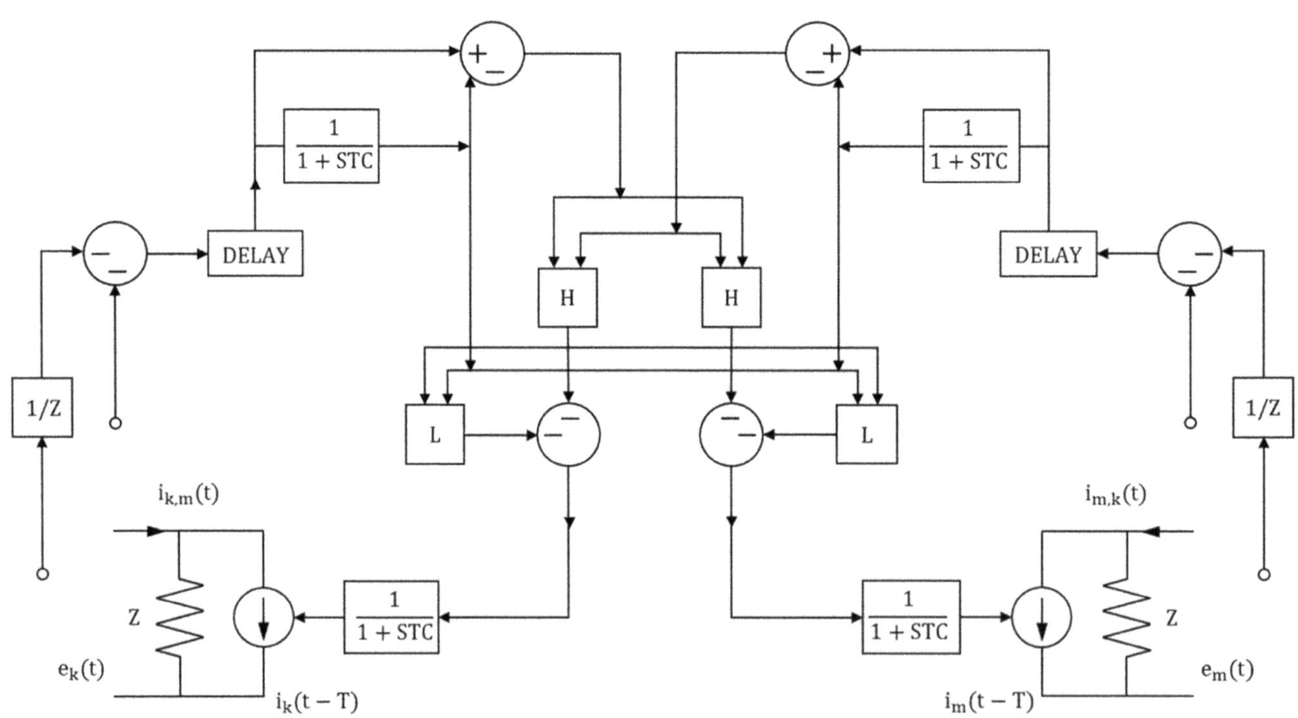

Figure 7.16 Block diagram of a single mode of the electromagnetic transient model distributed transmission line model

7.3.3 Multi-Conductor

So far, the mechanism of the EMT model's frequency-dependent distributed transmission line has been discussed considering one conductor or mode only. Generally, transmission lines consist of several mutually coupled phases or conductors. These have to be broken up into modes. There is one mode for each phase that is mutually coupled. Each mode can be treated as an independent single phase transmission line. This method allows unbalanced mutually coupled lines to be modeled. A modal transformation converts the modal response back into the EMT model network quantities. The modal transformation and modal quantities are obtained automatically by the line/cable constants routines from an eigenvalue analysis.

For unbalanced transmission lines, each mode will have different surge impedances and travel times. Usually there are two main types of modes: first, the ground mode or common mode or zero sequence mode. This mode is active whenever ground currents flow. Second, the remaining modes are known as metallic modes, differential modes, or positive and negative sequence modes. The ground mode has a longer travel time, higher surge impedance, and higher line resistance than the metallic mode(s). The important aspect of modal analysis is the modal transformation matrix. To summarize this procedure, a matrix diagonalization technique is undertaken so that

$$[V_{\mathrm{phase}}] = [T_e]\,[V_{mode}] \tag{7.18}$$

$$[I_{\mathrm{phase}}] = [T_i]\,[I_{mode}] \tag{7.19}$$

$$[\mathrm{T_e}]^T = [T_i]^{-1} \tag{7.20}$$

where

$[V_{\mathrm{mode}}]$ = Line end mode volts, $[I_{\mathrm{mode}}]$ = Line end mode current, $[V_{phase}]$ = Line end phase volts, $[I_{\mathrm{phase}}]$ = Line end phase current

$[T_e]$ = A square matrix transforming mode volts to phase volts

$[T_i]$ = A square matrix transforming mode current to phase current

If there is only a single conductor above a ground, then there is just a single mode. The path the current can circulate is along the conductor and back through the ground. This is particularly true if one end of the conductor is energized with a voltage source and the other is grounded as shown in Figure 7.17.

For a two-conductor transmission line such as a bipolar DC line, there are two paths return current can flow as shown in Figure 7.18.

For the ground mode, the current flows down both conductors and back through the ground. The current flows down one conductor and back on the other for the metallic mode. Each of these two modes can be

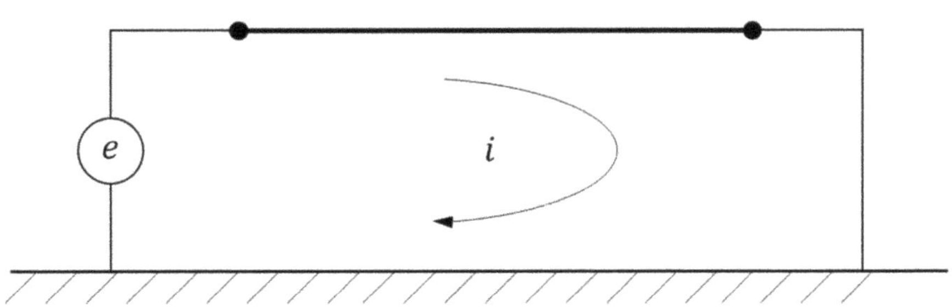

Figure 7.17 Single mode for a single conductor with ground path return

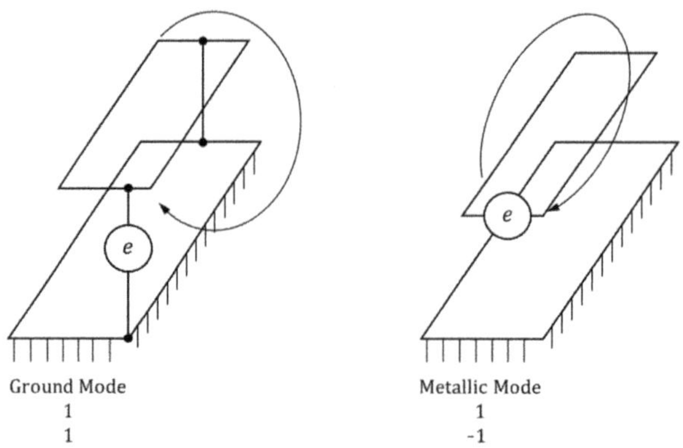

Figure 7.18 Two modes of a two conductor transmission line

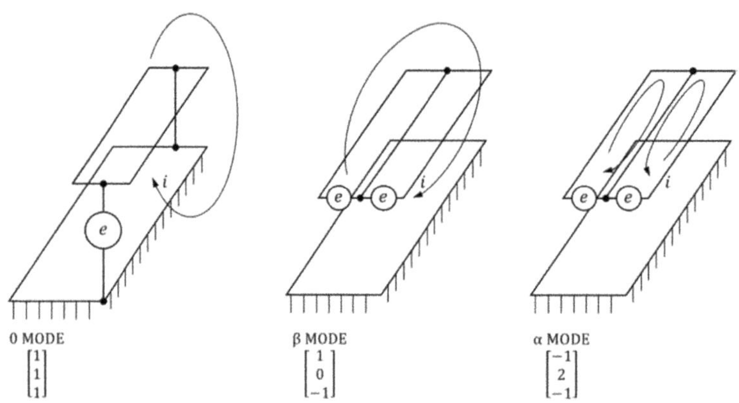

Figure 7.19 Three modes of a three conductor transmission line using Clark components

differentiated by defining the direction in which current flows as either a +1 or −1. The modal transformation matrix is then

$$[T_e] = \begin{bmatrix} 1 & 1 \\ 1 & -1 \end{bmatrix} \tag{7.21}$$

and when normalized becomes,

$$[T_e] = [T_i] = \begin{bmatrix} 0.7071 & 0.7071 \\ 0.7071 & -0.7071 \end{bmatrix} \tag{7.22}$$

It is worth noting that the transformation matrix for 1 and 2 conductor flat lines does not change with frequency. This is because the conductors are always balanced.

If a three-phase line is balanced, then the transformation is also constant with frequency. This is the case for ideally transposed lines, or a delta-configured line very high in the air. Under those conditions, the transformation becomes Clark components:

$$[T_e] = \begin{bmatrix} 1 & 1 & -1 \\ 1 & 0 & 2 \\ 1 & -1 & -1 \end{bmatrix} \tag{7.23}$$

The physical realization of the Clark transformation is shown in Figure 7.19.

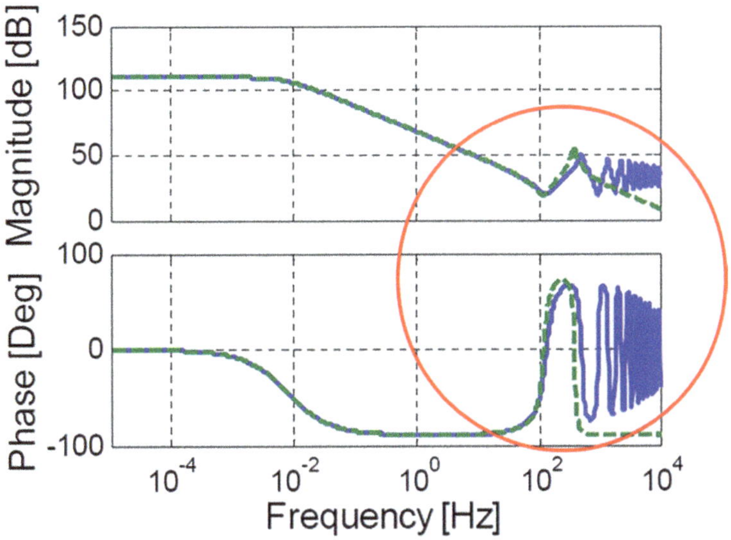

Figure 7.20 Frequency response result of Bergeron model

If a three-phase circuit is not usually perfectly balanced, then $[T_e] \neq [T_i]$ and Equation (7.20) determines the relation between the voltage and current transformation matrices. The transformation matrices are represented as real, even though in reality they may be frequency dependent and consist of both real and imaginary components. For overhead conductors, the imaginary components are small compared with the real components and it is reasonable approximation to neglect the imaginary parts.

The Bergeron model represents the line inductance and capacitance as distributed parameters, where as the cable resistance is represented as a lumped parameter. It is accurate for a certain frequency because all calculated parameters are calculated at this specific frequency. The Bergeron model is roughly equal to the PI model, but with infinite sections, so when the transmission line is long enough (transport delay longer than simulation time step), the Bergeron model is preferable to the PI model. The time domain implementation for the Bergeron model is shown in Figure 7.20.

7.3.4 Frequency–Dependent Model

The frequency-dependent line model is basically a distributed R-L-C traveling wave model that incorporates the frequency dependence of all parameters. Because the Bergeron model is adequate for studies that essentially only require the correct fundamental frequency impedance, the frequency-dependent line model should be used for all studies that require frequencies other than the fundamental to be represented accurately.

The frequency-dependent line model is based on the theory developed in except that the frequency dependence of the transformation matrix is also incorporated because of improvements in the line constants calculations. To arrive at the time domain formation of the line equations, it is convenient to first work in the frequency domain, where an exact solution for a given frequency is possible. Consider the following circuit of a transmission line as seen from the terminations.

For a given frequency, the voltages and currents at one end of the line may be represented in terms of the voltage and current at the other end by

$$V_k(\omega) = \cosh[\gamma(\omega) \cdot L]\, V_m(\omega) - Z_c(\omega) \cdot \sinh[\gamma(\omega) \cdot L]\, i_m(\omega) \tag{7.24}$$

Figure 7.21 Frequency domain line circuit

$$V_k(\omega) = \frac{\sinh[\gamma(\omega)\cdot L]}{Z_c(\omega)} V_m(\omega) - \cosh[\gamma(\omega)\cdot L]\, i_m(\omega) \tag{7.25}$$

where

$\gamma(\omega) = \sqrt{Y(\omega)Z(\omega)}$ is the propagation constant, $Z_c(\omega) = \sqrt{\frac{Z(\omega)}{Y(\omega)}}$ is the surge impedance, $Y(\omega) = G + j\omega C$ is the shunt admittance of the line, $Z(\omega) = R + j\omega L$ is the series impedance of the line

By introducing the forward and backward traveling wave functions F_k and B_k,

$$F_k(\omega) = V_k(\omega) + Z_c(\omega)\cdot i_k(\omega) \tag{7.26}$$

$$B_k(\omega) = V_k(\omega) - Z_c(\omega)\cdot i_k(\omega) \tag{7.27}$$

and similarly at node m

$$F_m(\omega) = V_m(\omega) + Z_c(\omega)\cdot i_m(\omega) \tag{7.28}$$

$$B_m(\omega) = V_m(\omega) - Z_c(\omega)\cdot i_m(\omega) \tag{7.29}$$

Removing $i_k(\omega)$ from Equation (7.26) using Equation (7.27) yields

$$F_k(\omega) = V_k(\omega) - B_k(\omega) \tag{7.30}$$

and similarly

$$F_m(\omega) = 2V_m(\omega) - B_m(\omega) \tag{7.31}$$

Equations (7.24) and (7.25) (and their equivalents at node m) can now be expressed in terms of the forward and backward traveling functions:

$$B_k(\omega) = A(\omega)\cdot[2V_m(\omega) - B_m(\omega)] \tag{7.32}$$

$$B_m(\omega) = A(\omega)\cdot[2V_k(\omega) - B_k(\omega)] \tag{7.33}$$

where

$$A(\omega) = \frac{1}{\cosh[\gamma(\omega)\cdot L]\sinh[\gamma(\omega)\cdot L]} = e^{-\gamma(\omega)L} \tag{7.34}$$

An (ω) is known as the propagation constant and is a complex number. The real part (a) is the attenuation constant, and the imaginary part (jß) is the phase constant. Equations (7.27) and (7.29) can be represented with the equivalent circuit as shown in Figure 7.21. Equations (7.31) and (7.32) relate the source $B_k(\omega)$ with the m side quantities $V_m(\omega)$ and $B_m(\omega)$:

$$B_k(\omega) = A(\omega)\cdot[2V_m(\omega) - B_m(\omega)] \tag{7.35}$$

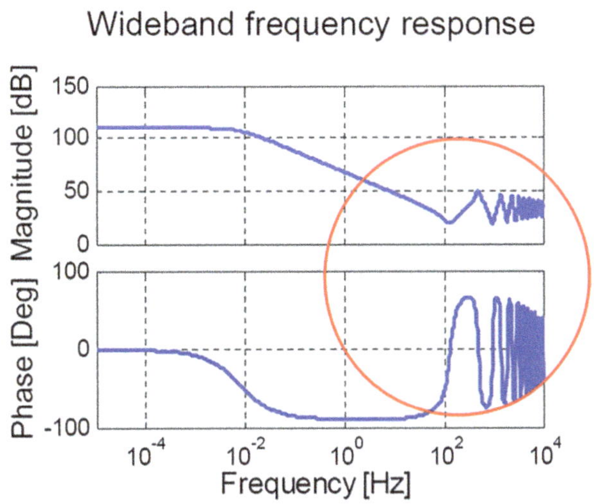

Figure 7.22 Frequency response result of frequency-dependent model

and similarly

$$B_m(\omega) = A(\omega) \cdot [2V_k(\omega) - B_k(\omega)] \tag{7.36}$$

The time domain implementation for the frequency-dependent model is shown in Figure 7.22.

References

1. U. N. Gnanarathna, A. M. Gole, and R. P. Jayasinghe, "Modeling of modular multilevel converter on electromagnetic transients programs," *IEEE Trans. Power Del.*, vol. 26, pp. 316–324, Jan. 2011.
2. S. Liu, et al., "Electromechanical transient modeling of modular multilevel converter based multi-terminal HVDC systems," *IEEE Trans. Power Syst.*, vol. 29, pp. 72–83, Jan. 2014.
3. A. D. Rajapakse, A. M. Gole, and P. L.Wilson, "Electromagnetic transients simulation models for accurate representation of switching losses and thermal performance in power electronic systems," *IEEE Trans. Power Del.*, vol. 20, no. 1, pp. 319–327, Jan. 2005.
4. Q. Tu, Z. Xu, Y. Chang, and L. Guan "Suppressing DC voltage ripples of MMC-HVDC under unbalanced grid conditions," *IEEE Trans. Power Del.*, vol. 27, no. 3, pp. 1332–1338, Jul. 2012.

8

Main Circuit Parameter Design

This chapter deals with the functional requirements of AC/DC converters for the system parameter design of HVDC grids. The robust and reliable operation of an high-voltage direct current (HVDC) grid in cooperation with the connected AC systems poses requirements for maximizing the active power transmission availability and reactive power support. The converter rating is a major aspect that defines the transfer capability in the DC-grid and the reactive power capability for the support and optimization of the interaction with the AC grid.

The main contents described in this chapter are the modulation index, submodules with IGBT (including gate drivers), loss calculation, DC busbars, arm reactors, and noise calculation.

Figure8.1 shows the relations and the requirements among HVDC functions for the HVDC system design. As shown in Figure 8.1, AC network conditions and control strategies are required, the ratings of the voltage-source converter (VSC) HVDC system are determined, and harmonic analysis, temporary over-voltage (TOV) analysis, and insulation coordination are studied.

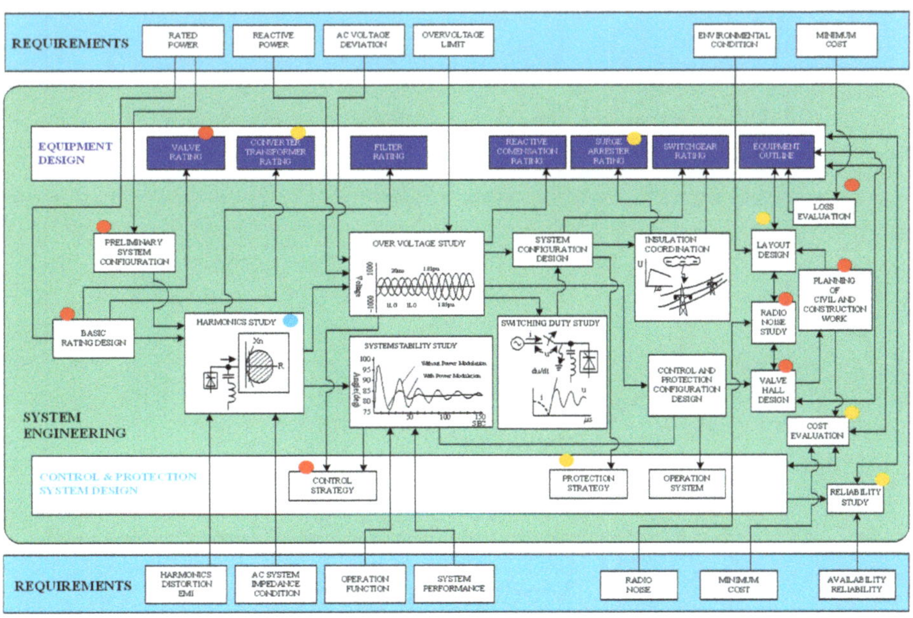

Figure 8.1 Relations and requirements among HVDC functions

8.1 Maximum Modulation Index

Modulation Index (MI) is defined as the ratio of the DC voltage of the converter to the AC reference voltage. The fact that MI is 1 means that the peak value of the AC reference voltage and the magnitude of the DC voltage are the same.

In a converter system, MI has an important meaning. The main consideration for MI in a two-level converter is to make the AC reference voltage maximize at a fixed DC voltage; methods include a space vector method, a min/max method, and a third harmonic injection method.

Unlike two-level converters, in the modular multi-level converter (MMC)-based VSC HVDC system, the factors affecting the MI are divided into internal factors (e.g., the influence of the second harmonic frequency flowing inside the MMC converter) and an external factor, which is the dropping of the DC voltage of the MMC converter when a fault occurs in the AC system in which the VSC HVDC converter is connected.

The conventional design for the MI of an MMC finds the minimum cost of IGBT (Insulated-Gate Bipolar Transistor) or, in other words, the minimum power loss point between power loss and MI. In practice, the range of MI in industrial converters have to have sufficient margin to allow the converter to operate under transient conditions (HVDC internal or external faults) as well as steady state conditions. Figure 8.2 shows the relationship between power loss and MI.

The condition for determining MI is to avoid the overmodulation conditions in the MMC converter. Overmodulation conditions in MMC converters are as follows:

- When the AC Network voltage suddenly changes (typically 10%)
- When the second harmonic circulating current flows inside the converter
- When the negative-sequence is compensated for by the unbalance of AC network voltage

Consequently, a new MI determination method that reflects the factors considered earlier is described in this chapter.

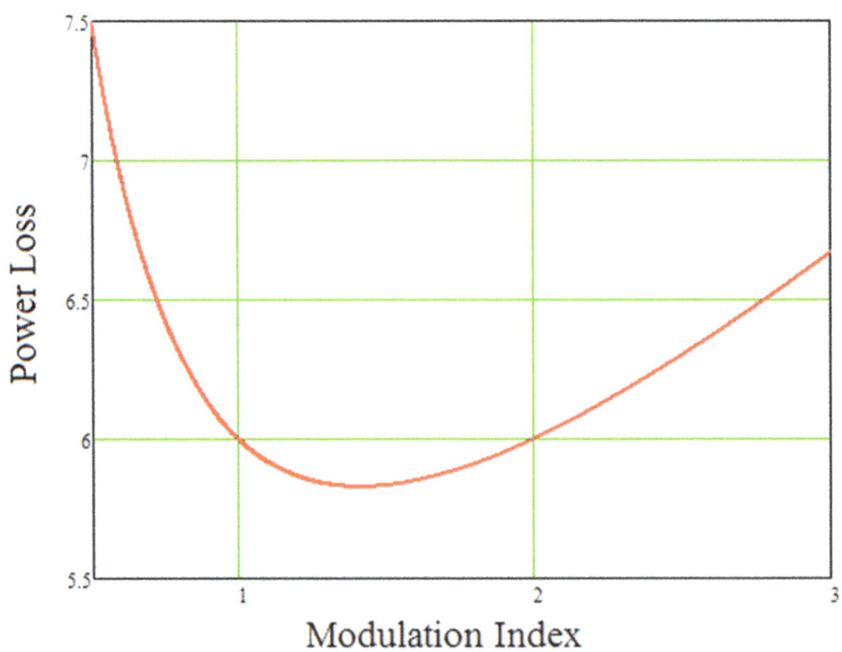

Figure 8.2 Relation between power loss and modulation index

8.1.1 Fundamentals of the Operation of MMC

Figure 8.3 shows the principle configuration of the MMC. The three-phase converter consists of three legs, which are composed of two arms. Each arm has series-connected N power submodules (SMs). Each power SM has a capacitor and two IGBTs. With these arrangements, the voltage of the SM can be switched to zero or to the capacitor voltage. Under this condition, the converter should be controlled such that the DC link voltage is constant and the capacitor voltages in all arms are balanced. From Figure 8.3, the arm voltage and currents for a phase can be expressed as in Equations (8.1)–(8.5):

$$i_{\text{ap}} = \frac{1}{3}I_{\text{DC}} + \frac{1}{2}i_{\text{a}} \tag{8.1a}$$

$$i_{\text{an}} = \frac{1}{3}I_{\text{DC}} - \frac{1}{2}i_{\text{a}} \tag{8.1b}$$

$$V_{\text{ap}} = n_{\text{ap}} \cdot N \cdot V_{\text{c\_ap}} \tag{8.2a}$$

$$V_{\text{an}} = n_{\text{an}} \cdot N \cdot V_{\text{c\_an}} \tag{8.2b}$$

$$\frac{dV_{\text{c\_ap}}}{dt} = \frac{1}{C_{\text{sub}}} \cdot n_{\text{ap}} \cdot i_{\text{ap}} \tag{8.3a}$$

$$\frac{dV_{\text{c\_an}}}{dt} = \frac{1}{C_{\text{sub}}} \cdot n_{\text{ap}} \cdot i_{\text{an}} \tag{8.3b}$$

$$v_{\text{a}} = \frac{V_{\text{an}} - V_{\text{ap}}}{2} - \frac{L_{\text{arm}}}{2} \cdot \frac{d\left(i_{\text{ap}} - i_{\text{an}}\right)}{dt} - \frac{R_{\text{arm}}}{2} \cdot \left(i_{\text{ap}} - i_{\text{an}}\right) \tag{8.4}$$

$$V_{\text{DC}} = \left(V_{\text{an}} + V_{\text{ap}}\right) L_{\text{arm}} \cdot \frac{d\left(i_{\text{ap}} + i_{\text{an}}\right)}{dt} + \frac{R_{\text{arm}}}{2} \cdot \left(i_{\text{ap}} + i_{\text{an}}\right) \tag{8.5}$$

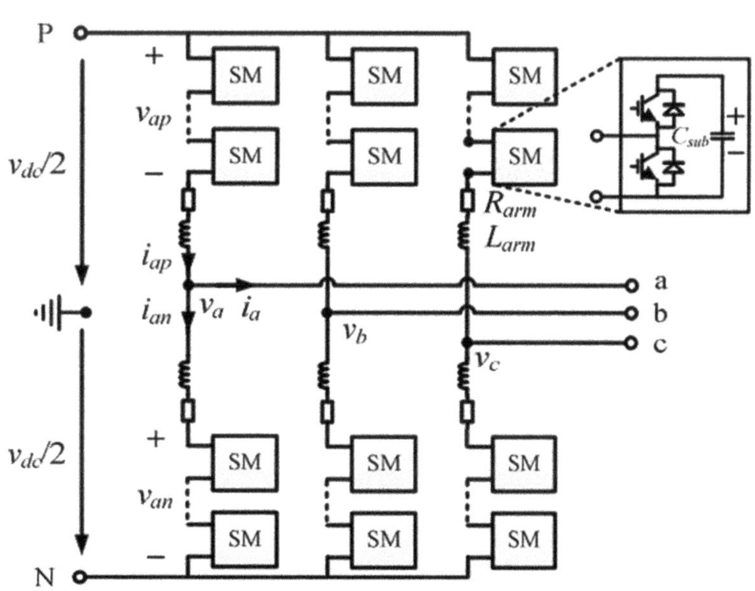

Figure 8.3 MMC HVDC system topology

The DC current I_{DC} is derived from the power balancing equation between the DC power and the active power at the output of the three-phase converter.

$$P = V_{\mathrm{DC}} \cdot I_{\mathrm{DC}} = 3 \cdot \frac{V_{\mathrm{a}} \cdot I_{\mathrm{a}}}{2} \cdot \cos \varnothing \tag{8.6}$$

With this I_{DC}, the minimum capacitor has been calculated in Equation (8.7):

$$C_{\mathrm{o}} = \frac{1}{2n \cdot \varepsilon \cdot \left(\overline{V_{\mathrm{c}}}\right)^2} \cdot \frac{2}{3} \cdot \frac{P_{\mathrm{s}}}{\omega_{\mathrm{N}} \cdot M} \cdot \left(1 - \left(\frac{M \cdot \cos\varphi}{2}\right)^2\right)^{\frac{3}{2}} \tag{8.7}$$

with the number of levels n, the relative voltage ripple ε, the average capacitor voltage $\overline{V_C}$, the output power P_s, the modulation index M, the angular frequency ω_N of the output voltage, and the displacement factor $\cos\varphi$.

From this formula, it can be seen that the amount of capacitive energy storage is directly proportional to the output power of the inverter. Furthermore, it is inversely proportional to the output frequency.

Equation (8.4) can be rewritten as in Equation (8.8):

$$\frac{L_{\mathrm{arm}}}{2}\frac{di_{\mathrm{a}}}{\mathrm{dt}} + \frac{R_{\mathrm{arm}}}{2} i_{\mathrm{a}} = -v_{\mathrm{a}} + N \cdot n_{\mathrm{c\_a\_dm}} \cdot v_{\mathrm{c\_a\_cm}} + N \cdot n_{\mathrm{a\_cm}} \cdot v_{\mathrm{c\_a\_dm}} \tag{8.8}$$

$$i_{\mathrm{a\_cm}} = \frac{i_{\mathrm{ap}} + i_{\mathrm{an}}}{2},\ i_{\mathrm{a}} = i_{\mathrm{ap}} - i_{\mathrm{an}} \tag{8.9a}$$

$$v_{\mathrm{c\_a\_cm}} = \frac{v_{\mathrm{c\_an}} + v_{\mathrm{c\_ap}}}{2},\ v_{\mathrm{c\_a\_dm}} = \frac{v_{\mathrm{c\_an}} - v_{\mathrm{c\_ap}}}{2} \tag{8.9b}$$

$$n_{\mathrm{a\_cm}} = \frac{n_{\mathrm{an}} + n_{\mathrm{ap}}}{2},\ n_{\mathrm{a\_dm}} = \frac{n_{an} - n_{\mathrm{ap}}}{2} \tag{8.9c}$$

where i_a is the AC current and i_{a_cm} is the common mode component of the arm currents.

Similarly, Equation (8.5) can be rewritten as in Equation (8.10):

$$2L_{\mathrm{arm}}\frac{di_{\mathrm{a\_cm}}}{dt} + 2R_{\mathrm{arm}} i_{\mathrm{a\_cm}} = v_{\mathrm{dc}} - 2\,N \cdot n_{\mathrm{a\_cm}} \cdot v_{\mathrm{c\_a\_cm}} - 2N \cdot n_{\mathrm{a\_dm}} \cdot v_{\mathrm{c\_a\_dm}} \tag{8.10}$$

8.1.2 MI in an MMC HVDC System

MI is the ratio of the AC voltage to half of the DC voltage. The methods to maximize the MI are as follows:

Min/Max Voltage Method

A zero-sequence voltage v_0 can be added to the phase-output voltage and is selected so that, at every time instant, the maximum and minimum values of the phase-voltage references are equal, but with opposite signs. The phase-voltage references become symmetrized, hence the name symmetrical references. This is obtained by selecting

$$v_0 = -\frac{\max\left(v_{\mathrm{sa}}^*, v_{\mathrm{sb}}^*, v_{\mathrm{sc}}^*\right) + \min\left(v_{\mathrm{sa}}^*, v_{\mathrm{sb}}^*, v_{\mathrm{sc}}^*\right)}{2} \tag{8.11}$$

The peak value of the total waveform resulting from symmetrical references is the same as for the third harmonic addition ($\sqrt{3}V_s/2$ given a peak value V_s for sinusoidal references).

Third Harmonic Injection

The third harmonic injection used in MMC control has significant advantages over sinusoidal modulation in terms of system power losses, SM capacitance, circulating current, and fault current. By injecting the third harmonics, the DC voltage utilization ratio is increased from 1/2 to $1/\sqrt{3}$. Compared with sinusoidal modulation, the maximum output-voltage peak with reference to the midpoint of the DC link is reduced from $1/2V_{DC}$ to $1/4\sqrt{3}V_{DC}$ as demonstrated in Figure 8.4, whereas the amplitude of the output fundamental voltage remains half of the DC voltage.

Only $1/2\sqrt{3}N$ of the SMs are utilized when third harmonic injection is used rather than sinusoidal modulation. In real application, the potentially used SM is the unit SM $\left(1/2\sqrt{3}N\right)$, which is quite close to $1/2\sqrt{3}N$ given that HVDC systems typically use hundreds of SMs per arm. The AC reference voltage with the third harmonic is shown in Equations (8.12) and (8.13):

$$v_{\text{as}}^* = A\cos(\omega t),\ v_{\text{sn}}^* = B\cos(3\omega t + \varphi) \tag{8.12}$$

$$\begin{aligned} v_{\text{an}}^* &= v_{\text{as}}^* + v_{\text{sn}}^* = A\cos(\omega t) + B\cos(3\omega t + \varphi) \\ \text{where, } B &= -\frac{A}{6}, \varphi = 0(\text{rad}) \end{aligned} \tag{8.13}$$

The additional advantages of the third harmonics injection are SM capacitance is reduced by 24% compared to that using sinusoidal modulation, which significantly reduces both the volume and the capital cost of SMs, and station converter conduction losses are reduced by 11% by injecting a third harmonic, thus reducing the capacity of the cooling system.

MI Calculation Considering Compensation Signal

From Figure 8.2, the following is an expression of the equivalent AC voltage:

$$v_{\text{ac}} = v_{\text{ac}}\cos(\omega t) = v_{\text{a}} + \frac{L_{\text{arm}}}{2}\frac{di_{\text{a}}}{\text{dt}} + \frac{R_{\text{arm}}}{2}i_{\text{a}} \tag{8.14}$$

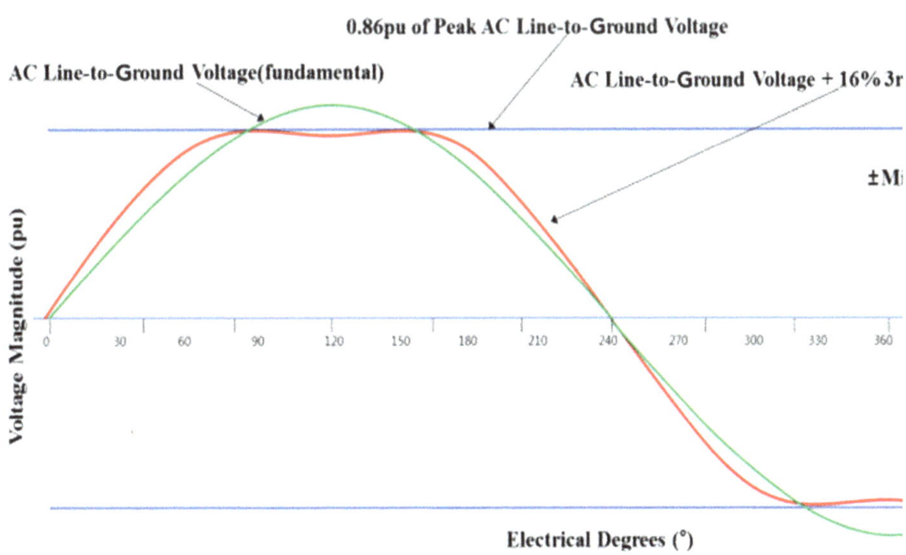

Figure 8.4 Third harmonic injection for MMC HVDC system topology

As shown in Equation (8.7), the maximum capacitor voltage ripple depends on the converter-operating range. Also, the maximum reactive power is usually defined as half of the maximum active power. So, the minimum power factor for an MMC operating at full apparent power is $\sqrt{3}/2$, and thus the maximum capacitor voltage ripple is obtained as in Equation (8.15):

$$V_{\text{Ripple}} = 0.73\frac{1}{C_{\text{sub}}}\frac{1}{4\omega}I_{\text{ac}} \tag{8.15}$$

where maximum M=1 is considered. The coefficient 0.73 is related to the maximum reactive power limitation. Thus, it is reasonable to neglect $\Delta v_{\text{c\_a\_cm}}$ in the analysis:

$$v_{\text{ac}}\cos(\omega t) = v_{\text{dc}}\cdot n_{\text{a\_dm}} + N\cdot n_{\text{a\_cm}}\cdot v_{\text{c\_a\_dm}} \tag{8.16}$$

$n_{\text{a\_dm}}$ is obtained as in Equation (8.17):

$$\begin{aligned} n_{\text{a\_dm}} &= \frac{v_{\text{ac}}cos(\omega t)}{v_{\text{dc}}} - \frac{N\cdot n_{\text{a\_cm}}\cdot v_{\text{c\_a\_dm}}}{v_{\text{dc}}} \\ &\approx \frac{M}{2}cos(\omega t) \end{aligned} \tag{8.17}$$

As the circulating current control regulates $i_{\text{a\_cm}}$ to its DC component, the left-hand side of Equation (8.10) should be equal to zero considering R_{arm} is small. That is,

$$2\Delta n_{\text{a\_cm}}v_{\text{dc}} + N\cdot\Delta v_{\text{c\_a\_am}} + 2N\cdot n_{\text{a\_dm}}\cdot v_{\text{c\_a\_dm}} = 0 \tag{8.18}$$

$$\begin{aligned} \Delta n_{\text{a\_cm}} = \frac{1}{2v_{\text{dc}}}\frac{N}{C_{\text{sub}}}\frac{M}{8\omega}I_{\text{ac}}\bigg\{&-\frac{3}{2}\sin(\varphi)\cos(2\omega t) \\ &+\left(\frac{3}{2}-\frac{M^2}{2}\right)\cos(\varphi)\sin(2\omega t)\bigg\} \end{aligned} \tag{8.19}$$

It is shown that the compensating component is a second harmonic and is related to the system parameters (C_{sub} and N) and operating conditions (M, I_{ac}, and ϕ). Based on Equation (8.15), the maximum relative SM capacitor voltage ripple can be defined as in Equation (8.20):

$$\varepsilon = 0.73\cdot\frac{N}{v_{\text{dc}}}\frac{1}{C_{\text{sub}}}\frac{1}{4\omega}I_{\text{ac}} \tag{8.20}$$

As ε can be considered as a fixed value, Equation (8.19) can be simplified as in Equation (8.21):

$$\Delta n_{\text{a\_cm}} = \varepsilon\frac{M}{2.92}\left\{-\frac{3}{2}\sin(\varphi)\cos(2\omega t) + \left(\frac{3}{2}-\frac{M^2}{2}\right)\cos(\varphi)\sin(2\omega t)\right\} \tag{8.21}$$

The insertion index for the lower arm (that for the upper arm is similar) can be obtained as

$$n_{\text{an}} = \frac{1}{2} + \frac{M}{2}\cos(\omega t) + \varepsilon\frac{M}{2.92}\left\{\frac{3}{2}\sin(\varphi)\cos(2\omega t) + \left(\frac{3}{2}-\frac{M^2}{2}\right)\cos(\varphi)\sin(2\omega t)\right\} \tag{8.22}$$

The insertion index should satisfy

$$0 \le n_{\text{an}} \le 1 \tag{8.23}$$

According to Equation (8.22), the maximum and minimum of the lower arm insertion index are obtained when $\cos\varphi$ is at its minimum value:

$$n_{\text{an}}(\max) = 0.5 + (0.5 + 0.26\varepsilon)\cdot M \tag{8.24a}$$

$$n_{\text{an}}(\min) = 0.5 - (0.5 + 0.26\varepsilon)\cdot M \tag{8.24b}$$

So, the limitation on the MI can be obtained as in Equation (8.25):

$$M \le \frac{1}{1 + 0.52\varepsilon} \tag{8.25}$$

Also, in the case of the third harmonic injection,

$$n_{\text{a\_dm}} = \frac{M}{2}\left[\cos(\omega t) - \frac{1}{6}\cos(3\omega t)\right] \tag{8.26}$$

With the similar calculation in Equation (8.15), the maximum relative capacitor voltage ripple can be obtained as in Equation (8.27):

$$\varepsilon = 0.68 \cdot \frac{N}{v_{\text{dc}}} \frac{1}{C_{\text{sub}}} \frac{1}{4\omega} I_{\text{ac}} \tag{8.27}$$

The compensating component can then be calculated as in Equation (8.28):

$$\begin{aligned}\Delta n_{\text{a\_cm}} = \varepsilon \frac{M}{2.72}\Big\{&\left(\frac{19}{12} - \frac{7M^2}{12}\right)\cos(\varphi)\sin(2\omega t)\\ &-\frac{17}{12}\sin(\varphi)\cos(2\omega t)\Big\}\end{aligned} \tag{8.28}$$

By calculating the maximum and minimum values of n_{an}, the limitation on the MI is obtained as in Equation (8.29):

$$M \le \frac{1}{0.87 + 0.70\varepsilon} \tag{8.29}$$

Considering the Internal Conditions of the MMC System

As seen in Equation (8.21), capacitor variation is related to the MI of an MMC system. The factors affecting the capacitor voltage ripple are capacitor, frequency, DC voltage, and AC current, assuming that the capacitor value is changed by aging, temperature, and sensor error. In practical systems, Δv_{dc_Error} and ΔI_{ac_Error} are 0.5%. These depend on sensor error, PT (potential transformer) and CT (current transformer). ΔC_{Error} is 5%. This value is determined by the aging curve of the capacitor characteristic. $\Delta\omega_{Error}$, which means frequency deviation during abnormal conditions of an AC network is normally 3%. ΔN_{Error} is the system error, including sampling time error and control error, and is normally below 1%.

Equation (8.20) is changed to Equation (8.30) by considering the MMC HVDC system conditions:

$$\varepsilon = 0.73 \cdot \frac{1}{2v_{\text{dc}}\left(1 + \Delta v_{\text{dc\_Error}}\right)} \cdot \frac{N(1 + \Delta N_{\text{Error}})}{C_{\text{sub}}(1 + \Delta C_{\text{Error}})} \cdot \frac{M}{8\omega(1 + \Delta\omega_{\text{Error}})} \cdot I_{\text{ac}} \cdot \left(1 + \Delta I_{\text{ac\_Error}}\right) \tag{8.30}$$

The MMC converter operates in the pure sinusoidal AC voltage. M in Equation (8.30) is Equation (8.20). In the case of the converter outputting the AC voltage with the third harmonic, M becomes Equation (8.29).

Considering the External Conditions of the MMC System

When calculating MI related to AC network conditions, Equation (8.25) is changed to Equation (8.31):

$$M \le \frac{\alpha_1}{1 + \alpha_2} \tag{8.31}$$

In Equation (8.31), α_2 is related to the MMC HVDC condition, and α_1 is close to AC network conditions. An investigation of the factors related to α_2 is shown in Equation (8.32). The AC negative-sequence voltage Δv_{ac}^{-} is below 2%, and The AC voltage variations Δv_{ac} and Δv_{Load} (these values are normally 5%–10%) and dead time ΔT_{Dead} are considered:

$$M \leq \frac{(1 + \Delta v_{\text{ac}}^{-} + \Delta T_{\text{Dead}} + \text{Max}\{\Delta v_{\text{ac}} + \Delta v_{\text{Load}}\})}{1 + 0.52\varepsilon} \tag{8.32}$$

In Equation (8.32), the dead time effect of the MMC system is meaningless in the case of high-level SMs, and the fact that the tap changer's effect Δv_{ac} is below 5%, has been considered. Ignoring the tap changer leads to a Δv_{ac} of under 10%.

Finally, the calculated equations are as follows:

Calculating α_2 Using Equation (8.31)

Inserting $\Delta v_{\text{dcError}}$, $\Delta I_{\text{ac\_Error}}$=0.5%, ΔC_{Error}=5%, $\Delta\omega_{\text{Error}}$ =3%, ΔN_{Error} =1%

Without Third Harmonic	With Third Harmonic
$M \leq \frac{\alpha_1}{1+0.577\varepsilon}$	$M \leq \frac{\alpha_1}{0.87+0.777\varepsilon}$

Calculating Maximum *M*

Case (1) α_1=12% (Δv_{ac}^{-}=2%, Δv_{ac}=10% (without tap changer))

Without Third Harmonic	With Third Harmonic
$M \leq \frac{0.88}{1+0.577\varepsilon}$	$M \leq \frac{0.88}{0.87+0.777\varepsilon}$
$\varepsilon = 0.1$	$\varepsilon = 0.1$
$M \leq 0.832$	$M \leq 0.929$

Case (2) α_1=7% (Δv_{ac}^{-}=2%, Δv_{ac}=5% (with tap changer))

Without Third Harmonic	With Third Harmonic
$M \leq \frac{0.93}{1+0.577\varepsilon}$	$M \leq \frac{0.93}{0.87+0.777\varepsilon}$
$\varepsilon = 0.1$	$\varepsilon = 0.1$
$M \leq 0.879$	$M \leq 0.982$

Figure 8.5 shows the simulation results including the above described contents (AC reference in steady state, 2nd harmonic, 3rd harmonic and external fault conditions). Figure 8.5(a) shows the MI signal and AC reference voltage, Figure 8.5(b) adds the compensation signal (circulation signal) to Figure 8.5(a), and Figure 8.5(c) shows the waveform injected with the third harmonic, in case of that a 15% voltage drop is observed. Figure 8.5(d) adds Figure 8.5(b) to Figure 8.5(c), Figure 8.5(e) shows the voltage variation of 10% in the AC network, and Figure 8.5(f) is the final simulation result. Figure 8.7 shows the waveform of Figure 8.6 undergoing an α-β transformation.

8.2 Rating Determination of MMC-Based VSC HVDC System

Figure 8.8 shows a P/Q diagram for when the VSC HVDC acts as an inverter. As in Figure 8.8, the rating of the VSC HVDC system is limited by the power factor and maximum current. Also, it is a standard practice for HVDC stations to provide reactive compensation to the AC line to improve the power throughput for the

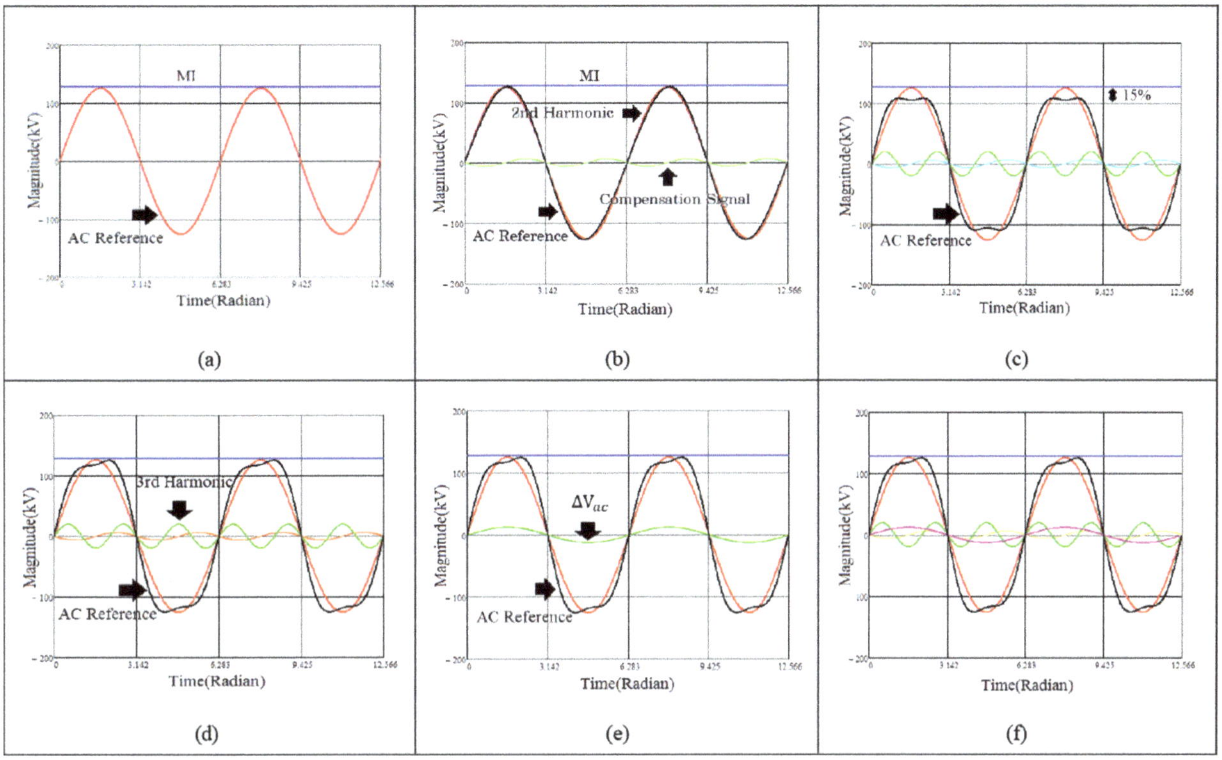

Figure 8.5 AC waveforms according to different MIs

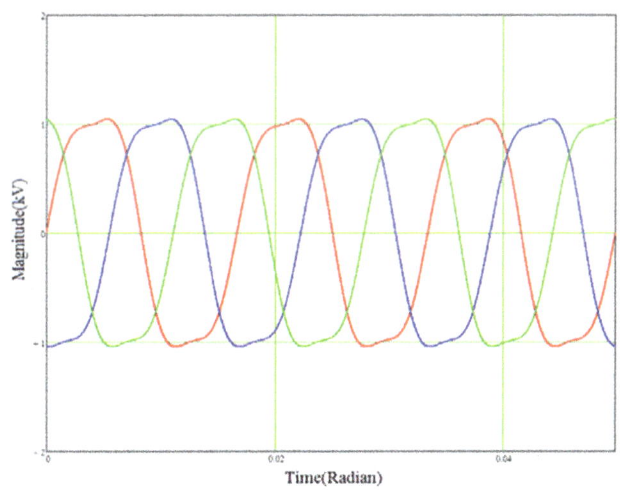

Figure 8.6 Three-phase output voltage considering AC network and system condition

station and eliminate the need for any additional VAR compensation. Figure 8.9 shows the limitation conditions for the VSC HVDC acting as an inverter.

As discussed earlier, the standard basis for operating AC converters with AC systems is controlling the AC-converter voltage so that it remains in a magnitude and phase relative to the AC supply, so that a differential voltage will appear across the reactance, coupling the converter to the power system and defining the magnitude and phase of the AC current flow. This approach means that the AC voltage must often be set using a

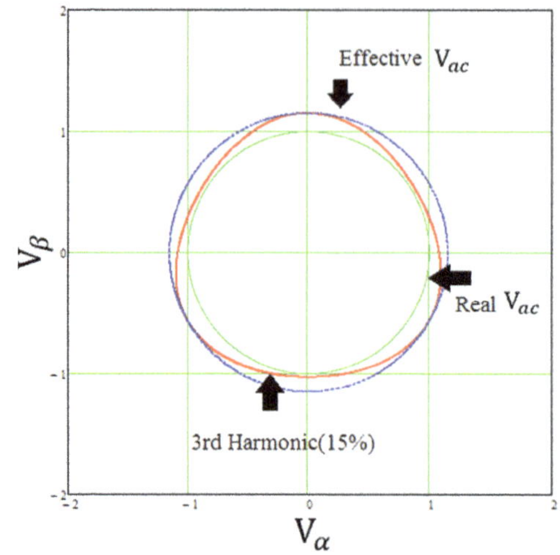

Figure 8.7 Power limit circle-real V_{ac} considering the third harmonic

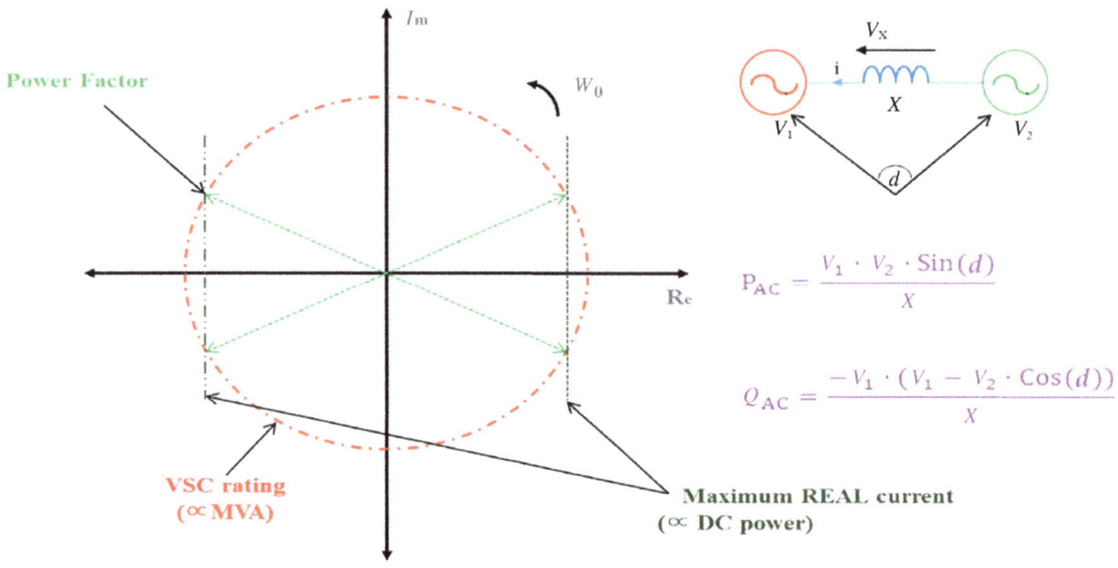

Figure 8.8 P/Q diagram for the VSC HVDC acting as an inverter

transformer, so that when normal variation in supply voltage potentially occurs, the AC-converter voltage has sufficient movement to ensure that the AC current requirements can be met.

Figure 8.10 presents a vector diagram in which the three-phase system is represented by single vectors, each having magnitude equal to the peak amplitude of a single-phase term. The AC voltage V_{ac} is used as a reference and the converter voltage V_C varies in magnitude and phase to control the voltage V_I, across the line reactor. The current vector I_C is at a right angle to the inductor voltage, at one end representing the current into the converter and at the other the current out of the AC system. The red circle shown in Figure 8.10 shows the line reactor voltage limit representing the maximum rated current for the system I_{Cm}, whereas the brown dashed arc indicates the maximum voltage for the converter, $V_{Cm} = V_{DC}$. For a converter phase, the AC current divides equally between the top and bottom arms. Also, the DC current that arises as a result of the power balance between the AC and DC systems flows for the rectifier station from the minus to the plus DC-link rail to maintain capacitance charge.

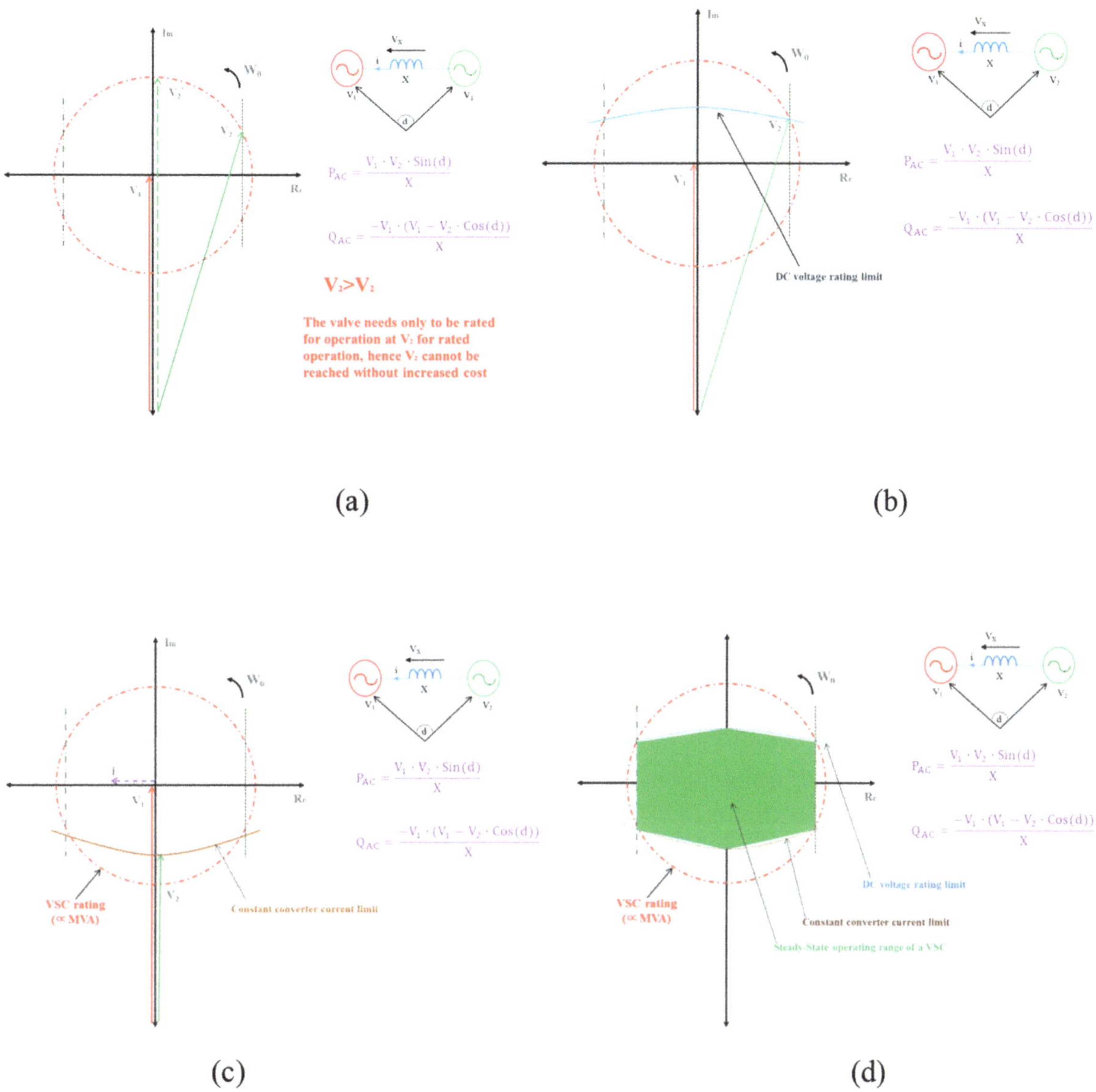

Figure 8.9 Limitation conditions for the VSC HVDC acting as an inverter.

The main variation to be considered is that of the AC supply voltage, which for this supply level is +6%/-10%. The DC voltage level will be regulated by the converter, but there is a high degree of noise because of the module switching process combined with the wide variation in capacitor voltage because of the normal charge/discharge process.

Since these all values are all below the proposed DC voltage, this design process is the appropriate, and the DC voltage needs to vary to permit the required system rating still must be considered. It is necessary to set a value for the converter capability required to achieve a given power factor under all operating conditions. There is very little guidance on this matter, and as the value shown in Figure 8.9 equates to ±0.92, which is within the range typically required for power system compensation, it will be taken as a design aim for the system. The relationship between the AC and DC systems is given by the power

$$W_{DC} = \sqrt{3}V_S \cdot I_S \cdot \mathrm{PF} = I_{DC} \cdot 2V_{DC} \tag{8.33}$$

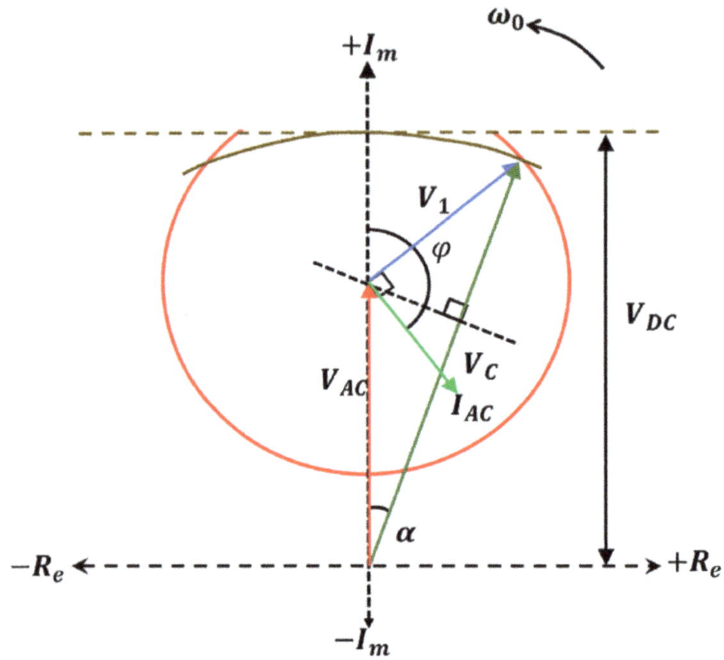

Figure 8.10 AC vector diagram

where W_{DC} is the DC power, I_S is the root-mean square (RMS) AC current, PF is the power factor, I_{DC} is the DC rated current, and V_{DC} is the nominal DC voltage.

As in Figure 8.10, as the angle "α" is the phase displacement between the converter vector and the AC supply vector and the angle "β" is the phase angle between the converter current and the converter voltage, the power factor is given by

$$\mathrm{PF} = \cos(\beta - \alpha) \tag{8.34}$$

By intersecting the converter voltage with a line that is normal to the vector that touches the peak of the AC supply voltage vector, two basic relationships can be derived:

$$V_C = V_{AC} \cdot \cos(\alpha) + V_I \cdot \sin(\beta) \tag{8.35}$$

$$V_{AC} \cdot \sin(\alpha) = V_I \cdot \cos(\beta) \tag{8.36}$$

Figure 8.11 shows the converter voltage for the value derived from Equations (8.35) and (8.36), showing how the peak value intersects and is limited by the current value of DC voltage.

Figure 8.12 shows the combination of power factor and current functions derived against angle β, highlighting how the operation principle of the current falls where the converter voltage is restricted by the DC-link voltage.

Combining the functions for P and Q gives the diagram with power factors of 0.95. These two sets of loci intersect with the *P/Q* locus for the converter with minimum AC voltage, therefore, demonstrating that the basic specification has been achieved. The diagram also shows that the locus for the maximum AC voltage presents the limiting factor.

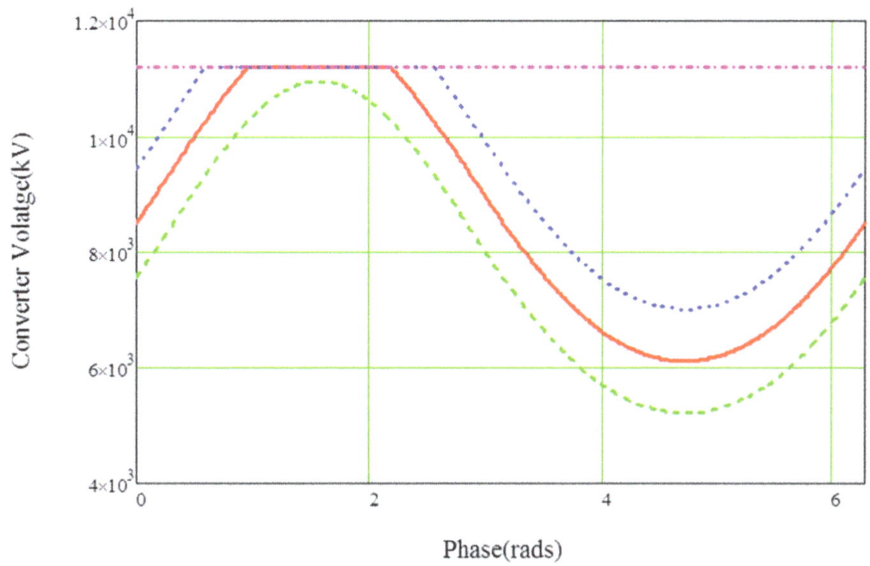

Figure 8.11 Peak converter voltage with different AC voltage ranges for the full range of angle β

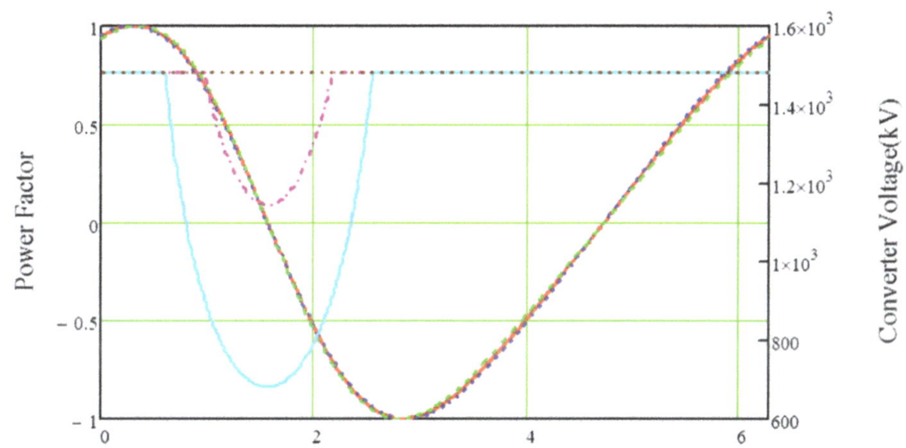

Figure 8.12 Variation of power factor and AC current with angle β

8.3 Determination of IGBT Rating

The rating for the submodule (SM) is based around components that are readily available from many sources. Although the commonly available semiconductor switch, the IGBT, is available in both plastic module packaging and press pack style, the latter is only sourced by ABB and is significantly more expensive. For SMs, the IGBT power capabilities are similar, so although the voltage can rise to 6.5kV, the current capability becomes too low for the application. The currents through the IGBTs are a combination of half the AC phase current and one third the total DC current, giving an RMS current in the region of 700 A and a peak in the region of 1.2 kA. For this rating, the 3.3-kV IGBT is most appropriate.

IGBTs are characterized by the voltage they are expected to be operated at, which is nominally half their voltage rating. It is important that the nominal operating voltage is less than or equal to this value because it is subject to a random failure mechanism whose rate increases rapidly as the voltage rating of the device is approached.

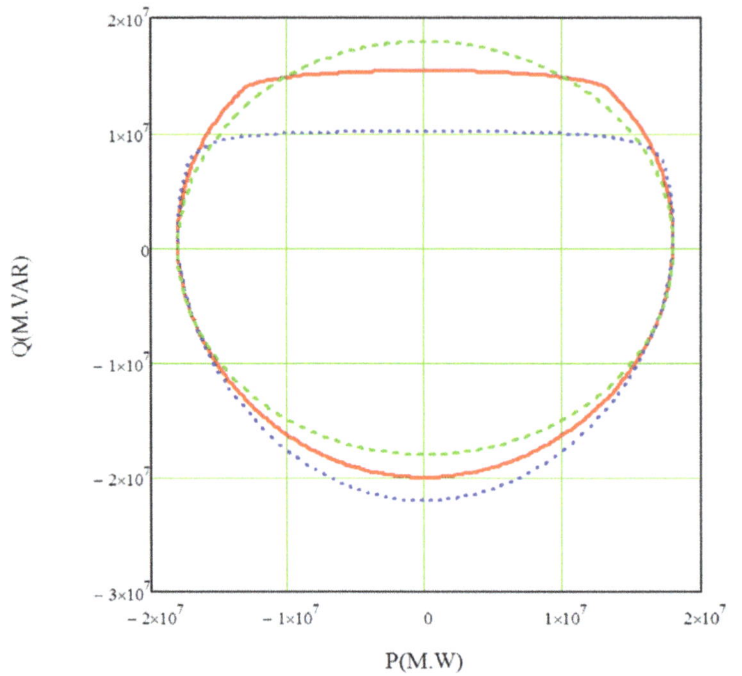

Figure 8.13 *P/Q* diagram for the proposed converter

The 3.3-kV IGBT is typically characterized as being between 1.8 and 2.0 kV. The capacitor for the module needs to be non-polarized and to have low equivalent series resistance (ESR) and low inductance, while having a capacitance value in the mF region. The most appropriate type is a metalized polypropylene (MPP) capacitor, which commonly has a maximum AC voltage rating of 1.4 kV, or a 2 kV peak. In order for the maximum peak variation in voltage expected over a power cycle to be nominally that of the AC capacitor rating, the mean voltage of operation needs to be in the range of 1.6 kV.

Voltage Stress of IGBTs

From the operation principle of the SM shown in Figure 8.14, when $IGBT_1$ is turned on and $IGBT_2$ is turned off, the collector-emitter voltage of $IGBT_2$ is the voltage of the SM capacitor. Similarly, when $IGBT_1$ is turned off and $IGBT_2$ is turned on, the collector–emitter voltage of $IGBT_1$ is the voltage of the SM capacitor. Thus, the RMS voltage $V_{T,rms}$ and the peak voltage $V_{T,peak}$ of $IGBT_1$ and $IGBT_2$ are equal to those of the SM capacitor. As the diode D_1 (D_2) is in parallel with $IGBT_1$ ($IGBT_2$), their voltage stress is the same. From the analysis earlier, the following two equations can be obtained:

$$V_{\mathrm{T,rms}} = V_{\mathrm{D,rms}} + V_{\mathrm{c,rms}} \tag{8.37}$$

$$V_{\mathrm{T,peak}} = V_{\mathrm{D,peak}} + V_{\mathrm{c,peak}} \tag{8.38}$$

Current Stress of IGBT

Figure 8.14 shows the topology diagram of the SM. i_{V1} is the current of $IGBT_1$ or D_1. i_{V2} is the current of $IGBT_2$ or D_2. It is obvious that i_{V1} is the capacitor current of the SM. That is,

$$i_{V1} = i_{c.pa}(t) = \frac{I_{vm}}{4}\sin(\omega t + \varphi a) + \frac{I_{r2m}}{2}\sin(2\omega t + \Theta a) + Q_2 \tag{8.39}$$

According to KCL, i_{V2} can be written as

$$i_{V2} = i_{pa} - i_{V1} = \frac{I_{dc}}{3} + \frac{I_{vm}}{4}\sin(\omega t + \varphi a) + \frac{I_{r2m}}{2}\sin(2\omega t + \Theta a) + Q_3 \tag{8.40}$$

where the second-order harmonic component of the capacitor voltage, Q_x (x = 1,2,3,...), represents the higher-order terms that contain third and above harmonic components.

When $i_{V1} > 0$, the current of $IGBT_1$ is zero and the current of D_1 is i_{V1}; when $i_{V1} < 0$, the current of $IGBT_1$ is i_{V1} and the current of D_1 is zero. Therefore, the currents of $IGBT_1$ and D_1 can be expressed as

$$i_{T1} = \begin{cases} 0, & i_{V1} \geq 0 \\ i_{V1} & i_{V1} < 0 \end{cases} \tag{8.41}$$

$$i_{D1} = \begin{cases} i_{V1}, & i_{V1} \geq 0 \\ 0, & i_{V1} < 0 \end{cases} \tag{8.42}$$

Similarly, when $i_{V2} > 0$, the current of $IGBT_2$ is i_{V2} and the current of D_2 is zero; when $i_{V2} < 0$, the current of $IGBT_2$ is zero and the current of D_2 is i_{V2}. Therefore, the currents of $IGBT_2$ and D_2 can be expressed as

$$i_{T2} = \begin{cases} i_{V2}, & i_{V2} \geq 0 \\ 0, & i_{V1} < 0 \end{cases} \tag{8.43}$$

$$i_{D2} = \begin{cases} 0, & i_{V2} \geq 0 \\ i_{V2}, & i_{V2} < 0 \end{cases} \tag{8.44}$$

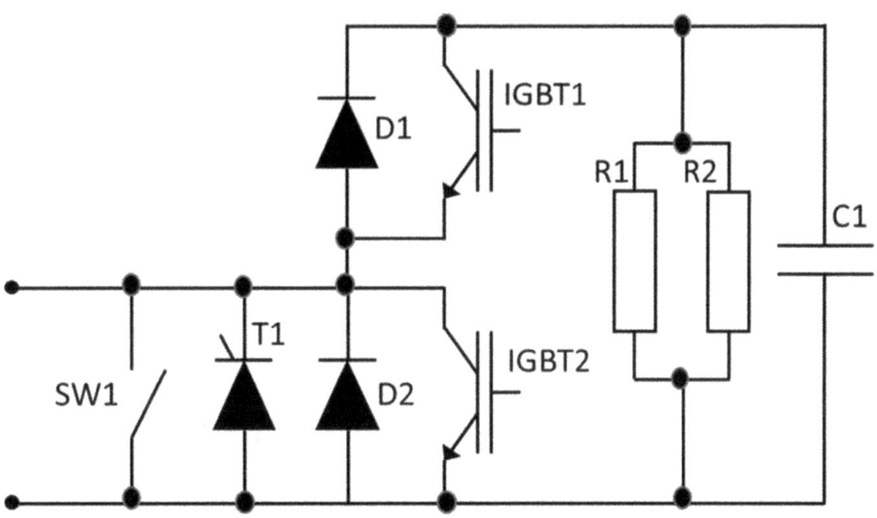

Figure 8.14 Submodule of MMC

The currents of i_{T1}, i_{D1}, i_{T2}, and i_{D2} can be calculated under four extreme operating conditions, including $P_{v,p.u.} = 1$, $Q_{v,p.u.} = 0$; $P_{v,p.u.} = -1$, $Q_{v,p.u.} = 0$; $P_{v,p.u.} = 0$, $Q_{v,p.u.} = 1$; $P_{v,p.u.} = 0$, $Q_{v,p.u.} = -1$.

The current stress varies greatly with different operating conditions. Specifically, the current stress of$IGBT_1$ and D^1 reaches the maximum value under the operating condition of $P_{v,p.u.} = 0$, $Q_{v,p.u.} = 1$. The maximum current stress of $IGBT_2$ appears in the $P_{v,p.u.} = 1$, $Q_{v,p.u.} = 0$ operating condition. The current stress of D^2 reaches its highest value in absorbing the full capacity of the active power-operating condition ($P_{v,p.u.} = -1$, $Q_{v,p.u.} = 0$). The maximum RMS currents of $IGBT_2$ and D^2 are larger than those of $IGBT_1$ and D^2 . However, the peak values of these four power electronic devices are almost the same. In addition, the RMS currents of the four power electronic devices are all smaller than half of the rated DC current I_{dcN}, and also smaller than the arm current rating I_{rN}.

8.4 Selection of Arm Reactor

There are three main roles of the arm reactor. First, as the arm reactor is the medium of power exchange between the converter and the AC system, it can smooth the injected current and suppress the negative-sequence current caused by an unbalanced grid. Second, the arm reactor can be used to suppress the circulating current, because the relationship of the arm reactor and the circulating current is complicated. Therefore, when selecting an arm reactor, priority should be given to preventing circulating current resonance. Third, the arm reactor can reduce the effects of faults arising inside or outside the converter. The phase reactance is a combination of transformer leakage reactance and arm reactance.

$$L_P = L_T + \frac{1}{2} \cdot L_V \tag{8.45}$$

where L_T is the transformer reactance and L_V is the arm reactance.

Transformer leakage reactance has a significant effect on the manufacturing cost of the interface transformer. Typically, the optimum reactance value to give the lowest price is about 14% of the equivalent base impedance. For HVDC, the main consideration is the possibility of a line-to-line fault occurring in the DC system. The arm reactance is essential in limiting the fault current that might otherwise develop.

With peak surge currents, potentially rising above the order of 10 kA, well beyond the capability of conventional IGBTs and diodes, additional bypass protection is necessary, so the selection of arm reactors will be used to limit this peak current to within the rating of this additional protection. The equivalent base impedance of the converter system to the AC supply, Z_C, may be defined as

$$Z_C = \frac{V_{ac}}{I_{ac}} = \frac{3}{2} \cdot \frac{V_{ac}^2 \cdot PF}{P} \tag{8.46}$$

It is essential that impedance of the phase reactance at power frequency should be a small proportion of this impedance, so relating this as a factor k_x of Z_C, the phase reactance is given as

$$X_P = k_x \cdot Z_C = k_x \cdot \frac{3}{2} \cdot \frac{V_{ac}^2 \cdot PF}{P} \tag{8.47}$$

The larger the k_x, the more the converter voltage will need to rise above the AC voltage. Typical values for k_x range from 10% to 30%, with most applications using values near 10% to provide simple control. VSC HVDCs, however, will require values greater than 20% to provide sufficient fault current limiting combined with optimal cost for the transformer.

Considering the current flow when a fault occurs between the DC sides, the phase current flows through the transformer reactance and a single-valve reactance, with the valve current flow being defined by the polarity of the AC voltage in relation to how the IGBTs in the valve will conduct or block current flows. Considering the frequency-domain expression, the peak current I_{acp} will be given as

$$I_{\text{acp}} = \frac{V_{\text{ac}}}{\omega_0 \cdot (L_{\text{T}} + L_{\text{V}})} + I_0 \tag{8.48}$$

where I_0 is the current flowing through the inductors at the instant the fault occurs and ω_0 is the power frequency in radians per second. Because L_T is shown by the notional 14% value noted earlier, this expression may be rearranged to give a value for valve reactance:

$$L_{\text{V}} = \frac{V_{\text{ac}}}{\omega_0 \cdot \left(I_{\text{acp}} - I_0\right)} - L_{\text{T}} \tag{8.49}$$

Thus, if I_0 is the maximum peak current flowing in normal operation, I_{ACp} is set to a value within the rating of the bypass protection, and V_{ac} is set to the nominal maximum operating value, the value of L_{V} can be defined.

For the MMC, the interface inductance Z_C contains two parts. One part is the inductance of the interface transformer Z_T, and the other part is the equivalent arm inductance $X_{L0}/2$. The reason for only counting half of the arm inductance is that two arm inductors in one phase unit are in parallel when looking from the AC side.

There are three aspects affecting the selection of Z_{C}. First the smaller the Z_{C}, the faster the output current can follow its reference value, and for reactive power loss, a smaller Z_{C} results in less consumption of the reactive power. Also, when considering the effects of suppressing the negative-sequence current, the larger the Z_{C} the better.

Consideration for the negative-sequence current is as follows: suppose the negative-sequence voltage of the AC system is V_{s}^{-} . When there is no negative-sequence current suppression controller, the MMC is short circuited as for negative-sequence current. Then the negative-sequence current I_{acp}^{-} flowing through the link inductor is

$$I_{\text{acp}}^{-} = \frac{V_{\text{s}}^{-}}{Z_{\text{C}}} = \frac{V_{\text{s}}^{-}}{(1/2)\sin(2\delta)} \tag{8.50}$$

$$\frac{I_{\text{v}}^{-}}{I_{\text{vN}}} = \frac{V_{\text{s}}^{-}}{(1/2)\sin(\delta)} \times \cos(\delta) = \frac{V_{\text{s}}^{-}}{\sin(\delta)} \tag{8.51}$$

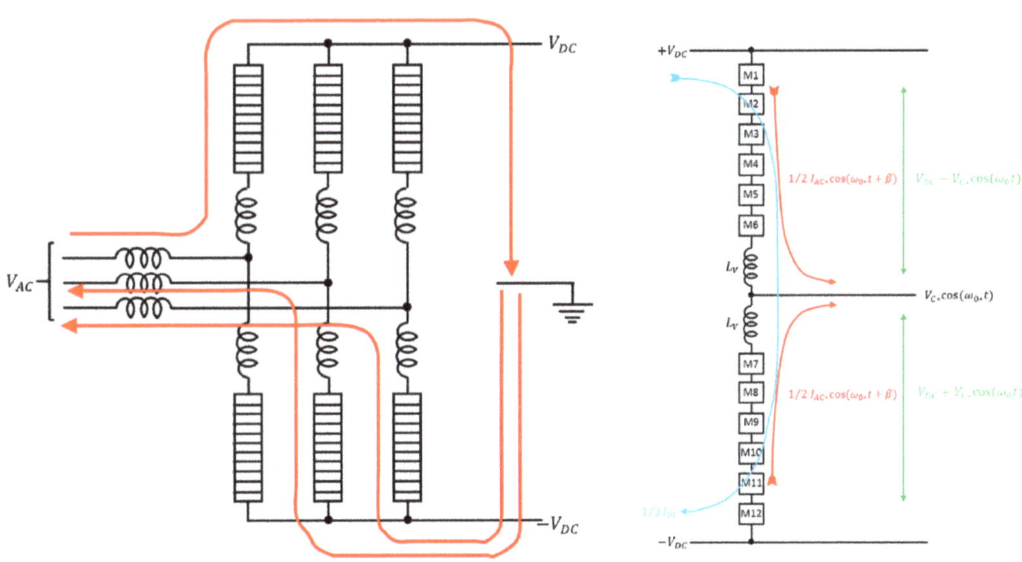

Figure 8.15 Current loop for MMC

Suppose that the negative-sequence voltage is 1.5% of the voltage rating. To enable the MMC to run for a long time, the negative-sequence current should not exceed 5–10% of the current rating. That is

$$\frac{V_s^-}{\sin(\delta)} = \frac{0.015}{\sin(\delta)} \leq 5\% \sim 10\% \tag{8.52}$$

Solving Equation (8.52), it leads to

$$\sin(\delta_N) \geq 0.15 \sim 0.3 \tag{8.53}$$

Inserting Equation (8.53) into Equation (8.38) yields

$$Z_C = \frac{1}{2}\sin(2\delta_N) \geq 0.15 \sim 0.29 \tag{8.54}$$

After the phase reactance Z_C is determined, the arm inductance X_{L0} *can* be easily obtained. The phenomenon of circulating current resonance can be easily understood by analytical expression of the amplitude of the second-order circulating current I_{Circle}. This expression is rewritten as

$$I_{\text{Circle}} = \frac{\sqrt{(F_1\cos(\varphi_{ac}) + F_2)^2 + (F_1\sin(\varphi))^2}}{F_3} \tag{8.55}$$

where, $F_1 = \frac{3f(1)I_{acp}}{8\pi\omega^2 L_0 C_0}$, $F_2 = -\frac{4f^2(1)I_{DC}}{3\pi^2 N\omega^2 L_0 C_0}$), $F_3 = 1 - \frac{N}{16\omega^2 L_0 C_0} - \frac{8f^2(1)}{3\pi^2 N\omega^2 L_0 C_0}$, and f(1) = πNm/8, ω is the network's angle frequency.

From Equation (8.55), it can be known that *F1* and *F2* are not equal to zero. Thus when $F_3 = 0$, the second-order circulating current will be infinite. This means the resonance does happen.

From Equation (8.55), the series resonance angle frequency ω_{res} of the phase unit is as follows:

$$\omega_{\text{res}} = \frac{1}{2}\sqrt{\frac{N}{L_0 C_0}} \tag{8.56}$$

Here, the second-order circulating current resonance angle frequency ω_{cir} is defined as

$$\omega_{\text{cir}} = \frac{2\omega}{\sqrt{1 + (128/3 \times (f^2(1)/\pi^2 N^2)}} \tag{8.57}$$

It can be proven that when the series resonance angle frequency ω_{res} of the phase unit equals the circulating current resonance angle frequency ω_{cir}, $F_3 = 0$. Then, the circulating current resonance occurs. Therefore, when selecting the arm inductance, the series resonance angle frequency ω_{res} should be far away from the circulating current resonance angle frequency ω_{cir}.

8.5 Selection of SM Capacitor Value

According to the model of the MMC, the analytical expression of SM capacitor $V_{\text{C,pa}}(t)$ is presented as Equation (8.58):

$$V_{\text{c,pa}}(t) = V_{\text{c}} + \Delta V_{\text{c,pa}}(t) \tag{8.58}$$

where $V_{\text{c}} = V_{\text{dc}}/N$ is the DC component and is the capacitor ripple.

To calculate the ripple of the capacitor voltage, the capacitor voltage ripple band is used, and the equation is given by

$$\varepsilon = \frac{\max|\Delta u_{c,pa}(t)|}{U_c} \tag{8.59}$$

On the basis of the relationship between the SM's energy storage and the capacitor voltage, another expression for the capacitor voltage ripple is as follows:

$$\varepsilon = \frac{1}{3}\frac{S_V}{MN\omega C_0 V_c}\left[1-\left(\frac{M\cdot\cos\varphi}{2}\right)^2\right]^{3/2} \tag{8.60}$$

where S_v and ϕ are the apparent power of the MMC and the angle of the power factor, respectively.

8.6 Redundancy Calculation of SMs for MMC Converter

The calculation of the required number of SMs for VSC HVDC multi-level converters depends on DC voltage and capacitor voltage. The system DC voltage is the combined value across all of the DC capacitors in the SM and, therefore, there must be sufficient modules to provide adequate DC capability. The number of SMs in the MMC depends on the selected IGBT devices. Three kinds of IGBT, such as 1.6, 1.8, and 2 kV were considered. The numbers of SMs required in the target system are shown in Table 8.1, assuming that the DC voltage is 600 kV.

However, the numbers included in SM levels should be considered for adequate redundant levels to satisfy maintenance periods, which depends on the reliability of each of the SM components. To estimate the number of additional modules required in each converter arm, an estimate of system reliability is made. This will then need to be refined by (a) substituting the values with values obtained after more calculation and (b) revising the circuit to reduce areas that are vulnerable.

Components such as the thyristor and the shorting switch shown by Figure 8.16 are only operated under exceptional circumstances and will not be included in the main calculation at this time. The main sources for component reliability are the measured values, Internet and maker's catalogues, and the failure rate data of the SM in the MMC that are presented in Table 8.2. The values are presented in units of "Failures in Time," which are defined as failures per billion hours (1 e-9 failures/hour).

As stated earlier, only the values shown in rows 3, 6, 7, 9–11, 13, and 14 in Table 8.2 are combined to give the failure rate of the module. If γ_0–γ_{11} represent the values from the "Total Failure Rate" column as elements in a vector, the expression becomes

$$\lambda_M = \lambda 0 + \lambda 3 + \lambda 4 + \lambda 6 + \lambda 7 + \lambda 8 + \lambda 10 + \lambda 11 = 1.762 \times 10^3\,(FIT)$$

The mean time to failure (MTTF) then becomes

$$\frac{1}{\lambda_M} = 64.744\,(\text{Year})$$

Table 8.1 Numbers of Submodules (SMs)

Module voltage	Numbers of SMs
1.6 kV	375
1.8 kV	334
2.0 kV	300

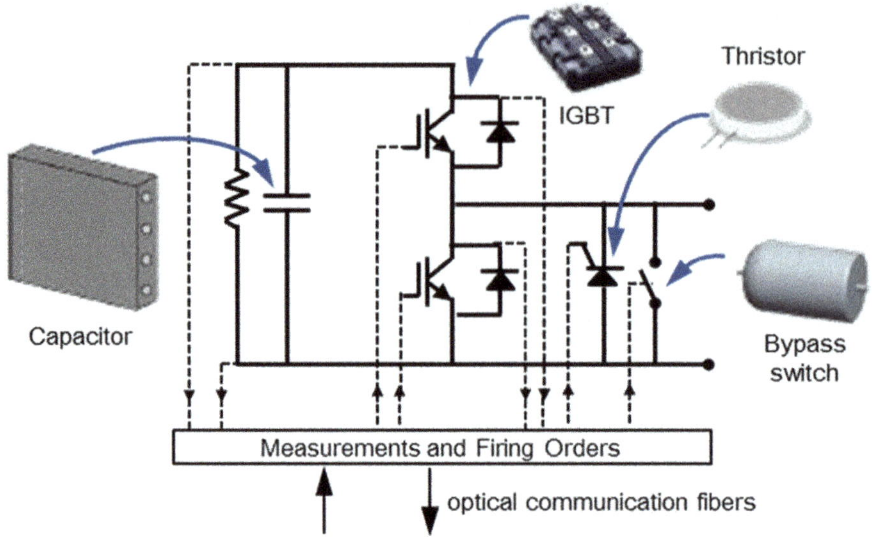

Figure 8.16 Main submodule components

Table 8.2 Failure Rate Data

Component	No.	Failure rate (FIT)	Total failure rate	Comments
Power circuit				
IGBT and gate drive	2	40	80	Traction data
Thyristor and gate drive	1	47	47	Traction data
Switch and actuator	1	1000	1000	Traction data
Film capacitor	1	10	10	Traction data
Power resistor	1	265	265	Traction data
Control panel				
PLD (Programmable Logic Device)	1	150	150	Traction data
Optical comms	2	100	200	Traction data
Crystal oscillator	1	13	13	Traction data
Power supply				
Transformer	2	22	44	Traction data
DC–DC converter	1	1000	1000	Traction data

As this reflects a random event, it is more meaningful to convert it to a value representing the availability of the module over a given time span, and through this obtain an estimate of the availability of the complete converter system over its life. This is commonly done by expression values in terms of "Unavailability, "Q:

$$Q(t) = 1 - e^{-\lambda t} \tag{8.63}$$

where t is the "time at risk." Thus, assuming full time operation, the proportion of the all modules in the target MMC HVDC arm to fail will be 4.5% over the 3-year maintenance period, whereas over the 30-year life of the equipment, 37% of the modules will have failed.

This equation can be adapted further to relate service life and maintenance period:

$$\mathrm{Q}\,(\mathrm{t}) = \frac{\lambda}{\lambda + \mu}\left(1 - e^{-(\lambda+\mu)\mathrm{t}}\right) \tag{8.64}$$

where t is the overall time at risk, in this case the equipment life, and μ is the maintenance rate, or 1/(3 years). As would be expected, the result is very similar to the above for the 3-year maintenance period, and 4.4% of the modules will fail.

To determine the availability of a complete inverter limb of n modules, in which m modules can be allowed to fail before the limb fails, the Binomial Failure Model needs to be used:

$$Q = \frac{N!}{k!\,(N-k)!} \sum_{i=k}^{N} \left(q^k \cdot \left(1 - q^{N-k}\right)\right) \tag{8.65}$$

$$_{N}C_{K} p^k q^{N-k} \approx \frac{1}{\sqrt{2\pi\,N\,p\,q}} e^{-\frac{(k-Np)^2}{2Npq}} \tag{8.66}$$

The unavailability of a single inverter arm in the target system has been designed to operate with a minimum of 334 modules. The unavailability of the system according to the redundant modules increment is shown in Figure 8.17.

To determine the number of redundant modules for the MMC, several scenarios are considered: maintenance periods are 1 year, 2 years, or 3 years, IGBT devices used are 1.6, 1.8, or 2 kV, and the FIT of DC/DC converter is changed from 1000 to 500.

Figure 8.18 shows the graph of MMC valve unavailability for several IGBT valves against a number of modules in the case of a 1-year maintenance period. In this case, the unavailability is 0.00568. To achieve an availability better than 99%, seven modules are required for a 1.6-kV IGBT device, six modules are required for a 1.8[kV] IGBT device, and five modules are required for a 2-kV IGBT device.

Figure 8.19 shows the graph of MMC valve unavailability for several IGBT valves against a number of modules in the case of a 2-year maintenance period. In this case, the unavailability is 0.0299. To achieve an availability better than 99%,22 modules are required for a 1.6-kV IGBT device, 20 modules are required for a 1.8-kV IGBT device, and 17 modules are required for a 2-kV IGBT device.

Figure 8.20 shows the graph of MMC valve unavailability for several IGBT valves against a number of modules in the case of a 3-year maintenance period. In this case, the unavailability is 0.044. To achievean availability better than 99%, 28 modules are required for a 1.6-kV IGBT device, 25 modules are required for a 1.8-kV IGBT device, and 20 modules are required for a 2-kV IGBT device.

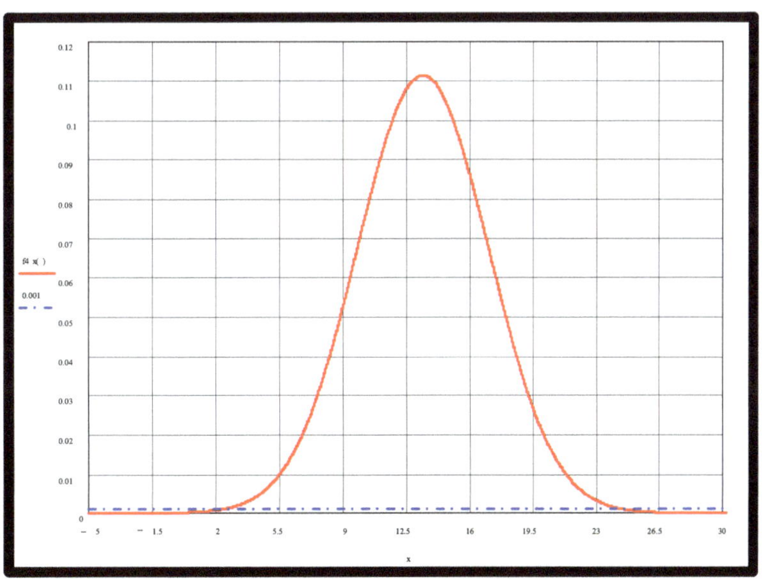

Figure 8.17 Unavailability function for the target system with 334 modules according to the redundant modules increment

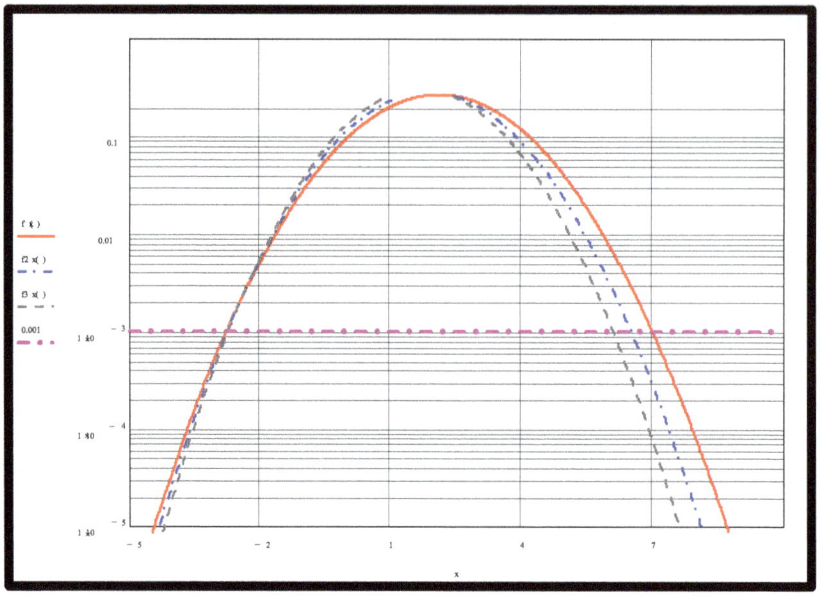

Figure 8.18 Unavailability function for MMC valve in the case of a 1-year maintenance period

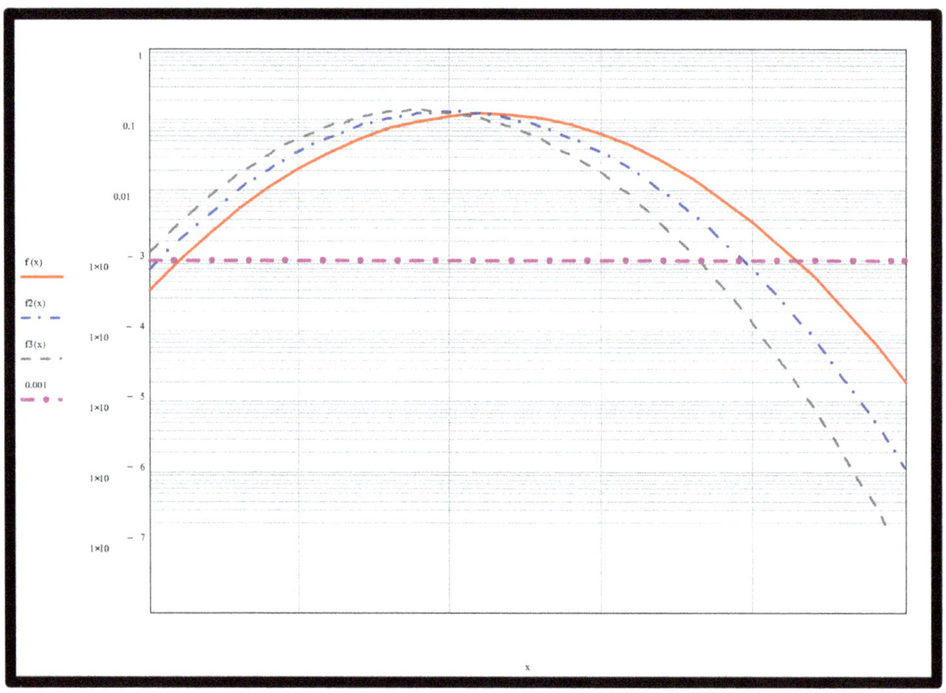

Figure 8.19 Unavailability function for MMC valve in the case of a 2-year maintenance period

The graph of MMC valve unavailability with the DC/DC converter failure rate reduced to 500 FIT is shown in Figure 8.21. In this case, the rate of IGBT is 2 kV with 300 modules. This shows the equivalent module requirement to be 7 modules for a 1-year maintenance interval, 12 modules for a 2-year maintenance interval, and 15 modules for a 3-year maintenance interval. Consequently, the number of MMC modules for maintenance and the range of mean capacitor voltage are given in Table 8.3.

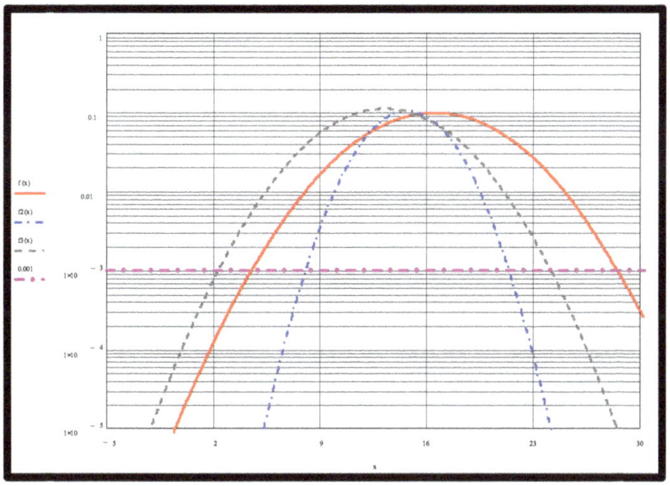

Figure 8.20 Unavailability function for MMC valve in the case of a 3-year maintenance period

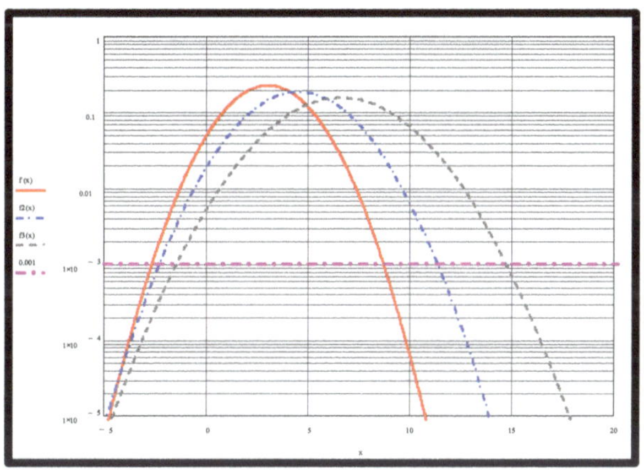

Figure 8.21 Unavailability function for MMC valve with the failure rate of the DC/DC converter = 500

Table 8.3 Additional Modules Against Maintenance Period and Module Voltage

		Additional cells maintenance period		
Module voltage	**No. of modules**	**1 year**	**2 year**	**3 year**
1.6 kV	**375**	7	22	28
1.8 kV	**334**	6	20	25
2.0 kV	**300**	5	17	20

8.7 Configuration and Valve Design of VSC HVDC System

Each VSC SM design consists of a number of series-connected SMs, each comprising a pair of IGBTs, as well as diodes, DC capacitors, and bleed resistors, along with control electronics and protection circuits. The mechanical support of the valves can be either floor-mounted or suspended. The capabilities of VSC topology permit reactive power control, multi-terminal schemes, and black start functionality. Figure 8.22 shows the typical arrangement of the half-bridge converter.

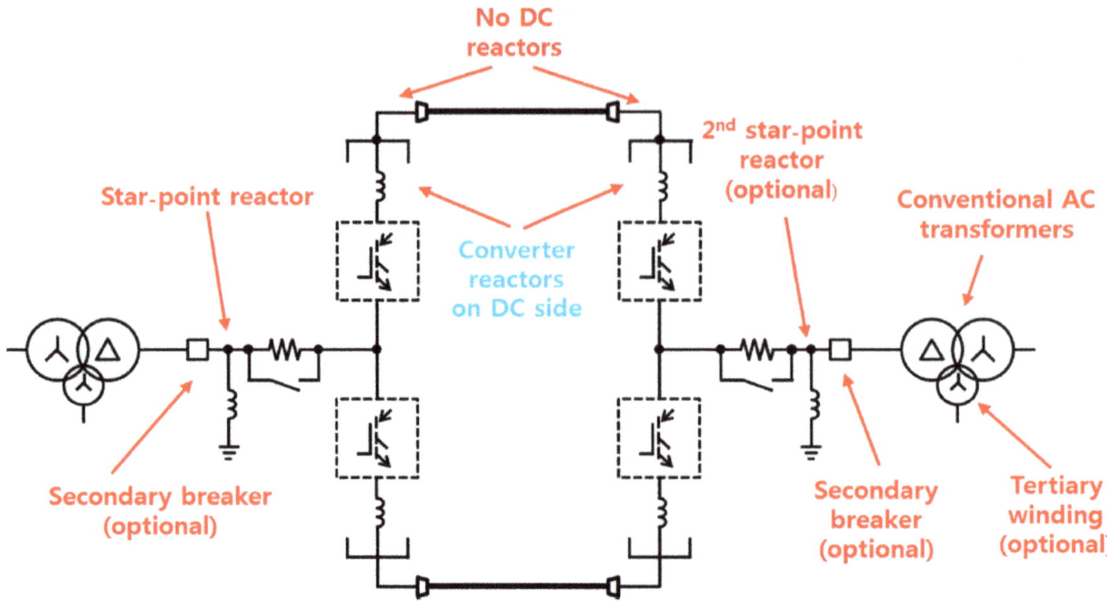

Figure 8.22 Typical point-to-point DC cables configuration

Monopolar Configuration

Figure 8.22 shows a typical configuration for point-to-point interconnectors with DC cables. The detailed configuration is dependent on project-specific requirements. Tertiary transformer winding is optional to supply station auxiliary power.

Back-to-Back Configuration

Figure 8.23 shows a configuration with the minimum space requirement for back-to-back. This detailed configuration depends on project-specific requirements. Tertiary transformer winding is optional to supply station auxiliary power.

Bipolar Configuration

Figure 8.24 shows a bipolar configuration that is used for DC overhead lines (OHLs) or hybrid DC circuits (DC circuits with OHL or with cable sections).

This bipolar configuration has a 50% redundancy upon single DC line and converter faults. Tertiary transformer winding is optional to supply station auxiliary power.

Parallel Configuration

Figure 8.25 shows a parallel configuration as an option for higher DC currents than a normal bipolar configuration. This parallel configuration has a 50% redundancy upon converter faults. Tertiary transformer winding is optional to supply station auxiliary power.

Valve General Description

Each valve within the VSC design consists of a number of series-connected SMs, and each of these SMs contains a pair of IGBTs, a pair of diodes, and the DC capacitor along with control electronics and protection circuits.

Each "valve" in the scheme consists of a number of series-connected SMs. These SMs are grouped into banks of eight known as "modules." Each of these modules is contained within a support structure that provides insulation, electrical connections between power modules, and a supply of coolant. A number of these modules are arranged to form a tier, on top of which further tiers are arranged.

SM Design

Every valve within a converter group consists of multiple SMs arranged in series. Each of these SMs is electrically identical to the next.

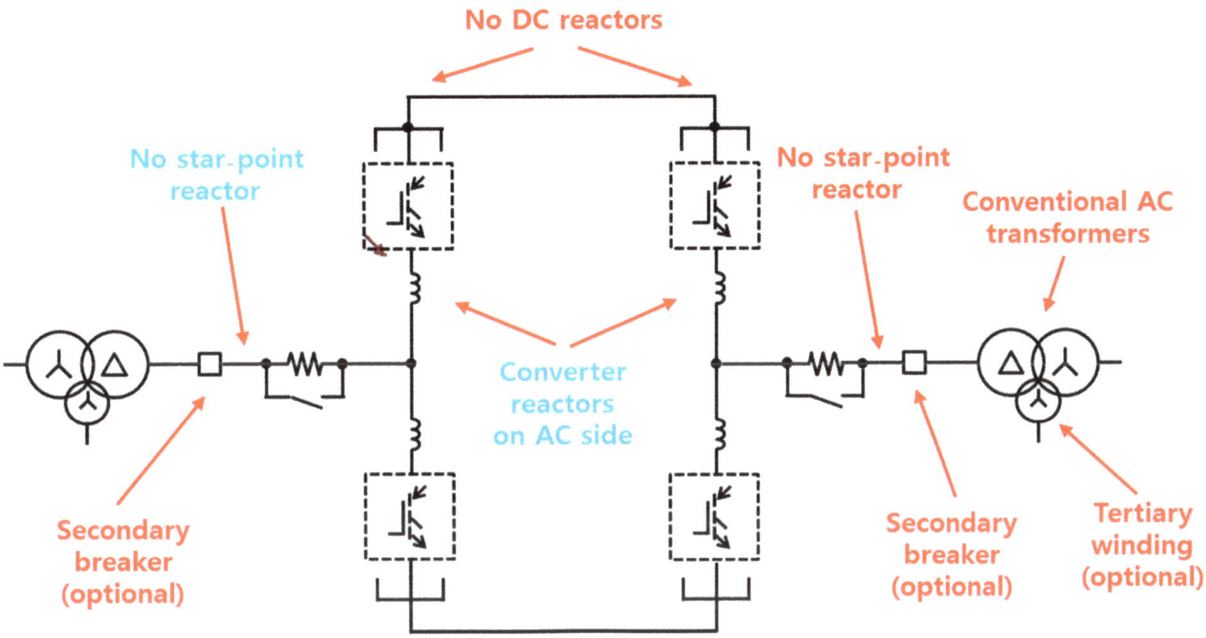

Figure 8.23 Typical back-to-back configuration

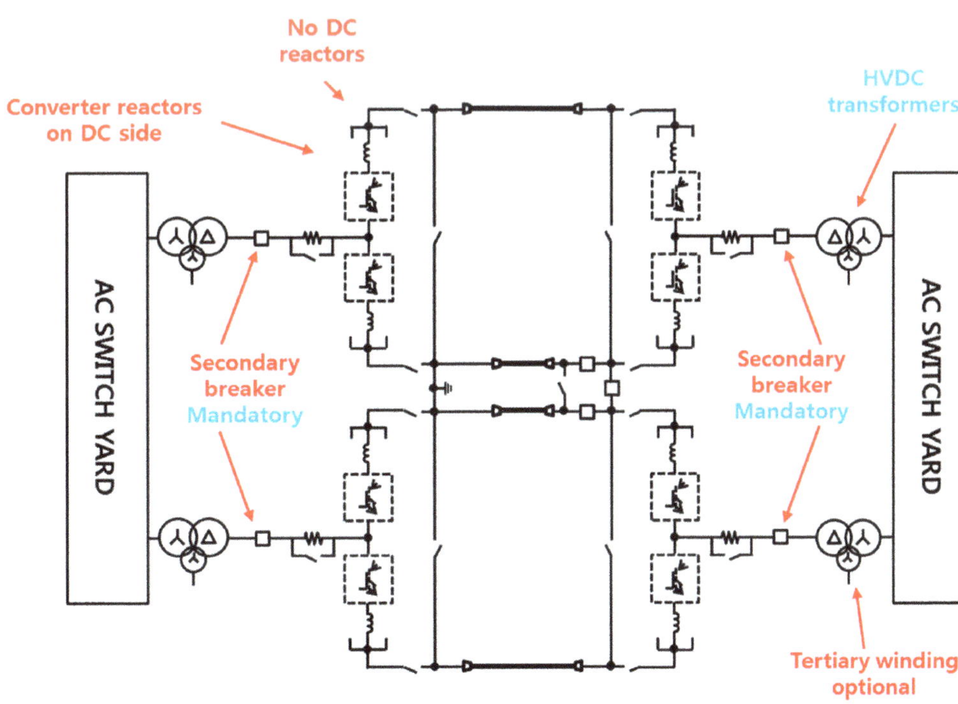

Figure 8.24 Bipolar configuration

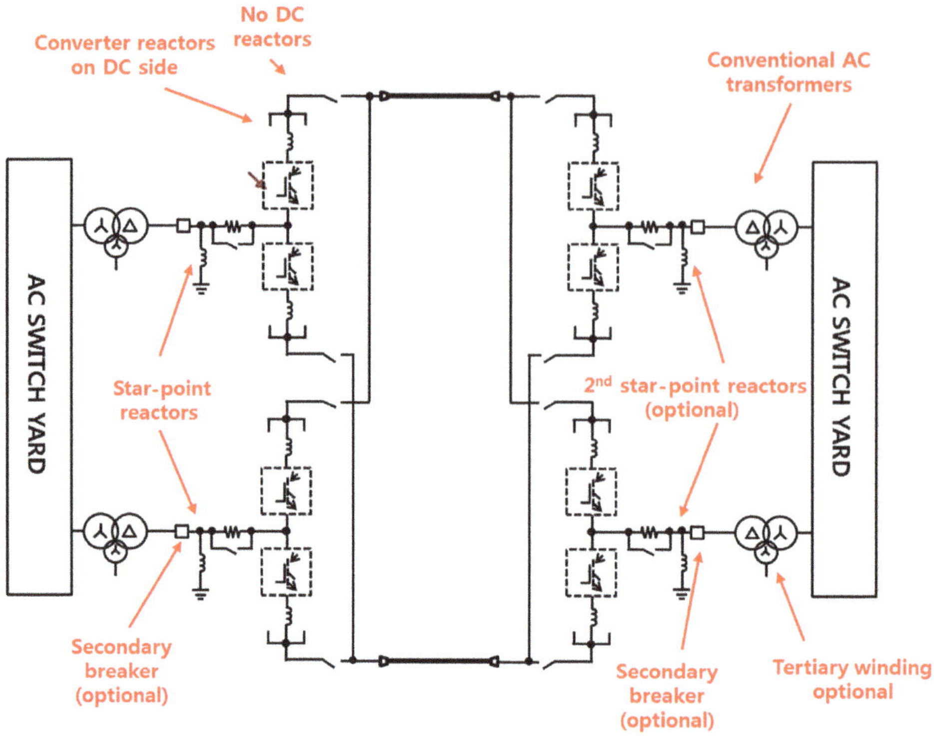

Figure 8.25 Parallel configuration

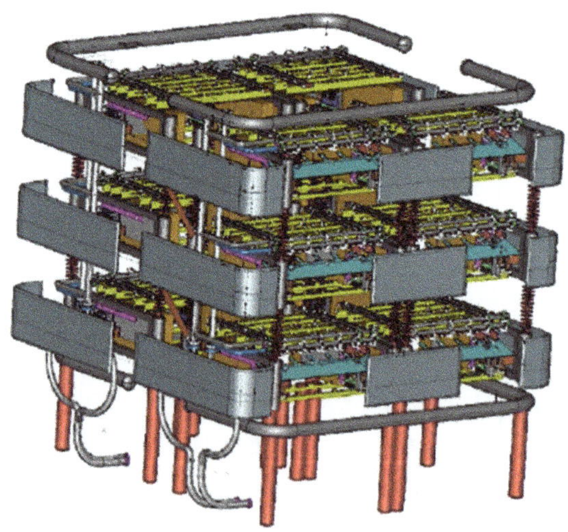

Figure 8.26 VSC HVDC valve [Courtesy of Siemens]

Figure 8.27 shows the major components that comprise each SM. Each of these components will be covered in more detail in the following sections. Not included in the earlier are the SM electronics. These control the turn-on and turnoff of the IGBTs, and also operate the thyristor and bypass switch (sw1) in the case of a level failure in order to short-circuit the SM.

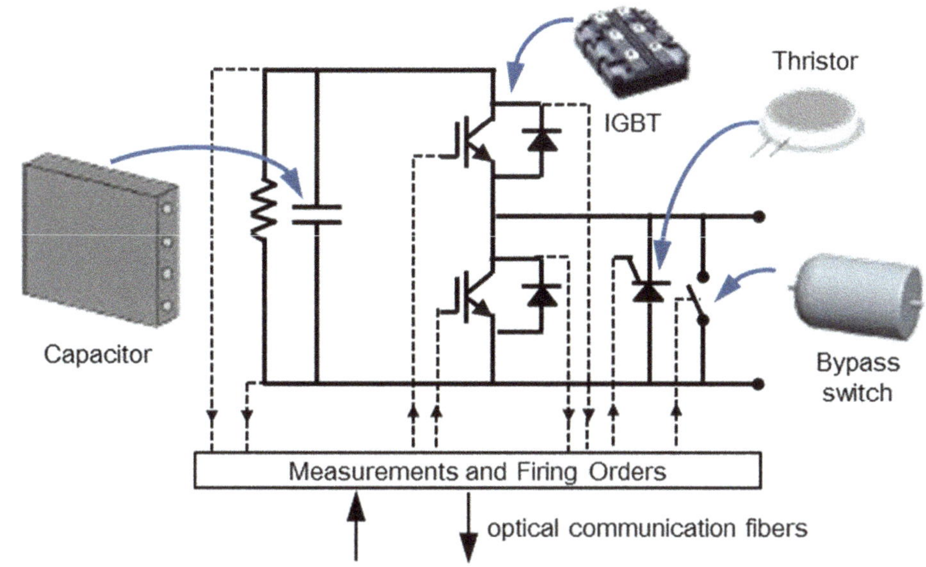

Figure 8.27 Submodule circuit

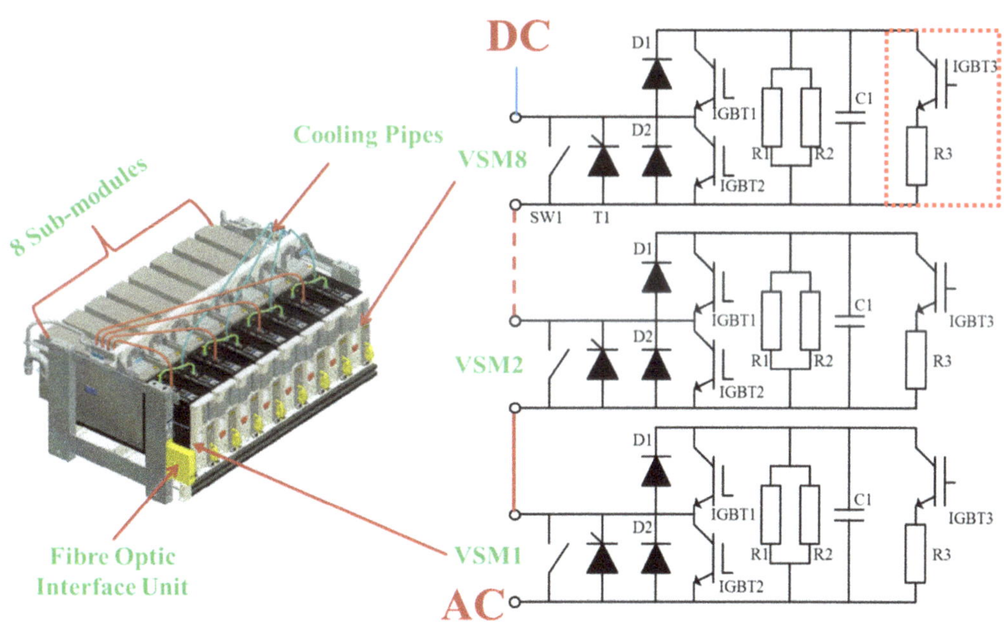

Figure 8.28 Power submodule

SM Electrical Components

IGBTs and Diodes

One IGBT used in VSC HVDC is a 3.3-kV IGBT made by Mitsubishi Electric. This IGBT is a new generation of 3.3-kV HVIGBT modules with Fine Planar MOS gate Light Punch Through HVIGBT (FP-LPT IGBT) and SR-HVDi (Soft reverse Recovery HV Diode). The advanced new HVIGBT improves the trade-off characteristics between CE (sat) and E_{off} while also keeping strong safe operating area (SOA) robustness. Also, reverse recovery current is reduced by the advanced new diode, and soft reverse recovery behavior maintains the robustness capability of conventional diode designs.

The voltage rating of the IGBT module and the number of cascaded-connected SMs depend on the DC-link voltage that needs to be supported. However, the IGBT module's voltage rating is also defined by the switching safe operating area (SSOA) voltage (VSSOA), which considers long-term stability against cosmic radiation. For improved reliability, and to avoid false triggering because of cosmic radiation, the maximum allowed DC-link voltage applied to the 3.3-kV IGBT module is set to 2.5 kV.

The margin between the maximum DC-link voltage and the maximum collector–emitter voltage CES of the IGBT module is because of voltage spikes caused by turn-off di/dt and stray inductance of the power circuit. One measure that should be taken to deal with the spike voltage problem during the turn-off of an IGBT module is designing the commutation circuit of the power section to be as low- inductance as possible. Also, an active clamping gate driver is another solution to limit spike voltage. Laminated bus bars and driver design are important technologies for lowering the turn-off spike voltage of IGBTs. Besides the external circuitry design, there are some crucial requirements in the IGBT module itself in terms of low power loss, high SOA robustness, low leakage current, and high reliability. A 3.3-kV and 1500-A IGBT module (also called the R-Series High-Voltage IGBT module) that could fulfill these requirements well is discussed in this chapter.

The new generation of Mitsubishi Electric's 3.3-kV R-Series adopts "Fine Planar MOS gate Light Punch Through IGBT (FP-LPT IGBT)" and "Soft Reverse Recovery Diode with High Robustness (SR-Diode)" structures that reduce the power losses and increase the rated current while maintaining the mechanical compatibility with the already existing H-Series. The new-generation chip technology provides benefits such as

- IGBT chip: Lower losses, controllability of di/dt and dv/dt or low EMI levels
- Diode chip: reduced reverse recovery current (Irr)
- Higher SOA robustness and a positive temperature coefficient (PTC) for easy parallel connection
- At rated current and 125°C, the new IGBT operates with a 30% lower VCE (sat) than a conventional IGBT
- The maximum operation temperature is extended from 125°C to 150°C while using minimum storage
- Temperature is further extended from −40°C to −55°C
- The new wire bond technology also improves the power-cycling capability of the module compared to conventional 3.3-kV IGBT modules

The cross-section structure of the 3.3-kV R-Series IGBT compared with a conventional 3.3-kV H-Series IGBT is shown in Figure 8.29. An advanced planar-gate design with a fine pattern MOS-structure is adopted in this R-Series IGBT module. Optimized carrier concentration with an additional N layer, a thinner N-draft layer, and an LPT structure are three key technologies by which low on-state losses are achieved.

Bypass Switch

The bypass switch shown in Figure 8.30 permits the SM to be short circuited, allowing a current path through a faulty device. It is actuated by the SM electronics and is designed to be fail-safe in situations where all power supplies in the SM fail.

Bypass Thyristor: A bypass thyristor is included so that in the case of a DC-side short-circuit current when the valve current exceeds a certain limit, the thyristor can be gated, protecting the free-wheeling diode (FWD) D_2 (in Figure 8.28). The thyristor itself is clamped between the input and output terminals of the SM, so that if the device fails it will fail closed circuit, thus bypassing the level.

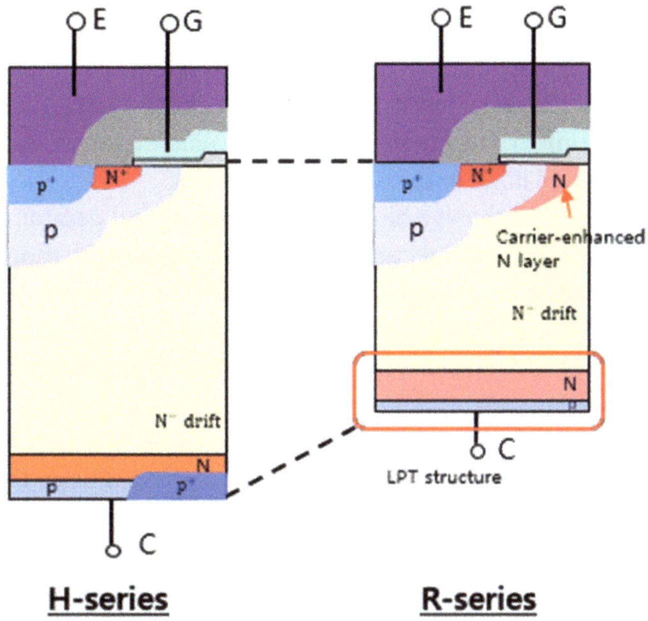

Figure 8.29 3.3-kV IGBT basic structure comparison of the new R-Series and the conventional H-Series

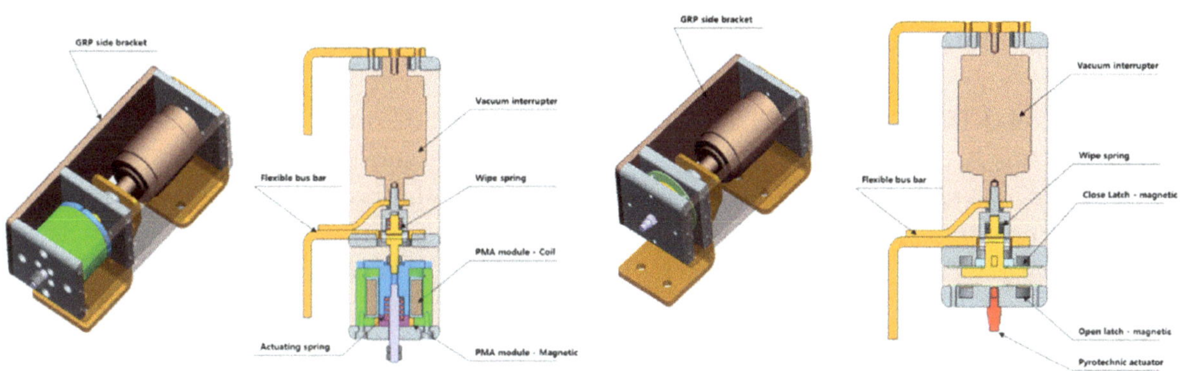

(a) PMA type (b) Pyrotechnic Type

Figure 8.30 Bypass switch

SM Electronics

Two other important components of the gate electronics are the IGBT gate drives. One operates as a master, connected directly to the main electronics and IGBT2, and the other as a slave, connected via an optical isolator to IGBT1. The gate drives generate the gate pulse that controls the state of the IGBTs. The following protections are provided by module electronics:

Fire-through Interlock: To minimize the possibility that both IGBTs are on at the same time, gate-voltage monitoring is used to inhabit either of the IGBTs while the other is on, as well as during the minimum dead time.

Minimum Dead Time: The electronics enforce a minimum dead time of 5 μs after an IGBT turnoff, during which the other device cannot be turned on.

Minimum On/Off Time: The electronics also enforce a minimum on/off time of 10 μs.

IGBT Protection

In many cases, it is desirable to implement short-circuit protection in the gate-drive circuit in order to provide the fast response required for protection against severe low-impedance short circuits. Typically, this protection has been provided by collector–emitter voltage sensing, or so-called desaturation detection. The operation of the desaturation detector will be explained in detail in the following:.

Desaturation Detection: Figure 8.31 shows a block diagram of a typical desaturation detector. In this circuit, a high-voltage fast recovery diode (D^1) is connected to the IGBT's collector to monitor the collector-to-emitter voltage. When the IGBT is in the off state, D^1 is reverse biased, and the (+) input of the comparator is pulled up to the positive gate-drive power supply, which is normally +15 V. When the IGBT turns on, the comparator's (+) input is pulled down by D^1 to the IGBT's CE (sat). The (-) input of the comparator is supplied with a fixed voltage V_{TRIP}, which is typically set at about 8 V. During normal switching, the comparator's output will be high when the IGBT is off and low when the IGBT is on. If the IGBT turns on into a short circuit, the high current will cause the collector–emitter voltage to rise above V_{TRIP}even though the gate of the IGBT is being driven on.

This abnormal presence of high CE when the IGBT is supposed to be on is often called ***desaturation***. Desaturation can be detected by a logical AND of the driver's input signal and the comparator output. When the output of the AND goes high, a short circuit is indicated. The output of the AND is used to command the IGBT to shut down in order to protect it from the short circuit. A delay (t_{TRIP}) must be provided after the comparator output to allow for the normal turn-on time of the IGBT. The t_{TRIP} delay is set so that the IGBT's V_{CE} has enough time to fall below V_{TRIP} during normal turn-on switching. If *tTRIP* is set too short, erroneous desaturation detection will occur. The maximum t_{TRIP} delay is limited by the IGBT's short-circuit-withstanding capability.

Real-Time Control (RTC) Detection: IGBT modules have a built in RTC circuit, and this limits short-circuit currents and maintains 10 μs of short-circuit durability. When excessive current flows through the device, the RTC circuit activates and reduces the gate voltage in order to limit the short-circuit current. Operation of the RTC circuit can be detected as shown in Figure 8.32. The RTC detector is similar to the desaturation detector except that the comparator's (-) input is connected to the gate of the IGBT module.

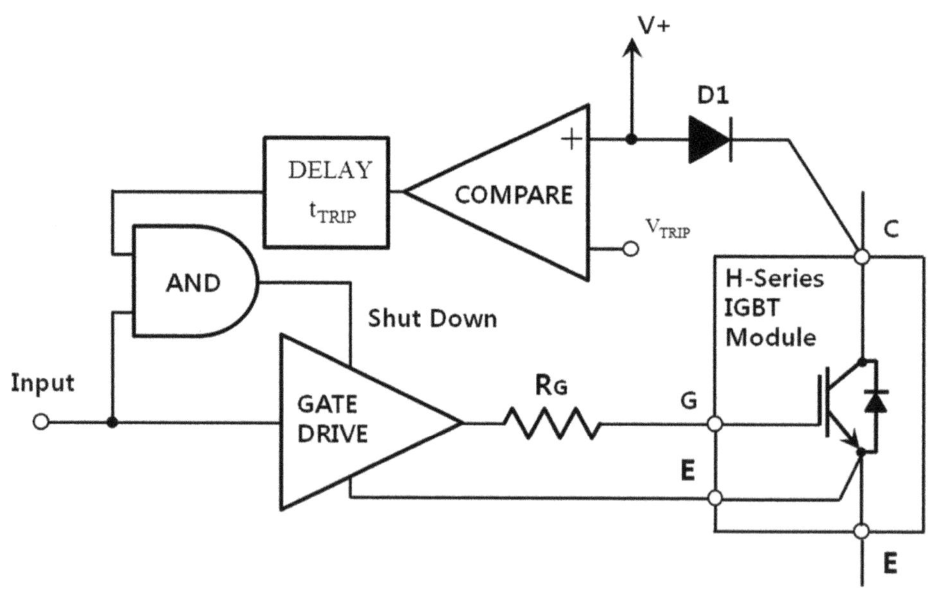

Figure 8.31 Desaturation detector

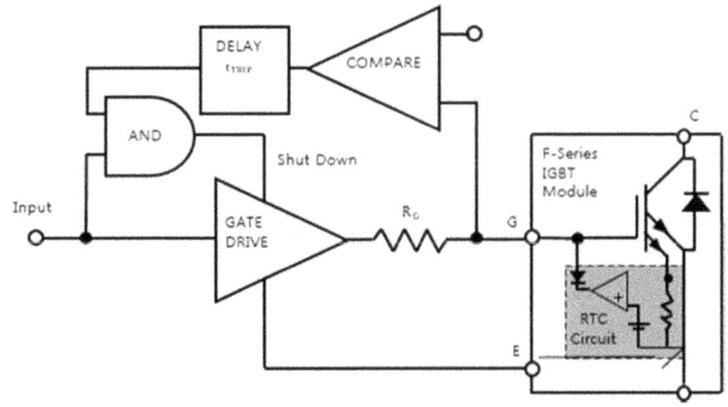

Figure 8.32 RTC detector

dV/dt Limit: Active control of the IGBTs is required to limit collector voltage dV/dt at turn-on and turnoff. This function is implemented as close as possible to the IGBT gate to maximize the availability of the protection.

Capacitor Over voltage Monitoring: The capacitor voltage is monitored, and if it exceeds the overvoltage threshold OV1, then the electronics communicate this problem to the controls. It is then the task of the controls to decide how to deal with this issue, either by switching the SM to HIGH or LOW depending on its current state. When the direction of the current through the valve is such that it would discharge the capacitor, the SM is put into the HIGH state; otherwise, it remains in the LOW state. If the voltage continues to rise and exceeds a second threshold, OV2, then the module electronics immediately switch the SM's state to LOW, and once again communicates this problem to the controls. This SM then remains in the LOW state until the capacitor voltage discharges to a safe level via its discharge resistors.

Capacitor Under-Voltage Monitoring: If the capacitor voltage drops below the under-voltage threshold UV1, an alarm will be sent to the controls. Once again, it is then the task of the control system to decide whether to set the module into a HIGH or LOW state. When the direction of the current through the valve is such that the current is in a direction that would charge the capacitor, the SM is put into a HIGH state; otherwise, it remains in the LOW state. If, although UV1 is true, the electronics detect that the PSU (Power Supply Unit) is low and the system is not in a shutdown state, the electronics will operate the bypass switch.

Gate-Drive Circuit Design Now that the basic requirements for a reliable, efficient gate drive have been reviewed, we can turn our attention to circuit design issues. The layout and component selection for the gate drive are critical for achieving the operating characteristics outlined earlier.

Maintaining dv/dt noise Immunity: IGBT gate-drive circuits are subjected to high common mode dv/dt. The driver circuit layout must minimize parasitic capacitances between adjacent drive circuits in order to prevent a C × *dv*/*dt* coupling of noise. The isolating interface for the gate-drive signals must be designed with appropriate noise immunity. If a pulse transformer is used, its interwinding capacitance must be small. If opto couplers are used, they must have an isolation that is designed for both high common mode voltage and transient noise immunity. Opto couplers should have a guaranteed minimum common mode transient noise immunity of 10 kV/μs specified at a common mode voltage (V_{CM}) of at least 1000 V. The layout of the isolating interface must minimize parasitic capacitance between the primary and secondary interfaces. The use of ground plane shield layers can be very helpful in controlling noise coupled through stray capacitances by the high dv/dt of the power circuit. If twisted pair gate-drive leads are used, the pairs should be kept separated from each other. If they must be bundled, shielded cables with the shield tied to the emitter potential of the IGBT being driven should be used. In general, minimizing gate-drive lead length will help to prevent noise coupling.

Minimum length wiring also helps to achieve the high peak-drive currents required for efficient switching. The best practice is to mount the driver circuit directly on the IGBT module.

Controlling On- and Off-Gate Voltages: Control of the steady-state on- and off-gate voltage is easily accomplished through appropriate regulation of the gate-drive power supply. However, during switching, and especially during the short-circuit operation miller effect, currents cause $i \times R$ voltage on the series gate resistor and (L × *di*/*dt*) voltage on gate-driver parasitic inductance. These voltages can add to the normal on-state gate voltage, causing a surge voltage on the gate. During switching, gate-voltage surges must be kept to less than the device's maximum gate-voltage rating (usually 20 V). Under short-circuit conditions, gate-voltage surges will cause degradation of the short-circuit-withstanding capability by allowing higher than normal currents to flow. To control gate-voltage surges, it is often desirable to implement gate-voltage clamping on the gate side of the series gate resistor. In order for the gate-voltage clamping circuit to be effective, it must be connected as close as possible to the gate and emitter terminals of the IGBT module. For a 15 V gate drive, 16-18 V zeners are typically used. An even more effective clamping circuit is shown in Figure 8.33. In this circuit, the gate voltage is a diode clamped to a local capacitor charged to the turn-on gate voltage. Gate-voltage clamping becomes critical with larger IGBT modules because miller-effect currents are more severe. Long gate-drive leads also aggravate gate-voltage surges, making gate-voltage clamping even more necessary.

Isolated Power Supplies: In most high-power applications, it is necessary to provide isolated power supplies that can float as needed to the emitter potential of the IGBT being driven. Isolated power supplies are required for high-side gates in single- and three-phase inverter circuits because the emitter potential of a high-side IGBT changes when a low-side IGBT is switched. Isolated power supplies are also recommended for low-side gate drive in high-power applications in order to avoid noise caused by (*L* x *di*/*dt*) voltages induced in the stray inductance of the negative DC bus. The gate-drive power supply should have high-voltage insulation designed to reliably withstand the high voltages in the power circuit. It is also critical that the power supplies have minimum capacitance between each other and the logic circuits in order to avoid *dv*/*dt*-induced noise. A common source of unwanted capacitance is adjacent or overlapping windings on the power supplies' isolation transformer.

Gate-Voltage Monitor: The gate voltages of the IGBTs are continuously monitored and compared to the current link state (HIGH/LOW). This permits the electronics to detect gate internal faults and short circuits. If the gate voltage does not tally with the expected link state, a "link state error" (LSE) is sent to the controls. If the controls then request a new link state that is successful, the error is cleared. If, however, the alarm persists, then the bypass switch will be activated.

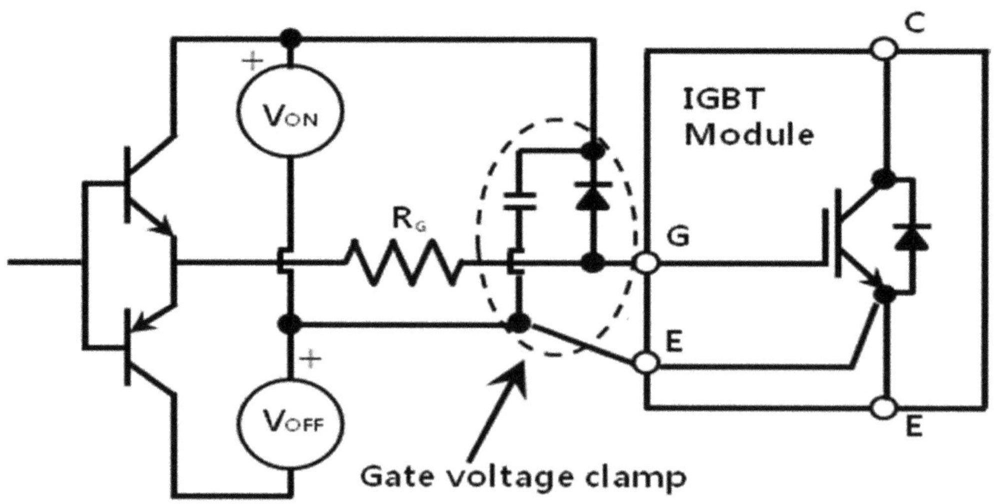

Figure 8.33 Gate-voltage clamp

8.8 HVDC Valve Loss Calculation

The initial VSC HVDCs based on two-level or three-level converter structures rely on a relatively high IGBT-switching frequency, typically 1–2 kHz, to obtain adequately low harmonic distortion. Therefore, high switching frequency leads to very high switching losses in the IGBTs and FWDs, whereas conduction losses are high in thyristor-based line commutated converter (LCC) HVDC systems. As a result, the power losses of such VSC HVDC stations have been in the range of 2%–4% per end, excluding the losses in the cable or line. This is much higher than for LCC stations, which are generally around 0.75% per end. In addition, the initial VSC HVDC required significant screening and filtering to eliminate both airborne and conducted radio frequency interference (RFI).

As a result, the overall losses per VSC station now stand at around 1% per end; still higher than for LCC, but close enough for the two technologies to be able to compete in some applications. The evaluation of losses is required to determine the component and equipment ratings and increase operation efficiency of the transmission system through optimization design for devices that produce large losses. The intention of this chapteris to present an analytical method for calculating the power losses of an MMC-based VSC HVDC converter.

Analysis of Losses of Converter

The MMC converter valves are composed of high power IGBT and anti-parallel FWDs in the same semiconductor package to ensure current capability in the opposite direction and to prevent the application of reverse voltage. Furthermore, the valves must include components that realize functions like voltage grading, monitor, control and heat sink, and so on. So, a complete IGBT valve unit consists of an IGBT, a gate driver, a voltage grading circuit, a monitoring and protecting circuit, and a current detecting circuit as shown in Figures 8.28 and 8.29.

The losses of IGBT SMs are categorized into several groups as follows:

- P_{V1} **IGBT conduction losses**
- P_{V2} **Diode conduction losses**
- P_{V3} **Other valve conduction losses**
- P_{V4} **DC-voltage dependent losses**
- P_{V5} **DC capacitor losses**
- P_{V6} **IGBT-switching losses**
- P_{V7} **Diode turn-off losses**
- P_{V8} **Snubber losses**
- P_{V9} **Valve electronics power consumption**

The dominant components of the valve losses are the semiconductor device conduction losses, P_{V1} and P_{V2}, and switching losses, P_{V6} and P_{V7}. Snubber losses (P_{V8}) can potentially be quite large in valves that use snubbers, but as most current VSC valves using IGBTs do not employ snubbers, this component can often be neglected. Before the loss calculation, the following assumptions have to be adopted:

- Neglect the switching time and dead time
- The semiconductor junction temperature keeps constant
- The converter operates in linear modulation
- Neglect the harmonic in the current of network
- The modulation frequency is much larger than the frequency of AC system

The relation between the collector–emitter voltage of IGBT (the forward voltage dropping of FWD) and the device current can be obtained using a linear equation based on the assumptions earlier.

IGBT conduction losses (P_{V1} and P_{V2}—IGBT and Diode Conduction Losses): One of the most important components of the VSC valve losses is the conduction loss in the FWDs and IGBTs. In the conducting state shown in Figure 8.34, the semiconductor device has an on-state voltage that varies nonlinearly with current and the on-state voltage, and is typically about 3 V at rated current.

The instantaneous power dissipation in the device is simply obtained by multiplying the on-state voltage by the current. This can then be averaged over a cycle to find the expected conduction losses in each device. It is possible to obtain a relatively accurate curve fit to the on-state voltage as a function of current, and thus to approximate very accurately the characteristics shown on the left side of Figure 8.34. In order to make the process of determining losses more transparent, an approximation is used whereby the on-state voltage is modeled as a piecewise-linear characteristic as shown on the right side of Figure 8.34.

$$\textbf{IGBT: } V_{ce(sat)}\,(I_T) = V_{0T} + R_{0T} \cdot I_T \tag{8.67}$$

$$\textbf{Diode: } V_F\,(I_D) = V_{0D} + R_{0D} \cdot I_D \tag{8.68}$$

The time-averaged conduction losses can then be expressed in terms of the mean and RMS currents in each device:

$$\textbf{IGBT: } P_{V1} = V_{0T} \cdot I_{Tav} + R_{0T} \cdot I^2_{Trms} \tag{8.69}$$

$$\textbf{Diode: } P_{V2} = V_{0D} \cdot I_{Dav} + R_{0D} \cdot I^2_{Drms} \tag{8.70}$$

The earlier approximation allows the conduction losses to be determined simply from a knowledge of $\boldsymbol{V_0}$ and $\boldsymbol{R_0}$ for each type of IGBT device, and from a knowledge of the mean and RMS currents in each device.

The conduction losses, which are generated because of volt drop across the IGBT or diode during conduction, are simply a product of the on-state volt drop and the current in the IGBT device as is given in Equation (8.71):

$$P_{cond} = \frac{1}{2\pi} \cdot \int_0^{2\pi} I_T(\omega t) \cdot V_{CE(sat)}(I_T) \cdot d(\omega t) \tag{8.71}$$

where $I_T(\omega t)$ represents the current in the device and $V_{CE(sat)}(I_T)$ the on-state volt drop across the device for a given current.

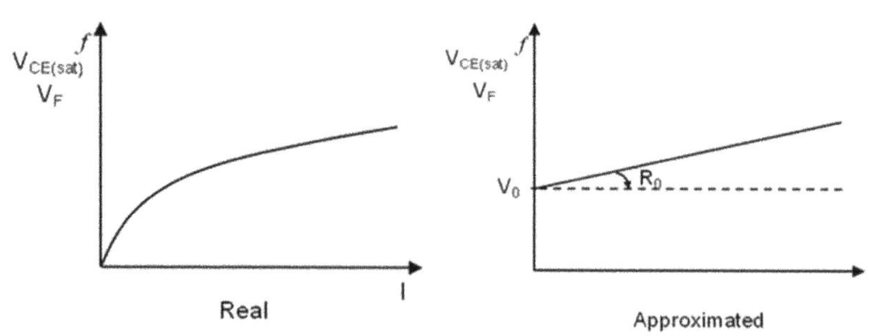

Figure 8.34 Typical IGBT and diode on-state characteristics: (Real (left) and piecewise-linear approximation (right))

Equation (8.72) is simplified using a linear approximation for $V_{CE(sat)}(I_T)$, using a threshold value and a slope angle to model the relationship. Using this simplified relationship Equation (8.72) can be rewritten as follows:

$$P_{\text{cond}} = V_0 I_{\text{av}} + R_0 I_{\text{rms}}^2 \tag{8.72}$$

where V_0 is the threshold voltage for the approximation of the relationship shown in Figure 8.34, and R_0 is the slope resistance shown in the same figure. Once V_0 and R_0 have been determined, the average and RMS current in the device are calculated. The average current is given as follows:

$$I_{\text{RMS}} = \sqrt{\frac{1}{2\pi}\int_0^{2\pi} I(\omega t)^2 \cdot d\,(\omega t)} \tag{8.73}$$

The RMS current is given by

$$I_{\text{RMS}} = \sqrt{\frac{1}{2\pi}\int_0^{2\pi} I(\omega t)^2 \cdot d(\omega t)} \tag{8.74}$$

To assess these values and obtain results, the conduction periods of each IGBT and diode in the SM need to be considered. Using the earlier equations and Figure 8.4, and considering the conduction periods, equations can be derived to describe the losses for each device. IGBT1 and D^1 only conduct when the link is in the output state, and IGBT2 and D^2 only conduct when the link is bypassed. Also, using conduction criteria, the appropriate limits can be selected in order to evaluate the integrals for the periods in which the components are conducting during a cycle.

Each converter arm contains series-connected capacitors, which can be either inserted or bypassed from the current path. Thus, the total arm voltage is synthesized by the summation of the inserted capacitor voltages. The low-level control continuously generates firing pulses for each individual SM based on the required voltage level at each time-step. In summary, the high-level control generates the reference signal for each converter arm, whereas the low-level control decides how many SMs to use and which ones should be inserted.

There are a variety of combinations that can result in the required arm voltage level. Assuming identical capacitor voltage level for all SMs, the SM selection scheme can be perceived as a scheme to select some items out of a collection. The binomial coefficient gives the number of available combinations for each required voltage level. The binomial coefficient for the selection of n_{ref} out of N SMs is calculated as

$$n_{\text{ref}} = \frac{N!}{n_{\text{ref}}!\,(N - n_{\text{ref}})\,!} \tag{8.75}$$

where N is the total number of SMs per arm and the insertion index n_{ref} is a reference signal that equals the normalized required voltage level of each converter arm. The main objective of an SM selection scheme is to insert the proper SMs at proper time instants.

As stated earlier, the behavior of a given SM is hard to predict because of the way the capacitor-balancing algorithm works, so it is not possible to say with certainty during which periods of a cycle an SM will be in the output or bypassed states. However, with power loss calculations, we are only interested in the average behavior rather than the instantaneous behavior. Since the number of SMs in a valve is very large, it is possible to use a statistical approach to calculate the average and RMS currents in each IGBT. This approach is based on determining the probability that a particular SM will be in the output state, as a function of time. $Pc(\omega t)$ is a term representing the probability that an SM is conducting, and therefore generating, voltage. At a given time in the cycle, each SM has an associated probability that it is conducting. In effect, $\text{Pc}(\omega t)$ is directly proportional to valve voltage. Figure 8.35 shows the conduction period, the voltage, and the current in the valve during rectifier and inverter operation. The point at which the current first passes the zero crossing can be calculated as

$$\theta = \cos^{-1}\left(-\frac{\lambda}{2}\right) \tag{8.76}$$

The MI, M, is a quantity that describes the factor by which the peak AC voltage and DC voltage differ. This can be seen in the following equation:

$$\lambda = \frac{2}{V_{DC}} \cdot \left(V_{AC} \cdot \frac{\sqrt{2}}{\sqrt{3}}\right) \tag{8.77}$$

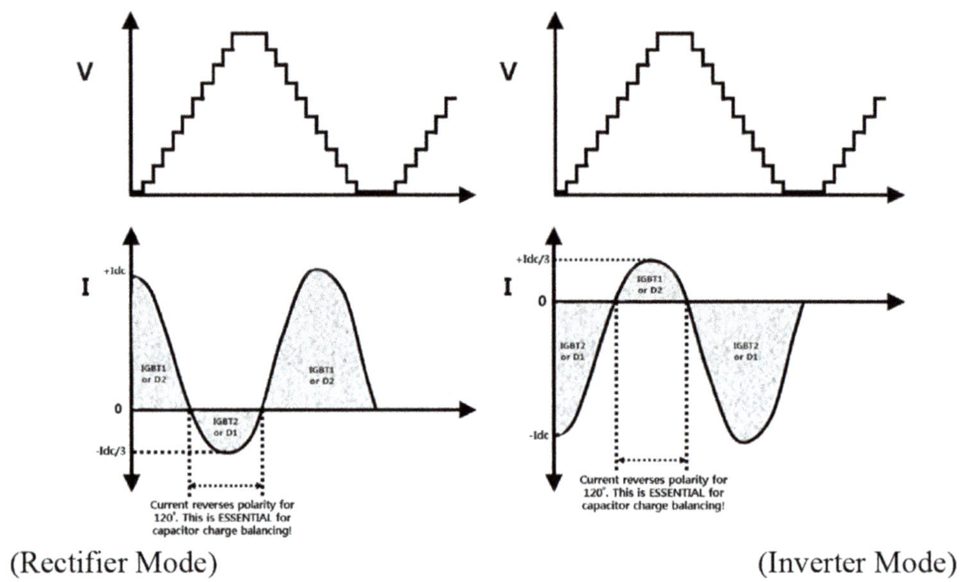

Figure 8.35 Submodule conduction operation

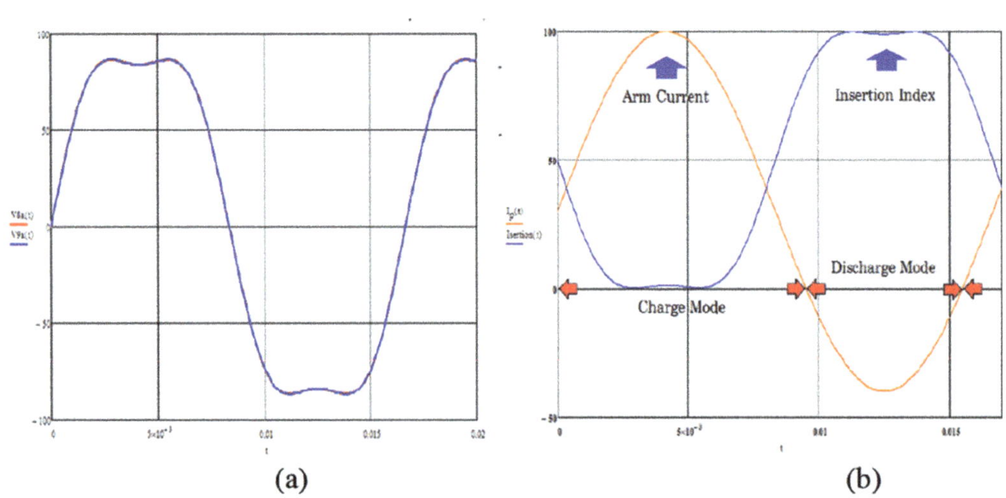

Figure 8.36 Conduction condition of IGBT and FWD

Therefore, the conduction loss is given as follows:

$$P_{con}(\omega t) = \frac{1}{2}(1 - \lambda \cdot \cos(\omega t)) \quad (8.78)$$

Equation (8.79) uses the fundamental voltage. This allows for the inclusion of third harmonic injection. Assuming unity power factor, with the current 180 degree out of phase with the voltage, the current becomes

$$I(\omega t) = \frac{1}{3}\left(1 + \frac{2}{\lambda}\cos(\omega t)\right) \quad (8.79)$$

For the upper components IGBT1 and D^2, the general equation for the average and RMS current is as follows:

$$I_{av} = \frac{1}{2\pi}\int_{\omega t1}^{\omega t2} I(\omega t)\, P_c(\omega t) \cdot d(\omega t) \quad (8.80)$$

For the lower components IGBT2 and D^1, to reflect the reversed voltage across the device, the equation becomes

$$I_{av} = \frac{1}{2\pi}\int_{\omega t1}^{\omega t2} I(\omega t)(1 - P_c(\omega t)) \cdot d\omega t \quad (8.81)$$

$$I_{RMS} = \sqrt{\frac{1}{2\pi}\int_{\omega t1}^{\omega t2} (I(\omega t))^2 (1 - P_c(\omega t)) \cdot d\omega t} \quad (8.82)$$

The next step is to use Equation (8.80) for I(ωt) and Equation (8.79) for Pc(ωt), and to evaluate the integrals for the following conduction periods. Using the earlier equations, the losses because of conduction can be calculated. Therefore, assuming the same IGBT and diodes are used in both positions in the SM, the per-SM losses for P_{V1} and P_{V2} can be summarized as follows:

Rectifier Mode

$$\begin{aligned} P_{V1} + P_{V2} = \tfrac{I_{DC}}{6\pi}&\left[\left(2\theta - 2\pi + \tfrac{4}{\lambda}\sqrt{1 - \tfrac{\lambda^2}{4}}\right).V_{0D} + \left(2\theta + \tfrac{4}{\lambda}\sqrt{1 - \tfrac{\lambda^2}{4}}\right).V_{0T}\right] + \\ \tfrac{I_{DC}^2}{18\pi}&\left[\left[\left(3 + \tfrac{3}{\lambda^2}\right).\pi - 4\theta - \left(\tfrac{16}{3\lambda} + \tfrac{2\lambda}{3}\right)\sqrt{1 - \tfrac{\lambda^2}{4}}\right].R_{0D} + \right. \\ &\left.\left[\left(\tfrac{2}{\lambda^2} - 1\right).\pi + 4\theta + \left(\tfrac{16}{3\lambda} + \tfrac{2\lambda}{3}\right)\sqrt{1 - \tfrac{\lambda^2}{4}}\right].R_{0T}\right] \end{aligned} \quad (8.83)$$

Inverter Mode

$$\begin{aligned} P_{V1} + P_{V2} = \tfrac{I_{DC}}{6\pi}&\left[\left(2\theta - 2\pi + \tfrac{4}{\lambda}\sqrt{1 - \tfrac{\lambda^2}{4}}\right).V_{0D} + \left(2\theta + \tfrac{4}{\lambda}\sqrt{1 - \tfrac{\lambda^2}{4}}\right).V_{0T}\right] + \\ \tfrac{I_{DC}^2}{18\pi}&\left[\left[\left(3 + \tfrac{3}{\lambda^2}\right).\pi - 4\theta - \left(\tfrac{16}{3\lambda} + \tfrac{2\lambda}{3}\right)\sqrt{1 - \tfrac{\lambda^2}{4}}\right].R_{0D} + \right. \\ &\left.\left[\left(\tfrac{2}{\lambda^2} - 1\right).\pi + 4\theta + \left(\tfrac{16}{3\lambda} + \tfrac{2\lambda}{3}\right)\sqrt{1 - \tfrac{\lambda^2}{4}}\right].R_{0T}\right] \end{aligned} \quad (8.84)$$

P_{V3}—Other Valve Conduction Losses: Other valve conduction losses are because of the resistance of conducting components in the valve other than the IGBTs and diodes, such as bus bars within the SM and

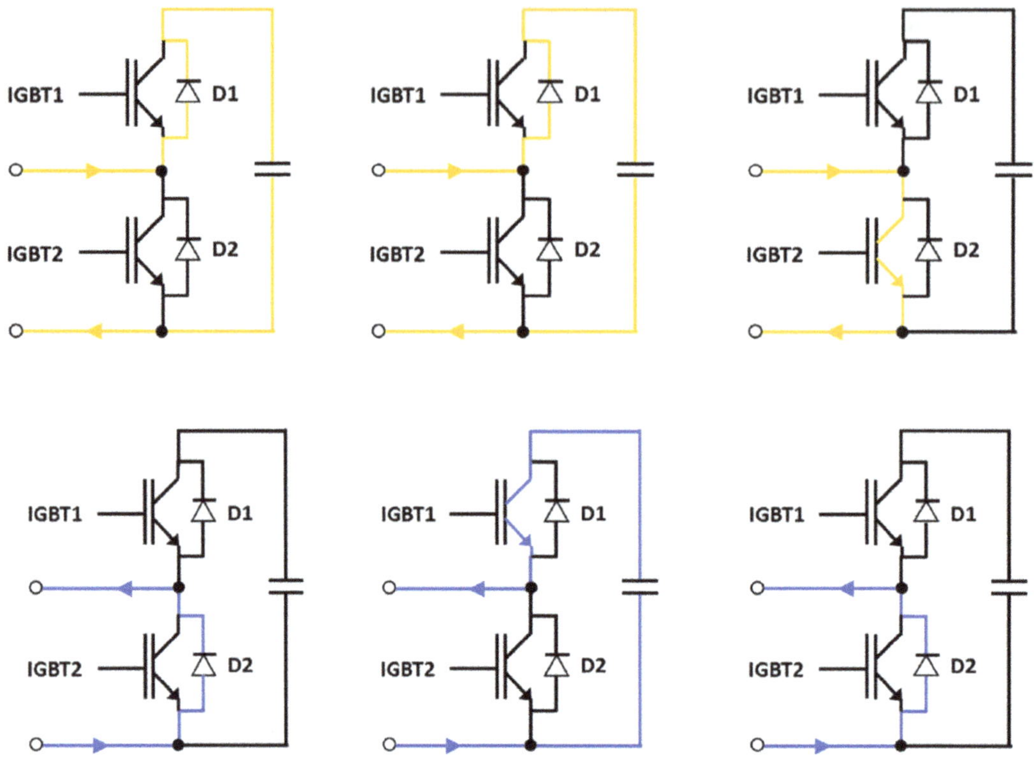

Figure 8.37 Conduction condition of IGBT and FWD

between SMs and tiers of the valve. To evaluate the losses from bus bar conduction within the valve, the arrangement of the current path needs to be understood. Figure 8.4 represents the current path through one valve. Each SM is connected to its neighbors using a copper-linking piece. This provides the value for Rs.

$$P_{V3} = I_{VRMS}^2 \cdot R_S \tag{8.85}$$

$\boldsymbol{P_{V4}}$**—DC-Voltage-Dependent Losses:** DC-voltage-dependent losses are because of shunt-resistive components in the valve. In the MMC converter, these are dominated by the "bleed" resistors connected across each SM DC capacitor. The loss through the SM bleed resistors is considered as this voltage. In the case of this specific SM design, with equal R_1 and R_2, their total resistance is simply R_1 / 2. Thus, P_{V4} is given by

$$P_{V4} = \frac{2}{R_1} \cdot \left(\frac{2 \cdot V_{DC}}{N_1} \right)^2 \tag{8.86}$$

$\boldsymbol{P_{V5}}$**—Losses in SM Capacitors:** Each SM capacitor carries a component of current and, as a consequence, contributes toward losses. These come from both dielectric and ohmic losses in the capacitor. These have been treated by considering the equivalent series resistance R_{ESR} of the capacitor:

$$P_{V5} = N_t \cdot I_{cRMS}^2 \cdot R_{ESR} \tag{8.87}$$

where N_t is the number of levels, I_{cRMS} is the RMS current flowing in the capacitor, and R_{ESR} is the effective series resistance of the capacitor. I_{cRMS} can be found by summing vectorially the RMS currents in IGBT1 and D^1, which upon simplification yields the following:

$$I_{cRMS} := \frac{Idc}{3 \cdot \sqrt{2}} \cdot \sqrt{\frac{2}{\lambda^2} - 1} \tag{8.88}$$

P_{V6} and P_{V7}—Switching Losses (Semiconductor switching losses): Every time an IGBT turns on or off, a small amount of energy is lost in the device, referred to as the switching energy. In addition, the turn-on of an IGBT is always accompanied by the turnoff of a diode elsewhere in the circuit. Diode turnoff also results in energy being lost. The IGBT turn-on and turn-off energies and the diode "recovery" energy are abbreviated as E_{on}, E_{off}, and E_{rec}, respectively. They depend on several parameters of the converter, but most importantly on the current in the device. The relationships between E_{on}, E_{off}, E_{rec}, and current may be conveniently represented as a linear relationship with current (IGBT switching energy) and a piece-wise linear representation (diode recovery energy):

- IGBT turn-on: $E_{on} = k_{on} \cdot I_T$
- IGBT turnoff: $E_{off} = k_{off} \cdot I_T$
- Diode turnoff(rec): $E_{rec} = k_{rec1} + k_{rec2} \cdot I_D$

If the waveform of the current in the valve can be represented mathematically, and the switching instants occur regularly in predictable instants (as is the case for a two-level converter), it is quite simple to calculate the total switching losses in the semiconductors by simply summing the switching energies in each cycle and multiplying the result by the switching frequency. The MMC topology, however, is more complex and a direct analytical approach is difficult.

Along with conduction losses, the switching losses in IGBTs and diodes account for one of the major sources of loss within a VSC valve. During switching, an IGBT device is subjected to both high current and high voltage simultaneously, and as a result, the IGBT incurs a high-power dissipation for a short time. The turn-on and turn-off switching losses depend on the value of the collector current during the on-state and the collector–emitter voltage during the off-state:

$$P_{V6} = N_t \cdot f \cdot \sum (E_{on}(V, I) + E_{off}(V, I)) \tag{8.89}$$

For diodes, the turn-on loss is negligible. The losses during turnoff are because of "Reverse Recovered Charge," Q_{rr}, which passes through the diode after turnoff, in a similar way to a thyristor. The product of Q_{rr} and SM DC capacitor voltage gives rise to a turn-off loss E_{rec}:

$$P_{V7} = N_t \cdot f \cdot \sum_{cycle} E_{rec}(V, I) \tag{8.90}$$

As with conduction losses, these relationships for E_{on}, E_{off}, and E_{rec} are approximated as linear relationships varying with current. E_{on} and E_{off} are given as follows:

$$\begin{aligned} E_{on} &= k_{on} \cdot I \\ E_{off} &= k_{off} \cdot I \end{aligned} \tag{8.91}$$

For the diode this is modeled as a straight line plus intercept:

$$E_{rec} = k_{rec1} + k_{rec2} \cdot I \tag{8.92}$$

Therefore, if a switching event takes place at a point ωt, then the losses incurred are as follows:

$$\text{IGBT } E_{swT}(\omega t) = (k_{on} + k_{off}) \cdot I(\omega t) \tag{8.93}$$

$$\text{Diode } E_{swD}(\omega t) = k_{rec1} + k_{rec2} \cdot I(\omega t) \tag{8.94}$$

Equations (8.93) and (8.94) can be considered as the "instantaneous switching loss" of the device. Although the instantaneous switching frequency will vary from cycle to cycle, the average switching frequency has been shown by MATLAB simulation to be approximately 3 times the fundamental frequency, that is, n_s=3. For the purposes of the following analysis, switching events are assumed to be equally distributed throughout the power frequency cycle. The quantity *fs* is defined as the average number of times that an SM will switch per radian. Hence:

$$f_s = \frac{n_s}{2 \cdot \pi} \tag{8.95}$$

Using this concept, we can calculate the average switching power of a particular device by multiplying the "instantaneous switching loss" by the switching frequency f_s and integrating over a period during which the device in question is conducting, then multiplying by the system frequency. If we use only the first half cycle, then we can simply multiply by twice the system frequency.

$\boldsymbol{P_{V8}}$**—Snubber Circuit Losses:** The $\mathbf{P_{V8}}$ term represents the losses in the snubber circuit of a generic VSC valve. As this circuit is not present in many companies' circuit topology, this loss is considered to be zero.

$\boldsymbol{P_{V9}}$**—Valve Electronic Losses:** To calculate the total valve electronic losses, the energy consumption of each SM is simply multiplied by the number of SMs:

$$P_{V6} = N_t \cdot P_{GU} \tag{8.96}$$

8.9 DC Bus Bar Design

Within each valve, the individual SMs are connected using an inter-SM connector. Two of the same connectors provide the connections between modules. Between tiers, the connection is made using hollow cylindrical bus bars. Along the full-tier lengths, these are separated into several sections in length and connected via a connector that permits expansion of the bus bars.

In hard-switched converters, the commutation loop stray inductance is responsible for DC bus overshoots and overvoltage spikes across power semiconductors during switching. As a consequence, semiconductor device losses increase. To limit losses, the allowable switching frequency has to be decreased. Furthermore, significant parasitic high-frequency oscillations appear whenever the loop inductance between DC link capacitors, bus bars, and power devices becomes larger.

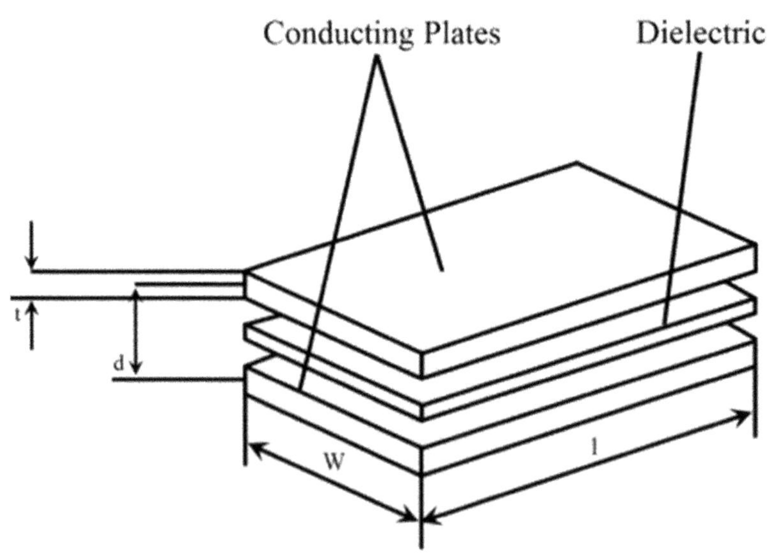

Figure 8.38 Physical representation of the bus bar, showing dimensional quantities

New generations of switching devices further amplify these voltage spikes, and higher *di/dt* transients occur. The high *di/dt* values produced during commutations by these new fast switching devices and diodes are currently the main source of conducted and radiated noise emissions, because the parasitic lay-out inductance can never be eliminated. The simplest method to limit improper high-dynamic overvoltages is the reduction of the distributed inductances throughout the converter and its components. Traditional methods of power distribution inside the power converter (using cable) are no longer compatible with new state-of-the-art packaging and high-speed switching technology. The advantages of using bus bars are well-known:

1. Low stray inductance
2. Low characteristic impedance

Bus Bar Model and Design

In literature, several planar bus bar designs have been presented for different applications. For each bus bar type, a suitable model to aid the designer was developed. The basic bus bar structure used in this study is composed of two parallel conducting plates of resistivity ρ [Ωm] separated by a dielectric material with relative permittivity ε_r [p.u.], relative permeability u_r [p.u.] and dielectric conductivity σ [1/ Ωm] (Figure 8.38).

The mechanical dimensions of the bus bar are length 1[m], width w[m], thickness of the plate t [m], and distance between the two plates d [m], as shown in Figure 8.38. The converter DC link spectral frequency content and the DC link path length (l) are important modeling considerations. The converter spectral content, or more specifically the highest frequency component, determines whether a lumped parameter model or a transmission line model is required to analyze the bus parasitic inductance. The highest critical frequency f_c is determined by the device fall time (t_f), according to

$$f_c = \frac{1}{2\pi t_f}[Hz] \tag{8.97}$$

The corresponding wavelength is defined as

$$\lambda = \frac{c}{f_c}[m] \tag{8.98}$$

Standing waves may occur whenever one-quarter of the wavelength $\lambda/4$ reaches the same dimensions as the bus bar. In this case, transmission line analysis would be required. Modern IGBT devices have typical fall times down to 40 ns, resulting in a maximum frequency of

$$f_c = \frac{1}{2\pi t_f} = \frac{1}{2\pi \cdot 40 \cdot 10^{-9}} = 4\text{MHz}$$

The correspondence wavelength is

$$\lambda = \frac{c}{f_c} = \frac{3 \cdot 10^8}{4 \cdot 10^6} = 75[\text{m}]$$

Thus, one-quarter wavelength $\lambda/4$ is about 19 m. One can conclude that a lumped stray inductance model is adequate for state-of-the-art power converters. The equivalent circuit is shown in Figure 8.39, constituted of four-lumped parameters describing the bus bar inductance L, capacitance C, and resistance R. The dielectric losses are represented by the conductance G. Simple relationships between physical dimensions and electrical parameters can be derived for the planar bus bar structure.

The capacitance of the bus bar arrangements depends on dielectric material and the physical dimensions of the system. Increased capacitance leads to lower characteristic impedance, improving signal suppression and

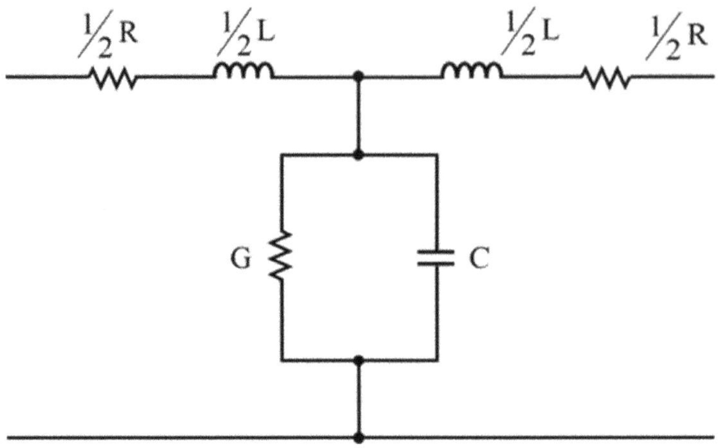

Figure 8.39 Electrical equivalent circuit to model power converter bus bar

noise elimination. It is, therefore, desirable to develop maximum capacitance between the conductors. This may be achieved by keeping the thickness of the dielectric as thin as possible with good manufacturing and design practices, as well as utilizing dielectrics that have a high relative permittivity:

$$C = \varepsilon_0 \varepsilon_r \frac{\Omega \cdot I_{\text{Busbar}}}{t_{\text{dielectric}}} \text{ [F]}$$

The shunt conductance depends on the physical dimensions of the bus bar and the dielectric conductivity of the insulation. The shunt conductance can vary with frequency and environmental changes (temperature, humidity):

$$G = \Sigma \frac{\Omega \cdot I}{t_{\text{dielectric}}} \text{ [S]}$$

At high frequencies, conductor current is not uniformly distributed over the entire cross-sectional area of the bus bar, but rather tends to be concentrated near the surface. Proximity effects between positive and negative bus bar plates force current to flow through the outermost surfaces. This effect influences bus bar stray inductance and resistance.

The most important parameter regarding electromagnetic compatibility (EMC) performance is the calculation of the bus bar inductance. Maintaining a low inductance results in a low characteristic impedance and greater noise attenuation. When designing a bus system for minimum inductance, the following aspects should be considered:

1. Inductance decreases by decreasing dielectric thickness (*d*) and by increasing conductor width (*w*)
2. Inductance decreases at higher frequency; two types of inductance can be considered:
3. An internal inductance (L_i), which is a result of flux linkages within a conductor or bus bar plates.
4. This inductance is frequency dependent because of the skin and proximity effects.
5. An external inductance (L_e), which is determined by the geometry of the two current-carrying conductors.
6. This inductance is frequency independent.

At high frequencies, the current distribution in a conductor tends to concentrate near the surface. The internal inductance is reduced. In many power electronic applications, it is usually sufficient to consider only the external inductance; however, at low frequencies, especially in the case of bus bars, the internal inductance L_i may become an important part of the total inductance. Its value can be calculated as

$$L_i = \frac{\mu_0 \mu_r}{8\pi} l\,[H] \tag{8.101}$$

An exact analytical solution to calculate the external inductance is not readily available in closed form. Formulas with increasing levels of complexity have been created up until now. Under certain conditions, it is possible to approximate the inductance value successfully with a simplified equation. A detailed calculation leads to the next equation for the external inductance (Equation (8.103)):

$$\mathrm{L}_e = \frac{\mu_0 \mu_\mathrm{r}}{\Pi} \mathrm{l} \left(\ln \frac{d}{t+w} + \frac{3}{2} + \Delta_\mathrm{k} + \Delta_e \right) [H]$$

where Δ_k and Δ_e are coefficients depending on the geometric parameters of the bus bar.

If $d \ll 2t$ and $d \ll t + w$, this formula can be simplified leading to

$$L_e = \frac{2\mu_0 \mu_r l}{\pi} ln \left(1 + \frac{t}{t+w} \right) [H] \tag{8.103}$$

If $t \ll w$, the previous equation can be simplified further

$$L_e = 2\mu_0 \mu_r \frac{t \cdot l}{\pi(t+w)} [H] \tag{8.104}$$

Bus Bar Resistance R

The DC resistance of a two conductor bus system is given by

$$R_{\mathrm{dc}} = \rho \frac{2 \cdot l}{w \cdot t}\,[\Omega] \tag{8.105}$$

For high frequency, the current can be assumed to be concentrated in an annulus at the wire surface of thickness equal to the skin depth

$$\delta = \sqrt{\frac{\rho}{\pi f \mu_0}} [m] \tag{8.106}$$

where $\rho[\Omega\ \mathrm{m}]$ is the resistivity of the conducting plates, u_0 $[H/m]$ is the absolute permeability, and f is the frequency.

For high-frequency operation and for $t > 2\delta$, the skin depth must be considered, and the formula for the AC resistance becomes R

$$R_{\mathrm{ac}} = \rho \frac{4 \cdot 1}{\mathrm{w} \cdot \delta}\ [\Omega] \tag{8.107}$$

Bus Bar Characteristic Impedance Z

The impedance of the bus bar must be kept as low as possible to reduce the electromagnetic interference (EMI). Increasing the capacitance and reducing the inductance is the most critical factor in eliminating noise. When the conductor resistance (R) and the insulation conductance (G) are negligible, the characteristic impedance of a two conductor bus bar is given by

$$Z = \sqrt{C}\ [\Omega] \tag{8.108}$$

8.10 MPP Capacitor

Traditional capacitors use oil and are susceptible to leaks. MPP capacitors, which are used in VSC HVDC, are non-oil and soft-cell DC capacitors with high capacity and reliability that maintain a minimal environmental footprint. The MPP capacitor is constructed of small modules, each equipped with self-healing metalized film, therefore, boosting the product's reliability. This film enables the capacitor to progressively reduce its capacity while reducing the risk for short circuiting and complete power outage in case of a failure.

Since MPP contains no oil, it is a safer alternative to traditional capacitors for converters used in hazardous environments such as on oil platforms. Its high-energy density capability allows for compact designs that can provide the same capacity as larger oil-based capacitors, thus saving space. The compact design allows it to be installed on the same support as the IGBT valves, thus reducing inductance and consequential stress on the IGBTs, resulting in more reliable converter operation. Also, this film enables the capacitor to progressively reduce its capacity and reduces the risk for short circuiting and complete power outage in case of a failure. MPP capacitors provide an added level of safety to end-of-life behavior that makes them especially suitable for a wide range of applications such as HVDC and SVC converters and motor drives where high reliability is required.

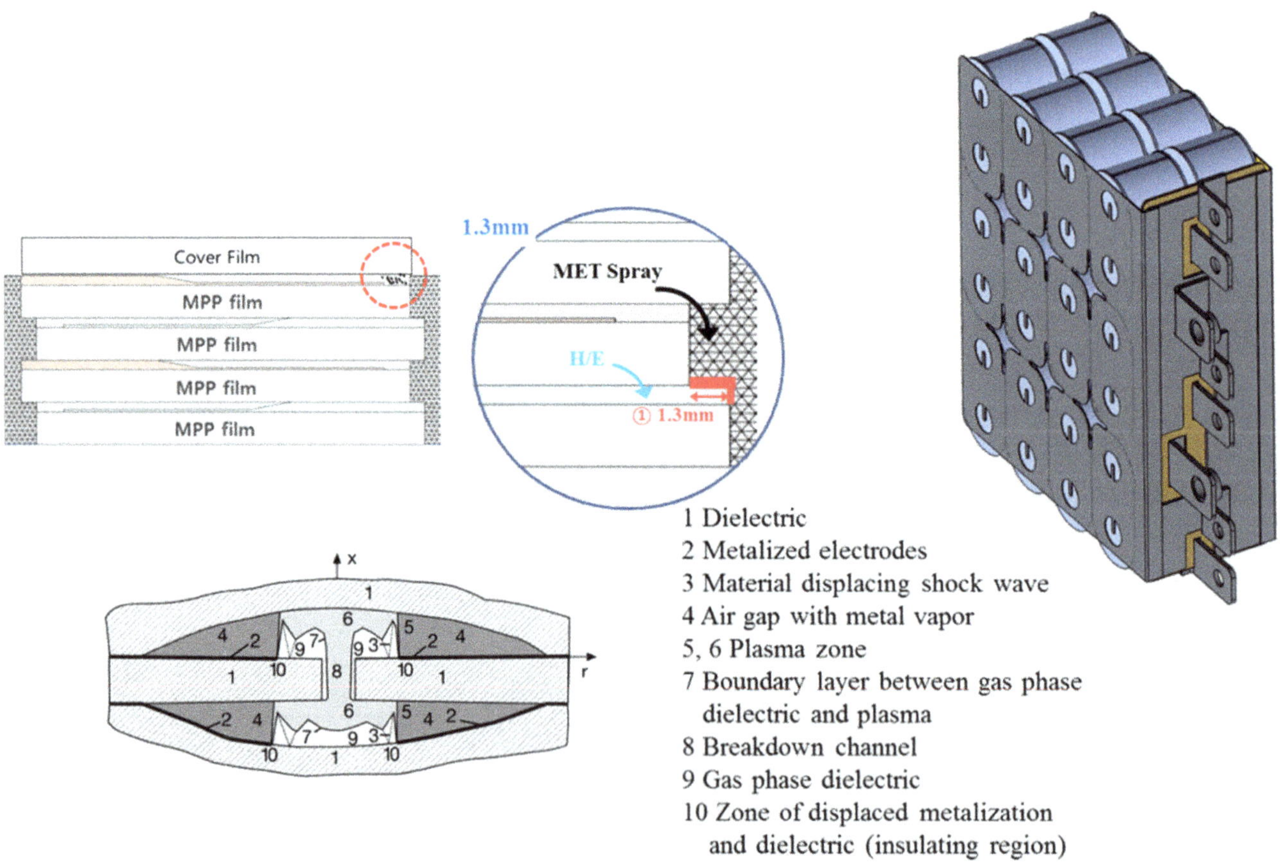

Figure 8.40 Metalized film capacitor

Capacitor Technologies

The following different power capacitor technologies are used in inverters:

- Electrolytic capacitors characterized by very big capacitance per volume unit, but with low-rated voltages and very important power losses because of ionic conductivity. In particular, the bigger the capacitance density, the lower the rated voltage.
- Film foil capacitors made of dielectric films between two plain aluminum foils. These capacitors can sustain very high currents.
- Metalized film capacitors that are made with dielectric films with a metallic coating on the surface. With this technology, the electric field stress may be much bigger than with film capacitors thanks to the metalization self-healing capability.

Today, the dielectric films that are used are mainly polypropylene (PP) or polyethylene terephthalate (PET). Paper (PA) was formerly used in film foil technology—either pure paper or mixed with PP (DM). In special applications, where high temperatures are required, polyethylene naphthalene (PEN) up to 125°C or polyphenylene sulfide (PPS) up to 150°C are used.

PET presents the following advantages over PP: a dielectric constant that is 50% bigger ($\varepsilon = 3.3$ vs. 2.2), which means 50% more capacitance in the same volume; a better mechanical resistance (which means a higher endurance to self-healing); and the possibility of manipulating thinner films, consequently leading to a smaller capacitance and a higher exploitation temperature (+10°C). The negative point is that the loss factor is 10 times larger, which means a 10-fold increase in temperature elevation for the same rated power. The nominal electrical field is about the same.

The capacitive elements must be dried to remove moisture, which would cause accelerated aging and bigger losses if left in the capacitor. In the case of power capacitors, the dried elements are either impregnated with vegetable oil or with gas (SF_6, N_2, etc.).

The dielectric films are either wound or stacked before being inserted in a plastic or metallic container. The best winding machines are required to produce active wound elements of reliable quality in the case of oil-free capacitors. One way of overcoming the difficulty of controlling the space ratio between the gas and the film in the winding curves is to wind the film on a large-diameter wheel and to cut the film layers to obtain a stack.

The plastic containers are not completely moisture tight—there is always some residual permeability in polymers. In the case of metalized films, this may lead to electrode corrosion when the capacitors are submitted to environmental conditions of high humidity.

The electric field stress in metalized film capacitors may be much larger than in film foil capacitors. This is obtained thanks to the ability of the electrodes to self-heal. If a breakdown occurs in the polymer, the current will increase through the defect and on the electrode near the defect. Closer to the defect, the current density will be big enough to evaporate the 100-nm metallic layer. If the capacitor is well designed, the phenomenon will stop when the diameter is large enough to insulate the defect and small enough not to damage the film. The electrode resistance (given in ohm/square) is the key parameter to define to achieve good self-healing behavior, with Joule losses as small as possible. A thick metalized layer will present a lower resistance, but higher energies will be involved during the self-healing process, leading to greater damage.

The typology of PP offers the capability of increasing the maximum voltage stress and maximum working temperature. The potential of this film is to reach an energy density of 2–2.5 J/cm in the future. The good reliability degree of PP film is less dependent on raw material and on the production lot than paper, which is generally used for these applications.

The parameters that should be taken into consideration for these kinds of applications are energy density, the number of shots with the required reversal voltage, peak current, ESR, dimensions, reliability, and cost.

Equivalent Series Resistance

The ESR can be defined as follows:

$$\mathrm{ESR} = \frac{\frac{1}{2} \cdot CV^2 - \int i \cdot v \cdot dt}{\int i^2 \cdot dt} \tag{8.109}$$

where C is the measured capacitance, V is the charge voltage on the capacitor before discharge, i is the current out of the capacitor during discharge, and v is the voltage across the capacitor during discharge.

A lower ESR value means better efficiency during the energy transfer from the capacitor to the load. The ESR is composed of two factors: the first one related to the dielectric losses and the second one because of the resistance of the electrodes.

The current is distributed linearly in the electrodes: it is at maximum value near the sprayed ends and at minimum near the free margin. Considering the metal thickness expressed as ohm/sq to be constant, it is possible to express the dissipated power W in the dielectric film in the following way:

$$W = \int_0^h dW = \int_0^h i^2 \cdot dR = \int_0^h \left(\frac{I}{h} \cdot x\right)^2 \cdot \frac{\mathrm{ohm/sq}}{L_\omega \cdot h} \cdot dx \tag{8.100}$$

where i is the capacitor current, L_ω is the winding length, h is the film height, and ohm/sq is the electrode resistivity.

Equation (8.111) allows the calculation of the equivalent electrode resistance:

$$R_S = \frac{W}{I^2} = \frac{\mathrm{ohm/sq}}{L_\omega} \cdot \frac{h}{3} \tag{8.111}$$

If the capacitor discharge is oscillatory, the dielectric factor can be calculated by way of the dissipation factor (DF) (10^{-4} for PP film). In this situation, the ESR can be expressed as follows:

$$\mathrm{ESR} = R_S + \frac{\mathrm{DF}}{2 \cdot \pi \cdot F \cdot C} \tag{8.112}$$

where F is the frequency of the oscillatory discharge.

MPP Capacitor: Self-Healing Characteristics

In metalized film capacitors, the electrodes are composed of very thin layers of metal evaporated under a vacuum onto the surface of the polymer film. Capacitors based on this technology have the ability to self-heal, following the application of voltage.

Defects, such as pinholes and microflaws in the polymer structure, can lead to a localized breakdown of the film when the AC or DC voltage is increased. The breakdown results in a discharge of a portion of the stored charge, with consequent localized high temperature. At the same time, the film is punctured and the thin metal layer at the defect site is rapidly vaporized. This phenomenon is called self-healing or clearing. It is well known that the polymer chemistry is important in determining the suitability of a self-healing metalized capacitor system. Polymer structures rich in oxygen clear better than those without oxygen. In materials with oxygen deficiency, the graphite formed by the decomposition of polymer during the breakdown arc accumulates at the clearing site. This produces conductive shorts, causing a catastrophic failure. The orientation change occurring within the polymer as a result of the applied stress is another important factor in the breakdown event. Polymers with highly polar molecules have great orientation even at a low-voltage level. The greater the orientation of the molecules, the greater the graphite residues that accumulate at the weak location during the clearings, resulting in a major failure.

As for PP, no polarization occurs because the molecule clears immediately as soon as the breakdown voltage of the weak spot is reached. Relatively little graphite accumulation is produced, resulting in less failure.

During the clearing of a weak spot, there is a loss of energy in the capacitor. This energy depends on the applied voltage, the thickness of the electrodes, the mechanical pressure on the individual capacitor layers, and the capacitance. The energy of a clearing in a capacitor can be expressed as follows:

$$E_{\text{discharge}} = \frac{k \cdot V^{4.7} \cdot C}{(\text{ohm/sq})^{1.8} \cdot \alpha(P)} \tag{8.113}$$

where $\alpha(P)$ is a function that relates the interlayer pressure to the clearing energy. A discharge energy reduction corresponds to a lower failure probability because the probability of a thermal runaway failure mode decreases. With a good choice of V, C, ohm/sq, and interlayer pressure, it is ideally possible to manufacture capacitors with very low discharge energy, having a limited decrease of capacitance during the capacitor's life.

Effect Because of the Applied Voltage: The loss of energy per discharge in the capacitor generally has a fifth-power dependence on the applied voltage as shown in Equation (8.114). To reduce the discharge energy to a minimum value, it is best to use a low thickness film (maintaining the same electric field stress) and increase the number of series elements; in this way, every capacitor will work at a lower-rated voltage. This solution is expensive and not always applicable.

Effect Because of the Capacitance: At fixed voltage, the discharge energy depends linearly on capacitance. Current capacitor technology uses metalized polymer film, mainly PP, containing a large number of individual capacitor elements in a specific segmented pattern, with small fuse links connecting neighboring elements together. This system permits the capacitor to improve its self-healing properties to isolate a defective area from its neighbors, thus clearing the faults. In this way, the capacitor can tolerate many clearings, each one with very low discharge energy, without becoming short circuited, but with more capacitance variation versus time.

Effects Because of Electrode Thickness: The dependence between dielectric strength and metal thickness is expressed as ohm/sq, as shown in Figure 8.1. If the metalization is made as thin as possible (from 5 to 200 ohm/sq as opposed to the 1–10 ohm/sq of the usual industry metalization thickness), it is possible to increase the film's dielectric strength and consequently the voltage breakdown. It is evident that reducing the electrode thickness results in an increase in the energy density of the capacitor.

A disadvantage connected to the use of film having high electrode resistivity is the increase of ESR. This problem can be solved using a metalization thickness variable along the film height. The equivalent electrode resistance for ohm/sq variable is expressed by

$$R_S = \frac{1}{L_\omega \cdot h^2} \cdot \int_0^h \text{ohm/sq}(x) \cdot x^2 \cdot dx \tag{8.114}$$

It is possible to select the function ohm/sq(x) in order to get low losses without reducing the voltage stress. The self-healing discharges during the "clearing" process produce a progressive loss of capacitance. The reduction of the conducting area around the discharge can produce an insulation of the area itself. This phenomenon is very similar to that which occurs in segmented film (self-segmentation).

Effects Because of Pressure: An important parameter to consider in self-healing is round capacitor elements, the mechanical pressure on the individual capacitor layers. It is difficult to maintain a fixed pressure: all elastic films add compressive radial forces to every wound wrap, and these forces add to the total compression on the film layers and core below. The clearing energy is dependent upon applied total mechanical pressure; the clearing energy in the outer turns is higher than that in the interior. In addition, the duration of the clearing is much shorter at high-pressure valves.

It is possible to obtain clearings at high-voltage levels, when the energy becomes larger, maintaining a high mechanical pressure. The effect is a lower reduction of capacitance during the capacitor's lifetime. At the same time, a higher pressure on the film layer increases the risk of a short circuit, since the discharge energy is completely absorbed by the dielectric, causing thermal damage in successive layers.

Flat elements have a lower risk of short circuiting (the self-healing capability results improved), but present a large decrease of capacitance during the capacitor's lifetime.

Capacitor Failure Modes Analysis

Most of the metalized film capacitors fail because the capacitance drops below the required tolerance. This normally occurs after the end of the expected lifetime given by the manufacturer. The capacitance drop is generally accompanied by an increase of the loss factor.

From a general point of view, the causes of capacitor failures may occur because of bad design, bad processes, or inappropriate application conditions. During the design phase, the following causes may lead to failure: the dielectric film is too thin, insulation distances are too small, the metalization layer is too thick or too thin, or the conductor is the wrong size.

During production, causes may include the following: poor mechanical tension control during the winding, bad drying (leaving too high a humidity content in the capacitor), or bad sealing. In application, the causes may be higher voltages, EMI, lightning, higher temperature, or a high-humidity environment.

The failure modes are a little more complicated to describe because different causes may lead to the same modes. Figure 8.1 gives a non-exhaustive summary of the possible failure modes that can occur in metalized film capacitors.

For example, bad space factor control of the dielectric films during the winding operation will cause electrode corona demetalization, which will lead to a fast capacitance drop and a loss of functionality of the capacitor.

A bad choice of the metalization resistance value or poor metalization control during the film-manufacturing process can lead to bad self-healing management, which may damage the dielectric film mechanically and produce heat that is transmitted locally to the next film layers. At this location, the dielectric strength of the film drops, and breakdown may occur. Consequently, chimneys of melted PP may appear through the winding. The formed channel is conductive, inducing a drop in insulation resistance and a leakage current that can generate enough heat to melt the PP and increase the internal pressure of the capacitor. Along with bad metalization resistance, the final consequence can, in the worst case, lead to fire ignition or even a capacitor explosion.

One of the main failure modes is often because of high currents, which increase the capacitor temperature and lead to a reduction of the breakdown voltage and, in the worse cases, even the melting of the capacitor. In this regard, the shape of the capacitor is very important. For high-power applications, it is important to build short elements in order to reduce the current path length and increase the number of parallel layers, and consequently reduce the heating. The current capability of a capacitor is specified through the series resistance R_s and the loss factor $tan\delta_s$ at different frequencies. The relation between the two factors in the high-frequency domain, where the effect of the insulation resistance is negligible, is given by the linear relation

$$\tan\delta_{\mathrm{s}} = \frac{Z_{\mathrm{R}}}{Z_{\mathrm{i}}} = R_{\mathrm{s}}\omega C \tag{8.115}$$

where C is the capacitance, $\omega = 2\pi f$ is the frequency, Z_R is the impedance real part, and Z_i the impedance imaginary part.

The presence of humidity in the capacitor, because of poor drying during the manufacturing process , because the moisture permeability of the material was too high, or because the humidity level where the capacitors are installed was too high, may lead to three failure modes with different effects and consequences.

The first is electrode corrosion, where the series resistance will slowly increase over time. The effect is a loss factor increase, because of the electrode thickness reduction, and a heat dissipation increase. The elevation of the temperature will accelerate the capacitance loss because of the reduction of dielectric strength with temperature, ending with loss of functionality of the capacitor.

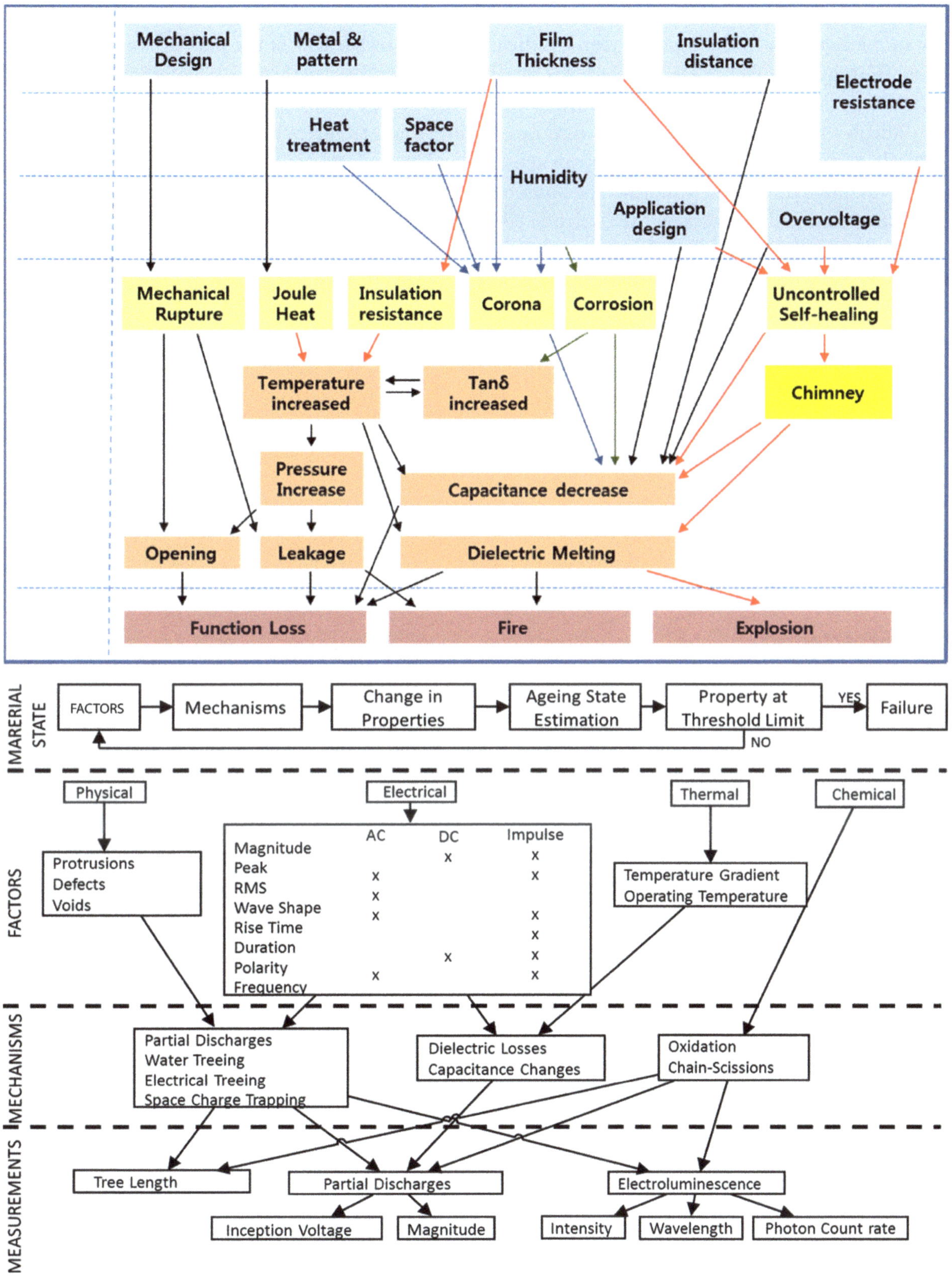

Figure 8.41 Metalized film capacitor failure modes with their causes, effects, and consequences

The second, today known as "corona," is either because of a decrease of the dielectric strength of the gas present in the capacitor in the gaps between the dielectric films or because of the poor space factor control of the films. The bigger the gap, the more severe the problem. The thickness of the gap is characterized by the space factor, which is the ratio of dielectric thickness to the total distance between the electrodes. This space factor is very difficult to control in curves of flat windings, leading manufacturers to build either round windings or stacks. Only performant winding machines can achieve good space factor control by managing the mechanical tension of the films during the winding. The consequence of this is a fast capacitance decrease because of the appearance of corona discharges on the electrode edges, that is, the locations where the electrical field is more intense because of the point effect. In the case of segmented metalization, the corona failure mode may also propagate from the non-metallic lines that separate the active electrode metallic areas.

The third failure mode is a reduction of the insulation resistance, which is the parallel resistance of the capacitor. A decrease in insulation resistance leads to an increase in current leakage from one electrode to the other. This phenomenon is present at lower frequencies. It may be measured via either the loss factor (tan δ) or the DC resistance R_p. The relation between the two parameters is given by the following relation (only true at very low frequencies):

$$tan\delta_{\mathrm{p}} = \frac{Z_{\mathrm{R}}}{Z_{\mathrm{i}}} = \frac{1}{R_{\mathrm{p}}\omega C} \quad (8.116)$$

This later failure mode may show runaway behavior. The more the insulation resistance decreases, the more heat is produced, and the more the temperature increases, which leads to a new insulation decrease. This phenomenon may end with the appearance of chimneys and the melting of the dielectric.

Lifetime Expectancy

The lifetime of a capacitor is the time to failure, where failure is defined as the lack of the ability of a component to fulfill its specified function. The failure modes are classified into two main categories: " early failures" and "wear out failures," which are reflected in the curve known as the "bathtub" curve: at the beginning of the component's existence, in its "infancy," the failure rate is rapidly decreasing. These "youth" failures are normally screened out by routine tests performed by the manufacturer. They occur because of design and process weaknesses that have not been detected by the design and process failure modes and effects analysis (FMEA) performed during the development. They are more likely to be caused by production process variations or by changes in material quality. The process variations are because of tool wear, operator change, and lack of formation. This early failure mode is not considered by the Weibull model theory. In normal operation, this failure process should not be observed in the field of applications. If it occurs, the capacitors are normally covered by a manufacturer's product warranty.

Once the "early failures" regime is past, the failure rate starts to follow a statistical prediction law that depends on several parameters that may be defined experimentally as a function of voltage, temperature, and environmental humidity. It has been shown that a Weibull statistic can provide a good prediction of the capacitor lifetime expectancy.

Weibull statistics can also be used to predict the capacitance evolution of a metalized capacitor under electrical, thermal, and humidity stresses. In such cases, the failure definition will be, for example, 1% capacitance loss, depending on the available resolution of the measurement device. The capacitance will be given straightforwardly by the survivor function. The Weibull reliable life that gives the expected lifetime of a capacitor for a given reliability level (the proportion of remaining working objects, or in our case capacitance) is

$$T_R = \frac{1}{\lambda_0}\{-\ln(R)\}^{1/p} \quad (8.117)$$

where λ_0 is the failure rate for the special case when 1/e, or 36.8%, of the samples still remain.

The speed of capacitance drop depends on the temperature, the voltage, and the humidity. Increases in these parameters are considered to be aging acceleration factors. They are determined experimentally based on the following theories.

Temperature: Capacitor aging as a function of the temperature follows an Arrhenius law, otherwise known as an exponential law.

$$t(\mathrm{T}) = t_{Tn} \exp\left(\frac{E_a}{k_B}\left(\frac{1}{T} - \frac{1}{T_n}\right)\right) \tag{8.118}$$

where t_{Tn} is the expected lifetime at a reference temperature, 70°C or 85°C, for example, k_B is the Boltzmann constant, and E_a is an activation energy. A relatively good fit of Epcos/Vishay factors may be obtained with the ratio $E_a/k_B = 7000$ K between 40°C and 70°C.

Voltage: Regarding voltage dependency, authors use either an inverse power law

$$t = t_{U_n}\left(\frac{U}{U_n}\right)^{-n} \tag{8.119}$$

or an exponential law

$$t = t_{U_n} exp\left(-\alpha\frac{(U - U_n)}{U_n}\right) \tag{8.120}$$

where t_{Un} is the expected lifetime at the nominal voltage or reference voltage and t/t_{Un} is the voltage acceleration factor.

Humidity: Humidity is a concern for capacitors contained in plastic because moisture can permeate through this type of material. Once inside the capacitor, the moisture has several effects: first, it decreases the electrical strength of the gas in the case of oil-free capacitors, leading to corona demetalization of the electrode, and second, it corrodes the electrode. When moisture is present in the dielectric film, the loss factor is increased because of the presence of water dipoles, and the insulation resistance is also reduced, leading to current leakage and generation of heat. The lifetime as a function of the humidity level can be estimated using the following relation:

$$t(RH) = t_{Hn}\left(\frac{RH_n}{RH}\right)^m \tag{8.121}$$

where t_{Hn} is the expected lifetime at a reference humidity level.

8.10.1 Arm Reactor (Air Core Reactor)

The arm reactors used in VSC HVDCs are air core reactors, which are devices that are used in distribution networks and transmission networks in both AC and DC systems for a variety of applications such as fault current limiting, power flow control, reactive compensation (shunt reactors), DC smoothing, and coupling reactors, and as the inductive part of tuned harmonic filters, among others. Despite the wide range of applications and sizes, air core reactors share a basic common structure.

As seen in Figure 8.42, an air core reactor consists of (1) one or more concentric cylindrical windings, which are electrically connected in parallel by welded connections to aluminum cross arms, commonly referred to as (2) spiders, which are located at the top and bottom of the coil. The individual windings are wound with one type of a variety of specialty cables or individual aluminum wires that are insulated with insulating tapes or films and encapsulated in a fiberglass epoxy composite. Each spider carries (3) a terminal for the electrical

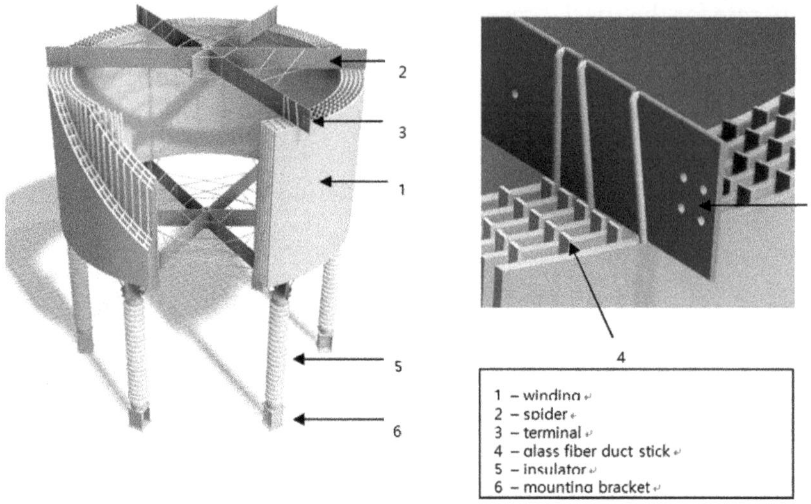

Figure 8.42 Air core reactor-basic construction

connections of the reactor. The individual concentric cylindrical windings are radially separated from each other by (4) glass fiber polyester resin composite duct sticks , which form the air ducts that are necessary for the cooling of the winding. Cooling is provided by the natural convection of ambient air that enters at the bottom end of the winding and exits at its top end. A protective paint or coating is applied to the surface of the fiberglass epoxy encapsulated winding to protect the epoxy from ultraviolet radiation and to improve tracking withstand and weather resistance. In the case of adverse pollution conditions at the site of installation, reactors for some applications may be equipped with a protective roof or so-called top hat (not shown). The reactor is mounted on several (5) base insulators and (6) mounting brackets. The rating of the insulators depends on the specific system requirements at the site of installation.

Magnetic Fields of Air Core Reactors

Since air core reactors do not have a magnetic core to constrain the magnetic field, the magnetic field is "broadcast" in and around the reactor. The strength of this "broadcasted field," more commonly referred to as a "stray field," depends on the power rating of the reactor. In general, the higher the rating of the reactor, the higher the strength of the stray field. This stray field impinges on both the reactor components themselves such as windings, spiders, corona electrodes, and supports, and also adjacent parts such as terminals, connectors, connecting buses or cables, bus supports, and generally any adjacent electrically conductive material.

Figure 8.43 illustrates the magnetic field distribution of a typical air core reactor; its magnitude is in milli Tesla [mT]. As can be seen from the magnetic field lines (Plot (d)), in the mid-plane as well as on the rotational axis (Plots (a) and (c)), the field is directed in the axial direction, whereas in all the other locations the field has components in both the axial and radial directions. In the vicinity of the winding ends of the reactor, the magnetic field is predominantly directed in a radial direction (Plot (b)).

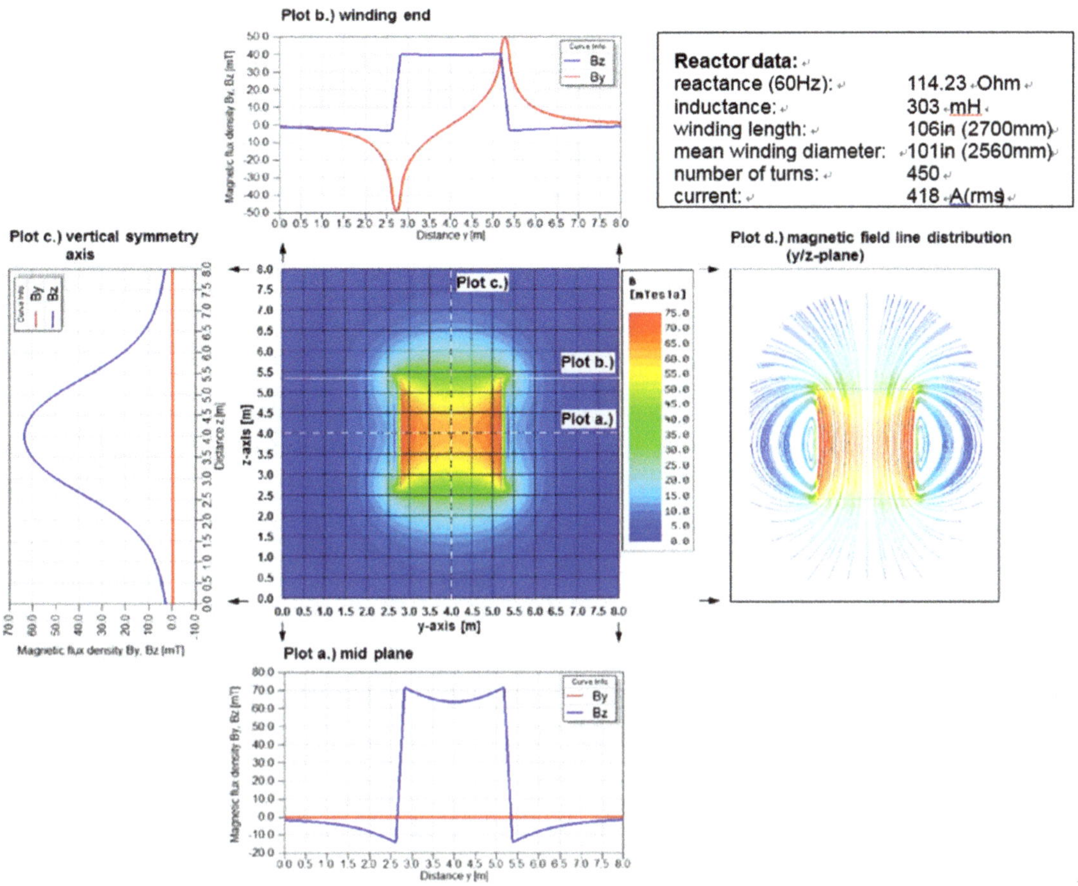

Figure 8.43 Magnetic field plot

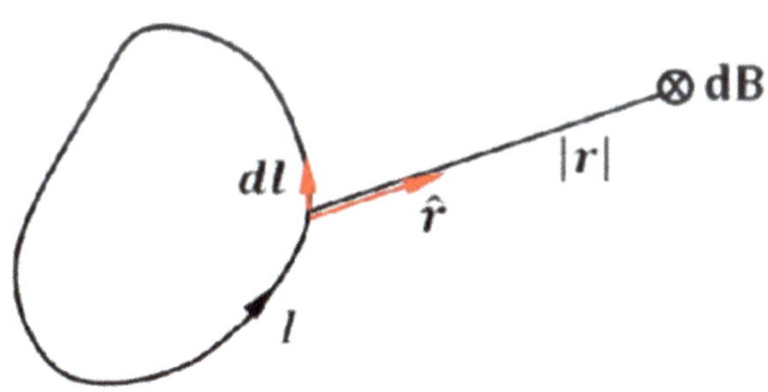

dB	magnetic flux density of a short current filament
μ_0	Physical constant $\left(4\pi \cdot 10^{-7}\ \mathrm{H/m}\right)$
I	electric current wire filament
dl	current -carrying wire filament
$\hat{r}$	Unit vector of vector **r**
$\lvert r\rvert$	Distance between current filament and field point

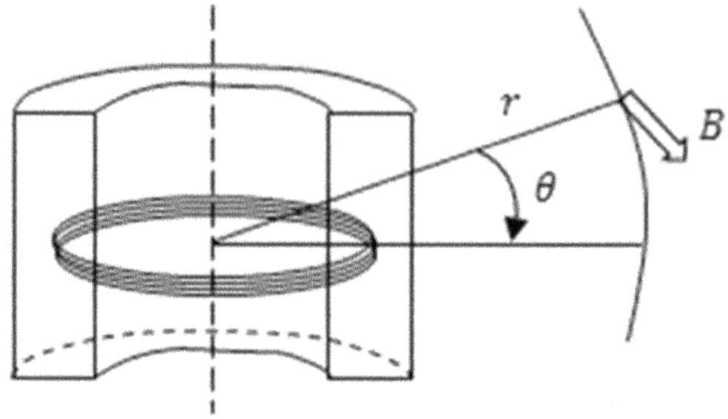

Figure 8.44 Loop, equivalent to a reactor winding

Biot–Savart's law Equation (8.122) may be applied to calculate the magnetic field of a cylindrical winding:

$$dB = \frac{\mu_0}{4\pi}\frac{Idl \times \hat{r}}{|r|^2} \tag{8.122}$$

The external magnetic field of an aircore reactor winding at a significant distance from the winding may be approximated by the field of a current loop as shown in Figure 8.43. This approximation holds for coils having a winding length shorter than about three times the winding diameter. The field produced by a current-carrying winding loop in a distance r of more than around three times the loop diameter may be approximated by Equation (8.122) and Equation (8.123).

$$|B| = \frac{\mu_o \cdot n \cdot l \cdot D^2}{8 \cdot r^3} f(\Theta) \tag{8.123}$$

$f(\Theta) = \sqrt{\sin^2(\Theta) + \frac{\cos^2(\Theta)}{4}}$ (8.124)	
$\lvert B \rvert$	Magnitude of the magnetic field
μ_0	Permeability in air constant ($4\pi \cdot 10^{-7}$ H/m)
n	No. of turns
I	current
D	Loop diameter (mean winding diameter)
r,Θ	coordinates
f(Θ)	Directivity function as per Equation (8.125)

Using (8.123) and (8.124) in the lateral direction $\Theta = 0, f(\Theta) = 0.5$, the magnitude of the magnetic flux density at moderate distances away from the reactor may be estimated by

$$|B| = \frac{\pi \cdot n \cdot I \cdot D^2}{4 \cdot r^3} 10^{-7} \text{ [Tesla]} \tag{8.125}$$

As shown in Equation (8.126), the field strength quickly drops off with increasing distance from the reactor, being inversely proportional to the cube of the distance.

In the case of air core reactors carrying alternating current, the stray magnetic field will induce eddy currents and associated eddy current losses in any electrically conductive parts that reside in the stray field, whether they be made of ferromagnetic or non-magnetic material. Where the stray field links closed conducting loops, very large closed loop currents may be induced, with associated steady-state heating and potentially large forces under short-circuit conditions. Finally, where the stray field interacts with current-carrying conductors such as cables, oscillatory vibration forces will be induced under steady-state conditions and correspondingly larger forces under short-circuit conditions.

Eddy Current Losses

To reach an estimation of the eddy losses in the structural elements of an air core reactor such as spiders and the terminals as well as the connection bus work, the following approach can be considered.

The terminals and spiders are considered as a plate with thickness d (profile thickness), with the assumption that the thickness is less than the so-called skin depth or penetration depth χ and the field direction is parallel to the plane of the plate. The penetration depth is calculated as follows:

$$\delta = \sqrt{\frac{2}{\omega\kappa\mu}} \tag{8.126}$$

where δ is penetration depth [m],
ω is 2πf, *and f is* in [Hz]
κ is specific conductance [S/m]
μ is permeability, [H/m]
The eddy losses can be calculated with some approximation as follows:

$$\mathrm{p} = \frac{B^2\gamma^3 h}{6\mu_0^2 2\kappa\delta} \text{ where } \gamma = \frac{d}{\delta} \tag{8.127}$$

where:
p is specific eddy losses [Watt/m]
B is magnetic flux density [T_{peak}] at the spider or terminal location
d is thickness of profile [m]
h is height of profile [m]
μ_0 is $4\pi \cdot 10^{-7}$ H/m
κ is specific conductance, [S/m]
δ *is* penetration depth [m]

The foregoing approximation is acceptably accurate for non-magnetic material geometries with $d < \chi$ at fundamental frequency. At high frequencies, or with ferromagnetic material, the reduced skin depth because of the higher frequency and/or high magnetic permeability results in significant errors if this approach is used. Hysteresis losses are also not accounted for in the above formulae.

Induced Currents and Losses in Closed Loops

Faraday's law of induction states that the electromotive force (EMF) induced in a closed loop is proportional to the negative of the time rate of change of the magnetic flux linked by the loop.

$$MP = -\frac{d\phi}{dt} \tag{8.128}$$

In the case of a magnetic field, the flux linking the loop is

$$\phi = \int_A B \cdot dA \tag{8.129}$$

where $dA = dA \cdot n$ is a vector normal to the small surface dA and **n** is a unit vector normal to this surface.

From Equations (8.129) and (8.130) it is clear that the EMF will depend on the size of the loop, and on its orientation relative to the field direction, being thus proportional to the component that is normal to the plane of the loop and the loop area. The loop current I, which results from the induced EMF, will be in accordance with Ohm's law, where the inductance and resistance of the loop are considered:

$$|I| = \left|\left(\frac{\text{EMP}}{Z_{loop}}\right)\right| \tag{8.130}$$

$$Z_{\text{loop}} = \sqrt{\left(R_{\text{loop}}^2 + X_{\text{loop}}^2\right)} \tag{8.131}$$

To analyze specific cases, the loop inductances and resistances must be calculated along with the linking flux and resulting EMF. Calculation of the flux linking a loop by formulas may be done only for simple geometries. More complex geometries typically require either purpose-built software or finite element modeling.

Short-Circuit Forces

The Lorentz force law states that the force on a current-carrying conductor in a time-varying magnetic field is

$$\vec{F} = I\vec{l}x\vec{B} \tag{8.132}$$

The direction of the force, in accordance with the well-known right-hand rule, is such that it is directed normally to the plane formed by the B-field vector and the current direction vector.

Another method for calculating the magnetic field forces is to determine the change in the magnetic energy that is produced by an imaginary change in the configuration of the current circuit. If any dimension x of a current-carrying circuit is changed by a small distance dx, magnetic field forces must be overcome (in addition to elastic stresses). If we denote the magnetic field force in the direction of x with F_x, a displacement of dx becomes a mechanical work done:

$$dW_1 = F_x dx \tag{8.133}$$

Assuming the current I is held constant in this change, for example, by providing a sufficiently high resistance in the circuit, the total flux of the current circuit increases with such a change in the amount:

$$\Phi = LI \tag{8.134}$$

$$d\Phi = IdL = I\frac{\partial L}{\partial x}dx \tag{8.135}$$

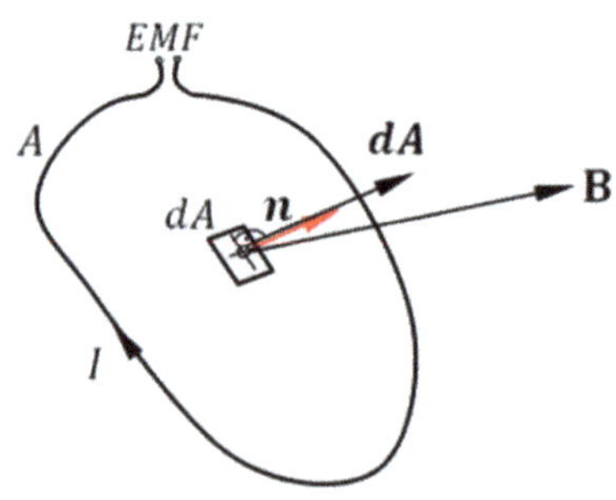

Figure 8.45 EMF and flux linkage

If the change takes place in time dt, then a self-induction voltage is obtained:

$$u_L = \frac{d\Phi}{dt} = I\frac{\partial L}{\partial x}\frac{dx}{dt} \tag{8.136}$$

This requires an electrical work for the current I during the time dt of

$$dW_2 = u_L I dt = I^2 \frac{\partial L}{\partial x} dx \tag{8.137}$$

Finally, the energy accumulated in the magnetic field is increased

$$W_m = \frac{1}{2} L I^2 \text{ by } dW_m = \frac{1}{2} I^2 \frac{\partial L}{\partial x} dx \tag{8.138}$$

As the energy expended must equal the energy obtained, it follows:

$$dW_2 = dW_m + dW_1 \tag{8.139}$$

After inserting the expressions Equations (8.133)–(8.135):

$$F_x = \frac{1}{2} I^2 \frac{\partial L}{\partial x} \tag{8.140}$$

The force is, therefore, always directed to increase the inductance. It can be calculated if the dependence of the inductance of the current circuit of x is known. In the case of two different circuits 1 and 2, the forces occurring between the circuits can be found through a similar method. The displacement dx of the two current circuits relative to one another results in a change in the mutual inductance by which, on the one hand, the voltages induced in the two current circuits and on the other hand, the change in the field energy. This follows the force acting in the direction of the displacement and brings the current loops into such a position that the mutual inductance becomes as high as possible:

$$F_x = I_1 I_2 \frac{\partial M}{\partial x} \tag{8.141}$$

In addition to the effects of the force on the current-carrying structures, it should also be noted that this force has a vibrational nature (in the case of AC systems). The mechanical frequency of this force is twice the excitation frequency (f_{mech}=120 Hz in case of a 60-Hz AC system). Thus, considerable noise emissions may also be triggered because of the vibration of the affected structures.

Clearance Guidelines for Air Core Reactors

The following points serve to summarize the basic installation clearance requirements associated with air core reactors. The clearance requirements are of three types:

- Electrical
- Ventilation
- Magnetic

Electrical Clearance

As the normal air core reactor has exposed live parts at essentially all points on its outer surface, provisions must be made for electrical clearance from the reactor surface to nearby grounded surfaces, as well as to the surface of other reactors or other live parts in adjacent phases or circuits. Standard electrical substation clearances to live parts are perfectly adequate. No special precautions over and above normal substation practices are required.

Ventilation Clearance

For a typical air core reactor that is mounted such that the cooling ducts are oriented in the vertical direction, adequate arrangements must be made to allow unimpeded entrance and exit of cooling air at the bottom and the top of the cooling ducts. In most cases, the ventilation clearance will be less than the magnetic clearance requirements, and as such is typically not a decisive factor in the installation arrangement

Magnetic Clearances

The magnetic clearance requirements for air core reactors are a consequence of the interaction of their stray magnetic fields with conductive parts in the vicinity of the reactor and the resulting currents that may be induced in those parts. The induced currents are of two types, and as such give rise to two types of clearance requirements:

- Eddy Currents: induced in nearby conductive parts, give rise to a minimum clearance requirement for metallic parts that do not form closed loops.
- Circulating Currents: caused by the coil flux linking a closed electrically conducting loop; examples of such loops are those formed by concrete reinforcement rebar, nearby fencing, nearby structural members in a building, inappropriately arranged grounding connections, or a combination of the above.

8.11 Energy Diverting Converter Topologies

Grid codes are also evolving to regulate the connection of such systems to national grids. Some of the most challenging requirements from a system design point of view are those concerning the fault ride-through (FRT) capability during onshore AC network faults (Figure 8.46). During a fault, the onshore VSC may see its capability to inject power to the local grid greatly reduced, and this might lead to an overvoltage situation in the HVDC interconnector because of the wind farm continuing to feed power to the interconnector offshore station.

Research in the area of the FRT capability of wind parks has been very active in the past few years. Several different solutions have been proposed to manage the control systems of the wind farm and HVDC interconnector to minimize delays in the wind farm's output power reduction and to add elements that protect the system until the power balance can be restored. The major propositions can be classified into two

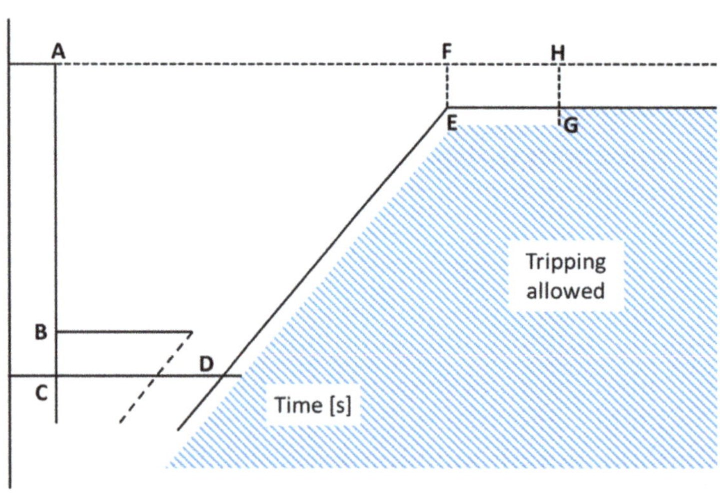

Figure 8.46 FRT requirements in different countries

Grid Code	BC	BD	AF	FE	AH	HG
Denmark	25%	0.1 s	0.75 s	25%	10 s	NA
Germany(*EON*)[1]	0%	0.15 s	0.15 s	30%	0.7 s	10%
Germany(*EON*)[2]	15%	0.625 s	3 s	10%	NA	NA
Ireland(EirGrid)	15%	0.625 s	3 s	10%	NA	NA
Spain	20%	0.5 s	1 s	20%	15 s	5%
UK(NG)	0%	0.14 s	1.2 s	20%	2.5%	15%

For generating unit with large symmetrical short-circuit current component
For generating unit with low symmetrical short-circuit current component

groups: (1) control strategies for a fast reduction of power from the wind farm and (2) methods to divert the excess energy, usually by dissipation in a dedicated circuit. The term **Energy Diverting Converter (EDC)** is used in this chapter to refer to the second group of solutions in general, while dynamic braking resistor (DBR) circuits can be considered a particular subgroup of EDC circuits in which DC link energy is diverted to a resistive device where it is dissipated as heat. DBR circuits connected across the HVDC link provide an additional level of protection against overvoltages in the HVDC line for a prudent and cautious industry where safety is paramount.

8.11.1 Dynamic Braking Resistor (DBR) circuits

At present, there are a number of HVDC manufacturers offering interconnection solutions for the offshore wind farm market. From an economical point of view, because of the high cost of adding extra equipment to an offshore station platform, the DBR circuit is assumed to be located in the onshore station next to the main VSC converter (Figure 8.47).

Figure 8.48 shows the different solutions. In the first solution, referred to as DBR circuit I and presented in Figure 8.48 (a), each switch consists of stacks of series-connected switches, is referred to as a valve, and operates as a DC chopper to apply full DC-link voltage across the braking resistor R_{diss}. The series-connected switches are commutated simultaneously, and PWM is used to regulate the amount of power dissipated. The second solution, presented in Figure 8.48(b) and referred to as DBR circuit II, has a modular structure where the braking resistor R_{diss} is distributed inside each of the individual cells. The third solution, shown in Figure 8.48(c) and referred

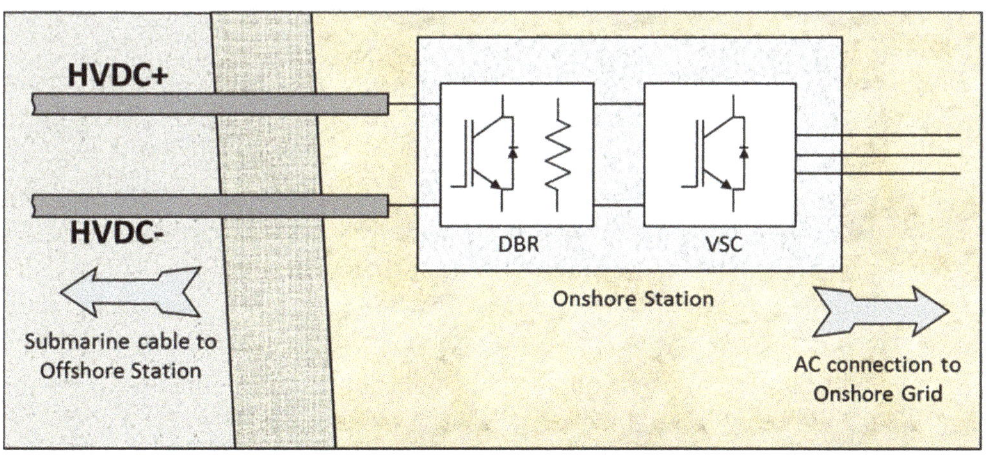

Figure 8.47 Location for the DBR circuit

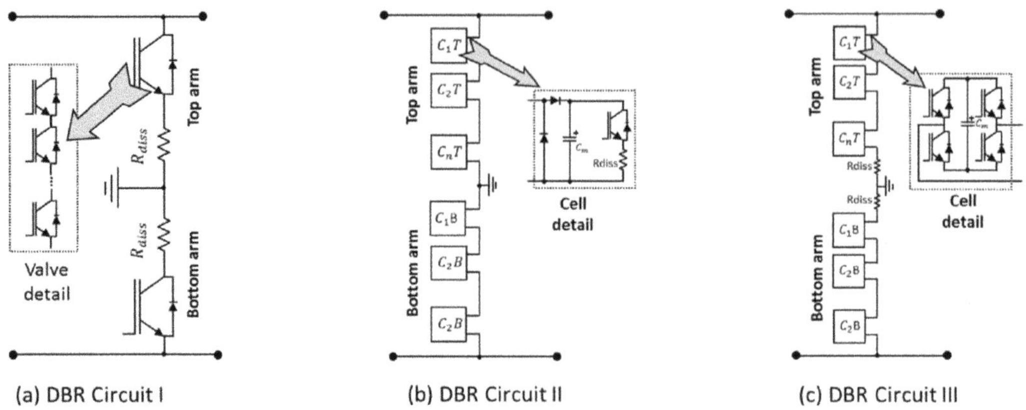

Figure 8.48 Different DBR solutions considered in this paper

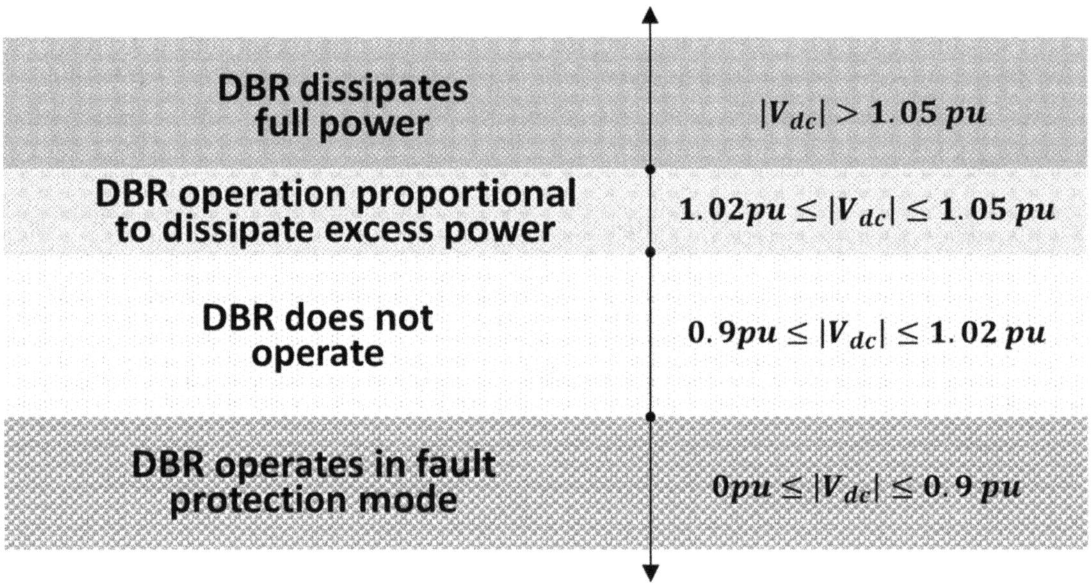

Figure 8.49 DBR operation regions

to as DBR circuit III, also presents a modular structure, but an external braking resistor, as for DBR circuit I, is used.

Figure 8.49 shows the different operation regions for the DBR converters. The control system keeps the DBR arms blocked when the DC-link voltage is between 0.9 and 1.02 p.u. Values greater than 1.02 p.u. are regarded as an overvoltage situation in this example, and the controller will modulate the DBR arms to dissipate the excess power in the resistive element. For values of V_{dc} greater than or equal to 1.05 p.u., the DBR arms will be modulated to dissipate the HVDC-link nominal power (P_{nom}) in the resistive element. When the DC voltage collapses below 0.9 p.u., as can be the case during DC faults, the DBR arm cells can either be blocked, or put into bypass mode, in both cases generating no voltage across the arm. It is important to note that during a pole-to-ground DC fault in the point-to-point HVDC link considered in this document, the DBR will operate to reduce the voltage of the non-faulted pole, which may suddenly see the full DC voltage appearing across it, by putting all the modules in bypass and applying all the voltage across the braking resistors.

One of the most important aspects in the design of the modular topologies is the sizing of the cell capacitance. This value determines the cell voltage ripple and, therefore, affects the operation of the whole converter.

Additionally, it has a significant impact on the size and cost of each of the converter cells. For this reason, it is important to choose the smallest value of capacitance that still offers a satisfactory performance of the cell in terms of voltage ripple during the circuit operation. In this study, the cell capacitance has been designed using the following expression:

$$C_{\text{cell}} = I_{\text{cell}} \cdot \frac{\Delta T}{\Delta V_{\text{cell}}} \tag{8.142}$$

where I_{cell} is the averaged current value through the cells during the DBR operation in full-power dissipation mode, ΔT is the execution period of the capacitor-balancing algorithm implemented in the control system, and ΔV_{cell} is the peak-to-peak value of the desired cell voltage ripple.

The value of the cell capacitor for the two modular topologies is calculated according to Equation (8.143), considering a balancing function period ΔT of 500 μs for DBR circuit II and 200 μs for DBR circuit III, and a maximum ripple ΔV_{dc} of 20% of the nominal cell voltage. For DBR circuits II and III, 12 cells are used per converter arm; this is 24 cells in total for the whole converter, resulting in a nominal cell voltage of around 1.05 kV. For the calculation of the resistive element value, the task is to dissipate nominal power when the full DC voltage of 1.05 p.u. is applied across it as shown in Equation (8.144). In the case of DBR circuit II, the distributed resistance inside each cell is calculated according to Equation (8.145), which is the total value of resistance required divided by the number of cells per converter arm:

$$R_{\text{diss}} = \frac{\left(\frac{v_{\text{dc nom}}}{2} \cdot 1.05\right)^2}{P_{\text{nom}}} \tag{8.143}$$

$$R_{\text{diss cell}} = \frac{\left(\frac{v_{\text{dc nom}}}{2} \cdot 1.05\right)^2}{P_{\text{nom}} \cdot N_{\text{cell arm}}} \tag{8.144}$$

Table 8.4 Shows the summary of Advantages and Disadvantages of the Three DBR circuits

	DBR Circuit I	DBR Circuit II	DBR Circuit III
Advantages	• Fewer semiconductors and compact in size • Simple lumped air-cooled resistor	• Low current ripple • Modular design	• Low dvdt didt • Modular design • Simple lumped air-cooled resistor
Disadvantages	• Use of series-connected devices technology • Large *dv*/*dt* and *di*/*dt*	• Large number of semiconductors • Complex distributed resistor cooling needed	• Large number of semiconductors

8.12 A Study of the Interference of HVDC

HVDC converters are complex systems with a range of sensitive control systems. HVDC converters also, by their basic operation, generate electromagnetic noise. In general, HVDC converters generate radiated noise within the range of 150 kHz to approximately 10 MHz. At higher frequencies, other sources such as corona are more significant. At lower frequencies, the radiation process is not efficient, and conducted noise is more significant.

High-frequency interference noise from HVDC converter stations might be generated by several mechanisms:

1. Voltage breakdowns in the converter valves during the firing process and the consequent current impulses in the valves
2. Corona discharges in the air at the surfaces of conductors that are similar to those in pure high-voltage AC systems
3. Discharges and sparking in highly stressed areas of insulators
4. Sparking at loose or imperfect contacts.

Voltage breakdowns in the earlier comments are called conducted noise, whereas the last three types can be summarized as background noise. Conducted noise, when compared with background noise, shows two notable characteristics:

1. Conducted noise is generally higher than background noise at AC and DC power transmission lines connected to HVDC converter stations.
2. The differences in level are greater at lower frequencies.

Based on the earlier principles, conducted noise in HVDC converter stations is completely unrelated to background noise.

The converter is the main high-frequency noise source. As in the prior conditions of high-frequency interference calculation, the model of the converter and correct noise-conducted channel are deemed as the main contributors to the accurate result of the noise calculation.

Studies are performed for the calculation of the radiated interference performance of a converter station as part of the design process. These studies are required to confirm that the valve hall screening is adequate to ensure the converter stations are within the limits specified. The purposes of RFI studies are to

- Outline and describe the sources of the radio interference and their methods of propagation
- Describe the method and design used to limit radio interference
- Describe the method proposed to meet the performance requirements

Several established documents such as the CIGRE *Guide on Power Line Carrier*, the EPRI report *Radio Interference from HVDC Converter Stations*, EL-3712, and CISPR 18 are used as background information in this report. To limit the emission of RFI from the converter station building, the walls, floor, and roof of the valve hall will be enclosed in a grounded Faraday cage made mostly from a wire mesh screen. Other options for screening include the use of a bonded metal building, the use of building reinforcement, or even sheet metal screening.

The high-frequency noise radiating from the converter station needs to be limited to an acceptable level. Some typical limits are as follows:

The main design requirement for RFI shielding is that it should limit the interference down to the specified limits. The radio frequency (RF) noise generated by the converter stations can be categorized as follows:

Location	Maximum level
Any point outside the boundary fence of the Station and at least 500 m from the nearest Bar which connects the valves to the converter transformers	A maximum of 100 μV/m. To be measured with quasi-peak detector at 1 MHz.

- Direct radiation from the valves
- Radiation from the converter station bus bars because of high-frequency currents circulating in the AC bus bars and associated stray capacitances and stray inductance
- RF noise because of corona from the high-voltage equipment and bus bars

The scheme can be divided into a number of discrete elements: the valves, the converter transformer, and the AC systems (AC harmonics filters and a simplified representation of the relevant AC system).

Noise Source Model

The conductedand radiated interference noise produced by the operation of VSC HVDC converters is caused by rapid changes in voltage and current during the turn-on and turn-off of the IGBT in the converter SM. Several hundreds of SMs per converter can act as individual sources of interference. Also, the superposition of all SMs switching events results in continuous interference noise emanating from the converters.

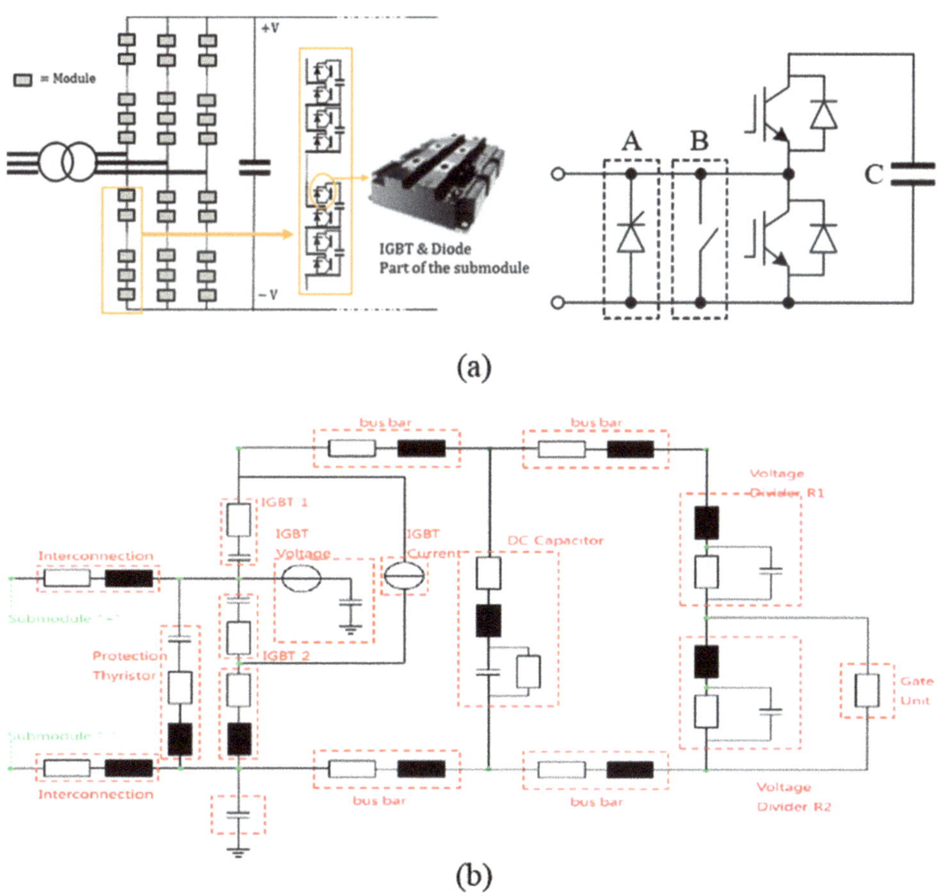

Figure 8.50 HF equivalent circuit of VSC HVDC SM

There are two events during valve operation that can cause high-frequency noise: turn-on (which is the sudden collapse of voltage across the valve as the IGBT "turns on" [the "breakdown voltage"], bringing about the discharge of parallel-connected capacitances) and turn off. For RFI purposes, only turn-on is significant, since turn-off is a comparatively slow event (the rate of change in the voltage is controlled by the valve-damping network), so its influence on the frequency range of interest is small.

The turn-on event is modeled mathematically as an exponentially decaying voltage leading to the equation for equivalent noise voltage and repeated as Equation ((8.145) below Figure 8.50):

$$V_{\text{eq}} \approx \frac{V_0 \cdot K_1}{\omega \cdot \sqrt{1 + (\omega \cdot T_0)^2}} \tag{8.145}$$

where:
V_{eq} = equivalent noise voltage at angular frequency ω
V_0 = breakdown voltage
$K_1 = 6320 \times A(n)$ for RF quasi peak measurement
$A(n)$ = pulse repetition rate weighting factor from ANSI C63.2 = 1.3
ω = angular frequency
T_0 = breakdown time

Valve-Generated Noise

In the frequency range of interest, the wavelength of the valve radiation is comparable with the valve dimensions. The resulting equation for electric field strength (E_θ) at distance r is

$$E_\theta = \frac{120 \cdot \pi \cdot I \cdot \sin(\theta)}{2 \cdot r \cdot \lambda} \int_{-\frac{h}{2}}^{\frac{h}{2}} \left[\cos\left(\frac{2 \cdot \pi \cdot x}{\lambda}\right) \times \text{Cos}\left(\omega \cdot \left(t - \frac{r + x \cdot \text{Cos}(\theta)}{c}\right)\right)\right] dx \tag{8.146}$$

where:
r = distance from valve
I = displacement current at wavelength λ in valve because of change in voltage
θ = angle between measurement point and valve orientation (0 degree vertically above valve, 90 degree level with valve)
h = height of valve
c = speed of light
λ = wavelength of interest
x = position on radiating structure

Radiated Noise Because of High-Frequency Bus Bar Currents

The turn-on event is brought about by a change of valve impedance from near open circuit to near short circuit. The sudden collapse in voltage brings about the discharge of any parallel-connected capacitances, generating oscillations in circuits that are affected by it. Precise modeling of this noise source to evaluate radiated noise is extremely laborious, requiring a great deal of information regarding the high-frequency impedances of components, their precise location, and their orientation in the substation. At a distance of approximately 40 m (assuming 30 m to the station perimeter) from the bus bars, it is possible, with little error, to use an equivalent bus bar loop as an antenna with appropriate current values. These high-frequency current values are derived

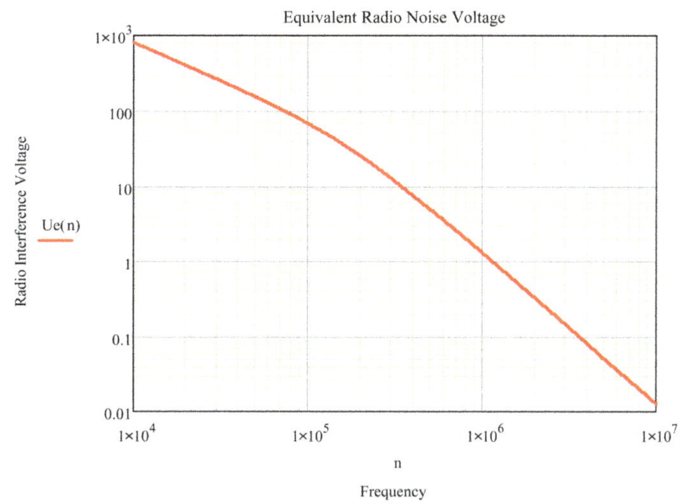

Figure 8.51 Equivalent noise voltage of IGBT

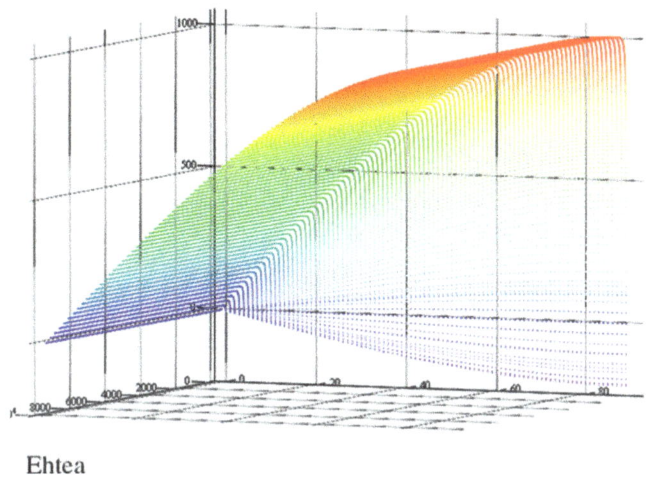

Figure 8.52 Valve-generated noise

from the aforementioned thyristor valve design studies by applying the calculated equivalent noise voltage to a detailed high-frequency digital model of the converter station.

Equivalent Circuit of VSC HVDC Converter Station

The converter station components represent the propagation paths for interference currents and voltages. Within the converter station design, the following components have to be considered in the high-frequency range:

- Converter SMs
- Station bus bars
- Converter hall bushings
- Converter reactors
- AC PLC measuring unit
- Converter transformers

Electromagnetic Compatibility –Immunity Issues

The complex electronic systems in converter stations need to be sufficiently immune to the interference from converter station operation and maintenance. More specifically, the installation needs to be sufficiently immune to the following phenomena:

- Electrostatic discharge from personnel
- Electric surges
- Bursts of conducted interference because of switchgear and relay operation
- Radiated interference from converter operation and also communications such as radio, television, and mobile phones; this requirement is particularly onerous for the equipment in close proximity to the converter valves
- Power supply interruptions and variations

Shielding for Radiated Noise

The radiated noise frequency range of interest is from 100 kHz to 50 MHz. The equations for the shielding effectiveness (SE) of a wire mesh are

For: g (wire gap in meters) > $150/f$ (in MHz)

$$\rightarrow \mathrm{SE}\,(\mathrm{dB} = 0) \tag{8.147}$$

For: g 150 / f and $r \gg g$

- Far filed – r (source-screen distance) > wavelength / $(2 \cdot \pi)$

$$\rightarrow \mathrm{SE}_E\,(\mathrm{dB}) = \mathrm{SE}_H\,(\mathrm{dB}) = 20 \cdot \log\left(150/\mathrm{g} \cdot \mathrm{f}\right) \tag{8.148}$$

- Near field – r < wavelength / $(2 \cdot \pi)$

$$\rightarrow \mathrm{SE}_E\,(\mathrm{dB}) = 20 \cdot \log\left(\pi \cdot \mathrm{r}/g\right) \tag{8.149}$$

$$\rightarrow \mathrm{SE}_E\,(dB) = 77 - 40 \cdot \log\left(f\right) - 20 \cdot \log\left(\mathrm{r} \cdot \mathrm{g}\right) \tag{8.150}$$

To ensure effective shielding, aperture losses must be minimized. The overall attenuation introduced by the building cannot be better than that of the worst aperture. Considerable attention is given to minimize the size and number of apertures in the building. All doors and windows are bonded to the building to form an effective part of the shield.

References

1. R. Picas, et al., "Improving capacitor voltage ripples and power losses of modular multilevel converters through discontinuous modulation," in *39th Annual Conference of the IEEE Industrial Electronics Society*, Vienna, Austria, 2013, pp. 6233–6238.
2. M. Saeedifard and R. Iravani, "Dynamic performance of a modular multilevel back-to-back HVDC system," *IEEE Trans. Power Del.*, vol. 25, pp. 2903–2912, 2010.
3. N. Flourentzou, V. G. Agelidis, and G. D. Demetriades, "VSC-based HVDC power transmission systems: An overview," *IEEE Trans. Power Electron.*, vol. 24, no. 3, pp. 592–602, 2009.
4. K. Friedrich, "Modern HVDC PLUS application of VSC in modular multilevel converter topology," in *Proceedings of the IEEE International Symposium on Industrial Electronics. (ISIE),* Bari, Italy, Jul. 2010, pp. 3807–3810.
5. S. Lu, Y. Liu, and J. D. L. Ree, "Harmonics generated from a dc biased transformer,"*IEEE Trans. Power Del.*, vol. 8, no. 2, pp. 725–731, Apr. 1993.
6. Y. You, et al., "Reactive power demand of transformers with DC bias," *IEEE Ind. Appl. Mag.*, vol. 2, no. 4, pp. 45–52, Jul. 1996.

7. Q. Tu, et al., "Parameter design principle of the arm inductor in modular multilevel converter based HVDC," in *Presented at the International Conference on Power System Technology*, Hangzhou, China, 2010.
8. H. Suzuki, et al., "Development and testing of prototype models for a high performance 300 MW self-commutated AC/DC converter," *IEEE Trans. Power Del.*, vol. 12, no. 4, Oct. 1997.
9. K. Sakamoto, et al., "Development of a control system for a high-performance self-commutated converter," *IEEE Trans. Power Del.*, vol. 13, no. 1, Jan. 1998.
10. T. Nakajima, et al., "A converter transformer with series-connected line-side windings for a DC link using voltage source converters," in *Proceedings of the IEEE Power Engineering Society Winter Meeting*, New York, USA, Feb. 1999.
11. R. A. Walling and A. H. Khan, "Characteristics of transformer exciting current during geomagnetic disturbances," *IEEE Trans. Power Del.*, vol. 6, no. 4, pp. 1707–1714, Oct. 1991.
12. Y. You, et al., "Reactive power demand of transformers with DC bias," *IEEE Ind. Appl. Mag.*, vol. 2, no. 4, pp. 45–52, Jul. 1996.
13. ABB, "HVDC light: It's time to connect," ABB, Ludvika, Sweden, Tech. Rep. POW-0038, 2012.
14. J. Candelaria and J.-D. Park, "VSC-HVDC system protection: A review of current methods," in *Presented at the IEEE PSCE*, Phoenix, AZ, Mar. 20–23, 2011.
15. A. Bidadfar, et al., "Power system stability analysis using feedback control system modeling including HVDC transmission links," *IEEE Trans. Power Syst.*, vol. 31, pp. 116–124, 2016.
16. J. Lago, J. Moia, and M. L. Heldwein, "Evaluation of power converters to implement bipolar dc active distribution networks–DC-DC converters," *IEEE Energy Conversion Congress and Exposition*, Phoenix, AZ, USA, pp. 985–990, 2011.
17. Y. Gu,W. Li, and X. He, "Analysis and control of bipolar LVDC grid with dc symmetrical component method," *IEEE Trans. on Power Syst.*, vol. 31, pp. 685–694, 2016.

9

Protection and Insulation of MMC

Figure 9.1 shows the relationship and range of power quality disturbance in the power system and the resulting stress placed on the device. As the high-voltage direct current (HVDC) system is also part of the electrical system, life can be shortened due to the fatigue and aging of the HVDC system because of the disturbance of this power quality. Therefore, it is necessary to protect the system by distinguishing the range of the power quality and the range of the electrical stress.

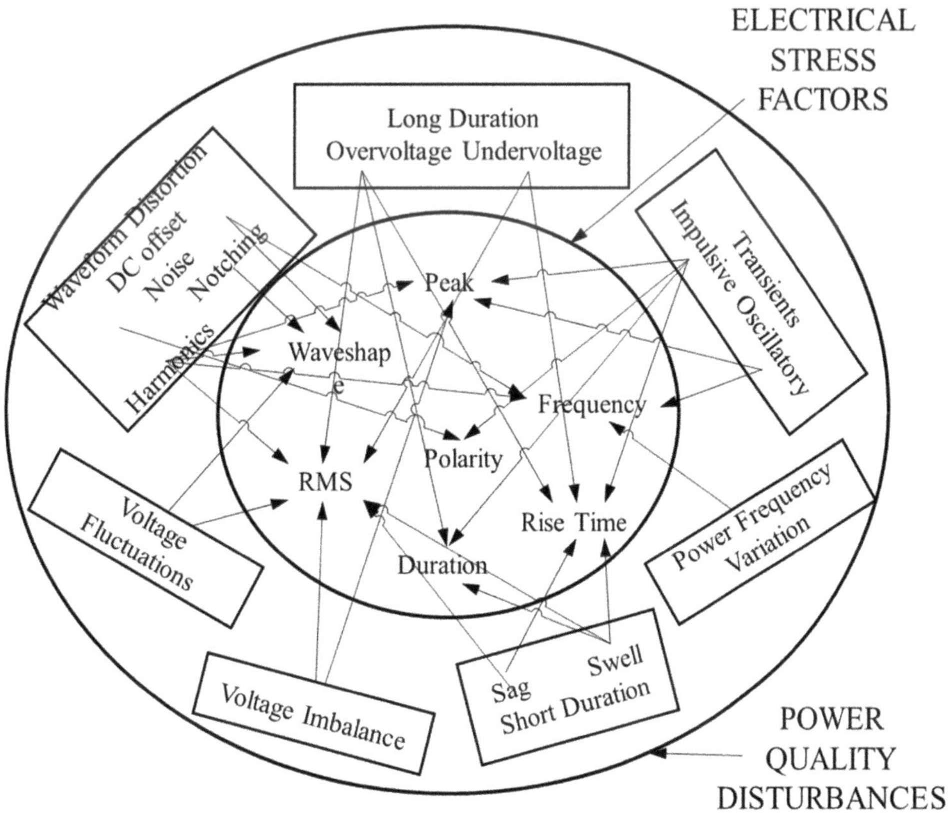

Figure 9.1 Power quality disturbance and range and effect of electrical stress in the power system

9.1 Fault Analysis of VSC HVDC [1] – [5]

Most of the research reported on DC fault analysis with voltage-sourced converter (VSC) configurations is based on numerical simulations without the theoretical basis provided by circuit analysis. The speed requirement for DC circuit breakers (CBs) can only be configured after identifying the critical time limits under various fault conditions. An AC side switchgear is not considered fast enough to cope with the rapid rise the fault current characteristics of freewheel diode conduction, which can damage power electronic devices within several milliseconds.

In addition, most work focuses on the DC short-circuit faults at the DC circuits. However, a cable short-circuit fault is potentially more common than a DC circuit fault, and the impact of a DC fault on the freewheel diodes in the VSC can be worse than that of a direct DC circuit short circuited because of the increased inductive component in the discharge path. Although underground cables are seldom short circuited in comparison to overhead lines, it is a critical condition that needs to be analyzed, particularly for switchgear relay and protection design. In contrast, ground faults are more common, but less serious. However, accurate fault location for effective protection coordination is required for high-impedance ground faults.

9.1.1 VSC DC-Cable Fault Analysis

According to circuit theory, DC bus faults are the same as DC-cable faults. The theoretical solution of the nonlinear system that represents the faulted network can be defined by different stages that assist in understanding system response. The characteristics of the DC fault current response are analyzed for accurate fault location.

***VSC DC-Cable Short-Circuit Fault Analysis*:** A DC short-circuit fault is the most serious condition for VSCs. IGBTs can be blocked for self-protection during faults, leaving reverse diodes exposed to overcurrent. Regardless of where the DC-cable short circuit occurs, it can be expressed by an equivalent circuit shown in Figure 9.2, where R and L are the π-model equivalent resistance and inductance of the positive and negative cables from the VSC to the location of the short circuit.

The cable grounding capacitor, which is dominated by the DC-link capacitance, is removed here. Multi-level VSCs are particularly promising for high-power conversion applications, such as diode neutral-point-clamped converters and flying capacitor converters. The main fault characteristics are the same as with the two-level converter in Figure 9.2, that is, a closed loop via the freewheel diodes is available for capacitance discharge.

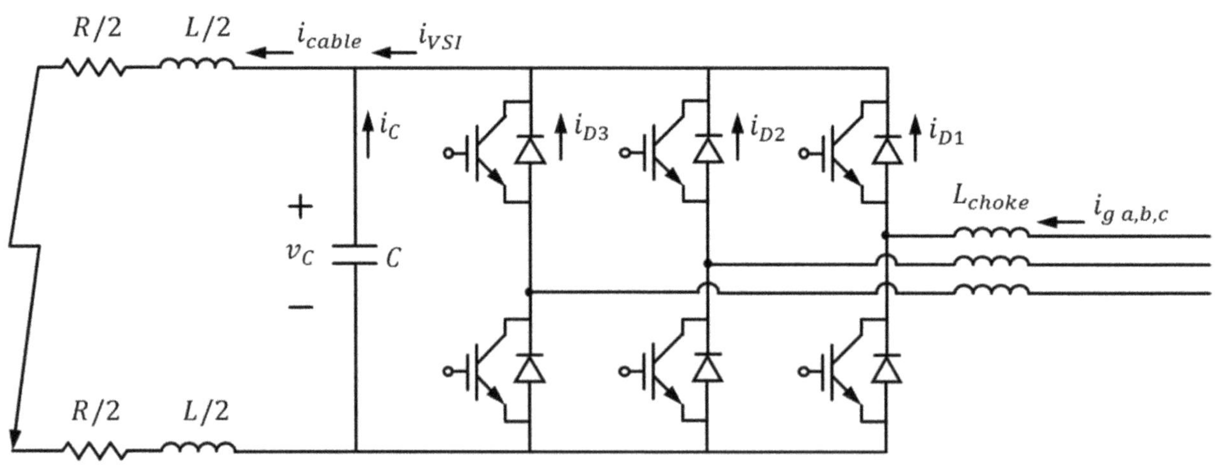

Figure 9.2 VSC with cable short-circuit fault condition

Therefore, the treatment of these multi-level converters is similar to the two-level system analyzed here. In the multi-level modular converter (MMC), the large DC-link capacitance is replaced with low-value cascaded capacitors. With appropriate fault control methods for the converter switches, the MMC can be tolerant to DC fault conditions.

To solve the complete response of this nonlinear circuit, the different stages of the fault, as it progresses, are analyzed individually. Expressions for the DC-link voltage and cable current are as follows.

Capacitor Discharge Stage (Natural Response): This stage occurs as the DC-link capacitor discharges, and the equivalent circuit is shown in Figure 9.3(a). Under the condition $R < 2\sqrt{L/C}$, the solution of the second-order circuit natural response gives an oscillation. Assuming that the fault happens at time t_0, the natural response (without inverter-side current i_{VS}) under the initial conditions of $v_C(t_0) = V_0$ and $i_{cable}(t_0) = I_0$ is

$$v_C = \frac{V_0\omega_0}{\omega}e^{-\delta t}\sin(\omega t + \beta) - \frac{I_0}{\omega C}e^{-\delta t}\sin(\omega t) \tag{9.1}$$

$$i_{cable} = C\frac{dv_C}{dt} = -\frac{I_0\omega_0}{\omega}e^{-\delta t}\sin(\omega t - \beta) + \frac{V_0}{\omega L}e^{-\delta t}\sin\omega t \tag{9.2}$$

where
$\delta = R/2L$, $\omega^2 = 1/LC - (R/2L)^2$, $\omega_0 = \sqrt{\delta^2 + \omega^2}$, $\beta = \arctan(\omega/\delta)$.

The time when the capacitor voltage drops to zero is as follows.

$$t_1 = t_0 + (\pi - \gamma)/\omega \tag{9.3}$$

where $\gamma = \arctan[(V_0\omega_0 C\sin\beta)/(V_0\omega_0 C\cos\beta - I_0)]$, t_0 = fault initial time.

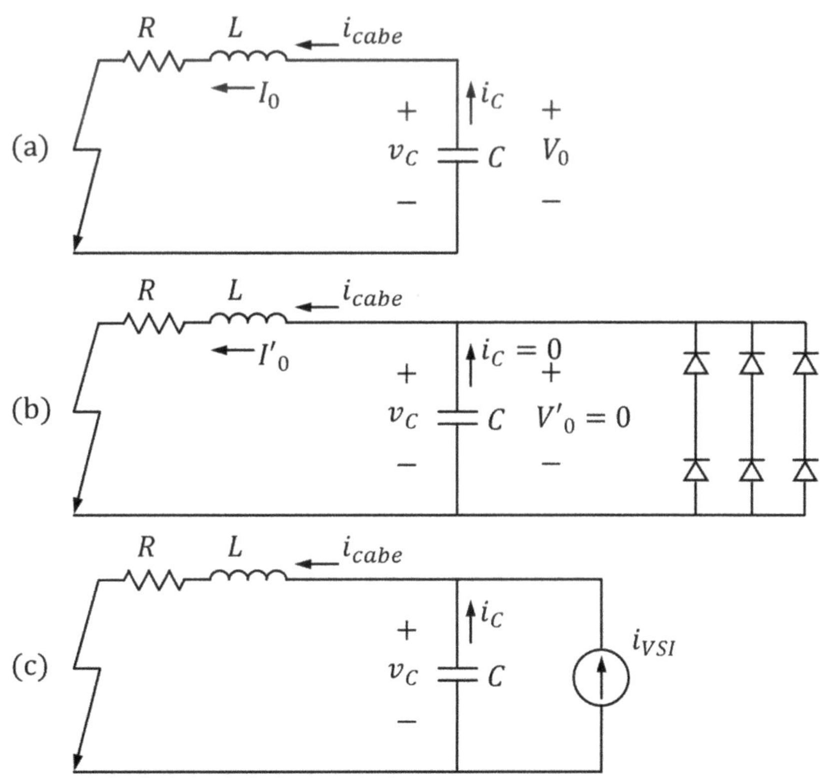

Figure 9.3 Equivalent circuit for VSI cable short-circuit fault. (a) Stage 1—Capacitor discharge; (b) Stage 2—Diode freewheel; (c) Stage 3—Grid current feeding

Diode Freewheel Stage (After $v_c = 0$; Natural Response): This stage is initiated as the cable current commutates to the VSI freewheel diodes when the DC-link voltage reaches zero and the cable inductance drives the current around the freewheel path. It is solved using the first-order equivalent circuit [Figure 9.3(b)]. The cable current has an initial value of $i_{\text{cable}} = I'_0$. The expression of the cable inductor current, where each phase-leg freewheel diode current carries a third of the current, is

$$i_{\text{cable}} = I'_0 e^{-(R/L)t} \; i_{D1} = i_{\text{cable}}/3 \tag{9.4}$$

This is the most challenging phase for VSI freewheel diodes because the free-wheel overcurrent is abrupt with a high initial value, which can rapidly damage the diodes.

Grid-Side Current Feeding Stage (Forced Response): During this stage, the DC-link capacitor and cable inductor have a forced current-source response (with i_{VSI} when the VSC IGBT gate control signals are blocked; v_C is not necessarily zero) shown in Figure 9.3(c). To calculate the fault current contribution from the converter, a three-phase short-circuit current equation is obtained by three-phase short-circuit analysis. For phase *a*, assume that the grid voltage after the fault is

$$v_{ga} = V_g \sin(\omega_s t + \alpha) \tag{9.5}$$

where V_g is the grid voltage amplitude, ω_s is the synchronous angular frequency, and α is the phase-a voltage angle at t_1.

The phase current is

$$\begin{aligned} i_{ga} &= I_g \sin(\omega_s t + \alpha - \varphi) + \left[I_{g|0|} \sin(\alpha - \varphi_0) - I_g \sin(\alpha - \varphi)\right] e^{-\frac{t}{\tau}} \\ &= I_g \sin(\omega_s t + \alpha - \varphi) + I_{gn} e^{-t/\tau} \end{aligned} \tag{9.6}$$

where $\varphi = \arctan\left[\omega_s \left(L_{\text{grid}} + L\right)/R\right], \tau = \left(L_{\text{grid}} + L\right)/R$, $I_{g|0|}$, and φ_0 are the initial grid current amplitude and phase angle, and L_{choke} is the grid-side choke inductance.

The positive i_{ga} current flows from diode D_1 to contribute to the i_{VSI}, along with those of i_{gb} and i_{gc}, so the total i_{VSI} is the summation of the positive three-phase short-circuit currents

$$i_{VSI} = i_{D1} + i_{D2} + i_{D3} = i_{ga,(>0)} + i_{gb,(>0)} + i_{gc,(>0)} \tag{9.7}$$

Here, only the phase-*a* part $i_{ga,(>0)}$ response is analyzed; phases *b* and *c* can be superimposed afterward. To analyze the most serious condition, the phase that has the highest current magnitude is determined (phase *a* in this case), with grid voltage angle being zero at the initiation of this stage of the fault. The cable currents are solved as

$$i_{\text{cable}} = A \sin(\omega_s t + \gamma) + Be^{-t/\tau} + \left[C_1 \omega_0 e^{-\delta t} \sin(\omega t + \beta)\right]/\omega + \left(C_2 e^{-\delta t} \sin \omega t\right)/\omega \tag{9.8}$$

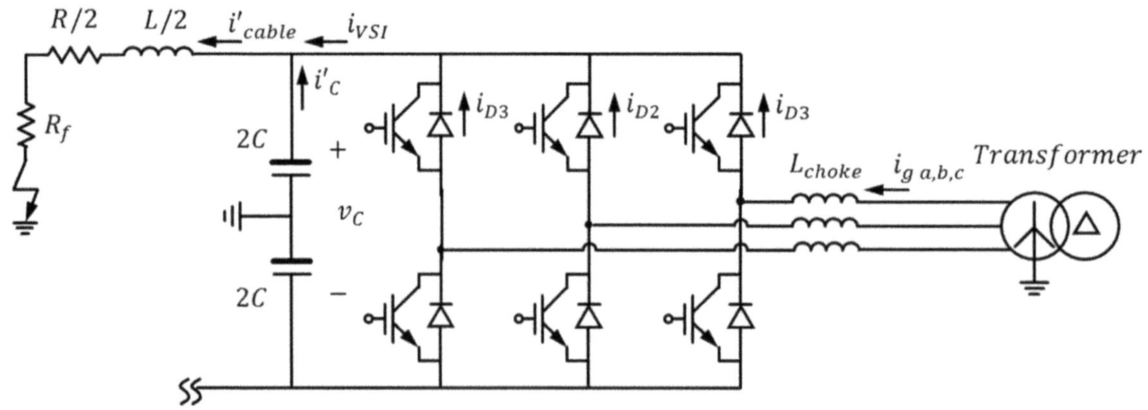

Figure 9.4 VSC with positive cable ground fault condition

$$v_C = Ri_{\text{cable}} + L\frac{di_{\text{cable}}}{dt} \tag{9.9}$$

where

$$A = I_g\left[\left(1-\omega_s^2 LC\right)^2 + (RC\omega_s)^2\right]^{-1/2}$$

$$\gamma = \alpha - \varepsilon - \theta$$

$$\theta = \arctan\left[(RC\omega_s)/\left(1-\omega_s^2 LC\right)\right]$$

$$B = I_{gn}\left[\tau^2/\left(\tau^2 - RC\tau + LC\right)\right]$$

$$C_1 = -(A\sin\gamma + B)\ C_2 = B/\tau - \omega_s A\cos\gamma$$

The most vulnerable components—the freewheel diodes—suffer during the freewheel phase, in which the current (in this model) can be ten times the nominal value (from 0.085 to 0.862 kA) and will rise rapidly (from equation (9.8) and equation(9.9)). The capacitor delivers the discharge current. This current could be eliminated by including a dedicated DC capacitor CB, adding capacitor overcurrent protection, or using high-speed fuses for distribution system capacitor banks. The DC cables should be protected from overcurrent, but they are more robust than the power-electronic devices because of the cable's thermal mass.

Influence of Fault Resistance: Usually, the fault resistance is such that $R < 2\sqrt{L/C}$, where the circuit will experience oscillation. In cases of short-circuit faults, fault resistances are generally small. Sometimes, however, a fault resistance does exist, and a large fault resistance R_f will make the condition $R_f + R > 2\sqrt{L/C}$. This is a first-order damped process. The DC-link voltage will not drop to zero, so no freewheel diode conduction occurs. Thus, the most critical phase can sometimes be avoided. The overcurrent protection relay time setting is not that critical in this case. The damped fault response will be shown with the cable ground fault scenario in which the ground resistance is always considerable.

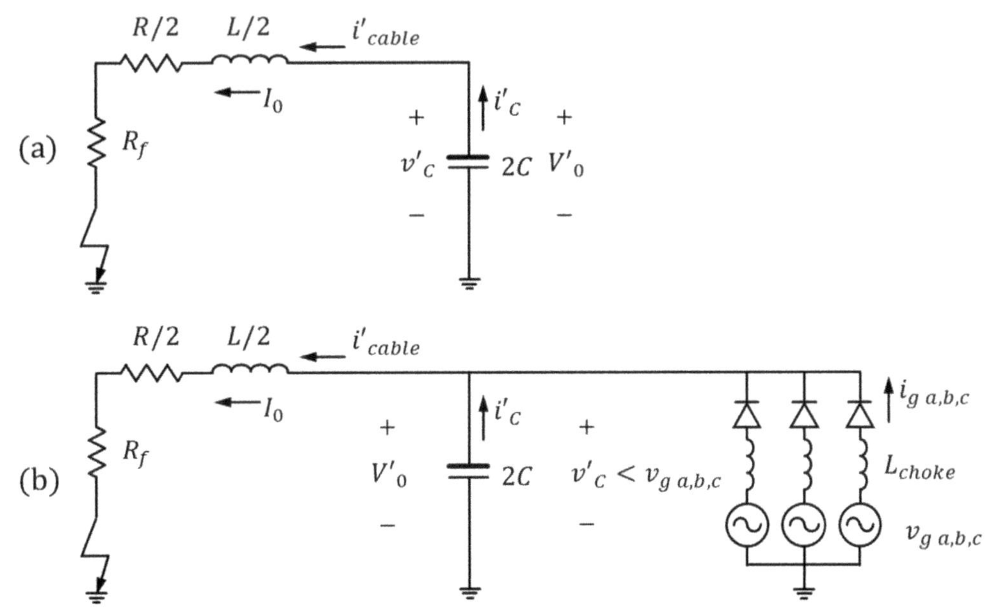

Figure 9.5 Equivalent circuit for VSI cable ground fault: (a) Stage 1—capacitor discharge; (b) Stage 2—grid current feeding

VSC DC-Cable Ground Fault Analysis: Ground fault analysis depends on the grounding of the DC system. Usually, the grounding points in a DC network include the neutral–ground link of the step-up transformer and the DC-link midpoint as shown in Figure 9.4. The DC-link grounding point is used to reduce imbalance between the positive and negative currents and voltages.

A ground fault will form a ground loop with the grounding points. An IGBT-blocked VSI will act like an uncontrolled rectifier with DC-link voltage modified to the rectified voltage, so fault current will flow through the freewheel diodes. This current depends on the impedance between the transformer and the ground fault point. The difference between faults on the positive and negative sides of the DC link is the direction of the current and the bridge diodes that conduct. The ground fault resistance cannot be ignored—it usually varies from amounts smaller than ohms to hundreds of ohms. The equivalent circuit is shown in Figure 9.5 for fault calculation, which is divided into three stages.

Capacitor Discharge Stage (Natural Response): This is the DC-link capacitor discharging stage as represented by Figure 9.5(a). Under the condition of $R_f + R/2 < 2\sqrt{L/4C}$, the solution of the second-order circuit natural response gives a non-oscillating discharge process. The DC-link voltage will not drop to zero, so no freewheel diode conduction occurs. Assuming that the fault happens at time t_0, the natural responses (without inverter-side current i_{VSI}) under initial conditions of $v'_C(t_0) = V_0$ and $i'_{\text{cable}}(t_0) = I_0$ are

$$v'_C = A_1 e^{p_1 t} + qA_2 e^{p_2 t} \tag{9.10}$$

$$i'_{\text{cable}} = 2C\frac{dv'_C}{dt} = A_1 p_1 e^{p_1 t} + A_2 p_2 e^{p_2 t} \tag{9.11}$$

where

$$p_{1,2} = -\frac{R_f + R/2}{L} \pm \sqrt{\left(\frac{R_f + R/2}{L}\right)^2 - \frac{1}{LC}}$$

$$A_1 = \frac{1}{p_2 - p_1} = \left(p_2 V_0 + \frac{I_0}{2C}\right)$$

$$A_2 = \frac{1}{p_2 - p_1}\left(-p_1 V_0 - \frac{I_0}{2C}\right)$$

This stage is an RLC circuit until the positive DC voltage drops to below any grid phase voltage. It is difficult to determine an analytical expression for the time t_1, when capacitor voltage drops below any grid phase voltage, but numerical methods can be used to find the time solution.

Grid-Side Current Feeding Stage (Forced Response): This transient phase can be expressed by third-order state space equations

$$\begin{pmatrix} \dot{v}'_C \\ \dot{i}'_{\text{cable}} \\ \dot{i}_{L_{\text{grid}}} \end{pmatrix} = \begin{pmatrix} 0 & -\frac{1}{2C} & \frac{1}{2C} \\ \frac{1}{L/2} & -\frac{R_f+R/2}{L/2} & 0 \\ -\frac{1}{L_{\text{grid}}} & 0 & 0 \end{pmatrix} \begin{pmatrix} v'_C \\ i'_{\text{cable}} \\ i_{L_{\text{grid}}} \end{pmatrix} + \begin{pmatrix} 0 \\ 0 \\ 1/L_{\text{grid}} \end{pmatrix} v_{ga,b,c} \tag{9.12}$$

where v'_C, i'_{cable}, and $i_{L_{\text{grid}}}$ are the state variables. The choke inductance can also include the transformer and its star-grounding inductance in the case of an arc-distinguishing coil connected in low- or medium-voltage situations. Equation (9.12) provides a theoretical method of determining the short-circuit fault response. However, it is difficult to derive analytical expressions for the voltages and currents during the fault, and so

the fault response must be numerically simulated. There are no particular effects on the diodes (unlike the free-wheel phase during short circuits).

It cannot be solved continuously because of the commutation between diodes. Therefore, for each diode conduction period, the state equations of Equation (9.12) need to be solved using the previous variable as the initial state for the present calculation.

Steady State: The steady-state equations can be determined. The total impedance is

$$Z = \left(R_f + R + j\omega_s L\right) \parallel \left(1/j\omega_s C\right) + j\omega_s L_{\text{choke}} = |Z|\angle\theta' \tag{9.13}$$

Then, the current through a diode is

$$i_{D1} = i_{ga,(>0)} = \frac{V_g\angle\alpha}{|Z|\angle\theta'} = \frac{V_g}{|Z|}\angle\alpha - \theta' \tag{9.14}$$

DC Fault Location: Traditional AC network distance protection uses impedance to represent the distance from the relay point to the fault point. This distance judgment is made with a mho characteristic or an impedance circle. However, for a DC system in a fault transient state, the frequency changes abruptly. No grid fundamental frequency impedance can be defined for distance protection. Therefore, an electrical circuit parameter evaluation can be used to represent the fault distance.

A. Short-Circuit Fault Location: A system short-circuit response is featured by the distance characteristics of overcurrent value and critical time for the freewheel effect. This critical time limit is when the DC-link voltage drops to zero and the free-wheel diodes conduct, assuming that there is no dedicated or effective DC-link capacitor protection to prevent the discharge. This critical time should be the upper limit before which both the main and backup DC circuit breakers operate. Using Equation (9.3), in respect to the distance x, the critical time is

$$t_c = t_1 - t_0 = [\pi - arctan\,(V_0 C\omega'/\,(V_0 C\delta - I_0))]\,\omega' \tag{9.15}$$

where

$$\omega' = \sqrt{\omega_0^2/x - \delta^2} \tag{9.16}$$

The total freewheel overcurrent is the cable current at this critical time.

Two DC voltage dividers are used for distance measurement and representation. The measurements and distance relationship are shown in Figure 9.6. The fault voltage at switchgear relay point (n) is where x^* is the real fault distance and R_f is the fault resistance.

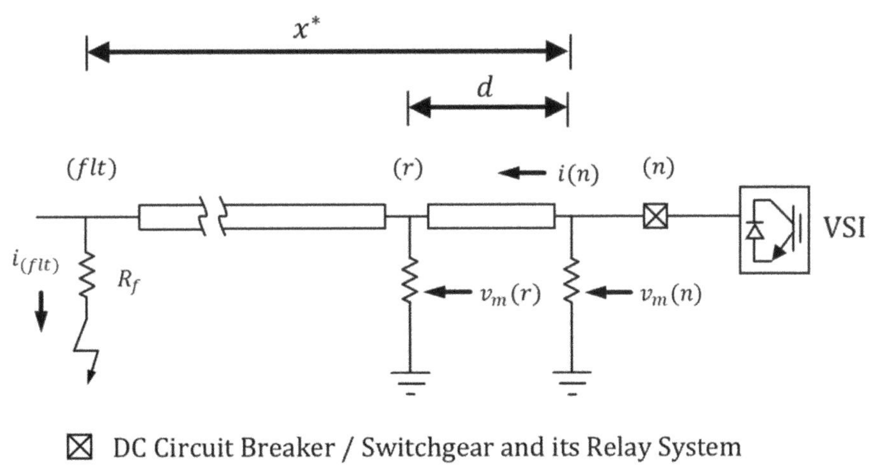

Figure 9.6 Distance evaluation with two voltage divider measurements

$$V_{(n)} = V_{(flt)} + x^* ri_{(n)} + x^* l \frac{di_{(n)}}{dt} = R_f i_{(flt)} + x^* \left(ri_{(n)} + l \frac{di_{(n)}}{dt} \right) \tag{9.17}$$

Another relay voltage sensor unit (r) is used as reference for the relative voltage calculation; it is located near the main relay point on the same section of the cable to avoid communications over distance. The measured values using voltage dividers are $Vm(n)$ = kvV(n) and $Vm(r) = V(r)$, where kv is the voltage divider ratio. The distance d between them is known, so the fault distance measured from this reference is

$$x = \frac{V_{(n)}}{V_{(n)} - V_{(r)}} d = \frac{V_{m(n)}}{V_{m(n)} - V_{m(r)}} d \tag{9.18}$$

where

$$V_{(n)} - V_{(r)} = d \left(ri_{(n)} + l \frac{di_{(n)}}{dt} \right) \tag{9.19}$$

For metallic grounding or short-circuit faults, $v(flt) = R_f i(flt)) = 0$, so the cable impedance is in proportion to the distance: the measured distance $x = x^*$. For high-resistance faults, which are more common in ground faults, the existence of R_f and the difference between $i(flt)$ and $i_{(n)}$ make the evaluation of x^* difficult. Equation (9.19) presents the real voltage drop between the two relay points, which reflects the real voltage drop excluding the influence of i(flt). However, R_f still cannot be exactly obtained even with the source side tripped. Usually, this kind of fault is not as serious as the metallic grounding or short circuit faults, and may not require fast time-response protection, which can be fulfilled by overcurrent setting.

B. Ground Fault Location and Ground Resistance Evaluation: In three-phase AC systems, distance protection uses symmetrical component analysis to avoid the aforementioned influence of fault resistance. However, in DC systems, this is not available. Ground faults are not as serious as the short circuit condition, as the grounding always comes with a large fault resistance; however, they occur more frequently. Moreover, a large fault resistance results in an inaccurate evaluation of distance for protection coordination. Generally, as resistance and distance increase, the fault overcurrent reduces, and the time during which the diodes are conducting increases.

Based on the aforementioned analysis, a new fault location approach for distance and ground resistance evaluation is proposed here for online applications. The results can also be used for offline maintenance and fault location without injecting signals into a faulty cable, or a prediction before the application of time-consuming location tracing methods. With the measurement values of $v'_{C,mea}$ and $i'_{\text{cable},mea}$, and the time when $v'_{C,mea}$ drops to below any phase value of the grid voltages $v_{g\,a,b,c} - t_{1,mea}$, the fault loop total resistance R_{total} and inductance L_{total} can be solved through

$$\begin{cases} R_{total} = R_f + r \cdot x \\ L_{total} = l \cdot x \end{cases} \tag{9.20}$$

9.1.2 Characteristics and Recovery Performance of VSC HVDC [5]–[9]

9.1.2.1 The Choice of DC Line and Fault Types

Normally, the DC transmission line of VSC HVDC can be composed by:

- DC cable
- DC overhead line

The choice between the use of cables and overhead lines is determined by environmental constraints, as well as an overall optimization considering the total capital cost and the reliability of the transmission system. Generally, the environmental impact of cable will be much smaller compared with that of an overhead line.

The common fault types for DC transmission lines are line-to-ground fault and line-to-line fault. Because faults of cables are usually caused by external mechanical stress, the faults are generally permanent, and require lengthy repair. The converters should be stopped immediately when a cable fault is detected. But for an overhead line, the faults are always caused by lightning strikes and pollution; faults along the line are likely to be temporary, which merely demands a fault restoration after the fault clearance.

Line-to-ground fault: As shown in Figure 9.7, the line-to-ground fault means insulation failure between one DC conductor and ground. For overhead lines, the fault is usually temporary, and is caused by lightning strikes and pollution. The primary fault consequence is the direct discharging of the capacitor on the faulty pole. The voltage of the healthy pole will rise to 2 p.u. because of the action of the DC voltage controller, which will impose an overvoltage on the healthy pole capacitor even if the restriction of the DC pole arrester is considered. For DC cables, the system does not need to be restored because of the permanence of the fault, but for overhead lines, restoration of the system should be considered.

Line-to-line fault: A line-to-line fault means insulation failure between two DC conductors. The capacitor will be rapidly discharged because of fault, while the AC system will be three-phase short-circuited through the fault point, as shown in Figure 9.8. Obviously, the valve will be over current stressed.

In the line-to-line fault condition, the AC system is still short-circuited through the VSC Free Wheeling Diodes (FWD), which means that it will continue to feed current into the fault even if the converter is blocked. To avoid this, besides blocking the converters, the DC line also needs to be isolated from the AC system by tripping AC breakers to enable the air insulation to deionize. Another method is to introduce DC breakers and open them when a fault is detected.

Protection and Recovery Demands

Line-to-ground fault

The required course of action for cable faults is to correctly detect the faults and then block and trip the converter for repairing. The course of action for overhead line faults is also to block and trip the converter, but a recovery process should be involved. However, as the capacitors are difficult to rebalance even after the fault is cleared, the rebalance of capacitors becomes the key point before system restart. Typically, the capacitor should first be discharged, followed by a normal start-up of the system. If the unbalance of the

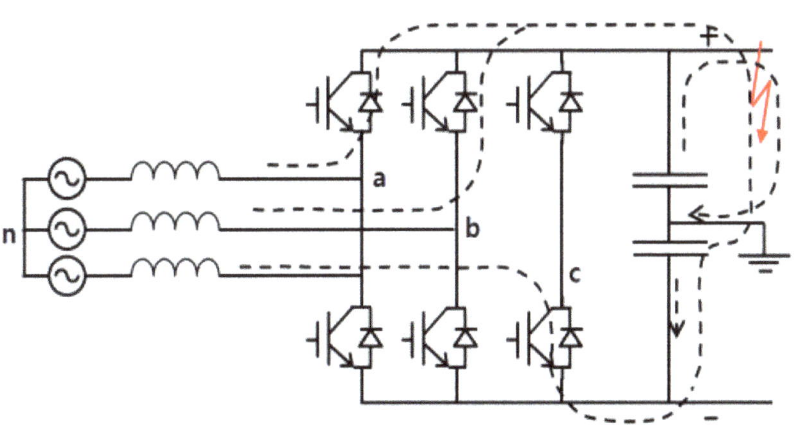

Figure 9.7 Discharging path under line to ground fault of bipolar system

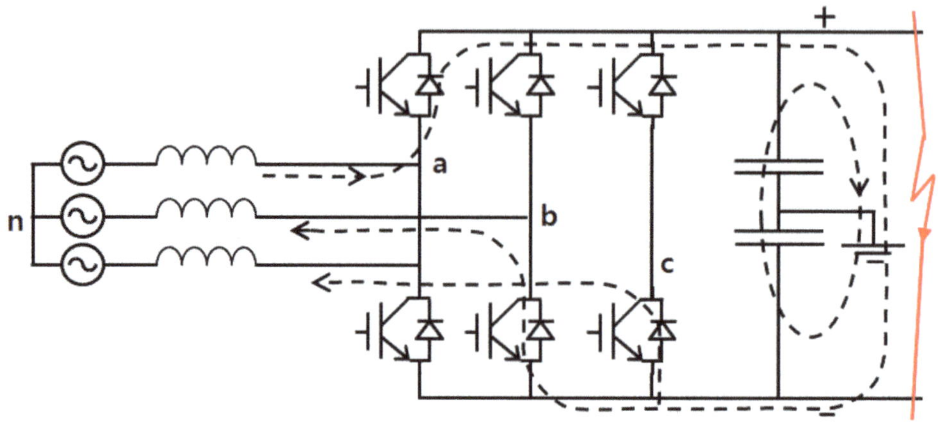

Figure 9.8 Discharging path under line-to-line fault of bipolar system

DC voltage disappears, then the system can be normally restarted; otherwise, the whole system should be stopped for a fault investigation.

If there is a convenient method to rebalance the capacitors, the converters do not need to be stopped because there is no current entering the fault point after the faulty pole capacitor is fully discharged. After the fault is cleared, the capacitors can be rebalanced automatically.

Line-to-line fault

Another requirement for cable faults is to first correctly detect the faults, then block and trip the converter for repairing. To avoid the over current stress of FWD after blocking, DC breakers can be introduced.

Overhead line faults also sends the the signals of the blocking and trip of the converter to converters. After a certain time, the AC CB with a closing resistor should be closed to charge the DC capacitors, and if the DC voltage can be re-established, the converters can be deblocked for normal operation. Otherwise, the whole system should be stopped for a fault investigation. DC breakers can also be introduced to avoid the overstress of FWD, but the restarting procedure of the converters should be reconsidered.

Solutions for Fault Recovery

The main issues in fault recovery are the rebalance of the capacitors and the design of the fast DC breakers.

Grounding by high-impedance branch

In a direct grounding system, the unbalance of the capacitor is caused by the discharging of the faulty pole and the DC voltage controller. In a high-impedance grounding system, the discharging current is very small so that the voltage of the capacitor can be maintained without any over current stress.

Change transformer secondary winding to Yn type

By changing the transformer secondary winding to Yn, the unbalance of the capacitors will rebalance automatically through the path shown in Figure 9.9. However, if there is no suppressing method, the balancing current will be too large for secure operation. According to this, a high resistor can be added to the neutral point of the transformer.

Monopolar scheme operation

The common arrangement of a monopolar scheme is shown in Figure 9.10. Obviously, the consequences of a line-to-ground fault of the monopolar system are similar to those of line-to-line faults; there is no unbalance during fault.

For an overhead line system, adopting a monopolar scheme has the following advantages:

- The system can adopt a DC current grounding return arrangement to save investment
- The system is generally grounded by the positive pole, so the total corona noise and fault probability can be reduced

Generally, the monopolar scheme is better for overhead line systems compared with the bipolar scheme.

Installation of equalization resistors

Equalization resistors, which are installed in parallel across the capacitors, will take a long time to rebalance the voltage because of the large resistance.

The design philosophy of fast DC breakers

The basic requirement of fast DC breakers is that they isolate the DC pole and converter punctually so as to avoid the AC system feeding current to the DC side by a free-wheeling diode (FWD). A representative design is shown in Figure 9.11.

By using an IGBT with an anti-paralleled diode, the normal operating current can pass through the IGBT on the sending end and the diode on the receiving end. Although the DC line is under fault condition, the discharging current will force the IGBT to turn off immediately to realize the pole isolation. This can avoid the AC breaker tripping, and the converter can restart faster.

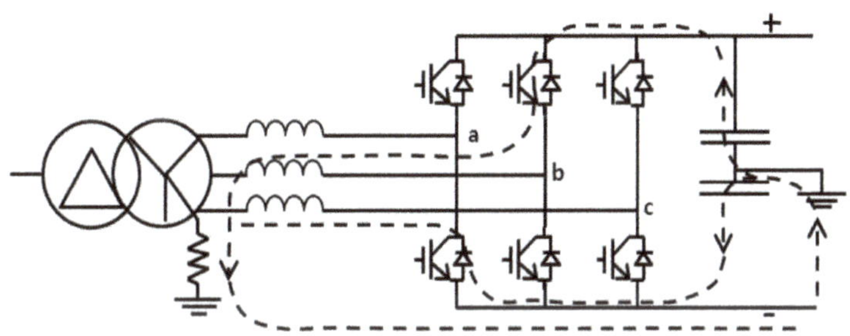

Figure 9.9 DC voltage rebalancing path under temporary line-to-ground fault

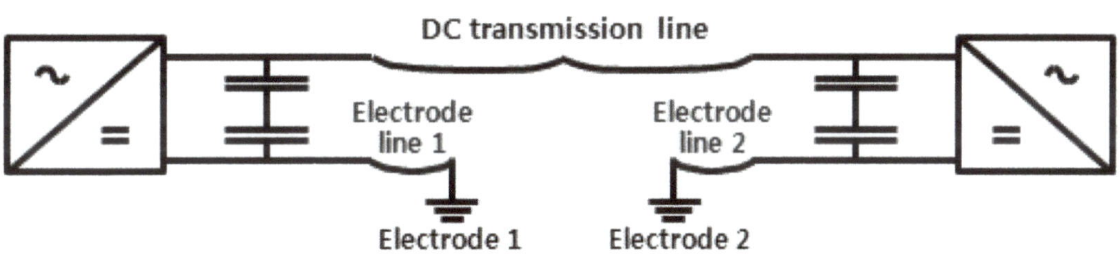

Figure 9.10 Arrangement of monopolar scheme

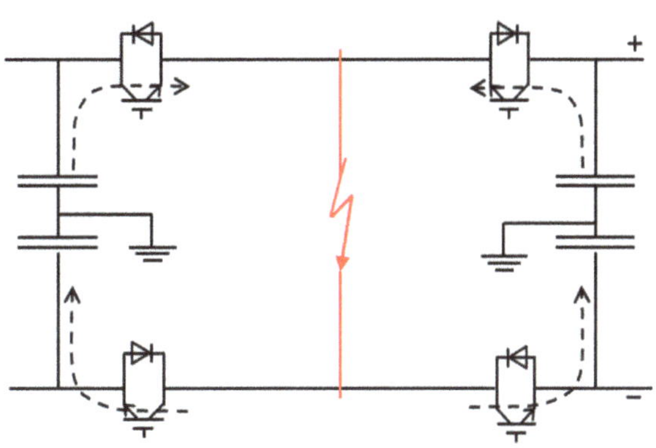

Figure 9.11 Discharging current direction in fast DC breaker under line-to-line fault

9.2. DC Protection Functions

The protection system needs to act as a background monitor to confirm the correct operation of the converter and to switch it off in the event of the measured operation going out of a given set of bounds. So, the system may be considered as divided into two categories:

- Those that monitor general operation and would be used within existing HVDC systems
- Those relevant to the converter operation, in particular with the associated module operation and capacitor state of charge

Figure 9.12 shows the DC protection types of the VSC HVDC system. The function of each protection function is as follows.

1. **DC Overcurrent:** This protects against DC over currents in the HVDC link, which can result from control failures such as failure of the control loop or the loss of the current measurement signal to the control system.
   Protection Task
   This protection function detects the following faults:
   - Secondary busbar solid line-to-ground faults
   - Secondary busbar high-impedance line-to-ground faults
   - Converter busbar solid line-to-ground faults
   - Converter busbar high-impedance line-to-ground faults

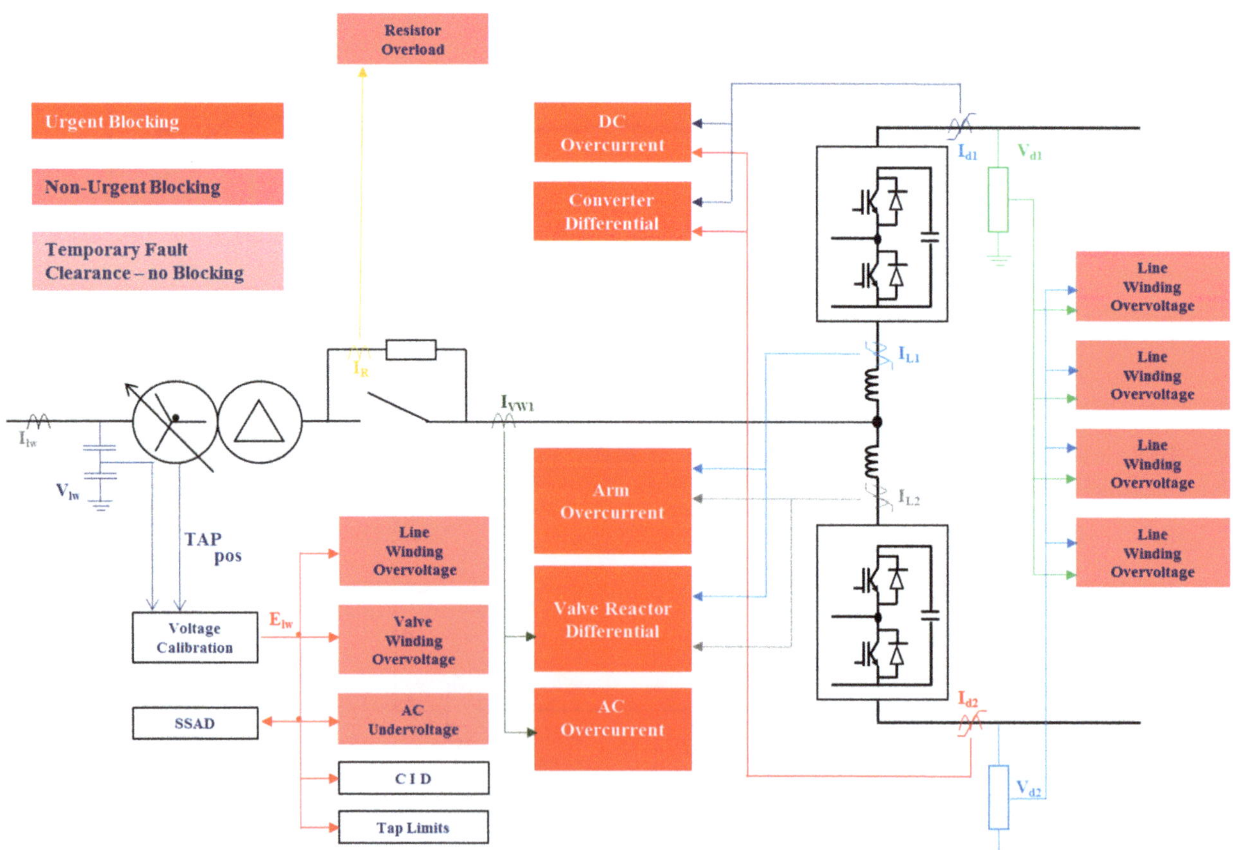

Figure 9.12 Protections of VSC HVDC

- DC busbar / terminal solid ground faults
- DC busbar / terminal high-impedance ground faults

Initiated Sequences

- Opening of AC CB
- Emergency DC Cable grounding

Protection Coordination
The overvoltage protection must be coordinated with the component overvoltage capability and the threshold of the HV-arresters. Protection must react for unallowable high-voltage conditions.

2. **Pole DC Differential**: This protects against faults to earth in the converter station HVDC main circuit. It monitors the difference between the currents measured in the DC connections. Any difference would be caused by a flashover to ground or the failure of a ZFCT.
   This protection function detects the following faults:
   - Secondary busbar solid two line-to-ground faults
   - Secondary busbar solid line-to-ground faults
   - Secondary busbar high-impedance two line-to-ground faults
   - Secondary busbar high-impedance line-to-ground faults
   - Converter busbar solid line-to-ground faults
   - Converter busbar high-impedance line-to-ground faults
   - DC busbar / terminal solid ground faults
   - DC busbar / terminal high-impedance ground faults

   Protection Coordination
   This protection function has to be coordinated with converter energization.
   Back-Up Function
   DC Undervoltage Protection, DC Overvoltage Protection, Power Differential Protection, Phase Current Imbalance, DC Current Differential Main Protection, DC Current Differential Backup Protection, Positive Busbar Differential Protection, Negative Busbar Differential Protection.

3. **Low DC Current:** This checks for the minimum expected DC current level when the converters are deblocked (which corresponds to the minimum power transfer level), and also provides protection against the failure of one converter side to deblock.

4. **Valve winding Overvoltage:** This protects the valve side equipment against the stresses caused by valve overvoltages.

5. **AC Undervoltage:** This detects undervoltage in the valve side windings and is used to inhibit the data back function in VBE on start up.

6. **AC Overcurrent:** This protects against AC overcurrent in any of the valve winding phases, which can result from phase to phase valve winding faults or control failure.
   Protection Task
   This protection detects the following faults:
   - Secondary busbar solid three phase-to-ground faults
   - Secondary busbar solid two phase-to-ground faults
   - Secondary busbar solid line-to-line faults

7. **Asymmetry:** This detects the presence of fundamental frequency and second harmonic voltages between the DC terminals of the pole caused by, for example, a valve misfire or a commutation failure.
   Protection Task:
   This protection function detects the following faults:
   - Secondary busbar solid two line-to-ground faults
   - Secondary busbar solid line-to-ground faults
   - Secondary busbar high-impedance two line-to-ground faults
   - Secondary busbar high-impedance line-to-ground faults
   - Converter busbar solid line-to-ground faults
   - Converter busbar high-impedance line-to-ground faults
   - DC busbar / terminal solid ground faults
   - DC busbar / terminal high-impedance ground faults

   Protection Coordination
   This protection function has to be coordinated with converter energization.
   Back-Up Function
   DC Undervoltage Protection, DC Overvoltage Protection, Group Power Differential Protection, Phase Current Imbalance, DC Current Differential Main Protection,
   DC Current Differential Backup Protection, Positive Busbar Differential Protection, Negative Busbar Differential Protection.

8. **Line winding Overvoltage:** This protects the converter against AC overvoltages on the AC busbars.

9. **Tap Limits:** The tap-changer limit signals are "Tap Down" and "Inhibit Raise," which act on the tap-changer to prevent long-term overvoltages.

10. **Earth Fault Detector:** The valve voltage is monitored by use of the capacitive taps of the transformer secondary bushings. When the relay detects an undervoltage, it is deemed to be an earth fault of the transformer secondary bushings. The signal is only used to inhibit the deblocking of the converter if an earth fault is detected.

11. **Sub-synchronous Activity Detector (SSAD):** The SSAD is used to detect oscillations in AC power at modulation frequencies below the AC system fundamental frequency. This is a commissioning tool only and is normally switched off. Its relevancy is uncertain. It is possible that the stability of the converter will interact with high Q resonances within the AC system.

12. **Converter Isolation Detector:** The Converter Isolation Detector is used to detect the condition known as "islanding." This is where the inverter becomes isolated from its AC system because of remote breaker(s) opening, but DC power transfer may continue feeding power into the isolated converter. This may result in an overvoltage that could cause damage to main circuit components without the response of this protection.

13. **Abnormal Neutral Area Voltage**
    Neutral-bus overvoltage (NBOV): This protection detects DC overvoltages on the neutral bus. The most likely cause of a NBOV is an open circuit in the electrode line or problems with the electrode causing an unusually high resistance. The LV bus voltage (Vn) is measured using a voltage divider.

14. **Common Neutral Area Protection:** This protection detects a short circuit in the common electrode line. It covers the common neutral bus area between the poles and the electrode line that is outside of the pole differential protection or filter protection zones.

9.2.1 Backup AC Protection behind MMC System [10] – [14]

The VSC HVDC system has different protection actions depending on whether the valve is configured with half-bridge or full-bridge, as well as the symmetry or asymmetry method.

As the half-bridge converter cannot interrupt the fault current when a DC fault occurs, it is necessary to prepare two AC breakers for primary protection and backup protection on the AC side. On the other hand, the primary protection for the full-bridge is the valve and the backup protection is the AC breaker.

Figure 9.13 shows this classification. Figure 9.13(a) uses the symmetry half-bridge method and requires two AC breakers on the AC side for primary protection and backup protection. Figure 9.13(b) shows the full-bridge type. The primary protection is the valve, and the backup protection is the AC breaker. Figure 9.13(c) shows the method using the asymmetry half-bridge type, with two AC breakers on the AC side for primary and backup protection. Figure 9.13(d) shows the asymmetry fulll-bridge type. The primary protection is the valve and the backup protection is the AC breaker.

The difference between the asymmetry scheme and the symmetry scheme is where the converter's ground is located. Since the asymmetry method grounds the negative line, a transformer that can withstand DC Bias should be used instead of a general transformer.

The relationship between the primary protection and back up protection of the AC breaker is as follows.

Local Backup Protection in AC Systems

The protection of an HVDC grid fulfills the same objectives as that of an AC system, for example, minimizing the impact of faulty elements on grid operation and preventing damage to system equipment. However, protection of an HVDC grid is more difficult than protection of the AC system as prospective DC fault currents rapidly reach high values. These currents should be quickly interrupted to prevent damage to the power-electronics components of the HVDC converters. Although recent advances in DC breaker technology have enabled fast fault current interruption, current interruption capabilities of existing prototypes are limited.

Consequently, fast protective relaying algorithms are required to promptly detect the fault and identify its location. Primary protection provides the fastest fault clearing for elements within the primary protection zone whereas backup protection must be provided in case of primary protection failure. The primary protection can fail due to the malfunction of breakers or relays.

The main requirement of HVDC grid backup protection is high operation speed. A delay in the backup operation will result in DC fault currents with high steady-state values that existing breakers with limited interruption capability will no longer be able to clear. However, a delay between primary and backup protective actions is required to prevent backup before primary actions.

The HVDC grid backup relaying algorithms proposed in literature do not provide a high operation speed. The proposed algorithm delays backup actions until the primary protection reduces the fault current to a near-zero value. This results in a considerable delay on backup actions as breaker series inductors also limit the rate of change of the current during fault current interruption. In low-speed HVDC, grid protection based on mechanical DC breakers is proposed. The fault clearing times of the primary and backup protections are 60 and 20 ms, respectively. The backup protection requires fault current limiting equipment, which results in additional costs.

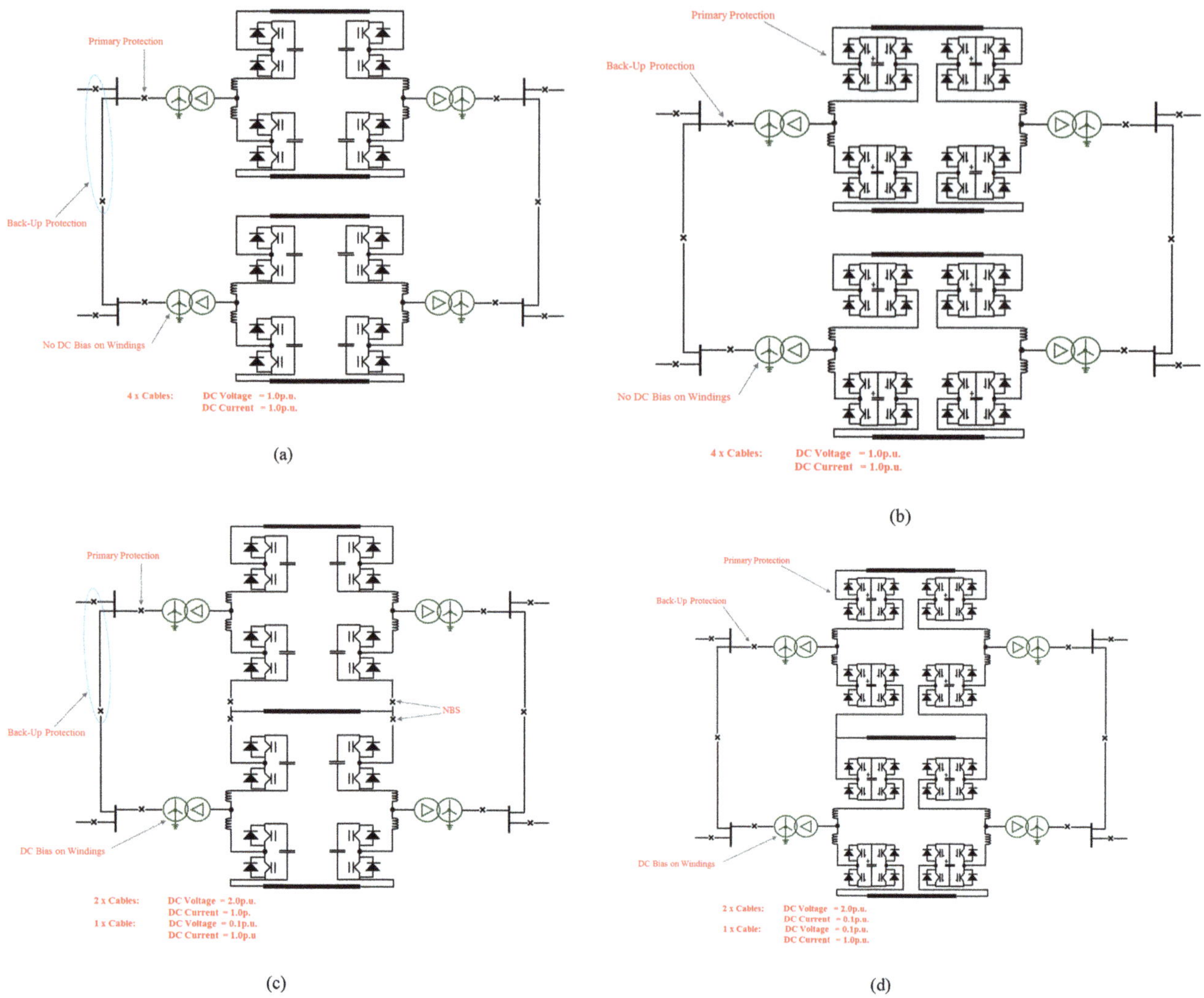

Figure 9.13 Configuration of VSC HVDC according to grounding types and valve types

In AC systems, the fault clearance time in case of primary protection failure is the sum of the primary fault clearance time, a delay between primary and backup actions, and the backup protection fault clearance time (Figure 9.14). The primary protection system detects the fault at t_p^d, which is the sum of time required for the fault wave to reach the relay (t_f), the processing time required to detect the fault, and the time that must pass before a trip signal is sent to the associated breaker. The associated breaker opens att_p^o and interrupts the fault current at t_p^c. The backup protection initiates the detection of primary protection failure at t_b^s, which is the sum of $t_p^c$and the delay Δt_{pb}. If primary protection failure is detected at this point, the backup protection trips the breakers on all lines adjacent to the faulted line. The adjacent breakers open at t_b^o and interrupt the fault current at t_b^c. The total time required to clear the fault in the case of primary protection failure is from t_f to t_{bc}.

Although in AC systems Δt_{pb} is used to provide a margin between primary and backup actions, the delay of backup actions in HVDC grids should be minimized. In AC backup protection, Δt_{pb} is typically set to multiple cycles of the fundamental frequency to avoid the potential tripping of multiple circuit breakers due to the misoperating of the backup protection. For HVDC grid protection, such a large margin would lead to impractical requirements on ratings of DC breakers and HVDC converters.

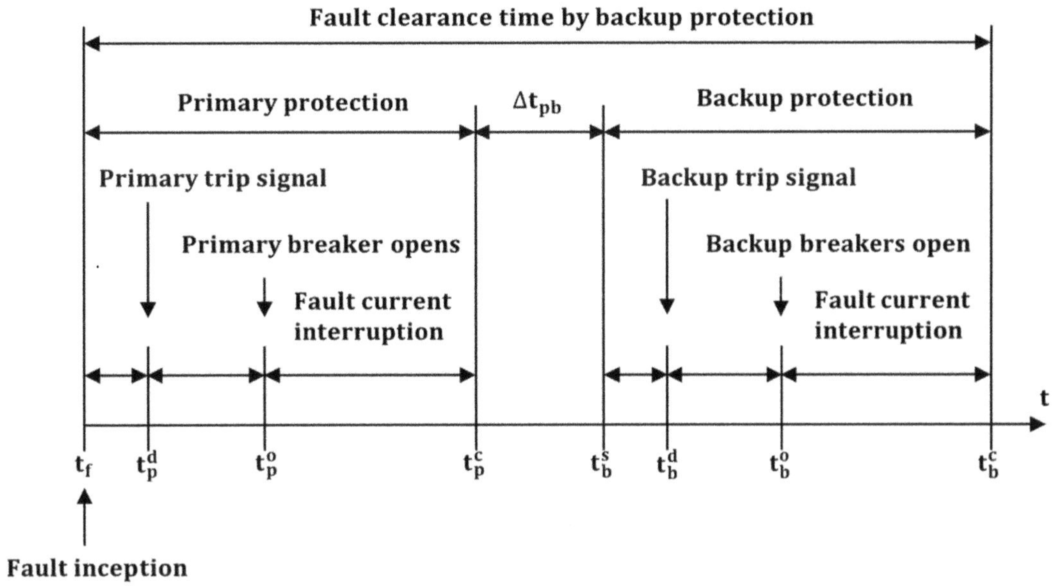

Figure 9.14 Primary and backup protection continua in AC systems

Loci of DC Fault Currents and Voltages with Primary and/or Backup Protection in Service

A short-circuit fault in an HVDC grid, in the initial stage, can be characterized by an increasing current and a decreasing voltage in the faulted pole. In the case of pole-to-ground faults in a high-impedance grounded system, the fault current in the faulted pole oscillates and returns to zero. In the case of a pole-to-ground fault in a low impedance grounded system or a pole-to-pole fault, the current increases to a high prospective steady-state value. Figure 9.15 provides an overview of the loci of the DC voltage and current for a cleared and an uncleared fault as a function of time (assuming a high prospective steady-state fault current). To simplify Figure 9.15, oscillations caused by electromagnetic effects were excluded. In Figure 9.15, points 1 and 2 correspond to fault inception and breaker trip instants ($t_f$and $t^o{}_p$), respectively. The solid line from point 1 to 2 shows the variations in current and voltage measured at the primary relay during the $[t_f t_p{}^o]$ interval as the fault is detected by the primary protection system and the primary breaker trip signal is generated. The dashed line between points 2 and 5 shows the loci of DC voltages and currents after the primary breaker trips and the faulty line is removed. As the fault is being cleared, the voltage increases from $U_f^{o,p}$ to U_f^0 while the current decreases from $I_f^{o,p}$ to 0.

In Figure 9.15, the solid line from point 2 to 4 shows the loci of DC voltages and currents measured at the primary relay after primary protection failure. For an uncleared fault, the DC voltage decreases from $U_f^{o,p}$ to $U_f^{s,}$ whereas the current increases until a steady-state value, I_f^s, is reached at point 4.

If the backup protection clears the fault after primary protection failure, the DC voltages and currents will follow the path from point 2 to 3 and then to point 6 (in case of relay failure) or 6' (in case of breaker failure). Based on the backup relaying algorithm, the backup protective actions begin at $t^o{}_p$ and continue through the $\left[t_p^o,\ t_b^s\right]$ interval. During the $\left[t_p^o,\ t_b^s\right]$ interval (from point 2 to 3) the current increases from $I_f^{o,p}$ to $I_f^{o,b}$. If the backup protection does not operate fast enough, $I_f^{o,b}$ will exceed the current interruption capability of the DC breakers and the fault cannot be cleared. For existing DC breaker prototypes, the maximum interruptible current is typically reached within a few milliseconds. The dotted line from point 3 to 6 shows DC currents and voltages at the primary relay as all transmission lines connected to the same bus as the faulty line are disconnected to clear the fault. At point $6\left(t_b^c\right)$, the backup protection clears the fault and the DC fault current becomes zero.

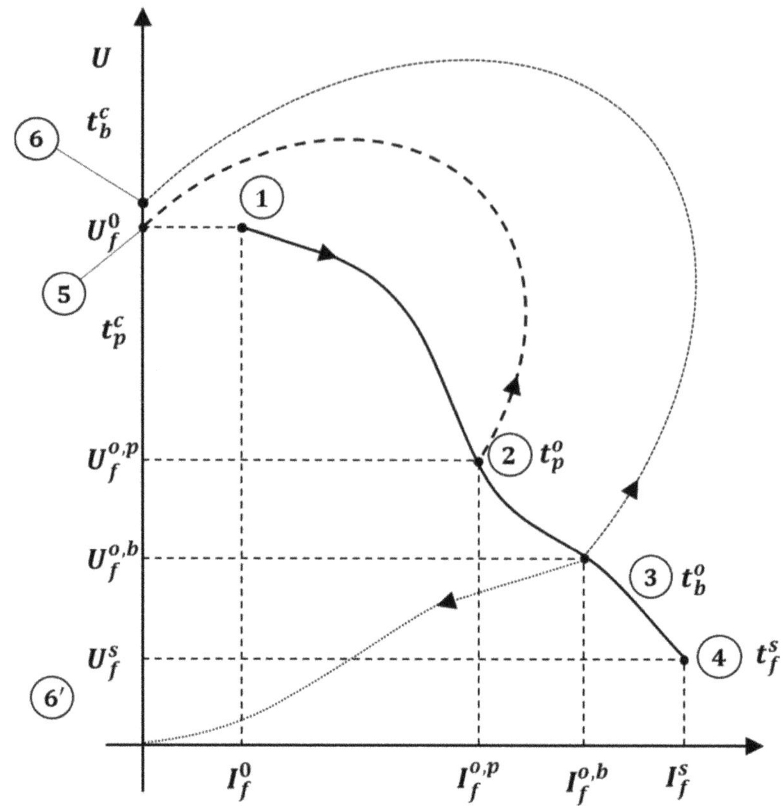

Figure 9.15 Conceptual sketch of loci of the DC fault currents and voltages at the primary and adjacent relay for a cleared and an uncleared fault

Backup Protection Principles

The proposed local backup protection algorithm achieves a high operation speed by detecting primary protection failure immediately after the instant at which the primary protection is expected to initiate fault current interruption (Figure 9.16). In Figure 9.16, similar to Figure 9.15-, t_f, $t^o{}_p$, and $t^o{}_p$ correspond to the fault inception, fault detection, and expected breaker opening instants. From the instant the primary protection is expected to initiate fault current interruption, $t^o{}_p$, the local backup relaying algorithm starts tracking uncleared faults. If the primary protection succeeds in interrupting the fault current, the backup relaying algorithm detects a cleared fault and no backup protective actions are carried out. If the primary protection fails, the backup relaying algorithm detects an uncleared fault at $\hat{t}^d{}_b$ and trips the adjacent breakers to interrupt the fault current. The adjacent breakers open at $\hat{t}^o{}_b$ and the backup protection clears the fault at $\hat{t}^c{}_b$.

Compared to algorithms based on an AC backup protection philosophy, the proposed algorithm poses less stringent requirements on ratings of HVDC grid components as the fault clearance time, and thus the system's exposure to abnormal DC voltages and currents, is shorter ($\hat{t}^c{}_b < t^c{}_b$). A comparison between Figures 9.14 and 9.16 show that the difference between $t^c{}_b$ and $\hat{t}^c{}_b$ is the sum of the fault current interruption time by the primary protection ($t^c{}_p - t^o{}_p$) and the time delay between primary and backup protective actions Δt_{pb}, that is, $t^c{}_p - t^o{}_p + \Delta t_{pb}$.

Figure 9.17 summarizes the backup relaying algorithm steps and the inputs and outputs of the backup protection system.

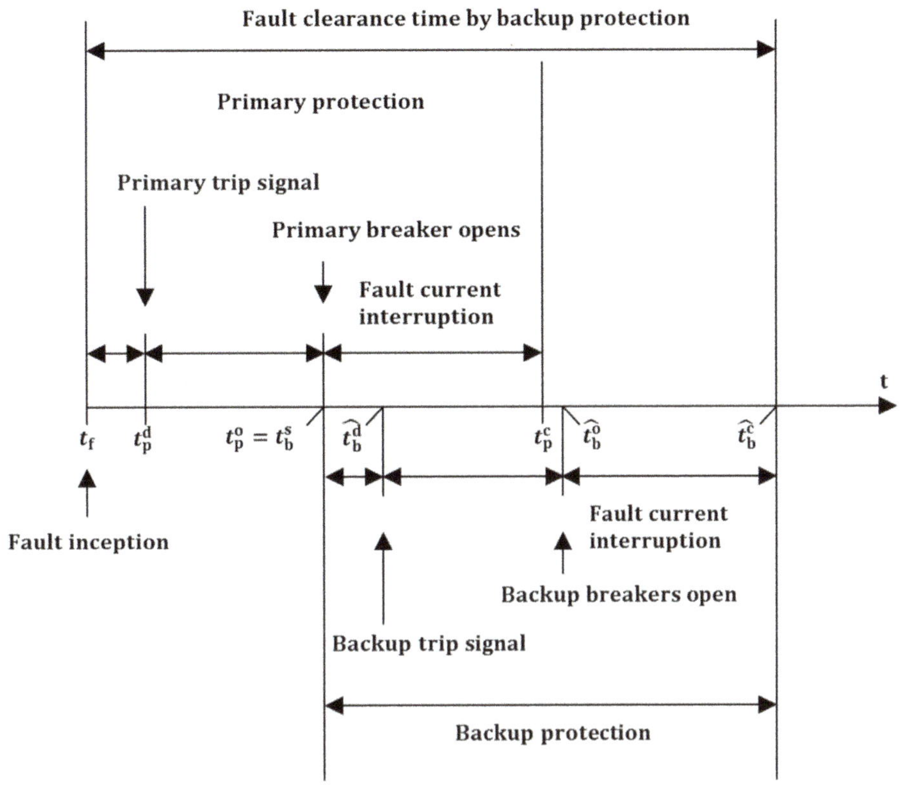

Figure 9.16 Primary and proposed local backup protection continua in HVDC grids

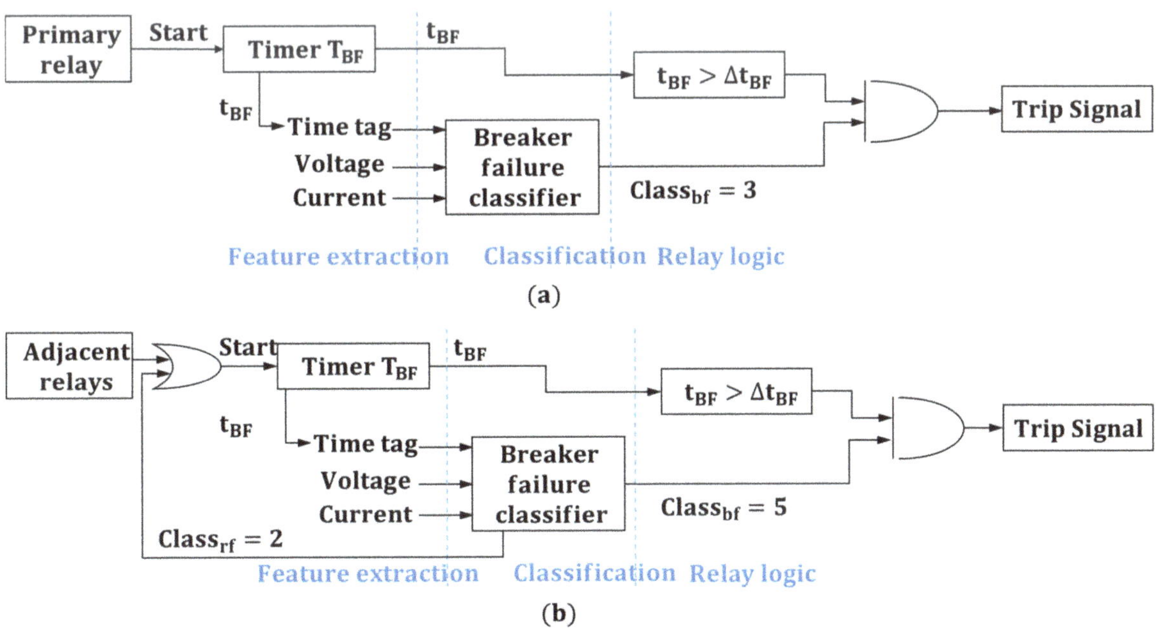

Figure 9.17 Block diagram of backup protection system. (a) Breaker failure subsystem, (b) Relay failure subsystem

9.3 Basic Insulation Coordination [7]

The insulation co-ordination process for a VSC HVDC converter station serves two purposes:

- To ensure that major items of expense (such as transformers and valves) are adequately protected from damage in the event of impulse overvoltages appearing on the system
- To optimize parameters such as BIL, BSL, and clearance to provide a reliable and economically viable converter station design

Before an arrester is determined, the voltage determination factor has to be considered. The voltage determination factor is a part of determining the rating of the arrester. The process is as follows:

The ratio between V_{1mA}, the arrester voltage at lmA discharge current, and V_{rmp}, the maximum repetitive operating voltage of the valve, is called "the voltage determination factor," which is defined by the next equation.

$$S = \frac{V_{rmp}}{V_{1mA}} \tag{9.21}$$

Figure 9.18(b) shows some of the results obtained so far from the lifetime test. The voltage application factor S should be less than 1.0 when the ZnO element's lifetime of more than 10 years is required at the environmental temperature of 60°C.

Figure 9.19 shows the surge arrester arrangement of the VSC converter station.

The following insulation withstand safety margins have been used:

	Valves*	Other Equipment (%)
Switching impulse safety margin (SSM)	25	5
Lightning impulse safety margin (LSM)	25	25

*As at the modular level the VSC valve has significant capacitance parallel to the valve, surge arrester is not applied across the valve.

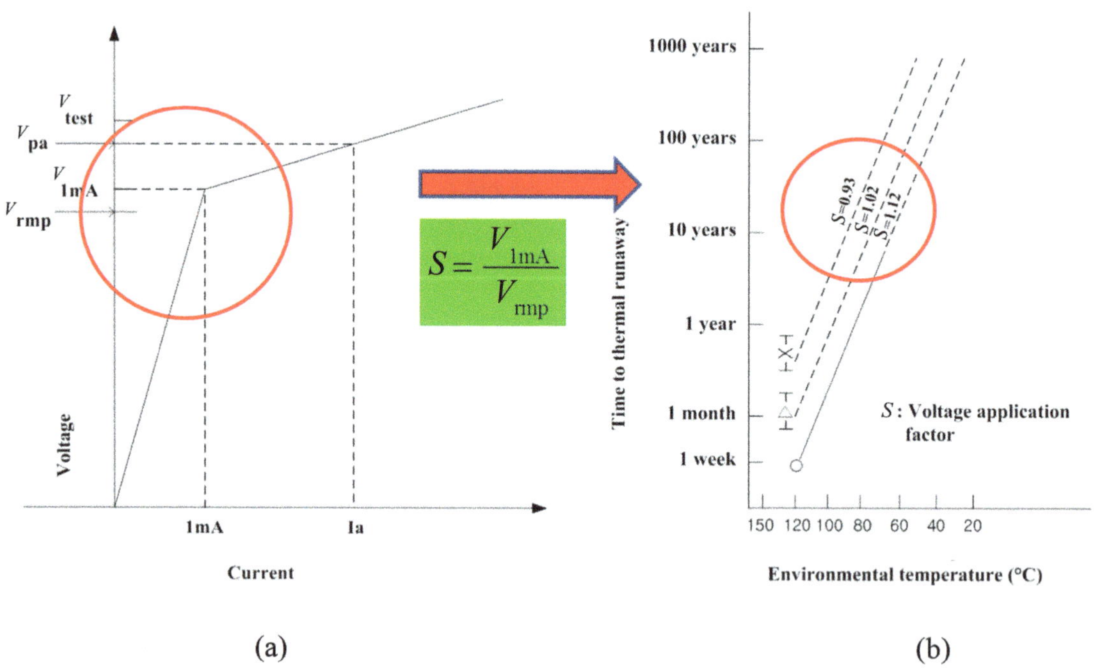

Figure 9.18 Surge arrester determination method

- **Overvoltages:**
  - Switching & Lightning
  - AC & DC
  - Limited by SAs

- **Surge Arresters:**
  - Critical components directly protected
  - Close to the equipment
  - Determined by different types of faults

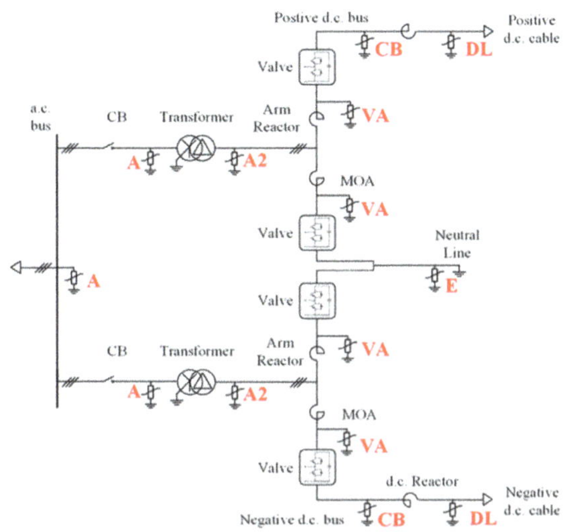

No.	Fault Events	A	A2	VA	CB	DL	E
S0	Clearance of earth fault of AC network	×	×				
S1	Earth fault between valve and arm reactor		×	×	×	×	×
S2	Earth fault between the secondary side of transformer and arm reactor		×	×	×	×	×
S3	Earth fault between arm and smoothing reactor			×	×	×	×
S4	Earth fault of DC line			×	×	×	×
S5	Loss of DC path			×	×	×	×
S6	Deblocking rectifier with blocked inverter			×	×	×	×
L1	Lightning from AC network	×	×	×	×	×	×

Figure 9.19 Surge arrester arrangement of VSC HVDC station

Atmospheric Correction Factor: The atmospheric correction factor $\mathbf{E_f}$ is given by:

$\mathbf{E_f = K_d * K_h}$

where $\mathbf{K_d}$ = air density correction factor, $\mathbf{K_h}$ = humidity correction factor.

The air density correction factor is given by

$$\mathbf{K_d} = \left[\frac{\mathbf{b}}{\mathbf{b_0}} * \frac{273 + \mathbf{t_0}}{273 + \mathbf{t}}\right]^{\mathbf{m}}$$

where

$\mathbf{t_0}$ = 20°C (standard reference temperature)

$\mathbf{b_0}$ =1.013*10s N/m^2 (standard reference pressure)

T = maximum ambient temperature

b = normal barometric pressure

m = 1

AIR CLEARANCES

DC Side: DC side clearances are calculated according to techniques described in ELECTRA 29, IEC60071, IEC60060, and considering the recommendations of ELECTRA 96. These techniques are outlined as following. The design probability of withstand to flashover for the calculation of clearance is considered as follows:

	Probability of withstand (%)	**Standard Deviations (SDs)**
DC switchyard	97.7	2
Valve hall	99.7	3

DC clearances are calculated using 50% flashover voltages and electrode configuration factors. Two different calculation techniques are used, depending on whether it is a phase-to-ground clearance or a phase-to-phase clearance (i.e., clearance between two conductors at different potentials).

50% Flashover Voltage: To achieve the specified probability of withstand, the corresponding 50% flashover voltage U_{50} is given by (σ is per unit value of the conventional deviation of an overvoltage distribution):

$$U_{50} = \frac{U_z}{1 - \sigma Z}$$

where

$\sigma = 0.06$ for switching, $\sigma = 0.03$ for lightning

Z = conventional deviation of the discharge probability function of a self-restoring insulation

Uz = required withstand level

Phase-to-Ground Air Clearance Calculation: Phase-to-ground clearances are influenced by both switching and lightning transients. The clearances resulting from switching and lightning transients are calculated by different techniques, and the larger is selected. For switching transients, the clearance distance is given by

$$d_{sw} = \left[\frac{U_{50s}}{E_f} * \frac{1}{K^*500}\right]^{1.667}$$

where

K = electrode configuration factor based on ELECTRA29 and IEC60071-2

E_f = Atmospheric correction factor

U_{50s} = 50% flashover voltage for switching impulse

For a lightning transient, the clearance distance is given by

$$d_{LW} = \frac{U_{50L}}{E_f} * \frac{1}{575}$$

where U_{50L} = 50% flashover voltage for lightning impulse

Phase-to-Phase Air Clearance Calculation: Phase-to-phase clearances are determined by switching transients only, as for a conductor-to-conductor clearance the effect of a lightning stroke will be of similar magnitude on both conductors, thus lessening the voltage developed between the two conductors. The phase-to-phase clearance is calculated by referring the corrected U_{50s} (i.e., U_{50s}/E_f) to Curve 2 in Figure 9.20, obtained from ELECTRA 29.

AC Side

AC side clearances are based on IEC60071-2. Where the BIL (Basic Lightning Impulse Insulation Level) and BSL (Basic Switching Impulse Insulation Level) determine different clearance requirements, the larger of the two is selected.

Phase-to-phase clearances are largely governed by switching withstand levels. In the lower voltage range (up to about 750KV BSL), phase-to-phase withstand requirements are the same as phase-to-ground.

CREEPAGES

DC Side Creepages The creepage level is calculated by applying a creepage distance specified by the client to the MCAV, which appears across the insulation for equipment.

Description	Rectifier/Inverter (mm/kV)
DC indoors (valve hall)	20
DC +AC outdoors	35
DC bushings (outdoor part)	35

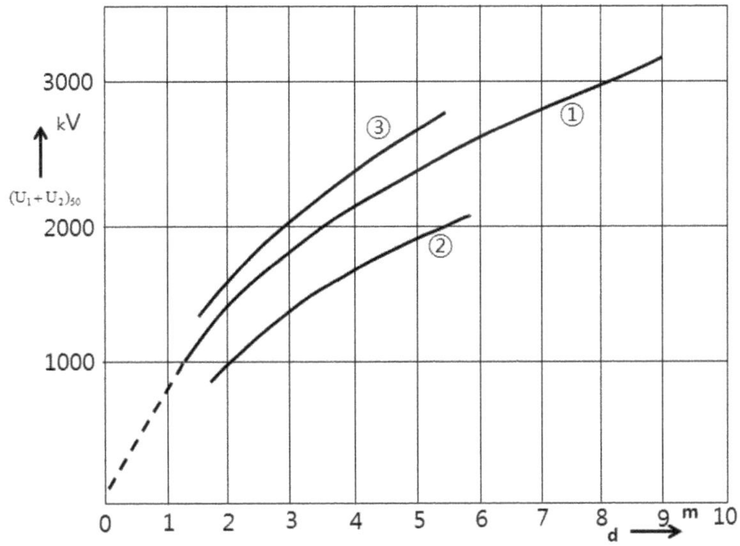

Figure 9.20 50% flashover voltage between phases versus the gap distance d $\alpha = U\_1/(U\_1 + U\_2) = 0.5$; $\Delta t = 0$; 1: ring electrode configuration, 2: rod-plate electrode configuration (Positive rod—negative plate), 3: rod-plate electrode configuration (Negative rod—positive plate)

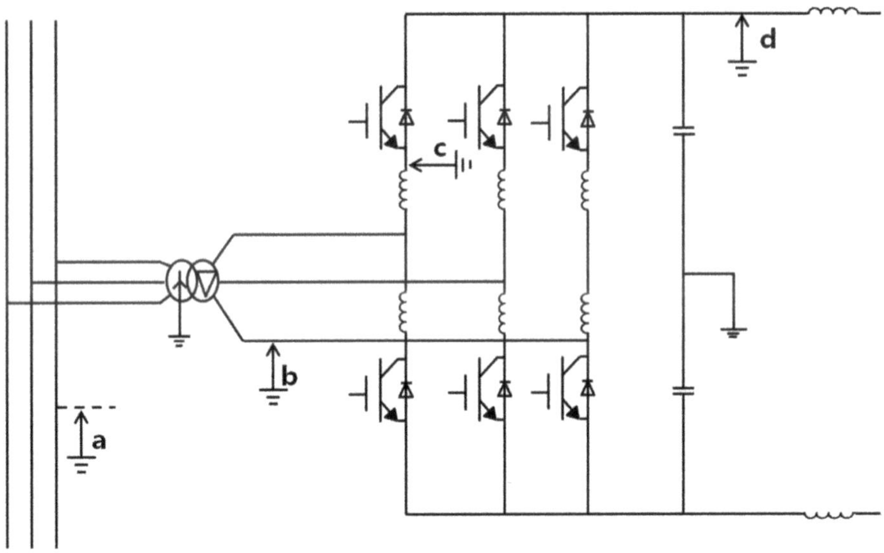

Figure 9.21 Insulation level locations

AC Side Creepages The creepage distance across insulation can be calculated by multiplying the maximum RMS value of the voltage across the insulation.

Description	Rectifier	Inverter (mm/kV)
AC	20	25

The creepage distance across insulation is calculated by multiplying the RMS phase-to-phase value of the voltage at the insulation point by the specified specific creepage distance.

Diameter Correction Factor

A correction factor (K_{di}) shall be applied to the calculated minimum creepage distances to account for the diameter of the insulator or bushing. The correction factor is as follows:

$$K_{di} = 1 \text{ for average diameter Dm} < 300 \text{ mm}$$
$$= 1.1 \text{ for } 300 < \text{Dm} < 500 \text{ mm}$$
$$= 1.2 \text{ for Dm} > 500 \text{ mm}$$

For all outdoor bushings, excluding capacitors, attached in a horizontal position (±30°), all creepage distances shall be increased by 20%.

9.4 Islanding [16]

The concepts of islanding and paralleling are relatively simple to understand. Basically, an islanding operation is operating an electric generating plant without an external voltage and frequency reference. Operating in parallel is the opposite of islanding, as an external voltage and frequency reference is provided. To better understand islanding and paralleling, it is important to understand how generator control systems operate.

When large and rapid load changes occur, as in the case of a loss of the grid, HVDC is stressed. The HVDC system may react to this stress by tripping the facility equipment. Detailed operator training must be created to ensure that the operators understand this capability. Also, HVDC may have to manually synchronize an operating unit to the grid. This requires continuous training reinforcement to ensure that the HVDC is properly synchronized to the grid.

Table 9.1 Electrode Configuration Factors and Clearance Calculation Methods

Clearance Reference (Referring to Figure 9.20)	Electrode Factor K	Method of Clearance Calculation
d	1.3 Conductor-Structure	Phase-Ground

Table 9.2 Altitude Correction Factors from IEC 60664

Altitude(m)	Normal barometric pressure (kPa)
0	101.3
500	95.0
1000	90.0
2000	80.0
3000	70.0
4000	62.5
5000	54.0
6000	47.0
7000	41.0
8000	35.5
9000	30.5
10000	26.5
15000	12.0
20000	5.5

9.4.1 Islanding Detection in Smart Grids [17] – [23]

The booming of the distributed generators industry combined with requirements for islanding detection has led to intensive research and development of methods for the identification of islanding conditions. These methods can be classified into three main categories:

- Passive
- Active
- Communication-based

A comprehensive report on methods and their classification has been realized. A brief description is given as follows:

Passive methods

Passive methods are based on the monitoring of grid variables by a dedicated algorithm residing in the control of a distributed generator or outside in a dedicated device. Most passive methods are looking for abnormal changes in frequency, voltage, or phase angle, but also in some particular harmonics or the total harmonic distortion (THD).

If the monitoring algorithm detects large or sudden changes of these variables at the point of connection with the utility grid, the inverter will trip. The most common passive methods are:

- Over/under voltage—monitors whether or not the grid voltage goes out of the limits established by the relevant standards (see Figure 9.22)
- Over/under frequency—monitors whether or not the grid frequency goes out of the limits imposed by the relevant standards (see Figure 9.22)
- Monitoring rate of change of frequency (ROCOF) and voltage (ROCOV)
- Phase monitoring—monitors fast jumps of grid voltage phase

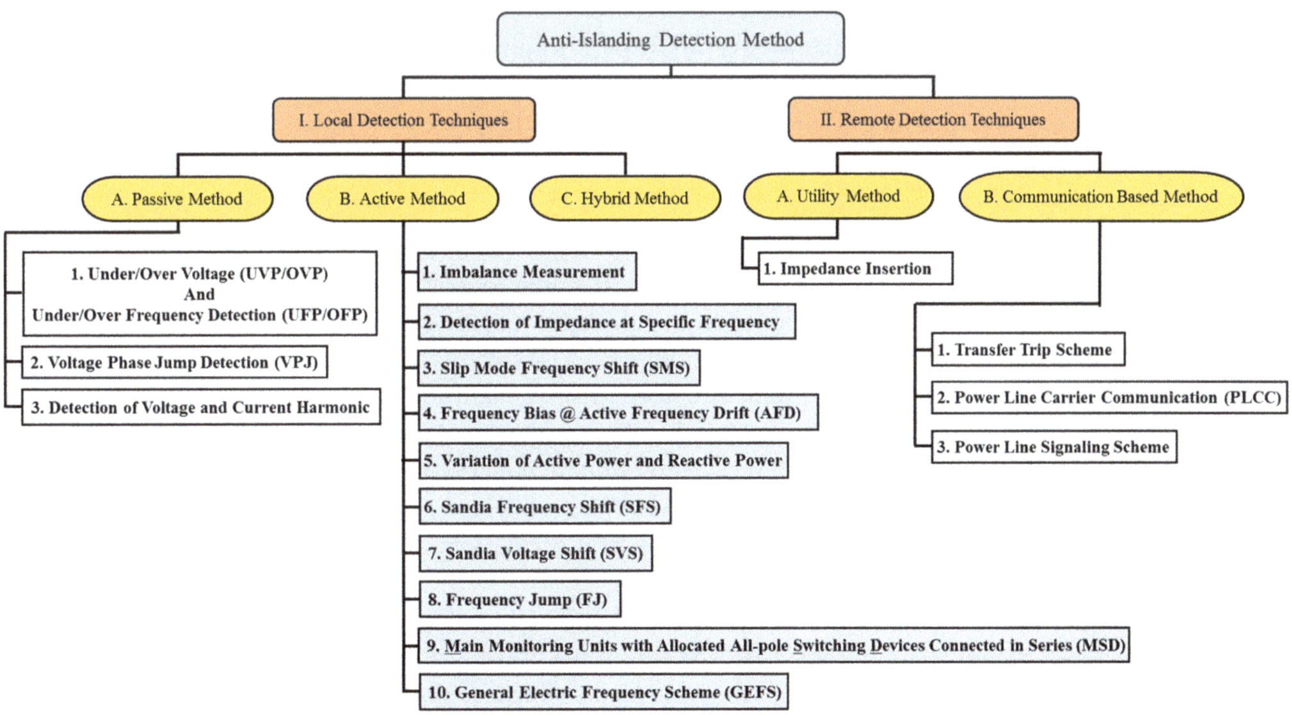

Figure 9.22 Classification of anti-islanding detection methods

- Voltage harmonic—monitors selective (3rd, 5th, etc.) or THD of grid voltage Passive methods are quite effective in a majority of situations that may occur in the grid; however, their non-detection zone (NDZ) does not cover situations when the power absorbed by the load matches the power generated by HVDC almost perfectly. In such cases, the variations in voltage, frequency, or phase angle are lower than specified in the standard because the network remains balanced even though the connection with the main grid has been lost. Therefore, the HVDC would not trip even though an island has been formed. One can use a combination of passive methods and multi-criteria decision making. The islanding detection algorithm uses passive methods including frequency and voltage monitoring and the most widely recognized methods like the ROCOF, Voltage Vector Shift (VVS), and Voltage Unbalance. However, passive methods are generally considered as insufficient anti-islanding protection.

Active methods

Active methods appeared as a necessity to minimize the non-detection zone of islanding detection methods in conditions where generation matches load. Normally, active methods inject a small disturbance to the utility grid and, based on the grid response, it is decided whether or not an island has formed. Disturbances in terms of shifts from normal values to grid voltage magnitude, frequency, or phase angle can be added by the HVDC and, in the case of a grid-connected situation, these disturbances should be corrected by the grid through the voltage and frequency control. However, if the voltage magnitude, frequency, or phase angle follow the shift introduced by the HVDC, it is most likely that the grid has been disconnected and an island has been formed. The most common active methods are:

- Positive feedback inside the HVDC control: the controller tries to alter grid variables such as frequency, phase, or voltage magnitude in order to perform
- Frequency jump or phase jump: the HVDC deliberately alters the frequency or phase of the injected current in order to change the frequency or phase of the voltage. If the grid frequency follows the inverter current, an island has formed and the HVDC inverter should disconnect.
- Frequency bias: similar to previous.
- Sandia Frequency Shift: similar to frequency jump but elaborated on by Sandia National Laboratories.
- Voltage positive feedback, Sandia Voltage Shift: tries to alter the voltage magnitude at the point of common coupling. If the grid follows the changes generated by the HVDC, the grid voltage will go out of the imposed operating ranges, and thus the inverter would trip.
- Injection of harmonics via HVDC: non-characteristic harmonics are injected by the HVDC and the grid response is registered. This method is also called detection of impedance at a certain frequency.
- Impedance detection: an active method that has been promoted by the requirements in the German standard. A current spike is periodically injected at the point of common coupling by a grid-tied power converter. Based on the voltage response to this disturbance, the grid impedance value is determined using Fourier transform. The influence of nonlinear loads connected close to the point of common coupling (PCC) is also addressed, and as a consequence, additional signal processing methods are necessary in order to obtain accurate results. The phase angle signal used to generate the reference for the current controller is slightly altered to be able to estimate the grid impedance based on grid reaction to the generated current. A high-frequency signal (600 Hz) is injected at zero crossing in order to determine the value of grid impedance. Active and reactive power oscillations are used to identify the value of grid impedance.

  Although active methods give better islanding identification, they also distort the delivered power in order to detect islanding conditions. Disturbances in the power grid are not suitable when a significant number of HVDCs are connected on the same feeder. Because the injected interharmonic generated by the HVDC is unique in the utility network, this particular method would allow the connection of more than one similar generation unit on the same grid because synchronization in respect to the interharmonic injection is possible.

However, only a finite number of inverters can be connected on the same feeder, and this number is directly dependent on the standard demands with regard to disconnection time and the number of injections necessary to obtain accurate impedance identification. A large penetration level of inverters using active methods not only decreases the quality of power in the grid but also increases the non-detection zone of all active methods.

Communication-based methods

Another category of methods for detecting islanding is based on communication between HVDC and the utility grid. Three main methods using communication are detailed:

- The power line used as a carrier for communication between the HVDC inverter and utility grid: A continuous signal is transmitted by a utility network via the power line. A receiver must be connected to the HVDC in order to detect the loss of this signal and hence determining islanding conditions
- The signal produced by the disconnector: This method assumes that the utility recloser is equipped with a transmitter, which communicates with HVDC when it opens
- The SCADA (Supervisory Control And Data Acquisition) based method: Uses the placement of voltage sensors at the location where HVDC is connected and the integration of those sensors into the SCADA system in order to monitor and warn the HVDC system to disconnect in the case of islanding

Table 9.3 draws on the main features of the main categories of islanding methods. Passive methods are the basic protection package of every distributed generator connected to a utility grid. Currently, active methods are preferred because of their low non-detection zone. However, one of the main drawbacks that may contribute to a shift from using active methods is their negative contribution to the quality of power in the grid. Future

Table 9.3 Comparison of Islanding Detection Methods

Method type	Strengths	Weaknesses	Grid friendly
Passive	– grid friendly – easy and cheap to implement	– NDZ larger compared to others	yes
Active	– low NDZ – some easy to implement	– can create power quality problems – can lead to nuisance trip – some difficult to implement – possible interaction between converters in the same grid	– suitable for a finite number of generators
Communication-based	– reliable – some easy to implement – theoretically no NDZ	– expensive to implement – need communication infrastructure – need involvement of utility	yes

availability of a communication network for the power grid combined with more interest from utilities to monitor the assets in the grid may facilitate a move toward the use of communication-based methods for islanding detection.

References

1. J. Yang, J. E. Fletcher, and J. O'Reilly, "Short-circuit and ground fault analyses and location in VSC based DC network cables," *IEEE Trans. Ind. Electron.*, vol. 59, Oct. 2012.
2. P. Bresesti et al "HVDC connection of offshore wind farms to the transmission system," *IEEE Trans. Energy Convers.*, vol. 22, no. 1, pp. 37–43, Mar. 2007.
3. M. J. Mousavi and K. L. Butler-Purry, "A novel condition assessment system for underground distribution applications," *IEEE Trans. Power Syst.*,vol. 24, no. 3, pp. 1115–1125, Aug. 2009.
4. E. W. Kimbark, "Transient overvoltages caused by monopolar ground fault on bipolar dc line: Theory and simulation," *IEEE Trans. Power App. Syst.*, vol. PAS-89, no. 4, pp. 584–592, Apr. 1970.
5. G. T. Wrate et al "Transient overvoltages on a three terminal dc transmission system due to monopolar ground faults," *IEEE Trans. Power Del.*, vol. 5, no. 2, pp. 1047–1053, Apr. 1990.
6. B. R. Andersen, "Wind farms interconnections," in *2006 International Conference on Power System Technology*, Chongqing, China.
7. J. Dixon, L. Moran, and J. Rodriguez, "Reactive power compensation technologies: State-of-the art review," *Proc. IEEE*, vol. 93, pp. 2144–2164.
8. T. Larsson, "Eagle pass hack-to-back tie: A dual purpose application ol voltage source converter technology," in *IEEE Power Engineering Society Summer Meeting*, Vancouver, BC, Canada, 2001.
9. M. Hyttinen, J.-O. Lamell, and T. F. Nestli, "New application of voltage source converter (VSC) HVDC to be installed on the gas platform Troll A," in *Cigre Session*, Paris, France, 2004.
10. J. Yang and J. C. Zheng, "Characteristics and recovery performance of VSC-HVDC DC transmission line fault," in *2010 Asia-Pacific Power and Energy Engineering Conference.*, Chengdu, China, 15, April, 2010.
11. J. Hafner and B. Jacobson, "Proactive hybrid HVDC breakers: A key innovation for reliable HVDC grids," in *Proceedings of the CIGRE Bologna Symposium*, Bologna, Italy, Sep. 13–15, 2011, p. 8.
12. C. Davidson, et al., "A new ultra-fast hvdc circuit breaker for meshed dc networks," in *Proceedings of the IET AC and DC Power Transmission*, Birmingham, Feb. 10–12, 2015, p. 7.
13. K. Tahata, et al., "HVDC circuit breakers for HVDC grid applications," in *Proceedings of the AORC-CIGRE*, Tokyo, Japan, May 27–29 2014, p. 9.
14. O. Cwikowski, et al., "Analysis and simulation of the proactive hybrid circuit breaker," in *Proceedings IEEE Power Electronics and Drive Systems*, Sydney, Australia, Jun. 9–12, 2015, pp. 4–11.
15. W. Leterme, S. P. Azad, and D. V. Hertem, "A local backup protection algorithm for HVDC grids," *IEEE Trans. Power Del.*, vol. 31, pp. 1767–1775, Aug. 2016.
16. Main Circuit Parameters of Transbay Cable project", Test Reports, Siemens, 2009.
17. P. L. Villenueve, "Converns generated by islanding," *IEEE Power Energy Mag.*, vol. 2, May-June 2004.
18. A. Timbus, A. Oudalov, and N. M. Carl, "Islanding detection in smart grids," in *2010 IEEE Energy Conversion Congress and Exposition.*, Atlanta, GA, USA, 12–16 Sep, 2010.
19. W. Bower and M. Ropp, "Evaluation of islanding detection methods for photovoltaic utility interactive power systems," March 2002. IEA-PVPS, http://www.iea-pvps.org, Online report T5-09, [Online]. Available at: http://www.oja-services.nl/ieapvps/products/download/rep5 09.pdf
20. F. De Mango, et al., "Overview of anti-islanding algorithms for PV systems. Part I: Passive methods," in *Proceedings of Power Electronics and Motion Control Conference*, Portoroz, Slovenia 2006, pp. 1878–1883.
21. S.-I. Jang and K.-H. Kim, "An islanding detection method for distributed generations using voltage unbalance and total harmonic distortion of current," *IEEE Trans. Power Delivery*, vol. 19, no. 2, pp. 745–752, 2004.
22. H. Zeineldin, et al., "Safe controlled islanding of inverter based distributed generation,"in *Proceedings of PESC'04*, vol. 4, Aachen, Germany, Germany, 2004, pp. 2515–2520.

23. J. B. Jeong, et al., "An improved method for anti-islanding by reactive power control," in *Proceedings of ICEMS'05*, vol. 2, 2005, pp. 965–970.
24. V. John, Z. Ye, and A. Kolwalkar, "Investigation of anti-islanding protection of power converter based distributed generators using frequency domain analysis," *IEEE Trans. Power Electron*, vol. 19, no. 5, pp. 1177–1183, 2004.

10

HVDC and AC Network

10.1 Grid Service

10.1.1 Different Stability Issues

Unlike the conventional line-commutated converter high-voltage direct current (LCC-HVDC) scheme, voltage source converter (VSC HVDC) is a new technology utilizing forced-commutated voltage source converters. VSC-HVDC can control both active and reactive power independently without commutation failures on the inverter side. It doesn't require reactive power compensators, which results in much smaller equipment sizes. VSC-HVDC can be applied to voltage support in the receiver systems, interconnection between asynchronous power systems, the grid connection of large wind farms or offshore wind farms, and subsea power transmission.

HVDC systems are responsive and can be controlled within tens of milliseconds. HVDC systems can also improve stability, especially the transient stability of a power system. The appearance of the VSC-HVDC, which has more advantages than conventional HVDC systems, could bring about a new approach for AC/DC hybrid transmission systems and multi-infeed HVDC transmission systems. If the receiving system is weak, voltage stability and commutation failure may become serious problems in conventional HVDC systems. A VSC-HVDC controls active and reactive power independently, and does not need reactive compensation in both the rectifier and inverter sides. It can also be used as static synchronous series compensation (SSSC) characteristics to damp the power system oscillations. A VSC-HVDC applies self-commutated solid-state devices, so there is no commutation failure issue. Figure 10.1 shows the performances and functions of a VSC HVDC as commented earlier.

For AC/DC interconnected transmission systems, the introduction of a VSC-HVDC can enhance the voltage support and improve the system stability. This chapter will discuss the effect of a VSC-HVDC in terms of the improvement of AC/DC interconnected systems. The performance of VSC-HVDC in enhancing both power angle stability and voltage stability in these systems is taken into consideration.

HVAC/DC hybrid transmission systems are very common nowadays. If AC transmission lines are paralleled with VSC-HVDC, there will be more advantages.

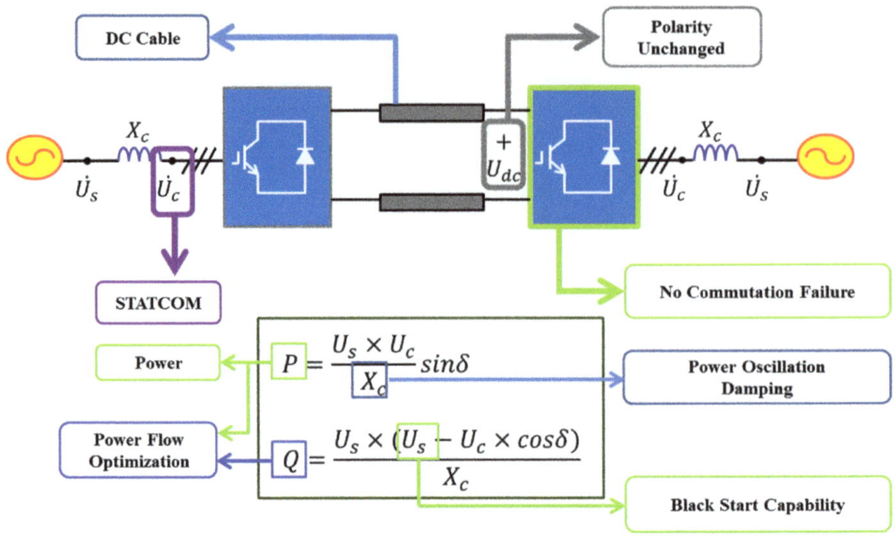

Figure 10.1 Performances and functions of VSC-HVDC

Different Stability Issues and Varying Grid Configuration

There are two different aspects to consider. First is the type of power system stability issue that the VSC transmission system can be exposed to. For example, two phenomena are voltage stability and rotor angle stability. Second is the grid configuration in which the VSC is configured to the AC system. Three basic types can be identified:

- The series connection of the VSC converter and the AC system
- The parallel connection of the VSC transmission system and the AC system
- The asynchronous infeed from the VSC converter into the AC system

Strategy for Series Connection and Voltage Instability

The topology of series-connected VSC HVDC is show in Figure 10.2. From Figure 10.2, the power flow equations for the receiving end are shown as a power circle diagram. The capability curve of the VSC transmission system shows the maximum transferable active power. The intersection of the capability curve of the VSC transmission and the receiving end circle will indicate the stable solution for that particular voltage level. If the sending end voltage drops, it is possible to immediately establish the new stable solution in the power circle plane by recalculating the figure.

Strategy for Parallel Connection and Voltage Instability

If the VSC transmission system is connected in parallel to an AC line, the control of the VSC transmission will have an impact on the AC flow. By varying the power factor of the DC transmission, the AC system will be utilized better. To enhance system operation capability, the best power factor operation for the VSC transmission must be determined when the system becomes overloaded. Figure 10.3 shows a parallel case with the associated power circle plane. If we begin studying the receiving end circle, we can see that the power flow on the AC line (following the arc with angle δ) can be found, to which the power flow via the VSC transmission is added (the vector within the smaller circle). In this example, the AC line requires some reactive power fed from the VSC system. In this figure, the MVA circle (the small one) is valid for VSC transmission. It is feasible that the MVA capacity is at its maximum point for the DC system, transferring more power over the combination is impossible. An increase of DC flow or AC flow (requiring more reactive power to keep AC voltage) would

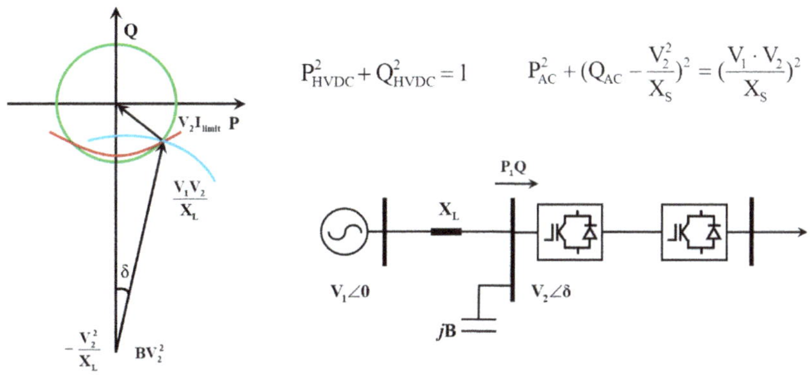

Figure 10.2 Receiving a power circle plane, for a VSC transmission in series with an AC line

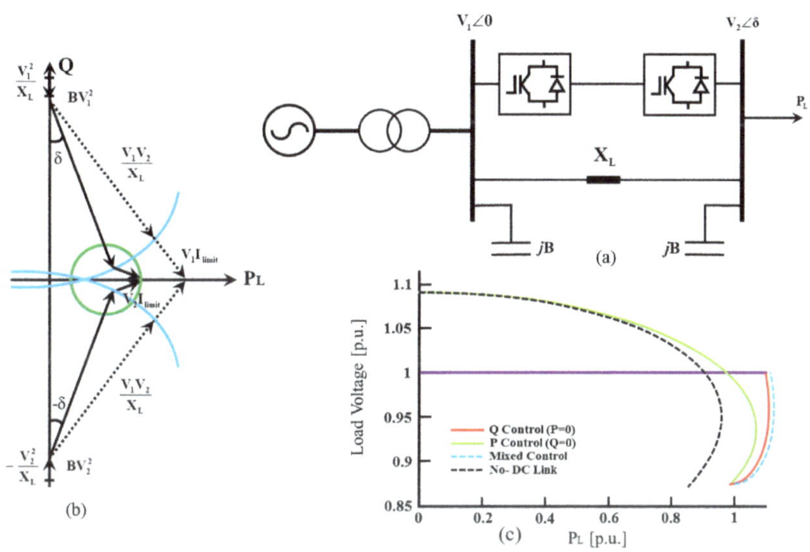

Figure 10.3 A parallel case (b) with a power circle plane, (a) to indicate "best" solution. PV curve in (c). Key grid parameters are XL = 0.5, B = 0.2, (SIL = 0.872 p.u.), Xt = 0.2 p.u., and converter size = 0.08 p.u. and converter size = 0.08 p.u.

violate the capability curve. If DC power transfer, is decreased and we are able to inject more reactive power, it is possible to transfer more active power over the combination. The best choice is made according to

$$P_{dc} = I_{limit} \cdot V_2 \sin(\delta_1 - \delta_2) \text{ and } Q_{dc} = I_{limit} \cdot V_2 \cos(\delta_1 - \delta_2) \tag{10.1}$$

where I_{limit} is the maximum steady state current allowed in he converter.

The associated PV curve is shown in Figure 10.3. It shows three different ways to utilize the capacity in the VSC transmission system. The first one is active power transfer only, the second is only reactive power generation in each end (STATCOM operation), and the third one is a mixture created from the best choices described earlier. The Points of Maximum Loadability (P_{max}) for the three possibilities are indicated in Table 10.1. The best choice in this example increases the point of maximum loadability to 149% of installed MVA capacity.

Table 10.1 Loadability as a Function of Different Control Strategies

Method	P_{max} [p.u.]	Gain [p.u.]	Gain/converter size (%)
No DC	0.8923		
P Control	0.9561	0.0638	-20
Q Control	0.9908	0.0985	+23
Mixed control	1.0117	0.1194	+49

Strategy for Asynchronous Infeed and Voltage Instability

Many HVDC links are connected between asynchronous grids operating with different frequencies. At the connection point, where active power is fed into the AC grid, VSC transmission can improve performance. If we investigate a Thevenin equivalent of the infeed shown in Figure 10.4, the qualitative behavior of changing power factors in the VSC transmission can also be studied. Compare the solid and dotted vectors in the diagram. By aligning the vectors through changing power factors maximum loadability is achieved.

The loadability P_L as a function of P_{dc} is shown to the left in Figure 10.5. The operating points are indicated in the schematic power circle planes to the right of Figure 10.5. Note point A, where active power (similar to classic HVDC) is transferred, and the optimum point B where the loadability has increased more than the installed MVA capacity in this numerical example. It means that from an SIL transmission level of 0.87 p.u., the power transmission capacity has been increased to 1.04 p.u., that is, a gain of 0.17 p.u., which is twice the installed converter capacity of 0.08 p.u. If active power is not available remotely the loadability increases to point C in the figure (assuming that maximum DC voltage has not limited reactive power output).

Two things are important to remember: firstly, the study has not taken the maximum DC voltage level limit into account, and secondly the power angle δ can be too high for certain parameter combinations. Nevertheless, it is clearly observed that the grid in a stressed situation is best supported by a mixture of active and reactive power. The gain will decrease with decreasing voltage. A critical bottleneck with a VSC-based transmission system in the remote end can be equipped with controls alleviating an N-1 contingency with up to two times its installed capacity.

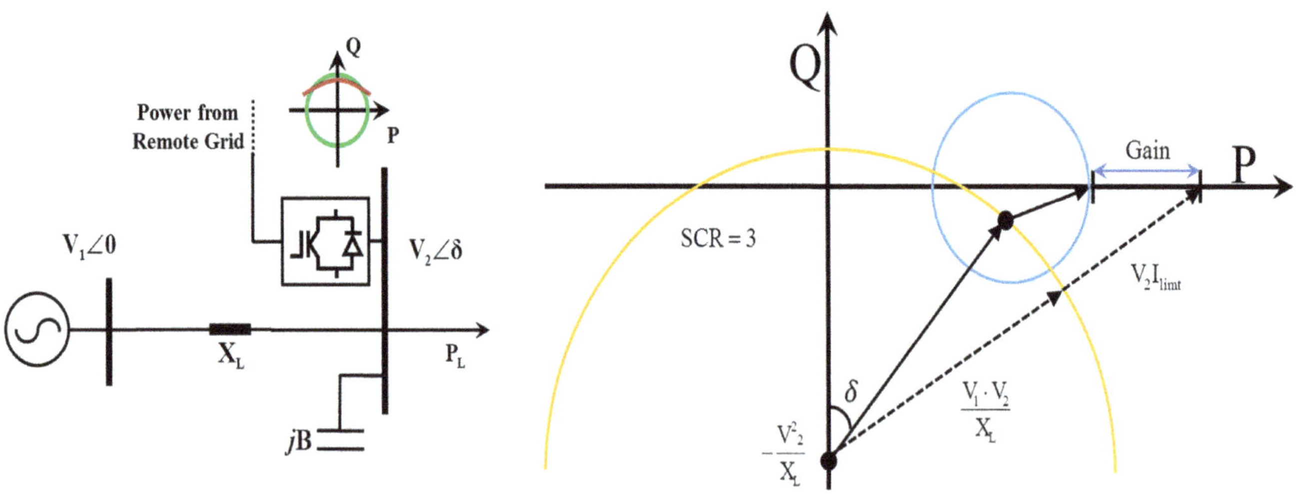

Figure 10.4 Asynchronous infeed and enhanced loadability shown in the diagram

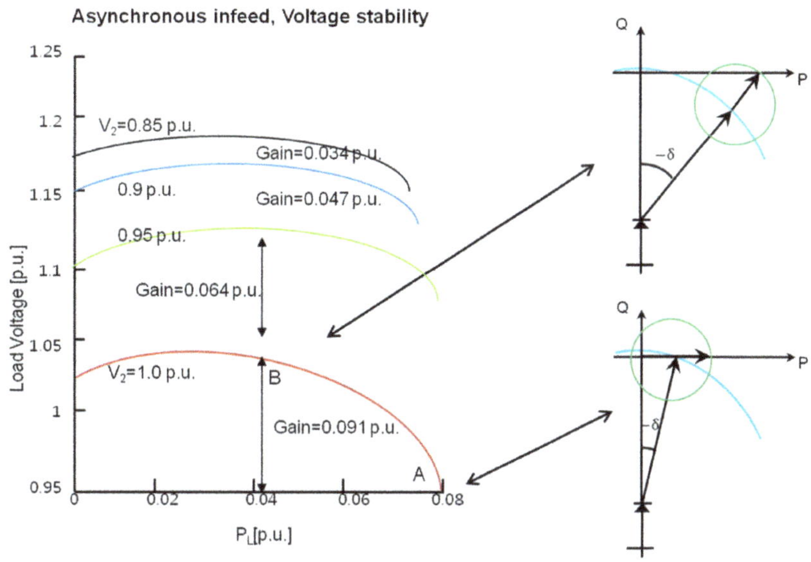

Figure 10.5 Loadability as a function of injected DC voltage. Key grid parameters are XL = 0.5, B = 0.2 (SIL = 0.872 p.u.), and converter size = 0.08 p.u.

10.2 AC/DC Interaction

An important procedure when designing a DC link is to analyze the AC/DC interactions. There are various interaction phenomena such as power instability, voltage instability, frequency instability problems, harmonic resonance, sub-synchronous torsional interaction (SSTI), and temporary over-voltages (TOVs). One of the important criteria in the design of DC links is the value of the permissible TOVs at the AC terminals of the converter stations. Much research has been done on the effect of interactions in a situation where the short-circuit capacity (SCC) of the AC system is not large enough compared to the capacity of the DC link [4].

10.2.1 System Strength Index for Single-Infeed HVDC Systems

Short-Circuit Ratio (SCR)

An AC system is labeled as "weak" when the impedance of the AC system relative to the DC power is high or the inertia of the AC system is low. In this case, the interaction phenomenon is evaluated to occur more frequently. Previous studies were conducted to show the strength of the AC system. Thus, to show the relationship between the impedance and the DC capacity of the AC system as a specific index, an index called SCR was introduced to evaluate the strength of the AC system. The SCR can be calculated as in Equation (10.2). The system strength according to SCR can be evaluated as shown in Table 10.2.

$$SCR = \frac{V_T^2/Z_s}{P_{dc}} \tag{10.2}$$

where Z_s is the magnitude of Thévenin equivalent impedance of the AC system, φ is the phase angle of Thévenin equivalent impedance of the AC system, V_T is the rated AC terminal voltage magnitude, $\frac{V_T^2}{Z_s}$ is the short-circuit level (SCL) of the AC system, P_{dc} is the rated DC power of the HVDC,

Table 10.2 Strength Classification According to the SCR [5]

SCR	Classification
3 < SCR	High
2 < SCR < 3	Low
SCR < 2	Very low

Effective Short-Circuit Ratio (ESCR)

In the case of LCC-HVDC, since the phase of the current lags behind the voltage and has the lagging power factor, it always consumes reactive power, and generally consumes 50%–60% of the active power output. Because of this feature, shunt capacitors are installed to compensate reactive power to the AC terminal as shown in Figure 10.6. The equivalent impedance seen from the AC terminal appears to be connected in parallel with the system impedance and the capacitors. The system impedance is generally inductive, so the equivalent impedance of the AC terminal considering the shunt capacitor is large. Therefore, ESCR is defined as an index of system strength when considering the influence of shunt capacitors in the existing SCR. The ESCR can be calculated by Equation (10.3) where Q_f is the rated reactive power output of the shunt capacitors.

$$ESCR = \frac{SCL - Q_f}{P_{dc}} \tag{10.3}$$

In addition to the previously mentioned shunt capacitors, there are various facilities for reactive power and voltage compensation for LCC-HVDC. Typically, there are synchronous compensators and static var compensators (SVCs), and the SCR calculation method considering them is as stated as follows:

The synchronous compensators have the effect of reinforcing the system by lowering the impedance of the AC system, so it should be included in the calculation of the SCR. However, the SVCs are slightly different from that. Thyristor-based high-speed control facilities such as the thyristor-controlled reactor (TCR) and thyristor-switched shunt capacitor (TSC) are excluded from the SCR calculation. On the contrary, facilities that directly affect the impedance of AC systems, such as mechanically switched reactors (MSRs), mechanically switched capacitors (MSCs), and fixed reactors (FRs), should be included in the SCR calculation.

Maximum Power Curve (MPC)

One of the concepts useful in characterizing AC/DC systems is the MPC. This is used to analyze power stability. This is similar to the method of analyzing PV stability using nose curves in the AC system. As shown in Figure 10.7, the relationship between the active power and current of the HVDC for a given system parameter is unique.

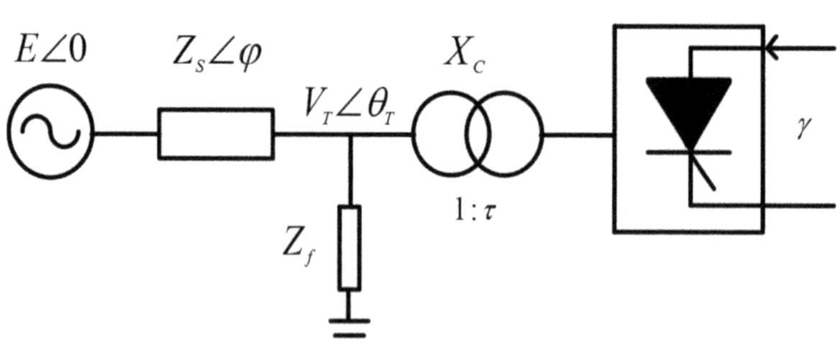

Figure 10.6 Single-infeed HVDC system

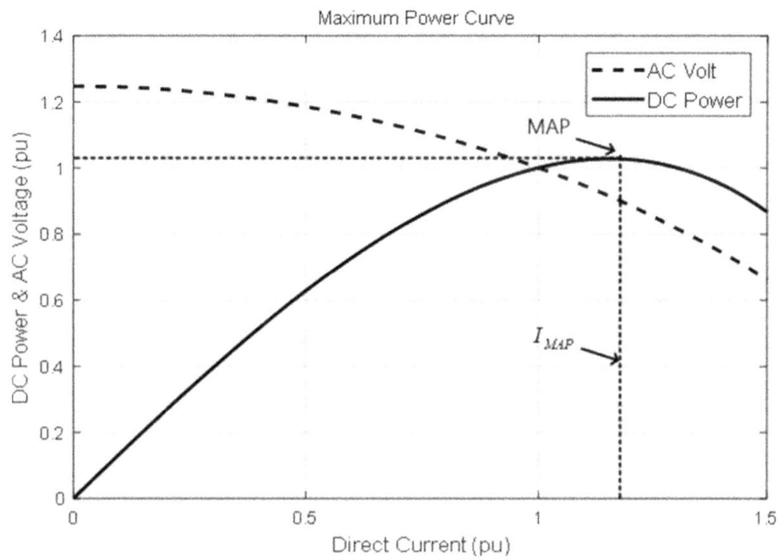

Figure 10.7 Maximum power curve (MPC) of a system (ESCR:2.0, γ:15°)

The method of deriving the MPC first defines the pre-disturbance steady-state operating point of the system. This assumes the nominal condition $V_T = I_d = P_d = 1.0\ (p.u.)$ as the operating point. To satisfy the nominal condition of the system, the power flow equations are solved for a given condition to calculate the phase of the AC system voltage (θ_T), tap ratio (τ), and AC terminal voltage (E).

The remaining parameters obtained from the power flow equations for the calculation of the operating point are assumed to be as follows: it is assumed that the DC current changes momentarily through control of the firing angle on the rectifier side, and that there is no change in the voltage of the AC system, extinction angle of the inverter, tap ratio, automatic voltage regulation (AVR), and value of shunt capacitors. Therefore, the change of other parameters according to the change of DC current can be calculated by solving the power flow equations shown in Figure 10.7.

Generally, the inverter station performs minimum extinction angle control to prevent commutation failure. The maximum DC voltage of the inverter station is obtained at the minimum extinction angle, so the maximum power for a specific DC current can be obtained. Since it is not possible to obtain a larger output curve than the power curve obtained at the minimum extinction angle, the curve derived from that is called the MPC.

As shown in Figure 10.7, an MPC has a maximum value called maximum available power (MAP). It can be seen that if the DC current is increased above the current in the MAP (I_{MAP}), the active power is relatively reduced. This is because as the DC current increases, the AC voltage decreases as the reactive power consumed by the HVDC converter increases.

The MPC is characterized by the sign of the slope of the graph based at the MAP. If the MAP exists in a region smaller than the operating point, it is operating in an unstable region. If the MAP exists in a region larger than the operating point, the MPC is operating in a stable region. The MAP condition can be mathematically expressed as Equation (10.4), and the linearization using the algebraic equation of Figure 10.6 can be obtained using Equation (10.5).

$$\frac{dP_d}{dI_d} = 0 \tag{10.4}$$

Table 10.3 shows the sensitivity according to the control mode of HVDC in Equation 10.5.

$$\frac{dP_d}{dI_d} = \frac{rV_T\cos(\gamma+\mu)\left[V_T^4ESCR^2 + b\left(V_T^2ESCR + Q_d\right) - Q_d^2 - P_d^2\right]}{\det[J1]}$$
$$r = \frac{V_d}{cos\gamma + \cos(\gamma+\mu)} \tag{10.5}$$
$$b = 2[Q_d - P_d\tan(\gamma+\mu)$$
$$\det[J1] = V_T^4ESCR^2 - Q_d^2 + V_T\frac{\partial Q_d}{\partial V_T}\left(V_T^2ESCR + Q_d\right) - P_d\left(P_d - V_T\frac{\partial P_d}{\partial V_T}\right)$$

where V_d is the DC voltage of the inverter, μ is the overlap angle of the inverter, and Q_d is the reactive power consumed by the inverter.

As mentioned earlier, MPC is a characteristic obtained in a given systematic situation, and exhibits different characteristics according to the ESCR. Figure 10.8 shows the MPCs according to various ESCRs. The results show that the larger the ESCR, the larger the MAP. Also, it can be seen that the MAP coincides with the operating point when the ESCR is between 1.5 and 2. If the ESCR is smaller than the value based on the specific ESCR, the

Table 10.3 Sensitivities According to the Control Mode of HVDC

Control mode	$\frac{\partial P_d}{\partial V_T}$	$\frac{\partial Q_d}{\partial V_T}$
Constant extinction angle (CEA)/CP	0	$\frac{2}{V_T}\left[Q_d - P_d\tan(\gamma+\mu)\right]$
CEA/CC	$\frac{2}{V_T}\left[\frac{P_d cos\gamma}{cos\gamma + \cos(\gamma+\mu)}\right]$	$\frac{2}{V_T}\left[Q_d - \frac{P_d\sin(\gamma+\mu)}{cos\gamma + \cos(\gamma+\mu)}\right]$
Constant U_d/CP	0	$\frac{2}{V_T}\left[Q_d + P_d\cot(2\gamma+\mu)\right]$

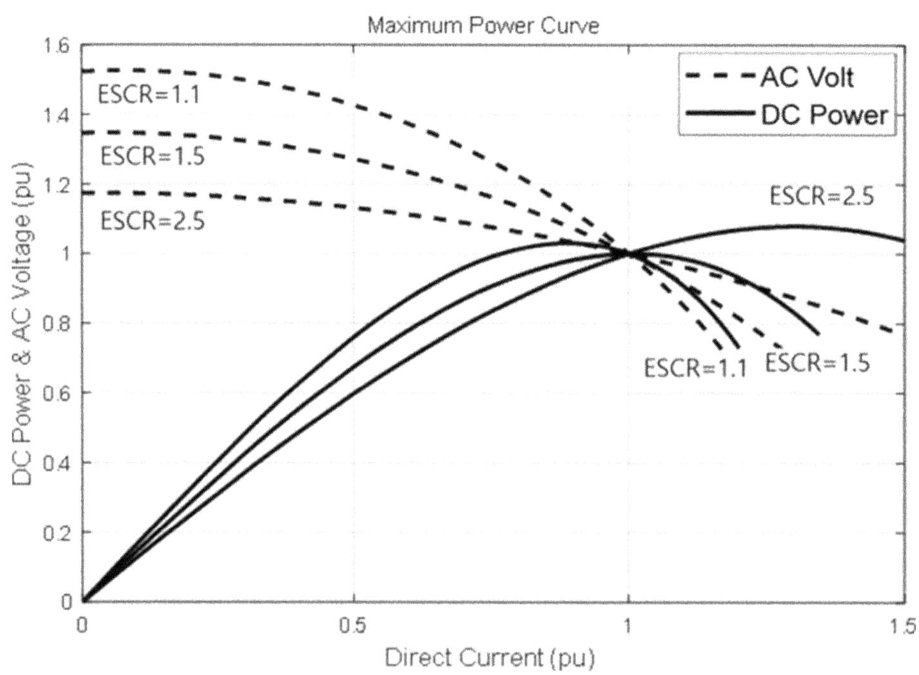

Figure 10.8 Variation of inverter AC terminal voltage and DC power

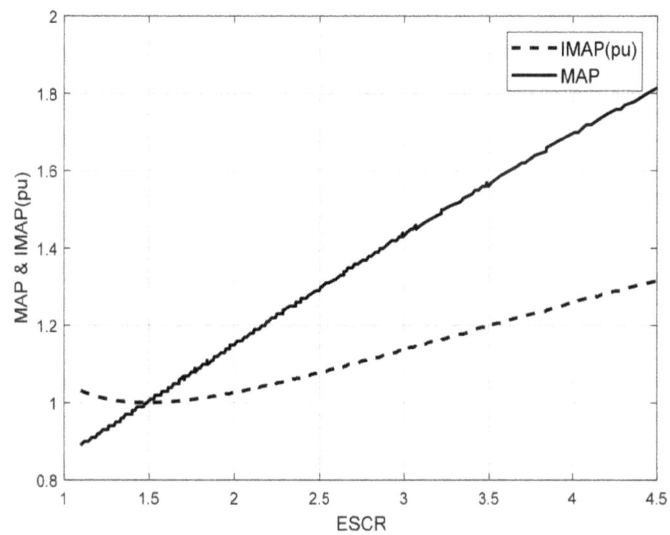

Figure 10.9 Variation of MAP and I_{MAP} with effective short-circuit ratio (ESCR)

operating point exists in the unstable region. If the ESCR is larger than the specific value, the operating point is formed in the stable region.

This specific ESCR is referred to as the critical effective short-circuit ratio (CESCR), and the MPC and MAP characteristics according to the various SCRs show a similar tendency to various ESCRs. For the SCR, the specific value is called the critical short-circuit ratio (CSCR). In general, CESCR has a value of about 1.5 and can be calculated as shown in Equation (10.6). The CSCR can be obtained by adding the reactive power output of the shunt capacitor in Equation (10.6).

$$CESCR = \frac{1}{V_T^2}[sin\varphi P_d \tan(\gamma + \mu) - Q_d + \sqrt{\left[\frac{P_d}{\cos(\gamma + \mu)}\right]^2 - \cos^2\varphi \left(P_d \tan(\gamma + \mu) - Q_d^2\right)} \tag{10.6}$$

10.3. System Strength Index for Multi-Infeed HVDC Systems

HVDC is widely used in electric power systems because of its advantages, such as long-distance bulk power transmission, asynchronous system connection, and submarine cable connection. Several HVDC systems are appearing in close proximity to meet growing demand, especially in developing countries. In other words, HVDC was added to the existing single-infeed HVDC system and changed into a multi-infeed HVDC system with multiple HVDCs.

Therefore, it is necessary to extend the multi-infeed HVDC system to the assessment of the interaction phenomena discussed in existing single-infeed HVDC systems. In a multi-infeed HVDC system, a number of converters are associated, resulting in a more complex interaction effect than a single-infeed HVDC system. In [6], important interaction phenomena in multi-infeed HVDC systems are classified into four types: TOV, commutation failure including fault recovery, harmonic problems, and voltage/power stability problems. It additionally examines whether SCR and ESCR can be used in multi-infeed situations in the existing single-infeed HVDC system, and proposes a new system strength index for multi-infeed HVDC systems.

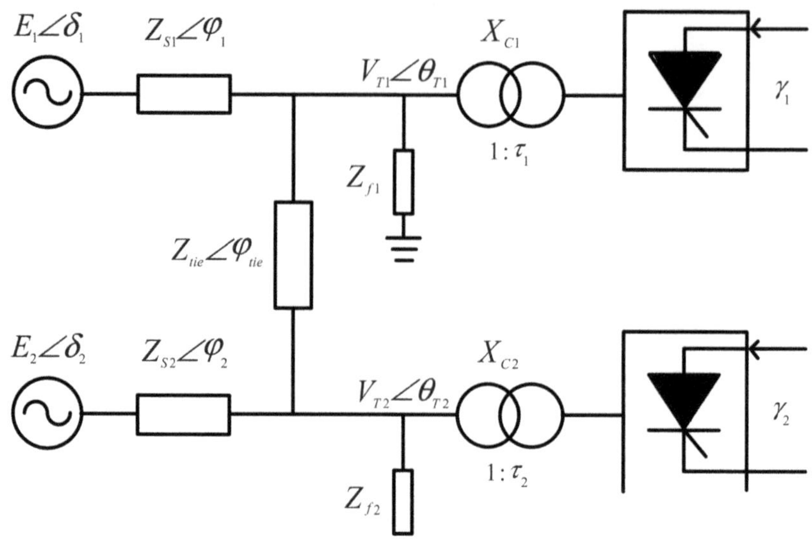

Figure 10.10 Multi-infeed HVDC system

10.3.1 Effective Short-Circuit Ratio (ESCR)

As we have seen in the previous section, SCR and ESCR have proven to be important indicators of robustness and interactions of single-infeed HVDC systems. Therefore, it is reasonable to use them in multi-infeed HVDC systems as well. If all of the HVDCs are connected to the same bus line, it is possible to calculate SCR and ESCR easily by equalizing with a single HVDC. However, if they is connected with electrical distance, it cannot be considered as such; therefore, it is necessary to consider this characteristic.

The method of calculating the SCR in a multi-infeed HVDC system assumes that other HVDC converters are blocked when calculating the SCC at the AC terminal under analysis. However, the compensating device, including the filter installed in the terminal, should be considered when calculating SCC. The SCR can be obtained by dividing the calculated SCC by the rated capacity of the HVDC to be analyzed. Also, ESCR can be calculated by subtracting the compensating capacity of the installed AC terminal, as in Equations (10.7) and (10.8).

$$SCR_i = \frac{SCC_i}{P_{d,i}} \tag{10.7}$$

$$ESCR_i = \frac{SCC_i - Q_{f,i}}{P_{d,i}} \tag{10.8}$$

where SCC_i is the SCC of ith AC terminal, $Q_{f,i}$ is the MVA rating of shunt capacitors and AC filters at ith AC terminal, and $P_{d,i}$ is the rated DC power of the ith HVDC.

However, if we look at the calculation process earlier, we can see that ESCR considers only the effects of the compensation facility installed in the AC terminals of other HVDCs and does not consider the output of other HVDCs. In general, the interaction effect of a specific AC terminal on a multi-infeed HVDC system appears to include not only the HVDC installed in the corresponding AC terminal, but also the influence of the HVDC installed on the other bus line. Therefore, the introduction of a new indicator is required.

10.3.2 Multi-infeed Interaction Factor (MIIF)

Among the parameters of the AC system, the AC terminal voltage at the inverter station is most relevant to the interaction phenomenon. Therefore, MIIF was introduced as an index to show correlation with the AC terminal voltage of other inverter stations.

The MIIF measurement method is to instantaneously apply a shunt reactor to drop the reference AC terminal voltage by about 1%, and then to measure the voltage drop at the other AC terminal and to define the ratio as in the MIIF Equation (10.9).

$$MIIF_{j,i} = \frac{\Delta V_{Tj}}{\Delta V_{Ti} = 1\%} \tag{10.9}$$

where V_{Tj} is the AC terminal voltage at jth bus and V_{Ti} is the AC terminal voltage at ith bus.

MIIF has a value between 0 and 1. The relationship between the two buses with an MIIF of 0 exists at an electrically infinite distance, and the voltage fluctuation of the AC terminal has no effect on the voltage of the other AC terminal. Conversely, if MIIF is 1, the two terminals are electrically connected, that is the two HVDCs are connected to the same bus. MIIF is generally asymmetrical and satisfies $MIIF_{i,j} \neq MIIF_{j,i}$. It is affected not only by the impedance between the two buses but also by the shunt impedance on each bus. The method defined by MIIF in [6] is a method of empirical measurement through dynamic simulation. However, it is difficult to use an empirical index for analytic methods such as stability analysis, and as such a method of deriving MIIF using an analytic method has been proposed in [7], [8] (Equation (10.10)).

$$MIIF_{j,i} = \frac{\Delta V_{Tj}}{\Delta V_{Ti}} = \frac{Z_{i,j}}{Z_{i,i}} \tag{10.10}$$

where Z is the impedance matrix, $Z_{i,j}$ is the (i,j)th entry of the impedance matrix, and $Z_{i,i}$ is the (i,i)th entry of the impedance matrix.

MIIF represents the voltage relationship between the corresponding AC terminal and the other AC terminal, which is the degree to which the HVDC installed on the other bus line affects the bus line. The degree of influence is determined by comparing the product of the HVDC active power and MIIF (e.g., $MIIF_{2,1} \times P_{d2}$), which is different from the active power of the reference HVDC(e.g., P_{d1}). If the product of the HVDC active power and MIIF on the other bus line is less than 15% of the active power of the reference HVDC, it is presumed that the mutual influence is small. If the result is more than 15% and less than 40%, interaction is likely to occur. If the result is more than 40%, interaction is extremely likely.

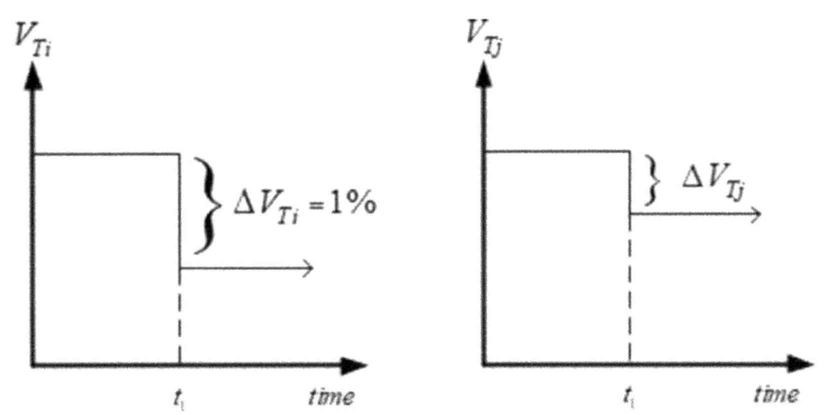

Figure 10.11 Determining the MIIF between two inverter AC terminals

10.3.3 Multi-infeed Effective Short-circuit Ratio (MIESCR)

In the definition of ESCR, the denominator represents the rated capacity of the HVDC installed in the target bus. Also, ESCR is a good indicator of power stability. For this reason, MIESCR, an extended form of ESCR, has also been introduced in multi-infeed HVDC systems (Equation (10.11)). Table 10.4 shows the typical megawatt-weighted related inverter AC voltage change matrix and Table 10.5 lists the indexes used for the analysis of single or multi-infeed HVDC systems and interaction phenomena types.

$$MIESCR_i = \frac{SCC_i - Q_{f,i}}{P_{d,i} + \sum_{j=1,\ j\neq i}^{n} MIIF_{j,i} \times P_{d,j}} \tag{10.11}$$

Table 10.4 Typical Megawatt-Weighted Relative Inverter AC Voltage Change Matrix

MIIF table		P_d	Megawatt-weighted relative inverter AC voltage change	
			Inverter 1	Inverter 2
Bus at which fixed reduction is applied	Inverter 1	P_{d1}	P_{d1}	$P_{d2} \times MIIF_{2,1}$
	Inverter 2	P_{d2}	$P_{d1} \times MIIF_{1,2}$	P_{d2}

Table 10.5 Comparison of Single- and Multi-Infeed Methodologies

	Single-infeed	Multi-infeed	Comments
Basic indicators	- Effective short-circuit ratio (ESCR)	- Multi-infeed effective short-circuit ratio (MIESCR) - MIIF - ESCR	
Fault recovery, Commutation failure	- ESCR - Detailed study	- MIESCR - Detailed study - Contour map - MIIF	The contour map is especially useful in multi-infeed situations to show common busses that result in c.f.
P/V stability	- Critical effective short-circuit ratio (CESCR) - ESCR - Detailed study	- Critical multi-infeed effective short-circuit ratio (CMIESCR) - MIESCR - Detailed study	Whether in a single- or multi-infeed situation, the applicable critical value may be similar.
TOV	- ESCR - Detailed study	- MIESCR - Detailed study	
Harmonics		- MIESCR - Detailed study - One-amp injection	Detailed studies could be a harmonic and/or EMT program

(continued)

Table 10.5 Comparison of Single- and Multi-Infeed Methodologies *(continued)*

	Single-infeed	Multi-infeed	Comments
Control interaction		- MIESCR - Detailed study	
Operating indicator		- MIESCR - MIIF - Detailed study	

10.3.4 MPC of Multi-infeed HVDC system

In a single-infeed HVDC system, ESCR is an important factor in determining the MAP, and we have seen in the previous section that I_{MAP} must be in a region larger than the operating point for the operating point to be in a stable region. It is also confirmed that the ESCR = CESCR (1.5), which matches the operation point and the MAP, should have a certain level of ESCR for stable operation of the HVDC. Therefore, we want to analyze the correlation between MIESCR, MPC, and MAP.

Figure 10.12 shows MPC according to the ESCR/MIESCR of HVDC1 when HVDC2 is operating in nominal values as shown in Figure 10.10. When ESCR1 is 1.9, ESCR1 is larger than a CRSCR of 1.5, so it can be seen that the HVDC is formed in an unstable region, though it should be operated in a stable one. If MIESCR1 is 1.50, HVDC1 is marginally stable, with I_{MAP} being equal to the operating point. Also, when MIESCR1 is 2.01, I_{MAP} is in a region larger than the operating point, and HVDC1 is operating in a stable region. It is shown that MIESCR is suitable for determining P/V stability in multi-infeed HVDC systems. To clearly determine this relation, the MAP and I_{MAP} of HVDC1 according to MIESCR1 are plotted in Figure 10.13, which shows that the system is marginally stable at an MIESCR of around 1.5.

In [6], the critical multi-infeed effective short-circuit ratio (CMIESCR) is slightly smaller than 1.5, but 1.5 is considered reasonable since a CMIESCR and MIESCR of 2.5 or more are recommended for stable operation of the HVDC.

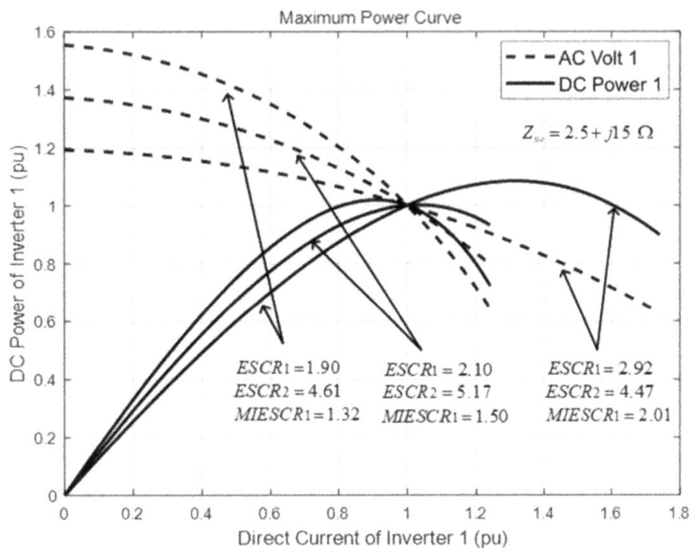

Figure 10.12 Variation of AC terminal voltage and DC power of inverter 1

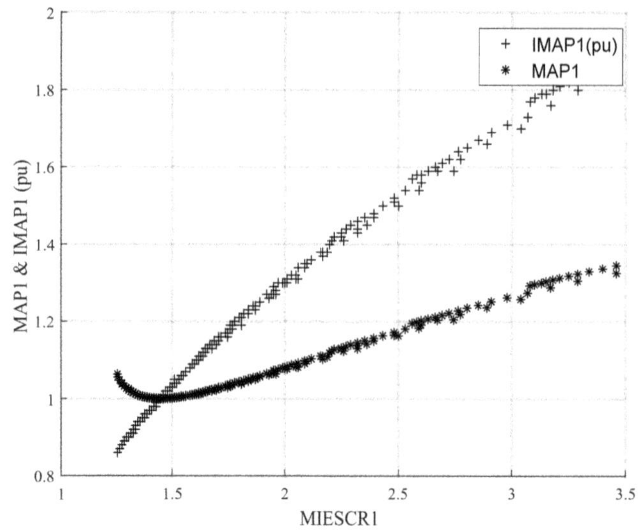

Figure 10.13 Plot of MAP and I_{MAP} with multi-infeed effective short-circuit ratio (MIESCR)

10.4. System Strength Index for LCC/VSC Multi-Infeed HVDC Systems

Because of the development of power electronics technology, the voltage source converters or VSC-HVDCs are being added to the existing HVDC systems. Since the VSC-HVDC uses an insulated bipolar transistor (IGBT) or a metal oxide semiconductor field effect transistor (MOSFET) with turn-on and turn-off capability, it generates a desired voltage and independently controls active power and reactive power [9]. Thanks to the many advantages of VSC-HVDC, there has been a shift to a hybrid multi-infeed HVDC system in which the VSC-HVDC is additionally installed in the existing power system and the LCC/VSC-HVDC is also present. In the previous section, we confirmed that ESCR and MIESCR are useful indexes for evaluating AC/DC interactions. However, since the relationship between existing interactions and indexes has been established for LCC-HVDC, this section discusses indexes that reflect the reactive power control capability of VSC-HVDC.

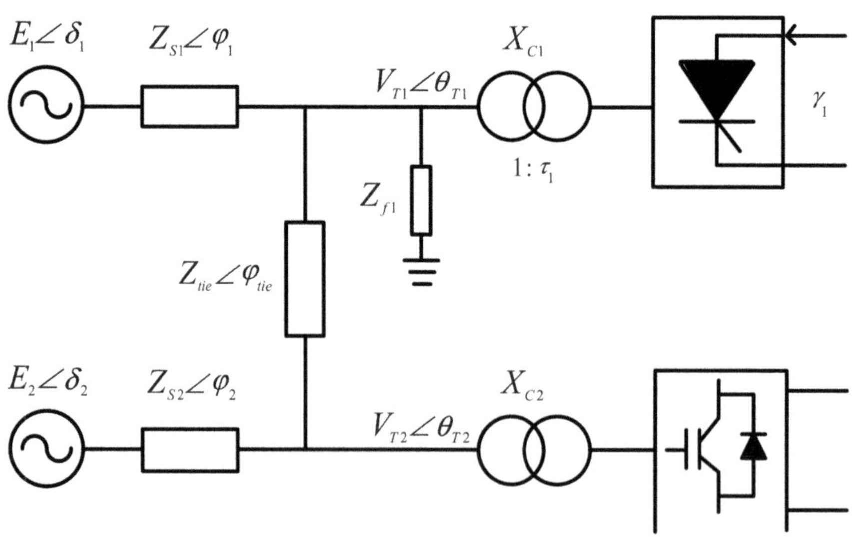

Figure 10.14 Multi-infeed HVDC system with LCC/VSC-HVDC

10.4.1 Capability Curve of VSC Transmission System

There are three main factors that limit the capability seen from a power system stability perspective. The first one is the maximum current through the IGBTs. This will give rise to a maximum MVA circle in the power plane where maximum current and actual AC voltage is multiplied. If the AC voltage decreases, so also will the MVA capability.

The second limit is the maximum DC voltage level. Reactive power is mainly dependent on the voltage difference between the AC voltage that the VSC can generate from the DC voltage and the grid AC voltage. If the grid AC voltage is high, the difference between the maximum DC voltage and the AC voltage will be low. The reactive power capability is then moderate, but increases with decreasing AC voltage. This makes sense from a stability point of view.

The third factor is the maximum DC current through the cable. The different limits are shown in Figure 10.15. For a decreasing AC voltage level, the maximum DC voltage level will vanish, and the maximum current level will decide the capability. The small bias in the Q-axis direction is because of the line reactor and filter capacitance within the VSC transmission system. Smaller adjustments in the following calculations will therefore be necessary when evaluating the qualitative results of a VSC transmission. Also, the tap changer on the converter transformers will play a role in certain cases. Note the similarities with this capability curve and the capability curve for a generator. Maximum DC voltage level corresponds to maximum field current in the rotor winding and IGBT current corresponds to armature current.

A VSC transmission system can virtually instantly take any working point within the capability chart. Instant active power flow reversals are also possible, since the VSC transmission system changes DC current direction and not DC voltage polarity. The XLPE-based cable technology used will handle such current reversals without any problems. Any limitations in the maximum power changing rates of the connecting AC grids must naturally be considered.

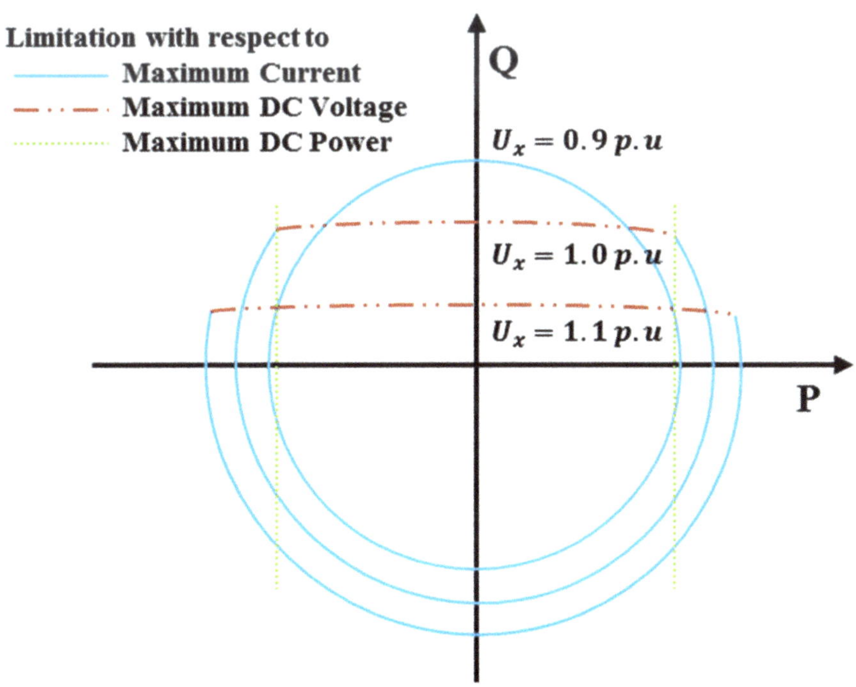

Figure 10.15 The capability curve of VSC

10.4.2 System Strength Index with VSC-HVDC

ESCR and MIESCR are the indexes for the strength of LCC-HVDC-based multi-infeed HVDC systems. Various previous studies have been carried out to define ESCR in a hybrid multi-infeed HVDC system with VSC-HVDC.

- Apparent increase in SCR (AISCR)
- Impedance-based ESCR (IESCR)
- Reactive power-based ESCR

10.4.3 Apparent Increase in SCR (AISCR)

As shown in Figure 10.16, a system with LCC-HVDC and VSC-HVDC installed in one AC terminal is assumed. First, we measure the relationship between the SCR and the MAP in a single LCC-HVDC system, where VSC-HVDC is blocked. Then, connect the VSC-HVDC to measure the MAP of the LCC-HVDC in a situation where the VSC-HVDC controls the active power injection and the AC terminal voltage. At this time, the measured MAP is substituted into the relation of SCR/MAP obtained in the single LCC-HVDC situation, and the SCR corresponding to that is searched (SCRa). The difference between the SCRa and the SCR of a given system is defined as AISCR, meaning the increase of SCR caused by VSC-HVDC.

Figure 10.17 is a graph of the SCR-dependent MAP of LCC-HVDC with VSC-HVDC blocked. In one example situation, the SCR of the system is 2.5, and the point "X" means the MAP at that time.

Figure 10.18 is a graph of the MAP of the LCC-HVDC according to the variation of the active power output of VSC-HVDC when the VSC-HVDC rating is 328MVA/300MW and the LCC-HVDC is the CIGRE Benchmark model. As the active power of the VSC-HVDC increases, the MAP of the LCC-HVDC decreases. When the active power is 315MW, it matches the MAP "X" in the single-infeed HVDC system. When the VCC-HVDC has a rated power of 300 MW, the MAP of the LCC-HVDC is the same as that of the single-infeed HVDC system where the SCR is 3.03. The VSC-HVDC has increased the SCR by 0.53 (= 3.03 – 2.5), which is the AISCR in this example situation.

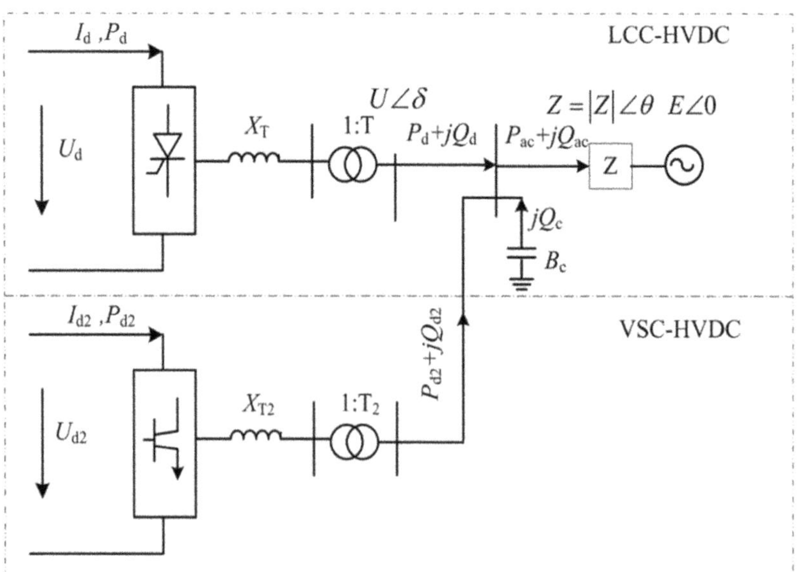

Figure 10.16 Schematic diagram of dual-infeed HVDC system

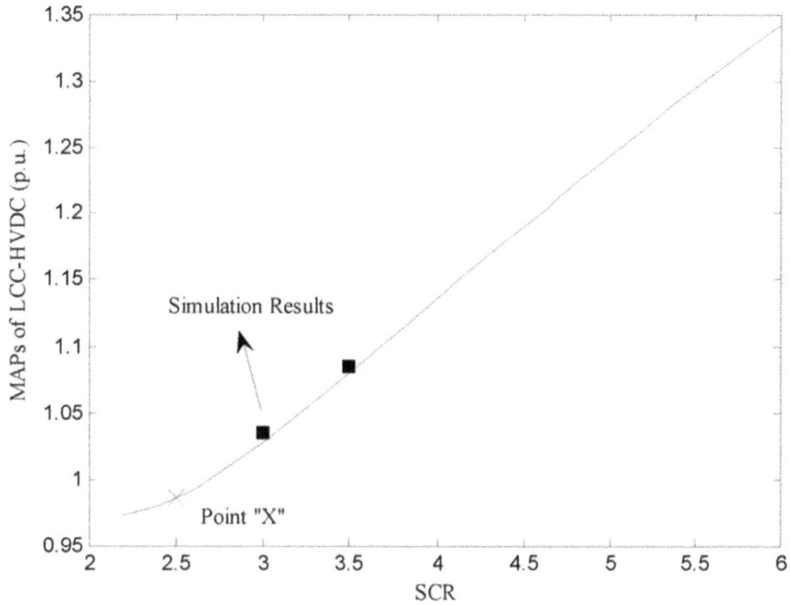

Figure 10.17 Maximum available power (MAP) and SCR relation of single LCC-HVDC

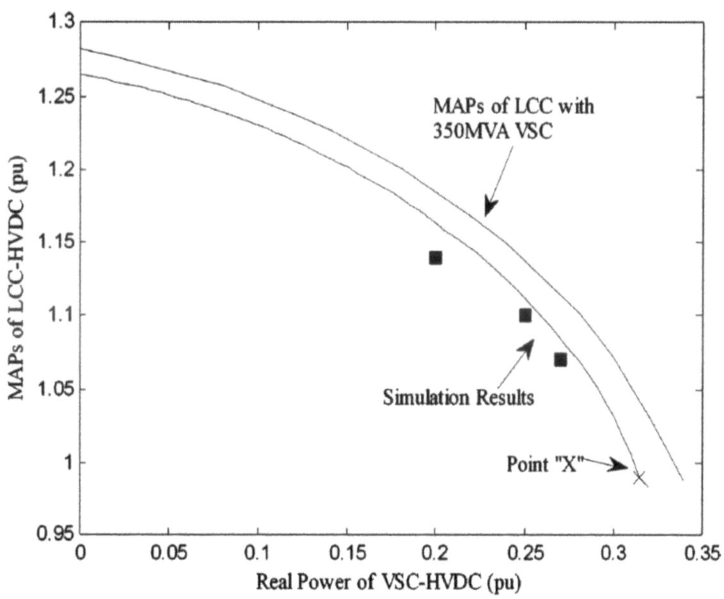

Figure 10.18 Maximum available powers (MAPs) of LCC-HVDC with varying VSC power loading

To verify the validity of AISCR, the interaction characteristics of a hybrid dual-infeed HVDC system and a single-infeed HVDC system with an SCR of 3.03 (= 2.5 + 0.53) were compared.

Table 10.6 shows that MAP, TOV, and CFII are relatively accurate compared with AISCR, but there is a difference in fault recovery time. Since AISCR is calculated through steady-state analysis, the analysis of the steady-state interaction effect shows relatively accurate results, and there is an inaccurate aspect to show transient phenomena such as fault recovery time. However, through the introduction and analysis of AISCR, it can be seen that when the VSC-HVDC is input, there is a reinforcement effect on the system such as AC voltage regulation.

Table 10.6 Steady-State and Transient Aspects Comparison

Scenarios / aspects		Hybrid dual-infeed	LCC-HVDC by itself SCR = 3.03
Maximum available power (MAP) of LCC (p.u.)		1.024	1.028
TOV (only block LCC (p.u.)		1.148	1.151
CFII of LCC converter (%)		16.5	16.3
Fault recovery time of LCC (ms)	SLGF	224	345
	3PGF	364	467

$$\text{where CFII (Commutation failure immunity index)} = \frac{\textit{Critical fault MVA}}{P_d} * 100\% \tag{10.12}$$

10.4.4 Impedance-based ESCR (IESCR)

The analysis of the interaction effect in the existing hybrid multi-infeed HVDC system mainly lacks the analysis of the effect of LCC-HVDC on VSC-HVDC (mainly by the influence of VSC-HVDC on LCC-HVDC). Also, it is known that CESCR is influenced by the control system of the HVDC. To reflect this, analysis using a method capable of reflecting control characteristics such as a small-signal model is required.

When calculating the ESCR in a multi-infeed HVDC system, it is important to calculate the equivalent impedance seen from the AC terminal. The important factor here is to calculate the impedance Z_t of Figure 10.19, while taking into account the effects of other converters. To do this, an IESCR (impedance-based ESCR) is proposed that calculates the ESCR by equalizing the other converter to the incremental impedance based on the small-signal model.

$$IESCR_i = \frac{\left(\frac{U_{i,n}^2}{Z_{equal}}\right)}{P_{di}} \tag{10.13}$$

where Z_{equal} is the impedance of the AC network seen from converter C_i's AC terminal.

In Figure 10.20, the VSC-HVDC is regarded as an equivalent impedance based on the small-signal model to calculate the system-equivalent impedance Z_{equal}, as viewed from the AC terminal of the LCC-HVDC.

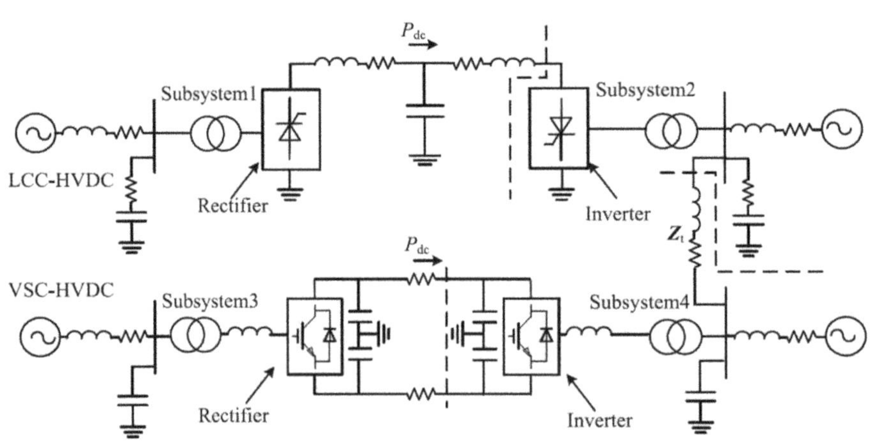

Figure 10.19 Schematic diagram of LCC/VSC dual-infeed HVDC system

At this time, the method of calculating the complex impedance of the VSC-HVDC (Z_{VSC}) is as follows: calculate the ratio of AC terminal voltage and current change of VSC-HVDC caused when the current command of LCC-HVDC (ΔI_{dcref}) changes slightly, as shown in Equation (10.16). Similarly, when calculating the complex impedance of the LCC-HVDC ((Z_{LCC}), the ratio of the AC terminal voltage to the current change of the LCC-HVDC is calculated when the power reference of the VSC-HVDC (ΔP_{VSC}) changes slightly, as shown in Equation (10.17).

$$Z_{equal} = \frac{1}{\frac{1}{Z_1} + \frac{1}{Z_{f1}} + \frac{1}{Z_B}} \tag{10.14}$$

$$Z_B = \frac{1}{\frac{1}{Z_1} + \frac{1}{Z_{f1}} + \frac{1}{Z_B}} \tag{10.15}$$

$$\Delta I_{dcref} \rightarrow Z_{VSC} = \frac{\Delta U_{VSC}}{\Delta I_{VSC}} \tag{10.16}$$

$$\Delta P_{VSC} \rightarrow Z_{LCC} = \frac{\Delta U_{LCC}}{\Delta I_{LCC}} \tag{10.17}$$

To verify the validity of IESCR, MPC analysis of LCC-HVDC was performed. As we have seen in the previous section, the LCC-HVDC has a CESCR where the MAP is equal to the operating point, and the slope of the MAP should be zero $= \frac{dP_d}{dI_d}$.

Figure 10.21 shows the LCC-HVDC value at the operating point according to the change in the impedance Z_1 in Figure 10.20. The dual-infeed system represents the LCC-VSC dual-infeed HVDC system, and the single-infeed system represents only the LCC-HVDC system as well as the PSCAD/EMTDC simulation results in each situation. At this time, the LCC-HVDC performs a 1000 MW rating and constant extinction angle (CEA) control, and the VSC-HVDC a 500 MW rating and AC terminal voltage control. In Figure 10.21, all cases have the same value. Also, as Z_1 becomes larger, the slope becomes smaller, since ESCR also becomes smaller. When

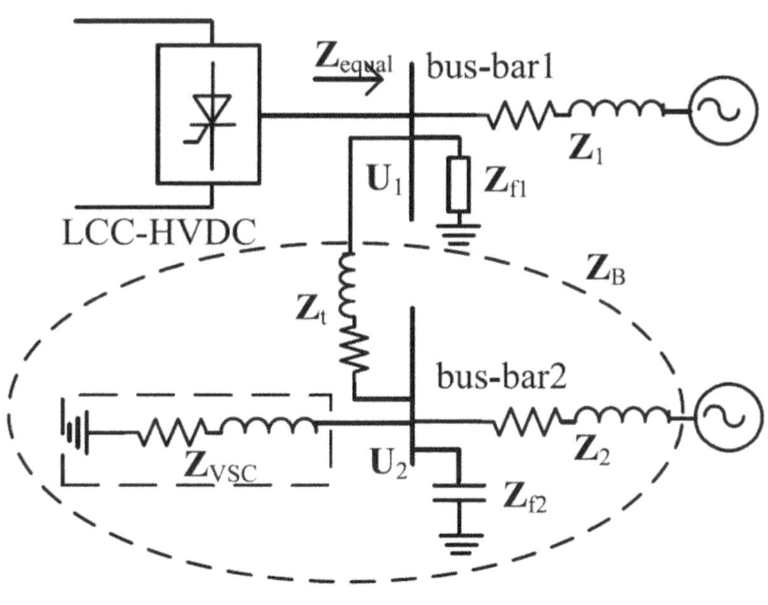

Figure 10.20 Equivalent circuit for LCC/VSC dual-infeed system

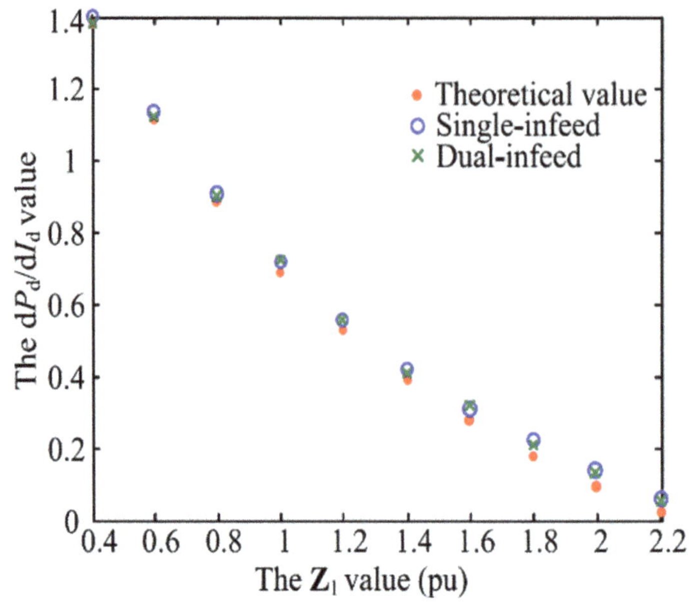

Figure 10.21 Comparison of $\frac{dP_d}{dI_d}$ values

Z_1 becomes about 2.2, the slope becomes zero, which means that CESCR is reached. This confirms that the results calculated based on the small signal model are reasonable.

In Table 10.7, the calculated Z_1 becomes CESCR according to the active power output of various VSC-HVDCs and the derived IESCR of LCC-HVDC accordingly. The results show that the IESCR corresponding to the CESCR is close to 1.5, confirming that IESCR plays the same role as the existing ESCR.

Table 10.7 Various IESCR of LCC-HVDC

$P_{VSC(p.u.)}$	$Z_{1(p.u.)}$	$Z_{VSC(p.u.)}$	$IESCR_{LCC}$
0	$2.56\angle75^\circ$	$0.042\angle179.3^\circ$	1.4864
0.1	$2.52\angle75^\circ$	$0.054\angle179.2^\circ$	1.4855
0.2	$2.48\angle75^\circ$	$0.067\angle178.9^\circ$	1.4832
0.3	$2.44\angle75^\circ$	$0.082\angle178.6^\circ$	1.4792
0.4	$2.36\angle75^\circ$	$0.100\angle178.2^\circ$	1.4776
0.5	$2.28\angle75^\circ$	$0.121\angle177.7^\circ$	1.4737

Figure 10.22 shows the change according to the control mode of VSC-HVDC. In the case of constant reactive power control mode, this means that the control is performed with unity power. As can be seen in both control modes, the IESCR decreases as the output of the VSC-HVDC inverter increases. This is in line with the decrease in MIESCR on the bus line as the output of another HVDC increases, as in the MIESCR definition in Equation (10.11).

Also, when the active power output of the VSC-HVDC is zero on the graph and the power factor control is performed, it is the same as the single LCC-HVDC situation since there is neither the active nor reactive power output of the VSC-HVDC. In this situation, when the VSC-HVDC performs an AC voltage control, it

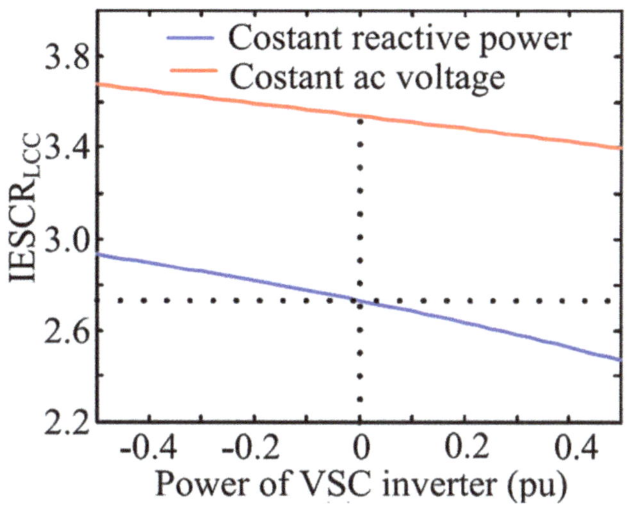

Figure 10.22 Dependence of impedance -based effective short-circuit ratio (IESCR) on control strategy

is operating in STATCOM mode. Comparing the IESCR of these two cases, the VSC-HVDC performs the AC voltage control, and the IESCR of the LCC-HVDC is improved by about 0.8.

10.4.5 Reactive power-based ESCR

In the previous section, we discussed the PQ capability curve of a VSC-HVDC. The PQ capability curve of a VSC- HVDC can be represented simply as a circle. Based on these characteristics, if the VSC-HVDC controls the active power output constantly, the reactive power equivalent to the remaining apparent power capacity can be flexibly used. In other words, it is possible to use reactive power up to the reactive power supply amount of the VSC-HVDC.

Using this concept, multiple HVDCs in a multi-infeed system composed of LCC-HVDCs can be represented as one equivalent HVDC as shown in Figure 10.25(a). On the other hand, when the VSC-HVDC system is equivalently synthesized in a way similar to the LCC-HVDC equivalent synthesis, it can be expressed as in

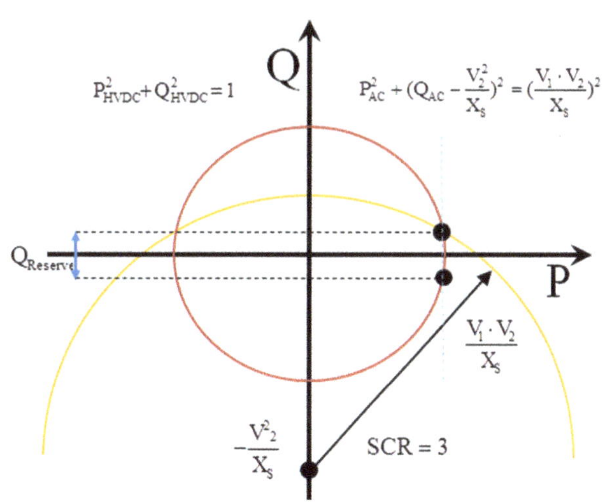

Figure 10.23 Capability curve of VSC-HVDC and AC system

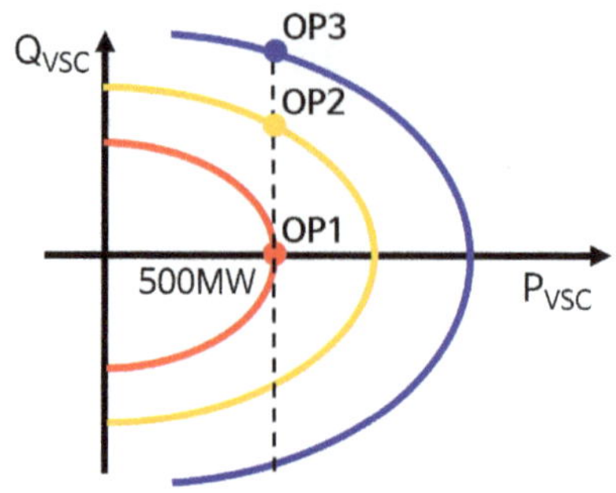

Figure 10.24 Variable PQ power curves according to reactive power supply availability

Figure 10.25(b). VSC-HVDC systems exhibit different aspects because they have reactive power supply capacities different from LCC-HVDCs.

Figure 10.26 shows the MPC of the LCC-HVDC in a multi-infeed system consisting of LCC-HVDC MPC, and LCC/VSC-HVDC in a multi-infeed system composed of LCC-HVDC for the same system.

From Figure 10.26(a), it can be confirmed that the AC voltage is continuously dropped because the LCC-HVDC has no reactive power control supply capability. On the other hand, in the case of Figure 10.26(b), it can be seen that the VSC-HVDC has the capability of maintaining the voltage within the capability range because of the reactive power supply capability of the VSC-HVDC. If the reactive power supply of the VSC-HVDC is larger, the voltage-holding period will also be widened, and the improved MAP will also become larger.

Figure 10.27 shows the change in the CMIESCR according to the amount of reactive power supply of a voltage type HVDC. The CMIESCR has a value of approximately 1.5, as mentioned earlier, which refers to the MIESCR when the MAP and the operating point coincide. If the existing MIESCR reflects the reactive power supply of the VSC-HVDC, it must have a constant CMIESCR regardless of Q_{VSC}. However, in Figure 10.27, it is clear that CMIESCR becomes smaller than 1.5 as Q_{VSC} increases. This means that the strength of the system increases in a positive correlation with Q_{VSC}.

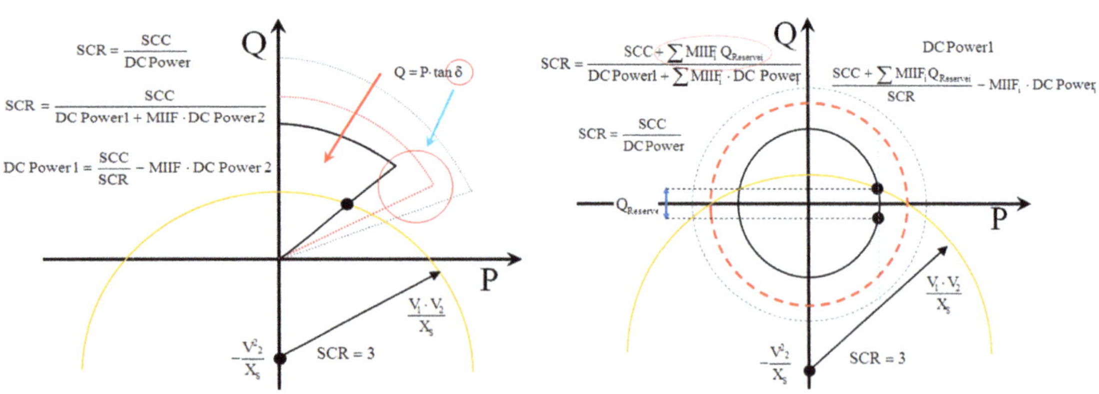

Figure 10.25 Capability curve synthesis for (a; left) LCC-HVDC; (b; right) VSC-HVDC-based multi-infeed system

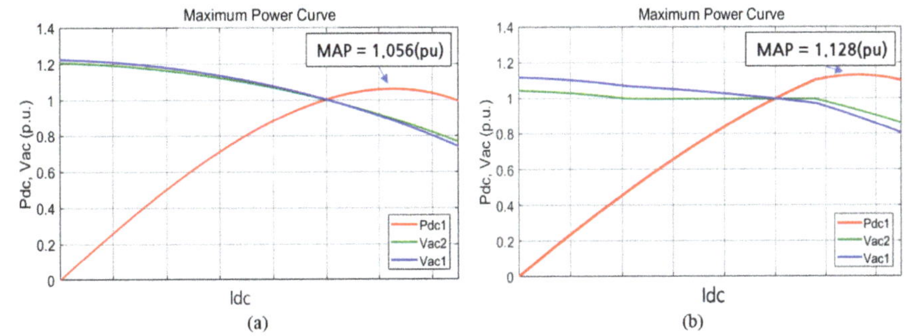

Figure 10.26 MPC for (a) LCC-HVDC-based multi-infeed system and (b) LCC/VSC-HVDC-based multi-infeed system

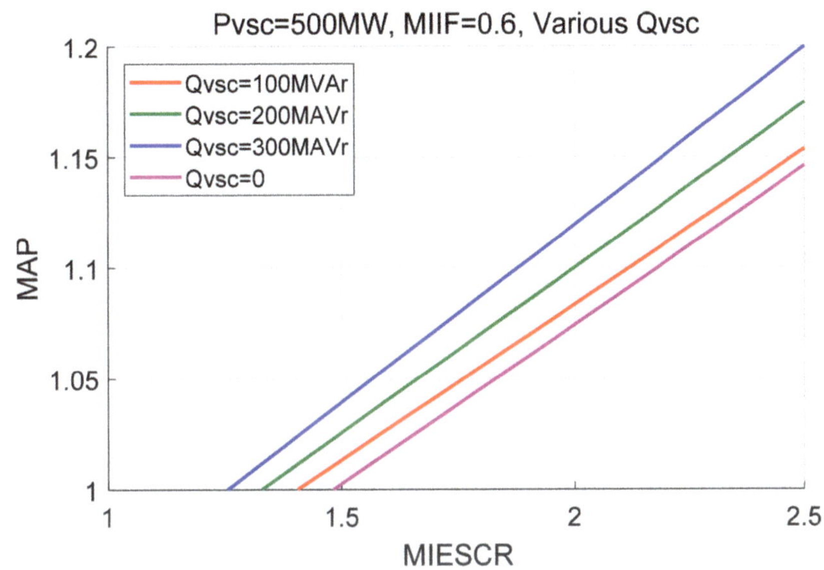

Figure 10.27 Variation of critical multi-infeed effective short-circuit ratio (CMIESCR) with VSC-HVDC availability of reactive power supply

10.5 Modeling of VSC HVDC for Power System Analysis

10.5.1 Steady-state Model of VSC HVDC

The ever growing need for transmission capacity in current day power systems has led to an increased interest in transmission based on voltage-source converter high-voltage direct current (VSC-HVDC) technology. One of the advantages of the VSC technology over conventional current source commutated (CSC) HVDCs is the relatively straightforward extension to multi-terminal (MTDC) configurations. The VSC-HVDC offers significant advantages over traditional AC grid reinforcements, and operation in an MTDC grid could facilitate the integration of renewable energy sources in the future. Recently, plans have even been suggested to build offshore grids based on DC technology or a DC supergrid to connect the growing share of renewable energy sources.

One of the outstanding research issues is the steady-state behavior of these integrated AC/DC systems. In the past, much effort has been spent on the joint solution of integrated AC/DC systems based on traditional CSC technology. Generally, the solution methods presented for CSC HVDC systems can be subdivided into unified and sequential methods. In the unified methods, the AC and DC system equations are solved together, whereas in the sequential methods, the AC and DC system equations are solved sequentially. The biggest advantage of the sequential methods over their unified counterparts is that the sequential methods can be implemented relatively easily as extensions to existing AC power flow programs, whereas a unified implementation requires the alteration of existing AC power flow algorithms. For AC power flow analysis, a steady-state model of VSC-HVDC is required to emulate the operating characteristics of VSC-HVDC systems.

10.5.2 Converter Model

In its most general format, a VSC-HVDC converter (Figure 10.28) can be represented by a controllable voltage source $\underline{U_C} = U_C\angle\delta_C$ behind the phase reactor with a complex impedance $\underline{Z_C} = R_C + jX_C$. As shown in Figure 10.28, the phase reactor is connected to the susceptance B_f which forms a part of the low pass filter. The filter bus is connected to the AC grid through a transformer, represented by its impedance $\underline{Z}_{th} = R_{th} + jX_{th}$. The equations for power flowing to the AC network can thus be written as

$$P_s = -U_s^2 G_{tf} + U_s U_f\left[G_{tf}\cos\left(\delta_s - \delta_f\right) + B_{tf}\sin\left(\delta_s - \delta_f\right)\right] \tag{10.18}$$

$$Q_s = U_s^2 B_{tf} + U_s U_f\left[G_{tf}\sin\left(\delta_s - \delta_f\right) - B_{tf}\cos\left(\delta_s - \delta_f\right)\right] \tag{10.19}$$

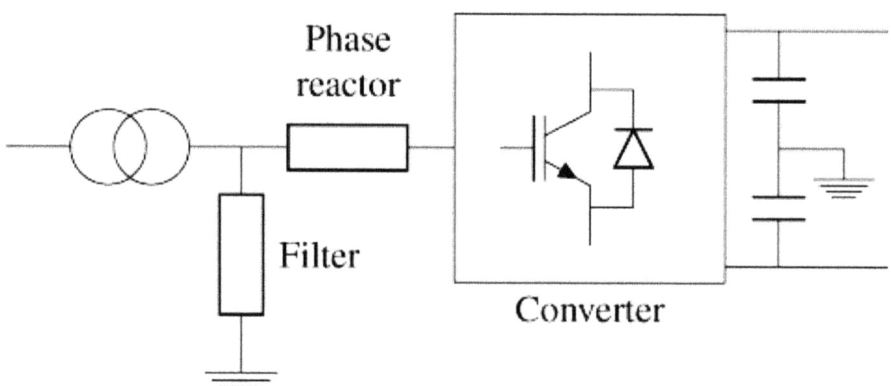

a) VSC HVDC Converter Station

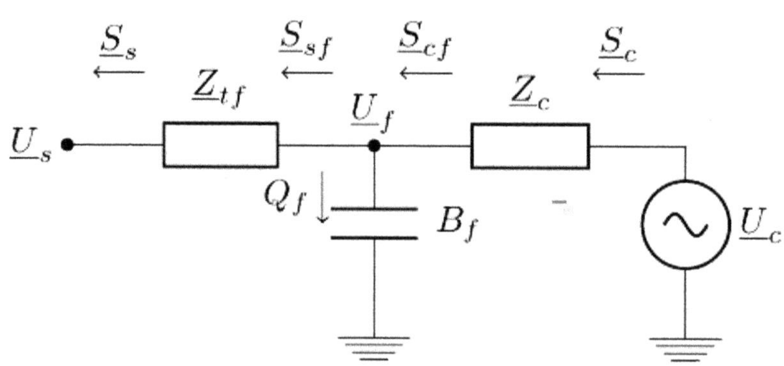

b) VSC HVDC Modeling

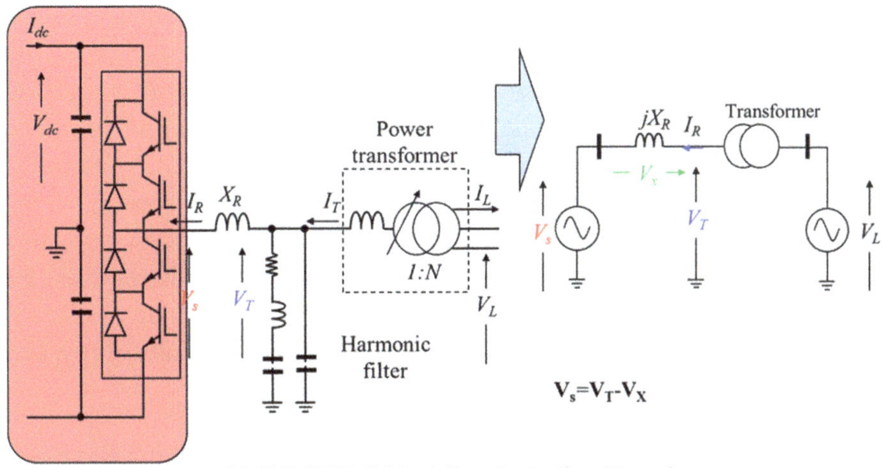

c) VSC HVDC Modeling including Transformer

Figure 10.28 VSC-HVDC modeling

with $\underline{U_S}$=$U_s\angle\delta_s$ and $\underline{U_f}$=$U_f\angle\delta_f$ representing the complex grid side and filter bus voltage respectively. The equations at the converter side can be written as

$$P_c = U_c^2 G_c - U_f U_c\left[G_c \cos\left(\delta_f - \delta_c\right) - B_c \sin\left(\delta_f - \delta_c\right)\right] \quad (10.20)$$

$$Q_c = -U_c^2 B_c + U_f U_c\left[G_c \sin\left(\delta_f - \delta_c\right) + B_c \cos\left(\delta_f - \delta_c\right)\right] \quad (10.21)$$

Assuming the AC filters to be lossless, the filter power reduces to its reactive part

$$Q_f = -U_f^2 B_f \quad (10.\ 22)$$

whereas the expressions for the filter side complex power flowing through the transformer are written as

$$P_{sf} = U_f^2 G_{tf} - U_f U_s\left[G_{tf} \cos\left(\delta_s - \delta_f\right) - B_{tf} \sin\left(\delta_s - \delta_f\right)\right] \quad (10.23)$$

$$Q_{sf} = -U_f^2 B_{tf} + U_f U_s\left[G_{tf} \sin\left(\delta_s - \delta_f\right) + B_{tf} \cos\left(\delta_s - \delta_f\right)\right] \quad (10.24)$$

and those flowing through the phase reactor side are

$$P_{cf} = U_f^2 G_c + U_f U_c\left[G_c \cos\left(\delta_f - \delta_c\right) + B_c \sin\left(\delta_f - \delta_c\right)\right] \quad (10.25)$$

$$Q_{cf} = U_f^2 B_c + U_f U_c\left[G_c \sin\left(\delta_f - \delta_c\right) - B_c \cos\left(\delta_f - \delta_c\right)\right] \quad (10.26)$$

In the absence of low pass filters or with the filters omitted, the phase reactor and the converter transformer can be lumped together, eliminating the dependence on the complex filter bus voltage in Equation (10.18) -(10.21). Similarly, if transformerless designs are considered, the filter bus and the AC grid bus coincide, simplifying the equations.

A VSC HVDC converter can independently control active and reactive power injection with respect to the AC system. In the power flow algorithm presented in this paper, the active power injection is modeled in two different ways:

1. Constant P: The converter has a constant active power P_s injection into the AC grid.
2. Constant U_{dc}: The algorithm adapts the active power injection P_s to obtain a constant DC bus voltage U_{dc}.

The actual steady-state behavior of a two-terminal VSC HVDC link can be modeled in a power flow algorithm by one constant P converter and one constant U_{dc} converter. The active power injection of the latter is unknown prior to the power flow and depends on the losses in the VSC HVDC system. This constant U_{dc} converter is referred to as the DC slack converter because of its similarity to the slack bus in AC power flow algorithms.

This two-terminal representation can be extended to model x different MTDC systems with a total of y DC converters. With one DC slack bus per MTDC system, this results in y–x converters controlling their active power output (constant P). The remaining DC slack converters, one per MTDC system, adjust their power to account for the DC system losses and to keep up the voltage of the DC grids. Alternatively, a voltage droop can be implemented to study the effect of distributing the DC voltage control function over the power flows.

The ability of the converter to independently control the reactive power results in two different representations in the power flow algorithm:

1. Constant Q: The converter has a constant reactive power injection Qs into the AC grid.
2. Constant U: The converter adapts the reactive power injection to obtain a constant AC bus voltage magnitude Us.

Commercial software models include the converter up to the filter bus, whereas the filter and transformer have to be added separately. The power and voltage set points are then defined with respect to the filter bus or the converter bus instead of the system bus. Although these alternative methods are easier to implement, they do not comply with current day practices in VSC-HVDC systems.

The model developed in this paper also includes the converter station losses using a generalized loss formula with the converter losses quadratically dependent on the converter current I_C:

$$P_{loss} = a + b \cdot I_c + c \cdot I_c^2 \tag{10.27}$$

with the per unit converter current magnitude given by

$$I_c = \frac{\sqrt{P_c^2 + Q_c^2}}{\sqrt{3U_c}} \tag{10.28}$$

This aggregated loss model was derived for the Södra Länken project, a VSC HVDC link with a rating of 600 MW and a DC voltage of ±300 kV. The model includes the losses in the different components of a VSC-HVDC converter substation. The loss data in this paper have been scaled down to the appropriate MVA base.

10.5.3 VSC-HVDC Converter Limits

To ensure an overall safe operation of the converter station, the steady-state working point must be situated within the PQ-capability chart of the converter, as depicted in Figure 10.29. The full lines include the effect of the filters, the dotted lines neglect the filters. The full circle forms the converter current limit. The arcs depict the upper and lower converter voltage limits.

With the apparent power injection written as

$$\underline{S}_s = \underline{U}_s \underline{I}_s^* \tag{10.29}$$

the system bus voltage equation, which is equal to

$$\underline{U}_s = \underline{U}_f - \underline{Z}_{tf} \underline{I}_s \tag{10.30}$$

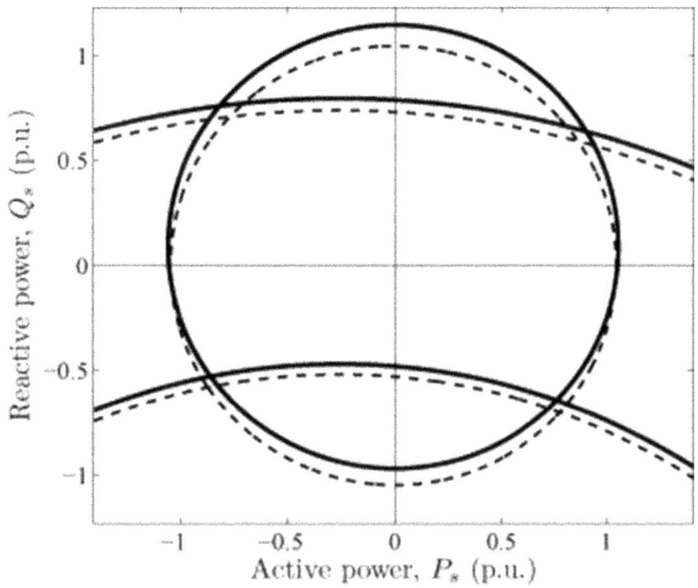

Figure 10.29 PQ-capability chart

can be reformulated and substituted in the equation for the current at the filter bus, yielding

$$\underline{I}_s = \underline{I}_c - \frac{\underline{U}_s + \underline{Z}_{tf}\,\underline{I}_s}{\underline{Z}_f} \tag{10.31}$$

Rewriting this equation in terms of the current $\underline{I}_s$ and substituting it in Equation (10.29) results in a closed expression of the converter current limit

$$\underline{S}_s = -\underline{U}_s^2\left(\frac{1}{\underline{Z}_f^* + \underline{Z}_{tf}^*}\right) + \underline{U}_s\underline{I}_{cm}^*\left(\frac{\underline{Z}_f^*}{\underline{Z}_f^* + \underline{Z}_{tf}^*}\right) \tag{10.32}$$

with $\underline{I}_{cm}$ the phasor of the maximum current

$$\underline{I}_{cm} = I_{c_{\max}}\angle\delta_{I_c} \tag{10.33}$$

With the filters omitted, Equation (10.32) simplifies to

$$\underline{S}_s = \underline{U}_s\underline{I}_{cm}^* \tag{10.34}$$

As shown in Figure 10.29, the circular current limit from Equation (10.33) slightly scales and shifts to a new center point as a result of the filter inclusion. The filter capacitance thus leads to a small bias in the Q-axis direction.

In VSC systems that use pulse width modulation (PWM), the modulation factor has an upper and lower limit in order to avoid overmodulation and the reappearance of harmonics. The lower reactive power limit is a constant at −0.5 p.u. With the modular multi-level converter (MMC) technology, the reactive power is limited by the current in the lagging region. The lower voltage limit can therefore be omitted in this case.

These limits are implemented as limits on the converter voltage. The voltage limits can be obtained by applying a Y-Δ transformation to the equivalent model from Figure 10.28. This is shown in Figure 10.29, with the equivalent impedances $\underline{Z}_1$, $\underline{Z}_2$, and $\underline{Z}_3$, respectively.

$$\underline{Z}_1 = \frac{\underline{Z}_{tf}\underline{Z}_c + \underline{Z}_c\underline{Z}_f + \underline{Z}_f\underline{Z}_{tf}}{\underline{Z}_c} \tag{10.35}$$

$$\underline{Z}_2 = \frac{\underline{Z}_{tf}\underline{Z}_c + \underline{Z}_c\underline{Z}_f + \underline{Z}_f\underline{Z}_{tf}}{\underline{Z}_f} \tag{10.36}$$

$$\underline{Z}_3 = \frac{\underline{Z}_{tf}\underline{Z}_c + \underline{Z}_c\underline{Z}_f + \underline{Z}_f\underline{Z}_{tf}}{\underline{Z}_{tf}} \tag{10.37}$$

Rewriting $\underline{I}_s^*$ in Equation (10.29) in terms of the currents $\underline{I}_1$ and $\underline{I}_2$, through $\underline{Z}_1$ and $\underline{Z}_2$ results in

$$\underline{S}_s = \underline{U}_s\left(\underline{I}_2^* - \underline{I}_1^*\right) \tag{10.38}$$

$$= \underline{U}_s\left(\frac{\underline{U}_c^* - \underline{U}_s^*}{\underline{Z}_2^*} - \frac{\underline{U}_s^*}{\underline{Z}_1^*}\right) \tag{10.39}$$

Rewriting this equation in terms of admittances results in the expression for $\underline{S}_s$ in terms of a converter voltage limit

$$\underline{S}_s = \underline{U}_s^2\left(\underline{Y}_1^* + \underline{Y}_2^*\right) + \underline{U}_s\underline{U}_{cm}^*\underline{Y}_2^* \tag{10.40}$$

with $\boldsymbol{\underline{U}_{cm}}$ being the phasor of the minimum or maximum converter voltage $\underline{\text{Ucmin}}$ or $\underline{\text{Ucmax}}$. $\underline{\text{Ucmax}}$ is the maximum voltage that can be generated by the converter while avoiding overmodulation. The lower voltage limit $\underline{\text{Ucmin}}$ in Figure 10.29 has been chosen to comply with the minimum reactive power limit. This limit can be omitted in an MMC or when no minimum reactive power limit is imposed on the converter.

Similar expressions can be found by applying l'Hôpital's rule to Equations (10.35)–(10.37) for $\underline{Z}_f$→inf or $\underline{Zt}_f$→0 when the filter or transformer are not included. The expression without the filter simplifies to

$$\underline{S}_s = -\underline{U}_s^2\left(\frac{1}{\underline{Z}_{tf}^* + \underline{Z}_c^*}\right) + \frac{\underline{U}_s\underline{U}_{cm}^*}{\underline{Z}_{tf}^* + \underline{Z}_c^*} \tag{10.41}$$

Similar to the current limit, the inclusion of the filter shifts the voltage limit circles from Equation (10.34) to a new center point, whereas their radii slightly change.

10.5.4 Dynamic Model of VSC-HVDC

In order to analyze an AC network connected to VSC-HVDC, detailed models of VSC HVDCs including harmonic characteristics, cell balancing, and so on are not required since the time constant of the AC network is much slower than that of a VSC controller. Therefore, to analyze an AC network connected to a VSC-HVDC, a dynamic model based on differential equations should be created. In this chapter, a dynamic model of a VSC-HVDC is represented.

10.5.5 Converter Model

VSC-HVDC converters are connected to the system through a phase reactor, represented here by an impedance Zpr = Rpr + *j*Xpr, that can also include the effect of the transformer (Figure 10.30).

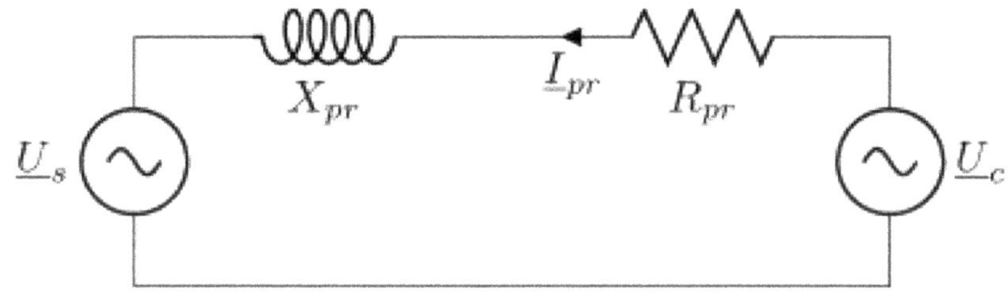

Figure 10.30 One-line diagram of the AC circuit

The basic equation of this circuit

$$u_c - u_s = L_{\text{pr}} \frac{di_{\text{pr}}}{dt} + R_{\text{pr}} i_{\text{pr}} \tag{10.42}$$

can be transformed to a power system synchronized rotating dq-reference frame

$$\frac{d\left(i_{\text{pr}}^{d} e^{j\omega t}\right)}{dt} = -\frac{R_{\text{pr}}}{L_{\text{pr}}} i_{\text{pr}}^{d} e^{j\omega t} + \omega i_{\text{pr}}^{q} e^{j\omega t} + \frac{1}{L_{\text{pr}}} \left(u_c^d - u_s^d\right) e^{j\omega t} \tag{10.43}$$

$$\frac{d\left(i_{\text{pr}}^{q} e^{j\omega t}\right)}{dt} = \frac{R_{\text{pr}}}{L_{\text{pr}}} i_{\text{pr}}^{q} e^{j\omega t} - \omega i_{\text{pr}}^{d} e^{j\omega t} + \frac{1}{L_{\text{pr}}} \left(u_c^q - u_s^q\right) e^{j\omega t} \tag{10.44}$$

The angle wt is provided by the phase-locked loop (PLL), arbitrarily assumed here to align system voltage with the q-axis. Dividing by e^{jwt}, the dynamic equations of the AC circuit in the dq-frame are finally obtained:

$$\frac{di_{\text{pr}}^{q}}{dt} = -\frac{R_{\text{pr}}}{L_{\text{pr}}} i_{\text{pr}}^{d} + \omega i_{\text{pr}}^{q} + \frac{1}{L_{\text{pr}}} \left(u_c^d - u_s^d\right) \tag{10.45}$$

$$\frac{di_{\text{pr}}^{q}}{dt} = -\frac{R_{\text{pr}}}{L_{\text{pr}}} i_{\text{pr}}^{q} - \omega i_{\text{pr}}^{d} + \frac{1}{L_{\text{pr}}} \left(u_c^q - u_s^q\right) \tag{10.46}$$

Most VSC converters have a cascaded control structure, composed of inner current control loops and outer controllers. The controllers themselves are usually PI controllers, "the de facto industry standard for current control," which guarantee high performance. In this paper, the emphasis is not placed on the control scheme. Therefore, a standard control scheme consisting of cascaded PI controllers is implemented here.

The converters described in Equations (10.45) and (10.46), obtained earlier can be controlled by two parallel control loops using PI controllers. The cross-coupling terms

$$\Delta u_c^d = \omega L_{\text{pr}} i_{\text{pr}}^{q} \tag{10.47}$$

$$\Delta u_c^q = -\omega L_{\text{pr}} i_{\text{pr}}^{d} \tag{10.48}$$

compensate for the cross coupling between the two control loops, introduced in Equations (10.47) and (10.48) by the transformation to a rotating reference frame. The reference voltage in the d and q components are then expressed as

$$u_{c\text{ref}}^{d} = u_s^d + \Delta u_c^d + \left(K_{p1} + \frac{K_{i1}}{s}\right)\left(i_{\text{pr}\text{ref}}^{d} - i_{\text{pr}}^{d}\right) \tag{10.49}$$

$$u_{c\text{ref}}^{q} = u_s^q + \Delta u_c^q + \left(K_{p1} + \frac{K_{i1}}{s}\right)\left(i_{\text{pr}\text{ref}}^{q} - i_{\text{pr}}^{q}\right) \tag{10.50}$$

After introducing two state variables, M_d and M_q, such that

$$\frac{dM_d}{dt} = K_{i1}\left(i^d_{\mathrm{pr}^{\mathrm{ref}}} - i^d_{\mathrm{pr}}\right) \tag{10.51}$$

$$\frac{dM_q}{dt} = K_{i1}\left(i^q_{\mathrm{pr}^{\mathrm{ref}}} - i^q_{\mathrm{pr}}\right) \tag{10.52}$$

the current controller equations become

$$u^d_{c^{\mathrm{ref}}} = u^d_s + \Delta u^d_c + K_{p1}\left(i^d_{\mathrm{pr}^{\mathrm{ref}}} - i^d_{\mathrm{pr}}\right) + M_d \tag{10.53}$$

$$u^q_{c^{\mathrm{ref}}} = u^q_s + \Delta u^q_c + K_{p1}\left(i^q_{\mathrm{pr}^{\mathrm{ref}}} - i^q_{\mathrm{pr}}\right) + M_q \tag{10.54}$$

The actual value of the voltage lags behind the reference because of the time-lag of the converter's power electronics. The relation between the actual value and the reference value can be represented by a time delay with time
constant $T\sigma$.

$$\frac{U^{dq}_c}{U^{dq}_{c^{\mathrm{ref}}}} = \frac{1}{T_\sigma s + 1} \tag{10.55}$$

or in the time-domain

$$\frac{du^d_c}{dt} = \frac{1}{T_\sigma}\left(u^d_{c^{\mathrm{ref}}} - u^d_c\right) \tag{10.56}$$

$$\frac{du^q_c}{dt} = \frac{1}{T_\sigma}\left(u^q_{c^{\mathrm{ref}}} - u^q_c\right) \tag{10.57}$$

Combining the time-lag in Equations (10.53) and (10.54) and the equations of the current controllers Equations (10.58) and (10.59) yields

$$\begin{aligned}\frac{du^d_c}{dt} = &-\frac{K_{p1}}{T_\sigma} i^d_{\mathrm{pr}} - \omega\frac{L_{\mathrm{pr}}}{T_\sigma} i^q_{\mathrm{pr}} - \frac{1}{T_\sigma} u^d_c \\ &+\frac{1}{T_\sigma} M_d + \frac{K_{p1}}{T_\sigma} i^d_{\mathrm{pr}^{\mathrm{ref}}} + \frac{1}{T_\sigma} u^d_s\end{aligned} \tag{10.58}$$

$$\begin{aligned}\frac{du^q_c}{dt} = &-\frac{K_{p1}}{T_\sigma} i^q_{\mathrm{pr}} + \omega\frac{L_{\mathrm{pr}}}{T_\sigma} i^d_{\mathrm{pr}} - \frac{1}{T_\sigma} u^q_c \\ &+\frac{1}{T_\sigma} M_q + \frac{K_{p1}}{T_\sigma} i^q_{\mathrm{pr}^{\mathrm{ref}}} + \frac{1}{T_\sigma} u^q_s\end{aligned} \tag{10.59}$$

10.5.6 Outer Controllers

Reactive Power Control and Voltage Control: In a VSC-HVDC system, every converter can independently control its reactive power injection in the power system. When the system voltage is aligned with the *q*-axis, the reactive power *Q* can be calculated as follows:

$$Q = u^q_s i^d_{\mathrm{pr}} \tag{10.60}$$

The d-axis current set point, $i^{d}_{\mathrm{pr}^{\mathrm{ref}}}$, is calculated from the reactive power set point Q_{ref}. A combination of an open loop and a PI controller is used to drive reactive power to its desired value, leading to the equation

$$i^{d}_{\mathrm{pr}^{\mathrm{ref}}} = \frac{Q_{\mathrm{ref}}}{u_s^q} + \left(K_{p2} + \frac{K_{i2}}{s}\right)(Q_{\mathrm{ref}} - Q) \tag{10.61}$$

We finally introduce state variable N^{d}, resulting in an additional differential equation for the PI controller

$$\frac{dN^d}{dt} = K_{i2}\left(Q_{\mathrm{ref}} - u_c^q i_{\mathrm{pr}}^d\right) \tag{10.62}$$

such that substitution of Equations (10.60) and (10.62) in Equation (10.61) leads to an expression for the d-axis current reference

$$i^{d}_{\mathrm{pr}^{\mathrm{ref}}} = \frac{Q_{\mathrm{ref}}}{u_s^q} + N^d + K_{p2}\left(Q_{\mathrm{ref}} - u_s^q i_{\mathrm{pr}}^d\right) \tag{10.63}$$

which can thereafter be substituted in Equation (10.58).

Instead of controlling reactive power, the AC system voltage can be controlled. The d-axis current reference can be calculated in a similar manner:

$$i^{d}_{\mathrm{pr}^{\mathrm{ref}}} = \left(K_{p3} + \frac{K_{i3}}{s}\right)\left(U_{s^{\mathrm{ref}}} - U_s\right) \tag{10.64}$$

with

$$U_s = \sqrt{\left(u_s^q\right)^2 + \left(u_s^d\right)^2} = u_s^q \tag{10.65}$$

After we introduce the state variable N^{d}, the following equations can be written down:

$$\frac{dN^d}{dt} = K_{i3}\left(U_{\mathrm{ref}} - u_s^q\right) \tag{10.66}$$

$$i^{d}_{\mathrm{pr}^{\mathrm{ref}}} = N^d + K_{p3}\left(U_{\mathrm{ref}} - u_s^q\right) \tag{10.67}$$

Equation (10.67) can be substituted in Equation (10.58).

Active Power Control: In a two-terminal VSC-HVDC system, one converter sets the active power set point. The variable i^{q}_{prref} is calculated from this active power set point. The other converter controls the DC voltage. We call it the "slack" converter, because it compensates the losses in the DC network and thus has a similar function when compared to a slack node in an AC grid. In general, an n converter VSC HVDC system has n-1 converters controlling active power and one controlling DC voltage. One must therefore differentiate between the slack converter and the other converters for the calculation of the q-axis current set point. As only n-1 converters can set an active power set point, n-1 q-axis reference currents can be calculated from the active power set point. As we will explain in Section IV, the remaining q-axis reference current can be obtained from the DC voltage control equation of the slack converter. Again, using a combination of an open loop and a PI controller, the q-axis reference current can be expressed as

$$i^{q}_{\mathrm{pr}^{\mathrm{ref}}} = \frac{P_{\mathrm{ref}}}{u_s^q} + \left(K_{p4} + \frac{K_{i4}}{s}\right)(P_{\mathrm{ref}} - P) \tag{10.68}$$

with

$$P = u_s^q i_{\mathrm{pr}}^q \tag{10.69}$$

Introducing a new state variable N^q leads to an extra equation

$$\frac{dN^q}{dt} = K_{i4}\left(P_{\mathrm{ref}} - u_s^q i_{\mathrm{pr}}^q\right) \tag{10.70}$$

After the substitution of Equations (10.70) and (10.69) in Equation (10.68), the expression for the q-axis reference current to be substituted in Equation (10.59) becomes

$$i^q_{\mathrm{pr}^{\mathrm{ref}}} = \frac{P_{\mathrm{ref}}}{u_s^q} + N^q + K_{p4}\left(P_{\mathrm{ref}} - u_s^q i_{\mathrm{pr}}^q\right) \tag{10.71}$$

All the equations of the converters have now been derived.

References

1. R. Shah, et al "Stability and control of mixed AC–DC systems with VSC-HVDC: A review," *IET Gener. Transm. Dis.*, vol. 12, no. 10, pp. 2207–2219, 2018.
2. Q. Zhong, et al., "Study of HVDC light for its enhancement of AC/DC interconnected transmission systems," in *2008 IEEE Power and Energy Society General Meeting-Conversion and Delivery of Electrical Energy in the 21st Century*, Pittsburgh, PA, USA, 2008.
3. F. B. Ajaei and R. Iravani, "Dynamic interactions of the MMC-HVDC grid and its host AC system due to AC-side disturbances," in *IEEE Trans. Power Del.*, vol. 31, no. 3, pp. 1289–1298, June 2016.
4. T. M. Haileselassie, et al., "Main grid frequency support strategy for VSC-HVDC connected wind farms with variable speed wind turbines," in *IEEE Trondheim PowerTech*, Trondheim, Norway, 2011.
5. "IEEE guide for planning DC links terminating at AC locations having low short-circuit capacities," IEEE std 1204–1997, IEEE press, 2003.
6. Y. Shao, "Voltage Stability Analysis of Multi-Infeed HVDC System Using Small-Signal Stability Assessment", in IEEE PES&T&D, 2010, New Orleans, LA, USA
7. G . Andersson, T . Adhikari, B . Davies, et al., "On voltage and power stability in AC/DC systems," *Cigre Working Group 14.05*, April, 2003.
8. E. Rahimi, "Voltage interactions and commutation failure phenomena in multi-infeed HVDC systems," Ph.D thesis, University of Manitoba, 2011.
9. E. Rahimi, et al., "Commutation failure analysis in multi-infeed HVDC systems," *IEEE Trans. Power Del.*, vol. 26, no. 1, pp. 378–384, Jan. 2011.
10. D. Aik and G. Andersson, "Analysis of voltage and power interactions in multi-infeed HVDC systems," *IEEE Trans. Power Del.*, vol. 28, no. 2, pp. 816–824, April 2013.
11. J. Beerten, S. Cole, and R. Belmans, "Generalized steady-state VSC MTDC model for sequential AC/DC power flow algorithms," *IEEE Trans. Power Syst.*, vol. 27, no. 2, pp. 821–829, May 2012.
12. Y. Liu and Z. Chen, "A flexible power control method of VSC-HVDC link for the enhancement of effective short-circuit ratio in a hybrid multi-infeed HVDC system," *IEEE Trans. Power Syst.*, vol. 28, no. 2, pp. 1568–1581, May 2013.
13. C. Guo, et al., "Analysis of dual-infeed HVDC with LCC-HVDC and VSC-HVDC," in *IEEE Trans. Power Del.*, vol. 27, no. 3, pp. 1529–1537, July 2012.
14. X. Ni, et al., "An improved measure of AC system strength for performance analysis of multi-infeed HVdc systems including VSC and LCC converters," *IEEE Trans. Power Del.*, vol. 33, no. 1, pp. 169–178, Feb. 2018.
15. C. Xinyu, A. Gole, "Analysis of Mixed Inverter/Rectifier Multi-Infeed HVDC System, in IEEE trans. on Power Delivery, Vol. 27, July, 2012.

11

DC Circuit Breaker, DC-to-DC Converter, and Multi-Terminal

High-voltage direct current (HVDC) systems have two main uses: First, to connect two AC networks with different frequencies or different frequency-control philosophies (back-to-back), and second, to transmit large amounts of power via long distances. For large distances, DC transmission lines have lower losses than AC connections. "Long" in this context means more than 800 km for overhead lines and more than 40 km for cables. There are two basic converter technologies in use today for HVDC power transmission: the "classical" current source commutated (CSC) technology based on thyristors and the more recent self-commutated voltage source converter (VCS) technology based on insulated-gate bipolar transistors (IGBTs).

Since thyristors are only turn-on devices, the active power flow of CSC systems is controlled by adjusting the turn-on (firing) and the extinction time instant (overlap) prior to commutation to another valve. Reactive power is consumed by the rectifier upon being sent, and by the inverter at the receiving end. This has to be compensated for with filters and additional capacitors on the AC sides. In particular, under transient conditions, the amount of reactive power consumed varies greatly. The power flow is unidirectional. The reversal of the power-flow direction requires a change in the polarity of the system, which could be problematic, in particular, for polymeric cable connections.

With advancements in power semiconductor devices and the availability of high-power transistors (IGBTs), it is also possible to use multi-level concepts for HVDC power transmission. The self-commutated VSC has a stiff DC voltage, and large capacitors are used.

VSC HVDC technology transmits active power, and can provide the required amount of reactive power at both the power sending and the power receiving end. However, the losses in one VSC terminal are about 1.6%, of which the converter valves contribute almost 70%. Also, in case of a DC-side fault, the diodes connected in parallel to the IGBT modules act as an uncontrolled rectifier, even if the IGBTs are blocked.

Short-circuit current is limited only by the AC system. A small DC-side inductance leads to a very high rate of rise in DC current. In addition, the DC capacitors discharge and add to the fault current. Thus, the DC and AC sides have to be decoupled very quickly to stop the fault current. Since the converter is no longer controllable in this fault case (in contrast to CSC HVDC systems), AC circuit breakers (CBs) are typically used. HVDC CBs could do the same, but they are not yet available. Because of this vulnerability to DC-side faults, VSC HVDC stations are preferably connected by cables rather than overhead lines, since cables are much less sensitive to environmental influences.

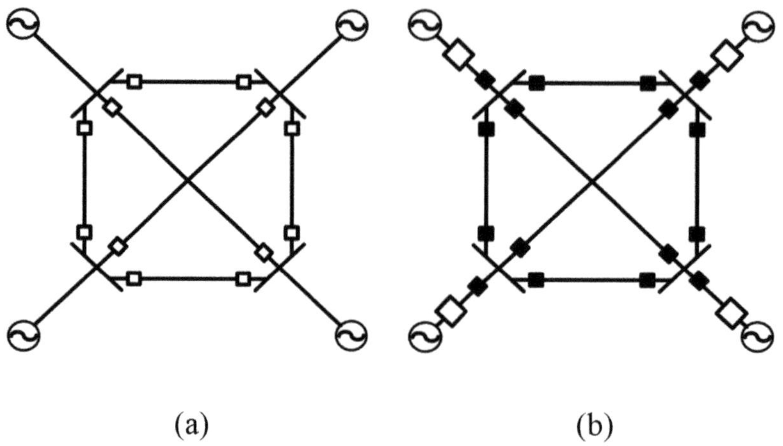

Figure 11.1 Sample comparison of the HVDC network configuration based on point-to-point systems (left) or with the use of HVDC CBs (right). Converter terminal □, HVDC CB: ■.

Most of today's HVDC transmission systems are point to point, connecting only two terminals. In principle, it is possible to connect a number of AC systems with many point-to-point connections in a so-called multi-infeed configuration shown in Figure 11.1(a).

However, there are several advantages if the lines are connected on the DC side, forming a true HVDC network shown in Figure 11.1(b). First, the number of converter stations could be reduced, since only one converter is needed per AC connection point. Not only does this significantly reduce the costs, but also the losses of the entire transmission system. In addition, each station can transmit (send or receive) power individually, and can even change from receiving to sending power without requiring that another station do the opposite. There is more redundancy, and power can be transmitted even if one line is lost. To ensure the reliable operation of a multi-terminal system, it must be possible to decouple faulty lines or stations from the system. In AC systems, CBs that are able to interrupt the full short-circuit current are used. For HVDCs, realized CBs have only limited ratings. Compared with AC CBs, they are much larger in size and considerably more expensive.

Thus, other concepts have been introduced in today's point-to-point systems. Most commonly, DC line faults are interrupted using converter control, which only works for CSC, or the CBs on the AC side of the converter stations are operated to deenergize the station and the line. In the case of a non-sustained fault, the stations and line can be reenergized. This operation sequence takes from at least a few hundred milliseconds up to a few seconds, but is well suited for point-to-point connections. Deenergizing the entire HVDC network has also been suggested, but it is surely limited to small systems with only three or four terminals.

The availability of HVDC CBs is thus a key enabling technology for the reliable and economical operation of multi-terminal HVDC systems. Supporting HVDC CBs with the fast action of terminal controls is undoubtedly advantageous and should be investigated further (in particular for VSC topologies).

There has been interest in multi-terminal HVDC systems since the 1960s, with a clear peak in the 1980s. Recently, the publications suggest VSC-based multi-terminal systems such as off-shore wind farms and urban distribution systems.

The question as to whether it is better to base an HVDC network on CSC or VSC cannot be answered unambiguously. Each technology has certain advantages and disadvantages. CSC HVDC is well established and has a higher power rating combined with lower losses, but a fault on the AC side can lead to commutation failures, which result in a collapse of the DC line voltage. A CSC-based network is thus vulnerable to AC-side faults. A VSC-based network, in turn, is vulnerable to DC-side faults, that is, any DC-side fault will result in

Table 11.1 Pros and Cons for Building a Multi-Terminal HVDC Based on CSC or VSC

	Advantage	Disadvantage
CSC-based	High voltage and power rating	Risk of commutation failures, collapse of DC line voltage
	Low losses	Strong AC-side requirement, consumes reactive power
	Low rate of rise of short-circuit current	Vulnerable to AC-side faults
VSC-based	Stiff voltage on DC side	Lower power rating
	Power reversal w/o voltage reversal	Higher losses
	More suitable for cables	Vulnerable to DC-side faults

a fault current with steeply increasing amplitude. Moreover, CSC requires relatively strong AC sources and consumes reactive power at every terminal location.

In contrast, a VSC-based network could help to strengthen regions with weak AC systems using its independently controllable active and reactive power. A reversal of power-flow direction for one terminal of the network is straightforward for VSC-based systems. In CSC-based systems, the voltage polarity would have to be changed, subsequently affecting the power-flow direction at all terminals connected to the common DC voltage line. The arguments for and against building a multi-terminal HVDC system based on CSC or VSC are summarized in Table 11.1.

11.1 HVDC Circuit Breakers

In today's point-to-point HVDC transmission systems, DC interrupters are used for several different switching duties. Neutral bus switches (NBSs), neutral bus ground switches (NBGSs), metal return transfer breakers (MRTBs), ground return transfer breakers (GRTBs), high-speed bypass switches (HSBSs) for parallel line switching, and isolation switches also exist. Interrupters to break DC short-circuit currents have only been realized in very limited numbers, and the maximum ratings are 250 kV, 8 kA or 500 kV, 4 kA, which are not more than 1.6 times the rated nominal current. The breaking time is to the order of 35 ms; for CSC-based systems, large inductances limit the rate of rise of fault current, and this time is sufficiently fast. However, the components of these breakers are very large, and more costly than AC CBs with comparable current and voltage ratings. In point-to-point HVDC transmission systems, the function of DC breakers has thus typically been substituted by deenergizing the converter stations (by acting with the station control or by operating the breakers on the AC side and opening isolation switches). In low-voltage (LV) and medium-voltage (MV) DC applications, short-circuit current interrupting breakers have been realized based on several different technologies (e.g., switching arcs or solid-state switches).

The difficulties in realizing HVDC CBs can be attributed to the demanding requirements placed on CBs in DC systems, which are quite different than those placed on AC CBs. One of the major differences is the absence of natural current zero crossings in DC systems. The breakers have to fulfill the basic requirements as follows:

1. Create a current zero crossing to interrupt the current
2. Dissipate the energy stored in the system inductance
3. Withstand the voltage response of the network after the current interruption

In particular, the first two requirements lead to a very strong interaction between the breaker and the DC system, an attribute that is also very different to AC CBs. In addition to these basic requirements, there are, depending on the breaker application, additional secondary requirements as follows:

- In VSC-based systems, the HVDC breaker has to be able to interrupt quickly, as was stated before
- The MV generated by the breaker must be low enough to comply with the insulation coordination of the DC system; this is particularly important for the switching of load currents where the network is at nominal voltage.

11.1.1 Basic Working Principles of HVDC CBs

The current in a DC circuit can be brought to zero by generating a counter voltage of similar amplitude to or larger amplitude than the system voltage. This counter voltage can be produced by inserting additional resistance or inductance in the current path. The energy of the DC system is dissipated across this device. The larger the counter voltage, the smaller the time needed to interrupt, but the larger the energy that is dissipated in the device. DC breakers with the current limiting and energy dissipating functions of an arc are commonly used in LV and MV applications. Some proposals for high-voltage (HV) systems have been made, but none of them have proven efficient and successful in real applications.

One alternative is to have several parallel paths in the breaker and to separate the requirements to different elements. The simplest is one nominal current path and one parallel path with a linear or nonlinear resistive element. The nominal current path typically consists of an interrupter with low ohmic losses in closed position, which is, so far, only possible with movable metallic contacts. Upon the opening of these contacts, an arc is established, and its arcing voltage is used to commutate the current to the resistive path, where the energy of the system is then dissipated. The advantage is that the interrupter in the nominal current path only needs to produce a voltage sufficient for commutation, not for counteracting the full system voltage. In addition, the breaker does not have to have a large energy dissipating function, which typically improves its interruption capability. If the commutation path only consists of a linear resistor, the arc voltage of the interrupter still has to be very high.

A gradual insertion of resistors or nonlinear resistors to limit the required commutation voltage is better. The commutation process can be eased by adding other elements, such as a capacitor that temporarily takes the current flow. More recent developments make use of actively controllable resistances of solid-state devices.

For most of the practically realized HVDC CBs, separate commutation and energy absorbing paths have been used, as sketched in Figure 11.2. The commutation path may then be a series resonance consisting of a capacitance C_C and inductance L_C so that current oscillation between the nominal and commutation paths can occur at the natural frequency $\omega_0^2 = 1/L_C C_C$. If the amplitude of the oscillating current i_n is larger than the system DC current I_0, a current zero crossing occurs in the nominal path, and the interrupter S_n can interrupt the current. Current I_0 continues to flow, charging the capacitor in the commutation path. If the capacitor voltage exceeds a given value, typically chosen to be the voltage capability of the breaker or the

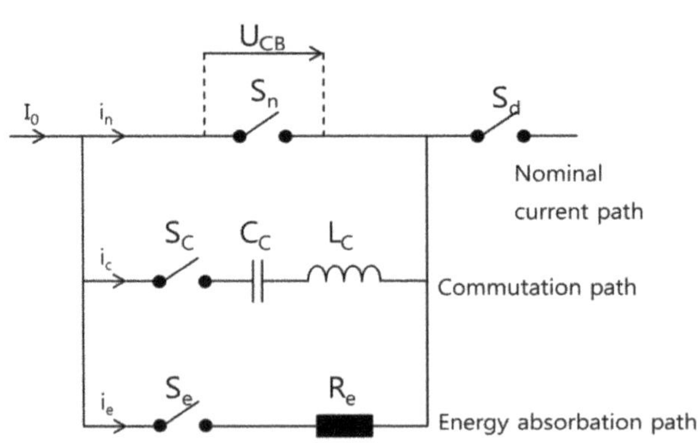

Figure 11.2 Basic arrangement of the HVDC Circuit Breaker

insulation coordination of the HVDC system, the energy absorption path acts, causing the system current I_0 to decrease. If the differential arc resistance dU/dI of the interrupter in the nominal path is negative, a current oscillation between the nominal and commutation paths with increasing amplitude occurs, started by the natural fluctuation in the arc voltage.

11.1.2 DC Circuit Breaker [1]–[4]

Mechanical Passive Resonance Circuit Breaker: The mechanical HVDC circuit breaker with a passive resonance circuit is an old technology, which was initially developed for CSC HVDC systems. Figure 11.3 shows a simplified diagram of a mechanical circuit breaker.

Typically, CB is an air blast circuit breaker with several interrupter units. During normal operation, current flows through the CB, while during interruption it is commutated into the commutation path. To understand the interruption process, it is necessary to analyze the current equation during the process. The differential equation during interruption can be written as follows:

$$L_c\frac{d^2 i_s}{dt^2} + \left(R_c + \frac{\partial u_{arc}}{\partial i_s}\right)\frac{di_s}{dt} + \frac{1}{C_c}i_s = \frac{I_0}{C_c} \tag{11.1}$$

An approximate solution for this equation is

$$i_s = I_0\left[1 + e^{\frac{1}{2L}\left(R_c + \frac{\partial u_{arc}}{\partial i_s}\right)t} \cdot \sin\omega_c t\right] \tag{11.2}$$

where $\omega_c = \sqrt{1/L_cC_c}$. If $\left(R_c + \frac{\partial u_{arc}}{\partial i_s}\right) < 0$, the i_s will oscillate with increasing amplitude. The first zero crossing of the current will be enough for a breaker to interrupt it.

Mechanical Active Resonance Circuit Breaker: In active mode, a current oscillation provided by the precharged commutation capacitor C_C will arise instantly and grow to oppose the current in the main *CB* when the current is commutated into the L^C branch. In some texts, this scheme is also introduced as a hybrid interruption method.

Generally, in these types of circuit breakers, thyristors are employed to act as commutator and disconnector switches. This concept is also known as the two-stage interruption method. There are several variants for this concept. Figure 11.4 depicts the first variant of mechanical active resonance circuit breakers.

In the first topology and under normal load conditions, only the main breaker (MB) S_1 is closed, whereas the other switches S_2, S_3, and S_4 are in open state. The capacitor C_C is precharged with a negative initial voltage V_{C0}. As soon as a fault current is detected, S_1 is opened and S_2 and S_3 are simultaneously closed. Then, the reverse current arising through the L^C branch opposes the fault current, and when it reaches an equal amount of fault current a zero crossing will happen and the current will be commutated to parallel branches. After charging

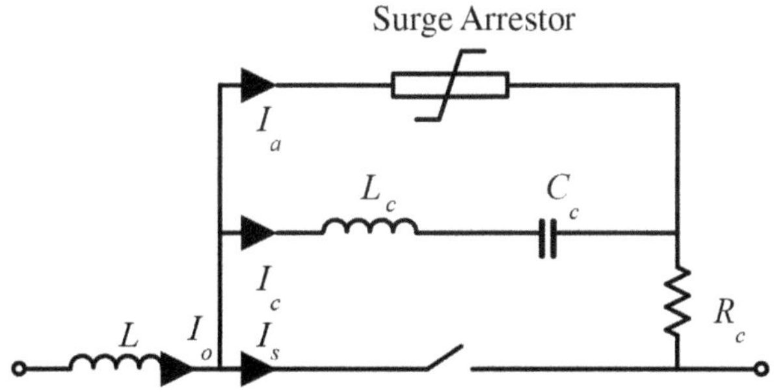

Figure 11.3 Mechanical HVDC circuit breaker

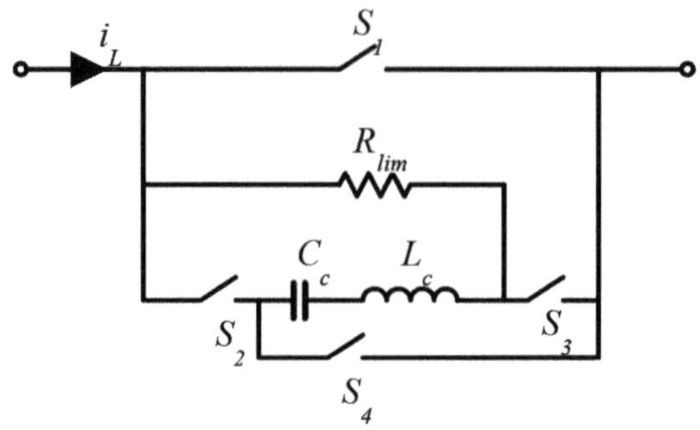

Figure 11.4 First variant of active resonance circuit breakers

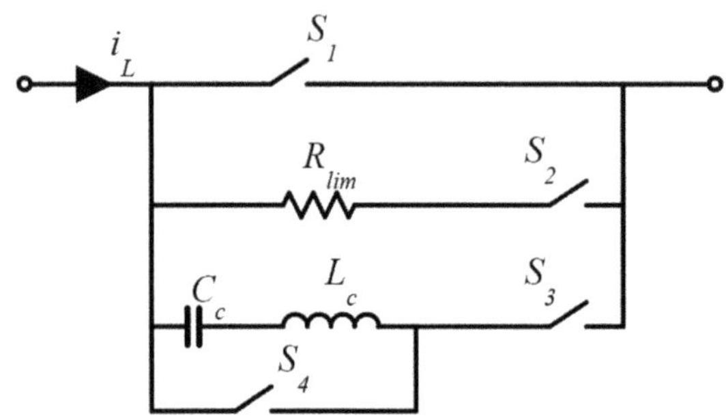

Figure 11.5 Variant 2 of active resonance circuit breakers

the C_C, the current inside S_2 will fall to zero, and it opens. The switch S_4 closed after discharging the capacitor C_C into the loop C_C-S_4-S_3-L_C-C_C. When a current zero occurred in S_3, it turns off, and the main current is commutated again into the path including R_{lim}-L_C -C_C –S_4. The switch S_4 opens when a zero current is created, resulting in a new energy balance in which the capacitor is fully charged.

In the second variant, it is possible to interrupt the nominal rated current by closing only the switch S_3. Figure 11.5 shows the second topology of conventional hybrid circuit breakers (HCB).

After fault detection in the second variant, the MB S_1 is opened and S_3 is closed to create a zero current inside the MB. Finally, S_3 will naturally open after the capacitor is fully charged.

Hybrid Technologies: Integrating controllable solid-state devices with a mechanical breaker or disconnector in a combined configuration is called a hybrid switching technique. Generally, within a HCB, the commutation path is introduced using solid-state switches, and only operates during the interruption process. All the switches are controlled by electronic circuits.

Recent developments in semiconductor switches and improvements in their characteristics such as breakdown voltage, conduction losses, switching time, and reliability bring about the possibility of using these devices as the main interrupters in circuit breakers. There are several possible topologies for HCB, but in practice, two main structures attract the most attention. Figure 11.6 shows the first basic hybrid topology.

In this topology, a fast mechanical breaker is equipped with a set of solid-state switches in the parallel path. This topology combines the low losses of a pure mechanical breaker with the fast switching response of a pure solid-state device. Since the arc chamber must only create sufficient voltage for commutation and not an artificial current zero crossing point, this topology is faster than conventional circuit breakers. Application of this topology has been developed for MV grids.

Another topology that has been introduced employs a fast solid-state device in the main path of the current and in series with a fast mechanical disconnector. The parallel path is built by a series connection of solid-state switches. The fast solid-state device in the main path can be an IGBT. This IGBT needs only to create a sufficiently HV for the commutation of the current to the parallel full IGBT breaker, so it has a lower rating than

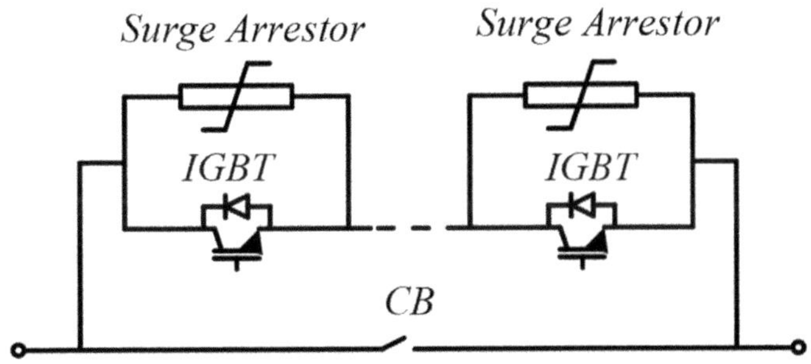

Figure 11.6 First basic hybrid DC circuit breaker

the parallel path breaker. Typically, it can be done through connection of a small number of IGBTs in series, so the conduction losses and voltage drop will be low enough.

During normal operation, the current will only flow through the bypass, and the current in the MB is zero. When a DC fault occurs, the auxiliary DC breaker immediately commutates the current to the main DC breaker, and the fast disconnector opens. With the mechanical switch in the open position, the main DC breaker breaks the current. The mechanical switch opens with zero current and LV stress. The fast disconnector will be exposed to recovery voltage defined by the protective level of the arrester banks first after being in open position while the main DC breaker opens. A simplified schematic of this topology is depicted in Figure 11.7.

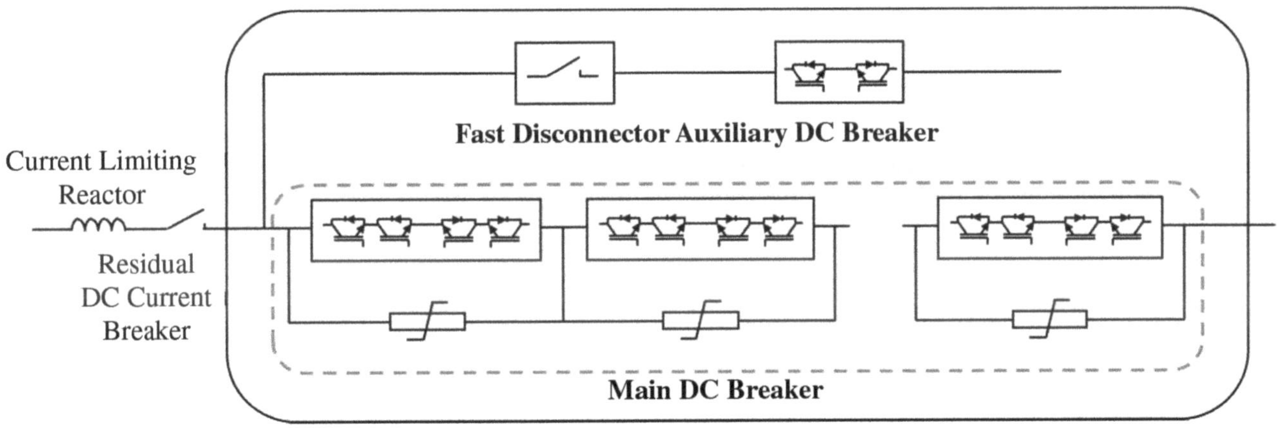

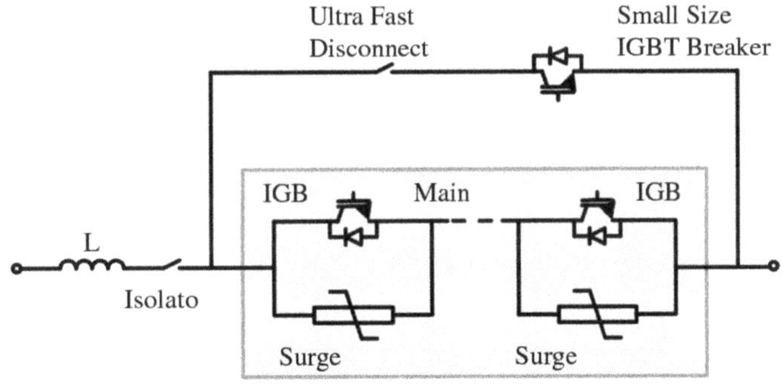

Figure 11.7 Simplified schematic of hybrid DC circuit breaker

Solid-State CB: The fast and ultrafast switching times of semiconductor devices make them a very strong candidate for DC fault interruption. A pure solid-state circuit breaker can be faster than all other topologies. The design of pure semiconductor-based CBs is possible through different combinations of solid-state switches and ancillary circuits. Considering proposed topologies in literature, there are two major topologies for this type of CB, and other structures can be classified as one of these topologies. Generally, in pure solid-state CBs, many IGBTs, integrated gate-commutated thyristors (IGCTs), or other semiconductor-based switches are connected in series and parallel to support the voltage and current of the system during normal and fault conditions. Research with the aim to optimize and improve the behaviors of solid-state CBs is ongoing, and new contributions have been reported.

CB Paralleling a Surge Arrestor: Figure 11.8 depicts a typical circuit configuration of a solid-state DC CB. The semiconductor switch T acts as the MB and a surge arrestor R_v is connected in parallel with it. During normal operation, T is in the on state and conducts the current from source to load. As soon as a short-circuit fault is detected, T will be turned off. Then, the load current commutates to the surge arrestor R_v. Surge voltage across T is limited to the clamping voltage of the surge arrestor R_v. The clamping voltage of the surge arrestor is assumed as $V_{dc} + V_{margin}$, and it is also assumed that the impedance of the fault point is negligible. When T turns off at time $t = 0$, V_{margin} is applied to L_{dc}, so the inductor current can be calculated as follows:

$$i_L = I_0 - \frac{V_{margin}}{L_{dc}} t \tag{11.3}$$

where I_0 is the amplitude of the fault current at $t = 0$. The time to turn off the fault current T_{open} is derived as follows:

$$T_{open} = \frac{L_{dc}}{V_{margin}} I_0 \tag{11.4}$$

and the energy absorbed in R_v can be given by

$$W_R = \left(\frac{V_{dc}}{V_{margin}} + 1 \right) \frac{1}{2} L_{dc} I_0^2 \tag{11.5}$$

where, V_{margin} is usually much smaller than V_{dc} in high-power applications in order to suppress the voltage across T and reduce T's conduction loss. In such cases, the term in the parentheses of Equation (11.4) becomes large, and W_R is much greater than the stored energy in L_{dc} at $t = 0$, which is $\frac{1}{2} L_{dc} I_0^2$. The more voltage across T is suppressed, the higher capacity the surge arrestor needs.

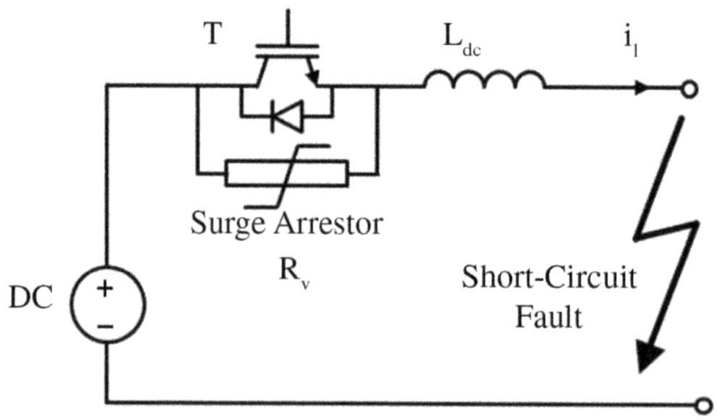

Figure 11.8 Typical configuration of solid-state DC circuit breaker

CB with Freewheeling Diode: As another topology of HCBs, a freewheeling diode D and a varistor R_v are connected across the DC line. The circuit reduces the energy absorbed in the breaker by suppressing the surge voltage across the valve device T. A circuit configuration of solid-state DC CBs using a freewheeling diode is shown in Figure 11.9.

During normal conditions, T remains on and leads the load current i_L. When T turns off at $t = 0$ after detecting a fault, the fault current i_L is commutated to D. Hence, the current flowing through V_{dc} immediately decreases to zero. The inductance L_{dc} is gradually demagnetized by R_v, and the fault current i_L decreases. When the clamping voltage of R_v is expressed in V_v, the surge voltage across T is $V_{dc}+ V_v$, which is the sum of the nominal DC voltage and the clamping voltage of R_v. Because V_{dc} does not supply any power after turning off T, the energy absorbed in the DC breaker at turn-off operation is equal to the energy stored in L_{dc} by the current I_0. So, the energy W_R is given by

$$W_R = \frac{1}{2} L_{dc} I_0^2 \tag{11.6}$$

Since W_R in Equation (11.6) is smaller than W_R in Equation (11.5), the rating and volume of R_v can be reduced by applying the freewheeling diode. Because the clamping voltage V_v does not affect W_R, the surge voltage across T can be suppressed without any increase in W_R by choosing a low value for V_v.

Voltage Rating: Nowadays, mechanical HVDC circuit breakers with nominal voltage up to 550 kV are available. HCBs have also been verified by experimental tests up to a voltage rating of 120 kV, and it is expected that the voltage rating will reach up to 320 kV. Pure semiconductor CBs are not available in HVs and power ratings and have only been designed and implemented for operation in MV applications. But, considering the developments in semiconductor devices, it is anticipated that an 800-kV voltage rating is achievable.

Current Rating: Mechanical HVDC CBs are able to interrupt currents up to 4 kA with a passive resonance system, while they can interrupt up to 8 kA with an active resonance circuit. For HCB topologies, the current interruption level of 9 kA has been proven experimentally, and in theory levels up to 16 kA are achievable. Considering the expected HV rating for pure semiconductor circuit breakers, a 5-kA current interruption rating is reasonable for them.

11.1.3 Fault Analysis of DC Circuit Breaker in the Modular Multi-Level Converter HVDC System [5]–[9]

Recently, a new fault analysis technique for DC circuit breaker has been published by Cwikowski and Alan Wood [5]. The contents of this paper are as follows.

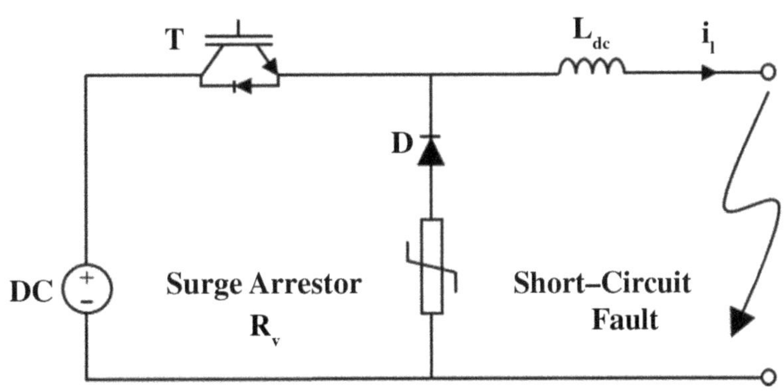

Figure 11.9 Circuit configuration of solid-state DC circuit breaker using a free-wheeling diode

Half-Bridge modular multi-Level converter (MMC) Pole-To-Pole Fault Analysis: The MMC is made from six arms consisting of stacks of submodules (SMs). Simplistically, from a protection point of view, the converter can be in one of two states: unblocked or blocked.

When the converter is unblocked, the control is capable of influencing the state variables through manipulation of the SMs. When the converter is blocked, the control is unable to influence the state variables. During the period the converter is blocked, the state variables are driven by the power system's forcing functions (AC and DC grid voltages, etc.).

Because of the converter's switching trajectory, the circuit cannot be described by a single equivalent circuit across the entire protection time. As such, the converter's fault response must be broken down into a sequence of stages. Between each stage within the sequence, the electrical circuit changes because of a controlled or uncontrolled switching of devices within the converter. The following sections describe key stages in the MMC's fault sequence. These stages can occur very rapidly, resulting in the converter's equivalent circuit changing many times before the circuit breaker is opened.

This analysis allows the fault currents to be described once the configuration of the converter is known. The analysis is kept general, and includes the influence of the cable voltage until the last step of the simplification. The resulting equations are, therefore, attempting to predict fault currents from a 0-km fault.

Stage A—Discharge: The converter's first stage in the fault sequence is the discharge stage. This stage starts from the moment the fault is seen by the converter and ends when the converter is blocked.

During this time, the MMC's SMs will discharge through the DC-side inductance, at a rate defined by the voltage that is presented across the DC-side inductor.

It is possible that before the converter is blocked, the SM sorting algorithm within the converter will change which SMs are inserted. This potentially exposes all SMs within the converter to some amount of discharge. Assume that significant discharge of the SMs is prohibited in order to maintain the ability to switch the IGBTs within each SM, and that the converter will attempt to maintain the DC link voltage during this discharge. Figure 11.10 shows the Thevenin equivalent circuit, and the fault current during this time can be estimated using

$$I_{DC_A}(t) = (1 + k_{TW})\frac{V_{DC}}{L_{Equ}}t + I_{0A} \tag{11.7}$$

$$L_{Equ} = 2L_{DC} + \frac{2}{3}L_a \tag{11.8}$$

V_{DC} is the DC link voltage at the moment the fault occurs. k_{TW} is a constant to compensate for the increased voltage across the equivalent DC inductance caused by the cable voltage; this value is bounded between 0 and 1, and proportionally relates the cable voltage to the DC link voltage. L_{Equ} is the Thevenin equivalent circuit

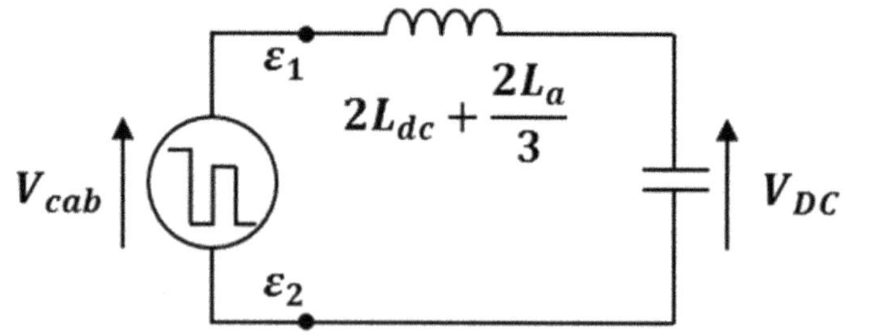

Figure 11.10 Stage A equivalent circuit diagram

inductance, given by Equation (11.8). The characteristic impedance of the cable is assumed to be a small relative to the DC-side reactor impedance.

An important note here is that when the arm current is positive at the moment the converter is blocked, the inserted capacitors will continue to discharge until the arm current has reached zero. This can result in a slight delay between the converter block signal being sent and the beginning of the second stage.

Stage B—Freewheeling: The second stage is the freewheeling stage. This stage starts after the converter is blocked and continues, as is explained in this section, until the moment a current zero crossing(s) occurs in a converter arm. The moment a zero crossing occurs in any of the converter's arms, the equivalent circuit changes, and another set of equations must be developed.

Blocking a converter prevents the SM voltage from driving the fault current. It also causes the fault current to commutate out of the IGBTs in the SMs and start to flow in the diodes of the lower switches. At this moment, all six arms act as diode stacks, and each leg of the converter carries a third of the DC line current. The converter layout is now represented by Figure 11.11.

As all six diodes are forward biased, this exposes all three phases of the AC grid to a reduction in load impedance. Thus, the AC grid will be exposed to an increase in current. However, this will not be a short circuit to the AC system, as the arm inductors provide some impedance. The AC grid is not capable of contributing to the DC fault current level at this stage, and this is explained as follows:

From Figure 11.12, providing the arm impedances match and the AC system voltages and impedances are balanced, each phase's contribution to the DC link voltage will have the same magnitude. This does not imply that the AC grid is subjected to a balanced set of currents. When the phase angle of these contributions is considered, these contributions will add up to zero. Thus, the AC grid cannot contribute to the DC fault while all six diodes are conducting. The AC grid voltages will influence the arm currents during this time; however, the cable voltage is capable of forcing a change in the DC line current during this time. The cable voltage can become negative because of inductive termination of the DC lines, which can increase the fault current. The fault current can be described using

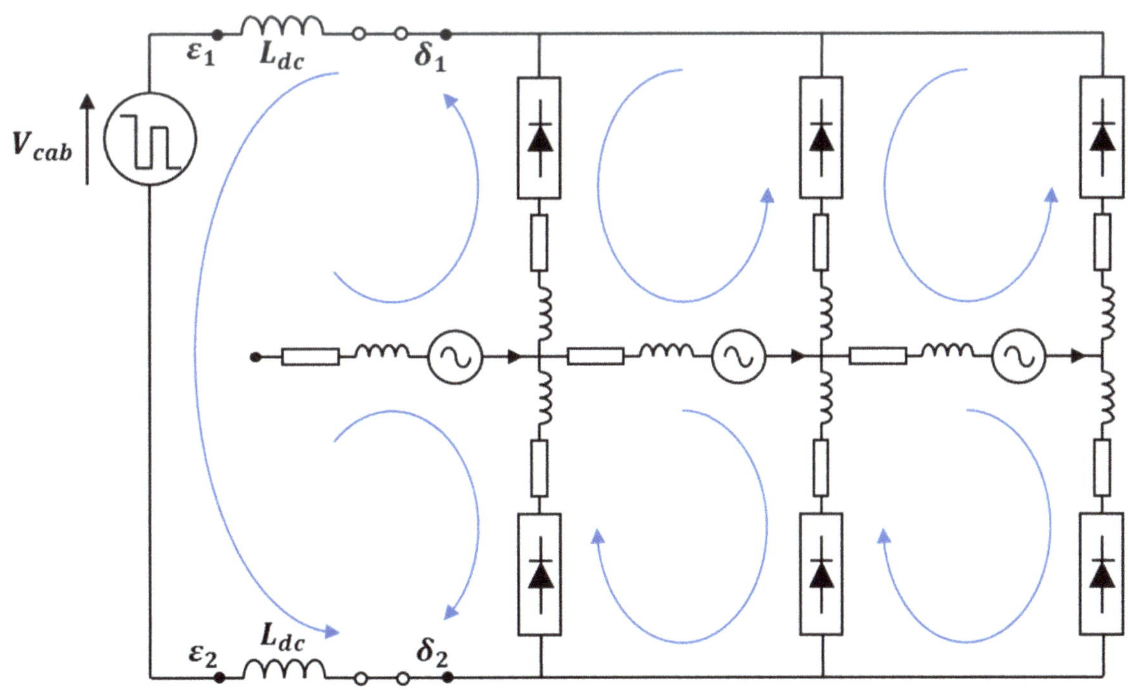

Figure 11.11 Stage B equivalent circuit diagram

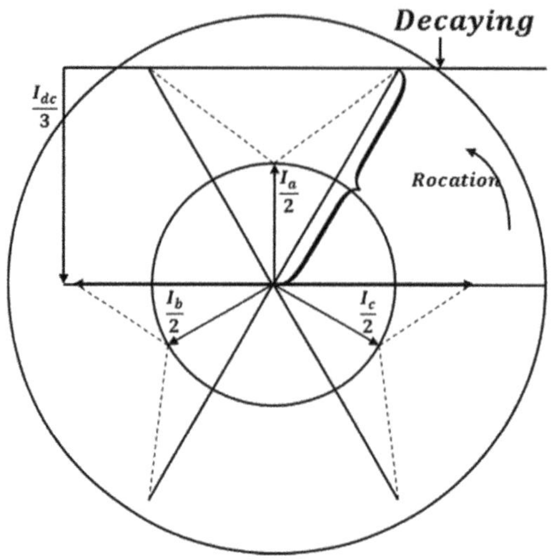

Figure 11.12 The dashed lines are the negatives of solid phase currents of the same color. The inner circle marks the magnitude of half the phase current. The outer circle marks the magnitude of the phase current

$$I_{DC_B}(t) = I_{02}e^{-\frac{R_{Equ}}{L_{Equ}}(t-t_B)} + I_{TW}(V_{cab}, t) \tag{11.9}$$

$$R_{Equ} = \frac{2R_a}{3} \tag{11.10}$$

which is made up of two components: an exponentially decaying initial current (I_{02}) and a travelling wave component I_{TW}. Although the freewheeling condition is maintained on the DC side of the converter, the AC grid will be subjected to an increase in current, which will flow through the converter's arms, but not into the DC circuit breaker. The AC grid current will circulate in all possible paths within the converter. However, the majority of the phase currents will flow in the loops shown in Figure 11.11. It can be seen that each phase-to-phase voltage increases the current in two arms and decreases the current in two other arms. Each phase superimposes its own contribution to each arm current on top of the DC component. Figure 11.12 shows a phasor diagram for the arm currents during this freewheeling stage. Each arm current is the sum of a DC component $\left(\frac{I_{DC}}{3}\right)$ and an AC component $\left(\tilde{I}_{xn}\right)$, as described by

$$I_{xn} = \frac{I_{dc}}{3} + \tilde{I}_{xn} = \frac{I_{02}}{3}e^{-\frac{R_{Equ}}{L_{Equ}}(t-t_B)} + \tilde{I}_{xn} \tag{11.11}$$

where, x denotes upper or lower arm positions and n denotes the phase leg. The AC component of each arm current $\left(\tilde{I}_{xn}\right)$ is the difference of two separate phase current contributions.

During Stage B, the AC component of each arm current grows while the DC component decays. This will eventually lead to a zero crossing in an arm, resulting in the converter moving to the next stage.

During the analysis of this stage, a generic problem was encountered that prohibited prediction of the time period the converter spends in Stage B, which is the main reason the time frames the converter spends in each stage cannot be directly predicted. The issue arises because of state variables being the sum of other state variables, which makes them dependent, one case being

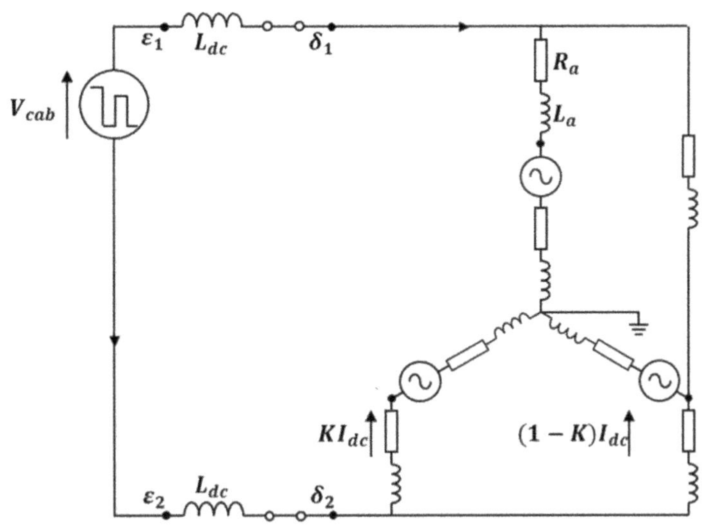

Figure 11.13 Stages C and D converter configuration diagram. The additional diode for Stage D is highlighted in red. For Stage C, this arm acts as an open circuit

$$I_{DC}(t) = I_{ua}(t) + I_{ub}(t) + I_{uc(t)} \tag{11.12}$$

This results in a singular matrix being formed during state space analysis, which cannot be inverted. This prevents prediction of the first zero crossings being made. If this problem can be overcome, then predictions for the time frame spent in each stage can be made without the need for simulation. The other stages of the converter response do not encounter this issue. This is known as a cut-set problem.

Stage C—Three-Diode Rectifier: Stage C starts the moment the three of the converter start conducting and ends at the next switching event. Figure 11.13 shows the converter configuration for Stage C when three of the converter diodes are conducting. The three conducting diodes are spread across both upper and lower arms.

Figure 11.13 can be reduced to the circuit in Figure 11.14, where the delta connection at the converter's terminals has been converted into an equivalent star connection. L_g' and R_g' are the AC grid impedances referred to the converter side of the transformer. This circuit can then be reduced into a Thevenin equivalent circuit shown in Figure 11.15, when the circuit resistance is assumed to yield much lower impedance than the inductors. V_{cab} is the cable's voltage at the point where it connects to the converter.

Using the Laplace transform given by

$$\mathcal{L}\{\sin(\omega t + \theta_n)\} = \frac{\sin(\theta_n)\, s + \cos(\theta_n)\, \omega}{\omega^2 + s^2} \tag{11.13}$$

the DC fault current can be described by

$$I_{DCc}(t) = \frac{\sqrt{3}|\hat{V}_{LL}|}{L_T \omega_{ac}} \left[1 - \cos(\omega_{ac} t + \theta_c)\right] + I_{0C} \tag{11.14}$$

where,

$$L_T = 2L_{dc} + \frac{3}{2}\left(L_a + L_g'\right) \tag{11.15}$$

$\hat{V}_{pnu}$ is the peak phase-to-neutral voltage that connects to the conducting upper arm during this period, and $\hat{V}_{LL}$ is the peak line-to-line voltage. The caret symbol (ˆ) is used to denote the peak throughout this paper. For the

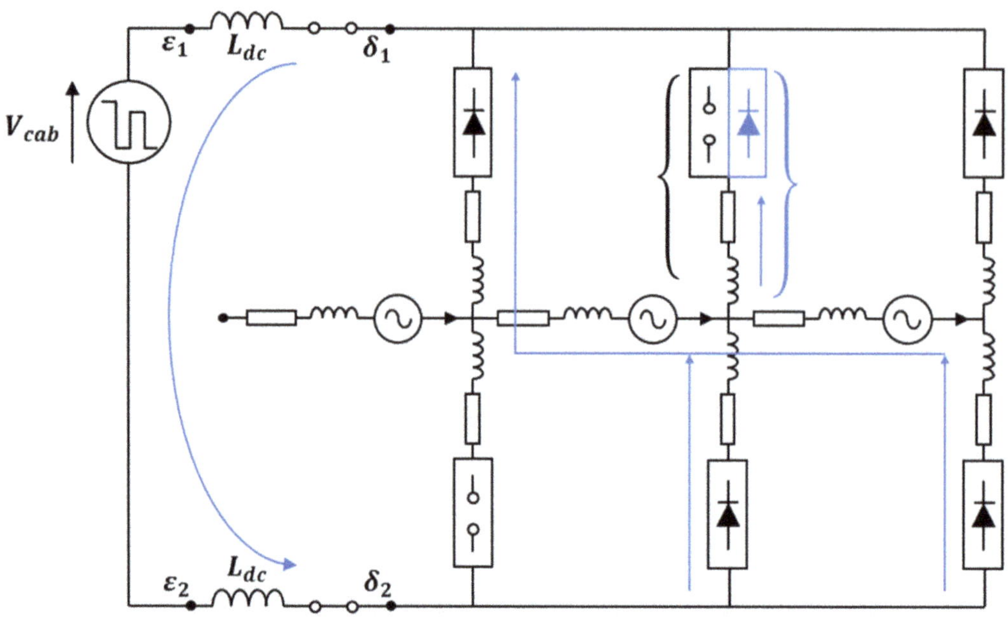

Figure 11.14 Reduced equivalent circuit for Stages C and D

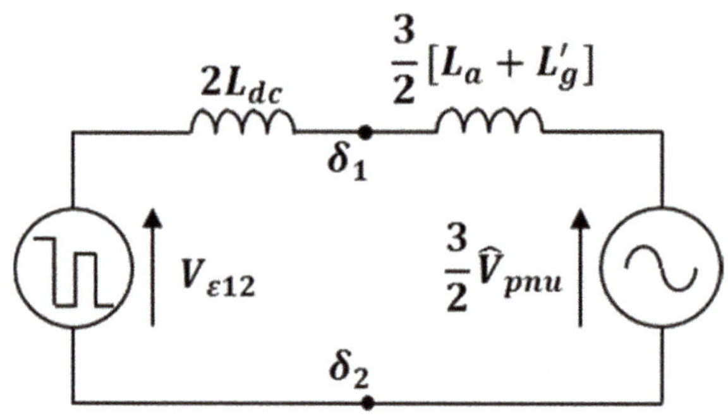

Figure 11.15 Thevenin equivalent circuit for Stage C

opposing diode configuration (two upper arms, one lower arm conducting), this voltage would be the phase to neutral voltage connected to the conducting lower arm. The cable voltage has been ignored in Equation (11.14) to provide a simplified result.

Stage D—Four-Diode Rectifier: Another possible pseudo-rectifier operation is when four of the diode stacks in the converter are conducting. It is assumed that the four diodes are spread evenly across the upper and lower arms. The converter configuration is given in Figure 11.13, with the additional upper arm diode marked in red. The equivalent circuit diagram is given in Figure 11.14 with the additional arm path highlighted.

Assuming low circuit resistance, and considering the voltages and impedances to the right of nodes δ_1 and δ_2, a reduced equivalent circuit can be found for this configuration (shown in Figure 11.16), indicating that only two of the phases are capable of providing a net contribution. Phase B, in this case, cannot, since the two contributions to voltage $V_{\delta 12}$ across the two impedances $L_a + L_g^{'}$ sum to zero; they act in opposing directions and have the same magnitude. This will hold true providing that the converter arm and AC system impedances are balanced. Figure 11.16 can then be reduced to the Thevenin equivalent circuit shown in Figure 11.17 whose inductance is given by

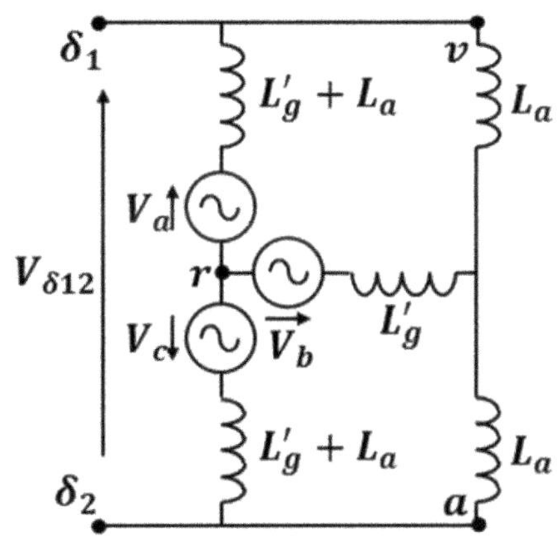

Figure 11.16 Reduced equivalent circuit for Stage D

$$L_{ThD} = 2L_a \left(\frac{L_a \left(L_a + L_g^{'}\right)\left(L_a + 2L_g^{'}\right)}{L_g^{'}\left(L_a + 2L_g^{'}\right) + L_a\left(2L_a + 3L_g^{'}\right)} \right) \tag{11.16}$$

and the Thevenin voltage is given by

$$V_{\delta 12} = \frac{|\hat{V}_{LN}|\sqrt{3}}{2} \frac{L_a}{L_a + L_g^{'}} \left[\cos\left(\omega_{ac} t \pm \frac{\pi}{3} + \theta_D\right)\right] \tag{11.17}$$

This reduction has been performed using a star-delta transformation on the nodes marked (τ, σ, υ) in Figure 11.16. Then, by subsuming impedance $L_a + L_g^{'}$ into one of the delta impedances, the circuit can be transformed back into a new star connection with reduced complexity.

Using the Laplace transform given in Equation (11.13), and assuming the cable voltage influence is small, current during Stage D can be described by

$$I_{DCD}(t) = \frac{L_a\sqrt{3}|\hat{V}_{LN}|}{2\left(L_{ThD} + 2L_{dc}\right)\left(L_a + L_g\right)}\left[1 - \cos\left(\omega_{ac} t + \theta_D\right)\right] + I_{0D} \tag{11.18}$$

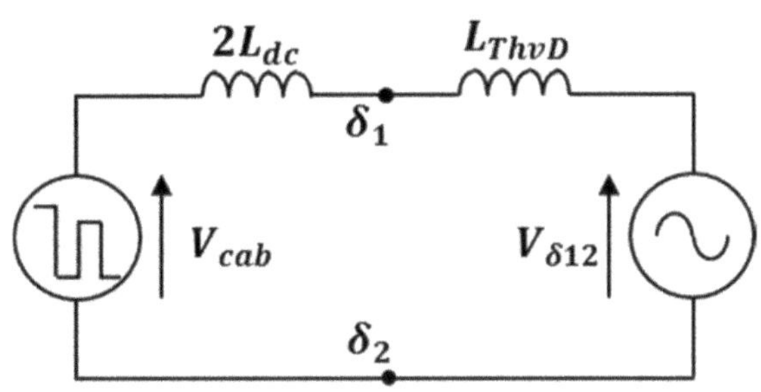

Figure 11.17 Thevenin equivalent circuit for Stage D

Figure 11.18 shows the equivalent circuit of final stage E, which is the moment the circuit breaker starts to generate the opposing voltage which starts to interrupt the fault current.

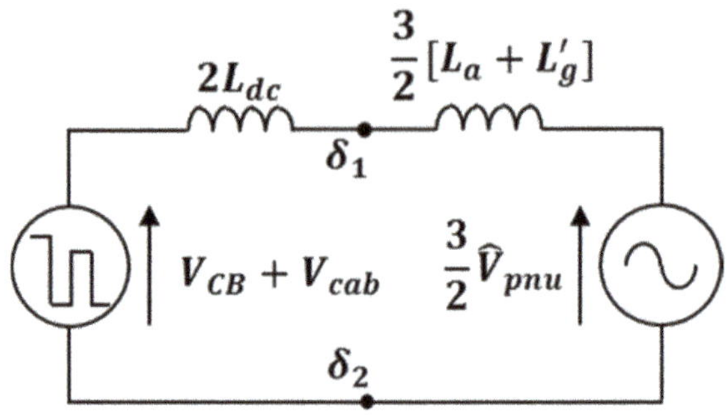

Figure 11.18 Stage E: Reduced equivalent circuit when opening in Stage C

11.1.4 HCB (Hybrid Circuit Breaker) with MMC [10]–[12]

In fact, it is not realistic to make DC breakers only with solid-state devices such as IGBTs. Therefore, a HCB (Hybrid Circuit Breaker) that combines mechanical switches and arresters with a solid-state device have been published [11]. The following is suggested by Chuanyue Li and so on, this method has many advantages when applied to a VSC HVDC multi-terminal.

An HCB is composed of two branches: the low-loss branch and the MB branch. The low-loss branch contains an ultrafast disconnector (UFD) and a load commutation switch (LCS) (see Figure 11.19). The UFD is a mechanical switch that can open within 2 ms to isolate the LCS from the main circuit. The LCS, which consists of a few IGBT modules, is designed to provide a low-loss current path for the load current. When a fault is detected, the LCS is switched OFF, and the fault current will be commutated into the MBs (Mechanical Breakers). The commutation time is normally 0.25 ms.

The MB branch is sectioned into several main break cells. Each cell contains one metal-oxide varistor (MOV) and one MB (see Figure 11.19). Each MB consists of a large number of IGBT modules. In IGBT modules, diodes are connected anti-parallel with IGBTs. The MBs stay ON state during normal conditions. The load current only

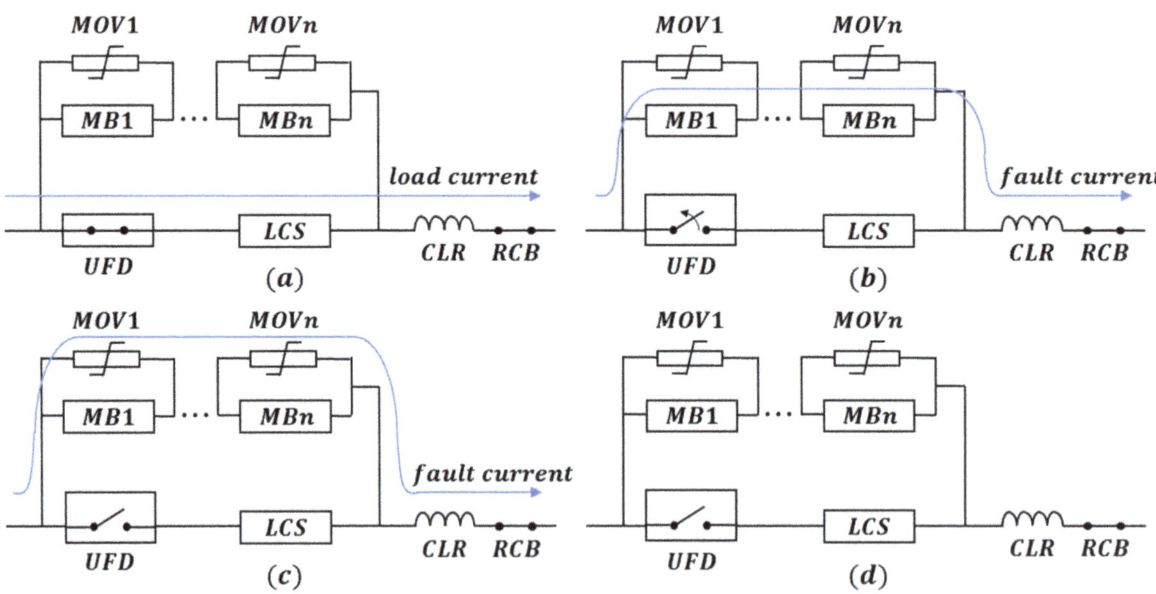

Figure 11.19 Fault interruption operation of an HCB: (a) $t_1 - t_3$; (b) $t_3 - t_4$; (c) $t_4 - t_5$; and (d) after t_5

flows through the low-loss branch because of the high ON-state resistance of MBs. The MOV is used to protect the MB from overvoltage and dissipate fault energy.

The HCB can break either unidirectional or bidirectional fault current depending on the connection of IGBT modules applied in the LCS and the MB.

The connection of the IGBT modules for unidirectional interruption or bidirectional interruption is shown in Figure 11.20. The number of IGBT modules in a bidirectional HCB is twice that of a unidirectional HCB.

A current limiting reactor (CLR) is used to limit the increasing speed of the fault current to no more than 3.5 kA/ms. The residual current breaker (RCB) is used to disconnect the HCB physically, typically within 1 s when the residual current is small enough.

The operation of the HCB is shown in Figure 11.21. When a DC fault occurs at t_1, the load current flowing through the low-loss branch will rise rapidly. The HCB will take some time to detect the fault ($t_1 - t_2$). Then, the LCS is switched OFF immediately at t_2, and thus the fault current starts to be commutated to the MBs. When the fault current is fully commutated into MBs at t_3, the UFD starts to open (see Figure 11.19(b)). Once the UFD opens at t_4, MBs are switched OFF to interrupt the fault current. The fault energy is dissipated by the MOVs and the fault current is, therefore, reduced to zero gradually (see Figure 11.19(c)). When the DC current drops to zero at t_5, the fault is interrupted by the HCB. If restore is not required, the RCB will open to disconnect the HCB physically (see Figure 11.19(d)).

A typical fault interruption process is given in Figure 11.21. For the LCS and the MB, their peak fault currents appear at t_3 and t_4, respectively. For the MOV, the fault energy is dissipated lasting fromt_4 to t_5, and the fault current via the MOV is, therefore, reduced gradually. During the fault energy dissipation, the voltage of the MB cell is limited at the protective level of the MOV most of the time, as shown in Figure 11.21.

After the fault is interrupted, the MB branch withstands the open-circuit DC grid voltage. The normal DC voltage of each cell is determined by the DC grid voltage dividing the number of cells. To achieve a short duration for fault current reduction, the protective level of the MOV is typically 1.5 times the normal DC voltage. Therefore, the voltage rating of the MB in each cell is 1.5 times the normal DC voltage.

An equivalent one-line diagram of a three-terminal MMC HVDC system, as shown in Figure 11.22, is used to explain the design of breaker parameters. The design will consider the peak currents of the MB and LCSs, and the maximum dissipated fault energy of the MOV under the most severe fault current. Thus, a DC bus fault F_l is used in the parameter design, along with F_b and a transmission line.

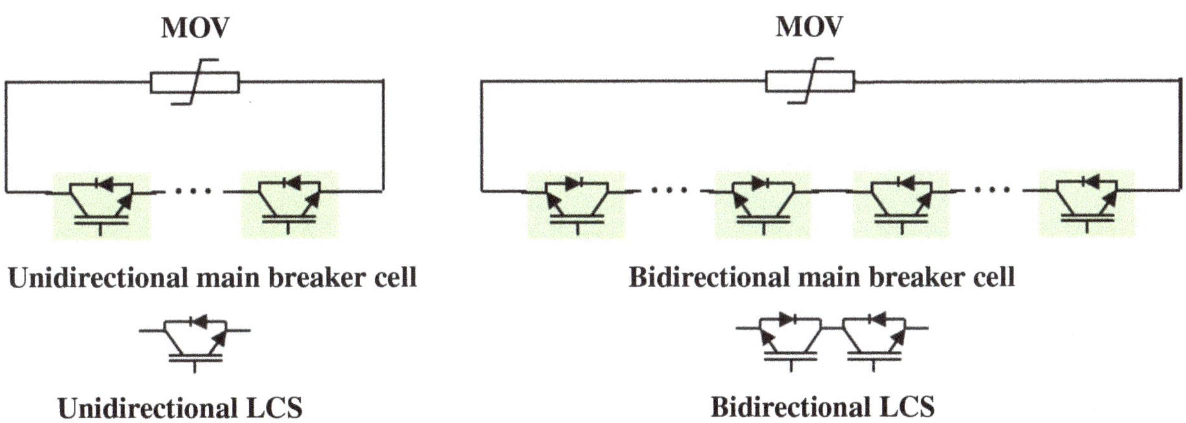

Figure 11.20 IGBT arrangement for the unidirectional or bidirectional interruption

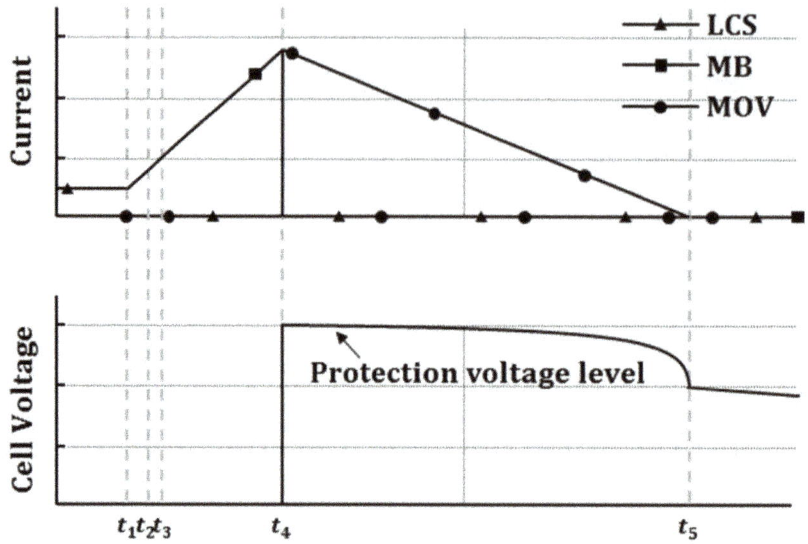

Figure 11.21 Typical fault interruption process in the hybrid DC circuit breaker

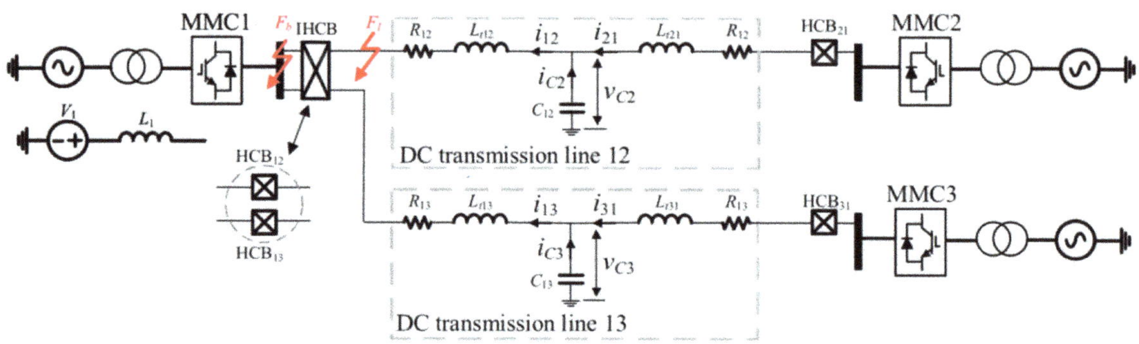

Figure 11.22 Equivalent one-line diagram of a three-terminal modular multi-level converter (MMC)-HVDC for the breaker's parameter analysis

The MMC converter is considered a constant voltage source with an inductor L_1 in series (see MMC1 in Figure 11.22). L_1 is the equivalent inductor of the arm inductors. Its value is one third the inductance of one-arm inductor, since arm inductors on three-phase units are in parallel. The transmission line is represented as a T-section RLC circuit. After a transmission line fault F_l occurs, the fault current flowing through the HCB is mainly contributed by MMC1, the fault current is expressed as

$$i_{fl2}(t) = I_{12} + \frac{V_1}{L_1 + L_{CLR12}} \times (t - t_1) \tag{11.19}$$

where V_1 is the DC voltage of MMC1, I_{12} is the prefaulted current of transmission Line 12, L_{CLR12} is the inductance of the CLR on Line 12, and L_1 is the equivalent arm inductor of MMC1.

For the HCB, the peak currents of the LCS and the MB appear at t_3 and t_4, respectively (see Figure 11.21), and are calculated as

$$I_{LCS12} = I_{12} + \frac{V_1}{L_1 + L_{CLR12}} \times (t_3 - t_1) \tag{11.20}$$

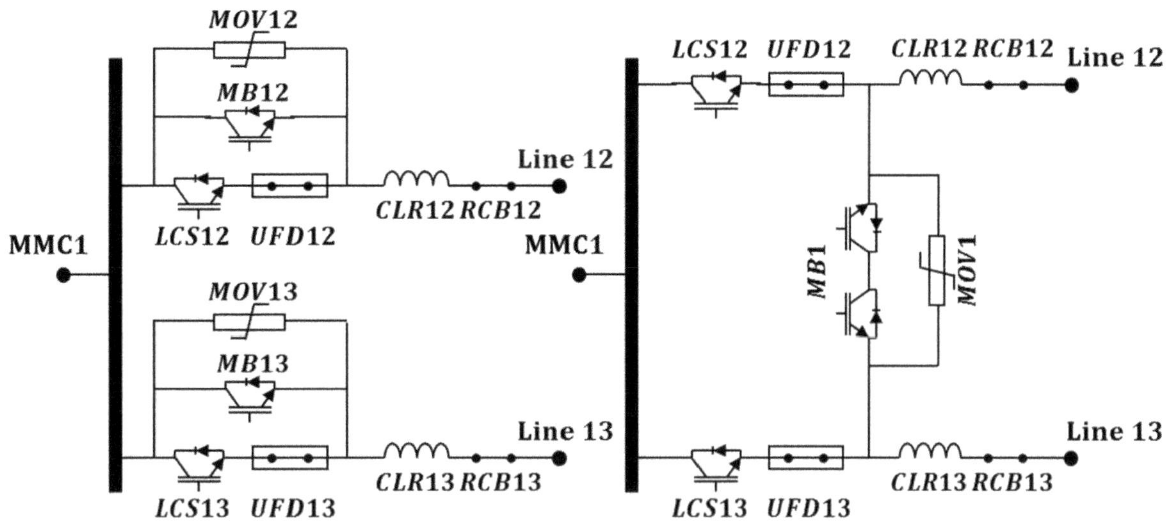

Figure 11.23 Comparison between two unidirectional HCBs and the unidirectional IHCB: (a) two unidirectional HCBs and (b) unidirectional IHCB

$$I_{MB12} = I_{12} + \frac{V_1}{L_1 + L_{CLR12}} \times (t_4 - t_1) \tag{11.21}$$

For the dissipated energy calculation of the MOV, its voltage is assumed to be constant at the protection level during $t_4 - t_5$, as shown in Figure 11.21. The maximum dissipated energy of the MOV in each cell is calculated as

$$E_{MOV12} = \left[I_{12} + \frac{V_1}{L_1 + L_{CLR12}} \times (t_4 - t_1)\right] \times V_{cell} \times \frac{t_5 - t_4}{2} \tag{11.22}$$

where V_{cell} is the voltage rating of the MB in each cell, and is also the protection level of the MOV.

Unidirectional Interlink Hybrid HVDC Circuit Breaker

DC line faults can be interrupted by using only unidirectional HCBs. The design of each HCB having one MB branch (see Figure 11.23(a)) can be improved by sharing an MB branch between two HCBs. Two unidirectional MB branches can be replaced by one interlink MB branch, through which 50% of MOVs is reduced, as shown in Figure 11.23(b). The fault current of Lines 12 or 13 can be commutated to this MB branch to interrupt; therefore, bidirectional MB cells are required.

The possibility of simultaneous faults occurring on different lines is extremely low, so this situation has not been taken into account for breaker design. The unidirectional IHCB (Interlink Hybrid HVDC Circuit Breaker) is also unable to interrupt such faults. The descriptions for the operation principle, parameter design, and MOV reduction are given in the following.

Taking the fault F_1 occurring on Line 12 as an example, the operation principle of the unidirectional IHCB is given in Figure 11.24. Before the fault, the load currents flow through low-loss branches. A fault F_1 occurs on Line 12 at t_1 and is detected at t_2, and the fault current is then commutated from LCS12 to the MBs (see Figure 11.24(a)). At t_3, the current of the low-loss branch becomes zero, and UFD12 starts to open, completing its action at t_4. Then, MB1 will open to interrupt the fault current. From $t_4 - t_5$, the fault energy is dissipated by the MOV1, and the fault current reduces to zero gradually (see Figure 11.24(c)). From $t_3 - t_5$, the LCS13 provides the path for the fault current flowing into the MB branch. After the fault is interrupted at t_5, if restore is not required, the RCB12 is opened to disconnect Line 12 physically and protect the MOV from

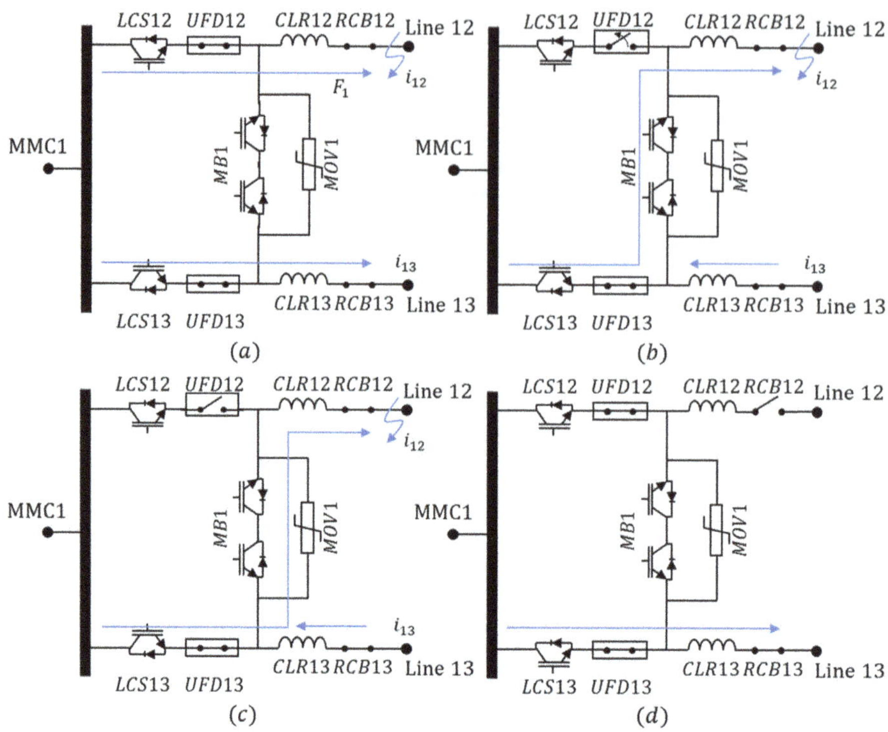

Figure 11.24 Operation of the unidirectional IHCB for the transmission line fault: (a) $t_1 - t_3$; (b) $t_3 - t_4$; (c) $t_4 - t_5$; and (d) after t_5

overload. The UFD12 and LCS12 will then reclose to be a part of the interlink MB branch to protect Line 13 from the line fault (see Figure 11.24(d)).

During the fault interruption $t_3 - t_5$ (see Figure 11.24(b) and (c)), LCS13 is the path for the fault current, which is commutated from LCS12 to the MBs, and the current i_3 of Line 13. Its peak current I_{LCS13} appears at t_4, when the fault current is interrupted by the MBs. The peak current of the LCS is

$$I_{LCS13} = \left[I_{12} + \frac{V_1}{L_1 + L_{CLR12}} \times (t_4 - t_1)\right] + i_{13}(t_4) \tag{11.23}$$

where $i_{13}(t_4)$ is assumed to be equal to the load current I_{13} of Line 13. The influence of the fault on Line 13's current i_{1i} is ignored because of quick fault current interrupting and the CLRs on Line 13.

The fault current keeps rising in MB1 until it is interrupted at t_4; the peak fault current of MB1 is

$$I_{\text{MB1}} = I_{12} + \frac{V_1}{L_1 + L_{CLR12}} \times (t_4 - t_1) \tag{11.24}$$

The MOVs in MB cells dissipated the fault energy evenly, and for one MOV in each cell, its maximum dissipated energy is expressed as

$$E_{\text{MOV}} = \left[I_{12} + \frac{V_1}{L_1 + L_{CLR12}} \times (t_4 - t_1)\right] \times V_{cell} \times \frac{t_5 - t_4}{2} \tag{11.25}$$

After the fault is interrupted, the MB branch of the IHCB withstands the open-circuit DC voltage (V_1) of MMC1. The voltage rating of the MB branch denoted as V_{MB} is 1.5 V_1. This interlink MB branch is composed of anti-series-connected IGBTs to achieve bidirectional interruption; the required numbers of IGBT modules and MOVs are

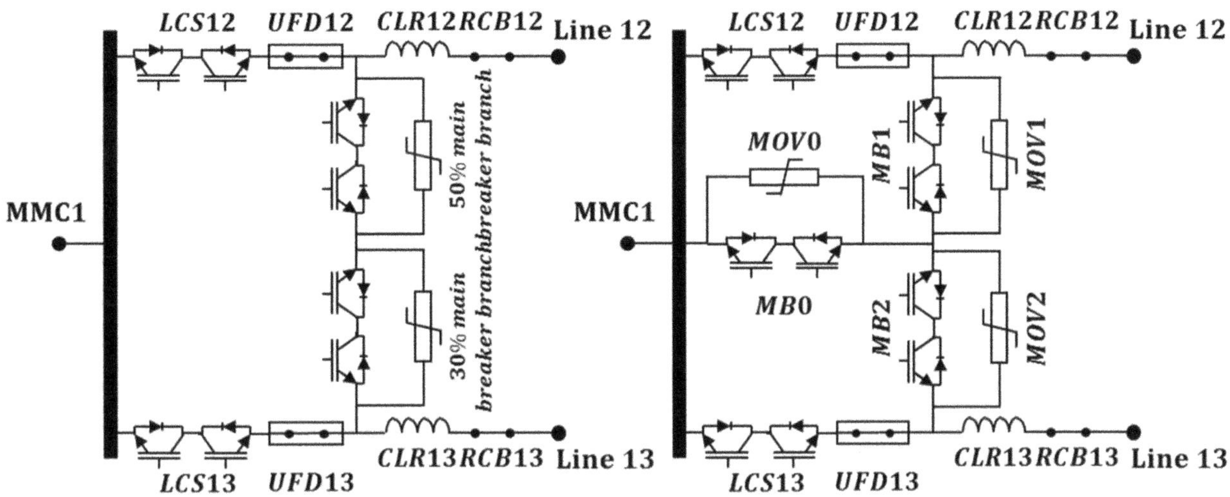

Figure 11.25 Bidirectional IHCB: (a) interlink main breaker branch and (b) Y-connected interlink main breaker branch

$$\begin{cases} n_{IGBT} = \dfrac{V_{MB}}{V_{IGBT}} \times 2 \\ n_{MOV} = \dfrac{V_{MB}}{V_{cell}} \end{cases} \tag{11.26}$$

where VIGBT are the voltage ratings of one IGBT module.

If two unidirectional HCBs are used, the voltage rating of the MB branch is same as that of the unidirectional IHCB, which is also $1.5V_1$. The numbers of IGBT modules and MOVs applied in two HCBs' MB branches are

$$\begin{cases} n_{IGBT} = \dfrac{V_{MB}}{V_{IGBT}} \times 2 \\ n_{MOV} = \dfrac{V_{MB}}{V_{cell}} \times 2 \end{cases} \tag{11.27}$$

Comparing Equations (11.26) and (11.27), the number of IGBT modules are the same for one unidirectional IHCB and two unidirectional HCBs. However, only 50% of MOVs are needed in the unidirectional IHCB.

Bidirectional Interlink Hybrid HVDC Circuit Breaker

The bidirectional HCB can be used to protect the system from not only line faults but also DC bus faults. When a DC bus fault occurs, the fault current of each line will be commuted from LCS to the MB branch before it is interrupted.

If a bidirectional IHCB would use the same MB branch as that of a unidirectional IHCB (see Figure 11.25(a)), then when a DC bus fault occurs, after both LSCs on Lines 12 and 13 are open, the LSCs would have to withstand high DC voltage, and the fault current cannot flow through the MB branch to the faulted bus. Therefore, one more branch is required to avoid overvoltage across LCSs, and to provide a path for the fault current commutation. A novel Y- connected interlink MB branch is thus proposed for the bidirectional IHCB to provide a path for the fault current after both LCSs open in order to interrupt the fault current. Its structure is shown in Figure 11.25(b); an additional MB branch (MB0 and MOV0) is added between the DC bus and the midpoint of the MB branch. The Y-connected MB branch also limits the number of IGBTs to be added.

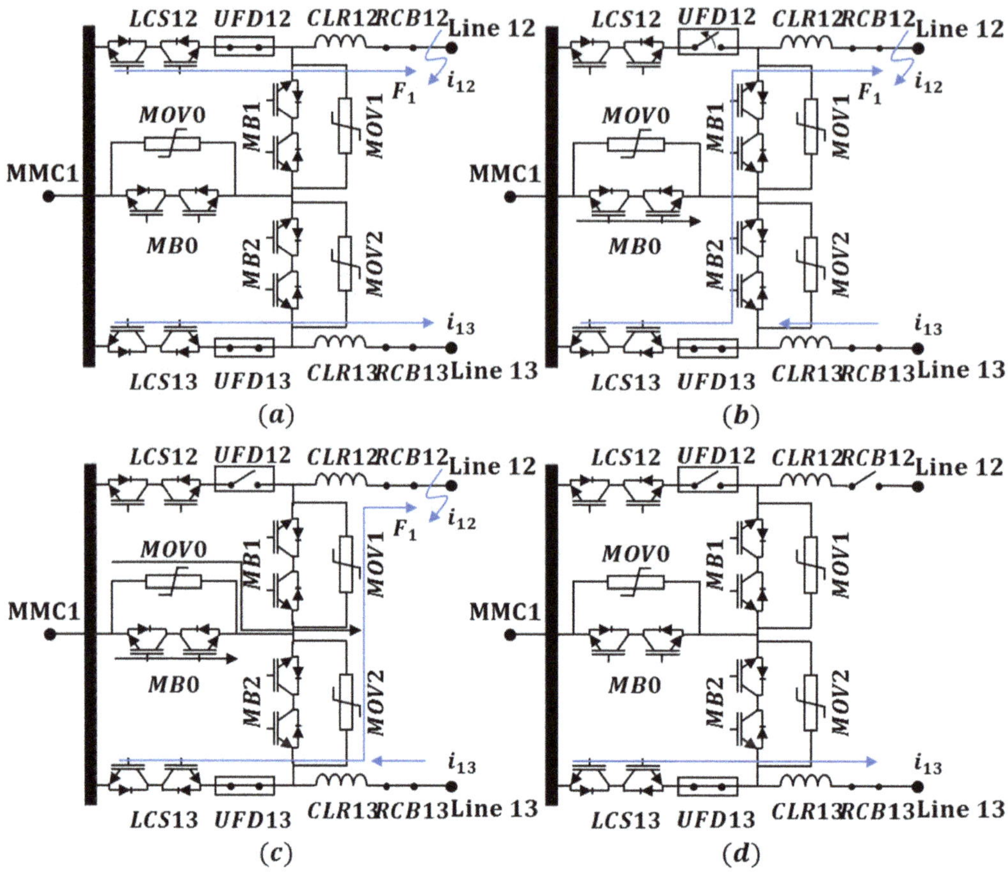

Figure 11.26 Operation of the bidirectional IHCB for the transmission line fault: (a) $t_1 - t_3$; (b) $t_3 - t_4$; (c) $t_4 - t_5$; and (d) after t_5

Compared to using bidirectional HCBs, the number of IGBTs and MOVs are reduced by 25% using the Y-connected branch. For either a line fault or a bus fault, the fault current will be commutated to the Y-connected interlink MB branch to interrupt.

The design of the bidirectional interlink hybrid DC circuit, which does not consider multiple faults occurring on different lines at the same time makes it very cost-effective. The presence of this fault situation, especially under an HVDC grid using cables for the off-shore application, is extremely rare. The descriptions of the operation principle, parameter design, and reduction of the numbers of IGBT modules and MOVs are given in the following.

To interrupt the fault at a transmission line, the operation of an IHCB is shown in Figure 11.26. Before the fault, the load currents flow through the low-loss branches. A fault occurs on Line 12 at t_1 (see Figure 11.26(a)). When it is detected at t_2, LCS12 is switched OFF, and the fault current starts to be commutated into Y-connected MBs. When the commutation process is finished at t_3, UFD12 starts to open, as shown in Figure 11.26(b). At t_4, when UFD12 is fully opened, the fault current will be interrupted by the Y-connected MBs, and the fault energy is dissipated over the MOVs 0, 1, and 2 (see Figure 11.26(c)). After the fault is interrupted at t_5, the RCB12 is opened to disconnect the Line 12 and protect the MOVs from overload. The MB branches 0 and 2 are still available to protect Line 13 (see Figure 11.26(d)).

To interrupt a fault occurring on the DC bus F_b, the operation of the bidirectional IHCB is shown in Figure 11.27. To test the breaker with the most severe fault, the load current is set to have the same direction as the fault current. Before the fault, the load current flows through the low-loss branch to the DC bus.

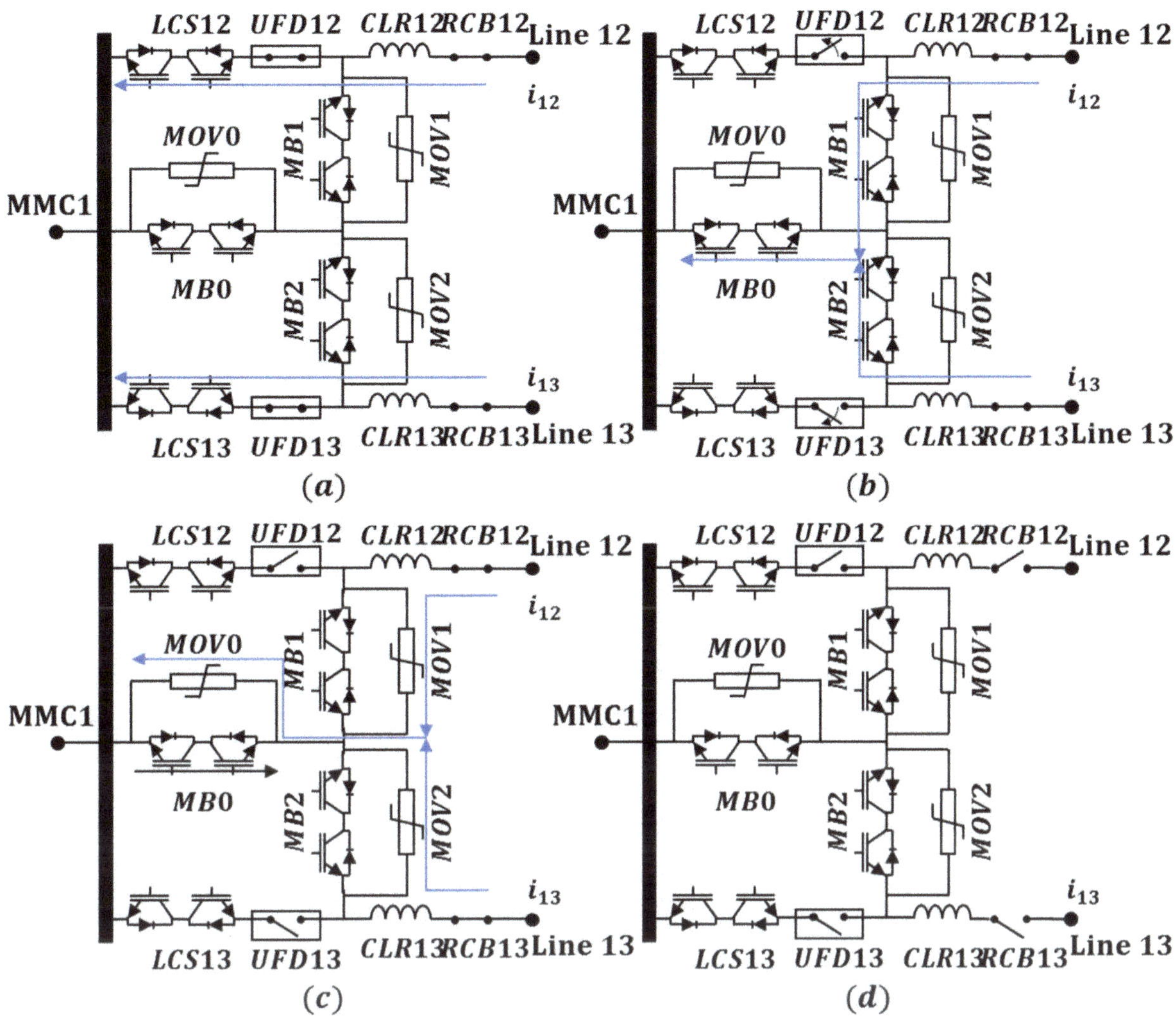

Figure 11.27 Operation of the bidirectional IHCB for the dc bus fault: (a) $t_1 - t_3$; (b) $t_3 - t_4$; (c) $t_4 - t_5$; and (d) after t_5

The DC bus fault occurs at t_1 (see Figure 11.27(a)). When it is detected at t_2, both LCS12 and LCS13 are switched OFF, and the fault currents of Lines 12 and 13 start to be commutated to the Y-connected MBs. From t_3, when the commutation process is finished, UFD12 and UFD13 start to open and complete the action at t_4 (see Figure 11.27(b)). From t_4, the fault current is interrupted by MBs 0, 1, and 2, and the fault energy will be dissipated by MOVs 0, 1, and 2 as shown in Figure 11.27(c). After the fault is interrupted at t_5, RCB12 and RCB13 open to disconnect the faulty DC bus physically (see Figure 11.27(d)). Line faults are used to determine the specifications for the bidirectional IHCB. Thus, a line fault, for example, F_1 in Figure 11.22, is selected.

During the fault interruption $t_3 - t_5$ (see Figure 11.26(b) and (c)), LCS13 is an alternate path for the fault current. The half fault current and the current i_{13} of Line 13 will pass through LCS13. Its peak current I_{LCS13} appears at t_4, when the fault current is interrupted by the MBs. The peak current of the LCS is shown in the following:

$$I_{LCS13} = i_{13}(t_4) + 0.5 \times \left[I_{12} + \frac{V_1}{L_1 + L_{CLR12}} \times (t_4 - t_1) \right] \tag{11.28}$$

where the influence of the fault on Line 12's current is ignored because of quick fault current interrupting and CLRs on Line 13. $i_{13}(t_4)$ is equal to the load current of Line 13. The MB1 withstands the whole fault current (see Figure 11.26(b)). Its peak current appears at t_4, when the fault current is interrupted. The peak current of the MB1 or 2 is shown as

$$I_{MB1} = I_{12} + \frac{V_1}{L_1 + L_{CLR12}} \times (t_4 - t_1) \tag{11.29}$$

The maximum dissipated energy of the MOV1 or MOV2 located on main are calculated as

$$E_{MOV1} = \left[I_{12} + \frac{V_1}{L_1 + L_{CLR12}} \times (t_4 - t_1)\right] \times V_{cell} \times \frac{t_5 - t_4}{2} \tag{11.30}$$

since MB branch 0 needs to withstand the high fault current from Lines 12 and 13 caused by the DC bus fault. The DC bus fault is used to determine the parameter of the MB branch 0. When a DC bus fault like Fb in Figure 11.22 occurs, the following equations of Lines 12 or 13 are obtained using the KVL and KCL laws

$$R_{1j} i_{j1}(t) + \left(L_{CLRj1} + L_{tj1} + L_j\right) \frac{di_{j1}(t)}{dt} + v_{Cj}(t) = V_j \tag{11.31}$$

$$\left(L_{CLR1j} + L_{t1j}\right) \frac{di_{1j}(t)}{dt} + R_{1j} i_{1j}(t) = v_{Cj}(t) \tag{11.32}$$

$$C_j \frac{dv_{cj}(t)}{dt} = -i_{Cj}(t) \tag{11.33}$$

$$i_{1j}(t) = i_{j1}(t) + i_{Cj}(t) \tag{11.34}$$

where $j = 2$ or 3 and V_j is the DC voltage of MMC*j*.

To simplify the fault current analysis, the total inductances on both sides of the capacitor C_{1j} are supposed to be equal. The values of L_{t1j} and L_{tj1} of the transmission lines should satisfy the relationship

$$L_{CLR1j} + L_{t1j} = L_{CLRj1} + L_{tj1} = L_{1j} \tag{11.35}$$

where L_{1j} denotes the equivalent inductance between the faulted DC bus and the capacitor C_{1j} and between the capacitor C_{1j} and MMCj.

The expression of v_c is obtained by substituting Equations (11.32)–(11.35) into Equation (11.31)

$$2v_{Cj}(t) + R_{1j} C_j \frac{dv_{Cj}(t)}{dt} + L_{1j} C_j \frac{d^2 v_{Cj}(t)}{dt^2} = V_j \tag{11.36}$$

Applying the Laplace transformation to Equation (11.36) yields

$$2v_{Cj}(s) + R_{1j} C_j \left[s v_c(s) - v_c(0)\right] + L_{1j} C_j \left[s^2 v_c(s) - s v_c(0)\right] = \frac{V_j}{s} \tag{11.37}$$

where, $v_c(0) = V_j$ is assumed because of the small value of R_{1j}.

The expression of $v_c(s)$ is obtained by arranging (19)

$$v_{cj}(s) = 0.5 V_j \left[\frac{1}{s} + \frac{C_j\left(L_{1j} s + R_{1j}\right)}{C_j L_{1j} s^2 + C_j R_{1j} s + 2}\right] \tag{11.38}$$

Applying the Laplace transformation to Equation (11.32) and considering Equation (11.38) yield

$$L_{1j}\left[si_{1j}(s) - i_{1j}(0)\right] + R_{1j}i_{1j}(s) = 0.5V_j\left[\frac{1}{s} + \frac{C_j\left(L_{1j}s + R_{1j}\right)}{C_jL_{1j}s^2 + C_jR_{1j}s + 2}\right] \tag{11.39}$$

The expression of $i_{1j}(s)$ is thus obtained from Equation (11.39)

$$i_{1j}(s) = 0.5iV\left[\frac{1}{R_{1j}}\left(\frac{1}{s} - \frac{1}{s+a}\right) + \frac{1}{L_{1j}b} \times \frac{b^2}{\left(s + \frac{a}{2}\right)^2 + b^2}\right] + i_{1j}(0) \times \frac{1}{s+a} \tag{11.40}$$

where, a $= \frac{R_{1j}}{L_{1j}}$, $b = \sqrt{\frac{2}{L_{1j}C_j} - \left(\frac{R_{1j}}{2L_{1j}}\right)^2}$, and $i_{1j}(0) = I_{1j}$, which is the rated current in normal condition.
The time-domain expression of i_{1j} is obtained through the inverse Laplace transform of Equation (11.40)

$$i_{1j}(t) = 0.5V_j\left[\frac{1}{R_{1j}}\left(1 - e^{-at}\right) + \frac{1}{L_{1j}b}e^{-\frac{a}{2}t}\sin(bt)\right] + I_{1j}e^{-at} \tag{11.41}$$

As shown in Equation (11.41), the fault current is determined by the rated current of both lines, the impedance of the transmission lines, the inductance of the CLRs, and the arm inductor of the converters. The fault current in the Y-connected MBs keeps rising until it is interrupted. The peak current in the MB 0 appears at t_4

$$I_{MB0} = \sum_{i=2}^{3} i_{1j}(t_4) \tag{11.42}$$

The maximum dissipated energy of the MOV0 of each cell in this subbranch is

$$E_{MOV0} = I_{MB0} \times V_{cell} \times \frac{t_5 - t_4}{2} \tag{11.43}$$

When a line fault is interrupted by the breaker, the open-circuit voltage is the DC voltage (V_1) of MMC1. When a bus fault is cleared by the breaker, the open-circuit voltage of $V_j - \Delta V_{1j}$, V_j is the DC voltage of MMCj, and ΔV_{1j} is the voltage drop of Line 1j. Only the rated DC voltage V_1 is considered to determine the voltage rating of the breaker as $V_1 > V_j - \Delta V_{1j}$.

For the bidirectional IHCB, two branches of the Y-connected MB branch share the open-circuit voltage V_1. The voltage rating of each MB branch is $V_{Y-MB} = V_{MB}/2 = 1.5\,V_1/2 = 0.75\,V_1$. The numbers of IGBT modules and MOVs required for an interlink MB branch are

$$\begin{cases} n_{IGBT} = \dfrac{0.5V_{MB}}{V_{IGBT}} \times 3 \times 2 \\ n_{MOV} = \dfrac{0.5V_{MB}}{V_{cell}} \times 3 \end{cases} \tag{11.44}$$

If two bidirectional HCBs are used, the open-circuit voltage for each MB branch is V_1. The voltage rating V_{MB} of each branch is $1.5V_1$. And the numbers of IGBT modules and MOVs are used in two MB branches.

11.2 DC/DC Converter [20]–[25]

Except for the conventional small-scale DC-DC converter (Chopper), the large-scale DC-DC converters that can be applied to HVDC system have attracted much attention from Smart Grid and Off shore wind power linkage.

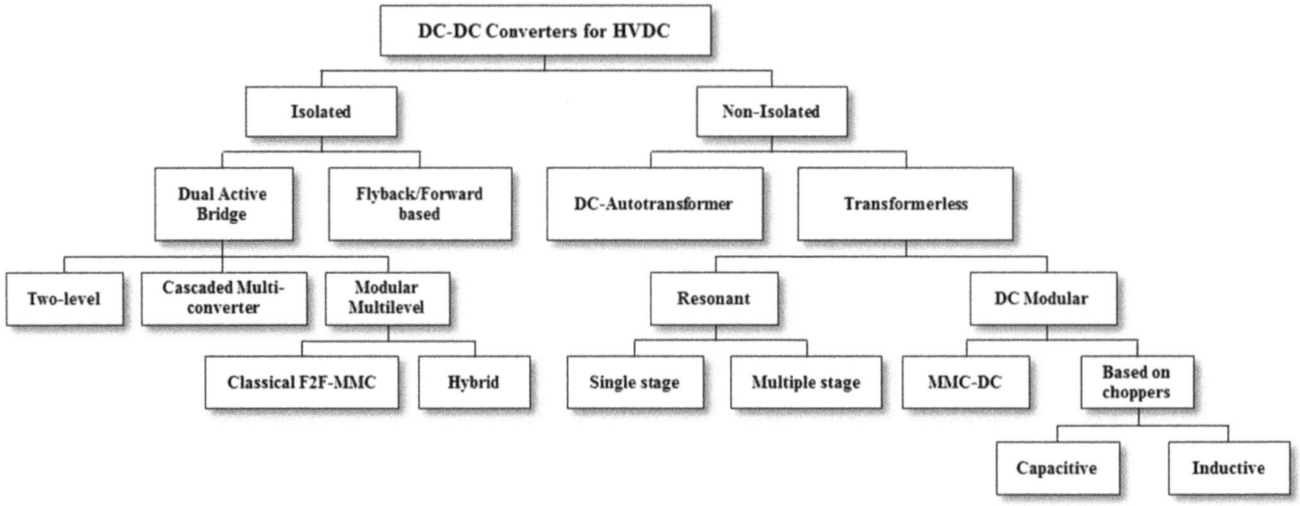

Figure 11.28 classification for DC-DC converters dedicated to HVDC applications

In a paper published on IEEE in 2019[20], a large-scale DC-DC converter is well summarized. The authors of this paper are Juan David paez and so on, and the contents are as follows.

DC–DC power converters dedicated to HVDC grids can provide more functionalities than just voltage stepping. For example, they can add power-flow control, DC grid voltage regulation, and fault isolation, as well as interface different DC transmission schemes such as monopolar or bipolar systems and serve as an interface between LCC and VSC systems (Figure 11.28).

Although several methods to achieve DC–DC conversion in LV and MV exist and are well known, the adaptation of these techniques to HV is not straightforward. HVDC converters require the association of several LV components, such as power semiconductor switches and/or LV converters, precluding the direct use of classical conversion methods. Table 11.2 shows the HVDC Applications Characteristics.

Given the differences in the LV and MV environments, an overview of the techniques specifically dedicated to HVDCs is needed. This allows the trends, challenges, and opportunities in this area to be identified. Reviewing the literature, there are few publications summarizing the different proposals. Nevertheless, their approach is oriented to specific circuits, thus making it difficult to identify a trend or general methods to build HVDC DC–DC structures.

11.2.1 Isolated Topologies

The isolated circuits published in literature rely on a DC-AC–DC conversion chain with two AC–DC conversion stages. Galvanic separation is done on the AC bus by magnetic coupling using an AC transformer or coupled inductors. Isolation may be needed for mainly safety and grounding reasons. In the first case, isolation is useful to prevent the apparition of HV on LV (LV) terminals, a very important aspect to consider, especially in high transformation ratio applications. Concerning grounding, isolated structures offer the possibility of having different grounding schemes on both DC grids to be interconnected.

Although those tasks can also be achieved by means other than galvanic isolation, using an isolated structure allows simplification of the design and safety assessment. Other advantages of these topologies are easy voltage adaptation using the AC transformer ratio and the inherent DC-fault blocking capability. In the case of a DC-fault, blocking both AC–DC conversion stages of the converter stops the fault current.

Converters that use a transformer can be considered as different versions of the dual active bridge (DAB). The proposed classification assumes that both DC–AC and AC–DC stages are identical, but combinations of

Table 11.2 HVDC Applications Characteristics

Parameter	Classification	Range
HV-side DC voltage $(V_{DC-HIGH})^*$	Medium High	$V_{DC} < 100\text{kV}$ $100\text{kV} \leq V_{DC}$
Transferred DC power	Low Medium High	$P_{DC} < 50MW$ $50MW \leq P_{DC} < 500MW$ $500MW \leq P_{DC}$
Transformation ratio $(V_{DC-HIGH}/V_{DC-LOW})$	Low Medium High	$1 < \text{Ratio} \leq 1.5$ $1.5 < \text{Ratio} \leq 5$ $5 < \text{Ratio}$

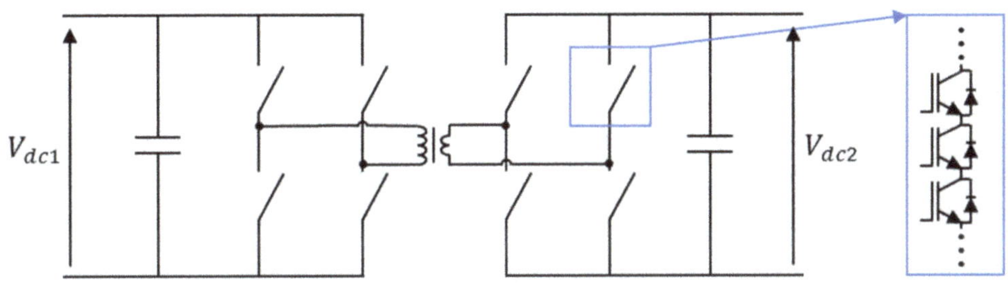

Figure 11.29 Two level high voltage DAB

different bridge types are also possible. The few circuits that do not fit into the DAB category are based on the Flyback/Forward principles. Figure 11.29 shows the isolated topology of DC-DC converter.

Two-Level HV DAB: The *Two-Level HV DAB* topology consists of two two-level VSCs interconnected with a transformer (Figure11.29). The converter standard operation is based on phase-shift modulation at a fixed duty cycle. Modified modulation schemes can be used to improve the harmonic behavior of the converter.

This converter presents several drawbacks for its utilization in HVDC. The high *dv/dt* imposed on the transformer, given the two-level operation, causes insulation and electromagnetic interference (EMI) issues. Furthermore, snubber circuits and/or active gate-driving are needed to avoid static and dynamic voltage sharing unbalances on the transistor valves, increasing switching losses.

Cascaded Multi-Converter DAB: In *Cascaded Multi-Converter DAB* circuits, the HV structure is built using low-power, LV DAB converters as elementary cells, avoiding the need for series-connected transistors since each cell only blocks a portion of the total DC voltage. In addition, since each cell only handles a fraction of the total power, the current rating of the semiconductors can be reduced.

The converter SMs can be associated in different manners (Figure 11.30). If the cell terminals are connected in parallel, the DC line current is distributed among all the SMs, whereas the DC line voltage is shared when using a series connection. The former configuration is advantageous when high DC currents are required, and the latter for HVs. Both schemes can be combined to meet the requirements of different HVDC applications. For the converter operation, each cell is controlled as a standard DAB, with the additional constraint of a balanced distribution of cell voltages and currents. Medium-frequency (MF) operation is possible in order to reduce the

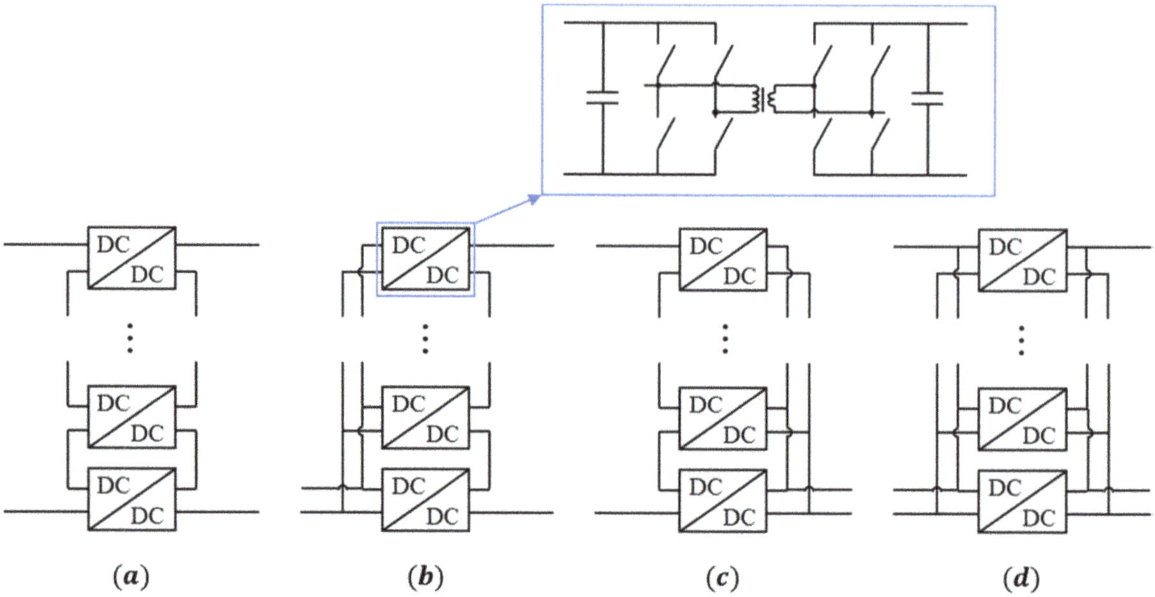

Figure 11.30 Cascaded multi-converter dual active bridge (DAB), with different possible configurations: (a) input-series output-series (ISOS); (b) input-parallel output-series (IPOS); (c) input-series output-parallel (ISOP); and (d) input-parallel output-parallel (IPOP)

size and weight of the passive components and the transformer. Additionally, if properly sized, the circuit can operate under soft-switching conditions, reducing losses.

The main advantages of this converter family are the modularity and scalability; the converter can be scaled in power and voltage using the same elementary cells. The different association variants allow high transformation ratios to be obtained at high power (input-parallel output-series [IPOS] scheme). However, the main issue of these structures is the high insulation requirement of the transformer, limiting their use to the MV range.

Modular Multi-Level DAB: *Modular multi-level DAB* converters are built by connecting two modular MMCs through an AC transformer in a **front-to-front (F2F)** structure [Figure 11.31a)]. Both MMCs are controlled to generate two phase-shifted AC waveforms at the transformer terminals to control the power transfer.

Generation of the AC voltages is done through the bypass or insertion of MMC-SM capacitors in each converter arm. The SMs can be implemented with various bridge types, giving different functionalities. The most common types are half-bridge (HB) SMs and full-bridge (FB) SMs. The stacks made of HBs generate unipolar voltages, whereas those made of FBs can generate positive and negative voltages. The second option can be useful to improve DC-fault management.

Since the SMs switch under hard-switching conditions at low frequency, the main source of losses in MMC-SMs-based converters is from semiconductor conduction.

Diverse modulation schemes can be used to optimize the operation or size of the converter. For example, the use of a trapezoidal modulation or MF operation has been proposed to reduce the size of the transformer, passive elements, and SM capacitors. Nevertheless, the design of a HV high-power AC transformer working with non-sinusoidal waveforms at MF is challenging. Additionally, a trade-off between size reduction and increase of switching losses has to be made.

Since the *Classical F2F-MMC* requires two fully rated MMCs, its size and cost are considerable. The *Hybrid* versions represent an alternative to improve these issues.

In these circuits, some SMs are replaced by transistor valves (Figure 11.31(b)), combining the advantages of the MMC and the two-level VSCs. The SMs act as wave-shapers, providing soft switching conditions to the valves, and serve to control the *dv*/*dt* on the transformer. The transistor valves control the current paths and can prevent simultaneous conduction of two same-leg converter arms, decreasing conduction losses. Because this

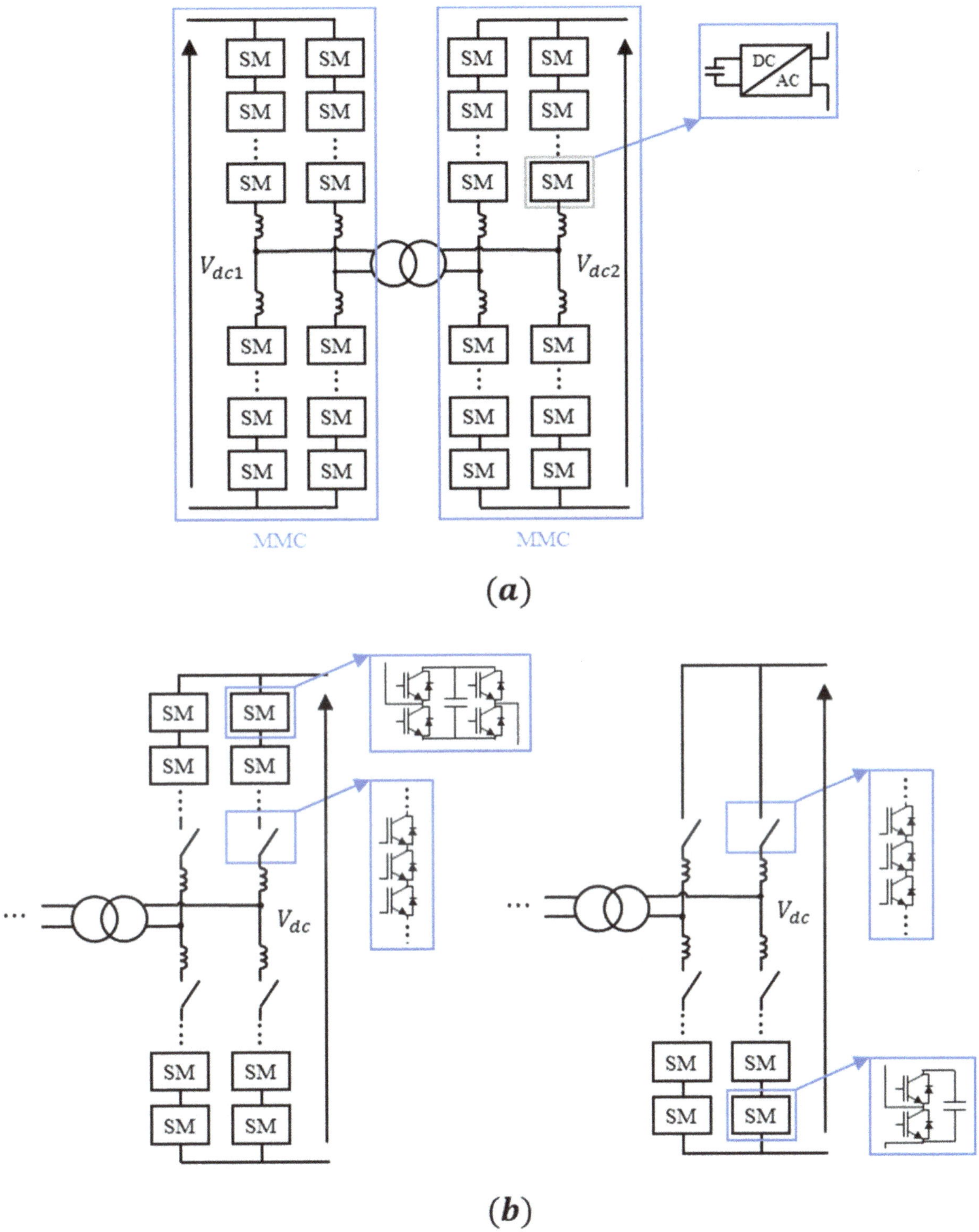

Figure 11.31 Modular multi-level DABs (F2F-MMCs): (a) classical F2F-MMC and (b) some hybrid F2F-MMCs: the AAC (Alternate Arm Converter) and the TAMC (Transition Arm Multilevel Converter)

operation is different to the classical MMC, each hybrid topology implements its own control and balancing methods.

Since all the *modular multi-level DAB* topologies present advantages of the MMC such as high modularity, reliability, and availability, these structures are suitable for HV and high-power applications at LV transformation ratios. Another advantage is the high level of technological maturity of the SMs and semiconductor valves. However, since two fully rated MMCs are needed, even in the *hybrid* versions, this could be a suboptimal solution.

Flyback/Forward Based: Only a few publications propose isolated converters different to the DAB solutions (Figure 11.32). They are based on the principles of flyback and forward converters. In some circuits, adaption to the HV of the classical structures is achieved using one central-coupled inductor and MMC-SMs to emulate HV switches.

Those structures have been proposed for high step-up voltage ratios. However, the high insulation requirements of the coupled inductors in the modular approaches, along with the high current requirements in the centralized inductor circuits, are challenging, potentially limiting these topologies to low-power applications.

11.2.2 Non-Isolated Topologies

The range of non-isolated topologies is larger than in the isolated case. A first classification is made based on the presence of a transformer in the circuit resulting in two families: *DC autotransformer* converters and *transformerless* converters.

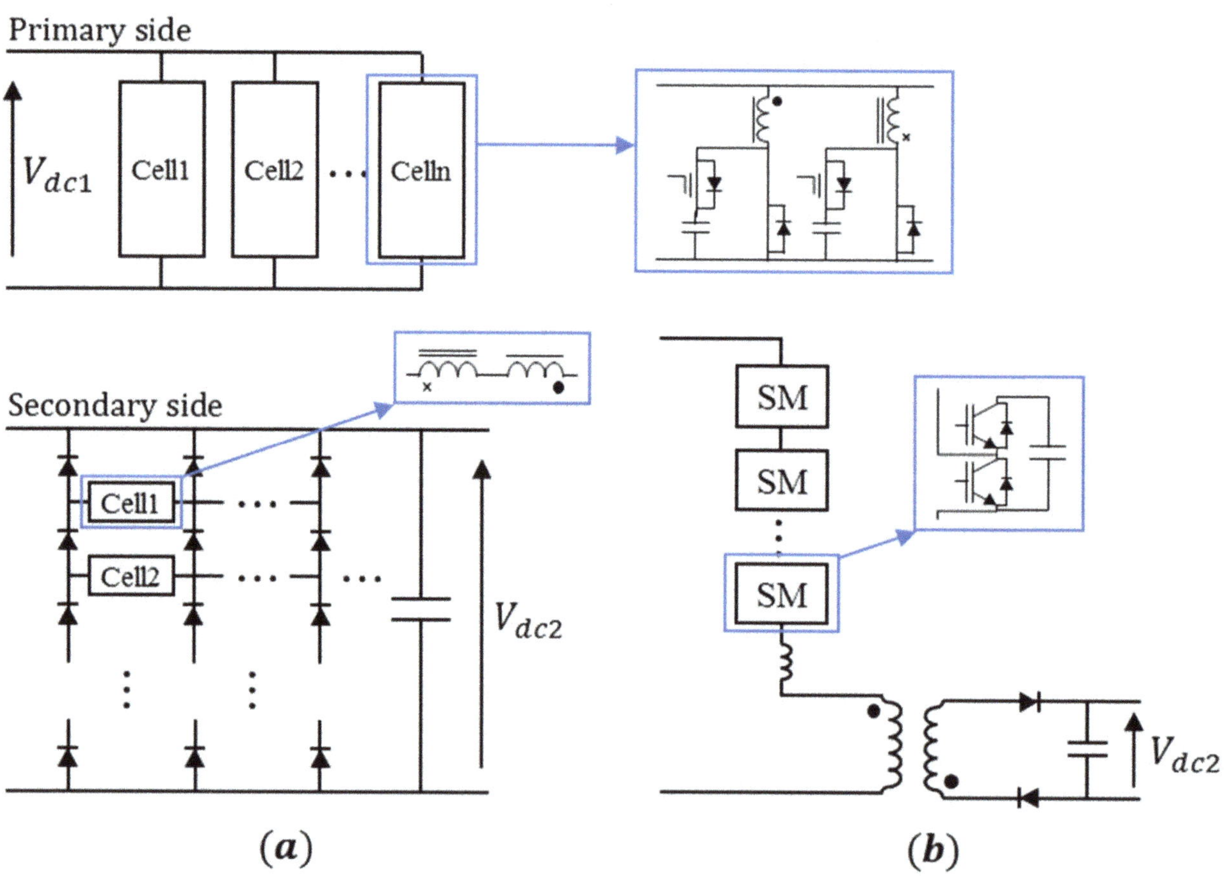

Figure 11.32 (a) Modular flyback-/forward-based converter and (b) centralized coupled inductor flyback-/forward-based converter

DC Autotransformer: The HVDC *autotransformer*-type converters (Figure 11.33) use two HV DC–AC and AC–DC structures connected in series on the DC ports and interconnected through an AC transformer on the AC side. This implementation is feasible with existing HVDC converter technology: VSC (two-level, three-level, or MMC), VSC-LCC, or VSC-diode rectifier.

This solution differs from those of the DAB family in that only a fraction of the power flows through the AC link. This feature contributes to decreasing power losses and the transformer rating. The voltage rating of each individual converter is also decreased thanks to their series connection. Nevertheless, these benefits are gradually lost as the voltage transformation ratio increases, so this structure is more appropriate for low- and medium-transformation ratios.

Resonant Topologies:

These converters are based on resonant LC tanks. The principle is to use the resonance as a step-up mechanism while achieving soft switching of the semiconductors at the same time. According to the number of resonant tanks, this family can be subdivided into single-stage, if only one is used, or multiple-stage, if several low-power tanks are sequentially activated.

Resonant Single-Stage: In these converters (Figure 11.34), the DC–DC conversion is done using a DC–AC–DC scheme without isolation. Semiconductor bridges, implemented with series-connected devices,

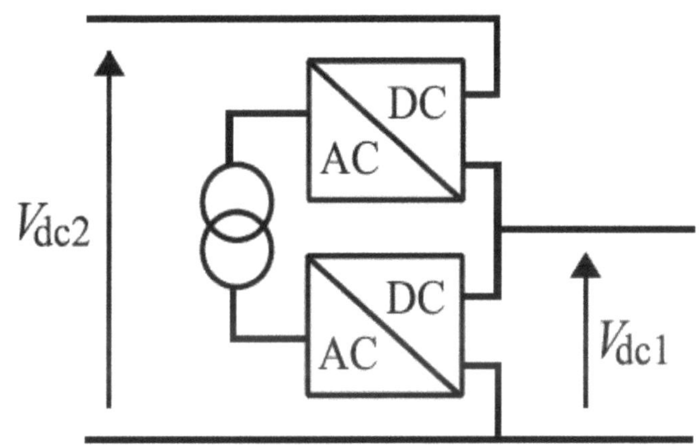

Figure 11.33 DC autotransformer

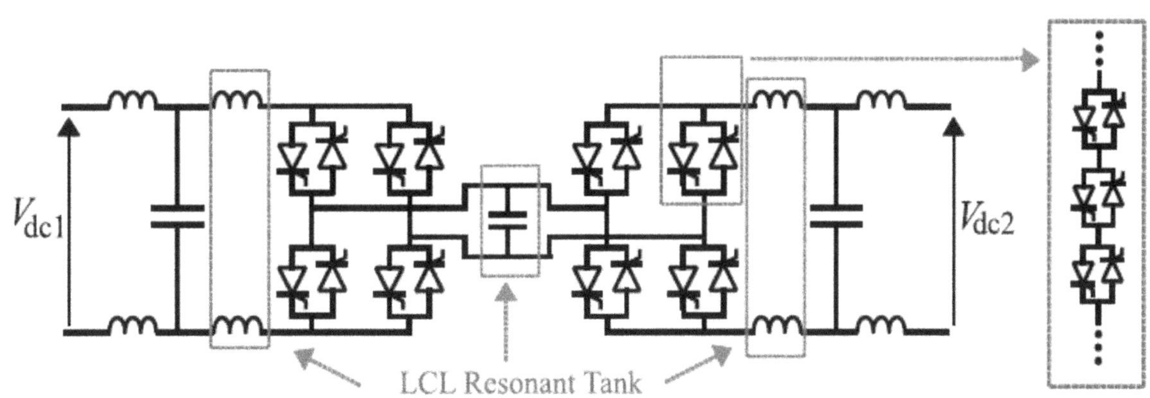

Figure 11.34 One single-stage resonant converter. LCL converter

are interconnected on their AC side through a main resonant tank. The topology in Figure 11.33 uses an LCL T network as the resonant element, whereas the converter uses an LC parallel tank followed by a voltage doubler.

All these converters were proposed for high transformation ratios. However, since only a central resonant tank is used, and because of the operation of the converter, the passive components are exposed to very high electrical stresses. Since all the power passes through just a few resonant elements, their current and voltage ratings could be very high. These issues could potentially limit the operation to the medium-power range at medium step transformation ratios.

Resonant Multiple-Stage: In this subfamily (Figure 11.35), the full-rated central resonant tank is replaced by several low-power resonant circuits, reducing the complexity of the design significantly and getting closer to a modular structure. In a sequential process, each tank is activated, and the power is consecutively transferred from one tank to the next one until it is delivered to the HV converter terminal. The inverse process is used for step-down operation.

In some topologies, the capacitors are charged connecting them in parallel across the LV DC bus, and the power is released to the HV bus connecting them in series. Some other circuits create a low AC voltage on the LV DC terminals, and the voltage is stepped up charging several capacitors. All *multiple-stage resonant converters* are proposed for high transformation ratios without having strong electrical stress on the resonant elements,

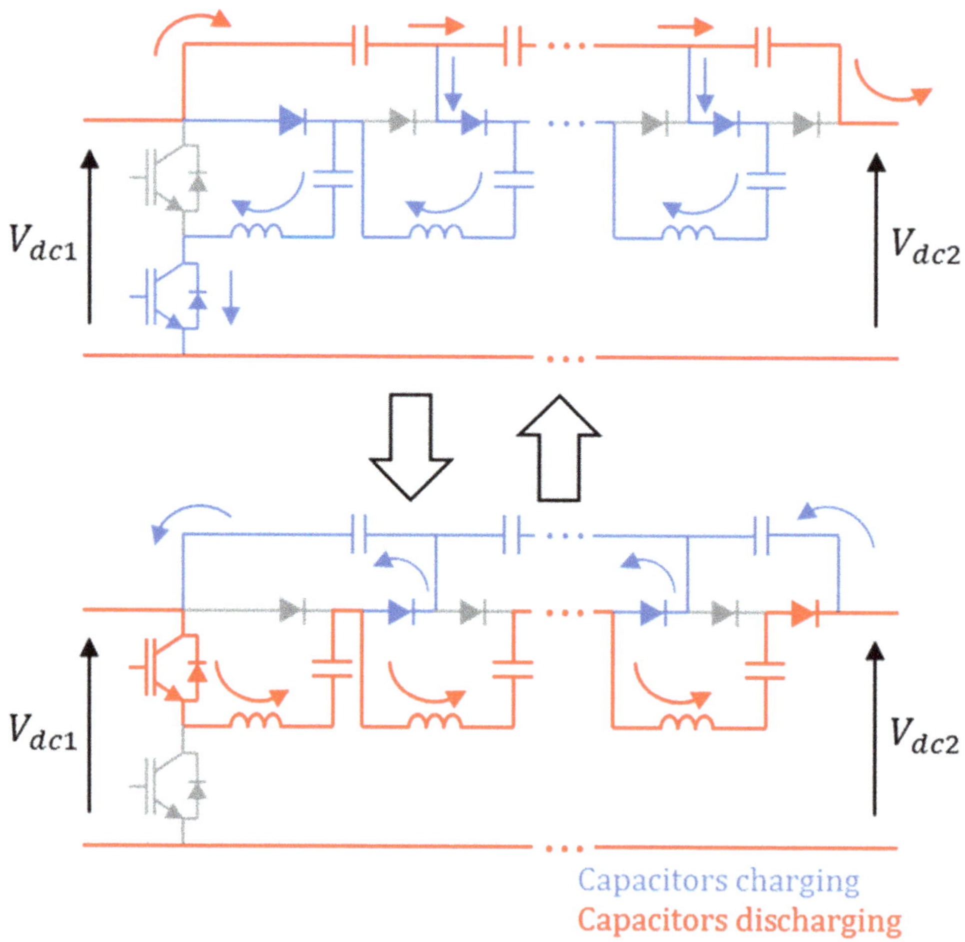

Figure 11.35 One multiple-stage resonant converter

as is the case for *single-stage* circuits. The main issue of these topologies is the uneven distribution of currents or voltages in the semiconductor devices, making it difficult to obtain a modular solution.

DC Modular Topologies

The circuits grouped in this subfamily adopt the fully modular approach of MMC to DC–DC conversion. Two trends are identified in literature: circuits whose operation is similar to the MMC and those whose operation principle is based on classical choppers.

DC-MMC: The so-called DC-MMC converters (Figure 11.36) use chains of MMC-SMs to generate currents and voltages at different frequencies. Two superposed current loops are created, one AC and one DC, and the operation principle is to transfer the power between these loops. The AC component is used for equilibrating the energy stored in the SMs, and the DC component is used for power transfer between the DC ports. To prevent the AC component from entering the DC buses, two techniques can be used: passive filters or control actions.

Concerning losses, these converters operate in hard-switching, but, like the MMC, at low switching frequency, so the main source of losses is semiconductor conduction. The increase of frequency for the AC current component can decrease the size of passives, but at the cost of increasing switching losses.

The DC-MMC structures have the advantages of modularity, high reliability, and scalability. They could be a solution for high-power and HV applications. Their main drawbacks are the rating of the filters, when they are required, and the potentially high AC circulating currents needed for the converter operation that could limit the use to the low or medium range of transformation ratios.

DC Modular Based on Choppers: Modular topologies based on the classical choppers have been proposed for HVDCs. The principle is to replace some of the switches on the original converters with MMC-SMs. This

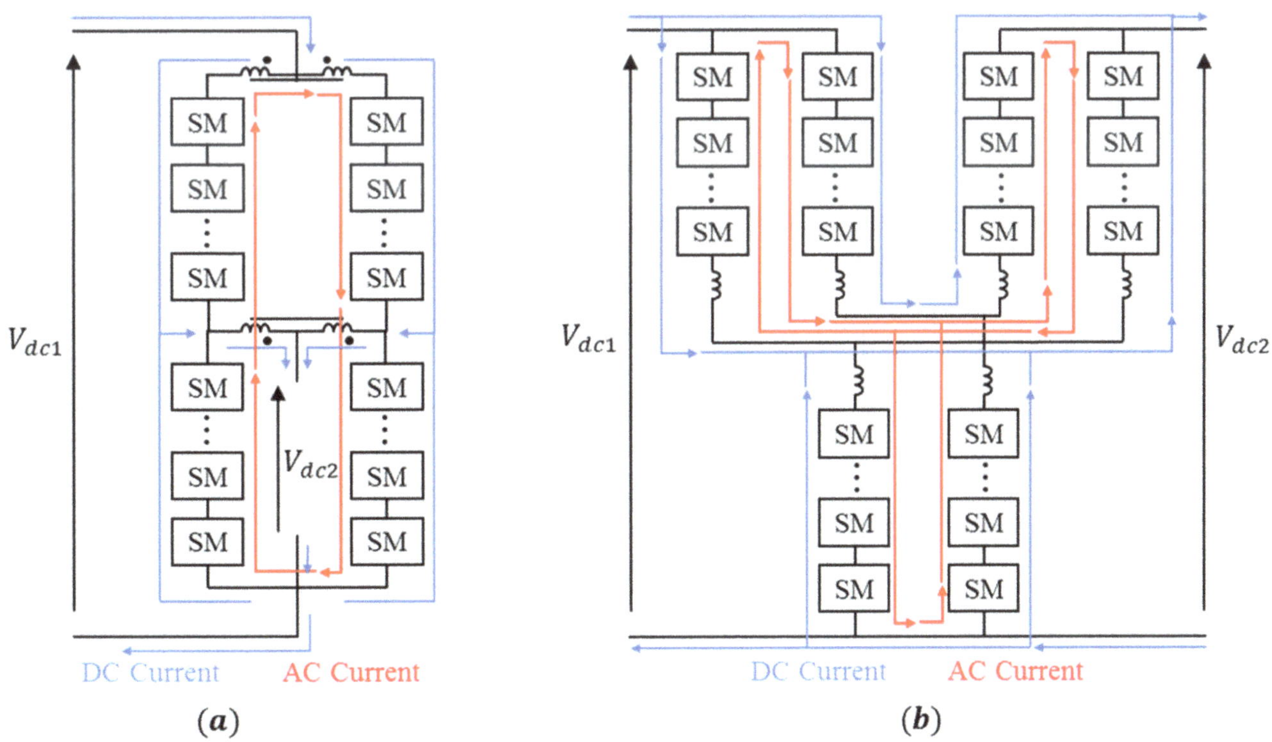

Figure 11.36 Some DC-modular multil-evel converters (MMCs): (a) AC current blocking with filters and (b) AC current blocking with control

replacement provides extended control features. An additional subdivision is proposed based on the employed energy storage mechanism:

Capacitive Accumulation Choppers: In these converters (Figure 11.37(a)), the MMC-SMs are used as a variable capacitor for storing energy. The principle is to charge and discharge the capacitors, switching the SM stacks connection between the input and output DC terminals. A dead time is required to accommodate the number of inserted SMs according to the voltage level of the DC terminal to be connected. The main converter valves are soft switched, whereas the SMs are hard-switched.

Inductive Accumulation Choppers: In these circuits (Figure 11.37(b)), the SMs control the charge and discharge of a central inductor, used as a storage element like in a classical chopper. The converter operation can be done by creating a staircase transition between the two states of the converter (inductor charging or discharging) using interleaved insertion of SMs to operate at resonance, or at resonant discontinuous-conduction modes. Soft switching of the SMs can be achieved, but this causes AC circulating currents to increase conduction losses. A compromise should be made to obtain an optimal operation point.

These converters present modularity, and are characterized by an HV transformation ratio at low-power. Their main drawback is the large energy rating of the central inductor.

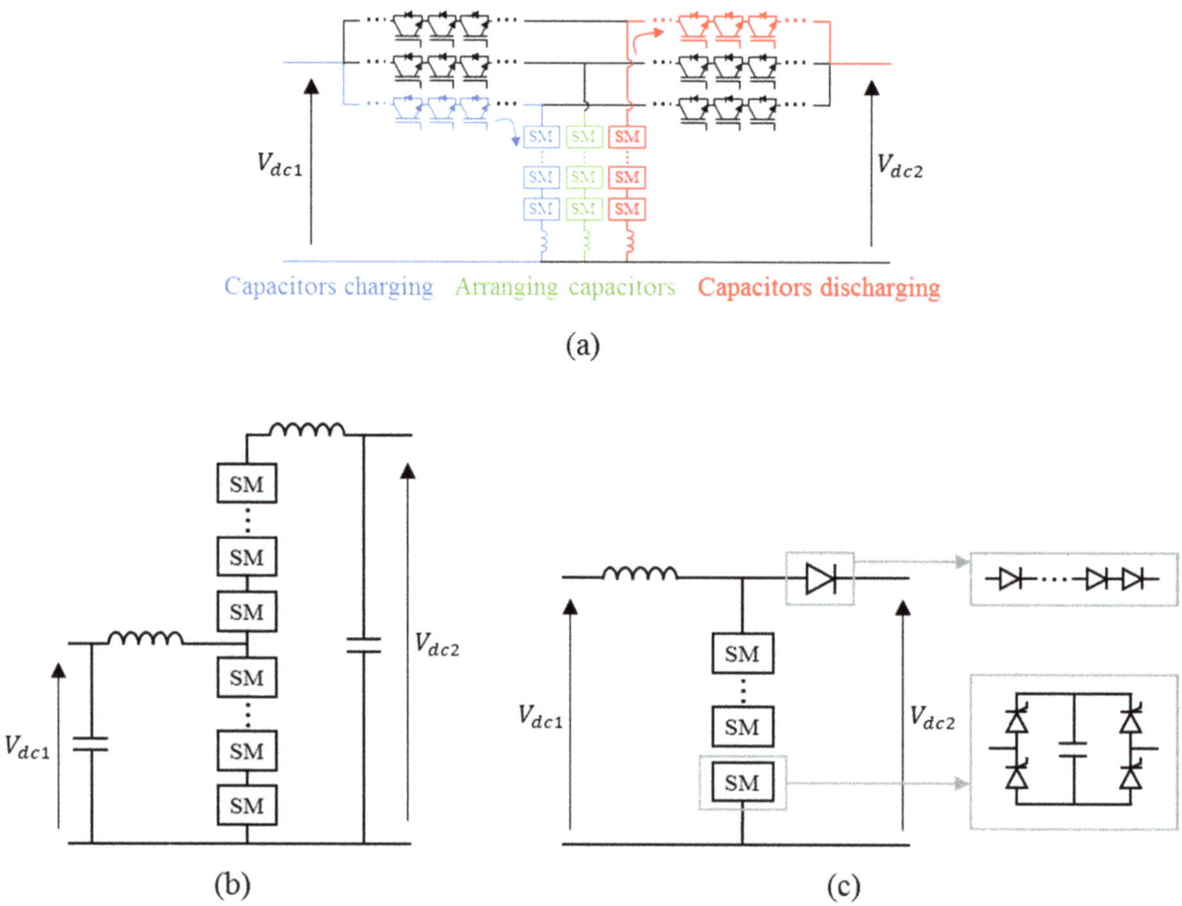

Figure 11.37 Some DC modular converters based on choppers: (a) capacitive accumulation and (b) inductive accumulation

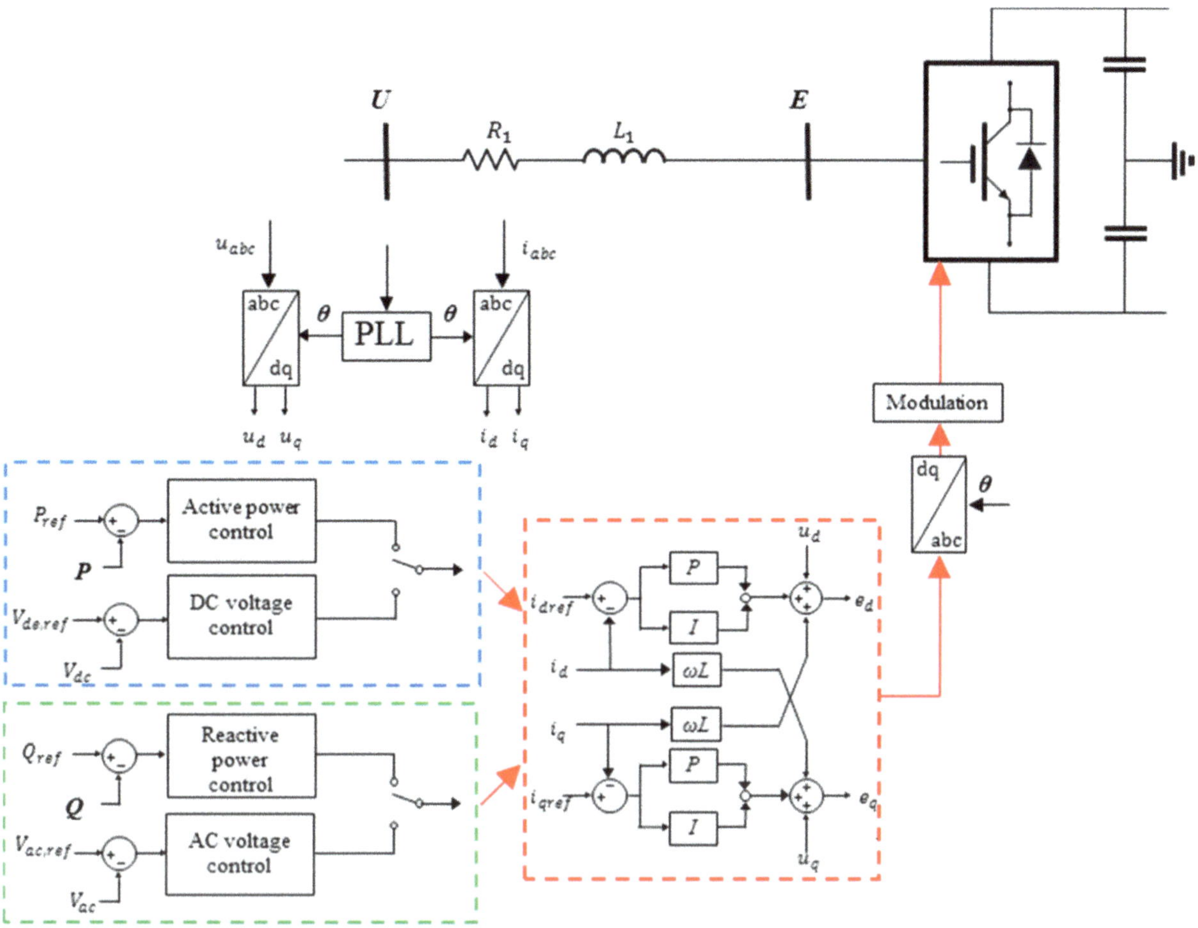

Figure 11.38 DC voltage and power control scheme of a VSC

11.3 Multi-Terminal HVDC [32]–[37]

11.3.1 Control Strategies of MMC-MTDC

To properly and efficiently control the VSC-Multi-Terminal HVDC systems, two control tasks should be considered: (1) DC voltage control and (2) AC-side auxiliary control. DC voltage control is used to stabilize DC grid operations. The DC voltage of each converter is related to the balance of active power and determines the power flow and sharing in the DC networks. On the other side, the VSC-based converter stations have the features of power decoupling and fast response, which could provide auxiliary support to the AC systems and potentially enhance performance and stability of the AC systems. The existing control strategies of the VSC-MTDC systems are applicable for the MMC-MTDC systems.

DC Voltage Control and Coordination Strategy

The commonly used control architecture of a VSC in an MTDC system is shown in Figure 11.38, which consists of inner and outer control loops and is also applicable for the MMC-based converter station. The DC voltage control is implemented by the outer control loop. Unlike the frequency of the AC system as a global control parameter, the DC voltage varies in the DC grid because of power-flow controlled by the difference between the bus voltages of the DC network. Based on this basic control architecture, various DC voltage control and coordination strategies have been developed.

Currently, there are primarily three methods for DC voltage control: master-slave control, voltage margin control, and voltage droop control. These methods can also be categorized as centralized and decentralized control methods. A brief introduction to these three DC voltage control methods is presented as follows:

Master-slave control: For this control, a master converter is selected and made responsible for controlling the voltage profile, whereas other converters control power. This method has been implemented in the Nan'ao five -terminal MTDC system. However, this method has instability issues, since only one master converter cannot ensure proper operations if that master converter fails. Additionally, the master converter must be connected to a very strong AC grid to ensure fast conditioning of the DC grid and avoid negative effects on the AC grid.

Voltage margin control: Voltage margin control is considered an improved version of master-slave voltage control, as shown in Figure 11.39. There are some reserved converters to ensure the capability of the DC voltage regulation when power limitation is exceeded. If the master converter is offline or reaches its power limitation, one of the reserved converters will immediately take over the role of DC voltage regulation. However, voltage margin control has similar instability issues to master-slave control because of only one regulating converter existing at one time. In worst case scenarios, master converter shifting will cause oscillation of the DC voltage and may lead to the collapse of the DC grid. Additionally, the dynamic response of the voltage margin control is slow because of its controller architecture.

Voltage droop control: Voltage droop control, as a decentralized strategy and as shown in Figure 11.40, is very different compared to the two aforementioned centralized control methods. In voltage droop control, DC voltage is controlled by multiple converters to maintain voltage stability and power balance. This approach originated from power-frequency droop control of the AC system. Compared to other methods, voltage droop control has higher reliability and does not cause oscillations.

Based on these three basic methods, some modified and improved strategies have been proposed. These emerging methods can be recognized as consisting of two categories. One category focuses on the design and optimization of the coefficients of voltage droop control, such as adaptive droop control for the available headroom of power limitation to improve stability, the integration of renewable generation and energy storage with trusted stability, and optimal DC power flow. The other category focuses on the modification and combination of these three basic methods for better performance.

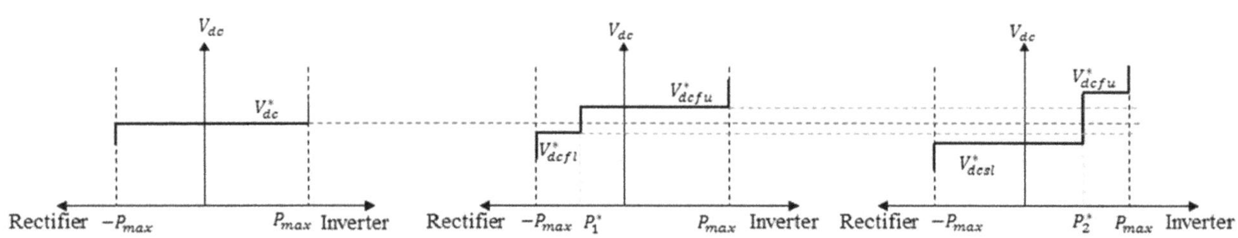

Figure 11.39 Characteristics of voltage margin control

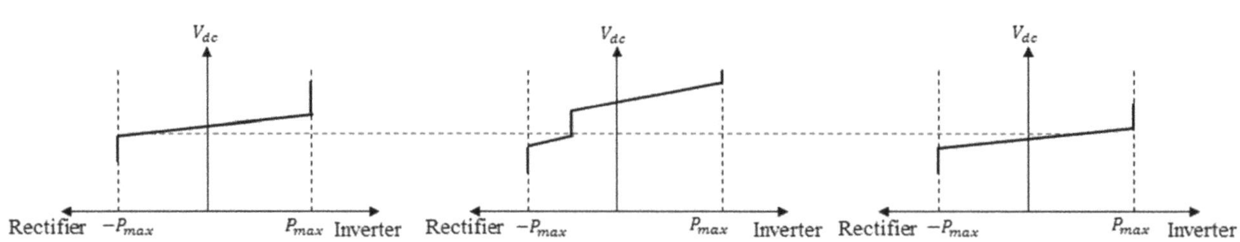

Figure 11.40 Characteristics of voltage droop control

11.3.2 Power Flow and Sharing Between Multiple Converters

The power flow and sharing of the VSC-MTDC systems are controlled by the DC voltage of each terminal, as shown in Figure 11.41. It is essential to consider the stability region, limitation, and optimal parameter determination of voltage control approaches to calculate the power flow and sharing.

11.3.3 Power Oscillation Damping and AC System Performance Enhancement

The MTDC systems can also offer additional control functions to improve system dynamic performance, such as power oscillation damping, transient stability, and fault recovery and subsynchronous damping enhancement.

Low-frequency interarea power oscillations are a common phenomenon in power systems and have caused a few wide-scale blackouts. Interarea oscillation is the main reason for power system separation and cascading failure. Neither the damper windings of the synchronous machines nor the modern digital electro-hydro control systems without global signal measurement can effectively attenuate interarea oscillations. Therefore, the attenuation of interarea power oscillations is important and has remained a challenge for a long time.

Power oscillation control can be integrated into the DC power control of the MTDC systems. The basic concept is to compensate for the oscillations of AC systems by exchanging extra power and modifying the DC power reference of Figure 11.38. The extra power utilized for damping the oscillation can be calculated based on the optimal control strategy with the following inputs: (1) generator rotor speed difference, (2) temporary frequency difference among converter stations when these stations are within the same AC network, and (3) instantaneous AC power flow at the converter stations.

Although the aforementioned control strategies for VSC-MTDC systems are applicable for MMC-MTDC systems as well, there is a lack of literature investigating the interaction of the internal dynamics of the MMCs, the DC networks, and the AC grids. Compared to the conventional two-level VSC, the MMC needs more complicated internal controllers to properly control the internal dynamics. Additionally, the internal energy storage of the MMC may provide a potentially enhanced capability to stabilize the AC systems as well as the DC networks.

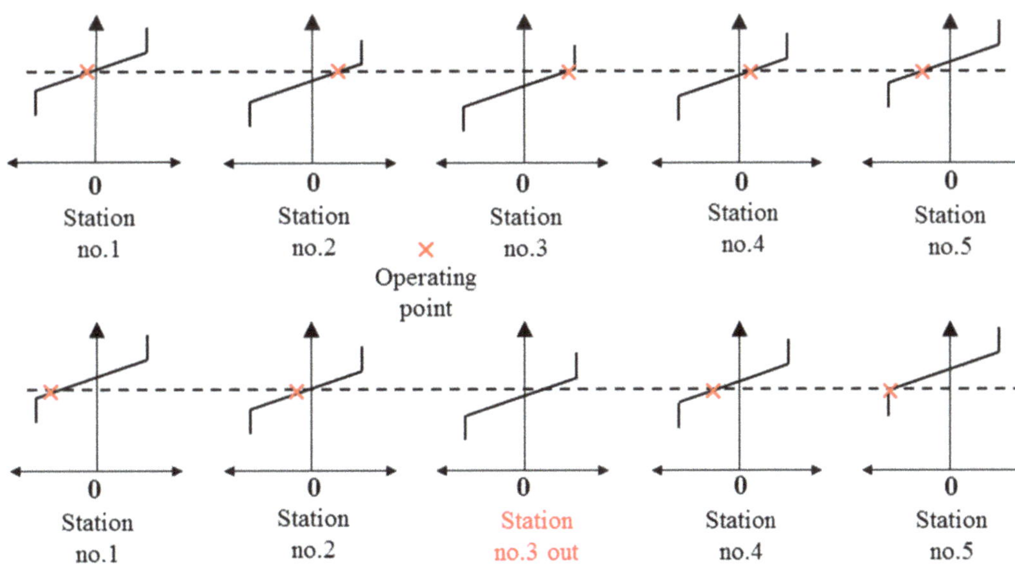

Figure 11.41 Characteristics of power sharing in a five-terminal MTDC system

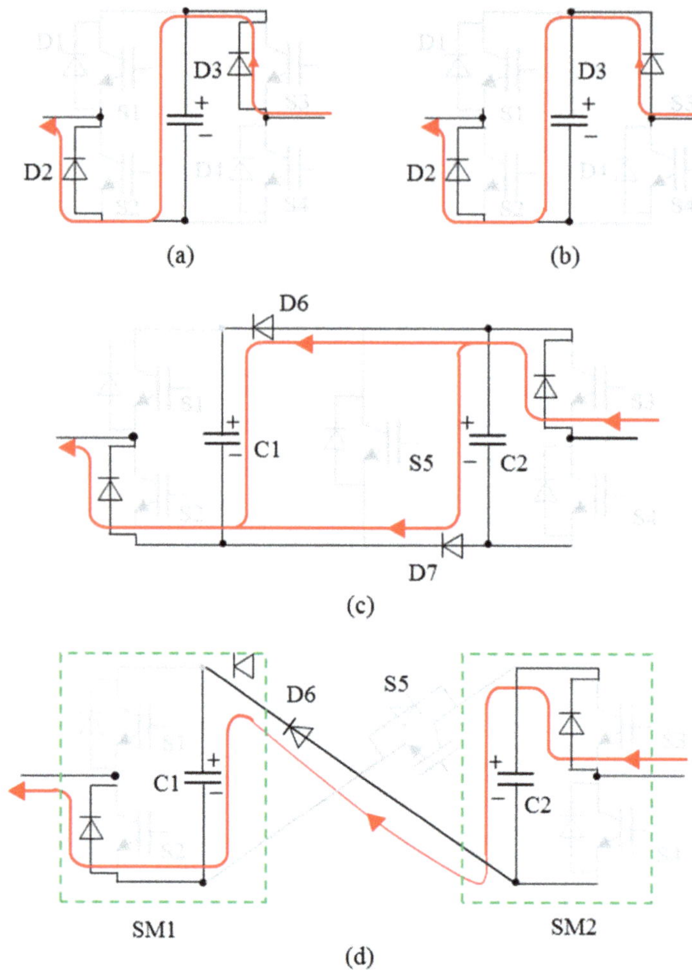

Figure 11.42 Fault current paths of various submodules (SMs): (a) full bridge (FB); (b) UFB; (c) CD; and (d) 3LCC SM circuits

However, although MMCs with fault-blocking SMs can block the DC fault current, the increased number of semiconductor switches leads to extra power losses. Figure 11.42 shows the fault current paths according to the kinds of SMs, Figure 11.43 shows the fault current path of the FB-MMC.

11.3.4 Coordination of the MMCs with the DC CBs

MMCs with fault-blocking SMs are not the most cost-effective solution because of their high cost and loss. To reliably isolate the faulty lines in the MTDC systems, DC CBs are still required. However, interrupting the DC fault current using only DC CBs is also not a cost-effective solution because of the large number of DC CBs required in meshed DC grids.

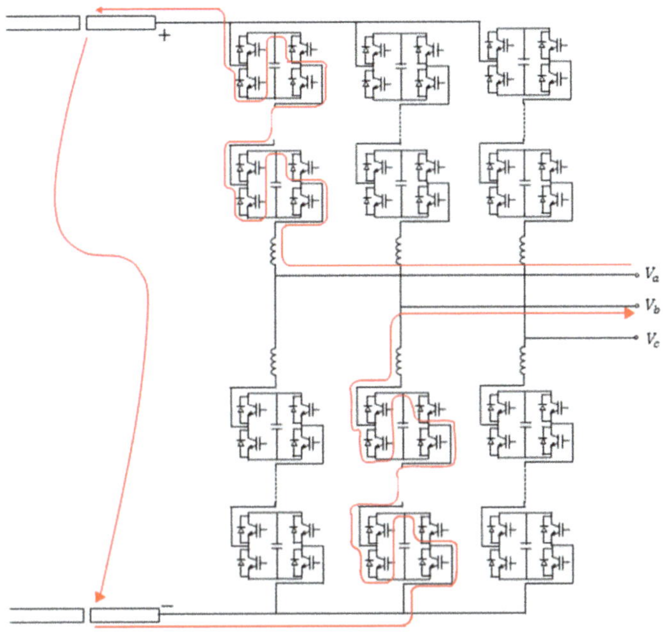

Figure 11.43 The fault current path of the FB-MMC

References

1. M. Ataollah, C. Adriano, L. Helder, S. Nuno, "A Review on HVDC Circuit Breakers" in proceeding of the 3rd Renewable Power Generation Conference (RPG 2014), Naples, Italy, 2014.
2. Franck, Christian M. "HVDC Circuit Breakers: A Review Identifying Future Research Needs." IEEE Trans. Power Delivery., vol. 26, pp. 998-1007, Jan. 2011.
3. J. Hafner and B. Jacobson, "Proactive hybrid HVDC breakers-A key innovation for reliable HVDC grids," in *Proceedings of the CIGRE Conference*, Bologna, Italy, Sep. 2011, pp. 1–8.
4. W. Grieshaber and L. Violleau, "Development and test of a 120 kV direct current circuit breaker," in *Proceedings of the GIGRE 2014*, Paris, 2014, pp. 1–11.
5. O. Cwikowski, et al "Operating DC circuit breakers with MMC," *IEEE Trans. Power Del.*, pp. 260–270, Feb. 2018. vol. 33, no. 1,
6. A. Hassanpoor, J. H¨afner, and B. Jacobson, "Technical assessment of load commutation switch in hybrid HVDC breaker," *IEEE Trans. Power Electron*, vol. 30, no. 10, pp. 5393–5400, Oct. 2015.
7. L. Chuanyue, et al., "The dc grid reliability and cost evaluation with Zhoushan five-terminal HVDC case study," in *Proceedings of 2015 50th International University Power Engineering Conference*, Stoke on Trent, UK, 2015, pp. 1–6.
8. W. Zhou, et al., "Development and test of a 200 kv full-bridge based hybrid HVDC breaker," in *Proceedings of 2015 17th European Conference on Power Electron and Applications*, Geneva, Switzerland, 2015, pp. 1–7.
9. R. Li, et al., "Hybrid cascaded modular multilevel converter with dc fault ride-through capability for the HVDC transmission system," *IEEE Trans. Power Del.*, vol. 30, no. 4, pp. 1853–1862, Aug. 2015.
10. V.-V. Nguyen, et al., "A novel topology of hybrid HVDC circuit breaker for VSC-HVDC Application." in Energies 2017, 10, 1675, pp. 1-15, Oct. 2017.
11. Chuanyue Li, etc, "Interlink Hybrid DC Circuit Breaker", IEEE Trans. on Industrial Electronics, Vol. 65, No. 11, Nov. 2018.
12. J. Hafner and B. Jacobson, "Proactive hybrid HVDC breakers—A key innovation for reliable HVDC grids," in *Proceedings of the Electric Power System of the Future—Integrating Supergrids and Microgrids International Symposium*, Bologna, Italy, pp. 1–9, 13–15 Sept. 2011.
13. G. Liu, et al., "Assembly HVDC breaker for HVDC grids with modular multilevel converters," *IEEE Trans. Power Electron*, vol. 32, pp. 931–941, 2017.
14. Han X. et al., "Transient characteristics under ground and short-circuit Faultss in a +-500 kV MMC-based HVDC system with hybrid DC circuit breakers," *IEEE Trans. Power Del.*, vol. 33, no. 3, pp. 1379–1387, June 2018.

15. N. Ahmed, et al., “Efficient modeling of an MMC-based multiterminal DC system employing hybrid HVDC breakers,” *IEEE Trans. Power Del.*, vol. 30, no. 4, pp. 1792–1801, Aug. 2015.
16. J. Häfner and B. Jacobson, “Device and method to break the current of a power transmission or distribution line and current limiting arrangement,” U.S. Patent 20 120 299 393, Nov. 29, 2012.
17. M. Callavik, et al., “The hybrid HVDC breaker—An innovation breakthrough enabling reliable HVDC grids,” *ABB Grid Syst.Tech Rep.*, pp. 1–10, Nov. 2012.
18. Y. Xue and Z. Xu, “On the bipolar MMC-HVDC topology suitable for bulk power overhead line transmission: Configuration, control, and dc fault analysis,” *IEEE Trans. Power Del.*, vol. 29, no. 6, pp. 2420–2429, Dec. 2014.
19. A. A. S. Emhemed and G. M. Burt, “An advanced protection scheme for enabling an LVDC last mile distribution network,” *IEEE Trans. Smart Grid.*, vol. 5, no. 5, pp. 2602–2609, Sep. 2014.
20. J. David Paezet al., “Overview of DC-DC converters dedicated to HVdc grids,” *IEEE Trans. Power Del.*, vol. 34, no. 1, pp. 119–128, Feb. 2019.
21. D. Jovcic, et al., “Feasibility of DC transmission networks,” in *Proceedings of the 2nd IEEE PES International Conference and Exhibition Innovative Smart Grid Technologies*, Manchester, UK, 2011, pp. 1–8.
22. E. highway2050 project, E. highway2050 project, “Europe's future secure and sustainable electricity infrastructure. e-Highway 2050 project results,” Nov. 2015.
23. N. Ahmed, et al., “HVDC SuperGrids with modular multilevel converters—The power transmission backbone of the future,” in *Proceeding of the 9th International Multi-Conference System Signals Devices*, Chemnitz, Germany, 2012, pp. 1–7.
24. J. Echeverrta, et al., “Multi-modular cascaded DC-DC converter for HVDC grid connection of large-scale photovoltaic power systems,” in *Proceedings of the 39th Annual Conference of the Industrial Electronics Society*, Vienna, Austria, 2013, pp. 6999–7005.
25. I. Etxeberria-Otadui, et al., “Medium voltage power electronics devices for distribution grids,” in *Advanced Solutions in Power Systems: HVDC, FACTS, and Artificial Intelligence*, Hoboken, NJ: Wiley, 2016, pp. 681–718.
26. J. Zhang, Z. Wang, and S. Shao, “A three-phase modular multilevel DC-DC converter for power electronic transformer applications,” IEEE Journal of Emerging and Selected Topics in Power Electronics, vol. 5, no. 1, March 2017, pp. 140–150.
27. C. P. Dick, A. Konig, and R. W. De Doncker, “Comparison of three-phase DC-DC converters vs. single-phase DC-DC converters,” in *Proceedings of the 7th International Conference on Power Electronics Drive Systems*, Bangkok, Thailand, Nov. 2007, pp. 217–224.
28. J. Xue, et al., “Single-phase vs. three-phase high density power transformers,” in *Proceedings of the IEEE Energy Conversion Congress and Exposition*, Atlanta, GA, USA, Sep. 2010, pp. 4368–4375.
29. H. Baars N. et al “A 80-kW isolated DC–DC converter for railway applications,” *IEEE Trans. Power Electron.*, vol. 30, no. 12, pp. 6639–6647, Dec. 2015.
30. H. van Hoek, M. Neubert, and R.W. De Doncker, “Enhanced modulation strategy for a three-phase dual active bridge—Boosting efficiency of an electric vehicle converter,” *IEEE Trans. Power Electron.*, vol. 28, no. 12, pp. 5499–5507, Dec. 2013.
31. H. Qin and J. W. Kimball, “Closed-loop control of DC–DC dual-active-bridge converters driving single-phase inverters,” *IEEE Trans. Power Electron.*, vol. 29, no. 2, pp. 1006–1017, Feb. 2014.
32. L. Zhang, et al., ‘Modeling, control and protectino of modular multilevel converter-based multi-terminal HVDC systems: A review,” *CSEE J Power Energy Syst.*, vol. 3, no. 4, pp. 340–352, Dec. 2017.
33. N. R. Chaudhuri, et al., “Modeling and stability analysis of MTDC grids for offshore wind farms: A case study on the North Sea benchmark system,” in *Proceedings of 2011 IEEE Power and Energy Society General Meeting*, Detroit, MI, USA, 2011, pp. 1–7.
34. D. Van Hertem and M. Ghandhari, “Multi-terminal VSC HVDC for the European supergrid: Obstacles,” Renew. Sust. Energ. Rev., vol. 14, no. 9, pp. 3156– 3163, Dec. 2010.
35. S. Rodrigues, et al., “Optimal power flow control of VSC-based multiterminal DC network for offshore wind integration in the North Sea,” *IEEE J. Emerg.Sel. Topics Power Electron.*, vol. 1, no. 4, pp. 260–268, Dec. 2013.
36. T. An, G. F. Tang, and W. N. Wang, “Research and application on multiterminal and DC grids based on VSC-HVDC technology in China,” *High Voltage*, vol. 2, no. 1, pp. 1–10, Mar. 2017.
37. A. Nami, et al., “Modular multilevel converters for HVDC applications: Review on converter cells and functionalities,” *IEEE Trans. Power Electron.*, vol. 30, no. 1, pp. 18–36, Jan. 2015.

12

Super Grid and the Future of HVDC

The development of electric power supply began more than 100 years ago. Residential areas and neighboring establishments were at first supplied with DC via short lines. At the end of the 19th century, AC transmission was introduced, using higher voltages to transmit power from remote power stations to consumers.

In Europe, 400 kV became the highest voltage level, in the Far East countries it was mostly 550 kV, and in America 550 and 765 kV. The 1150 kV voltage level was anticipated in some countries in the past, and some test lines have already been built.

Because of an increased demand for energy, as well as the construction of new generation plants, first built close and then at remote locations from the load centers, the size and complexity of power systems all over the world have grown. Power systems have been extended through the application of interconnections to neighboring systems in order to achieve technical and economic advantages. Large systems covering parts of or even whole continents came into existence to gain well-known advantages such as the possibility of using larger and more economical power plants, the reduction of reserve capacity in the systems, the utilization of the most efficient energy resources, and, potentially, the achievement of an increase in system reliability.

In the future of liberalized power markets, the following advantages will become even more important: pooling large power generation stations, sharing spinning reserve and using the most economic energy resources, and considering ecological constraints, building large nuclear and hydro power stations at suitable locations, harvesting solar energy from desert areas, and embedding big offshore wind farms.

It is, however, a crucial issue that, as the size of the interconnected systems increases (Figure 12.1), the advantages diminish. There are both technical and economical limitations to the interconnections if the energy has to be transmitted over extremely long distances through interconnected synchronous AC systems. These limitations are related to problems with low-frequency inter-area oscillations (Figure 12.1a), voltage quality, and load flow. This is the case in the UCTE system (the Union for the Coordination of Transmission of Electricity), where the 400 kV voltage level is in fact too low for large cross-border and inter-area power exchange. Bottlenecks have already been spotted (Figure 12.2) and, for an increase in power transfer, advanced solutions must be applied.

In deregulated markets, the loading of existing power systems will further increase, leading to bottlenecks and reliability problems. System enhancement will be essential to balance the load flow and to get more power out of the existing grid. Large blackouts in America and Europe confirmed clearly that the favorable close electrical coupling of the neighboring systems might also include the risk of uncontrollable cascading effects in large and heavily loaded synchronous AC systems.

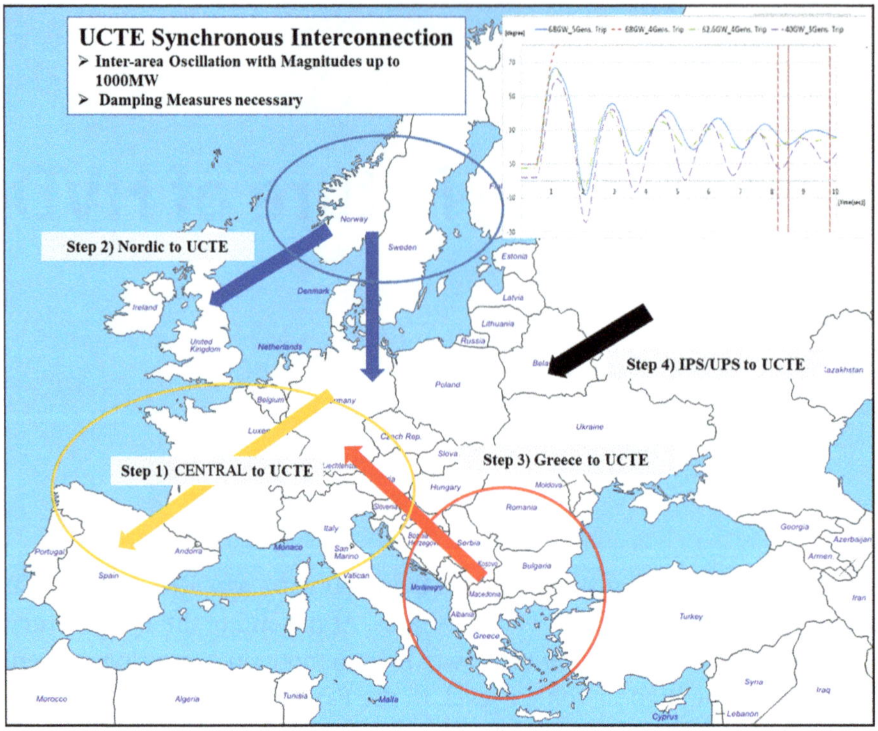

Figure 12.1 UCTE–Steps (a) and (b) for interconnection of Zones 1, 2, and further options (c)

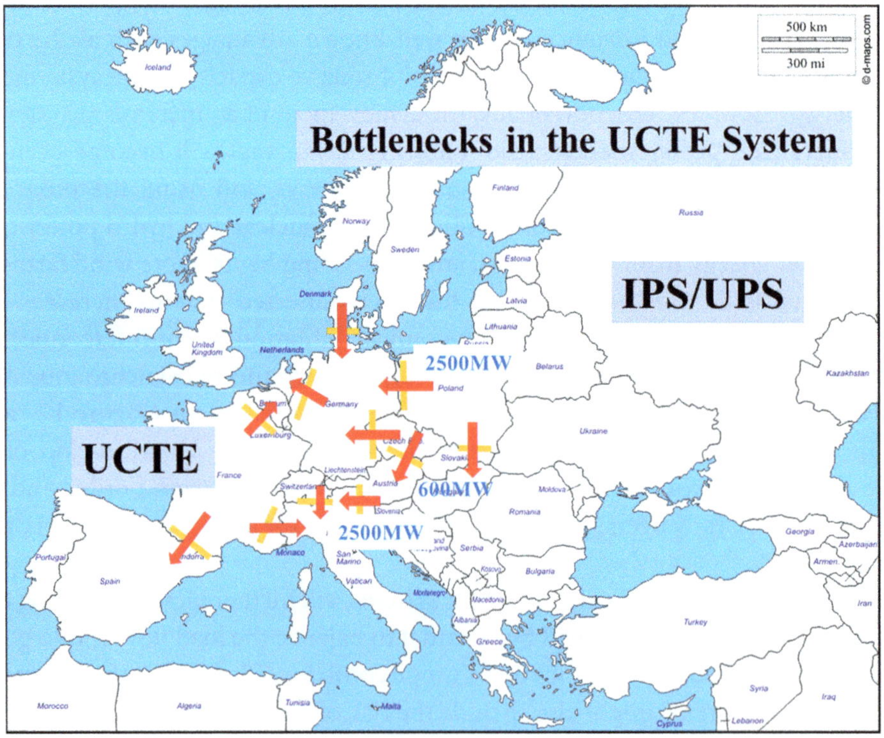

Figure 12.2 European power systems—Bottlenecks in UCTE

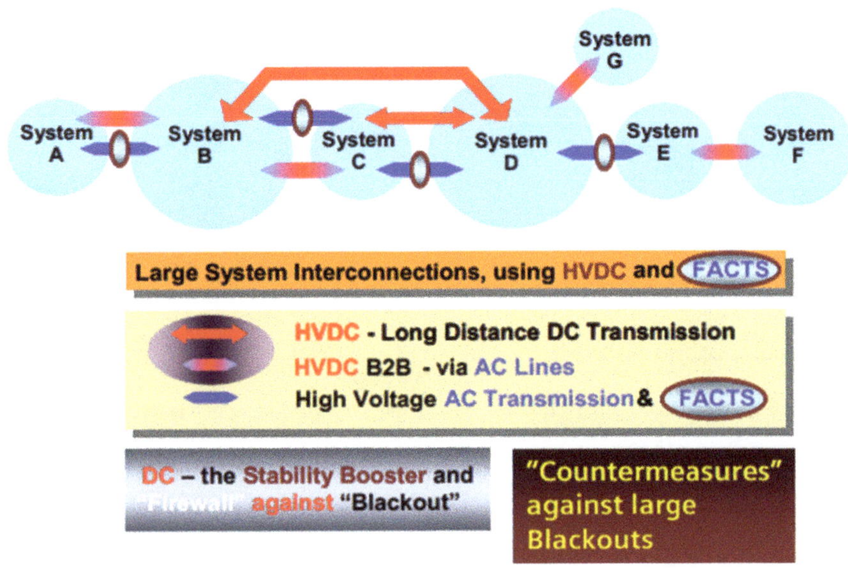

Figure 12.3 Large power system interconnections—Benefits of hybrid solutions

Therefore, strategies for the development of large power systems point clearly in the direction of hybrid transmissions consisting of DC and AC interconnections, including FACTS (Flexible AC Transmission System). Hybrid interconnected systems such as these offer significant advantages in terms of technology, cost savings, and reliability.

Figure 12.3 shows schematically a hybrid system with high-voltage direct current (HVDC) and FACTS. Power exchange in the neighboring areas of interconnected systems can be achieved by means of AC links, preferably with FACTS to increase transmission capacity and stability. The transmission of large power blocks over long distances should, however, be realized by means of HVDC links linked directly to locations at which power demand exists.

12.1 Super Grid and HVDC [1]–[17]

"A Super Grid" can be defined as a transmission solution that allows the massive integration of renewable energy sources and the connections of the existing grids while offering additional control. Also, **"Smart Grids"** can provide the vision and enhancement strategy for future electricity networks shown in Figure 12.4.

Features of a future "Smart Grid" such as this can be outlined as follows:

- **Flexible:** Fulfilling customers' needs while responding to the changes and challenges ahead
- **Accessible:** Granting connection access to all network users, particularly for RES (Renewable Energy Source) and high-efficiency local generation with zero or low carbon emissions
- **Reliable:** Assuring and improving security and quality of supply
- **Economic:** Providing the best value through innovation, efficient energy management, and "level playing field" competition and regulation

It is worthwhile to mention that the smart grid vision is applicable in the same way to system developments in other regions of the world. The key to achieving a Smart Grid performance is the use of HVDC using power electronics.

For interconnections of super grids and smart grids, innovative solutions to avoid congestion and improve stability will be essential. HVDC provides the features necessary to avoid technical problems in the power

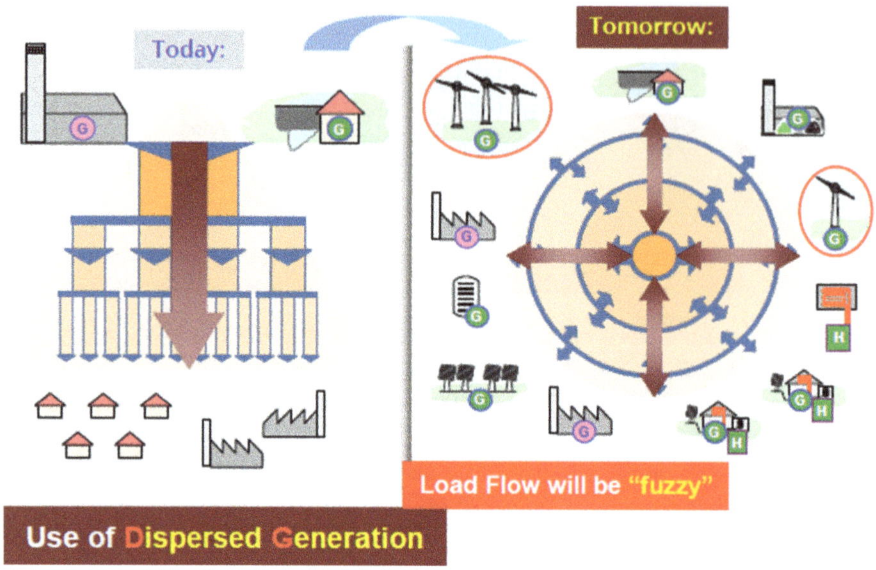

Figure 12.4 Regenerative energy sources and dispersed generation

systems. It also increases the transmission capacity and system stability very efficiently and helps prevent cascading disturbances.

HVDC can also be applied as a hybrid AC–DC solution in synchronous AC systems either as a **back-to-back** (B2B) for grid power flow control (elimination of congestion and loop flows) or as a long-distance **point-to-point** transmission.

Transmission Expansion Planning (TEP) in AC Systems: As shown in Figure 12.5, several HVDC systems have already been deployed in the North Sea, and several others are in the planning or commissioning phase. As can be inferred from their different voltage and power levels, years of commissioning, and technology used (Table 12.1), so far, they have been mostly defined and implemented independently from each other.

The gradual growth of the **North Sea Offshore Grid (NSOG)** from the interconnection of such preexisting point-to-point HVDC links will require expansion strategies that are substantially different from those traditionally applied in AC transmission systems. The main goal of **TEP** is to determine the installation plan for new facilities (i.e., type, number, location, and scheduled year) to meet the expected power demand while minimizing the cost and respecting technical, financial, and reliability constraints.

Based on the time horizon, TEP can be classified as ***static*** if the solution assumes that all the additions/modifications are implemented at the same time, whereas it is considered ***dynamic*** if subsequent time steps for the recommended expansion investments are also identified. TEP approaches also differ depending on input data uncertainty: examples of different approaches are the ***deterministic, non-deterministic,*** and ***solution methods***.

- Non-deterministic TEP captures the variety of operating conditions that the system may undergo (e.g., variability of generation and consumption input data), whereas deterministic TEP only relies on fixed scenarios.
- Solution methods can be based on ***mathematical optimization***, where the problem is formulated in terms of an objective function subject to a set of constraints, or *heuristic* approaches relying on the step-by-step solution of the TEP according to a predefined set of rules (evaluating different alternatives with or without user interaction). *Meta-heuristic* algorithms are also commonly used to integrate the benefits of the two previous methods. Comprehensive reviews of traditional approaches to TEP in AC systems are widely available.

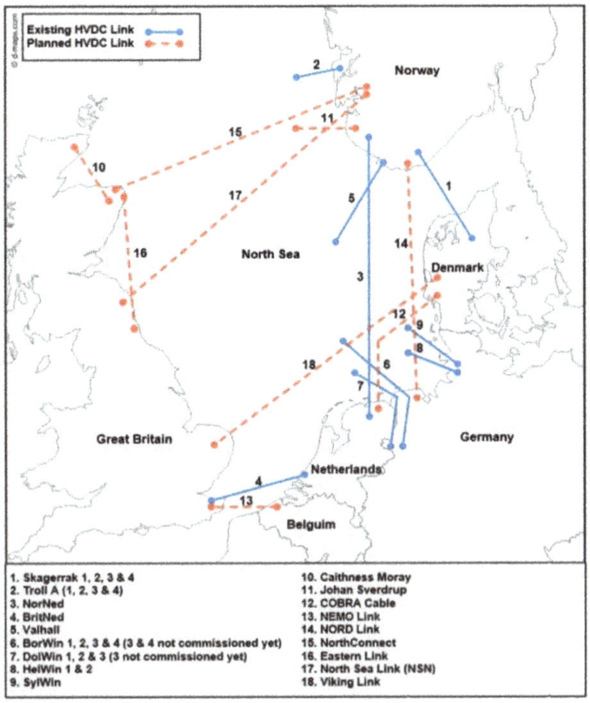

Figure 12.5 Existing and planned HVDC interconnections in the North Sea

Table 12.1 The North Sea HVDC Links

No.	Name	Country	Year of Commissioning	Power (MW)	DC Voltage (kV)	Transmission Length(km)	Converter Type
1	Skagmak 1&2	Norway – Denmark	1976 – 1977	500	250	113 (OHL) 127 (Submarine)	HVDC Classic
1	Skagmak 3	Norway – Denmark	1993	500	350	113 (OHL), 127 (Submarine)	HVDC Classic
1	Skagmak 4	Norway-Denmark	2014	700	500	104 (Under-ground) 140 (Submarine)	HVDC Light
2	Troll A (1&2)	Norway	2005	88		4 × 70 (Submarine)	HVDC Light
2	Troll A (3&4)	Norway	2015	100	±60	4 × 70 (Submarine)	HVDC Light
3	NorNed	Norway – Netherlands	2008	700	±450	2 × 580 (Submarine)	HVDC Classic
4	BritNed	Netherlands – GreatBri-tain	2011	1000	±450	260 (Submarine)	LCC

(continued)

Table 12.1 The North Sea HVDC Links *(continued)*

No.	Name	Country	Year of Commissioning	Power (MW)	DC Voltage (kV)	Transmission Length(km)	Converter Type
5	Valhall	Norway	2011	78	150	292 (Submarine)	HVDC Light
6	BorWin1	Germany	2015	400	±150	2 × 75 (Underground), 2 × 125 (Submarine)	HVDC Light
6	BorWin2	Germany	2015	800	±300	75 (Underground), 125 (Submarine)	HVDC PLUS (VSC MMC)
6	BorWin3	Germany	2019	900	±320	30 (Underground), 130 (Submarine)	HVDC PLUS (VSC MMC)
6	BorWin4	Germany	2020	900	600	2 × 172 (Underground), 2 × 122.4 (Submarine)	N.A.
7	DolWin1	Germany	2015	800	±320	2 × 90 (Underground), 2 × 75 (Submarine)	HVDC Light
7	DolWin2	Germany	2016	916	±320	2 × 90 (Underground), 2 × 45 (Submarine)	HVDC Light
7	DolWin3	Germany	2017	900	±320	79 (Underground), 83 (Submarine)	N.A.
8	HelWin1	Germany	2015	576	±250	45 (Underground), 85 (Submarine)	HVDC PLUS (VSC MMC)
8	HelWin2	Germany	2015	690	±320	45 (Underground), 85 (Submarine)	VSC
9	Sy1Win1	Germany	2015	864	±320	45 (Underground), 160 (Submarine)	VSC
10	Caithness – Moray	Great Britain	2018	1200	±320	160	HVDC Light
11	Johan Sverdrup	Norway	2019	100	±80	2 × 200 (Submarine)	HVDC Light
12	COBRA Cable	Denmark – Netherland	2019	700	±320	26 (Underground) 299 (Submarine)	VSC

(continued)

Table 12.1 The North Sea HVDC Links *(continued)*

13	NEMO Link	Belgium – Great Britain	2019	1000	±400	10 (Underground), 130 (Submarine)	LCC
14	NORD Link	Norway – Germany	2019	1400	±500	107 (Underground), 516 (Submarine)	HVDC Light
15	NorthConnect	Norway – Great Britain	2020	1400	±500	570 (Submarine)	N.A.
16	Eastern Link	Great Britain	2020	2000	400	305	N.A.
17	North Sea Link (NSN)	Norway – Great Britain	2021	1400	±525	730 (Submarine)	HVDC Light
18	Viking Link	Denmark – Great Britain	2022	1400	400	110 (Underground), 650 (Submarine)	VSC

Abbreviations: OHL: Overhead Line; LCC: Line Commutated Converter; VSC: Voltage Source Converter; MMC: Modular Multilevel Converter; N.A.: Not Available

In recent years, the deregulation of power markets determined an important change in expansion approaches. Although with centralized markets, the TEP's main goal is to meet the expected power demand while minimizing investment and operational costs, in the liberalized market where many players are active, different objectives and constraints are being considered.

TEP in AC/DC systems: The increased use of HVDC technology for bulk power transmission and offshore applications produced the need for a different approach to TEP. To properly quantify the benefits of HVDC interconnections and exploit their flexibility, it is crucial for expansion planning strategies to capture the HVDC operational characteristics that are substantially different from those of traditional AC systems.

Among the aspects that were most widely investigated in recent literature, there is the selection of the most convenient technological transmission alternative (HVDC vs. HVAC) for grid reinforcement, as well as the analyses of hybrid AC/DC grids, in particular with respect to interactions between the newly introduced HVDC system and preexisting HVAC grid, minimizing the cost of electricity by increasing inter-zonal HVAC and HVDC links for the interconnection of asynchronous networks using a heuristic approach based on linear programming optimization.

TEP in HVDC offshore grids: With the increase in distance from the shore and the installed capacity of offshore wind farms, HVDC transmission through submarine cables is the only viable option. The main challenge is not choosing between HVAC and HVDC, but proper planning of HVDC expansion in terms of topology, transmission capacity, and routing. Thus, in the application of HVDC transmission for offshore applications, planning objectives and constraints can be significantly different from those of onshore applications. However, the main objective is, in most cases, the maximization of social welfare while minimizing investment and operational costs. Although the use of HVDC in land-based applications mostly relates to grid reinforcement, offshore HVDC is normally justified by the need to maximize the exploitation of offshore renewables. Hence, the correct integration of information related to generation variability (because of the intermittent nature of the wind resource) becomes a mandatory requirement of TEP tools for offshore applications.

1. ***Role of HVDC topology optimization:*** It is worth noting that topology optimization is mostly the result of economic considerations. However, the technical challenges associated with the practical deployment of multi-terminal and meshed HVDC networks have triggered significant research on what the best topology (e.g., radial, ring, multi-terminal, and meshed) for the NSOG (North Seas Offshore Grid) will be from an operational standpoint. Such studies are often based on pure dynamic simulation techniques (and not optimization) and use different performance indicators, such as evaluation of steady-state losses and fault analyses, grid congestions and wind utilization, and the cost-benefit analysis of different topologies.

An actual optimization is carried out in order to identify the most economical topology, considering both HVAC and HVDC alternatives and optimizing the grid voltage level.

2. ***Role of HVDC controllability:*** The possibility of controlling the power flow through a transmission line is the main difference when comparing HVDC to HVAC systems. This implies that the HVDC grid cannot be fully captured by an impedance representation, as in traditional HVAC planning approaches, and that the controllable power flows have become additional variables in the optimization.

For this reason, several studies have been dedicated to optimizing the operating conditions of the NSOG, with optimal power flow from different perspectives, including steady-state optimization for the minimization of transmission losses and the maximization of social welfare, or the avoidance of critical conditions such as overcurrent and under-voltages. In other cases, such analyses have focused on additional services that can be provided by HVDC links under transient and contingency scenarios, such as (primary and secondary) reserve provision and increase in rotor angle stability.

Although such contributions do not explicitly target grid transmission expansion, the possibility of HVDC technology including controllers acting on different parameters and time scales (from slow controllers for frequency regulation to fast nonlinear controllers to counteract transient stability swings) calls for better integration of such features into expansion planning tools. A closer relation between ***strategic*** planning (targeting mid- and long-term optimization) and ***operational*** planning (working on a time horizon up to 5 years) is needed.

Use of HVDC technology for grid interconnection: The transmission of power in the range of a few hundred MW was introduced and continuously increased. Transmission ratings of 3 GW over large distances with only one bipolar DC line are state-of-the-art in many grids today. The world's first 800-kV DC project in China has a transmission rating of 5 GW, and further projects with 6 GW or even higher are at the planning stage.

In general, for transmission distances above 700 km, DC transmission is more economical than AC transmission (≥1000 MW). Power transmission of up to 600–800 MW over distances of about 300 km has already been achieved with submarine cables, and cable transmission lengths of up to about 1,000 km are at the planning stage. Because of these developments, HVDC became a mature and reliable technology. During the development of HVDC, different kinds of applications were carried out.

They are shown in Figure 12.6. The first commercial applications were HVDC sea cable transmissions, because using AC cable transmission over more than 80–120 km is technically not feasible because of reactive power limitations. Then, long-distance HVDC transmissions with overhead lines were built, as they are more economical than transmissions with AC lines. To interconnect systems operating at different frequencies, B2B schemes were applied. B2B converters can also be connected to long AC lines (Figure 12.6a). A further application of HVDC transmission that is very important for the future is its integration into the complex interconnected AC system (Figure 12.6c). The reasons for these hybrid solutions are lower transmission costs as well as the possibility of bypassing heavily loaded AC systems.

Typical configurations of HVDC are depicted in Figure 12.7. The major benefit of both the B2B and long-distance transmission varieties of the HVDC is their incorporated ability to block fault currents that serves as an automatic firewall for blackout prevention in case of cascading events, which is not possible with synchronous AC links.

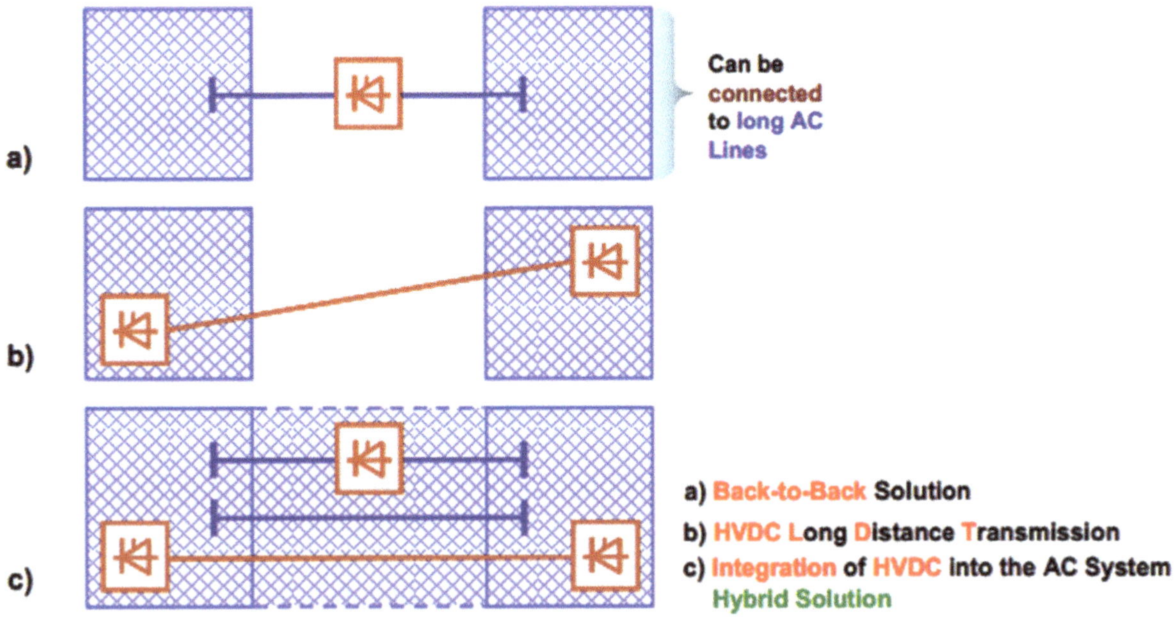

Figure 12.6 Types of HVDC transmissions

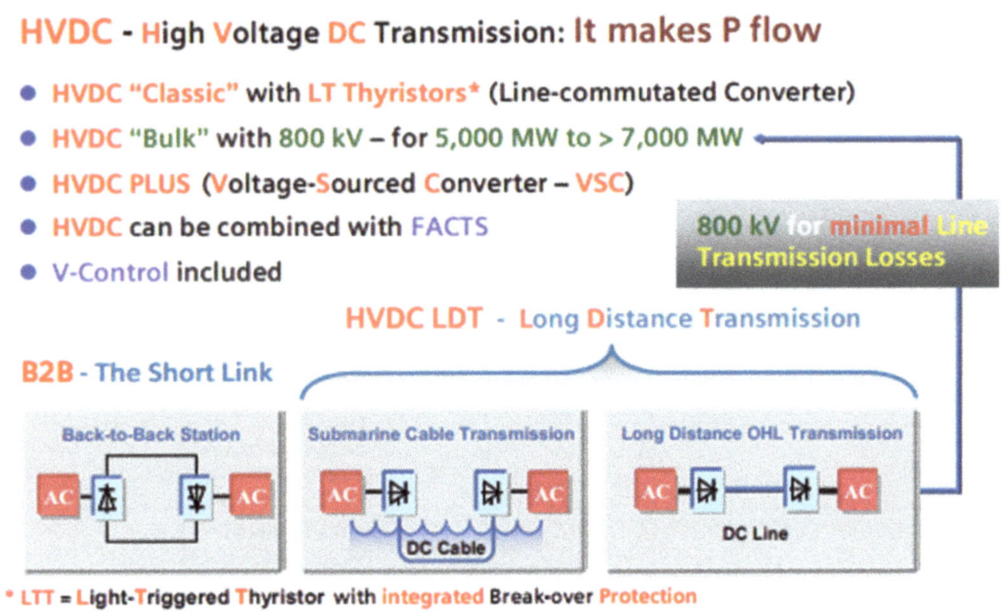

Figure 12.7 HVDC configurations and technologies

Security of supply—Blackouts: Figure 12.8 shows a summary of the United States–Canada Blackout on August 14, 2003. In Figure 12.8(a), a view of the affected area is given, and the satellite photos show the situation before and after the events. The figure indicates that Québec's system in Canada was not affected because of its DC interconnections to the United States, whereas Ontario (synchronous interconnection) fully "joined" the cascade. The reasons why Québec "survived" the blackout are very clear:

Figure 12.8(b) depicts the results of a study on the related transmission systems, published in May 2002. It can be seen that load flow in the system does not match the design criteria well, as shown by the "hot lines," represented in red. In the upper right-hand corner of the figure, one of the later blackout events with "giant" loop flows is attached. It occurred in the same area that had been under investigation 1 year before.

Figure 12.9 shows that the probability of large blackouts is much higher than calculated by mathematical modeling, particularly when the related number of power outages is very large. The reasons for this result are indicated in the figure. This means that, once the cascading sequence is started, it is mostly difficult or even impossible to stop it unless the direct causes are eliminated by means of investments in the grid and by the enhanced training of the system operators for better handling of emergency situations.

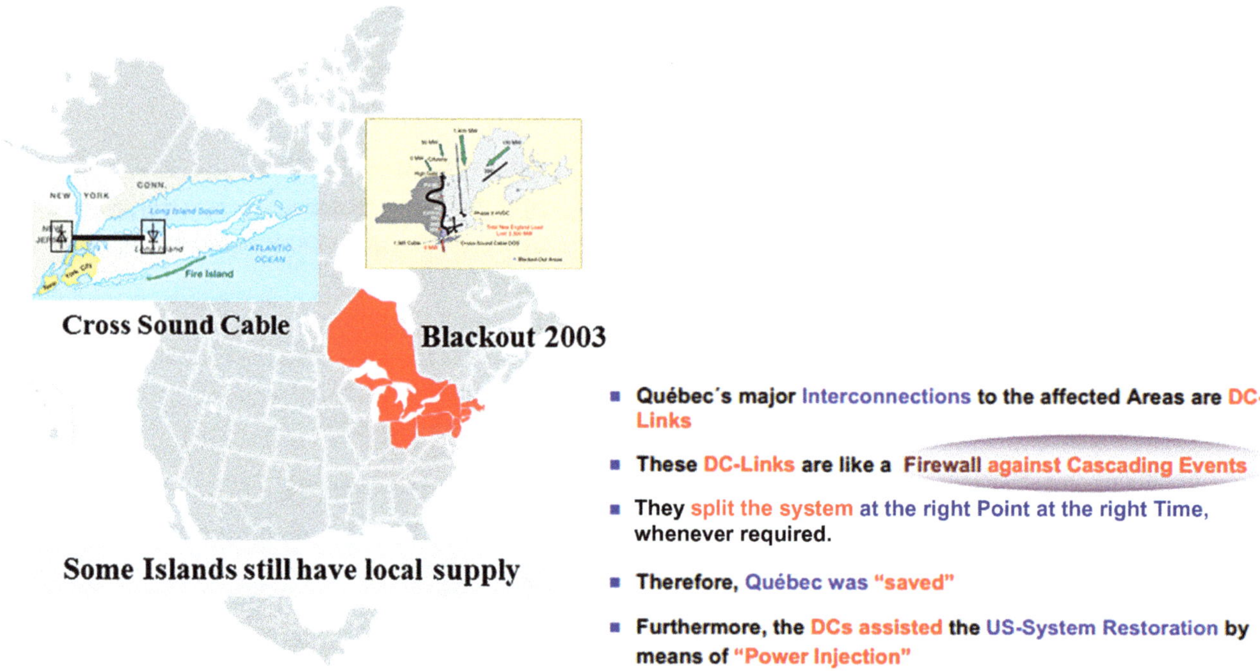

Figure 12.8 Blackouts 2003—The example of the United States: (a) The blackout area and a satellite view; (b) congestion and loop flows—forecasting studies and cascading events

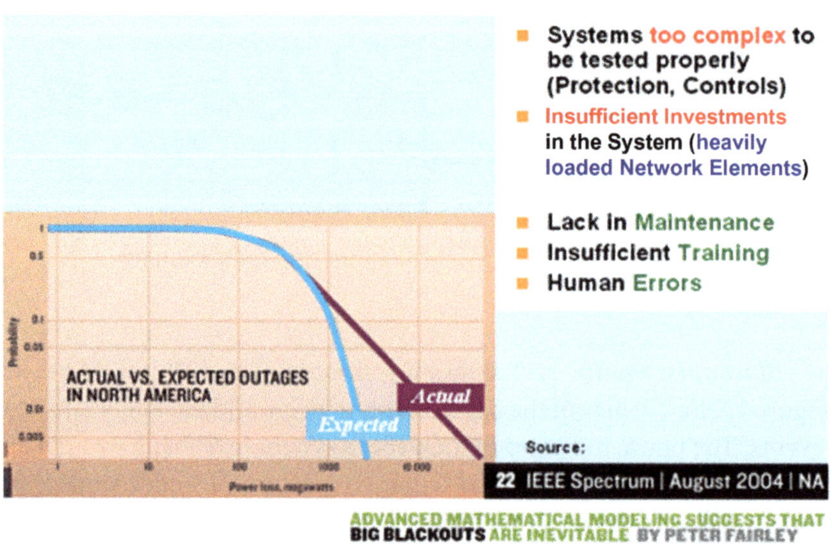

Figure 12.9 Reasons for high probability of large blackouts

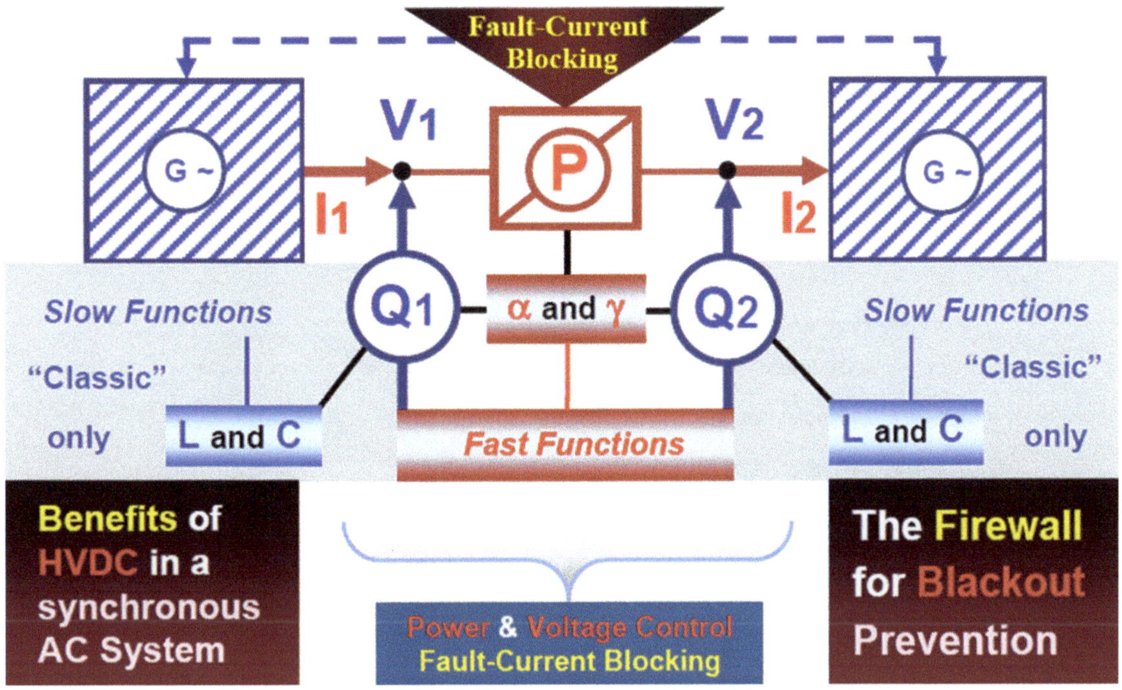

Figure 12.10 Benefits roles of HVDC

12.2 HVDC Systems in the World

12.2.1 HVDC Systems in Korea

The role of HVDC in power system expansion in the Republic of Korea: As South Korea's power network is isolated from those of neighboring countries, the electricity infrastructure in South Korea requires a high level of preparation before future grid technologies can be adopted. System plans must provide a practical solution for power system expansion that complies with strict reliability performance criteria, including contingencies (faults) on the 765-kV transmission system. System improvements also must provide a practical solution to the problem of how power can be exchanged between the mainland and Jeju Island, which is viewed as a "carbon-free" area because of the expected development of wind farms. Addressing these concerns led to the development of two upcoming HVDC projects with unique objectives.

The first project is a gigawatt-scale line-commutated converter high-voltage direct current (LCC-HVDC) system, which connects massive nuclear complexes with a metropolitan area. This project will replace an originally planned 765-kV transmission project that faced strong public opposition because of siting concerns. The HVDC project design offers environmental, public policy, and technical advantages while also meeting reliability performance standards.

The second project consists of a modular multi-level converter (MMC)-based voltage source converter (VSC)-HVDC connection between the mainland and Jeju Island. This project will deliver bidirectional power flow and coordinate with two existing LCC-HVDC ties.

The two new HVDC projects require complicated novel operating strategies. Therefore, comprehensive feasibility studies must be conducted that surpass the scope of feasibility studies commonly conducted for conventional technology. Many advanced planning and simulation methods and tools were considered by the Korea Electric Power Corporation (KEPCO) and partner universities in order to efficiently and accurately achieve practical and credible solutions.

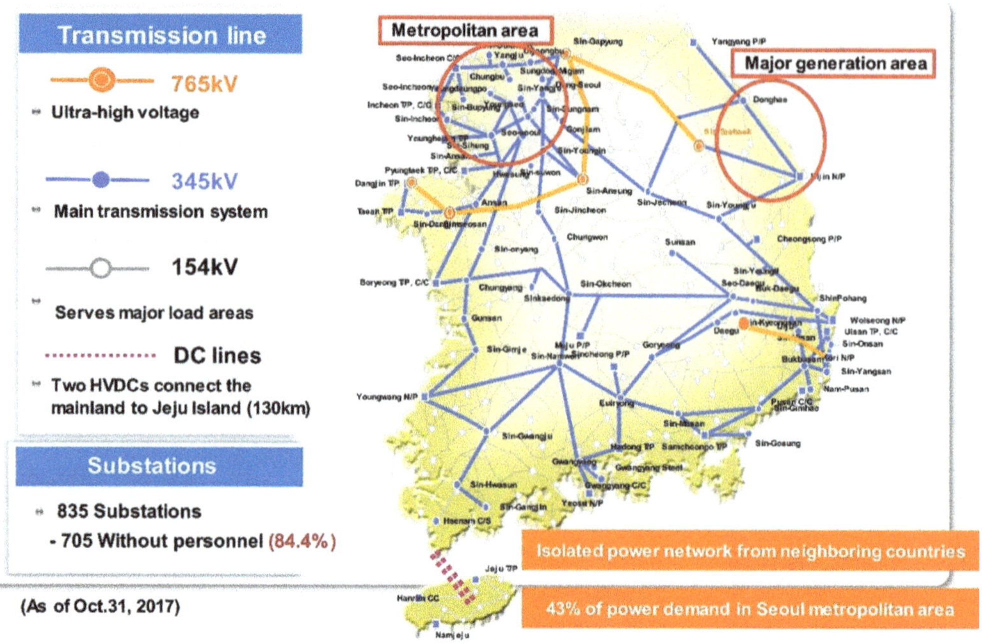

Figure 12.11 Electric power system of South Korea

Unique power system infrastructure of the Republic of Korea: The electric power system of the Republic of Korea shown in Figure 12.11 is totally isolated from neighboring countries because of political and geographical issues. The electricity consumption per capita in the Republic of Korea was 10,320 kWh/capita in 2017, which exceeds that of most of the European countries that have relatively larger land areas (see Figure 12.12(a)). Because of the large amount of power generation and consumption relative to the population (see Figure 12.12(b)), the generation and transmission facilities in the Republic of Korea are densely packed, with few opportunities for new greenfield sites. In addition, the ratio of peak load relative to generation capacity is high in the Republic of Korea, as shown in Figure 12.12(c).

The major generation complex (on the east coast) and load center (around the capital, Seoul, in the northwest of South Korea) are approximately 350 km apart, and are connected by several 345-kV lines and one 765-kV double-circuit, as shown in Figure 12.14. KEPCO plans to interconnect more bulk power plants in the major generation area, such as nuclear and coal-fired generation plants in the east, which will increase the regional generation capacity from 7 to 22 GW by 2022.

Currently, even when a power plant is constructed, it is difficult to build new transmission facilities because of social acceptance. The contingency study shows that a severe disturbance (*N-2*) resulting in the loss of two of the existing 765-kV lines would cause instability without tripping at least five nuclear generators (totaling approximately 6000 MW) through activation of a special protection scheme (SPS). However, the loss of 6000-MW generation would be too severe and could subsequently cause a blackout without grid reinforcements.

The introduction of more HVAC lines to the system was planned to prevent catastrophic consequences and comply with the KEPCO reliability performance criteria which states that no more than two nuclear generator units (totaling 3000 MW) may trip offline during the most critical contingency without load shedding. However, the HVAC transmission expansion option was strongly opposed by the public because of societal and environmental issues related to building new transmission towers and lines. As a result, KEPCO sought alternative solutions for expanding the transmission capacity that met reliability constraints and satisfied social requirements.

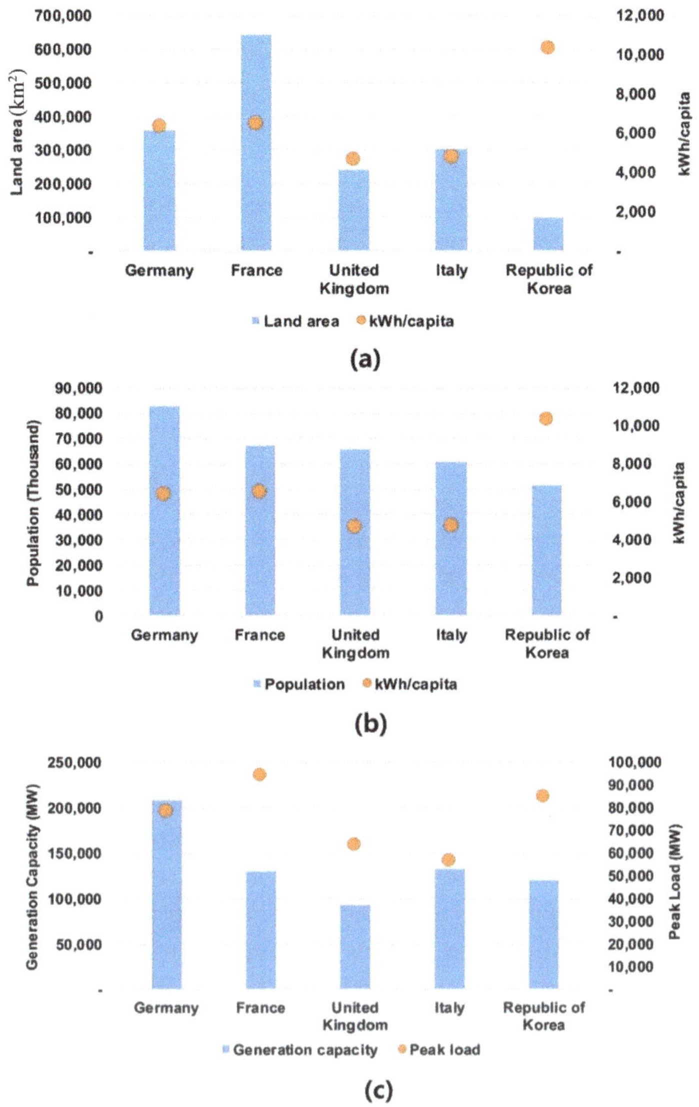

Figure 12.12 Comparison of (a) land area and kWh/capita; (b) population and kWh/capita; and (c) generation capacity and peak load of different countries

Figure 12.13 Photo of 765-kV double circuit line in Korea

Motivations for Alternatives: KEPCO's grid planners developed solutions that added 345-kV transmission facilities with the understanding that contingencies on 345-kV facilities would have less adverse effects on the system than 765-kV contingencies. However, study results showed that at least 24 additional 345-kV lines connecting the east and west of the country would be required to meet the reliability performance criteria for contingencies on the 765-kV lines. The problems with the 345-kV system expansion plans include the large number of lines needed; voltage stability concerns caused by the long-distance connections; and an increase in the short-circuit availability to unacceptably high levels throughout the system.

Against this backdrop, KEPCO sought alternatives that would meet reliability requirements, maximize the utilization of the existing system, and reduce the need for new transmission facility rights-of-way, which would avoid adverse environmental impact and meet social requirements. The plan called for LCC-HVDC transmission, static synchronous compensators (STATCOMs), and thyristor-controlled series capacitors (TCSCs).

Unique capabilities available with HVDC: There are many reasons for developing HVDC projects that drove the consideration of the LCC-HVDC project in the Republic of Korea. The proposed project connects large power plants (including plants that generate nuclear power) with distant load centers, keeps short-circuit availability within acceptable levels, requires minimal right-of-way, and uses fast power modulation and other controls to minimize the adverse effects of extremely high-impact contingencies. The LCC-HVDC project required multiple feasibility studies to ensure it could provide a solution to the many complex issues presented by the system in the Republic of Korea.

To meet the reliability performance standard when a critical 765-kV line contingency occurs, the project utilizes a sophisticated control strategy that rapidly ramps the DC power output in combination with other controls and an SPS. HVDC power modulation is a common industry application, especially for ramping down the HVDC system to protect its equipment in response to a contingency. However, this project also utilizes an overload capability for the HVDC system in coordination with the controls of FACTS to minimize generation rejection. Critical contingencies accelerate generators in the east, but they remain stable when HVDC transfers west are suddenly ramped up, which provides a cost-effective solution for these rare system events. Ramping the HVDC up also relieves potential thermal overloads of other transmission facilities. The plan still requires SPS actions to reconfigure the AC transmission system, reject some generation, and provide reactive power support.

Effects of combined LCC-HVDC and FACTS: The study of this HVDC project began in 2013 with a conventional feasibility study to estimate its size and terminal locations. The short-circuit ratio (SCR) and effective short-circuit ratio (ESCR) provide critical indices used to compare the strength of an AC system to acceptable HVDC expansion options. The short-circuit levels and SCR were examined for the most severe transmission outages for base case conditions with up to two generators normally out of service.

Using a transient stability analysis (TSA) tool, numerous credible operating scenarios were evaluated to identify the optimal connection points for the HVDC stations. Five locations on the 765-kV transmission lines were identified and simulated as the worst-case three-phase bus faults in the studies. However, KEPCO engineers performed TSAs of all credible contingencies throughout the system, particularly at the HVDC terminals. All results for a given plan must be stable to fulfill the intended role of the HVDC that must support continuous power delivery, even if there is a three-phase fault on the double-circuit 765- kV transmission lines.

However, this LCC-HVDC project needs more than the conventional approach normally used for planning HVDC facilities. So as to meet the high degree of accuracy required for studying the complex power system in the Republic of Korea, additional modeling, tools, studies, and analyses for this specific project were performed as follows:

- Maximum Power Curve analysis with STATCOM (MPCWS) for examining the initial overloading capability of HVDC based on different system strengths
- Equal area criterion (EAC) analysis for estimating the initial overloading amount of HVDC

- Feasibility study of a proposed TCSC with overload capability
- Feasibility study of a STATCOM to support the HVDC
- TSA using an advanced HVDC model
- Analysis to determine HVDC ramping rates for automatic operation and scheduling changes
- Electromagnetic transient (EMT) simulation analysis, including detailed models of the HVDC and FACTS

Analyses based on the EAC with HVDC and MPCWS turn out to be useful for analyzing the initial required overloading capability of the HVDC before refining the amounts using TSA analysis. Both analyses are different from conventional approaches, which can provide initial parameters for the detail study. The EAC analysis with HVDC can provide an initial overloading amount for the HVDC, which makes it possible to manage the deceleration of generators near the rectifier side of the HVDC. This analysis can estimate the level of overloading amount that secures enough stability margin, so that the number of generators required to trip by SPS can be within the reliability standard. The aggregation process of generators is required to perform such an analysis. An MPCWS analysis offers the expected overloading capability of a HVDC system under different system strengths and provides an initial rating of the STATCOM required to maintain the voltage on the inverter terminal in steady state. These procedures can prevent iterative tasks when performing further studies.

The use of advanced HVDC stability models, such as the CDC7T model in PSS®E, provides the most accurate results in planning studies. Through detailed modeling of the DC line and control algorithms, the model allows the simulation of HVDC performance to be as close as possible to the simulation results achieved with an EMT program. In addition, the stability model includes a controlled DC ramp-up rate, which is an important factor in determining the required overloading capacity of the HVDC for properly decelerating the generators after a contingency.

E-TRAN simulation provides detailed and accurate EMT studies after converting network data and dynamic models of TSA to those of EMT. EMT analysis is required to determine parameters such as DC voltage and current sensor time constants, the size of the DC smoothing reactor and line parameters, VDCOL (Voltage Dependant Current Order Limits) coordination, PI (Proportional and Integrated) controller values, and firing angle and extinction angle ranges. The analysis can tune controls and identify abnormal operation of the HVDC converter valves that would be observed in the field, but not seen by TSA simulation results. For example, commutation failures can be observed in EMT simulations, but not in the TSA. An examination of the EMT simulation results can inform the interpretation of TSA results and improve TSA models.

A feasibility study showed that the installation of FACTS resulted in improved HVDC performance and achieved better utilization of existing transmission assets. Series capacitors reduce line impedances that increase AC power flows, but a TCSC can also dynamically modulate the amount of series compensation and control line flows. The study examined a TCSC with dynamic overload capability to increase AC power transfer during transient conditions and reduce dependency on the HVDC overload capability. The analysis determined the TCSC firing angle, compensation level, and capacitor and inductor ratings. Studies also determined the dynamic-mode range of compensation and the expected operating ranges for normal and dynamic modes to avoid undesired resonances and enable estimates of the optimal firing angles for the TCSC.

The installation of a STATCOM would instantaneously provide needed reactive support and was selected as the best option for obtaining more stable LCC-HVDC performance. The location of the STATCOM electrically near the inverter side of the HVDC system must be determined prior to establishing the rating of the STATCOM. The proper reactive power support and dynamic capability of the STATCOM improves HVDC performance by avoiding commutation failures under severe contingencies and maintaining acceptable AC and DC voltages at the inverter terminal.

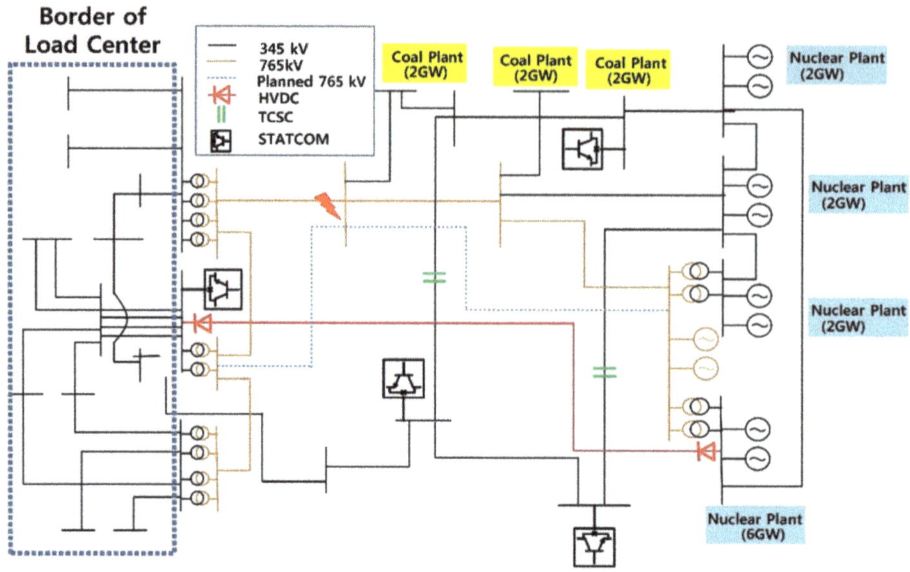

Figure 12.14 Simplified network from the generation complex to Seouls metropolitan area with HVDC and FACTS: The single transmission lines above are actually double-circuit lines

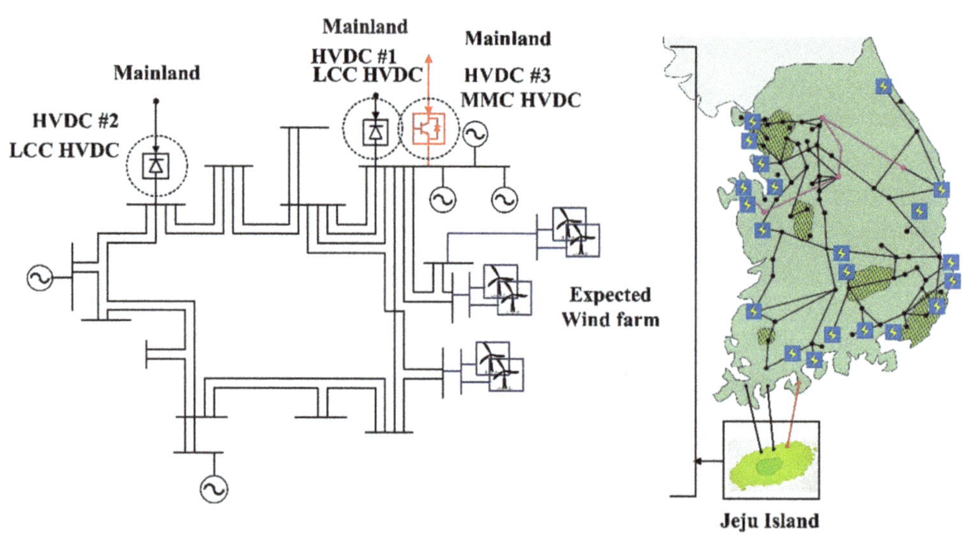

Figure 12.15 Simplified power system representation of Jeju island

Expected commissioning: Figure 12.14 shows a diagram for the planned power system in the area of interest. The east side is the main generation area, equipped with large-scale nuclear and coal plants; the west side of the system is the main load center of the country that includes Seoul, the capital city of the Republic of Korea. The project aims to stabilize the power system by coordinating the controls of the HVDC system with the FACTS, consistent with planning criteria that allows for up to two generators to trip offline upon occurrence of the most severe contingency: a three-phase fault on the double-circuit 765-kV transmission lines resulting in their loss. The addition of the HVDC system, the STATCOMs, and TCSCs replaces the original plan of building new 765-kV transmission lines. After reviewing various scenarios and strategies, two bipolar LCC-HVDC lines with a total rating of 8 GW have been planned. Two TCSCs, each with a rating of 500 MVAR, will support HVDC performance as well as maximize the AC power transfer capability under N-2 contingency conditions.

Jeju: Towards a zero-carbon island: A serious concern for KEPCO planners is the difficulty in performing accurate power system analysis with the increase in the use of power converters with HVDC and FACTS. The problem of choosing an appropriate simulation tool has arisen again with the Jeju HVDC No.3 project. Jeju Island is the largest island in Korea, and it connects to the mainland through HVDC lines, as shown in Figure 12.15. The Jeju electric power system has different characteristics compared to the Republic of Korea mainland power system. Owing to the abundant renewable energy sources, the Jeju special self-governing body declared their vision of becoming a "carbon-free island" by 2030.

The multi-infeed HVDC system presently in operation has had two LCC-HVDC lines (HVDC No.l: 300 MW, HVDC No.2: 400 MW) since 2010. The renewable power capacity on the island was 380 MW in 2017, and the Republic of Korea government aims to install further renewable energy sources totaling approximately 2 GW on Jeju by 2030. Although there are eight synchronous generators, three or four are typically in service, with minimal power generation during the off-peak load condition to support the minimum SCR required by the LCC-HVDC converters. The peak load of Jeju was approximately 930 MW in 2017. In the future, when Jeju's wind power penetration surpasses its peak load, the island will require the capability to transmit power back to the bulk power system, as shown in Figure 12.16. Therefore, HVDC No.3, which is in the planning stage, must be flexible and bidirectional. It is most likely to be an MMC-based VSC scheme.

Feasibility studies for Jeju HVDC No.3: Since 2010, KEPCO has conducted numerous feasibility studies for Jeju HVDC No.3. KEPCO first needs to determine the type of HVDC (LCC-HVDC or MMC-HVDC). Regarding power reversal capability, MMC-HVDC is more flexible than LCC-HVDC, and has less negative influence on the operation and reliability of the Jeju power system. The feasibility studies address key issues that include active power reserve, reactive power compensation, ESCR, multi-infeed effective short circuit ratio (MIESCR), effective inertial constant, the hosting capacity of renewables, and TSA. In particular, LCC-HVDC requires must-run synchronous generator operation to maintain a suitable short-circuit availability, and this constraint reduces the hosting capacity for the renewables. An EMT study determined the power reversal procedure and its influence on the Jeju power system. Although MMC-HVDC would not need any additional controls, LCC-HVDC requires many steps to reverse the power transfer. The application of LCC-HVDC requires the use of a mass-impregnated DC cable and a converter transformer with high insulation levels to cope with the DC voltage polarity change during the LCC-HVDC power reversal.

An economic feasibility study was conducted. When HVDC No.3 is installed and the penetration of the renewable energy source increases on Jeju, the island's system marginal price (SMP) can be significantly reduced; the SMP with the addition of the MMC-HVDC would be lower than that with the addition of an LCC-HVDC tie. This means that KEPCO can reduce the cost of purchasing power with the addition of the MMC-HVDC for HVDC No.3. However, the MMC converter station costs are higher than that for LCC. Therefore, KEPCO determined the type of HVDC based on the results of the economic and technical feasibility studies, and MMC-HVDC was chosen for HVDC No.3 because of the flexible power transfer capability.

Technical challenges for the Jeju power system: The future Jeju power system, which includes a high penetration of renewable energy sources, will encounter many technical challenges with multi-infeed (multiple HVDC system installations) and hybrid-infeed (different types of HVDC converter stations) HVDC systems located electrically close to each other in an area with low SCR and inertia. The massive penetration of wind power generation could threaten the stability performance of the Jeju power system. The two LCC-HVDC systems require reactive power compensation for power transfer. Overvoltage conditions may occur during ramp-down and recovery operations. Transient voltage depression may also occur because of the simultaneous recovery of the HVDC after an AC fault, a DC fault, or a commutation failure. Although the synchronous generators in Jeju provide AC voltage regulation, the number of must-run synchronous generators needs to be reduced in order to increase the penetration level of renewables. Research on the role allocation and coordination control of HVDC is also needed. In this complex operating situation, a feasibility study with hybrid simulation proved useful for KEPCO planners.

Outlook for a feasibility study of HVDC projects in the Republic of Korea: Currently, the Republic of Korea is going through a huge transition in its power system to increase public acceptance of power system expansion and address environmental concerns. Several HVDC projects are planned to be built in conjunction with numerous FACTSs to reinforce the power transfer capability and reliability, mitigate operational risks on the mainland, and integrate massive renewable energy growth on Jeju Island. In this situation, the planners have to carry out various feasibility studies with several tools and the weigh the results of the studies against each other, which should be resolved as existing power systems evolve into future power grids where power electronics-based components dominate networks.

The temporary overload performance of HVDC and FACTS should be considered as options to stabilize the power system and prevent cascading blackouts. This study, which was conducted using various tools, provides valuable insights into planning a power system with fast yet sophisticated converters; however, the accuracy of the results is not yet sufficient to reach a firm conclusion. The feasibility study of the MMC-HVDC, which connects the mainland and Jeju Island for flexible power transfer, also requires more than just the conventional approach. Since modeling at the TSA level cannot capture the unique and crucial dynamics and features of MMC in harmony with the two existing LCCs, the hybrid simulator should be useful in complementing the insights provided by the pure TSA-based system-wide feasibility studies. Advanced modeling with the hybrid simulator should help grid planners better define the grid requirements of the emerging equipment. Such modeling is also beneficial from the manufacturers' perspective because it provides clear design criteria, which avoids unnecessary conflicts after delivering the equipment. KEPCO planners: The project team will continue making efforts to improve their planning practices to meet the upcoming changes and challenges of the future power grid, and to fully exploit the opportunities.

12.2.2 HVDC Systems in India

In India, there are similar prospects for UHV DC (Ultra High Voltage Direct Current) system as there are in China because of a large extension of the grid. The road map for India's hybrid bulk power grid developments is depicted in Figure 12.17(a). India's energy growth is about 8%–9 % per annum, with an installed generation capacity of 124 GW in 2006 (92 GW peak load demand). The installed generation capacity is expected to increase to 333 GW by 2017.

Figure 12.17(b) depicts how the ideas of hybrid bulk power interconnections are reflected in China's UHV grid developments. The focus is on the interconnection of seven large inter-provincial grids in the Northern, Central, and Southern systems via three bulk power corridors, which will build up a redundant "backbone" for the whole grid. It is planned that each corridor will have about 20-GW transmission capacity, which shall be implemented with both AC and DC transmission lines with ratings of 4–10 GW each (at ±800-kV DC and 1000-kV AC as shown in the figure). Therefore, each corridor will have a setup with two to three systems for redundancy reasons. With these ideas, China envisages a total amount of about 900-GW installed generation capacity by 2020. For comparison, UCTE and IPS/UPS together sum up to 850 GW today.

12.2.3 HVDC Systems in China

Several VSC-HVDC projects have been constructed in China. At present, the 1000-MW ±320-kV Xiamen HVDC system (Figure 12.18) with a transmission distance of 10.8 km was built to increase the transmission capacity. It is integrated into a large AC interconnected system. Also, the ±500 kV/3000 MW MMC VSC converters are under development and will be used for the, Zhangbei HVDC grid project , for example. The

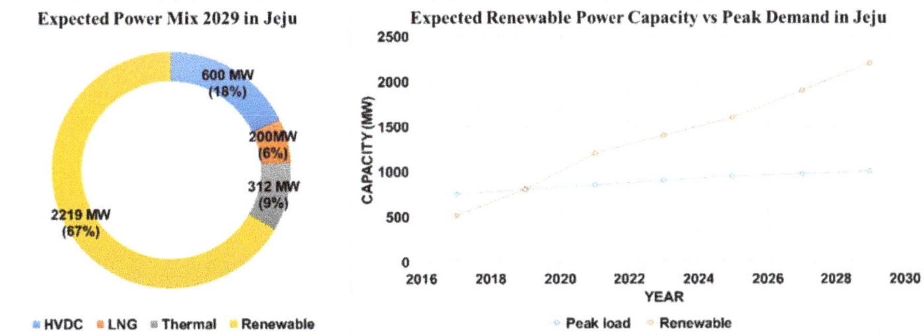

Figure 12.16 Future capacity (nameplate) of the Jeju Island power system

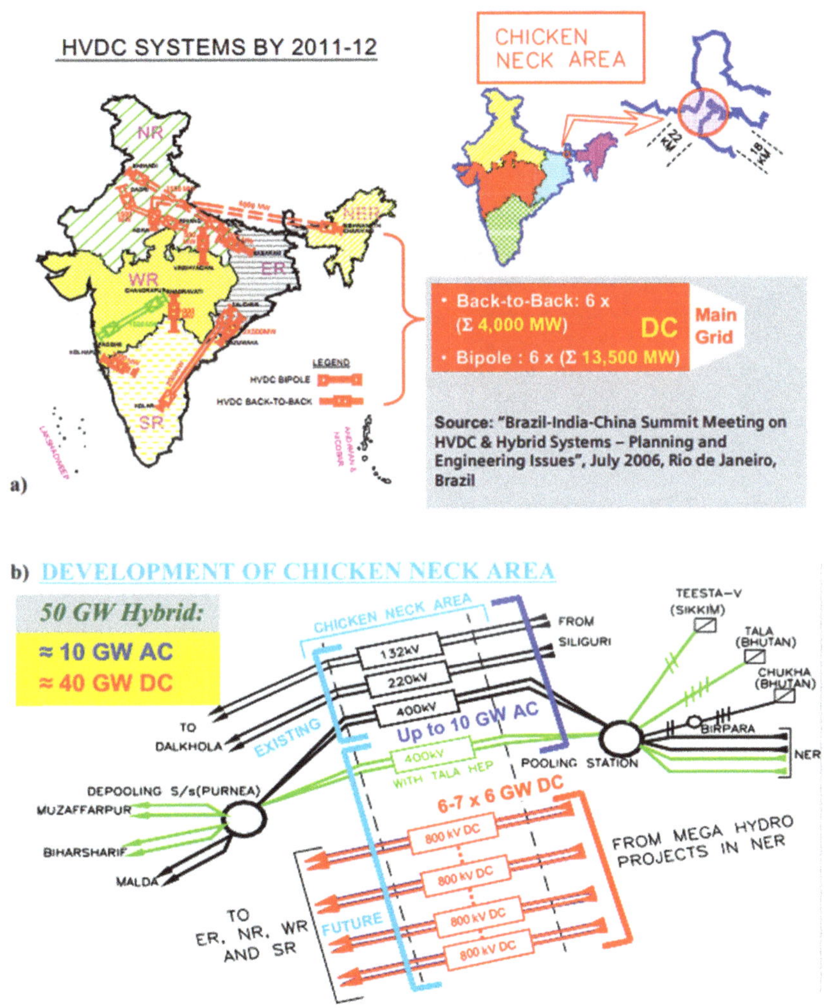

Figure 12.17 Grid developments in India: (a) System overview–including the "Chicken Neck Area"; (b) hybrid solutions in the Chicken Neck–including UHV DC

VSC-HVDC converters developed by Global Energy Interconnection Research Institute (GEIRI) are easily scalable with modular design, easy assembly and maintenance, have superior inflammability conforming to UL94 and excellent seismic performance (0.2 g typical), and use environmental friendly materials (halogen-free).

The hierarchical structure of a C&P system device developed by GEIRI is designed as a vertical containing an operator control layer, control layer and interface layer, and a horizontal performing DC control, DC protection, auxiliary power and equipment control. The C&P system device as shown in Figure 12.18 has a universal rack and modular design, employs an ETHERCAT bus and HDLC fiber optics for inter-rack communications, and has high reliability with multi-redundancy.

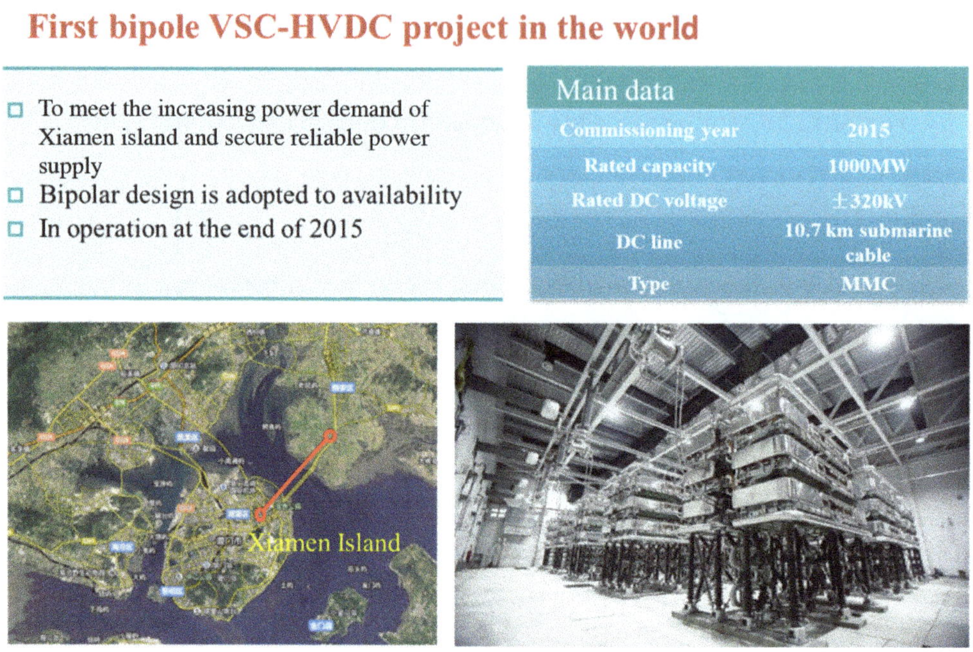

Figure 12.18 Highlights of the Xiamen HVDC transmission project

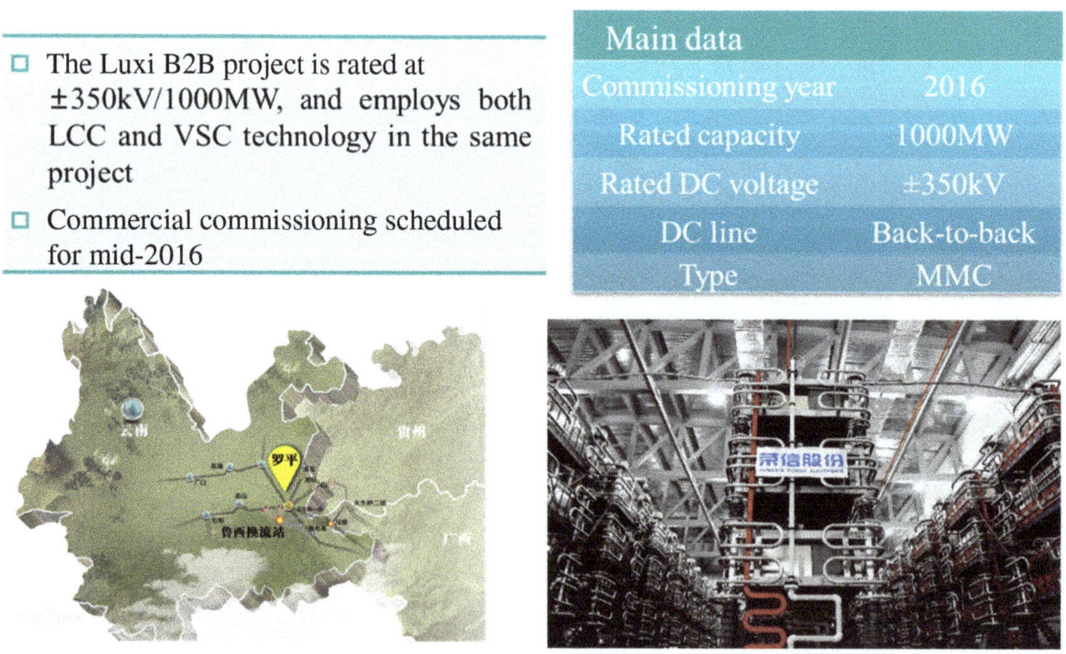

Figure 12.19 Highlights of the Luxi HVDC transmission project

- **Securing a reliable power supply to its five islands and integrating the offshore wind farm.**
- **Significant increased wind power integration and power supply capacity.**
- **A pilot project is planned to increase Zhoushan MTDC reliability and availability by using DC CBs scheduled for 2016.**

(Capacity: 400/300/100/100/100MW , DC voltage: ±200kV)

Figure 12.20 Highlights of the Zhoushan HVDC transmission project

- **To meet increasing demand and to eliminate the influence of the large capacity of wind power integration**
- **Commissioned for operation in 2015**

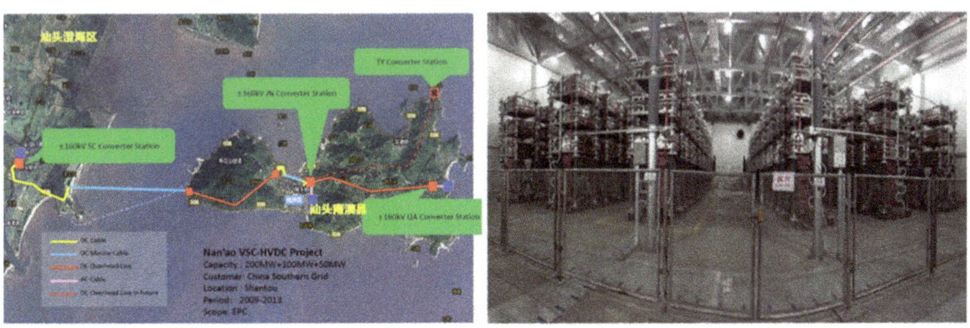

(Capacity: 200/100/50MW , DC voltage: ±160kV)

Figure 12.21 Highlights of the Nan'ao HVDC transmission project

12.2.4 HVDC Systems in Australia

Figure 12.22 gives an overview of the Basslink project in Australia, which transmits electric power from wind and hydro sources very cost-efficiently from George Town in Tasmania to Loy Yang in Victoria and back.

This happens by means of HVDC via a combination of a submarine cable (with the longest submarine cable in the world up to now being 295 km), subterranean cables (8 km for reasons of landscape protection), and overhead lines over a total transmission distance of 370 km. The nominal power is 500 MW at a DC voltage of

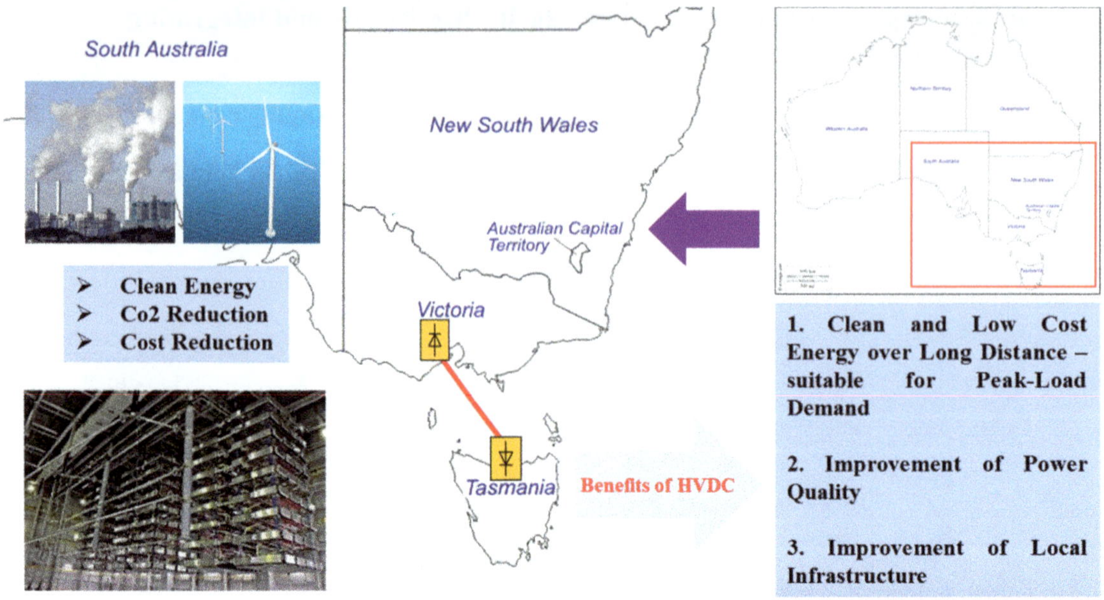

Figure 12.22 Basslink HVDC–sustainability of a "smart" and flexible grid

400 kV and a current of 1250 A. The overload capacity of the transmission system is 600 MW, 10 hours per day. Both Victoria and Tasmania profit from the interconnection of their networks.

During times of peak load, Tasmania delivers "green energy" from its hydro power stations to Victoria, whereas Tasmania can cover its base load demands through the use of Victoria's grid during dry seasons when the hydro-reservoirs are not sufficiently filled. Furthermore, the island of Tasmania receives access to the power market of the Australian continent.

Tasmania intends to install additional wind farms to increase its share in regenerative energy production. The figure shows that hydro power is perfectly suitable to be supplemented with the rather "fuzzy" wind energy—in terms of base load as well as through its ability to store energy for peak load demands. So far, the DC link can do much more to reduce CO_2 through the combined use of regenerative energies.

12.2.5 HVDC Systems in Germany

The idea of embedding huge amounts of wind energy in the German grid using HVDC, FACTS, and gas-insulated lines (GILs) is depicted in Figure 12.23. The goal is a significant CO_2 reduction through the replacement of conventional energy sources by renewable energies, mainly offshore wind farms. Power output of wind generation can vary fast in a wide range, depending on the weather conditions. Therefore, a sufficiently large amount of controlling power from the network is required to substitute the positive or negative deviation of actual wind power infeed from the scheduled wind power amount. Wind power infeed and the regional network load during a week of maximum load in the E.ON control area are plotted. The relation between consumption and supply in this control area is illustrated in the figure. In the northern areas of the German grid, the transmission capacity is already at its limits, especially during times with low load and high wind power generation.

An efficient alternative to the connection of offshore wind farms is the integration of HVDC long-distance transmission links into the synchronous AC system as shown in Figure 12.24.

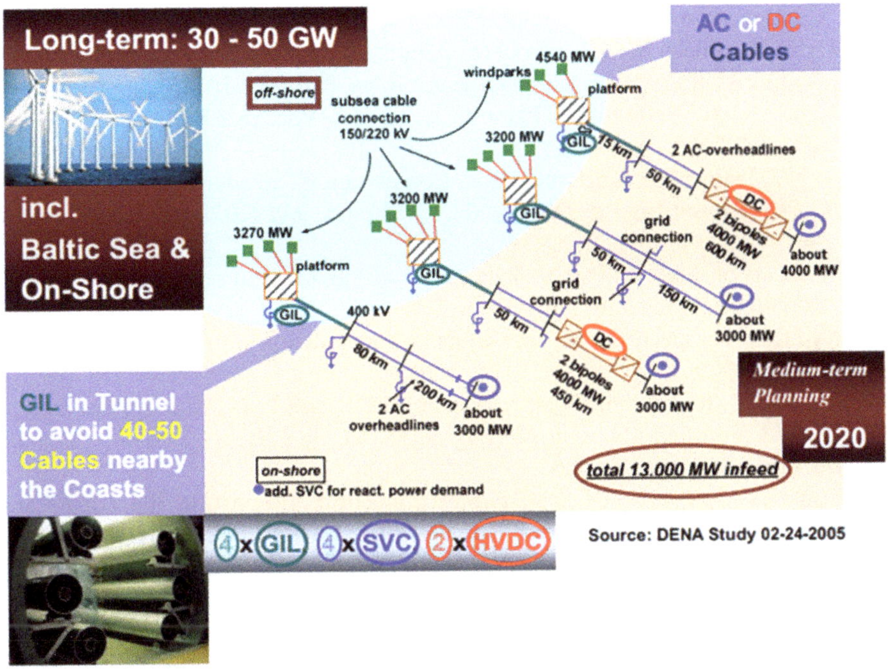

Figure 12.23 Integration of large offshore wind farms by means of GIL, FACTS, and HVDC-A smart grid solution

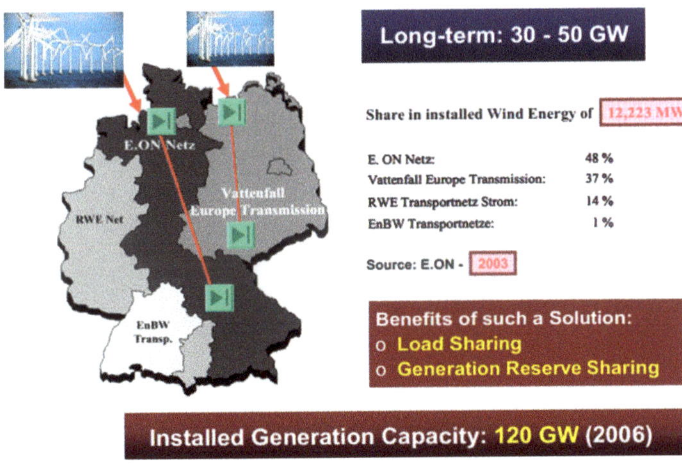

Figure 12.24 Use of HVDC long-distance transmission for wind farm grid access

References

1. A. Elahidoost and E. Tedeschi, "Expansion of offshore HVDC Grids," in *2017 IEEE 58th International Scientific Conference on Power and Electrical Engineering of Riga Technical University (RTUCON)*, Oct. 2017.
2. Windeurope, "Wind in Power-2016 European statistics," Feb. 2017. https://windeurope.org/wp-content/uploads/files/about-wind/statistics/WindEurope-Annual-Statistics-2016.pdf
3. F. Van Hulle, et al., "Powering Europe: wind energy and the electricity grid," in *A report by the European Wind Energy Association (EWEA)*, Nov. 2010, pp. 1–179.,
4. ENTSO-E, "10-year network development plan 2014," in *A report by European Network of Transmission System Operators for Electricity (ENTSO-E)*, 2014.
5. Greenpeace, "PowE[R] 2030: A European grid for 3/4 renewable electricity by 2030," in A report by Greenpeace, 2014.

6. G. Asplund, et al., "Continental overlay HVDC-grid," in *CIGRE B4-109,* 2010, pp. 1–9.
7. J. L. Rueda, et al., "Hybrid approach for power system operational planning with smart grid and small-signal stability enhancement considerations," *IEEE Trans. Smart Grid*, vol. 4, pp. 530–539, 2013.
8. M. Henderson, J. Gagnon, and D. Bertagnolli, "Planning issues for HVdc," in *2006 IEEE PES Power Systems Conference and Exposition*, Atlanta, GA, 2006, pp. 28–32.
9. A. H. Dominguez, et al., "Transmission network expansion planning considering HVAC/HVDC lines and technical losses," in *2016IEEE PES Transmission & Distribution Conference and Exposition-Latin America (PES T&D-LA)*, Morelia, 2016, pp. 1–6.
10. W. Breuer, et al., "Prospects of smart grid technologies for a sustainable and secure power supply," in *The 20th World Energy Congress & Exhibition*, Rome, Italy, Nov.11th to 15th, 2007.
11. J. M. Pérez de Andrés, et al., "Prospects of VSC converters for transmission system enhancement," in *Powergrid Europe 2007,* Madrid, Spain, June 26–28.
12. W. Breuer, M. Lemes, and D. Retzmann, "Perspectives of HVDC and FACTS for system interconnection and grid enhancement," in *Brazil-China-India Summit Meeting on HVDC and Hybrid Systems – Planning and Engineering Issues*, Rio de Janeiro, Brazil, July 16–18, 2006.
13. D. Povh, et al., "Advantages of large AC/DC system interconnections," Report B4-304, *CIGRE Session 2006*, Paris.
14. H. Waldhauer, "Grid reinforcement and SVC for full power operation of baltic cable HVDC Link," in *The 38th Meeting and Colloquium of Cigré Study Committee B4 "HVDC and Power Electronics,"* Technical Colloquium, Nuremberg, Germany, September 25, 2003.
15. N. G. Hingorani, "Flexible AC transmission," *IEEE Spectr.*, vol. 30, no. 4, pp. 40–45, Apr. 1993.
16. J. Dorn, H. Huang, and D. Retzmann, "Novel voltage-sourced converters for HVDC and FACTS applications," in *Cigre Symposium*, Osaka, Japan, Nov. 14, 2007.
17. Bamberger, Yves;Baptista, Joao, et al., "European Technology Platform Smart Grids, Vision and Strategy for Europe's Electricity Networks of the Future", Luxembourg, Belgium, 2006.

Index

R

S

T

U

V

W

Z

www.ingramcontent.com/pod-product-compliance
Ingram Content Group UK Ltd.
Pitfield, Milton Keynes, MK11 3LW, UK
UKHW051045150926
8624IPUK00006B/404

9 789811 212291